Journal of Drug Issues
Journal of Education for Business
Journal of Engineering for Industry
Journal of Environmental
  Education
Journal of Experimental Education
Journal of Experimental Social
  Psychology
Journal of Experimental Zoology
Journal of Financial Planning
Journal of Fish Biology
Journal of Geography
Journal of Gerontological Nursing
Journal of Hazardous Materials
Journal of Heat Transfer
Journal of Human Stress
Journal of Information Systems
Journal of Interdisciplinary History
Journal of Labor Research
Journal of Literacy Research
Journal of the Market Research
  Society
Journal of Marketing
Journal of Marriage and the Family
Journal of Materials Science
Journal of Modern History
Journal of Moral Education
Journal of Nursing Education
Journal of Nutrition
Journal of Occupational and
  Organizational Psychology
Journal of Performance of
  Constructed Facilities
Journal of Personal Selling & Sales
  Management
Journal of Personality
Journal of Psychology and Aging
Journal of Quality Technology
Journal of Quantitative
  Criminology
Journal of Research in Music
  Education
Journal of Sound and Vibration
Journal of Speech, Language, and
  Hearing Research
Journal of Sport & Social Issues

Journal of STAR Research
Journal of Studies in Alcohol
Journal of Teaching in Physical
  Education
Journal of Testing and Evaluation
Journal of Transportation
  Engineering
Journal of Travel Research
Journal of Zoology
Ladies Home Journal
Leisure Sciences
Lindleyana
Lisa Schwartz, M.D.
Lottery Buster
Marine Technology
Medical Interfaces
Memory & Cognition
Morbidity and Mortality Weekly
  Report
National Highway Traffic Safety
  Administration
National Wildlife
Nature
New Accountant
New England Journal of Medicine
New Jersey Governor's Council
New York Times
Newsday
Newsweek
Nuclear Science and Engineering
Orange County (CA) Register
Parade Magazine
Perception & Psychophysics
Physical Therapy
Presidential Studies Quarterly
Prison Journal
Proceedings of the Institute of
  Civil Engineers
Proceedings of the National Science
  Council, Republic of China
Prog. Oceanog.
Psychological Assessment
Psychological Reports
Psychological Science
Psychonomic Bulletin & Review
Public Health Reports

Quality
Quarterly Journal of Business and
  Economics
Quarterly Journal of Experimental
  Psychology
Real Estate Appraiser
Risk Management
San Francisco Chronicle
San Francisco Examiner
Science
Science News
Scientific American
Self
Shere Hite
Sleep
Sloan Management Review
Small Group Behavior
Small Group Research
Smithsonian
Social Science Quarterly
Special Interest Group on
  Computer-Human Interaction
  Bulletin
Sports Illustrated
Susan L. Farber
Tampa Tribune
Teaching Sociology
Technometrics
Time
Transportation Journal
University of South Florida
U.S. Army Corps of Engineers
U.S. Department of Agriculture
U.S. Department of Education
U.S. News & World Report
USA Today
Wall Street Journal
Washington Post
Wetlands
Women & Therapy
Zoological Science

Contents

Catriona MacGillivray
2007

Preface ix

Microsoft Excel Primer 1
    Overview 1
Getting Started with the User Interface 1
    EP.1.1   Getting Familiar with Mouse Operations 1
    EP.1.2   Getting Familiar with the Microsoft Excel Application Window 4
    EP.1.3   Getting Familiar with Menu Conventions and Dialog Boxes 5
Getting Started with Microsoft Excel 7
    EP.2.1   Configuring Microsoft Excel 7
    EP.2.2   Specifying Worksheet Locations 9
    EP.2.3   Using Formulas to Perform Calculations 9
    EP.2.4   Entering Data into Worksheets 10
    EP.2.5   Copying Cell Entries and Worksheets 11
    EP.2.6   Opening Workbooks 12
    EP.2.7   Printing Worksheets 13
    EP.2.8   Saving Workbooks 17
    EP.2.9   Enhancing the Appearance of a Worksheet 17
    EP.2.10 Getting Context-Sensitive Help in Microsoft Excel 19
Getting Started with Microsoft Excel Wizards and Add-Ins 21
    EP.3.1   Wizards 21
    EP.3.2   Using Add-Ins 23
    EP.3.3   Summary 26

Chapter 1   Introduction: Statistics and Data 27
 Real-World Application   Obedience to Authority—The Shocking Truth 27
    1.1   What Is Statistics? 28
    1.2   Types of Data 29
    1.3   Descriptive vs. Inferential Statistics 34
    1.4   Collecting Data 41
    1.5   Random Sampling 44
    1.6   Other Types of Samples (Optional) 48
    1.7   Ethical and Other Issues in Statistical Applications 50
 Real-World Revisited 55
Using Microsoft Excel 60
    1.E.1   Using Microsoft Excel to Select a Random Sample 60

Chapter 2   Exploring Data with Graphs and Tables 61
 Real-World Application   A Bad Moon Rising 61
    2.1   The Objective of Data Description 62
    2.2   Describing a Single Qualitative Variable: Frequency Tables, Bar Graphs, and Pie Charts 62

2.3   Describing a Single Quantitative Variable: Frequency Tables, Stem-and-Leaf Displays, and Histograms   72

2.4   Exploring the Relationship between Two Qualitative Variables: Cross-Classification Tables and Side-by-Side Bar Graphs   85

2.5   Exploring the Relationship between Two Quantitative Variables: Scatterplots   95

2.6   Proper Graphical Presentation   104

 Real-World Revisited   111

Using Microsoft Excel   125

2.E.1   Using Microsoft Excel to Describe a Single Qualitative Variable   125

2.E.2   Using Microsoft Excel to Describe a Single Quantitative Variable   126

2.E.3   Using Microsoft Excel to Explore the Relationship between Two Qualitative Variables   134

2.E.4   Using Microsoft Excel to Explore the Relationship between Two Quantitative Variables   141

Chapter 3   Exploring Data with Numerical Descriptive Measures   144

 Real-World Application   Bid-Collusion in the Highway Contracting Industry   144

3.1   Types of Numerical Descriptive Measures   145

3.2   Summation Notation   146

3.3   Measures of Central Tendency: Mean, Median, and Mode   148

3.4   Measures of Data Variation: Range, Variance, and Standard Deviation   157

3.5   Interpreting the Standard Deviation   159

3.6   Measures of Relative Standing: Percentiles and z Scores   173

3.7   Box-and-Whisker Plots   180

3.8   Methods for Detecting Outliers   184

3.9   A Measure of Association: Correlation   190

3.10   Numerical Descriptive Measures for Populations   198

 Real-World Revisited   198

Using Microsoft Excel   209

3.E.1   Using Microsoft Excel to Explore Data with Numerical Descriptive Measures   209

Chapter 4   Probability: Basic Concepts   212

 Real-World Application   Lottery Buster!   212

4.1   The Role of Probability in Statistics   213

4.2   Experiments, Events, and the Probability of an Event   213

4.3   Probability Rules for Mutually Exclusive Events   218

4.4   The Combinatorial Rule for Counting Simple Events (Optional)   230

4.5   Conditional Probability and Independence   233

4.6   The Additive and Multiplicative Laws of Probability (Optional)   241

 Real-World Revisited   250

Chapter 5   Discrete Probability Distributions   260

Real-World Application   Commitment to the Firm—Stayers vs. Leavers   260

     5.1   Random Variables   261

     5.2   Probability Models for Discrete Random Variables   262

     5.3   The Binomial Probability Distribution   270

     5.4   The Poisson Probability Distribution   283

     5.5   The Hypergeometric Probability Distribution (Optional)   288

Real-World Revisited   293

Using Microsoft Excel   299

     5.E.1   Using Microsoft Excel to Obtain the Expected Value and Variance of a Probability Distribution   299

     5.E.2   Using Microsoft Excel to Obtain Binomial Probabilities   300

     5.E.3   Using Microsoft Excel to Obtain Poisson Probabilities   301

     5.E.4   Using Microsoft Excel to Obtain Hypergeometric Probabilities   302

Chapter 6   Normal Probability Distributions   303

Real-World Application   Detecting Fire Insurance Fraud   303

     6.1   Probability Models for Continuous Random Variables   304

     6.2   The Normal Probability Distribution   305

     6.3   Descriptive Methods for Assessing Normality   321

     6.4   Sampling Distributions   330

     6.5   The Sampling Distribution of the Mean and the Central Limit Theorem   338

Real-World Revisited   347

Using Microsoft Excel   355

     6.E.1   Using Microsoft Excel to Obtain Normal Probabilities   355

     6.E.2   Using Microsoft Excel to Construct Normal Probability Plots   356

     6.E.3   Using Microsoft Excel to Simulate Sampling Distributions   357

Chapter 7   Estimation of Population Parameters Using Confidence Intervals: One Sample   361

Real-World Application   Scallops, Sampling, and the Law   361

     7.1   Point Estimators   362

     7.2   Estimation of a Population Mean: Normal (z) Statistic   362

     7.3   Estimation of a Population Mean: Student's t Statistic   375

     7.4   Estimation of a Population Proportion   383

     7.5   Choosing the Sample Size   390

     7.6   Estimation of a Population Variance (Optional)   395

Real-World Revisited   401

Using Microsoft Excel   408

     7.E.1   Using Microsoft Excel to Obtain the Confidence Interval Estimate for the Mean (σ known)   408

     7.E.2   Using Microsoft Excel to Obtain the Confidence Interval Estimate for the Mean (σ unknown)   409

7.E.3   Using Microsoft Excel to Obtain the Confidence Interval
Estimate for the Proportion   410

7.E.4   Using Microsoft Excel to Determine the Sample Size for
Estimating the Mean   411

7.E.5   Using Microsoft Excel to Determine the Sample Size for
Estimating the Proportion   413

7.E.6   Using Microsoft Excel to Obtain the Confidence Interval
Estimate for the Population Variance   414

Chapter 8   Testing Hypotheses about Population Parameters: One Sample   416

Real-World Application   Cellular Telephones—Road Hazards?   416

8.1   The Relationship between Hypothesis Tests
and Confidence Intervals   417

8.2   Hypothesis-Testing Methodology: Formulating Hypotheses   418

8.3   Hypothesis-Testing Methodology: Test Statistics
and Rejection Regions   425

8.4   Guidelines for Determining the Target Parameter   433

8.5   Testing a Population Mean   435

8.6   Reporting Test Results: p-Values   444

8.7   Testing a Population Proportion   452

8.8   Testing a Population Variance (Optional)   457

8.9   Potential Hypothesis-Testing Pitfalls and Ethical Issues   461

Real-World Revisited   464

Using Microsoft Excel   472

8.E.1   Using Microsoft Excel to Perform the Z Test of the Hypothesis
for the Mean (σ known)   472

8.E.2   Using Microsoft Excel to Perform the t Test of the Hypothesis
for the Mean (σ unknown)   473

8.E.3   Using Microsoft Excel to Perform the Z Test of the Hypothesis
for the Proportion   474

8.E.4   Using Microsoft Excel to Perform the Chi-Square Test of the
Hypothesis for the Variance   475

Chapter 9   Inferences about Population Parameters: Two Samples   478

Real-World Application   An IQ Comparison of Identical Twins Reared Apart   478

9.1   Determining the Target Parameter   479

9.2   Comparing Two Population Means: Independent Samples   480

9.3   Comparing Two Population Means: Matched Pairs   494

9.4   Comparing Two Population Proportions:
Independent Samples   504

9.5   Comparing Population Proportions: Contingency Tables   511

9.6   Comparing Two Population Variances (Optional)   523

Real-World Revisited   531

Using Microsoft Excel   543

9.E.1   Using Microsoft Excel to Compare Two Population Means:
Independent Samples   543

9.E.2   Using Microsoft Excel to Compare Two Population Means: Matched Pairs   545

9.E.3   Using Microsoft Excel to Compare Two Population Proportions   546

9.E.4   Using Microsoft Excel to Analyze Contingency Tables   548

9.E.5   Using Microsoft Excel to Perform the F Test for Differences in Two Variances   550

Chapter 10   Regression Analysis   554

Real-World Application   The S.O.B. Effect among College Administrators   554

10.1   Introduction to Regression Models   555
10.2   The Straight-Line Model: Simple Linear Regression   557
10.3   Estimating and Interpreting the Model Parameters   560
10.4   Model Assumptions   572
10.5   Measuring Variability around the Least Squares Line   574
10.6   Inferences about the Slope   576
10.7   Inferences about the Correlation Coefficient (Optional)   585
10.8   The Coefficient of Determination   592
10.9   Using the Model for Estimation and Prediction   596
10.10   Computations in Simple Linear Regression (Optional)   604
10.11   Residual Analysis: Checking the Assumptions (Optional)   612
10.12   Multiple Regression Models (Optional)   624
10.13   Pitfalls in Regression and Ethical Issues   641

Real-World Revisited   644

Using Microsoft Excel   661

10.E.1   Using Microsoft Excel to Generate Scatter Diagrams and a Regression Line   661
10.E.2   Using Microsoft Excel for Simple Linear Regression   663
10.E.3   Using Microsoft Excel to Test for the Existence of Correlation   665
10.E.4   Using Microsoft Excel for Multiple Regression (Optional)   667

Chapter 11   Analysis of Variance   669

Real-World Application   Reluctance to Transmit Bad News—The MUM Effect   669

11.1   Experimental Design   670
11.2   ANOVA Fundamentals   673
11.3   Completely Randomized Designs: One-Way ANOVA   675
11.4   Factorial Designs: Two-Way ANOVA   686
11.5   Follow-Up Analysis: Multiple Comparisons of Means   700
11.6   Checking ANOVA Assumptions   710
11.7   Calculation Formulas for ANOVA (Optional)   712

Real-World Revisited   718

Using Microsoft Excel   731

11.E.1   Using Microsoft Excel for a One-Way ANOVA   731
11.E.2   Using Microsoft Excel for a Two-Way ANOVA   732
11.E.3   Using Microsoft Excel for Multiple Comparisons   733

Chapter 12   Nonparametric Statistics   736

 Real-World Application   Do Women Really Understand the Benefit
of a Mammography?   736

12.1   Distribution-Free Tests   737
12.2   Testing for Location of a Single Population   738
12.3   Comparing Two Populations: Independent Random Samples   745
12.4   Comparing Two Populations: Matched-Pairs Design   754
12.5   Comparing Three or More Populations: Completely
       Randomized Design   762
12.6   Testing for Rank Correlation   769

Real-World Revisited   778

Using Microsoft Excel   786

12.E.1   Using Microsoft Excel for the Sign Test   786
12.E.2   Using Microsoft Excel to Perform the Wilcoxon Rank Sum Test
         for Differences in Two Medians   788
12.E.3   Using Microsoft Excel to Perform the Kruskal-Wallis H Test
         for Differences in c Medians   789

Appendix A   Review of Arithmetic and Algebra   791
Appendix B   Statistical Tables   799
Appendix C   Documentation for Microsoft Excel Diskette Files   819
Appendix D   Installation Instructions for the PHStat Add-In for Microsoft
               Excel and the Data Files on the CD-ROM   824
Answers to Self-Test Problems   825
Answers to Selected Problems   827
References   834
Excel Index   837
Subject Index   839

Preface

When planning this textbook, the authors focused on how advances in computer software have impacted the way in which data are analyzed. Today an increasing number of individuals use spreadsheet applications as the means to retrieve and analyze directly the data they need. Employers now are beginning to desire, if not demand, that their college-educated, entry-level employees have more than just a cursory awareness of spreadsheet applications.

These changes, along with the realization that current spreadsheet applications can perform the type of analyses once done only by specialized statistical software packages, have led us to develop *Practical Statistics by Example Using Microsoft® Excel.* Designed as an introductory text in statistics for students with a background in college algebra, our text contains the following features that distinguish it from the many other statistics texts available.

"BY EXAMPLE" INTRODUCTION OF CONCEPTS

Each new idea is introduced and illustrated by real data–based examples taken from a wide variety of disciplines and sources. These examples demonstrate how to solve various types of statistical problems encountered in the real world. We believe that students better understand definitions, generalizations, and concepts *after* seeing a real application. Each example is set off for easy identification and contains a full, detailed solution to the problem.

MICROSOFT EXCEL AS A TOOL FOR STATISTICAL ANALYSIS

The spreadsheet application Microsoft Excel is integrated throughout the entire text. This approach is fundamentally different from that of the many texts published and revised in the past twenty years. Since the advent of the computer revolution, statistics texts have struggled with the appropriate way to incorporate the use of statistical software packages (such as SAS, SPSS, and Minitab). A dilemma for faculty teaching this course has been how students could obtain access (often through site licenses and student versions) to the statistical software selected and how these packages could be used in the course. Often students are not familiar with these packages prior to the statistics course, and only a limited number may use them in subsequent courses or on the job. Thus, students may view them as one more hurdle to overcome in getting through the statistics course.

However, in the last several years, with the increasing functionality and power of spreadsheet applications, virtually all kinds of statistical analysis taught in the introductory course are supported by Excel or the statistics add-in provided with this text (PHStat). In addition to its possible use in a statistics course, students are usually exposed to a spreadsheet application such as Microsoft Excel in an introductory computer course or when they need to analyze data in advanced courses in their major area of study. Even if they are not familiar with Excel, they have almost certainly heard of this software, or Microsoft Office, and its use in the statistics course will give added relevancy to the course.

EMPHASIS ON CRITICAL THINKING AND INTERPRETATION OF EXCEL OUTPUT

Excel-generated graphs and output accompany almost every statistical technique presented, allowing instructors to focus on the statistical analysis of data and the interpretation of the results rather than the calculations required to obtain the results. Free from memorizing formulas and performing hand calculations, students are encouraged to develop critical thinking skills that will allow them to realize greater success in the workplace. Examples of hand calculations are provided for those instructors who desire flexibility in teaching the course.

TUTORIALS ON USING MICROSOFT EXCEL

The *Microsoft Excel Primer* provides basic instruction on using Windows and Microsoft Excel for the novice. Those with Excel experience may skip portions of the *Primer.*

Excel Tutorials appear at the end of pertinent chapters and give step-by-step instructions and screen shots for using Excel to perform the statistical techniques presented in the chapter.

All data sets for which Excel can be used are identified with a computer diskette icon (🖫) and the Excel Workbook file name is provided.

STATISTICS ADD-IN FOR MICROSOFT EXCEL: PHSTAT

The CD-ROM that accompanies the text includes a *statistics add-in for Microsoft Excel* to facilitate its use in introductory statistics courses. Although the off-the-shelf version of Microsoft Excel can perform statistical analysis for many of the topics in this text, the *PHStat* statistics add-in provides a custom menu of choices that lead to dialog boxes in which users make entries and selections to perform specific analyses. PHStat minimizes the work associated with setting up statistical solutions in Microsoft Excel by automating the creation of spreadsheets and charts. PHStat, along with Microsoft Excel's Data Analysis tool, now allows users to perform statistical analyses on most topics covered in an introductory statistics course. A complete list of topics on the PHStat custom menu are provided on page 25.

REAL-WORLD APPLICATION IN EACH CHAPTER

Each chapter opens with a real-world application to illustrate the material presented in the chapter and to provide a real-life context for learning statistics. The Real-World Application is revisited at the end of the chapter where a real data set is provided for analysis and questions are posed for the student to answer, either individually or in groups, based on the material learned in the chapter.

BUILT-IN STUDY GUIDE

Features in every chapter help students learn and retain new ideas:

Self-test questions appear immediately after important ideas have been introduced to test the student's comprehension of the concept and to help develop good study habits. Answers given at the end of the book allow the students to check their work.

Summary boxes are set off to provide step-by-step instructions for the statistical techniques presented.

Side boxes provide additional explanations of key ideas adjacent to where the concept is first referenced.

Each chapter ends with a list of *Key Terms, Formulas,* and *Symbols* with page references that guide students back to the text where they may review the element in context.

TOPICAL COVERAGE AT THE INTRODUCTORY LEVEL

This text includes all the topics covered in a basic introductory statistics course, including data collection (Chapter 1), descriptive statistics (Chapters 2 and 3), probability and probability distributions (Chapters 4–6), confidence intervals (Chapters 7 and 9), hypothesis tests (Chapters 8 and 9), regression (Chapter 10), analysis of variance (Chapter 11), and nonparametric statistics (Chapter 12). A minimal amount of probability is presented, allowing more time for instructors to teach statistical inference (Chapters 7–12).

Unique to this introductory text is a section on proper graphical presentation (Section 2.6), which promotes E. R. Tufte's principles of graphical excellence. Also, we have included a new, robust test for a population proportion (Section 8.7) that can be applied to both large and small samples.

SUPPLEMENTS FOR THE INSTRUCTOR

Each element in the package has been accuracy-checked to ensure clarity, adherence to the approaches presented in the main text, and freedom from computational, typographical, and statistical errors.

Instructor's Solutions Manual (by Mark Dummeldinger) (ISBN 0-13-020564-8). Complete solutions to all even-numbered problems are provided in this manual. Manual solutions are most frequently provided for the "Using the Tools" problems while a combination of hand and Excel solutions are presented for the "Applying the Concepts" problems. Solutions are also provided for the Real-World Application that begins and ends each chapter. Solutions to the odd-numbered problems are found in the *Student's Solutions Manual.*

Test Bank (by Mark Dummeldinger) (ISBN 0-13-020562-1). The *Test Bank* offers a full complement of more than 1,000 additional problems that correlate to exercises presented in the text. Microsoft Word™ files for this *Test Bank* are available from the publisher.

Windows PH Custom Test (ISBN 0-13-020635-0). Incorporates three levels of test creation: (1) selection of questions from a test bank; (2) addition of new questions with the ability to import text and graphics files from WordPerfect, Microsoft Word, and Wordstar; and (3) algorithmic generation of multiple questions from a single question template. PH Custom Test has a full-featured graphics editor supporting the complex formulas and graphics required by the statistics discipline.

Data Files. Data files for most problems and for the Real-World Applications are contained on the CD-ROM that is packaged with each copy of the text. When a given data set is referenced, a disk icon with the file name will appear in the text near the exercise. The data files may also be downloaded from the World Wide Web.

Web Site: http://www.prenhall.com/sincich Our Web site provides a central clearing house for information about the book for instructors and students. The data files, teaching tips, tips for students, and other useful information may be downloaded from the site. Extensive links to other useful and interesting sites and data sources are built in and updated frequently.

SUPPLEMENTS AVAILABLE FOR PURCHASE BY STUDENTS

Student's Solutions Manual (by Mark Dummeldinger) (ISBN 0-13-020564-8). Fully worked out solutions to all of the odd-numbered problems are provided in this manual. Careful attention has been paid to ensure that all methods of solution and notation are consistent with those used in the core text.

Text and SPSS 8.0 for Windows, Student Version Package (ISBN 0-13-021370-5). **Text and Minitab Rel. 12.0, Student Edition Package** (ISBN 0-13-021375-6). Both of these packages are designed specifically for hands-on classroom teaching and learning of data analysis, statistics, and research methods. Windows 95 and Windows 98 versions of the software allow users to take full advantage of the easy-to-use graphical user interface combined with the traditional power of these packages. Details on both products are available from the publisher.

For additional information about texts and other materials available from Prentice Hall, visit us on-line at **http://www.prenhall.com**

ACKNOWLEDGMENTS

We are extremely grateful to the many organizations and companies that granted us permission to use their data in developing problems and examples throughout the text. A complete list is provided on the accompanying page.

In addition, we would like to thank the Biometrika Trustees, American Cyanamid Company, *Annals of Mathematical Statistics,* and the Chemical Rubber Company for their kind permission to publish various tables in Appendix B.

This book reflects the efforts of a great many people over a number of years. Professor Emeritus William Mendenhall (University of Florida) and publisher Don Dellen (now deceased) were instrumental in developing and shaping earlier editions of *Statistics by Example,* upon which this text is partially based. Special thanks are due to our ancillary author, Mark Dummeldinger, and his typist, Kelly Barber. Priscilla Guthoni has done an excellent job of accuracy checking the manuscript and has helped us to ensure a clean answer appendix and solutions. The Prentice Hall staff of Ann Heath, Mindy McClard, Joanne Wendelken, Melody Marcus, Amy Lysik, Linda Behrens, Alan Fischer, and Maureen Eide and Elm Street Publishing Services's Martha Beyerlein, Barb Lange, Cathy Ferguson, and Sue Langguth helped greatly with all phases of the text development, production, and marketing effort. Pam Johnson did an outstanding job as copy editor, and

Faith Sincich prepared an immaculate manuscript. Finally, we would like to thank our wives and children for their patience, understanding, love, and assistance in making this book a reality. It is to them that we dedicate this book.

CORRESPONDENCE WITH THE AUTHORS

We have gone to great lengths to make this text both pedagogically sound and error free. If you have any suggestions, require clarification, or find potential errors, please contact Terry Sincich at **tsincich@coba.usf.edu** or David Levine at **dmlbb@cunyvm.cuny.edu**.

How to Use This Book

To the student: The following pages will demonstrate how to use this text in the most effective way—to make studying easier and to understand the connection between statistics and your world.

Chapter 2

Exploring Data with Graphs and Tables

CONTENTS

2.1 The Objective of Data Description

2.2 Describing a Single Qualitative Variable: Frequency Tables, Bar Graphs, and Pie Charts

2.3 Describing a Single Quantitative Variable: Frequency Tables, Stem-and-Leaf Displays, and Histograms

2.4 Exploring the Relationship between Two Qualitative Variables: Cross-Classification Tables and Side-by-Side Bar Graphs

2.5 Exploring the Relationship between Two Quantitative Variables: Scatterplots

2.6 Proper Graphical Presentation

EXCEL TUTORIAL

2.E.1 Using Microsoft Excel to Describe a Single Qualitative Variable

2.E.2 Using Microsoft Excel to Describe a Single Quantitative Variable

2.E.3 Using Microsoft Excel to Explore the Relationship between Two Qualitative Variables

2.E.4 Using Microsoft Excel to Explore the Relationship between Two Quantitative Variables

REAL-WORLD APPLICATION
A Bad Moon Rising

Is your behavior influenced by the phases of the moon? Despite the lack of supporting scientific evidence, many people still associate aberrant behavior with a full moon. To measure the degree to which people believe in lunar effects, a team of psychologists administered a questionnaire to a random sample of 157 college undergraduates. How can we use statistics to make sense of the data? Graphical methods that rapidly convey information contained in a data set are the topic of this chapter. In Real-World Revisited at the end of this chapter, we consider graphical methods to summarize the lunar-effects data.

61

Real-World Application

- Each chapter begins with a fascinating, real-world application.
- The application is revisited at the end of the chapter.
- The data for the analysis presented in the application are provided in an Excel Workbook.

Interesting Examples with Complete Solutions

- Examples, with complete solutions and explanations, illustrate every concept.
- Examples help to prepare for the section problem set.
- All examples are numbered for easy reference and the end of the solution is marked with a ❑ symbol.

Self-Test Problems

- At least one self-test problem is presented in each section to reinforce the concepts learned in the examples.
- Work through the solution carefully to prepare for the section problem set.
- Answers to the self-test questions are provided in the appendix.

Example 1.2   Classifying Variables

The U.S. Army Corps of Engineers recently conducted a study of contaminated fish inhabiting the Tennessee River (in Alabama) and its tributaries. A total of 144 fish were captured and the following variables measured for each:

1. Location of capture
2. Species
3. Length (centimeters)
4. Weight (grams)
5. DDT concentration (parts per million)

Classify each of the five variables measured as quantitative or qualitative.

Solution

The variables length, weight, and DDT are quantitative because they are all measured on a natural numerical scale: length in centimeters, weight in grams, and DDT in parts per million. In contrast, location and species cannot be measured quantitatively; they can only be classified (e.g., channel catfish, largemouth bass, and smallmouth buffalo for species). Consequently, data on location and species are qualitative. ❑

Self-Test 1.1

State whether each of the following variables measured on graduating high school students is quantitative or qualitative.

a. National Honor Society member or not
b. Scholastic Assessment Test (SAT) score
c. Number of colleges applied to
d. Part-time job status

Example 1.3   Classifying Variables

Refer to the *Good Housekeeping* (July 1996) study of shaving creams cited in Section 1.1. Recall that each of 30 women shaved one leg using a cream or gel and the other leg using soap or a beauty bar. At the end of the study, 20 of the 30 women stated their preference for shaving with creams and gels. Identify the experimental unit for this study and describe the variable of interest as quantitative or qualitative.

Solution

Since we are interested in the opinions of the 30 women who participated in the leg-shaving study, the ___ ___ental unit ___ woman who shaves her l___ The

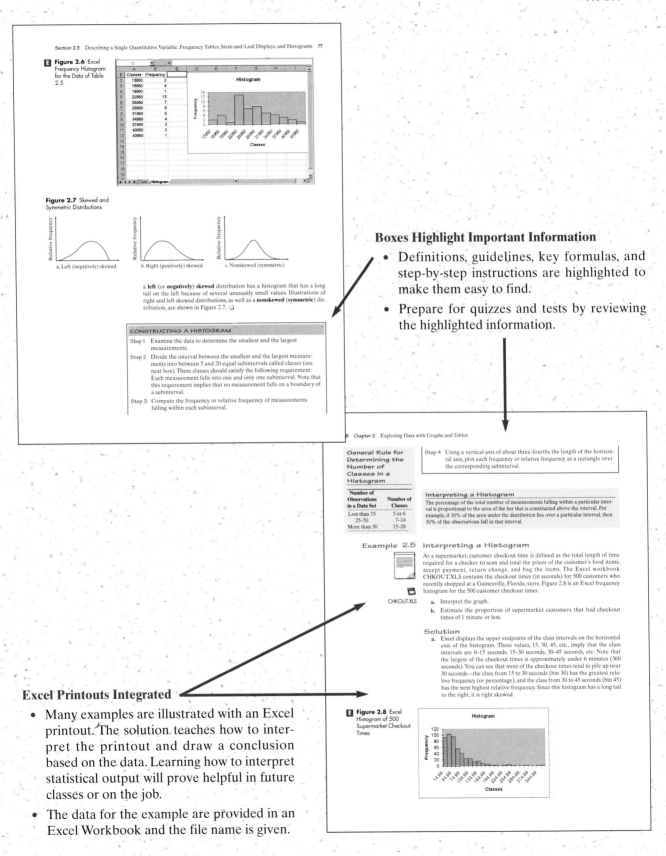

Figure 2.6 Excel Frequency Histogram for the Data of Table 2.5

	Classes	Frequency
1		
2	13960	2
3	16960	4
4	19960	1
5	22960	13
6	25960	7
7	28960	8
8	31960	6
9	34960	4
10	37960	3
11	40960	2
12	43960	1

Figure 2.7 Skewed and Symmetric Distributions

a. Left (negatively) skewed
b. Right (positively) skewed
c. Nonskewed (symmetric)

a **left** (or **negatively**) **skewed** distribution has a histogram that has a long tail on the left because of several unusually small values. Illustrations of right and left skewed distributions, as well as a **nonskewed (symmetric)** distribution, are shown in Figure 2.7. ☐

CONSTRUCTING A HISTOGRAM

Step 1 Examine the data to determine the smallest and the largest measurements.

Step 2 Divide the interval between the smallest and the largest measurements into between 5 and 20 equal subintervals called *classes* (see next box). These classes should satisfy the following requirement: Each measurement falls into one and only one subinterval. Note that this requirement implies that no measurement falls on a boundary of a subinterval.

Step 3 Compute the frequency or relative frequency of measurements falling within each subinterval.

Boxes Highlight Important Information

- Definitions, guidelines, key formulas, and step-by-step instructions are highlighted to make them easy to find.
- Prepare for quizzes and tests by reviewing the highlighted information.

8 Chapter 2 Exploring Data with Graphs and Tables

General Rule for Determining the Number of Classes in a Histogram

Number of Observations in a Data Set	Number of Classes
Less than 25	5 or 6
25–50	7–14
More than 50	15–20

Step 4 Using a vertical axis of about three-fourths the length of the horizontal axis, plot each frequency or relative frequency as a rectangle over the corresponding subinterval.

Interpreting a Histogram

The percentage of the total number of measurements falling within a particular interval is proportional to the area of the bar that is constructed above the interval. For example, if 30% of the area under the distribution lies over a particular interval, then 30% of the observations fall in that interval.

Example 2.5 Interpreting a Histogram

At a supermarket, customer checkout time is defined as the total length of time required for a checker to scan and total the prices of the customer's food items, accept payment, return change, and bag the items. The Excel workbook CHKOUT.XLS contains the checkout times (in seconds) for 500 customers who recently shopped at a Gainesville, Florida, store. Figure 2.8 is an Excel frequency histogram for the 500 customer checkout times.

CHKOUT.XLS

a. Interpret the graph.

b. Estimate the proportion of supermarket customers that had checkout times of 1 minute or less.

Solution

a. Excel displays the upper endpoints of the class intervals on the horizontal axis of the histogram. These values, 15, 30, 45, etc., imply that the class intervals are 0–15 seconds, 15–30 seconds, 30–45 seconds, etc. Note that the largest of the checkout times is approximately under 6 minutes (360 seconds). You can see that most of the checkout times tend to pile up near 30 seconds—the class from 15 to 30 seconds (bin 30) has the greatest relative frequency (or percentage), and the class from 30 to 45 seconds (bin 45) has the next highest relative frequency. Since this histogram has a long tail to the right, it is right skewed.

Figure 2.8 Excel Histogram of 500 Supermarket Checkout Times

Excel Printouts Integrated

- Many examples are illustrated with an Excel printout. The solution teaches how to interpret the printout and draw a conclusion based on the data. Learning how to interpret statistical output will prove helpful in future classes or on the job.
- The data for the example are provided in an Excel Workbook and the file name is given.

Lots of Problems for Practice

- Almost every section in the book is followed by a problem set divided into two parts:

Using the Tools has straightforward applications to test your mastery of definitions, concepts, and basic computation.

Applying the Concepts tests your understanding of concepts and requires you to apply statistical techniques in solving real-world problems. These problems help you develop your critical thinking skills.

(Illustration of textbook page)

Section 1.5   Random Sampling   47

PROBLEMS
1.26–1.33

Using the Tools

1.26 Consider a random sample from a population of size $N = 750$. To obtain the sample, you must assign each experimental unit in the population a code number. What number would you assign to:
  a. the first experimental unit on the list
  b. the twentieth experimental unit on the list
  c. the last experimental unit on the list

1.27 Starting in row 12, column 1 of the random number table, Table B.1, list the first random number for a random sample drawn from a population of size:
  a. $N = 300$ **b.** $N = 1,000$
  c. $N = 500$ **d.** $N = 50$

1.28 Use Table B.1 to draw a random sample of size $n = 10$ from a population of size $N = 70,000$.

1.29 Use Table B.1 to draw a random sample of size $n = 5$ from a population of size $N = 5,000$.

Applying the Concepts

FISH.XLS

1.30 Refer to the U.S. Army Corps of Engineers study of contaminated fish inhabiting the Tennessee River, Example 1.2, page 31. The data collected for all 144 fish captured are available in an Excel workbook. Select a random sample of size 10 from the population of fish weights.

CHKOUT.XLS

1.31 At a supermarket, customer checkout time is measured as the total length of time (in seconds) required for a checker to scan and total the prices of the customer's food items, accept payment, return change, and bag the items. We collected data on customer checkout times for 500 shoppers at an upscale supermarket in Gainesville, Florida. Randomly select a sample of 5 from the 500 checkout times in the data set.

1.32 A clinical psychologist is asked to view tapes in which each of six experimental subjects is discussing his or her recent dreams. Three of the six subjects have been previously classified as "high-anxiety" individuals, and the other three as "low-anxiety." The psychologist is told only that there are three of each type and is asked to select the three high-anxiety subjects.
  a. List the different samples of three subjects that may be selected by the psychologist.
  b. Do you think the sample chosen by the psychologist will be random? Explain.

1.33 A file clerk is assigned the task of selecting a random sample of 26 company accounts (from a total of 5,000) to be audited. The clerk is considering two sampling methods:

Applied Problems Extracted from Current Literature

- Almost all the applied problems contain data or information taken from newspaper articles, magazines, and journals published since 1990. Statistics are all around you.

(Illustration of textbook page)

4   Chapter 2   Exploring Data with Graphs and Tables

2.38 The *American Journal on Mental Retardation* (Jan. 1992) published a study of the social interactions of two groups of children. Independent random samples of 15 children who did and 15 children who did not display developmental delays (i.e., mild mental retardation) were taken in the experiment. After observing the children during "freeplay," the number of children who exhibited disruptive behavior (e.g., ignoring or rejecting other children, taking toys from another child) was recorded for each group. The data are summarized in the following table. Graphically portray the data with a side-by-side bar chart. Interpret the graph.

	Disruptive Behavior	Nondisruptive Behavior	TOTALS
With Developmental Delays	12	3	15
Without Developmental Delays	5	10	15
TOTALS	17	13	30

Source: Kopp, C. B., Baker, B., and Brown K. W. "Social skills and their correlates: Preschoolers with developmental delays." *American Journal on Mental Retardation,* Vol. 96, No. 4, Jan. 1992.

2.39 Since 1948, research psychologists have used the "water-level task" to test basic perceptual and conceptual skills. Subjects are shown a drawing of a glass tilted at a 45° angle and asked to assume the glass is filled with water. The task is to draw a line representing the surface of the water. *Psychological Science* (Mar. 1995) reported on the results of the water-level task given to 120 subjects. Each subject was classified by group and by performance on the test. A summary of the results is provided in the table.

| | GROUP | | | | | | |
| | FEMALES | | | MALES | | | |
Judged Line	Students	Waitresses	Housewives	Students	Bartenders	Bus Drivers	TOTALS
More than 5° below surface	0	0	1	1	1	1	4
More than 5° above surface	7	15	13	3	11	4	53
Within 5° of surface	13	5	6	16	8	15	63
TOTALS	20	20	20	20	20	20	120

Source: Hecht, H., and Proffitt, D. R. "The price of experience: Effects of experience on the water-level task." *Psychological Science,* Vol. 6, No. 2, Mar. 1995, p. 93 (Table 1).

  a. Use a graphical method to describe the overall results of the study (i.e., the performance responses for all 120 subjects).
  b. Construct graphs that the researchers could use to explore for group differences on the water-level task.
  c. Psychologists theorize that males do better than females, that younger adults do better than older adults, and that those experienced in handling liquid-filled containers do better than those who are not. Are these theories supported by the data?

2.40 Age-related macular degeneration (AMD), a progressive disease that deprives a person of central vision, is the leading cause of blindness in older adults. In a study published in the *Journal of the American Geriatrics Society* (Jan. 1998), researchers investigated the link between AMD incidence and alcohol consumption. Each individual in a national sample of 3,072 adults was asked about the most frequent alcohol type

Built-In Excel Manual

Microsoft Excel Primer

Microsoft Excel Primer

Overview

Spreadsheet programs such as Microsoft Excel allow users to create solutions that permit the interactive manipulation of data. First used in business, today the flexibility of modern spreadsheets makes them an everyday problem-solving tool for many, including students learning statistical problem solving in a first statistics course.

Of the many spreadsheet programs available, the authors have chosen to use Microsoft Excel in this text and not just for the obvious reasons that the program is widely available and incorporates the commonly used Microsoft Office user interface. Microsoft Excel also contains special statistical functions and procedures that aid in the analysis of data and can accept add-ins, preprogrammed procedures that extend the functionality of Excel. These features help construct statistical solutions in Excel and simplify its use. (Many Excel sections of this text use the Prentice Hall PHStat add-in (supplied on the CD-ROM that accompanies this text) to further streamline the use of Excel, as explained in the last section of this Primer.)

Microsoft Excel also allows users to create **workbooks**, collections of electronic **worksheets** and other information, including charts, that are combined into a single disk file. Workbooks facilitate the development of solutions that are consistent with the rules of good

- Designed for the newcomer to Excel, this *Primer* will acquaint you with the Excel user interface and basic functionality.

- More experienced users may skip portions of the *Primer*.

Using Microsoft Excel

2.E.1 Using Microsoft Excel to Describe a Single Qualitative Variable

Solution Summary:

Use the PHStat add-in to summarize and graph the data of a single qualitative variable.

Example: Intensity of men's cologne data of Table 2.1

Solution:

To generate a frequency table, bar chart, and pie chart from the intensity ratings of 17 men's cologne brands (Table 2.1 on page 63), do the following:

1. If the PHStat add-in has not been previously loaded, load the add-in using the instructions of Section EP.3.2. (You will also need to run the disk setup program using the procedure in Appendix D if you have never previously installed the files on the CD-ROM that accompanies this text.)
2. Open the cologne workbook (COLOGNE.XLS) and click the Data sheet tab.
3. Select PHStat | One-Way Tables & Charts.
4. In the One-Way Tables & Charts dialog box, enter the information and make the selections shown in Figure 2.E.1. Click the OK button.

Figure 2.E.1

The add-in produces on separate sheets a frequency table, bar chart, and pie chart, similar to the ones shown in Figures 2.E.2, 2.E.3, and 2.E.4.

Figure 2.E.2

	A	B
1	Intensity of Cologne	
2		
3	Count of Intensity	
4	Intensity	Total
5	Mild	2
6	Strong	10
7	Very strong	5
8	Grand Total	17

125

Excel Tutorials

- Each pertinent chapter ends with a tutorial that provides step-by-step guidelines and screen shots for using Excel to perform the statistical techniques presented.

PHStat CD-ROM

- PHStat, a statistics add-in for Excel, provides a custom menu of choices that lead to dialog boxes to help you perform statistical analyses more quickly and easily than "off-the-shelf" Excel permits.

- A complete list of topics in PHStat's custom menu are on page 25.

- All of the Excel workbooks used in examples and problems are on the CD.

Microsoft Excel Primer

Overview

Spreadsheet programs such as Microsoft Excel allow users to create solutions that permit the interactive manipulation of data. First used in business, today the flexibility of modern spreadsheets makes them an everyday problem-solving tool for many, including students learning statistical problem solving in a first statistics course.

   Of the many spreadsheet programs available, the authors have chosen to use Microsoft Excel in this text and not just for the obvious reasons that the program is widely available and incorporates the commonly used Microsoft Office user interface. Microsoft Excel also contains special statistical functions and procedures that aid in the analysis of data and can accept add-ins, preprogrammed procedures that extend the functionality of Excel. These features help construct statistical solutions in Excel and simplify its use. (Many Excel sections of this text use the Prentice Hall PHStat add-in (supplied on the CD-ROM that accompanies this text) to further streamline the use of Excel, as explained in the last section of this Primer.)

   Microsoft Excel also allows users to create **workbooks**, collections of electronic **worksheets** and other information, including charts, that are combined into a single disk file. Workbooks facilitate the development of solutions that are consistent with the rules of good application design. In this text, solutions generated follow a predictable pattern of placing a problem's data, calculations, and graphical outputs on separate sheets. These separate-sheet designs, in turn, enhance the reusability of the workbooks, provide easier opportunities for modifying the workbooks, and generally aid in the clarity of the presentation of the results.

   Although useful as a tool for learning statistics, readers should be aware that using Microsoft Excel is **not** an all-purpose substitute for using a standard statistical package. Very large data sets or data sets with unusual statistical properties can cause Excel, as well as any add-in code being used, to produce invalid results.

Getting Started with the User Interface

You operate Microsoft Excel through its user interface, typically by using a keyboard to enter values and using a pointing device, such as a mouse, to make selections and choices. This part of the Primer teaches the specifics of the Excel user interface. Learning the user interface first, before problem solving, helps you avoid getting bogged down with the operational details of the program later when you seek to focus on the statistical methods being used in a solution.

   Because Microsoft Excel uses the same user interface as other MS Office programs, you may wish to skip or skim this part if you have used a program such as Microsoft Word previously. Otherwise, use this section as a primer and reference guide to the user interface.

EP.1.1 Getting Familiar with Mouse Operations

Although almost all of the user interaction with the Microsoft Excel user interface can be done using some combination of keystrokes, many pointing or choosing tasks are more easily done with a pointer device such as a mouse, trackball, or touchpad. Moving a pointer device moves the **mouse pointer**, an onscreen graphic that in its most common form takes the shape of an arrow or an outlined plus sign (see Figure EP.1.1, a and b). Moving the mouse pointer over another object and pressing one of the buttons on the pointer device defines a mouse operation. Four pointer operations used throughout this text are defined as follows:

1. To **select** an on-screen object, move the mouse pointer directly over an object and press the left mouse button (or press the single button, if your mouse contains only

one button). If the onscreen object being selected is a button, **click** is often used as an alternative to the verb "select," as in the phrase "Click the OK button."

2. To **drag**, or move, an object, first move the mouse pointer over an object and then, while holding down the left (or single) mouse button, move the mouse. After the object has been dragged, release the mouse button.

3. To **double-click** an object, move the mouse pointer directly over an object and press the left mouse button twice in rapid succession.

4. To **right-click** an object, move the mouse pointer directly over an object and press the right mouse button.

The Mousing Practice workbook (Mousing Practice.XLS), supplied on the disk that accompanies this text, allows you to practice mouse operations.

E Figure EP.1.1a
Arrow Mouse Pointer

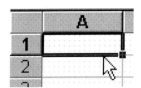

E Figure EP.1.1b
Outlined Plus Sign
Mouse Pointer

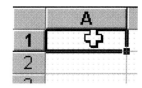

These mouse operations can be illustrated using the Windows desktop for the fictional Tadashi College. As configured, this desktop contains **icons**, graphics that represent a specific program application or document, a Microsoft Office **toolbar**, a set of icons in the form of onscreen **buttons** that simulate push buttons, and the Windows taskbar containing the system tray and the Start button (see Figure EP.1.2a).

Figure EP.1.2b shows the result of selecting the Microsoft Excel icon. Notice that when selected, an object is highlighted. Figure EP.1.2c shows the results of dragging the Microsoft Excel icon to a new desktop location. (Dragging an object around the screen can be useful to reduce clutter or to reveal objects that otherwise would be obscured.) Figure EP.1.2d shows the **shortcut menu** that appears as a result of right-clicking the icon. Contents of shortcut menus vary according to both the object right-clicked and the context in which the mouse operation was made.

E Figure EP.1.2a A
Typical Window
Desktop

2

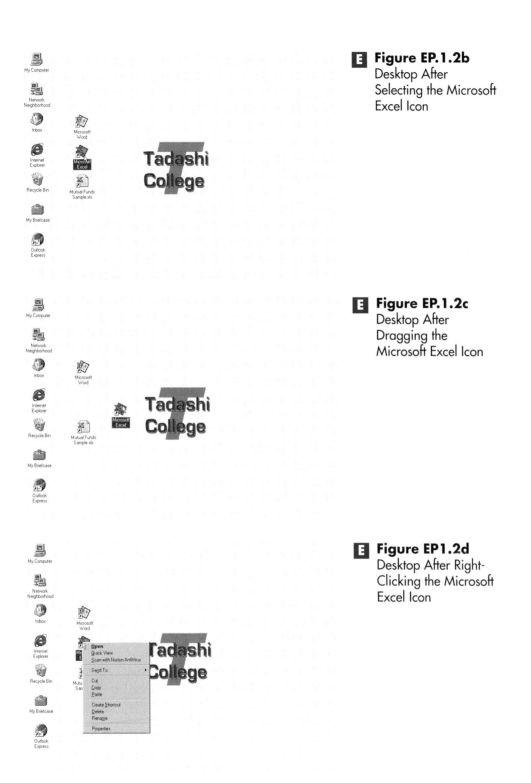

E **Figure EP.1.2b**
Desktop After
Selecting the Microsoft
Excel Icon

E **Figure EP.1.2c**
Desktop After
Dragging the
Microsoft Excel Icon

E **Figure EP1.2d**
Desktop After Right-
Clicking the Microsoft
Excel Icon

Double-clicking an icon loads and runs the program associated with the object. In the case of the Excel program and workbook files at the lower left, double-clicking would cause Windows to load Microsoft Excel and open the Excel application window (see Figure EP.1.3) in front of the Windows desktop.

EP.1.2 *Getting Familiar with the Microsoft Excel Application Window*

When Microsoft Excel program or workbook icons are double-clicked, Windows loads Excel and an Excel application window, similar to the one shown in Figure EP.1.3, appears.

E **Figure EP.1.3**
Microsoft Excel
Application Window

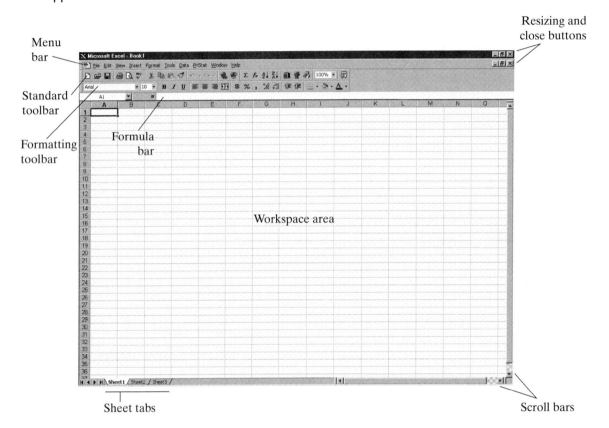

You can configure the placement of objects that appear in this window. In Figure EP.1.3, as in all illustrations of the Excel application window in this text, the standard and formatting toolbars and the formula bar have been configured to appear below the **menu bar**, the horizontal list of words that represent sets of command choices. The workbook displayed in the workspace area has been maximized, opened to cover the entire workspace area. Sets of **resizing** and **closing buttons** for both the Excel application window and the maximized workbook appear on the title bar and menu bar, respectively. **Scroll bars**, both horizontal and vertical, allow for the display of parts of the worksheet currently off screen (such as row 40 or column M). **Sheet tabs**, identifying the names of individual sheets, provide a means of "turning" to another sheet in the workbook. A **status bar** displays information about the current operation and the state of certain keyboard toggles. (Instructions to configure any Excel application window to look like the one illustrated in this text are given in Section EP.2.1.)

4

Because the workspace area is distinct from a workbook, it is possible to open two—or more—workbooks in one workspace when using Microsoft Excel. Should this occur, selecting Window from the menu bar will pull down a menu that permits switching between the opened workbooks.

EP.1.3 Getting Familiar with Menu Conventions and Dialog Boxes

When using Microsoft Excel, you will frequently want to open, or retrieve, files previously stored, save files containing new or updated information, and print sheets from an opened workbook. The File menu (see Figure EP.1.4), which appears when the word File is selected from the Excel menu bar, contains choices for all these tasks, as well as the Exit choice that closes the Excel application window and returns the user to the Windows desktop.

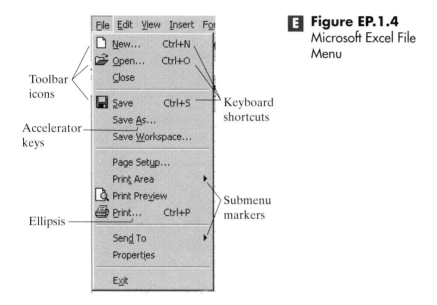

E Figure EP.1.4
Microsoft Excel File Menu

The File menu contains the conventions used in all menus. These conventions are:

- Using a triangular marker to denote that the menu choice leads to a second (sub)menu of choices.
- Displaying, at left, toolbar buttons that are equivalent to individual menu choices, if any.
- Listing, at right, keyboard shortcuts for individual menu choices, if any.
- Underlining "accelerator" letter keys that can be pressed to select individual choices.
- Using an ellipsis to mark choices that trigger the display of a **dialog box**, a special window that appears over the Excel application window, designed to accept user-supplied information or to report a status message.

Dialog boxes themselves contain any number of graphical objects, each with their own defined behaviors. Objects shown labeled in Figure EP.1.5, a and b, are explained on page 6.

5

E **Figure EP.1.5a**
Sample (File Open)
Dialog Box

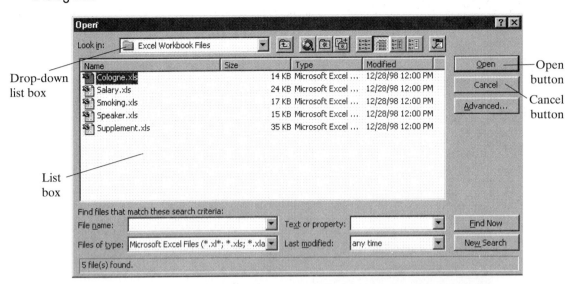

Drop-down
list box

List
box

Open
button

Cancel
button

E **Figure EP.1.5b**
Sample (File Print)
Dialog Box

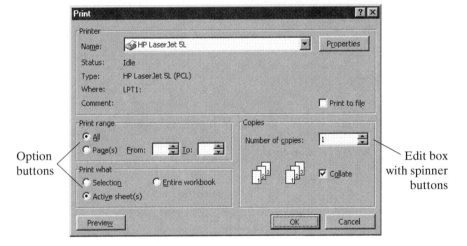

Option
buttons

Edit box
with spinner
buttons

Drop-down list boxes allow selection of an item from the list of choices that appears when the drop-down button, located on the right edge of the box, is clicked.

List boxes display a list of choices, in this case files or folders, for selection. When the list is too long to be seen in its entirety in the box, clicking the scroll buttons of these boxes reveals the rest of the list.

Edit boxes provide an area into which a value can be typed or edited. Edit boxes are often combined with either a drop-down list box or **spinner buttons** to provide an alternative to the typing of a value. (Pressing spinner buttons increases or decreases the numeric value that appears in the edit box.)

Option buttons present a set of mutually exclusive choices. Selecting an option button (also known as a radio button) always deselects, or clears, the other option buttons in the set, thereby allowing only one choice at any time.

Check boxes present a set of optional choices. Unlike option buttons, more than one check box in a set or none at all can be selected at any given time.

Clicking the **Open** or **OK buttons** causes Microsoft Excel to execute the operation with the current values and choices as shown in the dialog box.

Clicking the **Cancel button** closes a dialog box and cancels the operation (similar to pressing the Escape key).

From this point on, the authors will abbreviate menu selections by using the vertical slash character | to separate menu choices, as in File | Open, instead of the verbose "select the File menu and then select the Open choice." When directed to make a menu selection in this text, you can, of course, substitute a keyboard shortcut or accelerator key, as appropriate.

Getting Started with Microsoft Excel

This part of the Microsoft Excel Primer discusses the operational knowledge needed to use Excel effectively, beginning with the information needed in order to best set up Microsoft Excel for use with this text.

EP.2.1 Configuring Microsoft Excel

Microsoft Excel allows you to custom configure the display of the Excel application window. As a reader of this text, you may want to configure your application window to match as closely as is possible Figure EP.1.3, which illustrates how the application window appears throughout this text. To configure Microsoft Excel to match the authors' settings, load and run Excel and then do the following:

To display the formula and status bars and the standard and formatting toolbars:

1   Select View and if the Formula Bar choice is not checked, select it.

2   Select View and if the Status Bar choice is not checked, select it.

3   Select View | Toolbars and if the Standard choice is not checked, select it.

4   Select View | Toolbars and if the Formatting choice is not checked, select it.

Note: If a bar is not aligned to the top of the Excel application window, drag the bar towards the top of the screen until the dragged border of the bar changes to a slender band of small dots. The bar will then snap into place when the mouse button is released.

To standardize the display of the worksheet area:

1   Select Tools | Options.

2   In the Options dialog box:

     a   Select the View tab if another tab's options are visible (see Figure EP.2.1).

     b   Select the Gridlines, Zero values, Row & column headers, Horizontal scroll bar, Vertical scroll bar, Page breaks, Outline symbols, and Sheet tabs check boxes.

     c   Deselect (uncheck) the Formulas check box, if it has been selected.

     d   Click the OK button.

E **Figure EP.2.1** View Tab of the Tools Options Dialog Box

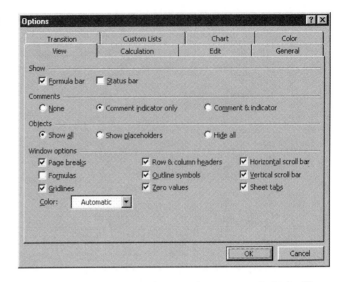

The Microsoft Excel examples in this text also assume various settings, some of which affect the look of illustrations such as Figure EP.1.3. To configure Microsoft Excel to match these settings, use the following procedures:

To verify calculation, edit, and general display settings:

[1] Select Tools | Options.

[2] In the Options dialog box:

  [a] Select the Calculation tab and verify that the Automatic option button of the Calculation group has been selected.

  [b] Select the Edit tab and verify that all check boxes except the Fixed decimal and Provide feedback with animation check boxes have been selected.

  [c] Select the General tab. Verify that the R1C1 reference style check box is deselected (unchecked) and that the Macro virus protection check box is selected. Enter 3 in the Sheets in new workbook: edit box. Select Arial (or similar font) from the Standard font: list box. Select 10 from the Size: drop-down list box. Change the values in the Default file location: and User name: edit boxes, if desired (see Figure EP.2.2).

  [d] Click the OK button.

E **Figure EP.2.2** General Tab of the Tools Options Dialog Box

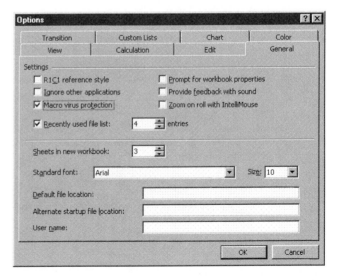

To include PHStat on the Excel menu bar as shown in Figure EP.1.3, you will need to load the PHStat add-in. (See Section EP.4.2 for further information on installing PHStat and other necessary add-ins.)

EP.2.2 Specifying Worksheet Locations

Worksheet locations into which entries are made are known as **cells**. Cells are formed by the intersections of the rows and columns on a worksheet. Individual entries can be referenced by specifying the (lettered) column and (numbered) row that forms the cell containing the entry. For example, we can refer to the contents of the cell in the first column and first row (the upper left corner cell) as cell A1, and talk about the cell in the second column and fourth row as cell B4.

Because a Microsoft Excel workbook can contain multiple worksheets, this column letter and row number format is, in certain contexts, insufficient to specify a particular cell. For such occasions, including certain dialog boxes encountered later in this text, the cell reference must be written in the form *Sheetname!ColumnRow*. Using this notation, one can distinguish between two similarly located cells of two different sheets in the workbook, for example, Data!A1 and Calculations!A1, the upper left corner cell of the Data and Calculations sheets, respectively. This notation is necessary when a cell reference is to a worksheet other than the one into which current entries are being made and is also required when entering cell references in certain Microsoft Excel dialog boxes.

Cell references can also be made to rectangular groups of adjacent cells in a worksheet. Such **cell ranges** are written using the cell references to the upper leftmost and lower rightmost cell in the block, using the form *UpperLeft:LowerRight*. For example, the cell range A1:B3 refers to the six-cell worksheet block containing the cells A1, B1, A2, B2, A3, and B3, and the range A1:A8 refers to the first eight cells in the first column of the worksheet. Ranges in the form *Sheetname!UpperLeft:LowerRight,* such as Data!A1:A8, are allowed and refer to cell ranges on other sheets.

EP.2.3 Using Formulas to Perform Calculations

Perhaps the most critical step in developing workbook-based solutions is translating calculations required into **formulas**, instructions to Microsoft Excel to perform arithmetic or other calculational tasks. Creating formulas requires knowledge of **arithmetic operators**, symbols used to express arithmetic operations in Excel, as well as a familiarity with Excel **functions**, preprogrammed calculations that perform common arithmetic, business, engineering, and statistical operations that might otherwise be hard or tedious to express using only the operators. (Microsoft Excel arithmetic operators are shown in Table EP.2.1.)

Table EP.2.1 Microsoft Excel Arithmetic Operators

Arithmetic Operator	Excel Operator
Addition	+
Subtraction	−
Multiplication	*
Division	/
Exponentiation (a number raised to a power)	^

Formulas always begin with the = (equals sign) symbol and contain operators combined with cell references in an algebraic-like notation. For example, the formula =A1−A2 calculates the difference between the values in cells A1 and A2 and the formula =B1+B2+B3+B4+B5+B6 calculates the sum of the values found in the first six cells of the second column of the current worksheet. (References to cells on other worksheets, as in the formula

=Data!A1+Data!A2, are also allowed.) Functions are used to simplify formulas. For example, the Sum function could be used in the formula =SUM(B1:B6) as an alternative to calculating the sum of the values found in the cell range B1:B6.

EP.2.4 Entering Data into Worksheets

To enter data into a worksheet, load Microsoft Excel and use the new, empty workbook that appears or, if no workbook appears, select File | New. Continue by giving each new worksheet a self-descriptive name and then by making the cell entries into that worksheet as called for by the design. Entries are made on a cell-by-cell basis and reflect the row and column contents presented in the designs. To accomplish these tasks, do the following:

To give a new or unused worksheet a self-descriptive name:

1. Insert a new worksheet by selecting Insert | Worksheet **or** select an unused (empty) worksheet in the workbook by clicking its sheet tab.
2. Double-click the sheet tab (see Figure EP.2.3) of the sheet to be renamed.
3. Type the new sheet name and press the Enter key. To simplify the entry of sheet names and formulas, avoid using spaces or special characters in the new name. (Two-word names, such as "Frequency Table", can be entered as "FrequencyTable".)

E Figure EP.2.3
Close-Up of Sheet Tabs

\ Sheet1 / Sheet2 \ **Sheet3** /

To make an entry into a cell:

1. Select the cell into which the entry is to be made. (When a cell is selected, three things occur: the column letter and row number of the selected cell are highlighted along the top and left borders of the worksheet, the selected cell's border is framed by a highlight, and the cell reference of the selected cell appears in the formula bar.)
2. Type the entry. The entry appears in the edit box of the formula bar as it is being typed and can be revised there (see Figure EP.2.4).
3. Complete the entry by pressing either the Enter key, which moves the cell highlight down one row, or the Tab key, which moves the cell highlight one column to the right. (Clicking the "check" button on the formula bar is equivalent to pressing the Enter key.)

E Figure EP.2.4
Microsoft Excel
Formula Bar

A1   ▼ **X** ✓ =   Cross-classification of Speaker Data

A very long entry may spill over into adjacent cells to the right. This does **not** mean, however, that those adjacent cells contain part of the entry. For example in Figure EP.2.5, the long worksheet title entered into cell A1 has spilled over to cells B1 and C1. Because of the spillover effect, casual visual inspection of the worksheet might lead one to believe that these two cells contained some part of the title. To demonstrate that these cells do not contain any part of the title, select them individually and examine the (blank) contents of the formula bar. This procedure of selecting cells and examining the contents of the formula bar is the *only* foolproof method for determining the contents of a particular cell.

E Figure EP.2.5 Long
Entry Spillover

	A	B	C
1	**Cross-classification of Speaker Data**		
2			

Correcting Errors

Errors made while making entries to implement a worksheet design can be corrected by doing one of the following:

- To start an entry over, press the Escape key (or click the "X" button on the formula bar) while typing the entry.

- To erase characters to the left of the cursor one character at a time, press the Backspace key.

- To erase characters to the right of the edit cursor one character at a time, press the Delete key.

- To correct an error in an entry, select the error by clicking at the start of the error and dragging the mouse pointer over the rest of the error. Then type the replacement text.

Corrections can be undone by selecting Edit | Undo. There is also an Edit | Redo command that allows the correction to be restored after it was undone.

Reviewing Entries

After all cell entries for a worksheet have been made, each entry should be reviewed for errors. One technique that makes this task easier is to switch the worksheet to formula view. Formula view causes Microsoft Excel to display the contents of each cell as they would appear in the formula bar edit box if selected. For formula entries, this means that the actual formulas, and not their results, are displayed on the screen. To switch to this view, do the following:

1   Select Tools | Options.

2   In the Options dialog box that appears, select the Formulas check box of the View tab. Click the OK button.

Some formulas, especially long formulas entered into narrow-width cells, may appear truncated after switching to formula view. To see the entire formula on the screen, adjust the width of the cell using the appropriate procedure of Section EP.2.9. With the workbook switched to formula view and with all column widths adjusted to display all formulas, the File | Print command, explained in Section EP.2.7, can be used to print out the worksheet for later review. To return the worksheet to normal view, select Tools | Options and deselect (leave unchecked) the formulas check box.

EP.2.5   Copying Cell Entries and Worksheets

The contents of a worksheet, whether they be an individual cell, cell range, or the entire worksheet itself, can be copied to simplify or speed the implementation of a worksheet design. Generally, copying involves first making a selection and then using the copy and paste commands. The specifics of copying cell entries and worksheets are given in the following:

To copy a single cell entry, or a range of entries:

1   Select the cell containing the entry to be copied (or select the cell range containing the entries to be copied by dragging the mouse pointer through all cells of the range).

2   Select Edit | Copy.

3   Select the cell to receive the copy (or select the first cell of the range to receive the copies).

4   Select Edit | Paste.

Copying formulas represents a special case of copying cell entries, as exact duplicates may or may not result depending on how cell references have been entered. Cell references entered in the form *ColumnletterRownumber,* such as in A1 or any of the cell references used in Section EP.2.2, the cell references are considered **relative cell references** and will be changed to reflect their new relative positions. For example, a cell C2 formula =A2+B2 when copied to cell C3 will be changed to =A3+B3 to reflect its new relative position one row down from the original. Likewise, a cell A5 formula =SUM(A1:A4) when copied to cell B5 will appear as =SUM(B1:B4).

When this automatic adjustment is not desired for a cell reference, you can enter an **absolute cell reference** in the form *$Columnletter$Rownumber.* For example, the formula =$A$2+$B$2 will always be unchanged regardless of the cell to which it is copied. Note the dollar sign symbol does not imply currency values; it is used solely to prevent the changing of a cell reference during a copy operation.

Formulas that mix relative and absolute cell references are allowed. For example, a cell C2 formula =A2/$B$10 when copied to cell C3 will be changed to =A3/$B$10. Such mixing of cell reference types often simplifies the implementation of worksheet formulas.

To copy an entire worksheet:

1   Select the worksheet to be copied by clicking on its sheet tab.

2   Select Edit | Move or Copy Sheet.

3   In the Move or Copy dialog box (see Figure EP.2.6):

   a   Select the Create a copy check box.

   b   Select (new book) from the To book: drop-down list if the copy of the worksheet is to be placed in a new workbook. If the copy of the worksheet is to be placed in an opened workbook, select the name of the workbook.

   c   If copying to an opened workbook, select the worksheet position for the copy by selecting a sheet from the Before sheet: list box.

   d   Click the OK button.

4   Rename the new worksheet, if necessary.

E **Figure EP.2.6** Move or Copy Dialog Box

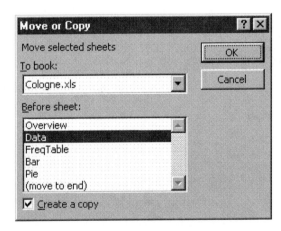

EP.2.6   Opening Workbooks

If a workbook has been previously implemented and saved, you can retrieve it using the following procedure:

1   Select File | Open.

2. In the Open dialog box (see Figure EP.2.7):

  a. Select the appropriate folder (also know as a directory) from the Look in: drop-down list box.

  b. If necessary, change the display of files in the scrollable files list box by selecting the appropriate format button. (Figure EP.2.7 shows the Preview button selected. Compare to Figure EP.1.5a in which the List files button has been selected.)

  c. Select the proper Files of type: value from the drop-down list. (In most situations, the default Microsoft Excel Files choice will be correct. For importing a text file as is explained in Section EP.3.1, this value must be changed to Text Files or All Files.)

  d. Select the file to be opened from the files list box. If the file does not appear, verify that steps 2a and 2c were done correctly.

  e. Click the OK button.

You should always remember to verify the contents of the file after it is retrieved.

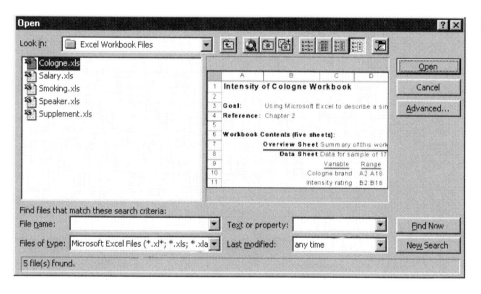

E ▶ **Figure EP.2.7**
Open Dialog Box
(Preview Button
Selected)

EP.2.7 Printing Worksheets

Printing worksheets allows you to review designs and results away from the computer screen. Although Microsoft Excel can print all of the sheets in a workbook in one operation, we suggest that you print each worksheet separately. To print a worksheet, do the following:

1. Select the worksheet to be printed, then select File | Print Preview to preview the output (see Figure EP.2.8). If errors are spotted, click the Close button, correct the errors, and reselect File | Print Preview. If the results displayed are not what is desired, click the Setup button and make the adjustments as necessary in the Page Setup dialog box (see Customizing Printouts at the end of this section).

2. Click the Print button from the print preview screen, or select File | Print if the preview has been closed.

Figure EP.2.8 Print Preview Screen

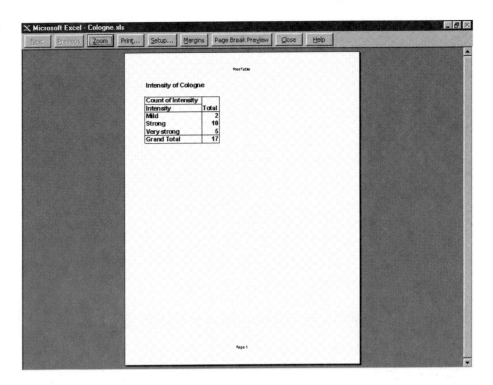

3 In the Print dialog box (see Figure EP.2.9):

a Select the printer to be used from the Name: drop-down list box.

b Select the All option button under the Print range heading.

c Select the Active sheet(s) option button under the Print what heading. (The authors recommend not selecting the Entire workbook option button.)

d Set the number of copies to the proper value.

e Click the OK button.

Figure EP.2.9 Print Dialog Box

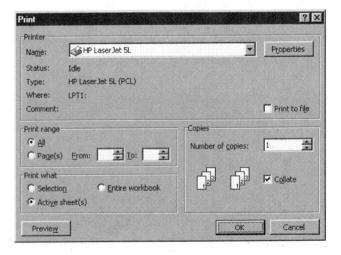

After printing ends, verify the contents of the printout. Most failures to print will result in the display of an informational dialog box. Click the OK button of that dialog box and correct any problems, if possible, before attempting to print the worksheet a second time.

Customizing Printouts

Printouts can be customized by selecting File | Page Setup (or clicking the Setup button in the Print Preview screen) to display the Page Setup dialog box. Settings found under each of the four tabs of this dialog box that are used to produce a proper-looking output are listed below. (Click a tab to display a tab's settings.)

- **Page tab settings**. Orientation option buttons control whether sheets are printed vertically (Portrait) or horizontally (Landscape). Use the landscape orientation for worksheets that have a greater width than length. The scaling choices allow you to either reduce or enlarge the printed size of a worksheet (Adjust option) or to specify the number of pages to fit on one printed page (Fit option). (See Figure EP.2.10a.)

- **Margins tab settings**. Edit boxes on this tab control the size of all margins on the printed page. Check boxes for centering the output are also found on this tab. (See Figure EP.2.10b.)

- **Header/Footer tab settings**. Choices here allow you to choose from many different styles of headers and footers that can be added to your printouts. You can also construct your own custom headers and footers. (See Figure EP.2.10c.)

- **Sheet tab settings**. Most useful are the Print check boxes that allow you to specify whether gridlines and row and column headings are included in the printouts of your worksheets. (See Figure EP.2.10d.)

After all settings have been made, click the OK button to close the dialog box and resume your work.

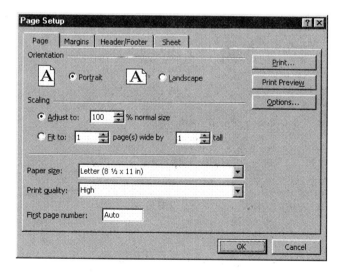

E **Figure EP.2.10a**
Page Tab of the Page
Setup Dialog Box

Figure EP.2.10b
Margins Tab of the Page Setup Dialog Box

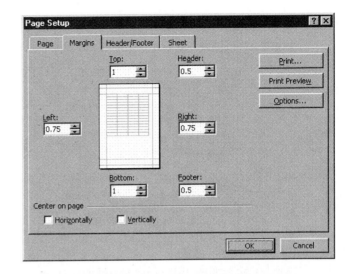

Figure EP.2.10c
Header/Footer Tab of the Page Setup Dialog Box

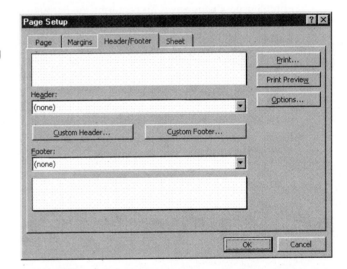

Figure EP.2.10d
Sheet Tab of the Page Setup Dialog Box

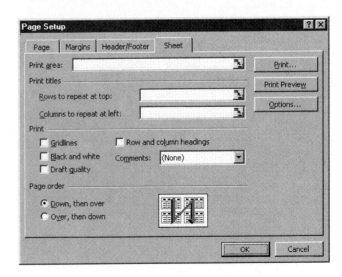

EP.2.8 Saving Workbooks

As you add worksheets and entries to a workbook, you should regularly save the workbook to disk. Saving a workbook assures its future availability and is protection against a sudden computer system failure such as those caused by component failures or loss of electric power. To save a workbook, do the following:

1. Select File | Save As. (Although many times File | Save can also be chosen, the authors recommend always using Save As to save files.)

2. In the Save As dialog box (see Figure EP.2.11):

   a. Select the folder (also know as a directory) to hold the file from the Save in: drop-down list box.

   b. Select the value from the Save as type: drop-down list. Typically, this value will and should be Microsoft Excel Workbook. However, Formatted Text (space delimited), Text (tab delimited), and Microsoft Excel 5.0/95 can be chosen when saving data that will be used by earlier versions of Excel or other programs.

   c. Enter the value for the name of the file in the File name: edit box

   d. Click the OK button.

   Saving the same workbook a second time, using a different name, is an easy way to create a duplicate of a workbook that can serve as a backup copy should some error render the original unusable.

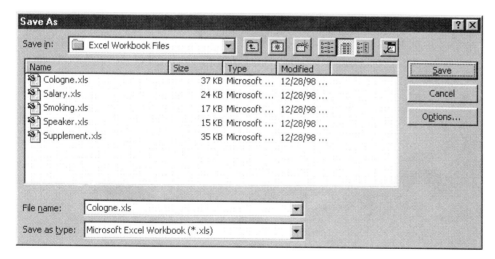

E Figure EP.2.11
Save As Dialog Box

   Note: Workbook files that Microsoft Excel identifies as "read-only" files cannot be saved under their original names, but can be saved using a different name.

EP.2.9 Enhancing the Appearance of a Worksheet

Any number of formatting selections can be used to enhance the appearance of a worksheet. Common formatting selections from the formatting toolbar (see Figure EP.2.12) that were used to enhance the worksheets shown in illustrations throughout this text are described briefly in the following procedures:

To display cell values in boldface type:

1. Select the cell (or cell range) containing the values to be boldfaced.
2. Click the Boldface button on the formatting toolbar.

To display cell values centered in a column:

1. Select the cell (or cell range) containing the values to be centered.
2. Click the Center button on the formatting toolbar. (The similar Align left and Align right buttons left or right justify values.)

To align the decimal point in a series of numeric entries:

1. Select the cell range containing the numeric entries to be aligned.
2. Click either the Increase Decimal or Decrease Decimal button on the formatting toolbar until the desired alignment is produced.

To display numeric values as percentages:

1. Select the cell range containing the numeric entries to be displayed as percentages.
2. Click the Percent button on the formatting toolbar.

To center a cell entry representing a title over a range of columns:

1. Select the cell row range that contains the entry to be used as the title and all the columns that the entry is to be centered over.
2. Click the Merge and Center button on the formatting toolbar.

E Figure EP.2.12
Microsoft Excel
Formatting Toolbar

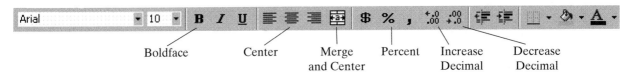

Boldface   Center   Merge   Percent   Increase   Decrease
                        and Center   Decimal   Decimal

To adjust the width of a column to fully display all displayed cell values in that column:

1. Select the column to be formatted by clicking the column heading (gray letter) of that column (see Figure EP.2.13).
2. Select Format | Column | AutoFit Selection.

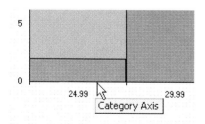
Additional formatting operations are available through the Format | Cells and Format | Auto-format dialog boxes. See the Microsoft Excel Help for further information on these topics.

EP.2.10 Getting Context-Sensitive Help in Microsoft Excel

Microsoft Excel offers the context-sensitive help methods of tool tips, question-mark buttons, and Office Assistant displays in addition to the standard command Help | Contents and Index.

Tool tips are brief identification messages displayed when the mouse pointer pauses and lingers over an on-screen object. Tool tips identify the commands associated with individual toolbar buttons, helping users to better learn these shortcuts. Tool tips also identify the name of an object when the mouse pointer lingers on an object. For example, Figure EP.2.14 shows how a tool tip confirms the location of the mouse pointer on the X-axis of a chart.

E **Figure EP.2.14**
Tool Tip Display for an
X-Axis of a Chart

When they appear in dialog boxes, **question-mark buttons** can be used to display an informational message about an object or part of a dialog box. To use this feature, do the following:

19

1. Click the question-mark button on the dialog box border. The mouse pointer changes from the standard mouse pointer to the special question-mark pointer (see Figure EP.2.15a).

2. Click over the dialog box object or part of interest. A help message for the object or part appears (see Figure EP.2.15b).

3. Click anywhere in the dialog box when finished. The help message disappears and the mouse pointer returns to its standard shape.

E Figure EP.2.15a
Mouse Pointer After the Question-Mark Button of a Dialog Box Has Been Selected

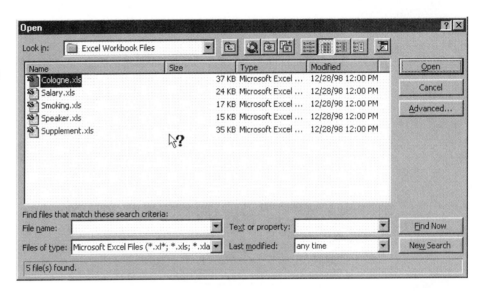

E Figure EP.2.15b
Question-Mark Help Message for Files List Box

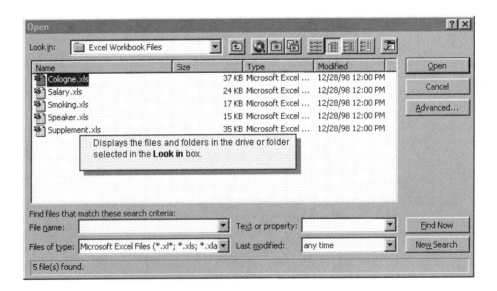

The **Office Assistant** is an animated cartoon character that stands by in its own window ready to offer help about Excel features. Figure EP.2.16 shows "Clippit," one of several identities that the Assistant can assume. The Assistant "talks" by using comic-like balloons that display information and offer choices. Figure EP.2.17 shows the contents of the balloon after a search was done on the word "help."

E **Figure EP.2.16**
The "Clippit" Office
Assistant Cartoon
Character

What would you like to do?

- Ways to get assistance while you work
- Connect to Microsoft technical resources
- Troubleshoot the Office Assistant
- Get Help without using the Office Assistant
- Get more help about maps
- ▼ See more...

help

● **Search**

● Tips   ● Options   ● Close

E **Figure EP.2.17**
Balloon Help for the
Search Phrase "Help"

The authors suggest hiding the Office Assistant to prevent its animations from distracting users and to prevent its window from obscuring the Excel application window. Hiding the Office Assistant places it in a stand-by mode, ready to reappear when certain program features, such as a wizard (see Section EP.3.1), are selected. To hide the Office Assistant, use the following procedure:

1   Right-click the Office Assistant animation.

2   Select Hide Assistant from the shortcut menu that appears.

If the Office Assistant does not appear when Microsoft Excel is loaded, the Assistant was either never installed or was previously hidden. To make the Assistant appear (or reappear), select Help | Microsoft Excel Help.

Getting Started with Microsoft Excel Wizards and Add-Ins

Wizards and add-ins are optional components that enhance Microsoft Excel by either giving the program additional functionality, such as the ability to perform a special statistical operation, or by simplifying certain tasks, such as the preparation of charts. This part of the Primer describes wizards and add-ins in general terms and looks at specific examples of both. (Later Excel sections introduce other wizards and add-ins as appropriate.)

EP.3.1 Wizards

Wizards are sets of linked dialog boxes that guide the user through the task of creating certain workbook objects. Users enter information and make selections in the linked boxes, and advance through the set by clicking a Next button (and ultimately, a Finish button to create the object; clicking a Cancel button cancels the task). As an example, consider the Microsoft Excel **Text Import Wizard** that assists in the importing, or transferring, of data from a **text file** into a worksheet. (A text file contains unlabeled and unformatted values that are separated by delimiters such as spaces, commas, or tab characters.).

To illustrate this wizard, assume that a set of student test scores has been stored in the text file STUDENTS.TXT (such a file is supplied on the disk that accompanies this text).

To import the data from this file into an Excel worksheet, the following procedure would be used:

1.   Select File | Open.

2.   In the Open dialog box (see Figure EP.2.7 on page 13):

      a   Select the folder that contains the STUDENTS.TXT file from the Look in: drop-down list.

      b   Select the Text Files (*.prn; *.txt; *.csv) choice from the Files of type: drop-down list. The All Files (*.*) choice can also be selected.

      c   Enter the name STUDENTS.TXT in the File name: edit box or select it from the files list box. (If the file does not appear in the files list box, verify that steps 2a and 2b were done correctly.)

      d   Click the OK button. This begins the three-step Text Import Wizard.

3.   In the Text Import Wizard Step 1 dialog box (see Figure EP.3.1a):

      a   Select the Delimited option button (because the data values for the variables in this file have been placed in fixed-width columns). Note that what the text calls variables are called fields in the dialog box.

      b   Click the Next button.

E   **Figure EP.3.1a** Text Import Wizard Step 1 Dialog Box

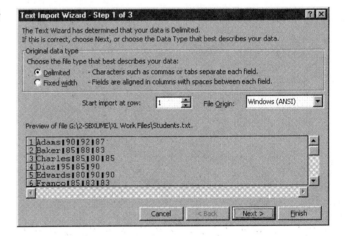

4.   In the Text Import Wizard Step 2 dialog box (see Figure EP.3.1b) click the Next button to accept the placement of the data from each line in the text file into columns.

E   **Figure EP.3.1b** Text Import Wizard Step 2 Dialog Box

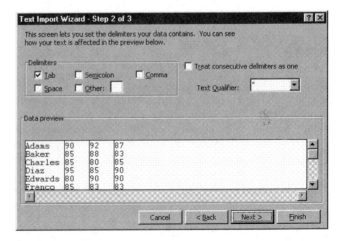

5 In the Text Import Wizard Step 3 dialog box (see Figure EP.3.1c):

  a Select the General option button under the Column data format heading.

  b Click the Finish button. The data of the text file is transferred to a new work-sheet, named after the text file name (STUDENTS in this example), in a new workbook.

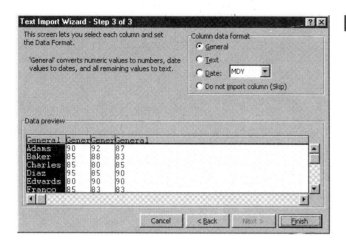

E **Figure EP.3.1c** Text Import Wizard Step 3 Dialog Box

Objects created by wizards should be reviewed for errors. Objects containing errors can be either edited, typically by right-clicking the object and selecting the appropriate choice from the shortcut menu, or deleted and then recreated by the wizard. (In this example, a good procedure would be to review the data just transferred and also, perhaps, insert column headings by first selecting any cell in row 1 and then selecting Insert I Rows and finally entering the actual column headings.)

EP.3.2 Using Add-Ins

As first mentioned in the overview to this Primer, **add-ins** are preprogrammed procedures that extend the functionality and simplify the use of the program. Add-ins appear as new Microsoft Excel menu choices and can be selected in the ways other menu items are. Some add-ins, such as the (Data) Analysis Toolpak used throughout this text, are supplied with Microsoft Excel; other add-ins, such as PHStat, the Prentice Hall statistical add-in included on the CD-ROM that accompanies this text, are obtained separately.

   Add-ins that are packaged with Microsoft Excel need to be permanently installed before they can be used. To permanently install such an add-in, do the following:

1 Select the add-in(s) for inclusion into Microsoft Excel during the Microsoft Excel (or Microsoft Office) setup process.

2 Run Microsoft Excel.

3 Select Tools I Add-Ins.

4 In the Add-Ins dialog box (see Figure EP.3.2):

  a Select the appropriate checkbox(es) in the Add-Ins available: list box.

  b Click the OK Button.

E Figure EP.3.2 The Add-Ins Dialog Box (The list of add-ins that appears on your system may be different from this list.)

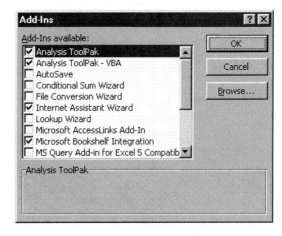

Once installed, the add-in will be automatically installed every time Microsoft Excel is run until its checkbox is deselected (unchecked). If a packaged add-in was not selected during the Microsoft Excel setup process, the Microsoft Excel setup can be rerun at another time to select it. For example, if you used the "Typical" option of the Microsoft Excel/Office setup program, the (Data) Analysis Toolpak add-in used in this text was not selected and you would need to run the Microsoft Excel/Office setup program a second time to include it.

Add-ins that are obtained separately are activated by the following similar procedure:

[1] Run the setup program for the add-in.

[2] Run Microsoft Excel.

[3] Select Tools | Add-Ins.

[4] In the Add-Ins dialog box:

[a] Select the appropriate checkbox in the Add-Ins available: list box. If there is no check box for the new add-in in the Add-Ins available: list, click the Browse button and locate and open the add-in.

[b] Click the OK Button.

Many separately obtained add-ins, including PHStat, can also be loaded temporarily using the procedure to open an Excel workbook file discussed in Section EP.2.6. (Selecting PHStat from the Prentice Hall Add-Ins Start menu loads the add-in using this method.) When loaded temporarily, add-ins may trigger the Microsoft macro virus dialog box that warns of the possibility of viruses (see Figure EP.3.3). Should this dialog box appear, click the Enable Macros button to allow a virus-free add-in (such as PHStat) to be loaded.

E Figure EP.3.3 Macro Virus Dialog Box

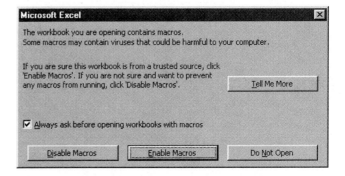

Add-ins loaded temporarily are not automatically installed the next time Microsoft Excel is run. Care should be taken never to temporarily load an add-in already permanently installed, as unpredictable results may occur.

Add-ins installed or loaded temporarily typically modify the menu bar of Microsoft Excel by adding new menu choices to a preexisting menu or by inserting a new menu of choices. For example, installing the (Data) Analysis ToolPak inserts the new choice Data Analysis to the Excel Tools menu (see Figure EP.3.4), while installing or loading the PHStat add-in inserts a PHStat menu similar to the one shown in Figure EP.3.5. These new choices can then be selected to use the add-in or one of its functions.

E Figure EP.3.4
Microsoft Excel Tools
Menu with the Data
Analysis Choice
Added

E Figure EP.3.5
PHStat Menu

To make full use of this text, you will want to install the Analysis ToolPak and Analysis ToolPak—VBA add-ins that are supplied with Microsoft Excel, and set up and install (or temporarily load) the Prentice Hall PHStat add-in included on the CD-ROM that accompanies this text. You must install the ToolPak add-ins to use PHStat, as PHStat will refuse to run if it detects that the ToolPak add-ins were not previously installed.

EP.3.3 Summary

The material in this Primer has been designed to prepare you to use Microsoft Excel in your study of statistics. In the Microsoft Excel sections of this text you will be learning about the Excel features that can be used to produce statistical analyses. Each section begins with a Solution Summary that outlines the steps necessary to perform a particular analysis. This is followed by at least one example that includes detailed instructions for using Microsoft Excel to perform the analysis. In many cases, the Prentice Hall PHStat add-in, supplied on the CD-ROM that accompanies this text, is used to quickly generate results.

Key Terms

Absolute cell reference 12	Edit box 6	Scroll bar 4
Add-ins 23	Formula 9	Select operation 1
Arithmetic operators 9	Function 9	Sheet tab 4
Buttons 2	Icons 2	Shortcut menu 2
Cancel button 7	List box 6	Spinner buttons 6
Cell ranges 9	Menu bar 4	Start button 2
Cells 9	Mouse pointer 1	Status bar 4
Check box 7	Office Assistant 20	Text files 21
Click mouse operation 2	Open/OK buttons 7	Text Import Wizard 21
Closing button 4	Option button 6	Toolbar 2
Dialog box 5	Question-mark button 19	Tool tips 19
Double-click operation 2	Relative cell reference 12	Wizards 21
Drag operation 2	Resizing button 4	Workbook 1
Drop-down list box 6	Right-click operation 2	Worksheets 1

Chapter 1

Introduction: Statistics and Data

CONTENTS

1.1 What Is Statistics?

1.2 Types of Data

1.3 Descriptive vs. Inferential Statistics

1.4 Collecting Data

1.5 Random Sampling

1.6 Other Types of Samples (Optional)

1.7 Ethical and Other Issues in Statistical Applications

EXCEL TUTORIAL

1.E.1 Using Microsoft Excel to Select a Random Sample

REAL-WORLD APPLICATION
Obedience to Authority—The Shocking Truth

In one of the most famous experiments in psychology, Dr. Stanley Milgram (1974) studied the factors that determine the extent to which people obey authority—even if that authority is pushing them to do something they are against. In Milgram's study, subjects playing the role of "teachers" were told to give electric shocks (up to 450 volts) to "learners" who answered questions incorrectly. The object of the experiment was to see how many volts a subject would be willing to give before refusing to comply with the request.

Imagine the massive and impossible task of testing everyone in the country to see how they would respond to the shock experiment. And if we could test everyone, how could we make sense out of the mass of information collected? This text describes how to use sampling and the science of statistics to solve practical problems encountered in today's world. Milgram's experiment is the focus of Real-World Revisited at the end of this chapter.

1.1 WHAT IS STATISTICS?

Consider the following recent items from the news media:

1. *USA Today,* Dec. 10, 1997—The biggest study ever of the health effects of alcohol concludes that a drink a day can cut your risk of death by 20%.... The researchers gave questionnaires to 490,000 men and women and then followed up nine years later, after 46,000 of them had died.... [However], the benefits decreased as people drank more. Among those who averaged four or five drinks a day, the risk of death among men was 10% lower, while among women it was 7% lower.

2. *Good Housekeeping,* July 1996—Do shaving creams and gels really work better than soap [when shaving your legs]? Yes. When *Good House-keeping* panelists put them to the test, two-thirds preferred creams and gels to shaving with soap or a beauty bar.... Using a new disposable shaver each time, the 30 women on our panel shaved one leg with a cream or gel, the other with soap or bar, switching legs midway through the study.

3. *Tampa Tribune,* Dec. 11, 1997—In a feat right from [the television show] *Star Trek,* scientists appear to have made photon light particles transfer to another place ... using a phenomenon called entanglement.... [Although] the process worked only 25 percent of the time ... it's possible to go up to 75 percent and scientists will shoot for that.

4. *New York Times,* Sept. 17, 1996—Millions of Americans routinely ignore one of mom's most important pieces of advice: Wash your hands after you go to the bathroom. This unsettling item of news was gathered in the only way possible—by actually watching what people do (or don't do) in public restrooms. The researchers—if that's what they should be called—hid in stalls or pretended to comb their hair while observing 6,333 men and women do their business in five cities.... Just 60% of those using restrooms in Penn Station (New York City) washed up afterward.

5. *Sports Illustrated,* Dec. 8, 1997—Results of a *Sports Illustrated* poll of 1,835 middle school and high school students from across the United States who said "yes" when asked whether they may be good enough to play professional basketball some day: African-American males—55%, White males—20%.

Every day we are inundated with bits of information—**data**—like those in the citations above, whether we are in the classroom, on the job, or at home. These data often guide the many decisions that we make during our lifetime. What decision-making tools will enable you to make sense of data? The answer is *statistics.*

A common misconception is that a statistician is simply a "number cruncher" or a person who calculates and summarizes numbers, like baseball batting averages or unemployment rates. Statistics involves numbers, but there is much more to it than that.

According to *The Random House College Dictionary* (1998 ed.), statistics is "the science that deals with the collection, classification, analysis, and interpretation of numerical facts or data." In short, statistics is the *science of data*—a science that will enable you to be proficient data producers and efficient data users.

Definition 1.1

Statistics is the science of data. It involves collecting, classifying, summarizing, organizing, analyzing, and interpreting data.

In this chapter we explore the different types of data and problems that you will encounter in your field of study and introduce you to some ideas on methods for collecting data. The various statistical methods for summarizing, analyzing, and interpreting those data are presented in the chapters that follow.

1.2 TYPES OF DATA

Data are obtained by measuring some characteristic or property of the objects (usually people or things) of interest to us. These objects upon which the measurements (or observations) are made are called *experimental units,* and the properties being measured are called *variables* (since, in virtually all studies of interest, the property varies from one observation to another).

A finer breakdown of data types into nominal, ordinal, interval, and ratio data is possible. *Nominal* data are qualitative data with categories that cannot be meaningfully ordered. *Ordinal* data are also qualitative data, but a distinct ranking of the groups from high to low exists. *Interval* and *ratio* data are two different types of quantitative data. For most statistical applications (and all the methods presented in this introductory text), it is sufficient to classify data as either quantitative or qualitative.

Definition 1.2

An **experimental unit** is an object (person or thing) upon which we collect data.

Definition 1.3

A **variable** is a characteristic (property) that differs or varies from one observation to the next.

All data (and, consequently, the variables we measure) are either *quantitative* or *qualitative* in nature. Quantitative data are data that can be measured on a numerical scale. In general, qualitative data take values that are nonnumerical; they can only be classified. The statistical tools that we use to analyze data depend on whether the data are quantitative or qualitative. Thus, it is important to be able to distinguish between the two types of data.

Definition 1.4

Quantitative data are observations measured on a natural numerical scale.

Definition 1.5

Nonnumerical data that can only be classified into one of a group of categories are **qualitative** (or **categorical**) **data**.

Example 1.1   Identifying Features of a Data Set

Many universities provide their students with a Career Resource Center. Periodically, the center will mail out questionnaires pertaining to the employment status and starting salary of students who recently graduated. A University

Table 1.1 Data for 10 University of Florida Graduates

Gender	Degree	Major	Job Type	Salary	Age	GPA
M	PhD	Law	Criminal	$38,000	31	3.97
M	Bachelor	Accounting	Auditing	$28,000	22	3.51
M	Bachelor	Ind/Systems	Sales	$43,000	25	2.90
F	Bachelor	Physical Therapy	Therapy	$27,600	34	3.05
M	Masters	Latin Amer. Studies	U.S. Army	$49,000	40	2.88
F	Bachelor	Medical Technology	Therapy	$21,100	24	3.15
F	Bachelor	Political Science	Ins. Claims	$27,500	26	2.74
M	Bachelor	Economics	Bank/Finance	$33,000	21	3.82
F	PhD	Dentistry	Dentist	$65,000	39	4.00
F	Bachelor	Nursing	Reg. Nurse	$30,000	25	4.00

of Florida (UF) questionnaire gathered the following information from each graduate who was contacted:

1. Gender
2. Degree
3. Major
4. Job type
5. Starting salary
6. Age
7. Overall grade point average

Data for 10 (fictitious) graduates from the data set are shown in Table 1.1.

a. For this data set, identify the experimental units and the variables measured.

b. Determine the type (quantitative or qualitative) of each variable measured.

Solution

a. Since the center collects data on each UF graduate surveyed, the UF graduates are the experimental units. The variables (properties) measured on each graduate are the seven items previously listed, i.e., Age, Grade point average, Gender, etc. These are called variables since their values vary from one graduate to another (i.e., they are not constant).

b. The first four variables listed are qualitative since the data they produce are values that are nonnumerical; they can only be classified into categories or groups. For example, Gender is either M (for male) or F (for female). Similarly, values for Major are accounting, nursing, political science, etc. We can classify a graduate according to Gender or Major, but we cannot represent Gender or Major as a meaningful numerical quantity.

The last three variables are quantitative since they are all measured on a natural numerical scale. Salary is measured in dollars, Age in years, and Grade point average on a 4-point scale. ❏

Caution: The values of a qualitative variable may be coded or recorded numerically. For example, Gender in Example 1.1 could have been coded 1 for males and 0 for females. Nevertheless, the data are qualitative since each observation can only be classified into one of a group of categories.

Example 1.2   Classifying Variables

The U.S. Army Corps of Engineers recently conducted a study of contaminated fish inhabiting the Tennessee River (in Alabama) and its tributaries. A total of 144 fish were captured and the following variables measured for each:

1. Location of capture
2. Species
3. Length (centimeters)
4. Weight (grams)
5. DDT concentration (parts per million)

Classify each of the five variables measured as quantitative or qualitative.

Solution

The variables length, weight, and DDT are quantitative because they are all measured on a natural numerical scale: length in centimeters, weight in grams, and DDT in parts per million. In contrast, location and species cannot be measured quantitatively; they can only be classified (e.g., channel catfish, largemouth bass, and smallmouth buffalo for species). Consequently, data on location and species are qualitative. ❑

Self-Test 1.1

State whether each of the following variables measured on graduating high school students is quantitative or qualitative.

a. National Honor Society member or not
b. Scholastic Assessment Test (SAT) score
c. Number of colleges applied to
d. Part-time job status

Example 1.3   Classifying Variables

Refer to the *Good Housekeeping* (July 1996) study of shaving creams cited in Section 1.1. Recall that each of 30 women shaved one leg using a cream or gel and the other leg using soap or a beauty bar. At the end of the study, 20 of the 30 women stated their preference for shaving with creams and gels. Identify the experimental unit for this study and describe the variable of interest as quantitative or qualitative.

Solution

Since we are interested in the opinions of the 30 women who participated in the leg-shaving study, the experimental unit is a woman who shaves her legs. The variable measured, shaving preference, can assume one of two categories in this experiment: (1) prefer a cream or gel, or (2) prefer soap or a beauty bar. Consequently, shaving preference is a qualitative variable. ❑

To summarize, knowing the type (quantitative or qualitative) of the data that you want to analyze is one of the keys to selecting the appropriate statistical method to use. A second key involves the concept of population and samples, as discussed in the next section.

PROBLEMS 1.1–1.9

Applying the Concepts

1.1 Determine whether each of the following variables is quantitative or qualitative.

a. Number of telephones per household

b. Whether there is a telephone line connected to a computer modem in the household

c. Whether there is a FAX machine in the household

d. Length (in minutes) of longest long-distance call made per month

e. Number of local calls made per month

1.2 Consider the following variables recorded in the Environmental Protection Agency Gas Mileage Guide for new automobiles. Determine the type, quantitative or qualitative, of each.

a. Manufacturer of automobile

b. Model name

c. Transmission type (automatic or manual)

d. Engine size

e. Number of cylinders

f. Estimated city gas miles per gallon

g. Estimated highway gas miles per gallon

1.3 Prior to 1980, only estates were permitted to own land in Hawaii and homeowners leased the land from the estate (a law that dates back to Hawaii's feudal period). The law now requires estates to sell land to homeowners at a fair market price. As part of a study to estimate the fair market value of its land (called the "leased fee" value), a large Hawaiian estate collected the data shown in the accompanying table for five properties.

Property	Leased Fee Value (thousands of dollars)	Lot Size (thousands of sq. ft.)	Neighborhood	Location of Lot
1	70.7	13.5	Cove	Cul-de-sac
2	52.6	9.6	Highlands	Interior
3	87.1	17.6	Cove	Corner
4	43.2	7.9	Highlands	Interior
5	144.3	13.8	Golf Course	Cul-de-sac

a. Identify the experimental units.

b. State whether each of the variables measured is quantitative or qualitative.

1.4   *Bioethics* describes ethical issues that arise in the life sciences. Examples include environmental pollution, wildlife conservation, animal rights, birth control (contraception/abortion), and herbicide/pesticide use. The *Journal of Moral Education* (Dec. 1997) examined the coverage of bioethics in biology textbooks for grades five to ten. Three questions of interest to the researchers were:

1. How much space of the textbook is dedicated to bioethics?
2. How often are catchwords such as "danger," "threatened," "polluted," "poisoned," and "protection/rescue" used in the textbook?
3. Does the textbook use pictures in discussion of bioethics?

a. Identify the experimental units in the study.

b. Identify the type of data (quantitative or qualitative) extracted from each question.

1.5   D. Pritchard and K. D. Hughes, mass communication researchers at the University of Wisconsin—Milwaukee, collected and analyzed data on 100 Milwaukee homicides (*Journal of Communication,* Summer 1997). The purpose of the research was to identify important determinants of the "newsworthiness" of a crime such as homicide. Several of the many variables measured for each homicide are listed below. Identify the type (quantitative or qualitative) of each variable.

a. Average length (in words) of news stories published

b. Number of news items published

c. Gender of the victim

d. Gender of the suspect

e. Whether or not the police banned the press from publishing crime-scene photos.

f. Per capita income of area in which crime occurred

g. Relationship between the victim and suspect

h. Age of victim

1.6   The *Journal of Performance of Constructed Facilities* (Feb. 1990) reported on the performance dimensions of water distribution networks in the Philadelphia area. For one part of the study, the following data were collected for a sample of water pipe sections:

1. Pipe diameter (inches)
2. Pipe material
3. Age (year of installation)
4. Location
5. Pipe length (feet)
6. Stability of surrounding soil (unstable, moderately stable, or stable)
7. Corrosiveness of surrounding soil (corrosive or noncorrosive)
8. Internal pressure (pounds per square inch)
9. Percentage of pipe covered with land cover
10. Breakage rate (number of times pipe had to be repaired due to breakage)

Identify each variable as quantitative or qualitative.

1.7   White-collar criminal offenders (e.g., those convicted of tax evasion, fraud, or embezzlement) often receive preferential treatment in the criminal justice system. A researcher studied the social and economic damage incurred by convicted white-collar criminals (*Criminology,* Nov. 1984). From the case histories of approximately seventy offenders, he collected data on several variables, listed here. Classify each of the variables as quantitative or qualitative.

a. Age in years

b. Length of sentence in years

c. Type of victim (business victim, government victim, individual victim, combination of victims)

d. Type of occupation (public/professional or private business)

e. Type of sentence (probation, work release, or incarceration)

f. Race

g. Recovery time in months (i.e., time from conviction to review of the file)

1.8   List five or more variables that your family physician considers while giving you a complete physical examination. State whether each is qualitative or quantitative.

1.9   Marketers are keenly interested in the factors that motivate coupon usage by consumers. Research reported in the *Journal of Consumer Marketing* (Spring 1988) surveyed 290 shoppers. Each shopper was asked to respond to the following questions:

a. Do you collect and redeem coupons?

b. Are you price-conscious while shopping?

c. On average, how much time per week do you spend clipping and collecting coupons?

Classify the responses to the questions as quantitative or qualitative variables.

1.3 DESCRIPTIVE VS. INFERENTIAL STATISTICS

When you examine a data set in the course of your studies, you will be doing so because the data characterize some phenomenon of interest to you. In statistics, the data set that is the target of your interest is called a *population.* This data set, which is typically large, exists in fact or is part of an ongoing operation and hence is conceptual. Some examples of phenomena and their corresponding populations are shown in Table 1.2.

Definition 1.6

A **population** is a collection (or set) of data that describe some phenomenon of interest to you.

If every measurement in the population is known and available, then statistical methodology can help you describe this set of data. In future chapters we will find graphical (Chapter 2) and numerical (Chapter 3) ways to make sense out of a large mass of data, i.e., to summarize and draw conclusions from it. The branch of statistics devoted to this application is called *descriptive statistics.*

Table 1.2 Some Typical Populations

Phenomenon	Experimental Units	Variable Measured	Type of Data	Population
a. White blood cell count of a hemophiliac	Hemophiliacs	White blood cell count	Quantitative	Set of white blood cell counts of all hemophiliacs
b. Grade point average (GPA) of a college freshman	College freshman	GPA	Quantitative	Set of grade point averages of all college freshman
c. July high temperature of a U.S. state	States	Highest temperature in July	Quantitative	Set of July high temperatures for all 50 U.S. states
d. Quality of items produced on an assembly line	Manufactured items	Defective status	Qualitative	Set of defective/nondefective measurements for all items manufactured over the recent past and future

> **Definition 1.7**
>
> The branch of statistics devoted to the organization, summarization, description, and presentation of data sets is called **descriptive statistics**.

Many populations are too large to measure each observation; others cannot be measured because they are conceptual. For example, population (a) in Table 1.2 cannot be measured since it would be impossible to identify all hemophiliacs. Even if we could identify them, it would be too costly and time-consuming to measure and record their white blood cell counts. Population (d) in Table 1.2 cannot be measured because it is part conceptual. Even though we may be able to record the quality measurements of all items manufactured over the recent past, we cannot measure quality in the future. Because of this problem, we are required to select a subset of values from a population, called a *sample*.

> **Definition 1.8**
>
> A **sample** is a subset of data selected from a population.

Example 1.4   Population and Sample

Refer to the starting salary data described in Example 1.1. Recall that the Career Resource Center (CRC) surveys all University of Florida (UF) graduates at the time of their graduation. However, the CRC estimates that only about half of the UF graduates return the questionnaire, and of these, only half indicate that they have secured a job as of the date of graduation. Thus, it is important to note that the data set collected by the CRC is not a complete listing of the annual starting salaries of all recent University of Florida graduates.

a. Suppose the target of your interest is the starting salary data collected by the CRC last year. Describe the target population.

b. Suppose the target of your interest is the first-year financial compensations of all recent University of Florida graduates. Is the data set collected by the CRC a population or a sample?

Solution

a. The measurements of particular interest to us are the starting salaries of UF graduates. Since the data set collected by the CRC includes only the starting salaries of last year's UF graduates who returned the CRC survey and secured a job, the target population consists of the collection of starting salaries of UF graduates *who returned the CRC questionnaire and had secured a job at graduation time last year*.

b. Since we are interested in the first-year financial compensations of *all* recent UF graduates, the target population is the set of starting salaries of all recent graduates—not only those who returned the CRC questionnaires but also those who failed to return the survey even though they had secured a job by the date of graduation. Since data collected by the CRC are only a subset of this target population, they represent a sample. ❏

Example 1.5   Population and Sample

Potential advertisers value television's well-known Nielsen ratings as a barometer of a TV show's popularity among viewers. The Nielsen rating of a certain TV program (e.g., NBC's hit medical drama *ER*) is an estimate of the proportion of viewers, expressed as a percentage, who tune their sets to the program on a given night at a given time. A typical Nielsen survey consists of 165 families selected nationwide who regularly watch television. Suppose we are interested in the Nielsen ratings for the latest episode of *ER* (currently shown on Thursday nights).

a. Identify the population of interest.   b. Describe the sample.

Solution

a. We want to know which TV program was watched by a family on Thursday during the *ER* time slot; consequently, the population of interest is the collection of programs watched by all families in the United States during this time slot. In this example, the qualitative measurements might be recorded as {family was tuned to *ER*} and {family was not tuned to *ER*}.

b. The sample—that is, the subset of the population—consists of the set of qualitative measurements defined in part **a** for each of the 165 families in the Nielsen survey. ❏

Self-Test

1.2

To evaluate the current status of the dental health of all U.S. school children, the American Dental Association conducted a survey. One thousand grade school children from across the country were selected and examined by a dentist; the number of cavities was recorded for each.

a. Identify the population of interest to the American Dental Association.

b. Identify the sample.

c. What is the variable of interest? Is it quantitative or qualitative?

In Chapters 2 and 3, we demonstrate several statistical methods for organizing, describing, summarizing, and presenting data sets similar to those of Examples 1.4 and 1.5. However, statistics may involve much more than data description. In succeeding chapters, we will discover statistical methods that enable us to infer the nature of the population from the information in the sample. The branch of statistics devoted to this application is called *inferential statistics*. In addition, this *statistical methodology provides measures of reliability for each inference obtained from a sample*. This is one of the major contributions of inferential statistics. Anyone can examine a sample and make a "guess" about the nature of the population. For example, we might estimate that the average grade point average of college freshmen is 2.35, or that the starting salary range of a college graduate is between $20,000 and $30,000. But statistical methodology enables us to go one step further. When the sample is selected in a specified way from the population, we can also say how accurate our estimate will be, i.e., how close the estimate of 2.35 will be to the true average GPA, or how confident we are that the typical starting salary falls between $20,000 and $30,000.

> ### Definition 1.9
> The branch of statistics concerned with using sample data to draw conclusions about a population is called **inferential statistics**. When proper sampling techniques are used, this methodology also provides a measure of **reliability** for the inference.

Example 1.6   Population, Sample, and Inference

Are state lottery winners who win big payoffs likely to quit their jobs within one year of winning? No, according to a study published in the *Journal of the Institute for Socioeconomic Studies* (Sept. 1985). The researcher mailed questionnaires to over 2,000 lottery winners who won at least $50,000 between 1975 and 1985. Of the 576 who responded, only 11% had quit their jobs during the first year after striking it rich. In this study, identify

    a. the population   **b.** the sample

    c. the inference made about the population

Solution

    a. The researcher is interested in state lottery winners who won at least $50,000 between 1975 and 1985. Consequently, the experimental units in the target population are all state lottery winners who won $50,000 or more during that decade. The measurement (or variable) of interest is the winner's job status—quit job within one year of winning or still working. Note that this is a qualitative variable.

    b. The sample consists of the collection of job status measurements (quit job or not) for the 576 lottery winners who responded to the questionnaire.

    c. Since 11% of the respondents in the sample had quit their jobs, the inference is that 11% of the state lottery winners in the population also quit their jobs within one year of winning the lottery. This leads to the conclusion reached by the researcher, namely, that most state lottery winners *do not* quit their jobs upon winning. Note that the researcher does not provide a measure of reliability for this inference. Enough information is available, however, to calculate such a measure. We will show in Chapter 7 that, with a high degree of "confidence," the estimate of 11% is within 2.6% of the true percentage. That is, the true percentage is no lower than 8.4% and no higher than 13.6%. ❑

The key facts to remember in this section are summarized in the accompanying boxes.

The Objective of Statistics

1. To describe data sets (populations or samples)
2. To use sample data to make inferences about a population

The Major Contribution of Descriptive Statistics

Conclusions about data can be made based on patterns revealed by descriptive statistical data.

The Major Contribution of Inferential Statistics

Statistical methodology allows us to provide a measure of reliability for every statistical inference based on a properly selected sample.

PROBLEMS 1.10–1.19

Applying the Concepts

1.10 A sociologist is interested in describing the socioeconomic status levels of all Florida households. Identify each of the following data sets as a population or sample.

a. Set of annual incomes of all Florida households

b. Set of job status measurements (employed or unemployed) for each of 50 households located in the city of Miami

c. Set of poverty levels for each of 1,000 households selected from across the state of Florida

d. Set of job type measurements (white collar or blue collar) for all Florida households

1.11 Identify whether the data sets analyzed in the following studies represent populations or samples.

a. Flavor quality ratings (poor, fair, good, very good, excellent) for the 16 best-selling regular and ice beer brands (*Consumer Reports,* June 1996)

b. Presence or absence of Hessian fly damage for 600 wheat plants selected from a large midwestern wheat field (*Journal of Agricultural, Biological, and Environmental Statistics,* June 1997)

c. Proposed orbit altitudes (kilometers) for the 14 satellite consortiums that have applied for FCC phone service licensing (*Forbes,* May 22, 1995)

d. Numbers of whale bones found at 35 prehistoric Thule Eskimo burial sites in the Canadian Arctic (*Journal of Archaeological Sciences,* Oct. 1997)

1.12 *Postpartum depression* is the term used to describe the usually short-lived period of emotional sensitivity that many women suffer following childbirth. Studies show that men, too, can suffer from postpartum depression. A developmental psychologist wants to estimate the proportion of fathers who suffer from postpartum blues. Fifty men who have recently fathered a child are interviewed and observed in the home, and the number experiencing some form of postpartum depression is recorded.

a. What is the population of interest to the developmental psychologist?

b. Describe the sample in this problem.

c. Suppose that 31 of the 50 men are diagnosed as having postpartum blues. The psychologist then estimates that 62% of all fathers experience postpartum depression. Do you believe that this estimate is equal to the proportion for the entire population? Explain.

1.13 As part of an undergraduate course in children's literature, a group of education majors read weekly to individually selected children. During the reading session, readers asked questions in an effort to draw the children into the literature. Three University of Colorado researchers analyzed all of the approximately 2,000 questions that readers asked and published the results in the *Journal of Literacy Research* (Dec. 1996). Each question was classified into one of four types:

1. Known information (e.g., "What's that?")
2. Opinion (e.g., "Why? What do you think about that?")
3. Conditional (e.g., "If you were in the story, what would you do?")
4. Connection (e.g., "Does that remind you of something in your life?")

a. How many variables are measured in this study?

b. What is the experimental unit?

c. Identify the type of data collected as quantitative or qualitative.

d. Would you classify the data as a population or a sample? Explain.

1.14 A new precooling method has been developed for preparing Florida vegetables for market. The system employs an air–water mixture designed to yield effective cooling with much less water flow than required for conventional hydrocooling. In an effort to compare the effectiveness of the two cooling systems, researchers divided 20 batches of green tomatoes into two groups of 10 each. One group was precooled with the new method, while the other was precooled with the conventional method. The total water flow (in gallons) required to effectively cool each batch was recorded.

a. Identify the populations of interest.

b. Describe the samples.

   c. How could the sample data be used to compare the cooling effectiveness of the two systems?

1.15 Tel Aviv University law professor A. Likhovski examined the history of the concept of "tyranny" in nineteenth-century American legal cases (*Journal of Interdisciplinary History,* Autumn 1997). Likhovski used "Lexis," a computer-searchable database containing all cases in which lawyers and/or judges used key words associated with tyranny (e.g., "tyranny," "despotism," "dictatorship," "autocracy," and "absolutism"). In the decade 1850–1859, there were a total of 3,854 U.S. federal cases. Likhovski's computer search found that 45 of these cases (1.17%) used one or more of these keywords.

   a. Identify the experimental unit in this analysis.

   b. For the decade 1850–1859, describe the target population.

   c. Describe the variable measured on each experimental unit and its type.

   d. Does the percentage, 1.17%, describe a population or a sample?

1.16 A panel of tobacco experts convened by the National Cancer Institute recommends that cigarette manufacturers put more descriptive labels on their cigarette packages, including a disclaimer that "light" brands are not really more healthful than "regular" (nonlight) brands (*Tampa Tribune,* Dec. 7, 1994). The panel's recommendations are based, in part, on data collected by the Federal Trade Commission (FTC). Each year, the FTC tests all domestic cigarette brands for carcinogens such as tar and nicotine. Suppose our goal is to compare the average nicotine content of all domestic light cigarette brands to the average nicotine content of all domestic regular cigarette brands. To do this, we record the nicotine contents (in milligrams) of 25 light cigarette brands and 25 regular cigarette brands.

   a. Describe the target populations. (Give the precise statistical definitions.)

   b. Describe the samples.

1.17 *Euthanasia,* the act of painlessly putting to death a person suffering from an incurable and painful disease or condition, has long been a dilemma of medical ethics. Recently, Dr. Jack Kervorkian's physician-assisted suicide efforts have made national headlines. Suppose you work for a major opinion pollster and you have been assigned the task of conducting a survey. The purpose of the survey is to estimate the proportion of American adults who support Dr. Jack Kervorkian's euthanasia movement.

   a. Clearly define the population of interest to the pollster.

   b. Do you think it is possible to obtain the entire population? Explain.

   c. Why should the sample you select for the survey be representative of the population?

1.18 Pesticides applied to an extensively grown crop can result in inadvertent air contamination. *Environmental Science & Technology* (Oct. 1993) reported on thion residues of the insecticide chlorpyrifos used on dormant orchards in the San Joaquin Valley, California. Surrounding air specimens were collected daily at an orchard site during an intensive period of spraying—a total of 13 days—and the thion level (ng/m^3) measured each day.

    a. Identify the population of interest to the researchers.

    b. Identify the sample.

1.19 The slope of a river delta region can sometimes be accurately predicted from knowledge of the typical size of stones found there. With this in mind, a geographer studying South America would like to estimate the average size of stones found in the delta region of the Amazon River. To obtain this estimate, the geographer collects 50 stones and measures the diameter of each.

    a. What is the population of interest to the geographer?

    b. Describe the sample in this problem.

    c. Suppose that the average diameter of the 50 stones is 7.2 inches. Do you believe that this sample average will equal the average for the population? Explain.

1.4 COLLECTING DATA

Once you decide on the type of data—quantitative or qualitative—appropriate for the problem at hand, you'll need to collect the data. Generally, you can obtain the data in four different ways:

    1. Data from a *published source*
    2. Data from a *designed experiment*
    3. Data from a *survey*
    4. Data collected *observationally*

Sometimes, the data set of interest has already been collected for you and is available in a **published source**, such as a book, journal, or newspaper. For example, you may want to examine and summarize the birth rates (i.e., number of births per 1,000 people) in the 50 states of the United States. You can find this data set (as well as numerous other data sets) at your library in the *Statistical Abstract of the United States,* published annually by the U.S. Department of Commerce. Some other examples of published data sources include *The Wall Street Journal* (financial data) and *The Sporting News* (sports information). Data from published sources may be obtained by computer over the *World Wide Web* (otherwise known as the *Internet*). Services such as *America Online* and software such as *Microsoft Internet Explorer* provide easy access to the online data.

A second method of collecting data involves conducting a **designed experiment**, in which the researcher exerts strict control over the units (people, objects, or events) in the study. For example, a recent medical study investigated the potential of aspirin in preventing heart attacks. Volunteer physicians were divided into two groups—the *treatment* group and the *control* group. In the treatment group, each physician took one aspirin tablet a day for one year, while each physician in the control group took an aspirin-free placebo (no drug) made to look like an aspirin tablet. The researchers, not the physicians under study, controlled who received the aspirin (the treatment) and who received the placebo. As you will learn in Chapter 11, a properly designed experiment allows you to extract more information from the data than is possible with an uncontrolled study.

Surveys are a third source of data. With a **survey**, the researcher samples a group of people, asks one or more questions, and records the responses. Probably the most familiar type of survey is a public opinion poll conducted by any one of a number of organizations (e.g., Harris, Gallup, Roper, and CNN). Another well-known survey is the Nielsen survey, which provides the major networks with

> With published data, we often make a distinction between the *primary source* and a *secondary source.* If the publisher is the original collector of the data, the source is primary. Otherwise, the data are secondary source data.

information on the most watched television programs. Surveys can be conducted through the mail, with telephone interviews, or with in-person interviews. Although in-person interviews are more expensive than mail or telephone surveys, they may be necessary when complex information must be collected.

Finally, observational studies can be employed to collect data. In an **observational study**, the researcher observes the experimental units in their natural setting and records the variable(s) of interest. For example, a psychologist might observe and record the level of "Type A" behavior of a sample of male managers. Similarly, a marine biologist may observe and record the length and gender of blue whales captured in Antarctica. Unlike a designed experiment, an observational study is one in which the researcher makes no attempt to control any aspect of the experimental units.

Example 1.7   Data Collection Methods

Identify the data collection method in each of the following studies:

a. The U.S. Army Corps of Engineers study of fish inhabiting the Tennessee River, Example 1.2 (page 31).

b. The *Good Housekeeping* study of leg shaving preferences of women, Example 1.3 (page 31).

c. The study of state lottery winners in Example 1.6 (page 37).

Solution

a. Recall that the U.S. Army Corps of Engineers captured 144 fish in their natural habitat—the Tennessee River—then recorded the values of several variables, e.g., capture location and species type. Consequently, the resulting data are observational in nature and the study is an observational study. The engineers made no attempt to control experimental variables such as weight and length.

b. The researchers in the *Good Housekeeping* study controlled the experimental units (women shavers) by requiring each woman shaver to shave one leg with a cream or gel and the other leg with soap or a beauty bar. Thus, it is a designed experiment. We learn in Chapter 7 that such a design attempts to remove an extraneous source of variation (the different shaving styles of women) in order to obtain a more reliable estimate of the proportion of women who prefer shaving their legs with a cream or gel.

c. The data collection method used by the researcher is a mail survey since questionnaires were mailed to lottery winners to elicit their job status (quit job within one year or not). ❑

Self-Test

1.3

All highway bridges in the United States are inspected for structural deficiency by the Federal Highway Administration (FHA). For each bridge, the variables measured include length of maximum span, number of vehicle lanes, and structural condition of bridge deck. Identify each of these variables as quantitative or qualitative, the data collection method, and whether or not the data represent a population or a sample.

Regardless of the data collection method employed, it is likely that the data will be a sample from some population. And if we wish to apply inferential statistics, we must obtain a *representative sample.*

Definition 1.10

A **representative sample** exhibits characteristics typical of those possessed by the target population.

For example, consider a political poll conducted during a presidential election year. Assume the pollster wants to estimate the percentage of all 120,000,000 registered voters in the United States who favor the incumbent president. The pollster would be unwise to base the estimate on survey data collected for a sample of voters from the incumbent's own state. Such an estimate would almost certainly be *biased* high.

The most common way to satisfy the representative sample requirement is to select a *random sample.* Random samples are the topic of the next section.

PROBLEMS 1.20–1.25

Year	DJIA
1990	2,634
1991	3,169
1992	3,330
1993	3,754
1994	3,834
1995	5,117
1996	6,448

Source: *The Dow Jones Investor's Handbook,* 1997.

Problem 1.20

Applying the Concepts

1.20   The Dow Jones Industrial Average (DJIA) for each of the past 7 years is listed in the table. Suppose you are interested in using these data to forecast the DJIA in the year 2000. Identify the data collection method.

1.21   Refer to the *Journal of Communication* (Summer 1997) study of 100 Milwaukee homicides, Problem 1.5 on page 33. Identify the data collection method.

1.22   Refer to the *Journal of Consumer Marketing* (Spring 1988) study of coupon usage by shoppers, Problem 1.9 on page 34. Identify the data collection method.

1.23   Refer to the study of a new precooling method for preparing vegetables for market, Problem 1.14 on page 39. Identify the data collection method.

1.24   The *Journal of Experimental Education* (Winter 1997) published a study that examined the benefit of repeated presentations of a lecture on student learning. The study participants were 109 college juniors and seniors enrolled in an educational psychology course where lectures are offered on videotape. The researchers were interested in how two variables, the number of times the videotaped lecture was presented (once, twice, or three times) and the type of organizational material presented before the lecture (conventional, linear, or matrix), impact lecture recall test scores. The students were randomly assigned to one of the $3 \times 3 = 9$ combinations of number of videotaped lectures and organization material; then the test scores of the nine groups were compared.

    a. Identify the experimental units for this study.

    b. Describe the variables measured and their type.

    c. Is this a designed study? Explain why or why not.

    d. Are the data collected in the study population data or sample data?

1.25 A study was conducted to explore the relation of self-esteem and positive inequity to on-the-job productivity (*Journal of Personality,* Dec. 1985). Eighty students enrolled in an industrial psychology course at a private New England university participated. All students were asked to complete a proofreading task and were compensated for their work on an hourly basis. Half the students were assigned to be overpaid for each hour they worked (positive inequity condition), while the other half were given fair compensation (equity condition). The results of the study revealed that individuals of high self-esteem were more productive (i.e., completed more of the task) in the positive inequity condition than in the equity condition. However, the reverse was true for individuals of low self-esteem.

    a. Identify the data collection method for this study.

    b. In this study, we can envision four experimental conditions: (1) high self-esteem/positive inequity, (2) high self-esteem/equity, (3) low self-esteem/positive inequity, and (4) low self-esteem/equity. Describe the population corresponding to the four conditions. (Recall that the variable of interest is productivity, measured as amount of the task completed.)

    c. Identify the samples. Assume that 20 students were assigned to each of the four experimental conditions.

    d. Do you think the samples adequately represent the populations described in part **a**?

    e. Do you think the results of the study were obtained by analyzing the data in the populations, or were they derived from sample information?

1.5 RANDOM SAMPLING

A careful analysis of most data sets will reveal that they are samples from larger data sets that are really the object of interest. Consequently, most applications of modern statistics involve sampling and using information in the sample to make inferences about the population. For these applications, it is essential that we obtain a representative sample.

    Assume we want to select a sample of size n from a population with N experimental units. The most common way to satisfy the requirement of representative sample is to select the sample in such a way that every different sample of size n has an equal probability (or chance) of selection from the population.[*] This procedure is called *random sampling* and the resulting sample is called a (*simple*) *random sample*. (Throughout this text, a "random sample" refers to a simple random sample as defined below.)

*A more formal definition of "probability" will be provided in Chapter 4.

> ### Definition 1.11
>
> A (**simple**) **random sample** of n experimental units is one selected from the population in such a way that every different sample of size n has an equal chance of selection.

How can a random sample be generated? If the population is not too large, each observation may be recorded on a piece of paper and placed in a suitable container. After the collection of papers is thoroughly mixed, the researcher can remove n pieces of paper from the container; the elements named on these n pieces of paper are the ones to be included in the sample. Lottery officials utilize such a technique in generating the winning numbers for the state of Florida's weekly 6/49 Lotto game. Forty-nine white Ping-Pong balls (the population), each identified with a number from 1 to 49 in black numerals, are placed into a clear plastic drum and mixed by blowing air into the container. The Ping-Pong balls bounce at random until a total of six balls "pop" into a tube attached to the drum. The numbers on the six balls (the random sample) are the winning Lotto numbers.

This method of random sampling is fairly easy to implement if the population is relatively small. It is not feasible, however, when the population consists of a large number of observations. Since it is also very difficult to achieve a thorough mixing, the procedure provides only an approximation to random sampling. Most scientific studies, however, rely on **random number generators** to automatically generate the random sample. Random number generators can be found in a table of random numbers (such as Table B.1 in Appendix B) and in statistical software (such as Microsoft Excel). We illustrate the use of these random number generators in the following example.

Example 1.8   Selecting a Random Sample

Suppose we want to update the study of state lottery winners, Example 1.6. Assume there were 1,800 people who won at least $1 million in state lotteries last year and we are interested in knowing what proportion of these 1,800 winners quit their jobs within a year of winning. The target population, then, consists of the job status of these 1,800 state lottery winners. Use the random numbers in Table B.1 to generate a random sample of size $n = 5$ from the population of size $N = 1,800$.

Solution

Begin by numbering (arbitrarily) the state lottery winners in the population from 1 to 1,800. A portion of Table B.1 is reproduced in Table 1.3. The steps for obtaining a random sample using the table are outlined in the box below. The five random numbers are shown in Table 1.4. The state lottery winners assigned these numbers comprise our random sample of size 5. ❑

USING A TABLE OF RANDOM NUMBERS TO GENERATE A RANDOM SAMPLE OF SIZE n FROM A POPULATION OF N ELEMENTS

Step 1   The elements in the population are numbered from 1 to $N = 1,800$. This labeling implies that we will obtain random numbers of four digits from the table, selecting only those numbers with values less than or equal to 1,800. Note that Table 1.3 gives 5-digit random numbers in each column. Consequently, we'll use only the first four digits of each random number.

Step 2   Arbitrarily, let's begin in row 1, column 1 of the table. The random number entry given there is 10480. Using the first four digits, the

Table 1.3 Reproduction of a Portion of Table B.1

		COLUMN				
	1	**2**	**3**	**4**	**5**	**6**
Row 1	10480	15011	01536	02011	81647	91646
2	22368	46573	25595	85393	30995	89198
3	24130	48360	22527	97265	76393	64809
4	42167	93093	06243	61680	07856	16376
5	37570	39975	81837	16656	06121	91782
6	77921	06907	11008	42751	27756	53498
7	99562	72905	56420	69994	98872	31016
8	96301	91977	05463	07972	18876	20922
9	89579	14342	63661	10281	17453	18103
10	85475	36857	53342	53988	53060	59533
11	28918	69578	88231	33276	70997	79936
12	63553	40961	48235	03427	49626	69445
13	09429	93969	52636	92737	88974	33488
14	10365	61129	87529	85689	48237	52267
15	07119	97336	71048	08178	77233	13916

Table 1.4 Random Sample of Five Selected from a Population of 1,800

Random Number
1048
1501
0153
0201
0624

random number generated is 1048 (highlighted in Table 1.3). Thus, we'll choose the state lottery winner numbered 1048 as our first element of the random sample.

Step 3   Proceeding horizontally to the right across the columns (this choice of direction is arbitrary), the next entry in the table is 15011. Therefore, the lottery winner numbered 1501 represents the second element in the sample. Continuing in this manner, the remaining elements to be included in the sample are those numbered 0153, 0201, 8164 (skip), 9164 (skip), 2236 (skip), 4657 (skip), …, (proceeding to row 4), 4216 (skip), 9309 (skip), and 0624.

Self-Test
1.4

Use the random number table, Table B.1, to select a random sample of size $n = 3$ from a population of size $N = 100$.

[*Note:* In the Excel tutorial at the end of this chapter we show you how to obtain a printout of 50 random numbers using Excel.]

Although random sampling represents one of the simplest of the multitude of sampling techniques available for research, most of the statistical techniques presented in this introductory text assume that a random sample (or a sample that closely approximates a random sample) has been collected. We consider a few of the other, more sophisticated, sampling methods in optional Section 1.6.

PROBLEMS 1.26–1.33

Using the Tools

1.26 Consider a random sample from a population of size $N = 750$. To obtain the sample, you must assign each experimental unit in the population a code number. What number would you assign to:

   a. the first experimental unit on the list

   b. the twentieth experimental unit on the list

   c. the last experimental unit on the list

1.27 Starting in row 12, column 1 of the random number table, Table B.1, list the first random number for a random sample drawn from a population of size:

   a. $N = 300$ **b.** $N = 1,000$

   c. $N = 500$ **d.** $N = 50$

1.28 Use Table B.1 to draw a random sample of size $n = 10$ from a population of size $N = 700$.

1.29 Use Table B.1 to draw a random sample of size $n = 5$ from a population of size $N = 5,000$.

Applying the Concepts

FISH.XLS

1.30 Refer to the U.S. Army Corps of Engineers study of contaminated fish inhabiting the Tennessee River, Example 1.2, page 31. The data collected for all 144 fish captured are available in an Excel workbook. Select a random sample of size 10 from the population of fish weights.

CHKOUT.XLS

1.31 At a supermarket, customer checkout time is measured as the total length of time (in seconds) required for a checker to scan and total the prices of the customer's food items, accept payment, return change, and bag the items. We collected data on customer checkout times for 500 shoppers at an upscale supermarket in Gainesville, Florida. Randomly select a sample of 5 from the 500 checkout times in the data set.

1.32 A clinical psychologist is asked to view tapes in which each of six experimental subjects is discussing his or her recent dreams. Three of the six subjects have been previously classified as "high-anxiety" individuals, and the other three as "low-anxiety." The psychologist is told only that there are three of each type and is asked to select the three high-anxiety subjects.

   a. List the different samples of three subjects that may be selected by the psychologist.

   b. Do you think the sample chosen by the psychologist will be random? Explain.

1.33 A file clerk is assigned the task of selecting a random sample of 26 company accounts (from a total of 5,000) to be audited. The clerk is considering two sampling methods:

Method A: Organize the 5,000 company accounts in alphabetical order (according to the first letter of the client's last name). Then randomly select one account card for each of the 26 letters of the alphabet.

Method B: Assign each company account a 4-digit number from 0001 to 5000. Using a computer random number generator, choose 26 4-digit numbers (from 0001 to 5000) and match the numbers with the corresponding company account.

Which of the two methods would you recommend to the file clerk? Which sampling method could possibly yield a biased sample?

1.6   OTHER TYPES OF SAMPLES (OPTIONAL)

Selecting a random sample can be difficult and costly. For example, suppose we want to sample the opinion of all residents age 18 or over in a certain community on the issue of mandatory testing for AIDS in the workplace. The first obstacle to the sample selection is acquiring a *frame*, a complete listing of all the experimental units in the population. In this example, the frame is all residents age 18 or over in the community. The second obstacle is contacting the residents who were selected in the sample to obtain their opinions.

Definition 1.12

A **frame** is a list of all experimental units in the population.

To reduce the difficulty and costs associated with acquiring a frame and selecting the sample, and to increase the precision of the sample information, experts trained in the art of *survey sampling* have devised some modifications to the simple random sampling procedure. In this section we discuss three alternative methods of sampling: *stratified random sampling, cluster sampling,* and *systematic sampling.*

Example 1.9   Stratified Random Sample

Suppose we wish to sample the opinions of all heads of household in a state on some issue (e.g., mandatory AIDS testing) and further suppose that the state contains ten counties (see Figure 1.1). It might be difficult to obtain a frame listing all households within the state, but suppose we know that each county possesses a frame—namely, the households listed on its tax roll. How could we proceed?

Figure 1.1 Map of a Fictitious State with 10 Counties

Solution

Instead of combining the 10 frames and selecting a random sample from the whole state, it would be easier and less costly to select a random sample of heads of household from each county. This would enable us to obtain sample opinions not only for each county (so that particular counties could be compared) but also for the entire state. This method of sampling is called *stratified random sampling* because the population is partitioned into a number of *strata* (in this example, counties), and random samples are then selected from among the elements in each *stratum*.

Stratification is advantageous because it allows you to acquire sample information on the individual strata (e.g., counties) as well as on the entire population (e.g., state). It is often less costly to select the sample and, because the variability of the responses within a stratum (county) is usually less than the variability in responses between strata, stratification may provide more accurate estimates of strata and population parameters. However, if the target population is stratified incorrectly, stratified sampling may produce worse results than simple random sampling. ❏

Definition 1.13

A **stratified random sample** is obtained by partitioning the sampling units in the population into nonoverlapping subpopulations called *strata*. Random samples are then selected from each stratum.

Example 1.10   Cluster Sampling

Suppose we want to sample the opinions of the heads of household in a city but we know that many households, for one reason or another, are not listed on the tax roll. How could we proceed?

Solution

Since the tax roll cannot be used as a frame, we could construct a frame by numbering each of the city blocks on a map. The list of all blocks in the city would then provide a frame for selecting a random sample of city blocks, each representing a *cluster* of households. After the random sample of clusters is selected, interviewers are sent out to contact and interview all heads of household within each cluster (block) that appears in the sample. The information contained in the clusters is ultimately combined to make inferences about the population of heads of household within the city. This method of sampling is commonly known as *cluster sampling*. The advantages of cluster sampling are clear: It is often easy to form a frame consisting of clusters, and it is less costly to conduct interviews within clusters than it is to interview individual elements selected at random from the population. Cluster sampling, however, will produce unreliable results if improper clusters are selected. ❏

Definition 1.14

A **cluster sample** is a random sample in which the sampling units consist of clusters of elements.

Example 1.11   Systematic Sampling

Suppose our objective is to sample the opinions of all persons who use the downtown area of the city concerning a prospective increase in bus fares. Since it would be difficult, if not impossible, to construct a frame consisting of all people who use the downtown area, how should we proceed?

Solution

The sample could be collected by *systematically* questioning every fifth person (or every 10th person or, in general, every kth person) encountered at a particular street location. This technique, known as **systematic sampling**, is very valuable when choosing large samples. Instead of generating many random numbers, you can order the sampling frame and select every nth element.

Although this type of sampling is easy and inexpensive, it is difficult to assess the accuracy of estimates based on a systematic sample. Also, systematic sampling produces poor results if a periodic or cyclical pattern exists in the data. (For example, if prices are sampled every 30th day at a retail outlet, the prices will tend to be too low because of end-of-the-month clearance sales.) ❑

> **Definition 1.15**
>
> A **systematic sample** is obtained by systematically selecting every kth element in the population when the elements are ordered from 1 to N.

1.7 ETHICAL AND OTHER ISSUES IN STATISTICAL APPLICATIONS

According to H. G. Wells, author of such science-fiction classics as *The War of the Worlds* and *The Time Machine,* "Statistical thinking will one day be as necessary for effective citizenship as the ability to read and write." Written more than a hundred years ago, Wells' prediction is proving true. Every day we read about the results of surveys or opinion polls in our newspaper, or we hear some titillating commentary on our radio or television. Clearly, advances in information technology have led to a proliferation of data-driven studies. Not all this information, unfortunately, is good, meaningful, or important. For example, a recent segment of the popular ABC television program *20/20* exposed the truth behind a "survey" that claimed that today's teachers worry most about being assaulted by their students. The original source of this information was not a survey at all, but a Texas oilman who simply published his opinions in a conservative newsletter.[*] It is essential that we learn to evaluate critically what we read and hear and discount studies that lack objectivity or credibility.

When evaluating information that is based on a survey, we should first determine whether or not the survey results were obtained from a random sample. Surveys that employ nonrandom sampling methods are subject to serious, perhaps unintentional biases that may render the results meaningless. Even when surveys employ random sampling, they are subject to potential errors. Three of these problems are due to *selection bias, nonresponse error,* and *measurement error,* as the following examples illustrate.

*"Fact or Fiction? Exposés of So-Called Surveys." Mar. 31, 1995, segment of *20/20,* ABC Television.

Example 1.12   Selection Bias

One of the most infamous examples of improper sampling was conducted in 1936 by *Literary Digest* magazine to determine the winner of the Landon–Roosevelt presidential election. The poll, which predicted Landon to be the winner, was conducted by sending ballots to a random sample of persons selected from among the names listed in the telephone directories, magazine subscription lists, club membership lists, and automobile registrations of that year. In the actual election, Landon won in Maine and Vermont but lost in the remaining 46 states. The *Literary Digest*'s erroneous forecast is believed to be the major reason for its eventual failure. What was the cause of the *Literary Digest*'s erroneous forecast?

Solution

In 1936, the United States was still suffering from the Great Depression. During this period of economic hardship, the majority of the voting population could not afford such amenities as telephones, magazine subscriptions, club memberships, and automobiles. Consequently, the majority of voters in the election were excluded from being selected in the sample because they were not on the lists compiled by *Literary Digest*—only the rich participated in the poll.

Recall, from Definition 1.12, that a frame is a listing of all possible experimental units in the population. **Selection bias** results from the exclusion of certain experimental units from the frame so that they have no chance of being selected in the random sample. The *Literary Digest* poll, then, suffered from selection bias, leading to an erroneous forecast of the presidential winner. ❑

> **Definition 1.16**
>
> **Selection bias** results when a subset of the experimental units in the population is excluded from the population frame so that they have no chance of being selected in the sample.

Example 1.13   Nonresponse Bias

Refer to the survey of state lottery winners, first presented in Example 1.6 (page 37). Recall that the researcher mailed job status questionnaires to over 2,000 lottery winners, and 576 (only 29%) responded. Why might the results of the mail survey be biased?

Solution

In mail surveys, not everyone mailed a questionnaire will be willing to respond. Busy people, a particular and unique social class, are often too preoccupied to complete the survey's questionnaire. Or, as in the study of state lottery winners, those surveyed may be totally disinterested in the survey question (remember, they have just struck it rich in the lottery) and, therefore, less likely to respond. *Nonresponse error* results from the failure to collect data on all experimental units (i.e., state lottery winners) in the sample. This type of error results in nonresponse bias. Methods for coping with nonresponse (based primarily on re-sampling the nonresponders) are available and are discussed in the references. ❑

Definition 1.17

Nonresponse error occurs when those conducting a survey or study are unable to obtain data on all experimental units in the sample.

Example 1.14   Misleading Survey Results

The Holocaust—the Nazi Germany extermination of the Jews during World War II—has been well documented through film, books, and interviews with the concentration camp survivors. But a recent Roper poll found that one in five U.S. residents doubts the Holocaust really occurred. A national sample of over 1,000 adults and high school students were asked, "Does it seem possible or does it seem impossible to you that the Nazi extermination of the Jews never happened?" Twenty-two percent of the adults and 20% of the high school students responded "Yes" (*New York Times,* May 16, 1994). Why might the results of this survey be misleading?

Solution

A good questionnaire is designed with the intent of extracting meaningful and valid data. Do responses to the survey questions actually measure the phenomenon of interest to the researcher? Clearly, the pollsters in this example want to know the proportion of U.S. residents who believe that the Holocaust actually occurred. But the question, "Does it seem possible or does it seem impossible to you that the Nazi extermination of the Jews never happened?", is poorly worded. It is unclear whether answering "Yes" to the question means that you believe the Holocaust was real or whether you believe it never happened! This results in *measurement error,* since the pollster is unsure of what the question is really measuring. In addition to ambiguous questions, other common sources of measurement error are "leading questions" (i.e., the questions are not objectively presented in a neutral manner) and the perceived obligation on the part of the respondent to "please" the interviewer by answering questions in a certain manner. ☐

Definition 1.18

Measurement error refers to inaccuracies in the values of the data recorded. In surveys, this error may be due to ambiguous or leading questions and the interviewer's effect on the respondent.

A fourth source of error in survey research is *sampling error.* This error reflects the chance differences that occur from one sample to another when selecting random samples from the target population. For example, surveys or polls often include a statement regarding the margin of error or precision of the results—"the results of this poll are expected to be within ±4 percentage points of the actual value." In all the inferential statistical procedures presented in this text, we show how to measure and interpret sampling error.

Definition 1.19

Sampling error results from the chance differences that occur from sample to sample, when random samples are selected from the population. It measures the margin of error or precision of a sample estimate, i.e., how close the sample estimate is to the true population value.

Self-Test
1.5

An editor of a magazine for Harley-Davidson bikers argued that motor-cyclists should not be required by law to wear helmets (*New York Times,* June 17, 1995). The editor cited a survey of 2,500 bikers at a rally that found 98% opposed the law. Identify the potential sources of error in this survey.

We conclude this section with some comments about ethics in surveys and other statistical studies. Eric Miller, editor of the newsletter *Research Alert,* recently stated that "there's been a slow sliding in ethics. The scary part is that people make decisions based on this stuff. It may be an invisible crime, but it's not a victimless one." As we stated earlier, not all surveys are good, meaningful, or important—and not all survey research is ethical. However, there is a distinction between poor survey design and **unethical survey design**. The key is *intent.*

For example, selection bias becomes an ethical issue only if particular groups or individuals are purposely excluded from the population frame so that the survey's results will be more likely to favor the position of the survey's sponsor. Similarly, nonresponse error becomes an ethical issue if there are individuals that are less likely to respond to a particular survey format and the researcher knowingly designs the survey with this in mind. Any time the survey designer purposely chooses loaded, lead-in questions that will guide responses in a particular direction, or advises the interviewer through mannerism and tone to guide the responses, measurement error becomes an ethical issue. Finally, if the researcher purposely presents a survey's results without reference to sampling error in order to promote a certain viewpoint, ethics can be called into question.

The ideas presented in this section will, hopefully, cause you to critically evaluate the results of surveys or other studies presented in the media. Examine the purpose of the study, why it was conducted, and for whom, and then discard the results if the study is found to be lacking in objectivity or credibility, or is unethical.

PROBLEMS
1.34–1.38

Applying the Concepts

1.34 To gauge consumer preferences for a new product, companies sometimes employ *simulated test marketing* over conventional sampling methods. In simulated test marketing, a consumer is recruited at a shopping center and given a free sample of a new product to try at home. Later, the consumer rates the product in a telephone interview. The telephone responses are used by the test-marketing firm to predict potential sales volume. Researchers have found that simulated marketing tests identify potential product failures reasonably well, but not potential successful products. Why might the sampling procedure yield a sample of consumer preferences that underestimates sales volume of a successful new product?

1.35 Researchers who publish in professional business journals are typically university professors. Consequently, the data upon which the research is based are sometimes obtained by using students in an experimental setting. For example, the *Academy of Management Journal* (Oct. 1993) published a study on the relationship between employee commitment and employee turnover at an aerospace firm. A major part of the study involved measuring and analyzing the commitment of employees (and former employees) of the firm. A secondary portion of the study, however, utilized commitment data measured on a sample of students in an undergraduate management class. In general, comment on the use of students as experimental units in business studies.

1.36 The California Department of Health Services, Licensing and Certification (CDHSL&C) recently commissioned a survey of skilled nursing facilities in the state. The survey's purpose was to examine the infection prevention and control programs implemented at the nursing facilities (*Journal of Gerontological Nursing,* Nov. 1997). A survey was mailed to the managers of all 1,454 skilled nursing facilities in the state. (The survey was mailed only once and no follow-up reminders were sent.) A total of 444 surveys were returned, for a response rate of 30.5%. Several of the survey questions are listed below.

1. Is your facility a profit or nonprofit facility?
2. Is your facility a free-standing or hospital-based facility?
3. How many licensed beds are at your facility?
4. How many private rooms are available at your facility?
5. How many hours per week do you spend in infection prevention and control activities?
6. Do you culture new admissions for infectious diseases?
7. Do you admit patients infected with vancomycin-resistant enterococcus (VRE)?
8. On a scale of 1–4 (where 1 = not important and 4 = very important), rate the importance of teaching health care workers, patients, and visitors to wash hands at your facility.

a. Identify the type, quantitative or qualitative, of the data extracted from each survey question.
b. Identify the population of interest to the CDHSL&C.
c. What are the experimental units in this survey?
d. Explain why the survey data are a sample.
e. Is the sample randomly selected from the population?
f. What type of error might this survey suffer from?

1.37 Many opinion surveys are conducted by mail. In such a sampling procedure, a random sample of persons is selected from among a list of people who are supposed to constitute a target population (e.g., purchasers of a product). Each is sent a questionnaire and is requested to complete and return the questionnaire to the pollster. Why might this type of survey yield a sample that would produce biased inferences?

1.38 A popular ratings gimmick for television networks is to conduct a telephone survey of its viewers concerning some newsworthy issue. The TV network usually provides two or more "900" telephone numbers (each corresponding to a different opinion). The viewer chooses which number to call and is charged for the call. Because the calls are electronically monitored, the results can be shown "live" almost immediately after the polling period ends. Several years ago, the cable news network CNN provided phone numbers for viewers of its popular *Larry King Live* talk-show to voice their opinion on the guilt or innocence of O. J. Simpson, the former NFL star accused (and since acquitted) of murdering his ex-wife Nicole Brown and Ronald Goldman. Explain why the sample of opinions collected by this TV survey could have produced biased results.

REAL-WORLD REVISITED
Obedience to Authority—The Shocking Truth

A basic principle upon which our society is organized is obedience to authority. Psychologists and sociologists agree that without obedience, our society would soon be very chaotic. But what if a person is asked to obey orders that appear to be evil or malicious? Could we expect that person to disobey? In a classic psychology study, Stanley Milgram (1974) conducted a series of experiments designed to isolate the psychological factors that influence a person's behavior in such a setting.

The basic set-up of the experiment was as follows. Two people were brought into a room at an appointed time; one person was assigned to play the role of the "learner" and the other was assigned to play the role of the "teacher." The learner's task was to learn a list of word pairs, for example, box-boat. After reviewing the entire list with the learner, the teacher (the subject of the experiment) went through the list one-by-one, giving the first member of the word pair (e.g., box). The learner was then asked to respond with the corresponding second member of the pair (e.g., boat). If the learner answered correctly, the teacher went on to the next word pair. However, if the learner gave the wrong answer, the teacher was instructed to deliver an electric shock by depressing one of 30 levers, ranging from 15 volts to 450 volts in increments of 15 volts. In addition to the voltage, the first 28 levers were labeled in groups of four as follows: Slight shock, Moderate shock, Strong shock, Very strong shock, Intense shock, Extreme intensity shock, and Danger: Severe shock. The final two levers were simply labeled "XXX." Teachers were instructed to start with the lowest-level shock and to increase the level of shock with each error until the entire list was learned correctly. [*Note:* Unknown to the teacher, the learner was an actor who conspired with the experimenter. The shocks were imaginary, but the learner responded in a manner that led the teacher to believe that, in fact, they were real.]

If the teacher raised questions about the experiment and whether it should be continued, the experimenter would respond with one of the following four statements:

Statement 1: Please continue or Please go on.

Statement 2: The experiment requires that you continue.

Statement 3: It is absolutely certain that you continue.

Statement 4: You have no other choice; you must go on.

Only if the teacher refused to continue after all four statements had been made was the experiment stopped. One variable measured by Milgram was how far (in the sequence of shocks) the teacher proceeded before the experiment was terminated. For this study, we will refer to this variable as the shock level (measured in volts).

Milgram manipulated the basic experiment in several ways to study the impact of different situational factors on obedience. Teachers (i.e., the subjects) were randomly assigned to one of five different experimental settings:

Setting 1: *Predicted behavior.* Subjects (i.e., teachers) had the experimental set-up described to them and were asked to predict how far they would go (although they did not actually perform the experiment).

Setting 2: *Remote.* The learner could not be heard or seen but pounded on the walls at 300 volts and ceased responding and pounding at 315 volts.

Setting 3: *Voice feedback.* The learner could not be seen but vocal protest could be clearly heard.

Setting 4: *Proximity.* The learner was placed in the same room so that he could be both seen and heard.

Setting 5: *Touch proximity.* The learner received a shock only when his hand rested on a shock plate. At 150 volts, the learner demanded to be released from the experiment and refused to put his hand on the shock plate. The teacher was instructed to take the learner's hand and force it onto the plate.

What were the results of Milgram's experiment? Surprisingly, two-thirds of all the teachers in the study continued to obey the experimenter and applied shock levels of 400 or more volts despite the pleas of the learners. From this, Milgram concluded that a high level of obedience to authority exists in our society.

Study Questions
a. Identify the response variable measured in Milgram's study and its type (quantitative or qualitative).
b. Identify the population(s) of interest to Milgram.
c. Identify the data collection method.
d. Describe how Milgram could have obtained a sample representative of each population identified in part b. Do you think he did this?
e. Select any two of the five experimental settings utilized by Milgram. Suppose you were to compare the shock levels under one setting to the shock levels under the other setting. Explain the dangers of drawing definitive conclusions from such a comparison.
f. Discuss other potential problems in Milgram's study.

Key Terms

Starred (*) terms are from the optional section of this chapter.

Cluster sample*   49
Data   28
Descriptive statistics   35
Designed experiment   41

Experimental unit   29
Frame*   48
Inferential statistics   37
Measurement error   52

Nonresponse error   52
Observational study   42
Population   34
Published source   41

Qualitative (categorical) data   29
Quantitative data   29
Random number generator   45
Random sample   44
Reliability   37

Representative sample   43
Sample   35
Sampling error   52
Selection bias   51
Statistics   29

Stratified random sample*   49
Survey   41
Systematic sample*   50
Unethical survey design   53
Variable   29

Supplementary Problems 1.39–1.46

Using the Tools

1.39   Use a random number table or a computer to generate a random sample of 10 observations from a population with 40,000 elements.

1.40   Use a random number table or a computer to generate a random sample of 100 observations from a population with 350,000 elements.

Applying the Concepts

1.41   State whether each of the following variables is quantitative or qualitative.

  a.   Number of acres in a plot of land

  b.   Mode of transportation (to and from work) for a city employee

  c.   Type of residential water-heating system

  d.   Time required for postoperative pain to be relieved in surgery patients

1.42   Classify each of the following variables as quantitative or qualitative.

  a.   Political affiliation of a chief executive whose firm is listed in the *Fortune* 500

  b.   Geographical region with the highest unemployment rate in the United States

  c.   Gas mileage attained by an automobile

  d.   Fee charged by an attorney to handle an uncontested divorce

  e.   Highest educational degree attained by members of the faculty at a community college

1.43   Hundreds of sea turtle hatchlings, instinctively following the bright lights of condominiums, wandered to their deaths across a coastal highway in Florida (*Tampa Tribune,* Sept. 16, 1990). This incident led researchers to begin experimenting with special low-pressure sodium lights. On one night, 60 turtle hatchlings were released on a dark beach and their direction of travel noted. The next night, the special lights were installed and the same 60 hatchlings were released. Finally, on the third night, tar paper was placed over the sodium lights. Consequently, the direction of travel was recorded for each hatchling under three experimental conditions—darkness, sodium lights, and sodium lights covered with tar paper.

  a.   Identify the population of interest to the researchers.

  b.   Identify the sample.

  c.   What type of data were collected, quantitative or qualitative?

  d.   What data collection method was employed?

1.44   When Nissan first introduced its new Infiniti luxury cars, its television ad campaign was renowned for a novel gimmick: The automobiles were nowhere in sight. The Infiniti ads, which depicted lushly photographed trees, boulders, lightning bolts, and ocean waves (but no cars) were found by a nationwide Gallup telephone poll of 1,000 consumers to be one of the best-recalled commercials on television (*Time,* Jan. 22, 1990).

a.   Identify the experimental units in this study.

b.   Describe the population of interest to the pollsters.

c.   Identify the sample.

d.   What is the inference made by the Gallup poll?

e.   Describe the data collection method used by the pollsters.

f.   What types of errors might bias the study results? Explain.

1.45   A Brandeis University researcher experimented with giving propranolol, one of the class of heart drugs called beta blockers, to nervous high school students prior to taking their SATs (*Newsweek,* Nov. 16, 1987). In theory, the same calming effect that beta blockers provide heart patients could also be used to reduce anxiety in test takers. To test this theory, the researcher selected 22 high school juniors who had not performed as well on the SAT as they should have based on IQ and other academic evaluations. One hour before the students repeated the test in their senior year, each was administered a dosage of a beta blocker. [*Note:* Typically, students who retake the test without special preparations will increase their scores by an average of 38 points. These 22 students improved their scores by an average of 120 points!]

a.   Identify the experimental units in this study.

b.   Identify the measured variable.

c.   Describe the population of interest to the researcher. (Give the precise statistical definition.)

d.   Identify the data collection method used.

e.   Describe the sample.

f.   Based on the sample results, what inference would you make about the use of beta blockers to increase SAT scores? (In Chapter 9 we show you how to assess the reliability of this type of inference.)

g.   The director of research and development for the College Board (sponsors of the SAT) warns that "the findings have to be taken with a great deal of caution" and that they should not be interpreted to mean "that someone has discovered the magic pill that will unlock the SAT for thousands of teenagers who believe they do not do as well as they should because they're nervous." Give several reasons for issuing such a warning.

1.46   In her book, *Women and Love: A Cultural Revolution in Progress* (Knopf Press, 1988), Shere Hite reveals some startling statistics describing how women she surveyed feel about contemporary relationships: 84% of women are not emotionally satisfied with their relationship; 95% of women report "emotional and psychological harassment" from their men; 70% of women married 5 years or more are having extramarital affairs; and only 13% of women married more than 2 years are "in love." Hite conducted the survey by mailing out 100,000 questionnaires to organizations such as church groups, women's voting/political groups, women's rights organizations, and counseling centers for women, and asked the groups to circulate the questionnaires to their members. Included were volunteer respondents who wrote in for copies of the survey. Each questionnaire consisted of 127 open-ended questions, many with numerous subquestions and follow-ups. Hite's instructions read: "It is not necessary to answer every question! Feel free to skip around and answer those questions you choose." Approximately 4,500 completed questionnaires were returned for a response rate of 4.5%, and they form the data set from which these percentages were determined.

a. Identify the population of interest to Shere Hite. What are the experimental units?

b. Identify the variables of interest to Hite. Are they quantitative or qualitative variables?

c. Describe Hite's data collection method.

d. Hite claims that the 4,500 women surveyed are a representative sample of all women in the United States, and therefore the survey results imply that vast numbers of women are "suffering a lot of pain in their love relationships with men." Do you agree?

e. Discuss the difficulty in obtaining a random sample of women across the United States to take part in a survey similar to the one conducted by Shere Hite.

1.E.1   Using Microsoft Excel to Select a Random Sample

Solution Summary:

Use the PHStat add-in to select a random sample.

 Example: Random Sample of $n = 50$.

Solution:

To select a random sample of $n = 50$ from a population of 1,800 state lottery winners, do the following:

1.   If the PHStat add-in has not been previously loaded, load the add-in using the instructions of Section EP.3.2. (You will also need to run the disk setup program using the procedure in Appendix C if you have never previously installed the files on the disk that accompanies this text.)

2.   Select File | New to open a new workbook.

3.   Select PHStat | Data Preparation | Random Sample Generator.

4.   In the Random Sample Generator, enter the information and make the selections show in Figure 1.E.1. Click the OK button.

E Figure 1.E.1
Random Sample
Generator Dialog Box

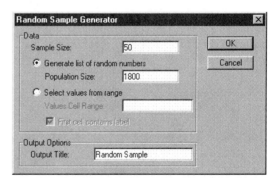

Using the settings of Figure 1.E.1, the add-in produces a randomly generated list of 50 different numbers.

Chapter 2

Exploring Data with Graphs and Tables

CONTENTS

2.1 The Objective of Data Description

2.2 Describing a Single Qualitative Variable: Frequency Tables, Bar Graphs, and Pie Charts

2.3 Describing a Single Quantitative Variable: Frequency Tables, Stem-and-Leaf Displays, and Histograms

2.4 Exploring the Relationship between Two Qualitative Variables: Cross-Classification Tables and Side-by-Side Bar Graphs

2.5 Exploring the Relationship between Two Quantitative Variables: Scatterplots

2.6 Proper Graphical Presentation

EXCEL TUTORIAL

2.E.1 Using Microsoft Excel to Describe a Single Qualitative Variable

2.E.2 Using Microsoft Excel to Describe a Single Quantitative Variable

2.E.3 Using Microsoft Excel to Explore the Relationship between Two Qualitative Variables

2.E.4 Using Microsoft Excel to Explore the Relationship between Two Quantitative Variables

REAL-WORLD APPLICATION
A Bad Moon Rising

Is your behavior influenced by the phases of the moon? Despite the lack of supporting scientific evidence, many people still associate aberrant behavior with a full moon. To measure the degree to which people believe in lunar effects, a team of psychologists administered a questionnaire to a random sample of 157 college undergraduates. How can we use statistics to make sense of the data? Graphical methods that rapidly convey information contained in a data set are the topic of this chapter. In Real-World Revisited at the end of this chapter, we consider graphical methods to summarize the lunar-effects data.

2.1 THE OBJECTIVE OF DATA DESCRIPTION

In Chapters 2 and 3 we demonstrate how to explore data using descriptive methods. The objective of data description is to summarize the characteristics of a data set, identify any patterns in the data, and to present that information in a convenient form. In this chapter we will show you how to construct charts, graphs, and tables that convey the nature of a data set. The procedure that we will use to accomplish this objective in a particular situation depends on the type of data, qualitative or quantitative, that you want to describe, and the number of variables measured. In the next chapter we present numerical descriptive measures of a data set.

2.2 DESCRIBING A SINGLE QUALITATIVE VARIABLE: FREQUENCY TABLES, BAR GRAPHS, AND PIE CHARTS

Consider the data collected from measuring one qualitative variable. For example, suppose we observe the rank of a university professor, where rank is recorded as assistant, associate, or full professor. Consequently, each value of rank falls into one of three different categories or *classes*.

Definition 2.1

A **class** is one of the categories into which the qualitative data can be classified.

The summary information we seek about a qualitative variable is either the number of observations falling in each class (called a *frequency*), the proportion of the total number of observations falling in each class (called a *relative frequency*), or the *percentage* of the total number of observations falling in each class.

Definition 2.2

The **class frequency** for a particular class is the number of observations falling in that class.

Definition 2.3

The **class relative frequency** for a particular class is equal to the class frequency divided by the total number of observations.

$$\text{Class relative frequency} = \frac{\text{Class frequency}}{\text{Total number of observations}}$$

Definition 2.4

A **class percentage** is the relative frequency for the class multiplied by 100.

$$\text{Class percentage} = (\text{Class relative frequency}) \times 100$$

Table 2.1 Intensity Ratings of 17 Men's Cologne Brands

Fragrance Brand	Intensity Rating
Aramis	Strong
Brut	Very strong
Drakkar Noir	Very strong
Egoïste	Strong
English Leather	Mild
Escape for Men	Strong
Eternity for Men	Very strong
Gravity	Strong
Lancer	Strong
Obsession for Men	Strong
Old Spice	Strong
Polo	Very strong
Preferred Stock	Strong
Realm for Men	Strong
Safari For Men	Very strong
Stetson	Mild
Tribute	Strong

Data Source: Consumer Reports, Dec. 1993, p. 773.

Class frequencies, relative frequencies, or percentages can be displayed in table form or graphically. **Bar graphs** and **pie charts** are two of the most widely used graphical methods for describing qualitative data. We illustrate in Example 2.1.

Example 2.1   Bar Graphs and Pie Charts

The popular magazine *Consumer Reports* recently evaluated 17 men's cologne (fragrance) brands. A number of characteristics were rated, including intensity of the cologne. The qualitative variable, intensity, was measured as mild, strong, or very strong. The intensity ratings for the 17 brands are listed in Table 2.1.

a. Construct a frequency table for the data.

b. Portray the data in a bar graph.

c. Portray the data in a pie chart.

Solution

a. Examining Table 2.1, we observe that 5 brands are rated very strong, 10 are rated strong, and 2 are rated mild. These numbers—5, 10, and 2—represent the class frequencies for the three classes of intensity rating and are shown in the accompanying frequency table, Table 2.2.

Table 2.2 Frequency Table for Cologne Intensity Rating Data

Class	Frequency	Relative Frequency	Percentage
Very strong	5	.294	29.4
Strong	10	.588	58.8
Mild	2	.118	11.8
TOTALS	17	1.000	100.0

E **Figure 2.1** Excel Frequency Bar Graph for Cologne Intensity Rating

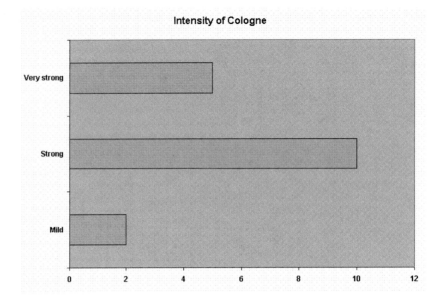

Intensity of Cologne

Class relative frequencies and percentages are also shown in Table 2.2. Using Definition 2.3, the relative frequencies for the three classes are:

$$\text{Very strong: } \frac{5}{17} = .294$$

$$\text{Strong: } \frac{10}{17} = .588$$

$$\text{Mild: } \frac{2}{17} = .118$$

Class percentages are obtained by multiplying each of the class frequencies by 100.

b. A bar graph displays the class frequencies, relative frequencies, or percentages. Figure 2.1 is a horizontal frequency bar graph for the cologne data created using Microsoft Excel. The figure contains a rectangle, or bar, for each intensity class; the length of the bar is proportional to the class frequency. (Optionally, the bar heights can be proportional to the class relative frequencies or percentages.) The bar graph makes it easy to see that the majority of the 17 men's cologne brands had a strong intensity rating. [*Note:* Some statistical software packages reverse the axes and display the bars in a vertical fashion. See the step box on how to construct a vertical bar graph.]

Qualitative variables can be either *nominal* or *ordinal* variables. The values of an ordinal variable can be naturally ordered—like "mild," "strong," and "very strong" in Example 2.1. The values of a nominal variable (e.g., "male" and "female" for gender) have no natural ordering. The bars in an ordinal variable bar graph are usually arranged according to their natural ordering as in Figure 2.1.

c. A pie chart conveys the same information as a bar chart. Typically, class relative frequencies are shown on a pie chart. Figure 2.2 is a pie chart for the cologne data, also created using Excel. Note that the pie is divided into three slices, one for each of the three classes. The size (angle) of each slice is proportional to the class relative frequency. For example, since a circle spans 360°, the slice assigned to a strong intensity rating is 59% of 360°, or .59(360) = 212°. It is common to show the percentage of measurements in each class on the pie chart as indicated. ❏

E **Figure 2.2** Excel Pie Chart for Cologne Intensity Ratings

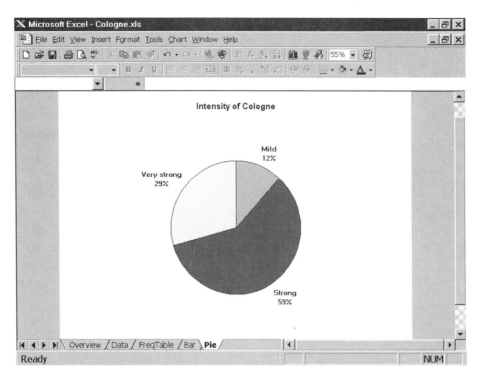

CONSTRUCTING A VERTICAL BAR GRAPH

Step 1   Summarize the data in a frequency table. The table should contain the frequency and relative frequency for each class (or category) of the qualitative variable.

Step 2   Draw horizontal and vertical axes on graph paper. The vertical axis can represent either class frequency or class relative frequency. The classes (or categories) of the qualitative variable should be marked under the horizontal axis.

Step 3   Draw in the bars for each class (or category). The height of the bar should be proportional to either the class frequency or class relative frequency.

CONSTRUCTING A PIE CHART

Step 1   Summarize the data in a frequency table. The table should contain the frequency and relative frequency for each class (or category) of the qualitative variable.

Step 2   Draw a circle (360°) on a sheet of paper.

Step 3   Calculate the size (angle) of each class (or category) by multiplying the class relative frequency by 360°.

Step 4   Draw the sections of the pie using the angles computed in step 3. Label each section by its class (or category) name. You may want to mark the class relative frequency (or percentage) on each slice.

E **Figure 2.3** Excel Pie
Chart for NCAP
Passenger Star Rating

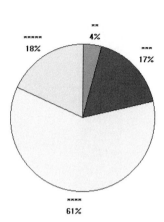

Pie Chart for Drivstar

Example 2.2   Interpreting a Pie Chart

CRASH.XLS

Each year the National Highway Traffic Safety Administration (NHTSA) crash tests new car models to determine how well they protect the driver and front-seat passenger in a head-on collision. The *New Car Assessment Program* (NCAP) uses two dummies of average human size that contain instruments designed to measure the forces and impacts that occur during a head-on crash into a fixed barrier at 35 mph. The NCAP test results for 98 cars (model year 1997) are stored in the Excel workbook CRASH.XLS. (See Appendix C for the file layout of this data set.) One of the variables reported by the NCAP is Driver Star Rating, which ranges from one star (*) to five stars (*****). The more stars in the rating, the better the level of crash protection in a head-on collision. Summarize the Driver Star Rating data for the 98 cars in the NCAP report. Interpret the results.

Solution

The variable, Driver Star Rating, is qualitative in nature; thus, we can use either a bar graph or pie chart to summarize its values. Using Excel, we produced the pie chart displayed in Figure 2.3. Each slice of the Excel pie chart represents one of the categories of Driver Star Rating. Note that the number and percentage of the 98 cars falling into each category are shown on the pie chart. We see that 18% (or 18 cars) had the highest level of crash protection (five stars), 61% (or 59 cars) were rated four stars, 17% (or 17 cars) were rated three stars, and 4% (or 4 cars) were rated two stars. None of the cars test crashed were rated one star (the lowest level of protection). ❑

Self-Test
2.1

Fetal alcohol syndrome is a group of abnormalities found in children born to chronic alcoholic mothers. Each of 60 children diagnosed as having the syndrome was examined for the abnormality of the most serious nature, with the results illustrated by the bar graph in Figure 2.4. Discuss the information provided by the graph. Which abnormality occurs most often as the child's most serious problem?

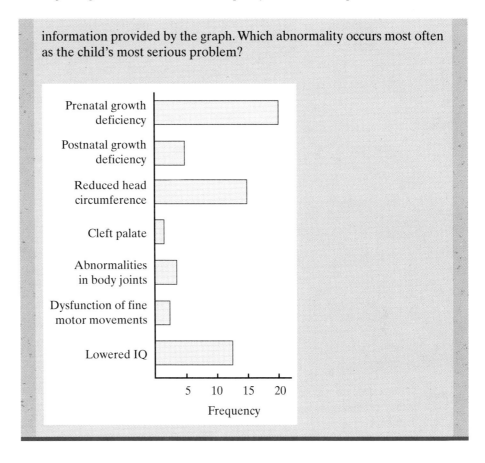

Figure 2.4 Frequency of Fetal Alcohol Syndrome Abnormalities

PROBLEMS 2.1–2.14

Using the Tools

Category	Frequency
A	13
B	28
C	9

Problem 2.1

2.1 Suppose that a qualitative variable has three categories with the frequency of occurrence shown at left.

a. Compute the percentage of values in each category.

b. Construct a bar graph.

c. Construct a pie chart.

2.2 Complete the following table.

Grade on Statistics Exam	Frequency	Relative Frequency	Percentage
A	16	.08	
B	36		
C	90		
D	30		
F	28		
TOTALS		1.00	100

Category	Relative Frequency
A	12
B	29
C	35
D	24

Problem 2.3

2.3 Suppose that a qualitative variable has four categories with the relative frequencies shown at left.

a. Construct a bar graph.   **b.** Construct a pie chart.

2.4 A qualitative variable with four classes (W, X, Y, and Z) is measured for each of 25 experimental units sampled from the population. The data are listed in the table.

Experimental Unit	Class	Experimental Unit	Class	Experimental Unit	Class
1	Y	10	Y	18	Y
2	Y	11	X	19	Y
3	W	12	Y	20	X
4	Z	13	X	21	W
5	X	14	X	22	Y
6	X	15	Z	23	W
7	Y	16	X	24	X
8	W	17	Y	25	Y
9	Z				

a. Compute the frequency of each class.

b. Compute the relative frequency of each class.

c. Construct a relative frequency bar graph for the data.

Applying the Concepts

2.5 The pie chart below describes the fate of the (estimated) 240 million automobile tires that are scrapped in the U.S. each year.

a. Interpret the pie chart.

b. Convert the pie chart into a relative frequency bar graph.

c. Convert the pie chart into a frequency bar graph.

2.6 *Sports Illustrated* (Dec. 8, 1997) polled 1,835 middle school and high school students from across the U.S. on how race impacts sports participation. Students were asked to agree or disagree with the statement: "African-American players have become so dominant in sports such as football and basketball that many whites feel they can't compete at the same level as blacks." The results are summarized in the accompanying pie charts. Interpret the results.

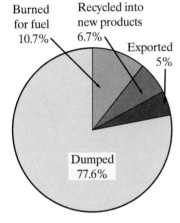

Burned for fuel 10.7%   Recycled into new products 6.7%   Exported 5%   Dumped 77.6%

Source: U.S. Environmental Protection Agency and National Solid Waste Management Association.

Problem 2.5

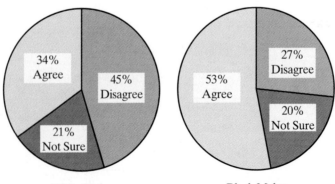

White Males — 34% Agree, 45% Disagree, 21% Not Sure

Black Males — 53% Agree, 27% Disagree, 20% Not Sure

Source: SI poll of 1,835 middle school and high school students from across the U.S.

2.7   According to the American Optometric Association (AOA), about 10% of the U.S. population have normal visual acuity, while 60% suffer from farsightedness and 30% suffer from nearsightedness.

a.   Given this information, construct a pie chart to describe the distribution of visual acuity problems in the United States.

b.   Of those that have visual acuity problems (nearsighted or farsighted), the AOA has found that many do not wear any sort of corrective lenses. Use the information provided in the table to construct a relative frequency bar chart for the four categories shown.

Vision Problem Category	Percent
Nearsighted, corrected	27
Nearsighted, uncorrected	7
Farsighted, corrected	27
Farsighted, uncorrected	39
TOTAL	100

Source: American Optometric Association.

c.   What percentage of those with vision impairments do not wear any corrective lenses?

2.8   Researchers at the University of Florida Center for Solid and Hazardous Waste Management inspected roadside trash at four sites in each of 67 Florida counties (*Focus,* Spring 1995). Each piece of trash was classified according to type. A percentage breakdown of the large litter pieces found is provided in the table. Portray the data graphically with a pie chart.

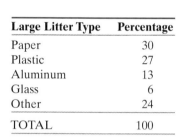

Large Litter Type	Percentage
Paper	30
Plastic	27
Aluminum	13
Glass	6
Other	24
TOTAL	100

Problem 2.8

2.9   Archaeologists use the distribution of artifacts found at an archaeological site to make inferences about the economy and lifestyle of ancient inhabitants. Recent archaeological digs at Motupore Island in Papua, New Guinea, have revealed that shell bead manufacturing was a significant part of the island's prehistoric economy (*Archaeol. Oceania,* Apr. 1997). The site was excavated for artifacts related to the manufacture of beads: stone drillpoints, shell beads, and grinding slabs. The distribution of these artifacts at the excavation site is shown in the accompanying table.

Type of Artifact	Number Discovered
Drillpoint	566
Bead	385
Grinding slab	43

Source: Allen, J., et al. "Identifying specialization, production, and exchange in the archaeological record: The case of shell bead manufacture on Motupore Island, Papua." *Archaeol. Oceania,* Vol.32, No. 1, Apr. 1997, p. 27 (Table 1).

a.   Identify the experimental unit and the variable measured for this study.

b.   Portray the artifact distribution with a graph.

c.   Identify the artifact found most often at the site. What percentage of the time does it occur?

2.10   Researchers at York University and Queen's University in Canada examined the formation of melodic expectancies in a melody-completion task (*Perception & Psychophysics,* Oct. 1997). Fifty subjects with a high training in music (i.e., 5 or more years of formal training and currently playing an instrument) and fifty subjects with little musical training participated in the study. After listening to a short (2-note) melody on a piano keyboard, each subject was asked to produce a melody that followed naturally from the two notes. The pitch of the initial note selected by the subject was recorded. The frequencies of the continuation notes selected are shown in the table.

TRAINED LISTENERS		UNTRAINED LISTENERS	
Note	Frequency	Note	Frequency
D_3	1	D_3	1
G_3	1	F_3	1
$A_3^{\#}$	1	G_3	2
C_4	12	B_3	1
$C_4^{\#}$	1	C_4	15
D_4	1	D_4	1
E_4	15	$D_4^{\#}$	2
F_4	7	E_4	15
G_4	8	F_4	3
A_4	3	G_4	6
		B_4	2
		C_5	1
TOTALS	50		50

Source: Thompson, W. F., et al. "Expectancies generated by melodic intervals: Evaluation of principles of melodic implication in a melody-completion task." *Perception & Psychophysics,* Vol. 59, No. 7, Oct. 1997, p. 1076.

   a.   Use a relative frequency bar graph to describe the notes selected by the trained listeners.

   b.   Repeat part **a** for the untrained listeners.

   c.   Compare the two bar graphs. What can you conclude from the data? Do the untrained listeners select certain notes with the same relative frequency as the trained listeners?

2.11   At the start of the 1995 National Football League (NFL) season, Jerry Jones, owner of the Dallas Cowboys, argued that since his team was more popular than other teams and generated the most revenues from sales of team products, it deserved a proportionally higher share in earnings which, under NFL policy, are distributed equally to all 30 teams. The summary table on page 71 displays the percentage market share of sales of licensed products by the various NFL teams.

   a.   Construct a bar graph for the data.

   b.   Construct a pie chart for the data.

   c.   Which of these two graphs do you prefer for presentation purposes? Why?

   d.   Use one of the graphs to support or refute Jerry Jones' claim.

2.12   Refer to the *Journal of Literacy Research* (Dec. 1996) study of teachers who read weekly to children, Problem 1.13, page 39. The teachers asked

Team	Percent Sales of Licensed Products
Carolina	5.0
Dallas	20.8
Green Bay	5.1
Kansas City	4.8
Miami	6.6
New England	3.6
New York Giants	3.2
Oakland	4.7
Pittsburgh	4.7
San Francisco	11.3
Other 20 teams	30.2
TOTAL	100.0

Problem 2.11

approximately 2,000 questions during three different sessions with the children. Each question was classified according to one of four different types: known information questions, opinion questions, connection questions, and conditional questions. (See Problem 1.13 for an example of each.) The percentage of questions falling into each category during each session is listed in the table.

Question Type	First Session	Second Session	Third Session
Known-information	28%	23%	15%
Opinion	49%	51%	56%
Connection	15%	18%	22%
Conditional	8%	8%	7%
TOTALS	100%	100%	100%

Source: Wolf, S. A., et al. "What's after 'What's that?': Preservice teachers learning to ask literary questions." *Journal of Literacy Research,* Vol. 28, No. 4, Dec. 1996, p. 466 (Figure 2).

a. For each reading session, construct a bar graph to describe the distribution of question types.

b. Do you detect any trends in the bar graphs, part **a**?

2.13 Consider a study of aphasia published in the *Journal of Communication Disorders* (Mar. 1995). Aphasia is the "impairment or loss of the faculty of using or understanding spoken or written language." Three types of aphasia have been identified by researchers: Broca's, conduction, and anomic. They wanted to determine whether one type of aphasia occurs more often than any other, and, if so, how often. Consequently, they measured aphasia type for a sample of 22 adult aphasics. The accompanying table gives the type of aphasia diagnosed for each aphasic in the sample. Analyze the data for researchers and present the results in graphical form.

Type of Aphasia	Number of Subjects	Proportion
Broca's	5	.227
Conduction	7	.318
Anomic	10	.455
TOTALS	22	1.000

2.14 *Child Welfare* (Jan./Feb. 1992) reported on a study of 124 adoptees who recently reunited with their biological mothers and relatives. One aspect of the study dealt with how often the adoptees experienced "quest" feelings (i.e., the desire to meet or learn about their biological parents) before their reunion. The accompanying table reports the frequency of quest feelings experienced by the 124 adoptees at three different ages: before age 10, between 10 and 17, and 18 or older.

a. For each category, summarize the data on frequency of quest feelings with a bar graph.

b. Compare and contrast the three graphs, part **a**. What inference can you make?

| | AGE CATEGORY | | |
Frequency of Quest Feelings	Before 10	10–17	18 or Older
Very often, or somewhat often	19.8%	47.4%	87.5%
Occasionally	16.5	27.8	12.5
Almost never	47.2	22.7	0
Uncertain	16.5	2.1	0
TOTALS	100.0%	100.0%	100.0%

Source: Sachdev, P. "Adoption reunion and after: A study of the search process and experience of adoptees." *Child Welfare,* Vol. 71, No. 1, Jan./Feb. 1992, p. 59 (Table 2).

2.3 DESCRIBING A SINGLE QUANTITATIVE VARIABLE: FREQUENCY TABLES, STEM-AND-LEAF DISPLAYS, AND HISTOGRAMS

Stem-and-leaf displays and **histograms** are the most popular graphical methods for describing quantitative data sets. Like the bar graphs and pie charts of Section 2.2, they may show class frequencies, class relative frequencies, or class percentages. The difference is that the classes do not represent categories of a qualitative variable; instead, they are formed by grouping the numerical values of the quantitative variable that you want to describe.

For small data sets (say, 30 or fewer observations) with measurements with only a few digits, stem-and-leaf displays can be constructed quickly by hand. Histograms, on the other hand, are better suited to the description of large data sets, and they permit greater flexibility in the choice of the classes. Of course, we can generate any of these graphs with computer software.

Example 2.3   Stem-and-Leaf Display

Postmortem interval (PMI) is defined as the elapsed time between death and an autopsy. Knowledge of PMI is considered essential when conducting research on human cadavers. Table 2.3 gives the PMI (in hours) for a sample of 22 human brain specimens obtained at autopsy.

a. Summarize the quantitative data with a stem-and-leaf display.

b. Interpret the graph.

Table 2.3 Postmortem Intervals for 22 Human Brain Specimens

PMI.XLS

5.5	14.5	6.0	5.5	5.3	5.8	11.0	6.1	7.0
14.5	10.4	4.6	4.3	7.2	10.5	6.5	3.3	7.0
4.1	6.2	10.4	4.9					

Source: Hayes, T. L., and Lewis, D. A. "Anatomical specialization of the anterior motor speech area: Hemispheric differences in magnopyramidal neurons." *Brain and Language,* Vol. 49, No. 3, June 1995, p. 292 (Table 1).

Solution

a. In a stem-and-leaf display, each quantitative measurement is broken into a *stem* and a *leaf.* One or more of the digits will make up the stem, while the remaining digits (or digit) will be the leaf. The stems represent the classes in a graph; the leaves reflect the number of measurements in each class.

   Figure 2.5 is a stem-and-leaf display for the PMI data, produced using the PHStat Add-in for Excel. The different stems are listed vertically to the left of the line in Figure 2.5. You can see that these stems, 3, 4, 5, 6, ..., 13, and 14, represent the digit(s) to the left of the decimal point. The leaf of each number in the data set (i.e., the digit to the right of the decimal) is placed in the row of the display corresponding to the number's stem. For example, for the PMI value of 7.0 hours, the leaf 0 is placed in stem row 7. Similarly, for the PMI of 4.9 hours, the leaf 9 is placed in stem row 4. The usual convention is to list the leaves of each stem in increasing order, as shown in Figure 2.5.

b. You can see that the stem-and-leaf display in Figure 2.5 partitions the data set into 12 classes corresponding to the 12 stems listed. The class corresponding to the stem 3 would contain all PMI values from 3.0 to 3.9 hours; the class corresponding to the stem 4 would contain all PMI values from 4.0 to 4.9 hours; etc. The number of leaves in each class gives the class frequency. Thus, a stem-and-leaf display makes it easy to calculate class frequencies and relative frequencies.

   The frequencies and relative frequencies for the 12 PMI classes (stems) are shown in Table 2.4.

   Note that the 22 PMI values are scattered over the interval from 3.0 hours to less than 14.9 hours, with most falling between stems 4 and 11.

E **Figure 2.5** Stem-and-Leaf Display for PMI Data Obtained from the PHStat Add-in for Excel

	A	B	C	D	E
1				Stem-and-Leaf Display	
2				for PMI	
3				Stem unit: 1	
4					
5	Statistics			3	3
6	Sample Size	22		4	1 3 6 9
7	Mean	7.3		5	3 5 5 8
8	Median	6.15		6	0 1 2 5
9	Std. Deviation	3.184935		7	0 0 2
10	Minimum	3.3		8	
11	Maximum	14.5		9	
12				10	4 4 5
13				11	0
14				12	
15				13	
16				14	5 5

Table 2.4 Frequency Table for Stem-and-Leaf Display of PMI Data

PMI Stem Class	Frequency	Relative Frequency
3	1	1/22
4	4	4/22
5	4	4/22
6	4	4/22
7	3	3/22
8	0	0/22
9	0	0/22
10	3	3/22
11	1	1/22
12	0	0/22
13	0	0/22
14	2	2/22
TOTALS	22	1

Summing the relative frequencies for the stems 4, 5, 6, 7, 8, 9, 10 and 11, we obtain

$$\frac{4}{22} + \frac{4}{22} + \frac{4}{22} + \frac{3}{22} + \frac{0}{22} + \frac{0}{22} + \frac{3}{22} + \frac{1}{22} = \frac{19}{22} = .86$$

Thus, 86% of the 22 postmortem intervals range between 4 hours and 11 hours. ❑

CONSTRUCTING A STEM-AND-LEAF DISPLAY

Step 1 Define the stems and leaves so that the resulting graph is meaningful. [*Hint:* Choose the stem so that the total number of stems will range between 5 and 15, approximately.]

Step 2 List the stems in order in a column, starting with the smallest stem and ending with the largest.

Step 3 Proceed through the data set, placing the leaf for each observation in the appropriate stem row. (You may want to place the leaves of each stem in increasing order.)

**Self-Test
2.2**

Consider the sample data shown here.

26	34	21	32	42	36	28	38	17	39	22	12
56	39	25	41	30	23	27	19				

a. Using the first digit as a stem, list the stem possibilities in order.

b. Place the leaf for each observation in the appropriate stem row to form a stem-and-leaf display.

c. Compute the relative frequencies for each stem.

SAL50.XLS

Table 2.5 Starting Salaries for a Sample of College Graduates

$30,000	$20,600	$32,400	$14,200	$15,700
25,200	26,100	30,300	20,400	38,900
22,800	20,400	17,400	30,900	29,400
15,300	24,400	32,700	23,200	23,100
34,600	31,400	22,100	26,700	33,100
26,600	36,700	22,800	21,400	16,800
20,700	43,900	25,200	26,700	40,500
11,000	36,600	28,200	26,300	20,100
21,500	20,500	25,600	24,900	20,000
20,400	37,500	28,700	13,200	28,300

Example 2.4 Frequency Histogram

Table 2.5 gives the starting salaries for 50 graduates of the University of South Florida who returned a mail questionnaire shortly after obtaining a job.

a. Construct a frequency histogram for the data.

b. Interpret the resulting figure.

Solution

a. *Step 1* The first step in constructing the frequency histogram for this sample is to define the **class intervals** (categories) into which the data will fall. To do this, we need to know the smallest and largest starting salaries in the data set. These salaries are $11,000 and $43,900, respectively. Since we want the smallest salary to fall in the lowest class interval and the largest salary to fall in the highest class interval, the class intervals must span starting salaries ranging from $11,000 to $43,900.

Step 2 The second step is to choose the **class interval width**; this will depend on how many intervals we want to use to span the starting salary range and whether we want to use equal or unequal interval widths. For this example, we will use equal class interval widths (the most popular choice) and 11 class intervals. (See the General Rule in the box on page 78 on choosing the number of class intervals.)

Note that the starting salary range is equal to

$$\text{Range} = \text{Largest measurement} - \text{Smallest measurement}$$

$$= \$43,900 - \$11,000$$

$$= \$32,900$$

Since we chose to use 11 class intervals, the class interval width should approximately equal

$$\text{Class interval width} = \frac{\text{Range}}{\text{Number of class intervals}}$$

$$= \frac{32,900}{11} = 2,990.9$$

$$= \$3,000$$

Table 2.6 Tabulation of Data for the Starting Salaries of Table 2.5

Class	Class Interval	Tally	Class Frequency	Class Relative Frequency
1	10,950–13,950	||	2	.04
2	13,950–16,950	||||	4	.08
3	16,950–19,950	|	1	.02
4	19,950–22,950	₩₩ ₩₩ |||	13	.26
5	22,950–25,950	₩₩ ||	7	.14
6	25,950–28,950	₩₩ |||	8	.16
7	28,950–31,950	₩₩	5	.10
8	31,950–34,950	||||	4	.08
9	34,950–37,950	|||	3	.06
10	37,950–40,950	||	2	.04
11	40,950–43,950	|	1	.02
TOTALS			50	1.00

We shall start the first class slightly below the smallest observation ($11,000) and choose the starting point so that no observation can fall on a **class boundary**. Since starting salaries are recorded to the nearest hundred dollars, we can do this by choosing the lower class boundary of the first class interval to be $10,950. [*Note:* We could just as easily have chosen $10,955, $10,975, $10,990, or any one of many other points below and near $11,000.] Then the class intervals will be $10,950 to $13,950, $13,950 to $16,950, and so on. The 11 class intervals are shown in the second column of Table 2.6.

Step 3 The third step in constructing a histogram is to obtain each class frequency—i.e., the number of observations falling within each class. This is usually done using the computer. For illustration purposes, we performed this task by examining each starting salary in Table 2.5 and recording by tally (as shown in the third column of Table 2.6) the class in which it falls. The tally for each class gives the class frequencies shown in column 4 of Table 2.6. (Optionally, class relative frequencies or percentages can also be calculated. Relative frequencies are shown in the fifth column of Table 2.6.)

Step 4 The final step is to produce the graph using computer software. The bars in the histogram will have heights proportional to the class frequency, the class relative frequency, or the class percentage. Since we desire a frequency histogram, the bar heights will be proportional to the class frequencies. The resulting histogram, generated by Excel, is shown in Figure 2.6. Note that Excel displays the upper endpoints of each class interval (called *bins*) on the histogram.

b. A histogram, like a stem-and-leaf display, conveys a visual picture of the quantitative data. In particular, a histogram will identify a range of values where most of the data fall. You can see that most of the 50 graduates in the sample had starting salaries between $19,950 and $34,950. The five classes (bars) associated with this interval have frequencies of 13, 7, 8, 5, and 4, respectively. These bars are highlighted on Figure 2.6. The sum of these frequencies is 37. Thus, 37/50 = .74, or 74%, of the graduates in the sample had starting salaries between $19,950 and $34,950.

Note also that none of the starting salaries was less than $10,950, but several high salaries caused the distribution to have a long tail on the right. We say that such a distribution is **right skewed** (or **positively skewed**). Similarly

E **Figure 2.6** Excel
Frequency Histogram
for the Data of Table
2.5

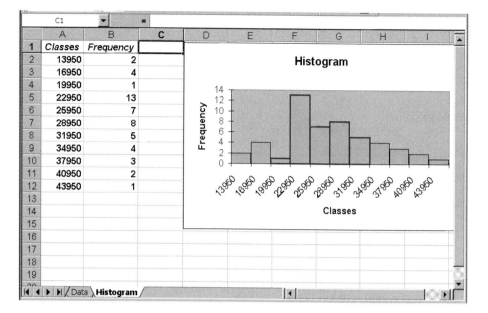

Figure 2.7 Skewed and
Symmetric Distributions

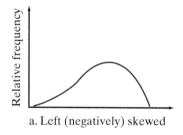

a. Left (negatively) skewed

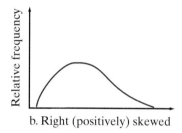

b. Right (positively) skewed

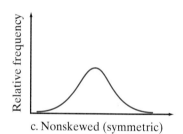

c. Nonskewed (symmetric)

a **left** (or **negatively**) **skewed** distribution has a histogram that has a long
tail on the left because of several unusually small values. Illustrations of
right and left skewed distributions, as well as a **nonskewed** (**symmetric**) distribution, are shown in Figure 2.7. ❑

CONSTRUCTING A HISTOGRAM

Step 1   Examine the data to determine the smallest and the largest
measurements.

Step 2   Divide the interval between the smallest and the largest measurements into between 5 and 20 equal subintervals called *classes* (see
next box). These classes should satisfy the following requirement:
Each measurement falls into one and only one subinterval. Note that
this requirement implies that no measurement falls on a boundary of
a subinterval.

Step 3   Compute the frequency or relative frequency of measurements
falling within each subinterval.

General Rule for Determining the Number of Classes in a Histogram	
Number of Observations in a Data Set	Number of Classes
Less than 25	5 or 6
25–50	7–14
More than 50	15–20

Step 4   Using a vertical axis of about three-fourths the length of the horizontal axis, plot each frequency or relative frequency as a rectangle over the corresponding subinterval.

Interpreting a Histogram

The percentage of the total number of measurements falling within a particular interval is proportional to the area of the bar that is constructed above the interval. For example, if 30% of the area under the distribution lies over a particular interval, then 30% of the observations fall in that interval.

Example 2.5   Interpreting a Histogram

CHKOUT.XLS

At a supermarket, customer checkout time is defined as the total length of time required for a checker to scan and total the prices of the customer's food items, accept payment, return change, and bag the items. The Excel workbook CHKOUT.XLS contains the checkout times (in seconds) for 500 customers who recently shopped at a Gainesville, Florida, store. Figure 2.8 is an Excel frequency histogram for the 500 customer checkout times.

a.   Interpret the graph.

b.   Estimate the proportion of supermarket customers that had checkout times of 1 minute or less.

Solution

a.   Excel displays the upper endpoints of the class intervals on the horizontal axis of the histogram. These values, 15, 30, 45, etc., imply that the class intervals are 0–15 seconds, 15–30 seconds, 30–45 seconds, etc. Note that the largest of the checkout times is approximately under 6 minutes (360 seconds). You can see that most of the checkout times tend to pile up near 30 seconds—the class from 15 to 30 seconds (bin 30) has the greatest relative frequency (or percentage), and the class from 30 to 45 seconds (bin 45) has the next highest relative frequency. Since this histogram has a long tail to the right, it is right skewed.

E Figure 2.8 Excel Histogram of 500 Supermarket Checkout Times

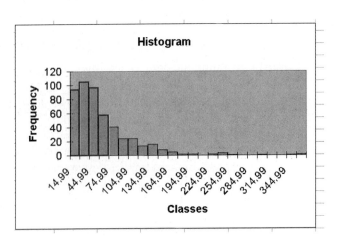

b. The four bars that fall in the interval from 0 to 60 seconds (bins 15, 30, 45, and 60) are highlighted in Figure 2.8. Based on the bar heights, we estimated the frequencies for these four intervals to be 95, 105, 100, and 60, respectively. The sum of these estimated frequencies, 360, represents approximately 360/500 = .70 of the total area of the bars for the complete distribution. This tells us that approximately 70% of the 500 customer checkout times were 60 seconds or less. ❑

Self-Test

2.3

In a *Developmental Psychology* (Mar. 1986) study on the sexual maturation of college students, 75 male college undergraduates were asked to report the age (in years) when they began to shave regularly. The frequency histogram of age at regular shaving is shown in Figure 2.9.

Figure 2.9 Frequency Histogram of Shaving Age

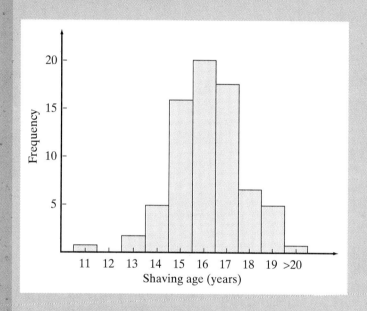

a. Approximately how many of the 75 male students began to shave regularly at age 16?

b. Approximately what proportion of the 75 male students began to shave regularly at age 16?

PROBLEMS
2.15–2.29

Using the Tools

2.15 Construct a stem-and-leaf display for the following sample of 20 quantitative measurements:

8.1	10.5	6.6	6.7	3.3	5.1	8.3	7.5	6.6	4.0	6.9	9.0
10.2	5.0	4.2	6.1	6.3	7.2	5.4	8.7				

2.16   Consider the following sample of 15 quantitative measurements:

35	40	50	35	40	40	45	35	30	40	30	35
45	50	40									

   a.   Using the digit in the tens place as a stem, list the stems in order.

   b.   Identify the leaf for each measurement.

   c.   Place the leaves for each measurement in the appropriate stem row to form a stem-and-leaf display.

2.17   Consider the following sample data:

213	228	241	268	234	303	274	316	319	320	227	226
224	267	303	266	265	237	288	291	285	270	254	215

   a.   Using the first two digits of each number as a stem, list the stems in order.

   b.   Place the leaf for each observation in the appropriate stem row to form a stem-and-leaf display.

2.18   Consider the sample data shown here. Using the first digit as a stem, construct a stem-and-leaf display.

5.9	5.3	1.6	7.4	8.6	1.2	2.1	4.0	7.3	8.4	8.9	6.7			
4.5	6.3	7.6	9.7	3.5	1.1	4.3	3.3	8.4	1.6	8.2	6.5	1.1	5.0	9.4	6.4

2.19   Refer to the sample data of Problem 2.18.

   a.   Find the difference between the largest and smallest measurements.

   b.   Divide the difference obtained in part **a** by 5 to determine the approximate class interval width for five class intervals.

   c.   Specify upper and lower boundaries for each of the five class intervals.

   d.   Construct a relative frequency distribution for the data.

2.20   A sample of 20 measurements follows.

26	34	21	32	32	22	12	26	39	25	36	28
38	17	39	31	30	23	27	19				

   a.   Using a class interval width of 5, give the upper and lower boundaries for six class intervals, where the lower boundary of the first class is 10.5.

   b.   Determine the relative frequency for each of the six classes specified in part **a**.

   c.   Construct a relative frequency distribution using the results of part **b**.

Applying the Concepts

2.21   The following data are the retail prices for a random sample of 22 VCR (video cassette recorder) models:

VCRPRIC.XLS

350	300	340	220	320	450	270	265	210	250	180	300
190	170	190	170	170	200	180	220	200	250		

Data Source: Adapted from "VCRs," *Consumer Reports,* Nov. 1996, pp. 36–38.

   a.   Form a stem-and-leaf display for the data.

   b.   Are you more likely to find a VCR for under $200 or over $300?

   c.   Does there seem to be a concentration of prices around or near any specific dollar amount? Explain.

2.22   The number of calories per 12-ounce serving for each of 22 beer brands is provided in the table.

  a.   Construct a stem-and-leaf display for the data.

  b.   Locate the data values for the six light beers on the stem-and-leaf display. Do light beers really have a lower caloric content than regular beers?

BEERCAL.XLS

Beer	Calories	Beer	Calories
Old Milwaukee	145	Miller High Life	143
Stroh's	142	Pabst Blue Ribbon	144
Red Dog	147	Milwaukee's Best	133
Budweiser	148	Miller Genuine Draft	143
Icehouse	149	Rolling Rock	143
Molson Ice	155	Michelob Light	134
Michelob	159	Bud Light	110
Bud Ice	148	Natural Light	110
Busch	143	Coors Light	105
Coors Original	137	Miller Light	96
Gennessee Cream Ale	153	Amstel Light	95

Source: Consumer Reports, June 1996, Vol. 61, No. 6, p. 16.

2.23   Over half of the nearly 60,000 members of the U.S. Chess Federation (USCF) have official chess ratings. A player with a rating of 1,100 or less is a "beginner"; "average" players' ratings range between 1,100 and 1,900; "experts" range between 2,000 and 2,200; "masters" range between 2,200 and 2,400; and "grand masters" have ratings higher than 2,400. (Gary Kasparov, the reigning world champion from Russia, has a chess rating of 2,900.) The accompanying graph, extracted from *Scientific American* (Oct. 1990), illustrates the distribution of the ratings of the 35,000 rated members of the USCF.

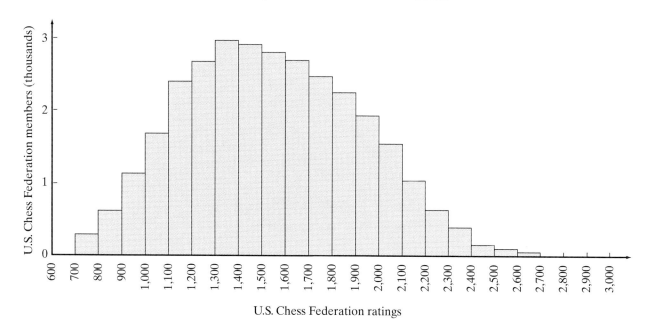

  a.   What type of graph is portrayed?

    b.   Use the graph to estimate the number of USCF grand masters.

    c.   Are the data skewed? Explain.

2.24   Under a voluntary cooperative inspection program, all passenger cruise ships arriving at U.S. ports are subject to unannounced inspection. The purpose of these inspections is to achieve levels of sanitation that will minimize the potential for gastrointestinal disease outbreaks on these ships. Ships are rated on a 0- to 100-point scale depending on how well they meet the Centers for Disease Control sanitation standards. In general, the lower the score, the lower the level of sanitation. The accompanying table lists the sanitation inspection scores for 91 international cruise ships during a recent year.

SANIT.XLS

Ship	Score	Ship	Score	Ship	Score
Americana	89	Hanseatic Renaissance	82	Seabourn Pride	99
Amerikanis	97	Holiday	91	Seabourn Spirit	92
Azure Seas	83	Horizon	94	Seabreeze I	96
Britanis	93	Island Princess	87	Seaward	89
Caribbean Prince	84	Jubilee	89	Sky Princess	97
Caribe I	90	Mardi Gras	92	Society Explorer	66
Carla C	90	Meridian	95	Song of America	95
Carnivale	92	Nantucket Clipper	89	Song of Flower	99
Celebration	95	New Horeham II	95	Song of Norway	92
Club Med I	94	Nieuw Amsterdam	97	Southward	89
Costa Classica	91	Noordam	92	Sovereign of the Seas	93
Costa Marina	91	Nordic Empress	93	Star Princess	94
Costa Riviera	91	Nordic Prince	92	Starship Atlantic	87
Crown Monarch	94	Norway	84	Starship Majestic	94
Crown Odyssey	88	Pacific Princess	88	Starship Oceanic	97
Crown Princess	88	Pacific Star	70	Starward	96
Crystal Harmony	99	Queen Elizabeth 2	98	Stella Solaris	94
Cunard Countess	96	Regent Sea	87	Sun Viking	90
Cunard Princess	89	Regent Star	74	Sunward	95
Daphne	86	Regent Sun	95	Triton	86
Dawn Princess	86	Rotterdam	92	Topicale	93
Discovery I	93	Royal Princess	93	Universe	92
Dolphin IV	96	Royal Viking Sun	86	Victoria	96
Ecstasy	94	Sagafjord	89	Viking Princess	90
Emerald Seas	95	Scandinavian Dawn	87	Viking Serenade	96
Enchanted Isle	86	Scandinavian Song	90	Vistafjord	94
Enchanted Seas	96	Scandinavian Sun	89	Westerdam	91
Fair Princess	87	Sea Bird	86	Wind Spirit	96
Fantasy	97	Sea Goddess I	97	Yorktown Clipper	92
Festival	94	Sea Lion	91		
Golden Odyssey	89	Sea Princess	88		

Source: Center of Environmental Health and Injury Control, Miami, Florida. (reported in *Tampa Tribune,* May 17, 1992).

    a.   A stem-and-leaf display of the data is shown at the top of page 83. Identify the stems and leaves of the graph.

    b.   Locate the inspection score of 70 (Pacific Star) on the stem-and-leaf display.

    c.   A score of 86 or higher at the time of inspection indicates the ship is providing an accepted standard of sanitation. Use the stem-and-leaf display in part **a** to estimate the proportion of ships that have an accepted

E	D	E
1	Stem-and-Leaf Display	
2	for Sanitation Inspection Scores	
3	Stem unit: 10	
4		
5	6	6
6	7	0 4
7	8	2 3 4 4 6 6 6 6 6 6 7 7 7 7 7 8 8 8 8 9 9 9 9 9 9 9 9 9
8	9	0 0 0 0 0 1 1 1 1 1 1 2 2 2 2 2 2 2 2 2 3 3 3 3 3 3 4 4 4 4 4 4 4 4 4 5 5 5 5 5 5 5 6 6 6 6 6 6 6 7 7 7 7 7 8 9 9 9

Problem 2.24a

sanitation standard. Does this answer influence your decision to take a cruise in the future?

2.25 Access the Excel workbook that contains the data on contaminated fish in the Tennessee River (Alabama) collected by the U.S. Army Corps of Engineers. Consider the 50 DDT measurements corresponding to fish specimens identified on the data set by observations numbered 51–100.

FISH.XLS

a. Use Excel to construct a frequency histogram for the 50 DDT values. Use 10 classes to span the range.

b. Repeat part **a**, but use only three classes to span the range. Compare the result with the frequency histogram you constructed in part **a**. Which is more informative? Why do an inadequate number of classes limit the information conveyed by a frequency histogram?

c. Repeat part **a**, but use 25 classes. Comment on the information provided by this histogram and compare it with the result of part **a**.

2.26 "Deep hole" drilling is a family of drilling processes used when the ratio of hole depth to hole diameter exceeds 10. Successful deep hole drilling depends on the satisfactory discharge of the drill chip. An experiment was conducted to investigate the performance of deep hole drilling when chip congestion exists (*Journal of Engineering for Industry,* May 1993). An analysis of drill chip congestion was performed using data generated via computer simulation. The simulated distribution of the length (in millimeters) of 50 drill chips is displayed here in a frequency histogram.

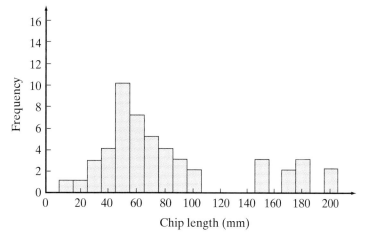

Source: Chin, Jih-Hua, et al. "The computer simulation and experimental analysis of chip monitoring for deep hole drilling." *Journal of Engineering for Industry, Transactions of the ASME,* Vol. 115, May 1993, p. 187 (Figure 12).

a. Convert the frequency histogram into a relative frequency histogram.

b. Based on the graph in part **a**, would you expect to observe a drill chip with a length of at least 190 mm? Explain.

2.27 Computer software that allows students to view electronic atlases and produce their own interactive maps has become popular in the geography classroom. A study was conducted to determine whether using computer software has a positive influence on learning geography (*Journal of Geography,* May/June 1997). One group of seventh graders (43 students) used computer resources (electronic atlases, encyclopedias, and interactive maps) to research the regions of Africa. Another group of seventh graders (22 students) studied the same African regions using notes from a presentation and teacher-produced worksheets. At the end of the study period, all students were given a test on the African regions. The test scores are listed in the table.

GEO.XLS

Group 1 (Computer Resources) Student Scores										
41	53	44	41	66	91	69	44	31	75	66	69
75	53	44	78	91	28	69	78	72	53	72	63
66	75	97	84	63	91	59	84	75	78	88	59
97	69	75	69	91	91	84					

Group 2 (No Computer) Student Scores										
56	59	9	59	25	66	31	44	47	50	63	19
66	53	44	66	50	66	19	53	78	78		

Source: Linn, S. E. "The effectiveness of interactive maps in the classroom: A selected example in studying Africa." *Journal of Geography,* Vol. 96, No. 3, May/June 1997, p. 167 (Table 1).

a. Use a graphical technique to describe the test scores for the first group of geography students.

b. Repeat part **a** for the second group of geography students.

c. Compare the graphs. What do you learn from this comparison?

2.28 Many Vietnam veterans have dangerously high levels of the dioxin 2,3,7,8-TCDD in their blood as a result of their exposure to the defoliant Agent Orange. A study published in *Chemosphere* (Vol. 20, 1990) reported on the TCDD levels of 20 Massachusetts Vietnam veterans who were possibly exposed to Agent Orange. The amounts of TCDD (measured in parts per trillion) in blood plasma drawn from each veteran are shown in the table. Use a graphical technique to describe the distribution of TCDD levels. Identify the type of skewness in the data.

TCDD.XLS

TCDD Levels in Plasma									
2.5	1.8	6.9	1.8	3.5	2.5	1.6	36.0	6.8	3.1
20.0	3.3	4.7	3.1	4.1	7.2	4.6	3.0	2.1	2.0

Source: Schecter, A. et al. "Partitioning of 2,3,7,8-chlorinated dibenzo-p-dioxins and dibenzofurans between adipose tissue and plasma lipid of 20 Massachusetts Vietnam veterans." *Chemosphere,* Vol. 20, Nos. 7–9, 1990, pp. 954–955 (Table I).

2.29 Nondestructive evaluation (NDE) describes the process of evaluating or inspecting components without causing any permanent physical change to the components. NDE was applied in a system for inspecting fluorescent-penetrant specimens that are prone to synthetic cracks (*Technometrics,* May 1996). The crack size (in inches) of each in a sample of 58 fluorescent-penetrant specimens was measured as well as whether or not the specimen was previously determined to be flawed. The crack sizes are reported in the table. [*Note:* Specimens judged to have a flaw are marked with an asterisk (*).]

a. Summarize the data on crack sizes with a stem-and-leaf display.

b. Locate the specimens judged to have a flaw on the graph. Do flawed specimens tend to have larger crack sizes?

CRACK.XLS

Crack Sizes (in inches) of 58 Specimens							
.003	.004	.012	.014	.021	.022*	.023	.024
.026*	.026*	.030*	.030	.031*	.034	.034*	.041
.041	.042*	.042	.043	.043*	.044*	.045	.046*
.046*	.052*	.055*	.057	.058*	.060*	.060*	.063
.070*	.071*	.073*	.073*	.074	.076	.078*	.079*
.079*	.083*	.090*	.095*	.095*	.096*	.100*	.102*
.103*	.105*	.114*	.119*	.120*	.130*	.160*	.306*
.328*	.440*						

Source: Olin, B. D. and Meeker, W. Q. "Applications of statistical methods to nondestructive evaluation." *Technometrics,* Vol. 38, No. 2, May 1996, p. 101 (Table 1).

2.4 EXPLORING THE RELATIONSHIP BETWEEN TWO QUALITATIVE VARIABLES: CROSS-CLASSIFICATION TABLES AND SIDE-BY-SIDE BAR GRAPHS

Experimental units in a study are often simultaneously categorized according to two qualitative variables. For example, every ten years the U.S. Census Bureau collects demographic data on each American citizen. Two of the many variables measured are gender and marital status. In this section, we demonstrate how to describe two categorical variables (e.g., gender and marital status) simultaneously with *cross-classification tables* and *side-by-side bar graphs.* These methods will enable us to explore a possible pattern or relationship between the two qualitative variables.

Example 2.6   Cross-Classification Table

In group discussions, do men and women interrupt the speaker equally often? This was the research question investigated in the *American Sociological Review* (June 1989). Undergraduate sociology students (half male and half female) were organized into discussion groups and their interactions recorded on video. Each time the speaker was interrupted, the researchers recorded two variables: (1) the gender of the speaker and (2) the gender of the interrupter. Hypothetical data for a sample of 40 interruptions are listed in Table 2.7. Use a cross-classification table to summarize the data in Table 2.7.

Solution

The two qualitative variables of interest, speaker gender and interrupter gender, each has two categories: male (M) and female (F). In order to investigate the

Table 2.7 Data on Speaker and Interrupter Gender for 40 Interruptions

Interruption	Speaker	Interrupter	Interruption	Speaker	Interrupter
1	M	M	21	F	M
2	M	M	22	F	F
3	F	F	23	F	F
4	M	F	24	M	M
5	F	M	25	F	F
6	M	F	26	M	M
7	M	M	27	F	M
8	F	M	28	F	F
9	F	M	29	M	M
10	F	M	30	M	M
11	F	F	31	F	F
12	M	M	32	M	M
13	F	M	33	F	M
14	F	M	34	M	F
15	M	M	35	M	M
16	M	M	36	M	M
17	F	F	37	F	M
18	M	F	38	F	F
19	M	M	39	F	F
20	M	F	40	M	M

Table 2.8 Cross-Classification Table for Data of Table 2.7

		INTERRUPTER		
		Male	Female	TOTALS
Speaker	Male	15	5	20
	Female	10	10	20
	TOTALS	25	15	40

relationship between the two variables, we determine the frequency of interruptions for each of the 2 × 2 = 4 *combined* categories. That is, we count the number of observations that fall into the four speaker/interrupter categories: M/M, M/F, F/M, and F/F. These four categories and their frequencies are displayed in Table 2.8. The summary table is known as a **cross-classification table**, since the values of one variable are classified across the values of the other variable.

For example, the number 15 in the upper-left cell of the table is a count of the number of male speakers who were interrupted by a male. Similarly, the number 5 in the upper-right cell is the number of male speakers who were interrupted by a female. You can see that the sum of the counts in the four categories equals 40, the total number of interruptions in the sample. ❑

To explore a possible relationship between speaker gender and interrupter gender, it is useful to compute the relative frequency for each cell of the cross-classification table. One way to obtain these relative frequencies is to divide each cell frequency by the corresponding total for the row in which the cell appears. Alternatively, the relative frequencies can be obtained by dividing the cell frequencies by either their column totals or their overall total (in this case, 40). The choice of divisor will depend on the objective of the analysis, as the next example illustrates.

Example 2.7   Side-by-Side Bar Graphs

Refer to Example 2.6. The researchers want to explore the nature of the relationship between speaker gender and interrupter gender. Specifically, they want to know the gender that is more likely to interrupt a male speaker and the gender that is more likely to interrupt a female speaker. To do this, they require relative frequencies of male interrupters and female interrupters for each speaker gender.

a. Compute these relative frequencies and display them in a table.

b. Portray the results, part **a**, graphically. Interpret the graph.

Solution

a. First, we consider male speakers (i.e., focus on the first row of Table 2.8). Of the 20 male speakers in the study, 15 were interrupted by a male. Thus, the relative frequency for the male speaker/male interrupter cell is 15/20 = .75. Our interpretation is that 75% of the male speakers in the study are interrupted by males. Similarly, 5 of the 20 male speakers were interrupted by females. The associated relative frequency is 5/20 = .25; this implies that 25% of the male speakers were interrupted by a female.

Now, consider the 20 female speakers in the study (i.e., focus on the second row of Table 2.8). Since 10 were interrupted by males and 10 were interrupted by females, the associated relative frequencies are 10/20 = .5 and 10/20 = .5. In other words, half of the female speakers in the study were interrupted by males and half were interrupted by females.

All four of these relative frequencies are shown in Table 2.9. Note that the relative frequencies in each row sum to 1.

b. The relative frequencies of Table 2.9 can be visually displayed using bar graphs. Since the cross-classification table contains two rows (male speaker and female speaker), we form one bar graph for male speakers and one for female speakers, and then place them side-by-side for comparison purposes. A **side-by-side horizontal bar graph**, generated with Excel, is shown in Figure 2.10.

Figure 2.10 clearly shows the dramatic difference in proportions of male and female interrupters for the two speaker genders. Males interrupt three times as much as females (.75 vs. .25) when a male is speaking. However, when a female is speaking, females are just as likely to interrupt as males (.50 vs. .50). Consequently, it appears that males are more likely to interrupt than females, but only when the speaker is a male. ❑

Table 2.9 Relative Frequencies for Table 2.8 Based on Row Totals

		GENDER OF INTERRUPTER		
		Male	Female	TOTALS
GENDER OF	Male	.75	.25	1.00
SPEAKER	Female	.50	.50	1.00

E **Figure 2.10** Excel Side-by-Side Bar Graph for Data Summarized in Table 2.9

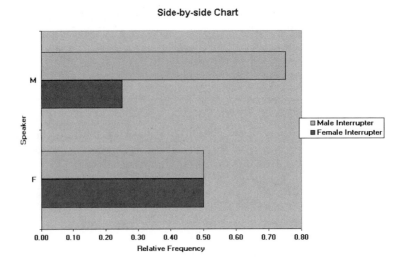

Side-by-side Chart

Self-Test 2.4

Each of 274 introductory psychology students (80 males and 194 females) was asked to indicate when he/she reached puberty in relation to others of the same gender (*Developmental Psychology,* Mar. 1986). One of five response categories were possible: much earlier, earlier, same time, later, and much later. The proportions responding in each of the categories for males and females are reported in Table 2.10.

Table 2.10 Summary for *Developmental Psychology* Study

Time Reached Puberty in Relation to Others	Males	Females
Much earlier	.02	.05
Earlier	.22	.20
Same time	.56	.57
Later	.20	.17
Much later	.00	.01
TOTALS	1.00	1.00

Data Source: Sanders, B. and Soares, M. P. "Sexual maturation and spatial ability in college students." *Developmental Psychology*, Vol. 22, No. 2, pp. 199–203 (Figure 1).

a. Interpret the value .22 shown in the table.

b. Construct a side-by-side bar chart for the data. Interpret the graph.

Example 2.8   Interpreting a Side-by-Side Bar Graph

A study published in the *Journal of Education for Business* (Jan./Feb. 1991) used a mail survey to ask a sample of 215 personnel directors, "Would you give hiring preference to applicants knowledgeable in foreign languages?" Responses were either "Yes," "No," or "Neutral." The directors were also classified according to type of business (foreign or domestic). A summary of the responses is shown in

Table 2.11 Cross-Classification Table for Example 2.8

| Firm type | FOREIGN LANGUAGE HIRING PREFERENCE | | | |
	Yes	Neutral	No	TOTALS
U.S.	50	57	19	126
Foreign	60	22	7	89
TOTALS	110	79	26	215

Source: Cornick, M. F., et al. "The value of foreign language skills for accounting and business majors." *Journal of Education for Business,* Jan./Feb. 1991, p. 162 (Table 2).

the cross-classification table, Table 2.11. Use a graphical method to explore a possible relationship between firm type and foreign language hiring preference.

Solution
The first step is to decide how to calculate relative frequencies for the qualitative data. One choice is to divide the cell frequencies shown in Table 2.11 by their respective row totals. (Remember, you could also choose to divide by column totals or by the overall total.) For example, the relative frequency for the U.S./Yes cell is 50/126 = .397. Similarly, the relative frequency for the Foreign/No cell is 7/89 = .079. These computations are shown in Table 2.12 and displayed graphically in the Excel side-by-side bar graphs, Figure 2.11.

Table 2.12 Relative Frequencies for Table 2.11 Based on Row Totals

| FIRM TYPE | FOREIGN LANGUAGE HIRING PREFERENCE | | | |
	Yes	Neutral	No	TOTALS
U.S.	50/126 = .397	57/126 = .452	19/126 = .151	1.000
Foreign	60/89 = .674	22/89 = .247	7/89 = .079	1.000

Figure 2.11 Excel Side-by-Side Bar Graphs for Data in Table 2.12

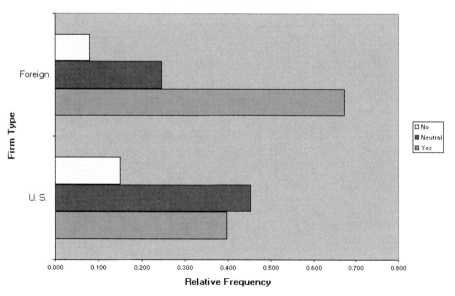

From Figure 2.11 we notice that the Yes bar for foreign firms is nearly twice as long as the Yes bar for U.S. firms. Numerically, these relative frequencies (from Table 2.12) are .674 and .397, respectively. That is, for foreign firms, about 67% of the personnel directors prefer hiring business majors with foreign language skills. In contrast, only about 40% of U.S. personnel directors prefer hiring business majors with foreign language skills. Consequently, it appears that the foreign language skill hiring preferences of the personnel directors in the sample does depend on whether the firm is foreign or domestic. ❏

Cross-classification tables and side-by-side bar graphs are useful methods for describing the simultaneous occurrence of two qualitative variables, and can be used to make inferences about whether or not the two variables are related. However, no measure of reliability can be attached to such an inference without a more sophisticated numerical analysis. In Chapter 9 we present a formal inferential method for analyzing cross-classification (two-way contingency) tables.

PROBLEMS 2.30–2.40

Using the Tools

2.30 Two qualitative variables were measured for each of 10 experimental units. The data are listed in the accompanying table.

Experimental Unit	Variable 1	Variable 2
1	Yes	A
2	Yes	B
3	No	B
4	Yes	A
5	Yes	B
6	No	B
7	No	A
8	No	A
9	No	B
10	No	A

a. List the possible outcomes of the experiment (i.e., the different combined categories of the two qualitative variables).

b. Find the frequency of each category. Present the results in the form of a cross-classification table.

c. Using the totals for each category of Variable 2, compute the relative frequencies for the different classes of Variable 1.

d. Using the totals for each category of Variable 1, compute the relative frequencies for the different classes of Variable 2.

e. Present the results, part **c**, in the form of a side-by-side bar graph.

f. Present the results, part **d**, in the form of a side-by-side bar graph.

2.31 Two qualitative variables were measured for each of 12 experimental units. The data are listed in the table on page 91.

Experimental Unit	Variable 1	Variable 2
1	X	A
2	X	B
3	Z	B
4	Y	A
5	Y	A
6	X	B
7	Z	A
8	Z	B
9	X	B
10	Y	A
11	Y	B
12	Y	B

    a. List the possible outcomes of the experiment (i.e., the different combined categories of the two qualitative variables).

    b. Find the frequency of each category. Present the results in the form of a cross-classification table.

    c. Using the totals for each category of Variable 2, compute the relative frequencies for the different classes of Variable 1.

    d. Using the totals for each category of Variable 1, compute the relative frequencies for the different classes of Variable 2.

    e. Present the results, part **c**, in the form of a side-by-side bar graph.

    f. Present the results, part **d**, in the form of a side-by-side bar graph.

2.32 Consider the cross-classification table shown below.

		COLOR			
		Red	**White**	**Blue**	**TOTALS**
Age	Old	10	21	3	34
	New	2	14	0	16
	TOTALS	12	35	3	50

    a. How many experimental units are included in the analysis?

    b. How many experimental units are old and white?

    c. How many experimental units are new and blue?

    d. Use row totals to calculate the relative frequencies for the three colors.

    e. Use column totals to calculate the relative frequencies for the two ages.

    f. Use the overall total to calculate the relative frequency for each cell of the table.

    g. Present the results, part **d**, in a side-by-side bar chart.

    h. Present the results, part **e**, in a side-by-side bar chart.

2.33 Consider the cross-classification table shown on page 92.

    a. How many experimental units are included in the analysis?

    b. How many experimental units are located in the north?

    c. How many experimental units of type B are located in the west?

    d. Use row totals to calculate relative frequencies for the four locations.

		LOCATION				
		North	East	South	West	TOTALS
	A	20	40	80	50	190
TYPE	B	50	40	10	25	125
	C	30	20	10	25	85
	TOTALS	100	100	100	100	400

   e. Use column totals to calculate relative frequencies for the three types.

   f. Use the overall total to calculate relative frequencies for each cell of the table.

   g. Present the results, part **d**, in a side-by-side bar chart.

   h. Present the results, part **e**, in a side-by-side bar chart.

Applying the Concepts

2.34 The faculty council of a large university would like to determine the opinion of students on a proposed trimester academic calendar. A random sample of 150 students is selected and each student is classified according to graduate status and opinion on trimester. The results are shown in the accompanying table.

	GRADUATE STATUS	
OPINION	**Undergraduates**	**Graduates**
Favor trimester	63	27
Oppose trimester	37	23
TOTALS	100	50

   a. Identify the two qualitative variables measured in this study. List the classes for each variable.

   b. Construct a table of column percentages.

   c. Construct a side-by-side bar chart to visually highlight the results in part **b**.

   d. Interpret the bar chart.

2.35 *Industrial Marketing Management* (Feb. 1993) published a study of humor in trade magazine advertisements. Each in a sample of 665 ads was classified according to nationality (British, German, American) of the trade magazine in which they appear and whether or not they were considered humorous by a panel of judges. The number of ads falling into each of the categories is provided in the accompanying table.

		HUMOROUS	
		Yes	No
	British	52	151
NATIONALITY	German	44	148
	American	56	214

Source: McCullough, L. S., and Taylor, R. K. "Humor in American, British, and German ads." *Industrial Marketing Management,* Vol. 22, No. 1, Feb. 1993, p. 22 (Table 3).

a. Identify the two qualitative variables measured in this study. List the classes for each variable.

b. Construct a table of row percentages.

c. Construct a side-by-side bar chart to visually highlight the results in part **b**.

d. Interpret the bar chart.

2.36 The victory of Bill Clinton in the 1996 presidential election was attributed to improved economic conditions and low unemployment. Suppose that a survey of 800 adults taken soon after the 1996 election resulted in the following cross-classification of financial condition with education level:

| | EDUCATION LEVEL | | | |
	High School Degree or Lower	Some College	College Degree or Higher	TOTALS
FINANCIAL CONDITIONS				
Worse off now than before	91	39	18	148
No difference	104	73	31	208
Better off now than before	235	48	161	444
TOTALS	430	160	210	800

a. Construct a table of column percentages.

b. Construct a side-by-side bar chart to visually highlight the results in part **a**.

c. Based on the results of parts **a** and **b**, do you think there is a clear difference in the current financial condition as compared to before based on level of education?

2.37 A group of cardiac physicians has been studying a new drug designed to reduce blood loss in coronary artery bypass operations. Data were collected for 114 coronary artery bypass patients—half who received a dosage of the new drug and half who received a placebo (i.e., no drug). In addition to monitoring who received the drug, the physicians also recorded the type of complication (if any) experienced by the patient. A summary table for the data on drug administered and complication type is shown below. [*Note:* The data for this study are real. For confidentiality reasons, the drug name and physician group shall remain anonymous.]

| | | DRUG | | |
		Yes	No	TOTALS
Complication Type	Redo surgery	7	5	12
	Post-op infection	7	4	11
	Both	3	1	4
	None	40	47	87
	TOTALS	57	57	114

a. Identify the two qualitative variables measured in this study. List the classes for each variable.

b. Construct a table of column percentages.

c. Display the results, part **b**, in a side-by-side bar graph.

d. Do you believe that patients who are administered the drug have more complications than those who receive a placebo? Explain.

2.38 The *American Journal on Mental Retardation* (Jan. 1992) published a study of the social interactions of two groups of children. Independent random samples of 15 children who did and 15 children who did not display developmental delays (i.e., mild mental retardation) were taken in the experiment. After observing the children during "freeplay," the number of children who exhibited disruptive behavior (e.g., ignoring or rejecting other children, taking toys from another child) was recorded for each group. The data are summarized in the following table. Graphically portray the data with a side-by-side bar chart. Interpret the graph.

	Disruptive Behavior	Nondisruptive Behavior	TOTALS
With Developmental Delays	12	3	15
Without Developmental Delays	5	10	15
TOTALS	17	13	30

Source: Kopp, C. B., Baker, B., and Brown K. W. "Social skills and their correlates: Preschoolers with developmental delays." *American Journal on Mental Retardation,* Vol. 96, No. 4, Jan. 1992.

2.39 Since 1948, research psychologists have used the "water-level task" to test basic perceptual and conceptual skills. Subjects are shown a drawing of a glass tilted at a 45° angle and asked to assume the glass is filled with water. The task is to draw a line representing the surface of the water. *Psychological Science* (Mar. 1995) reported on the results of the water-level task given to 120 subjects. Each subject was classified by group and by performance on the test. A summary of the results is provided in the table.

	GROUP						
	FEMALES			MALES			
Judged Line	Students	Waitresses	Housewives	Students	Bartenders	Bus Drivers	TOTALS
More than 5° below surface	0	0	1	1	1	1	4
More than 5° above surface	7	15	13	3	11	4	53
Within 5° of surface	13	5	6	16	8	15	63
TOTALS	20	20	20	20	20	20	120

Source: Hecht, H., and Proffitt, D. R. "The price of experience: Effects of experience on the water-level task." *Psychological Science,* Vol. 6, No. 2, Mar. 1995, p. 93 (Table 1).

a. Use a graphical method to describe the overall results of the study (i.e., the performance responses for all 120 subjects).

b. Construct graphs that the researchers could use to explore for group differences on the water-level task.

c. Psychologists theorize that males do better than females, that younger adults do better than older adults, and that those experienced in handling liquid-filled containers do better than those who are not. Are these theories supported by the data?

2.40 Age-related macular degeneration (AMD), a progressive disease that deprives a person of central vision, is the leading cause of blindness in older adults. In a study published in the *Journal of the American Geriatrics Society* (Jan. 1998), researchers investigated the link between AMD incidence and alcohol consumption. Each individual in a national sample of 3,072 adults was asked about the most frequent alcohol type (beer, wine, liquor, or none) consumed in the past year. In addition, each adult was clinically tested for AMD. The results are displayed in the accompanying table.

Most Frequent Alcohol Type Consumed	Total Number	Number Diagnosed with AMD*
None	965	87
Beer	430	30
Wine	1,284	51
Liquor	393	20
TOTALS	3,072	188

*Numbers based on percentages given in Figure 1.

Source: Obisean, T. O., et al. "Moderate wine consumption is associated with decreased odds of developing age-related macular degeneration in NHANES-1." *Journal of the American Geriatrics Society,* Vol. 46, No. 1, Jan. 1998, p. 3 (Figure 1).

a. Identify the two qualitative variables measured in this study.

b. Summarize the data on the two variables using a cross-classification table.

c. Construct a bar graph to describe the most frequent alcohol type consumed by all 3,072 adults in the study.

d. Construct a pie chart to describe the incidence of AMD among all 3,072 adults in the study.

e. Construct a side-by-side bar graph to compare the incidence of AMD among the four alcohol type response groups. Interpret the graph.

2.5 EXPLORING THE RELATIONSHIP BETWEEN TWO QUANTITATIVE VARIABLES: SCATTERPLOTS

In situations where two quantitative variables are measured on the experimental unit, it may be important to examine the nature of the relationship between the two variables. For example, a realtor may be interested in the relationship between the asking price and sale price of a home; an educator may be interested in relating a student's SAT score and high school GPA; a psychologist might want to know whether or not a person's annual salary is related to his or her IQ. In this section, we show how to graphically describe the relationship between two quantitative variables using a simple two-dimensional graph, called a *scatterplot.*

Example 2.9   Constructing a Scatterplot

Annually, the Federal Trade Commission (FTC) collects data on domestic cigarette brands. Smoking machines are used to "smoke" cigarettes to a certain length, and the residual "dry" particulate matter is tested for the amounts of various hazardous substances. The variables of interest to the FTC are carbon monoxide (CO) content and amount of nicotine, both measured in milligrams. Data collected on these two variables for 20 cigarette brands are listed in Table 2.13.

a. Construct a scatterplot for the data.   **b.** Interpret the graph.

Solution

a. A **scatterplot** is a two-dimensional graph, with a vertical axis and a horizontal axis as shown in the Excel graph, Figure 2.12. The values of one of

SMOLING.XLS

Table 2.13 Carbon Monoxide–Nicotine Data for Example 2.9

Brand	CO	Nicotine	Brand	CO	Nicotine
1	15	0.9	11	6	0.5
2	6	0.4	12	22	1.2
3	13	1.3	13	15	0.8
4	12	0.8	14	17	1.3
5	12	1.2	15	13	1.0
6	9	0.7	16	11	0.7
7	13	1.0	17	14	1.1
8	16	1.1	18	14	1.4
9	13	1.1	19	12	0.9
10	12	0.9	20	8	0.4

the quantitative variables of interest are located on the vertical axis while the other variable's values are located on the horizontal axis of the graph. Note that we have selected (arbitrarily) to locate CO content on the vertical axis and nicotine amount on the horizontal axis. Each of the 20 pairs of observations on CO and nicotine listed in Table 2.13 are in Figure 2.12. (For example, the point corresponding to a CO content of 15 mg and a nicotine amount of .9 mg for brand 1 is identified on the scatterplot.)

b. Scatterplots often reveal a pattern or trend that clearly indicates how the quantitative variables are related. From Figure 2.12 you can see that small values of CO content tend to be associated with small nicotine amounts, and large values of CO tend to be associated with large nicotine amounts. Stated another way, CO content tends to *increase* as nicotine amount *increases*. Since the two variables tend to move in the same direction, we say that there is a **positive association** between CO content and nicotine amount.

In contrast, a **negative association** occurs between two variables when one tends to *decrease* as the other *increases*. If no distinct pattern is revealed by the scatterplot, we say there is *little or no association* between the two

E Figure 2.12 Excel Scatterplot of CO Ranking versus Nicotine Content for Example 2.9

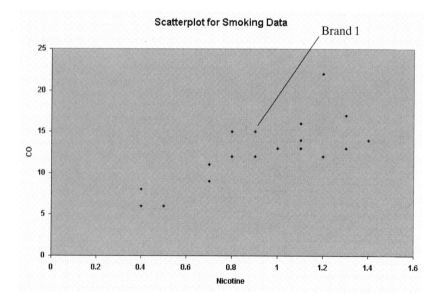

Figure 2.13 Scatterplots for Three Data Sets

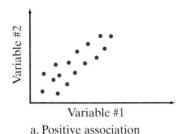

Variable #1
a. Positive association

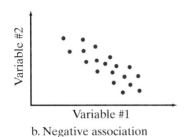

Variable #1
b. Negative association

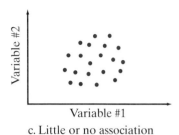

Variable #1
c. Little or no association

variables. That is, as one variable increases (or decreases), there is no definite trend in the values of the other variable. Graphs showing each of these three types of association are displayed in Figure 2.13. ❑

CONSTRUCTING A SCATTERPLOT

Step 1   Draw the horizontal and vertical axes for a two-dimensional graph.

Step 2   Choose one variable to plot on the horizontal axis (call this variable #1) and one to plot on the vertical axis (call this variable #2).

Step 3   Examine the data for variable #1 to determine the largest and smallest amounts. Mark these values (and the internal values between them) in equal intervals on the horizontal axis.

Step 4   Examine the data for variable #2 to determine the largest and smallest amounts. Mark these values (and the internal values between them) in equal intervals on the vertical axis.

Step 5   Plot each pair of observations on the two variables in the data set by marking the values with dots on the graph.

Self-Test
2.5

Consider the data on two quantitative variables for $n = 12$ experimental units shown in Table 2.14.

Table 2.14 Data for Two Quantitative Variables

Experimental Unit	Variable #1	Variable #2
1	−1	7
2	0	8
3	1	5
4	2	2
5	−3	10
6	1	6
7	−2	8
8	4	1
9	5	0
10	3	0
11	0	7
12	2	5

continued

Construct a scatterplot for the data. What type of relationship is revealed?

Example 2.10   Interpreting Scatterplots

Refer to Example 2.2 (page 66) and the data on 98 automobiles crash-tested for the 1997 New Car Assessment Program (NCAP). Several quantitative variables measured and published in the report are: (1) car weights in pounds (WEIGHT), (2) driver's severity of head injury score (DRIVHEAD), (3) driver's rate of chest deceleration score (DRIVCHEST), and (4) overall driver injury rating (DRIV-STAR). Use scatterplots to examine the relationship between each of the following pairs of quantitative variables. Do the graphs reveal any trends?

   a. WEIGHT vs. DRIVHEAD

   b. DRIVHEAD vs. DRIVCHEST

   c. DRIVHEAD vs. DRIVSTAR

Solution

Figure 2.14 shows three Excel scatterplots for the NCAP data: (a) WEIGHT vs. DRIVHEAD, (b) DRIVHEAD vs. DRIVCHEST, and (c) DRIVHEAD vs. DRIVSTAR.

Figure 2.14a shows little or no association between driver's head injury score and car weight. Figure 2.14b reveals a fairly strong positive association between severity of driver's head injury and chest deceleration. In contrast, Figure 2.14c shows a negative relationship between the driver's head injury score and overall injury rating; the higher the head injury severity score, the lower the overall injury rating. ❑

E **Figure 2.14** Excel Scatterplots for the NCAP Data

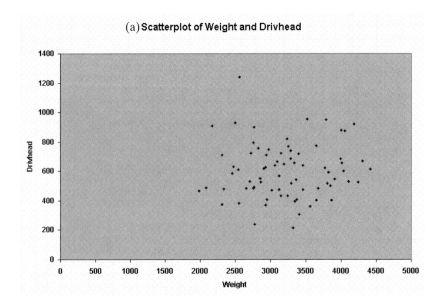

(a) Scatterplot of Weight and Drivhead

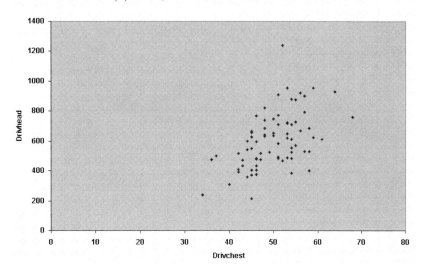

(b) Scatterplot of Drivchest and Drivhead

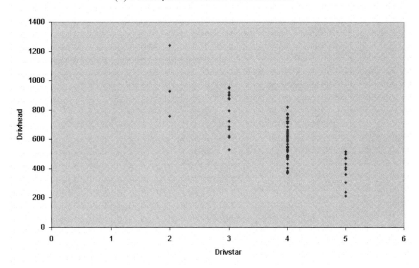

(c) Scatterplot of Drivstar and Drivhead

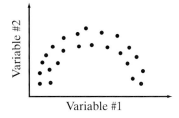

Figure 2.15 A Curvilinear Relationship

Comment: The positive and negative relationships displayed in Figures 2.12–2.14 show a general straight-line (or *linear*) relationship between the quantitative variables plotted. A scatterplot may reveal a nonlinear (or *curvilinear*) relationship similar to the one shown in Figure 2.15. In this case, we cannot label the relationship as positive or as negative. Later in this text (Chapter 10), we develop mathematical "models" for the different types of relationships that may exist between two quantitative variables.

PROBLEMS
2.41–2.50

Using the Tools

2.41 Construct a scatterplot for each of the following sets of data. Determine the nature of the relationship between the two variables.

Variable #1	Variable #2
−1	−1
0	1
1	2
2	4
3	5
4	8
5	9

a.

Variable #1	Variable #2
−1	4
0	1
1	5
2	2
3	1
4	7
5	3

b.

Variable #1	Variable #2
−1	11
0	8
1	7
2	3
3	1
4	0
5	−2

c.

2.42 Two quantitative variables are measured for each of 11 experimental units. The data are shown in the accompanying table.

Variable #1	7	5	8	3	6	10	12	4	9	15	18
Variable #2	21	15	24	9	18	30	36	12	27	45	54

   a. Graph the data using a scatterplot.
   b. Do you detect a relationship between the variables? Explain.

2.43 Two quantitative variables are measured for each of 10 experimental units. The data are shown in the accompanying table.

Variable #1	100	200	250	400	350	300	450	500	100	250
Variable #2	6	20	10	4	12	30	35	15	18	5

   a. Graph the data using a scatterplot.
   b. Do you detect a relationship between the variables? Explain.

2.44 For each scatterplot, determine the nature of the relationship (positive, negative, or none) between the two variables plotted.

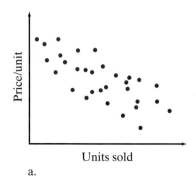

a.

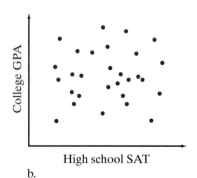

b.

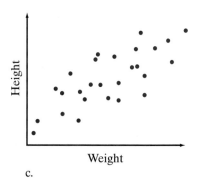

c.

Applying the Concepts

2.45 The following data represent the approximate retail price (in dollars) and the energy cost per year (in dollars) of nine large side-by-side refrigerators.

REFRIG.XLS

Brand	Price ($)	Energy Cost per Year ($)
KitchenAid Superba KSRS25QF	1,600	73
Kenmore (Sears) 5757	1,200	73
Whirlpool ED25DQXD	1,550	78
Amana SRD25S3	1,350	85
Kenmore (Sears) 5647	1,700	93
GE Profile TPX24PRY	1,700	93
Frigidaire Gallery FRS26ZGE	1,500	95
Maytag RSW2400EA	1,400	96
GE TFX25ZRY	1,200	94

Data Source: Adapted from "The Kings of Cool," *Consumer Reports,* Jan. 1998, p. 52.

a. With energy cost on the horizontal axis and price on the vertical axis, construct a scatterplot for the data.

b. Does there appear to be a relationship between price and energy cost? If so, is the relationship positive or negative?

c. Would you expect the higher-priced refrigerators to have greater energy efficiency? Is this borne out by the data?

2.46 Two processes for hydraulic drilling of rock are dry drilling and wet drilling. In a dry hole, compressed air is forced down the drill rods in order to flush the cuttings and drive the hammer; in a wet hole, water is forced down. An experiment was conducted to determine whether the time it takes to dry drill a distance of five feet in rock increases with depth (*The American Statistician,* Feb. 1991). The results for one portion of the experiment are shown in the accompanying table.

DRILL.XLS

Depth at Which Drilling Begins (Feet)	Time to Drill 5 Feet (Minutes)
0	4.90
25	7.41
50	6.19
75	5.57
100	5.17
125	6.89
150	7.05
175	7.11
200	6.19
225	8.28
250	4.84
275	8.29
300	8.91
325	8.54
350	11.79
375	12.12
395	11.02

Source: Penner, R., and Watts, D. G. "Mining information." *The American Statistician,* Vol. 45, No. 1, Feb. 1991, p. 6 (Table 1).

a. With depth on the horizontal axis and drill time on the vertical axis, construct a scatterplot for the data.

b. Identify the nature of the relationship between the two variables.

2.47 The following data represent the average charge (in dollars and cents per minute) and the amount of minutes expended (in billions) for all telephone calls placed from the United States to 20 different countries during 1996.

TELEPHON.XLS

Country	Charge per Minute (In Dollars)	Minutes Expended (In Billions)
Canada	0.34	3.049
Mexico	0.85	2.012
Britain	0.73	1.025
Germany	0.88	0.662
Japan	1.00	0.576
Dominican Republic	0.84	0.410
France	0.81	0.364
South Korea	1.09	0.319
Hong Kong	0.90	0.317
Philippines	1.29	0.297
India	1.38	0.287
Brazil	0.96	0.284
Italy	1.00	0.279
Taiwan	0.97	0.273
Colombia	1.00	0.257
China	1.47	0.232
Israel	1.16	0.214
Australia	1.01	0.201
Jamaica	1.03	0.188
Netherlands	0.78	0.167

Source: The New York Times, Feb. 17, 1997.

a. With charge per minute on the horizontal axis and minutes expended on the vertical axis, set up a scatterplot for the data.

b. Does there appear to be a relationship between charge per minute and expended minutes? If so, is the relationship positive or negative?

c. One might expect that the higher the charge per minute, the lower the number of minutes that would be used. Does the scatterplot reflect this expected relationship? Explain.

2.48 Neurologists have found that the hippocampus, a structure of the brain, plays an important role in short-term memory. Research published in the *American Journal of Psychiatry* (July 1995) attempted to establish a link between hippocampal volume and short-term verbal memory of patients with combat-related posttraumatic stress disorder (PTSD). A sample of 21 Vietnam veterans with a history of combat-related PTSD participated in the study. Magnetic resonance imaging was used to measure the volume of the right hippocampus (in cubic millimeters) of each subject. The verbal memory retention of each subject was also measured by the percent retention subscale of the Wechsler Memory Scale. The data for the 21 patients is plotted in the scatterplot on page 103. The researchers "hypothesized that smaller hippocampal volume would be associated with deficits in short-term verbal memory in patients with PTSD." Does the scatterplot provide visual evidence to support this theory?

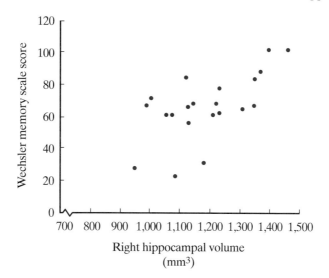

2.49 The physical layout of a warehouse must be carefully designed to prevent vehicle congestion and optimize response time. Optimal design of an automated warehouse was studied in the *Journal of Engineering for Industry* (Aug. 1993). The layout assumes that vehicles do not block each other when they travel within the warehouse, i.e., that there is no congestion. The validity of this assumption was checked by simulating warehouse operations. In each simulation, the number of vehicles was varied and the congestion time (total time one vehicle blocked another) was recorded. The data are shown in the accompanying table. Of interest to the researchers is the relationship between congestion time and number of vehicles.

WAREHOUS.XLS

Number of Vehicles	Congestion Time (Minutes)	Number of Vehicles	Congestion Time (Minutes)
1	0	9	.02
2	0	10	.04
3	.02	11	.04
4	.01	12	.04
5	.01	13	.03
6	.01	14	.04
7	.03	15	.05
8	.03		

Source: Pandit, R., and Palekar, U. S. "Response time considerations for optimal warehouse layout design." *Journal of Engineering for Industry*, Transactions of the ASME, Vol. 115, Aug. 1993, p. 326 (Table 2).

a. Construct a scatterplot for the data.

b. Identify the nature of the relationship between the two variables.

2.50 On January 28, 1986, the space shuttle *Challenger* exploded and seven American astronauts died. Experts agreed that the disaster was caused by two leaky rubber O-rings. Prior to the flight, some rocket engineers had theorized that the O-rings would not seal properly due to cold weather. (The predicted temperature for the launch was 26°F to 29°F.)

The engineers based their theory on data collected on O-ring damage and temperature for all 23 previous launches of the space shuttle. These data are given in the accompanying table. [*Note:* O-ring damage index is a measure of the total number of incidents of O-ring erosion, heating, and blow-by.]

a. Graph the data in a scatterplot. Do you detect a trend?

b. Based on the graph, would you have permitted *Challenger* to launch on the date of the explosion? Explain.

ORING.XLS

Flight Number	Temperature (°F)	O-Ring Damage Index
1	66	0
2	70	4
3	69	0
5	68	0
6	67	0
7	72	0
8	73	0
9	70	0
41-B	57	4
41-C	63	2
41-D	70	4
41-G	78	0
51-A	67	0
51-B	75	0
51-C	53	11
51-D	67	0
51-F	81	0
51-G	70	0
51-I	67	0
51-J	79	0
61-A	75	4
61-B	76	0
61-C	58	4

Note: Data for flight number 4 is omitted due to an unknown O-ring condition.

Primary Sources: Report of the Presidential Commission on the Space Shuttle Challenger Accident, Washington, D.C., 1986, Vol. II (pp. H1–H3) and Volume IV (p. 664). *Post-Challenger Evaluation of Space Shuttle Risk Assessment and Management,* Washington, D.C., 1988, pp. 135–136.

Secondary Source: Tufte, E. R., *Visual and Statistical Thinking: Displays of Evidence for Making Decisions.* Graphics Press, 1997.

2.6 PROPER GRAPHICAL PRESENTATION

According to the Chinese sage Confucius, "a picture is worth a thousand words." The previous sections demonstrated that well-designed graphical displays are powerful tools for presenting data. However, these pictures (e.g., bar graphs, pie charts, histograms, scatterplots) are susceptible to distortion, whether unintentional or as a result of unethical practices. The proliferation of graphics software and spreadsheet applications in recent years has compounded the problem. With a simple "point and click," almost anyone can generate a graph. But if the graph is not presented clearly and carefully, the true message given by the data can become clouded and distorted.

In this section, we present some of the ideas of Professor Edward R. Tufte, who has written a series of books devoted to proper methods of graphical design. We start with Tufte's five principles of **graphical excellence**,* listed in the box.

*Tufte, E. R. *Visual Display of Quantitative Information.* Cheshire, Conn.: Graphics Press, 1983.

Principles of Graphical Excellence

1. Graphical excellence is a well-designed presentation of data that provides substance, statistics, and design.
2. Graphical excellence communicates complex ideas with clarity, precision, and efficiency.
3. Graphical excellence gives the viewer the largest number of ideas in the shortest time with the least ink.
4. Graphical excellence almost always involves several dimensions.
5. Graphical excellence requires telling the truth about the data.

The excellence of a graph is usually compromised when the data are not presented clearly or when the data's message is distorted. To present clear graphs, Tufte suggests maximizing *data-ink* and minimizing *chart junk*. The following examples illustrate these ideas.

Definition 2.5

Data-ink is the amount of ink used in a graph that is devoted to non-redundant display of the data.

Definition 2.6

Chart junk is decoration in a graph that is non-data-ink.

Example 2.11   Chart Junk

The graph shown in Figure 2.16 was recently published in *Newsday,* a New York daily newspaper. Figure 2.16 employs a form of scatterplot to show the trend in annual milk prices in New York City metropolitan area from 1987 to 1997. Comment on the graphical excellence of Figure 2.16.

Figure 2.16 Plot of New York City Milk Prices
Source: "Pricing Milk" by Steve Madden, *Newsday,* Feb. 8, 1998. Reprinted with permission © Newsday, Inc., 1998.

Solution

In this graph, milk prices per gallon are plotted on the vertical axis and years (1987 through 1997) are plotted on the horizontal axis. Note that the years are not clearly marked. (Each vertical grid line represents a different year.) Also, two different milk prices are being plotted. The points at the top of the graph (connected by a straight line) represent annual price per gallon at New York City supermarkets.

The point of the graph is to compare the selling prices of milk at supermarkets to the selling prices established by dairy farmers. The graph, however, suffers from an overabundance of chart junk. The amount of ink used to display the decorative cow is too large relative to the data-ink. (Tufte refers to this type of graph as "The Duck." The chart junk shown in Figure 2.16 is, of course, a variation of "The Duck"—we would call it "A Cow.") ❏

Avoiding distortions of data is a central feature of graphical excellence. A graph does not distort if its visual representation is consistent with its numerical representation. One way to measure the amount of distortion in a graph is with the *lie factor*.

Definition 2.7

The **lie factor** in a graph is the ratio of the size of the effect shown in the graph to the actual size of the effect in the data.

$$\text{Lie factor} = \frac{\text{Size of effect shown in graph}}{\text{Size of effect in the data}}$$

Example 2.12   Lie Factor

Consider Figure 2.17, a graph extracted from the *New York Times*. The figure is a horizontal bar graph showing the oyster catch (in millions of bushels) in Chesapeake Bay for six time periods: 1890–1899, 1930–1939, 1962, 1972, 1982, and 1992. Comment on the graphical excellence of Figure 2.17. Compute the lie factor to support your observation.

Solution

Figure 2.17 uses oyster icons rather than rectangles to represent the "bars" in the bar graph. According to the numerical values shown at the end of the icon, 20 million bushels of oysters were caught during the 1890s and 4 million bushels were caught in 1962. Thus, according to the data, five times as many oysters were caught in Chesapeake Bay during the 1890s than during 1962 (i.e., an actual effect size of 5). The oyster icons, however, do not appear to be sized proportionally. In fact the 1890s icon, measuring lengthwise, is only about twice the size of the 1962 icon (i.e., an effect size of 2). Consequently, the lie factor is

$$\text{Lie factor} = \frac{\text{Visual ratio of 1890s catch to 1962 catch}}{\text{Actual ratio of 1890s catch to 1962 catch}} = \frac{2}{5} = .4$$

In other words, the decrease in the oyster catch from the 1890s to 1962 shown on the graph is only about 40% of the actual decrease conveyed by the data. The oyster icons shown in Figure 2.17 could also be considered chart junk, resulting in distortion of the visual impact of the declining oyster catch in Chesapeake Bay. ❏

Figure 2.17 Oyster Catch in Chesapeake Bay for Six Time Periods
Source: The New York Times, October 17, 1993, p. 26. Reprinted with permission.

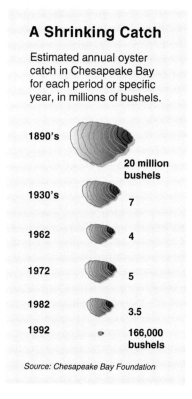

A Shrinking Catch

Estimated annual oyster catch in Chesapeake Bay for each period or specific year, in millions of bushels.

1890's   20 million bushels

1930's   7

1962   4

1972   5

1982   3.5

1992   166,000 bushels

Source: Chesapeake Bay Foundation

Self-Test
2.6

Figure 2.18 Bar Graph for American History Book Reviews

Figure 2.18 is a relative frequency bar graph for data collected on book reviews in American history. Identify one improper or misleading feature of the graph.

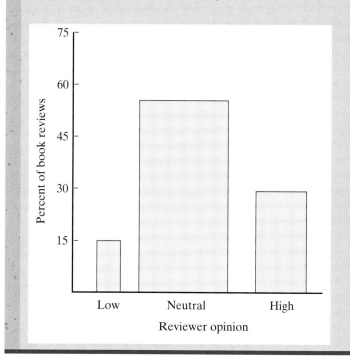

In addition to chart junk, another common way to distort a graph is to change the scale on the vertical axis, the horizontal axis, or both. This occurs most often when the zero point is not displayed or when a broken axis line—called a **scale break**—is used.

Example 2.13   Scale Break

Refer to the Chesapeake Bay oyster catch data, Example 2.12. Figure 2.19 is a bar graph comparing the 1890s catch to the 1962 catch. Identify the distortion in this graph.

Solution

Figure 2.19 uses a scale break on the vertical axis, effectively diminishing the perceived decline in number of oysters caught. We know from the data (displayed in Figure 2.17) that 20 million bushels were caught during the 1890s and 4 million were caught in 1962, a difference of 16 million bushels of oysters. Due to the scale break, however, the bar height for the 1890s does not appear to be 16 million bushels more than the bar height for 1962. ❑

According to Tufte, for many people the first word that comes to mind when they think about statistical charts is "lie." Too many graphics distort the underlying data, making it hard for the reader to learn the truth. Ethical considerations arise when we are deciding what data to present in tabular and chart format and what not to present. It is vitally important when presenting data to document both good and bad results. When making oral presentations and presenting written reports, it is essential that the results be given in a fair, objective, and neutral manner. Thus, we must try to distinguish between poor data presentation and unethical presentation. Again, as in our discussion of ethical considerations in data collection (Section 1.7), the key is *intent*. Often, when fancy tables and chart junk are presented or pertinent information is omitted, it is simply done out of ignorance. However, unethical behavior occurs when an individual willfully hides the facts by distorting a table or chart or by failing to report pertinent findings.

Figure 2.19 Bar Graph for Chesapeake Bay Oyster Catch: 1890s and 1962

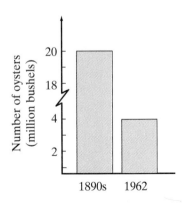

PROBLEMS
2.51–2.57

Applying the Concepts

2.51 Bring to class a graphic from a newspaper or magazine that you believe to be a poorly drawn representation of some *quantitative* variable. Be prepared to submit the graph to the instructor with comments as to why you feel it is inappropriate. Also, be prepared to present and comment on this in class.

2.52 Bring to class a graphic from a newspaper or magazine that you believe to be a poorly drawn representation of some *qualitative* variable. Be prepared to submit the graph to the instructor with comments as to why you feel it is inappropriate. Also, be prepared to present and comment on this in class.

2.53 Bring to class a graphic from a newspaper or magazine that you believe contains too much *chart junk* that may cloud the message given by the data. Be prepared to submit the graph to the instructor with comments as to why you feel it is inappropriate. Also, be prepared to present and comment on this in class.

2.54 Comment on the graphical excellence of the pie charts illustrated in the following graphic recently extracted from the newspaper.

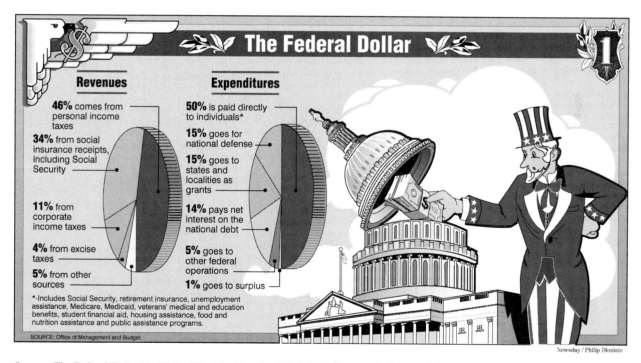

Source: "The Federal Dollar" by Philip Dionisio, *Newsday,* Feb. 3, 1998. Reprinted with permission © Newsday, Inc., 1998.

2.55   Comment on the graphical excellence of the bar graphs illustrated in the following graphic recently extracted from *The New York Times* (Dec. 1, 1997).

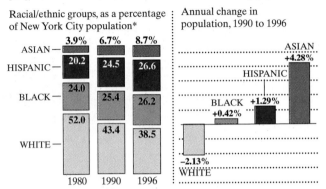

DEMOGRAPHICS

Today's New Yorkers

An analysis of Federal Census Bureau data shows that Hispanic people continue to increase their share of New York City population.

Racial/ethnic groups, as a percentage of New York City population*

Annual change in population, 1990 to 1996

*Hispanic people can be of any race. Statistics for Asian, black, and white groups do not include Hispanic New Yorkers.

Source: Taub Urban Research Center, New York University

Source: The New York Times, December 1, 1997, p. B3. Reprinted with permission.

2.56   Comment on the graphical excellence of the bar graph illustrated in the following graphic extracted from *Self* (Oct. 1992).

6. What are your favorite Olympic events?

Gymnastics 30%

Figure Skating 27%

Track and Field 17%

Skiing 13%

Basketball 11%

Swimming 11%

Data Source: Self, October 1992.

2.57 Comment on the graphical excellence of the bar graph illustrated in the following graphic extracted from *Fortune* (Sept. 18, 1995).

Top 20 Industries for Jobs
JANUARY **1990** THROUGH JUNE **1995**

Industry	899 thousands of jobs created
Temp and full-time employment agencies	899
Restaurants and bars	738
Local government administration	359
Hospitals	345
Recreation (health clubs, casinos, etc.)	344
Home health care	339
Nursing and personal care	310
Medical doctors' offices	290
Computer software	255
State government administration	228
Trucking and parcel delivery	222
Business services (security, drafting, etc.)	204
Residential care (rehab centers, etc.)	195
Management consulting and public relations	179
Social services	166
Offices of optometrists, podiatrists, etc.	144
Child day care	136
Grocery stores	133
Motion picture production	127
Elementary and secondary schools	106

FORTUNE CHART / SOURCE: NUALA BECK & ASSOCIATES

REAL-WORLD REVISITED
A Bad Moon Rising

Is our behavior influenced by the different phases of the moon? To the scientific mind, the idea that lunar phases influence behavior is matter-of-factly passed off as superstition or gobbledygook. In fact, several scientific studies have found no evidence of a link between human behavior and phases of the moon. Nevertheless, many of us at one time or another have associated aberrant behavior with a full moon. Whether or not our behavior is actually affected by the moon, studies have indicated that a significant percentage of people *believe* in lunar effects.

To quantify the degree to which people believe in this phenomenon, psychologists J. Rotton and I. W. Kelly (1985) developed a Belief in Lunar Effects (BILE) Scale. Questionnaires were administered to a sample of 157 undergraduates (54 male and 103 female). In addition to providing demographic and background information, students were asked to respond to nine statements relating lunar phases with human behavior. The nine items are listed here.

1. Lunar phases play an important role in human affairs.

2. There is some truth to the idea that "crazies" come out when the moon is full.

3. Some people behave strangely when the moon is full.

*4. I have never felt that the moon affects my behavior.

5. It is a good idea to stay at home when the moon is full.

6. My own behavior is affected by phases of the moon.

7. A full moon can trigger violence and aggression.

*8. There is absolutely no relationship between phases of the moon and behavior.

*9. Only superstitious people believe that a full moon influences behavior.

Respondents indicated degree of agreement with each statement by choosing numbers between 1 (strong disagreement) and 9 (strong agreement). Total scores were then obtained by summing the nine item scores for each respondent. [*Note:* For items marked with an asterisk (*), scores were reversed, i.e., a response of "1" was scored a "9," a response of "2" was scored an "8", etc.] On the BILE scale, this total score measures the degree to which a person "believes in lunar effects on behavior," with a maximum value of 81 indicating strong agreement and a minimum value of 9 indicating strong disagreement.

The BILE scores for all 157 respondents are listed in Table 2.15. Use the methods of this chapter to summarize and describe the distribution of BILE scores. [*Note:* The data in Table 2.15 are saved in the Excel workbook BILE.XLS.]

BILE.XLS

Table 2.15 Belief in Lunar Effects (BILE) Scale Scores for 157 Subjects

21	43	72	27	50	48	30	45	41	22	50	36	29	60	13
48	34	19	51	25	37	58	34	71	46	42	18	44	24	51
16	55	15	59	24	45	51	53	30	45	39	61	39	46	36
60	19	46	14	54	32	27	65	56	43	34	16	26	37	45
46	38	18	38	61	41	44	13	32	52	29	56	36	59	32
31	56	28	43	29	33	37	20	56	26	40	22	48	18	43
23	35	51	16	45	27	40	44	30	61	20	69	20	44	21
44	18	59	29	24	51	24	47	42	38	36	18	52	17	39
25	53	45	31	48	64	52	27	45	48	51	39	11	43	34
36	61	15	41	12	22	45	45	56	33	34	41	29	16	49
50	12	30	17	39	20	49								

Source: Rotton, J. and Kelly, I. W. "A scale for assessing belief in lunar effects: Reliability and concurrent validity." *Psychological Reports,* Vol. 57, 1985, pp. 239–245.

One of the objectives of the researchers was to determine whether a large proportion of respondents have either an overall strong disagreement or strong agreement with the notion that the moon affects human behavior.

Key Terms

Bar graph 65	Class boundary 76	Class interval width 75
Chart junk 105	Class frequency 62	Class percentage 62
Class 62	Class interval 75	Class relative frequency 62

Cross-classification table 86
Data-ink 105
Frequency 62
Graphical excellence 105
Histogram 77
Left (negatively) skewed 77
Lie factor 106

Negative association 96
Nonskewed (symmetric) 77
Pie chart 65
Positive association 96
Right (positively) skewed 76
Scale break 108
Scatterplot 95

Side-by-side horizontal
 bar chart 87
Skewed distribution 77
Stem-and-leaf display 74
Symmetric 77

Supplementary Problems 2.58–2.76

Applying the Concepts

2.58 The accompanying data indicate fat and cholesterol information concerning popular protein foods (fresh red meats, poultry, and fish). For the data relating to the amount of calories, protein, the percentage of calories from fat and saturated fat, and the cholesterol for the popular protein foods:

a. Construct the stem-and-leaf display.

b. Construct the frequency distribution and the percentage distribution.

c. Construct scatterplots for each pair of variables.

d. What conclusions can you reach concerning the amount of calories of these foods?

PROTEIN.XLS

Food	Calories	Protein (g)	Fat	Saturated Fat	Cholesterol (mg)
Beef, ground, extra lean	250	25	58	23	82
Beef, ground, regular	287	23	66	26	87
Beef, round	184	28	24	12	82
Brisket	263	28	54	21	91
Flank steak	244	28	51	22	71
Lamb leg roast	191	28	38	16	89
Lamb loin chop, broiled	215	30	42	17	94
Liver, fried	217	27	36	12	482
Pork loin roast	240	27	52	18	90
Sirloin	208	30	37	15	89
Spareribs	397	29	67	27	121
Veal cutlet, fried	183	33	42	20	127
Veal rib roast	175	26	37	15	131
Chicken, with skin, roasted	239	27	51	14	88
Chicken, no skin, roast	190	29	37	10	89
Turkey, light meat, no skin	157	30	18	6	69
Clams	98	16	6	0	39
Cod	98	22	8	1	74
Flounder	99	21	12	2	54
Mackerel	199	27	77	20	100
Ocean perch	110	23	13	3	53
Salmon	182	27	24	5	93
Scallops	112	23	8	1	56
Shrimp	116	24	15	2	156
Tuna	181	32	41	10	48

Source: United States Department of Agriculture.

e. What conclusions can you reach concerning the amount of protein of these foods?

f. What conclusions can you reach concerning the percentage of calories from fat and saturated fat of these foods?

g. Are there any foods that seem different from the others in terms of these variables? Explain.

h. Based on part **c**, what conclusions can you reach about the relationship among these five variables?

2.59 Consider an investigation of the natural survival rate of the green sea turtle (an endangered species). A total of 350 sea turtle nests at Tortuguero beach located on the Caribbean coast of Costa Rica were marked and monitored. The fate of each of the 350 marked nests was of prime interest; results are given in the table. Construct a bar graph for the data. Interpret your results.

Nest Fate	Number
Undisturbed, young emerged (successful hatching)	148
Disturbed by predators, some young emerged	18
Destroyed by animal predators	122
Washed out by surf	20
Lost to human predators	23
Dead although undisturbed	19
TOTAL	350

Data Source: Flower, L. E. "Hatching success and nest predation in the green sea turtle, *Chelonia mydas,* at Tortuguero, Costa Rica." *Ecology,* Oct. 1979, 60, pp. 946–955.

2.60 Electrical engineers recognize that high neutral current in computer power systems is a potential problem. To determine the extent of the problem, a survey of the computer power system load currents at 146 U.S. sites was taken (*IEEE Transactions on Industry Applications,* July/Aug. 1990). A relative frequency histogram for the load capacities (measured as a percentage) of the 146 sites in the sample is shown here.

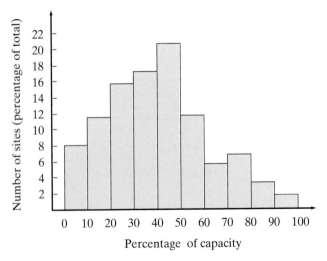

Source: Gruzs, T. M. "A survey of neutral currents in three-phase computer power systems." *IEEE Transactions on Industry Applications,* Vol. 26, No. 4, July/Aug. 1990, p. 722 (Figure 6).

a. Approximately what proportion of the 146 computer sites had a load capacity between 20% and 30%?

b. Approximately what proportion of the 146 computer sites had a load capacity of 50% or more?

2.61 The U.S. Department of Education takes corrective and punitive actions against colleges and universities with high student-loan default rates. Those schools with default rates above 60% face suspension from the government's massive student-loan program, whereas schools with default rates between 40% and 60% are mandated to reduce their default rates by 5% a year or face a similar penalty. A list of 66 colleges and universities in Florida and their student-loan default rates is provided in the accompanying table.

DEFAULT.XLS

College/University	Default Rate	College/University	Default Rate
Florida College of Business	76.2	Brevard Community College	9.4
Fort Lauderdale College	48.5	College of Boca Raton	9.1
Florida Career College	48.3	Florida International University	8.7
United College	46.8	Santa Fe Community College	8.6
Florida Memorial College	46.2	Edison Community College	8.5
Bethune Cookman College	43.0	Palm Beach Junior College	8.0
Edward Waters College	38.3	Eckerd College	7.9
Florida College of Medical and Dental Careers	32.6	University of Tampa	7.6
International Fine Arts College	26.5	Lakeland College of Business	7.2
Tampa College	23.9	Pensacola Junior College	6.8
Miami Technical College	23.3	University of Miami	6.7
Tallahassee Community College	20.6	Florida Institute of Technology	6.7
Charron Williams College	20.2	University of West Florida	6.3
Florida Community College	19.1	Palm Beach Atlantic College	6.0
Miami-Dade Community College	19.0	University of Central Florida	5.7
Broward Community College	18.4	Seminole Community College	5.6
Daytona Beach Community College	16.9	Polk Community College	5.6
Lake Sumter Community College	16.7	Phillips Junior College	5.6
Florida Technical College	16.6	Nova University	5.5
Florida A. & M. University	15.8	Rollins College	5.5
Prospect Hall College	15.1	St. Leo College	5.5
Hillsborough Community College	14.4	Gulf Coast Community College	5.4
Pasco-Hernando Community College	13.5	Southern College	5.3
Orlando College	13.5	Flagler College	4.7
Jones College	13.1	Florida Atlantic University	4.4
Webber College	11.8	University of South Florida	4.2
Warner Southern College	11.8	Manatee Junior College	4.1
Central Florida Community College	11.8	Florida State University	4.0
Indian River Community College	11.8	University of North Florida	3.9
St. Petersburg Community College	11.3	Barry University	3.1
Valencia Community College	10.8	University of Florida	3.1
Florida Southern College	10.3	Stetson University	2.9
Lake City Community College	9.8	Jacksonville University	1.5

a. Construct a relative frequency histogram for the data using 12 classes to span the range.

b. Repeat part **a**, but use only three classes to span the range. Compare with the relative frequency distribution you constructed in part **a**. Which is more informative? Why do an inadequate number of classes limit the information conveyed by the relative frequency distribution?

c. Repeat part **a**, but use 25 classes. Comment on the information provided by this graph as compared with that of part **a**.

d. Refer to the histogram, part **a**. Estimate the proportion of Florida colleges and universities with a default rate of 40% or higher.

e. Note that Florida College of Business has a default rate nearly 30% higher than the next highest rate. Omit the value for Florida College of Business from the data set and reconstruct the histogram.

f. Compare the histograms constructed in parts **a** and **e**. Which graph is more informative? Explain.

2.62 The *Journal of Consumer Marketing* (Summer 1992) reported on a study of company response to letters of consumer complaints. Marketing students at a large midwest public university "were asked to write letters of complaint to companies whose products legitimately caused them to be dissatisfied." Of the 750 students in the class, 286 wrote letters of complaint. The table shows the type of response received from the companies and number of each type.

Type of Response	Number
Letter and product replacement	42
Letter and good coupon	36
Letter and cents-off coupon	29
Letter and refund check	23
Letter, refund check, and coupon	13
Letter only	76
No Response	67
TOTAL	286

Source: Clark, G. L., Kaminski, P. F., and Rink, D. R. "Consumer complaints: Advice on how companies should respond based on an empirical study." *Journal of Consumer Marketing,* Vol. 9, No. 3, Summer 1992, p. 8 (Table 1).

a. Display the study results in a relative frequency bar graph.

b. What percentage of consumers received a response to their letter of complaint?

c. What conclusions can you reach about the survey responses?

2.63 Reporting in the *New England Journal of Medicine* (Mar. 18, 1991), the Centers for Disease Control (CDC) confirmed what many former cigarette smokers have learned from experience: People who quit smoking tend to gain weight. The CDC's research team reviewed data on 1,885 smokers and 768 former smokers who were studied over a 13-year period. Weight gain over the study period was classified as slight (3 kilograms or less), moderate (3–8 kilograms), significant (8–13 kilograms), and major (more than 13 kilograms). The smokers/quitters were also classified according to gender to compare male versus female weight gain. The percentages of men and women in the four weight-gain categories are provided in the table on page 117.

	QUITTERS		SMOKERS	
Weight Gain	**Men**	**Women**	**Men**	**Women**
Slight	55	50	66	63
Moderate	22	26	24	23
Significant	14	10	8	9
Major	9	14	2	5
TOTALS	100	100	100	100

Source: Time, Mar. 25, 1991, p. 55.

a. Describe the data with the appropriate graphical technique. Construct one graph for each column of the table.

b. Compare the four graphs, part **a**. Do quitters tend to gain more weight than smokers? Do female quitters tend to gain more weight than male quitters?

2.64 An investigation of the waiting time experienced by customers at a large Boston bank was conducted (*Sloan Management Review,* Winter 1991). With the aid of video cameras, the researchers recorded the actual time each of 227 customers waited in line for a bank teller. As customers finished their transactions, they were interviewed by the researchers and asked about perceived waiting time. In addition to perceived and actual waiting time, the researchers asked customers to describe their wait in line. Customers generally fell into three categories: "watchers" who enjoy observing people and events at the bank, "impatients" who could think of nothing more boring than waiting in line, and "neutrals" who fell somewhere in the middle. The accompanying table gives the breakdown, in percentages, of the 277 customers falling into the three categories of watchers, impatients, and neutrals.

SUMMARY OF CUSTOMER INTEREST LEVEL

Category	Percentage
Watchers	21
Impatients	45
Neutrals	34

Source: Katz, K. L., Larson, B. M., and Larson, R. C. "Prescription for the waiting-in-line blues: Entertain, enlighten, and engage." *Sloan Management Review,* Winter 1991, p. 49 (Fig. 5).

a. Are the data quantitative or qualitative?

b. Describe the data summarized in the table graphically. Interpret the graph.

2.65 Refer to the *Sloan Management Review* (Winter 1991) study in Problem 2.64.

a. Consider the data set consisting of the differences between the perceived and actual waiting times of the 277 customers. Are the data qualitative or quantitative?

b. A graph describing the data of part **a** is shown on page 118. What type of graph is displayed? Interpret the graph. In particular, obtain the approximate percentage of customers who overestimated their waiting time in line at the Bank of Boston.

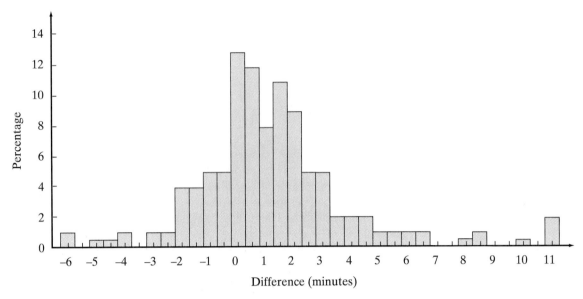

Difference (minutes)

Source: Katz, K. L., Larson, B. M., and Larson, R. C. "Prescription for the waiting-in-line blues: Entertain, enlighten, and engage." *Sloan Management Review,* Winter 1991, p. 48 (Fig. 3).

Problem 2.65

2.66 The *Journal of Modern History* (Dec. 1997) published a study of radio broadcasting in Germany during the 1920s. The aim of the research was to determine whether the introduction of the modern mass medium of radio broadcasting impacted German cultural distinctions. In 1929, the German radio station *Deutsche Stunde in Bayern* broadcast 3,578 hours of programming. The journal article classified each hour of programming into one of several categories. A summary of the number of broadcasting hours in each category is displayed in the accompanying table.

Broadcast Category	Number of Programming Hours
Music	1,943
Lectures	576
News	494
Plays/Literature	279
Programming for women	82
Programming for children	61
General reports	25
Sports reports	21
Miscellaneous	97
TOTAL	3,578

Source: Führer, K. C. "A medium of modernity? Broadcasting in Weimer Germany, 1923–1932." *The Journal of Modern History,* Vol. 69, No. 4, Dec. 1997, p. 743 (adapted from Table 3).

a. Summarize the data using a graphical method.

b. Identify the broadcast category that occurred most often.

c. What proportion of programming time was dedicated to lectures?

2.67 Chemical researchers experimented with a new copper-selenium metal compound and reported the results in *Dalton Transactions* (Dec. 1997). The chemists varied the ratio of copper to selenium in each metal cluster, then used x-ray crystallography to measure the bond length (pm) of each cluster. The bond lengths for 43 metal clusters are listed in the accompanying table. [*Note:* Bond lengths marked with an asterisk (*) represent copper-phosphorus metal compounds.]

BOND.XLS

256.1	259.0	256.0	264.7	223.8*	262.5
239.0	238.9	237.4	222.7*	265.9	240.2
239.7	238.3	223.3*	273.8	246.4	246.0
247.4	290.6	260.9	255.6	253.1	283.7
261.4	271.9	267.9	238.7	237.6	240.0
226.1*	241.1	240.3	240.8	226.3*	246.5
248.3	246.0	226.0*	250.3	250.1	254.6
225.0*					

Source: Deveson, A., et al. "Syntheses and structures of four new copper (I)-selenium clusters: Size dependence of the cluster on the reaction conditions." *Dalton Transactions,* No. 23, Dec. 1997, p. 4492 (Table 1).

a. Graphically summarize the data for all 43 metal clusters.

b. What type of skewness exists in the data?

c. If possible, locate the bond lengths for the copper-phosphorus metal compounds on the graph, part **a**. What do you observe?

2.68 A study was conducted to determine whether a student's final grade in an introductory sociology course was linearly related to his or her performance on the verbal ability test administered before college entrance. The verbal test scores and final grades for a random sample of 10 students are shown in the table.

SOCIOLOGY.XLS

Student	Verbal Ability Test Score	Final Sociology Grade
1	39	65
2	43	78
3	21	52
4	64	82
5	57	92
6	47	89
7	28	73
8	75	98
9	34	56
10	52	75

a. Construct a scatterplot for the data.

b. Describe the relationship between verbal ability test score and final sociology grade.

2.69 The data on page 120 represent the retail price (in dollars) and the printing speed (in number of pages per minute of double-spaced black text with standard margins) for a sample of 19 computer printers.

a. With price on the horizontal axis and text speed on the vertical axis, set up a scatterplot.

b. Does there appear to be a relationship between price and text speed? If so, is the relationship positive or negative?

c. One might expect that the higher the price, the higher the text speed. Does the scatterplot reflect this expected relationship? Explain.

PRINTER.XLS

Brand	Price (In Dollars)	Text Speed (Pages per Minute)
Hewlett-Packard DeskJet 855Cse	500	3.0
Hewlett-Packard DeskJet 682C	300	2.5
Hewlett-Packard DeskJet 600C	250	2.6
Epson Stylus ColorII	230	2.5
Hewlett-Packard DeskWriter 600	250	3.0
Canon BJC-610	430	1.3
Canon BJC-210	150	2.9
Apple Color StyleWriter 1500	280	3.1
Canon BJC-4100	230	1.7
Apple Color StyleWriter 2500	380	3.4
Epson Stylus ColorIIs	190	0.7
Lexmark 2070 Jetprinter	350	2.1
Lexmark 1020 Jetprinter	150	1.3
NEC SuperScript 860	500	7.9
Panasonic KX-P6500	450	5.5
Hewlett-Packard LaserJet 5L	480	4.2
Texas Instruments MicroLaser Win/4	380	4.2
Canon LBP-460	350	4.1
Okidata OL600e	400	3.9

Data Source: Adapted from "Computer printers," *Consumer Reports,* Oct. 1996, pp. 60–61.

Problem 2.69

2.70   The owner of a restaurant serving Continental-style entrees was interested in studying patterns of demand for dessert by patrons for the Friday-to-Sunday weekend time period. She decided that two other variables were to be studied along with whether a dessert was ordered: the gender of the individual and whether a beef entree was ordered. The data for 600 patrons were tabulated and are summarized in the accompanying tables.

		GENDER		
		Male	**Female**	**TOTALS**
DESSERT	Yes	96	40	136
ORDERED	No	224	240	464
	TOTALS	320	280	600

		BEEF ENTREE		
		Male	**Female**	**TOTALS**
DESSERT	Yes	71	65	136
ORDERED	No	116	348	464
	TOTALS	187	413	600

For each of the two cross-classification tables,

a.   Construct a table of row percentages.

b.   Construct a table of column percentages.

c.   Construct a table of total percentages.

d.   Which type of percentage (row, column, or total) do you think is most informative for the gender/dessert ordered table? Beef entree/dessert ordered table?

e. What conclusions concerning the pattern of dessert ordering can the owner of the restaurant reach?

2.71 The accompanying table gives the breakdown in the percentages of current Californians who speak foreign languages and how well they speak English.

Main Language Spoken	Percentage
English	54.2
Spanish	
Speak English very well	13.4
Speak English well	6.3
Speak little English	9.4
TOTAL	29.2
Asian languages (various)	
Speak English very well	4.5
Speak English well	3.0
Speak little English	2.6
TOTAL	10.1
Other	
Speak English very well	4.4
Speak English well	1.4
Speak little English	0.7
TOTAL	6.5

Source: U.S. Census Bureau, *San Francisco Chronicle,* May 13, 1992.

a. Use a graphical method to summarize the main languages spoken by Californians.

b. Use a graphical method to summarize how well Spanish-speaking Californians speak English.

2.72 In the book *Identical Twins Reared Apart,* author Susan Farber offers a survey and reanalysis of all published cases of identical (monozygotic) twins separated and reared apart. One chapter deals extensively with the evaluation and analysis of IQ tests conducted on the twins. A portion of the data for 19 pairs of twins is reproduced in the table below.

IDTWIN.XLS

	IQ				IQ	
Pair	Twin A	Twin B		Pair	Twin A	Twin B
1	99	101		11	115	105
2	85	84		12	90	88
3	92	116		13	91	90
4	94	95		14	66	78
5	105	106		15	85	97
6	92	77		16	102	94
7	122	127		17	89	106
8	96	77		18	102	96
9	89	93		19	88	79
10	116	109				

Data Source: Adapted from *Identical Twins Reared Apart,* by Susan L. Farber. Basic Books, Inc., 1981.

a. Classify the variable of interest, IQ score, as quantitative or qualitative.

b. What type of graphical method is appropriate for summarizing the data in the table?

c. Construct an appropriate graph to summarize the data for the 38 twins in the sample.

d. Use the graph to find the percentage of twins in the sample with IQ scores above 100.

e. What would be the advantages and disadvantages of using a stem-and-leaf display, rather than a relative frequency distribution, to describe the IQ data?

2.73 A study was conducted to obtain information on the background levels of the toxic substance polychlorinated biphenyl (PCB) in soil samples in the United Kingdom. The accompanying table contains the measured PCB levels of soil samples taken at 14 rural and 15 urban locations in the United Kingdom. (PCB concentration is measured in .0001 gram per kilogram of soil.) From these preliminary results, the researchers reported "a significant difference between (the PCB levels) for rural areas … and for urban areas."

PCBUK.XLS

Rural				
3.5	9.0	9.8	23.0	8.2
8.1	1.0	15.0	1.5	9.7
1.8	5.3	1.6	12.0	

Urban				
24.0	21.0	22.0	94.0	18.0
29.0	11.0	13.0	141.0	12.0
16.0	49.0	107.0	11.0	18.0

Data Source: Badsha, K. and Eduljee, G. "PCB in the U.K. environment—A preliminary survey." *Chemoshpere,* Vol. 15, No. 2, Feb 1986, p. 213 (Table 1), Pergamon Press, Ltd.

a. Construct a stem-and-leaf display for the PCB levels of rural soil samples.

b. Construct a stem-and-leaf display for the PCB levels of urban soil samples.

c. Combine the data for rural and urban soil samples and construct a stem-and-leaf display. Identify each of the urban PCB levels on the display with a circle. Does the graph support the researchers' conclusions?

2.74 The average New York state household produces about 44,600 pounds of carbon dioxide each year. The graphic on page 123, extracted from *The New York Times,* is intended to describe the amount of carbon dioxide emissions attributed to each household item. Comment on the excellence of the graph.

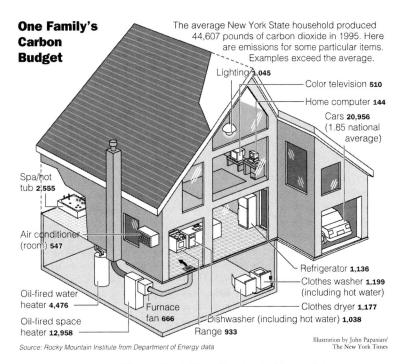

One Family's Carbon Budget

The average New York State household produced 44,607 pounds of carbon dioxide in 1995. Here are emissions for some particular items. Examples exceed the average.

Lighting **1,045**

Color television **510**

Home computer **144**

Cars **20,956** (1.85 national average)

Spa/hot tub **2,555**

Air conditioner (room) **547**

Oil-fired water heater **4,476**

Oil-fired space heater **12,958**

Furnace fan **666**

Refrigerator **1,136**

Clothes washer **1,199** (including hot water)

Clothes dryer **1,177**

Dishwasher (including hot water) **1,038**

Range **933**

Illustration by John Papasian/ The New York Times

Source: Rocky Mountain Institute from Department of Energy data

Source: John Papasian/*The New York Times.* Reprinted with permission.

Problem 2.74

2.75   The following figure appeared in an advertisement for satellite television in a recent issue of *Sports Illustrated.* Comment on the graphical excellence of the figure.

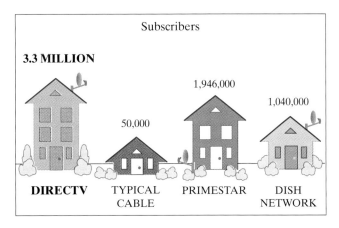

Subscribers

3.3 MILLION

1,946,000

1,040,000

50,000

DIRECTV   TYPICAL CABLE   PRIMESTAR   DISH NETWORK

Data Source: Media Business Corp., 12/31/97.

2.76   Teenage pregnancy and childbearing have become pressing social concerns in the United States. Data from individual 1989 and 1990 vital statistics records for births in New Jersey, by race of mother, are summarized in the accompanying table.

		NUMBER OF NEW JERSEY BIRTHS	
		White	**Black**
	<15	162	292
	15–17	3,084	3,551
	18–19	6,655	5,180
Maternal	20–24	29,057	12,699
Age Group	25–29	55,517	10,944
	30–34	49,023	6,588
	35–39	16,558	2,172
	≥40	2,273	376
	TOTALS	162,329	41,802

Source: Reichman, N. E., and Pagnini, D. L. "Maternal age and birth outcomes: Data from New Jersey." *Family Planning Perspectives,* Vol. 29, No. 6, Nov./Dec. 1997, p. 269 (Table 1).

a.   For white mothers, construct a table giving the percentages of births in each age group.

b.   Repeat part **a** for black mothers.

c.   Use a side-by-side bar graph to compare the distributions of New Jersey births for white and black mothers. Interpret the results.

2.E.1 Using Microsoft Excel to Describe a Single Qualitative Variable

Solution Summary:

Use the PHStat add-in to summarize and graph the data of a single qualitative variable.

 Example: Intensity of men's cologne data of Table 2.1

Solution:

To generate a frequency table, bar chart, and pie chart from the intensity ratings of 17 men's cologne brands (Table 2.1 on page 63), do the following:

1 If the PHStat add-in has not been previously loaded, load the add-in using the instructions of Section EP.3.2. (You will also need to run the disk setup program using the procedure in Appendix D if you have never previously installed the files on the CD-ROM that accompanies this text.)

2 Open the cologne workbook (COLOGNE.XLS) and click the Data sheet tab.

3 Select PHStat | One-Way Tables & Charts.

4 In the One-Way Tables & Charts dialog box, enter the information and make the selections shown in Figure 2.E.1. Click the OK button.

E Figure 2.E.1

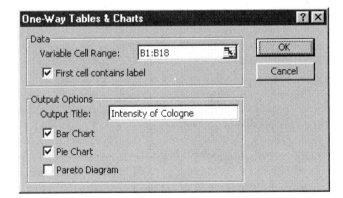

The add-in produces on separate sheets a frequency table, bar chart, and pie chart, similar to the ones shown in Figures 2.E.2, 2.E.3, and 2.E.4.

E Figure 2.E.2

	A	B
1	Intensity of Cologne	
2		
3	Count of Intensity	
4	Intensity	Total
5	Mild	2
6	Strong	10
7	Very strong	5
8	Grand Total	17

E **Figure 2.E.3**

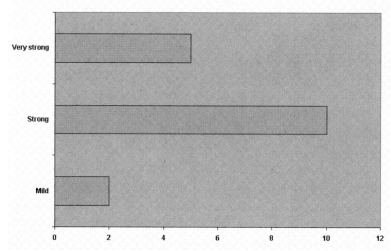

E **Figure 2.E.4**

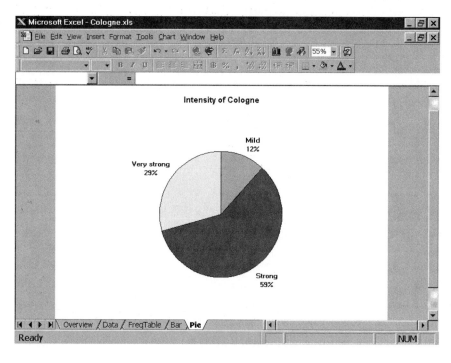

2.E.2 Using Microsoft Excel to Describe a Single Quantitative Variable

Solution Summary:

A. Use the PHStat add-in to construct a stem-and-leaf display.

B. Use the Data Analysis Histogram tool to construct a frequency table and frequency histogram.

C. Follow the "Correcting the Data Analysis Histogram Output" procedure to revise the frequency table and histogram.

D. Use the Microsoft Excel Chart Wizard to construct a relative frequency histogram.

Example: Starting Salaries Data of Table 2.5

Part A Solution: Using the PHStat add-in to construct a stem-and-leaf display

To construct a stem-and-leaf display from the starting salaries for 50 graduates of the University of South Florida (Table 2.5 on page 75), do the following:

1. If the PHStat add-in has not been previously loaded, load the add-in using the instructions of Section EP.3.2. (You will also need to run the disk setup program using the procedure in Appendix D if you have never previously installed the files on the disk that accompanies this text.)
2. Open the Salaries workbook (SAL50.XLS) and click the Data sheet tab.
3. Select PHStat | Stem-and-Leaf Display.
4. In the Stem-and-Leaf Display dialog box, enter the information and make the selections shown in Figure 2.E.5. Click the OK button.

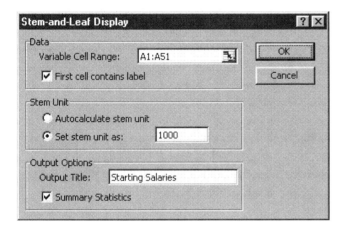

E **Figure 2.E.5**

E **Figure 2.E.6**

	A	B	C	D	E	F
1				Stem-and-Leaf Display for Starting Salaries		
2				Stem unit: 1000		
3	Statistics			11	0	
4	n	50		12		
5	Mean	25708		13	2	
6	Median	25200		14	2	
7	Std. dev.	7337.9316		15	3 7	
8	Minimum	11000		16	8	
9	Maximum	43900		17	4	
10				18		
11				19		
12				20	0 1 4 4 4 5 6 7	
13				21	4 5	
14				22	1 8 8	
15				23	1 2	
16				24	4 9	
17				25	2 2 6	
18				26	1 3 6 7 7	
19				27		
20				28	2 3 7	
21				29	4	
22				30	0 3 9	
23				31	4	
24				32	4 7	
25				33	1	
26				34	6	
27				35		
28				36	6 7	
29				37	5	
30				38	9	
31				39		
32				40	5	
33				41		
34				42		
35				43	9	

Table 2.E.1 Class
Maximum Equivalents of
Table 2.6 Ranges

Original Class Interval Range	Entered As:
10950–13949	13949
13950–16949	16949
16950–19949	19949
19950–22949	22949
22950–25949	25949
25950–28949	28949
28950–31949	31949
31950–34949	34949
34950–37949	37949
37950–40949	40949
40950–43949	43949

The add-in constructs on a new worksheet a stem-and-leaf display similar to the one shown in Figure 2.E.6. This chart is **not** dynamically changeable, so changes made to the underlying data would require repeating the procedure in order to produce a new chart.

Part B Solution: Using the Data Analysis Histogram tool to construct a frequency table and frequency histogram

Using the Data Analysis Histogram tool requires a preliminary step in which the data to define class intervals are entered in the worksheet containing the data to be summarized. Unlike the common way of defining a class interval by its range, as is done in Table 2.6 on page 76, Microsoft Excel requires that class intervals be specified as an ordered list of class maximum values. Table 2.E.1 shows how the ranges of Table 2.6 are entered as an ordered list.

To construct a frequency table and a frequency histogram from the starting salary data, do the following:

1️⃣ Open the Salaries workbook (SAL50.XLS) and click the Data sheet tab. Note that the class maximum values of Table 2.E.1 have been entered to the right of the salary data under the heading Classes in column C.

2️⃣ Select Tools | Data Analysis. If the Data Analysis does not appear as a choice on the Tools menu, the Microsoft Excel Analysis ToolPak add-in has not been previously installed. Install this add-in using the procedure discussed in Section EP.2.1 before continuing.

3️⃣ Select Histogram from the Analysis Tools list box in the Data Analysis dialog box (see Figure 2.E.7). Click the OK button.

E Figure 2.E.7

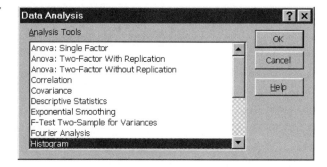

4️⃣ Enter the information and make the selections shown in Figure 2.E.8 in the Histogram dialog box. Click OK.

E Figure 2.E.8

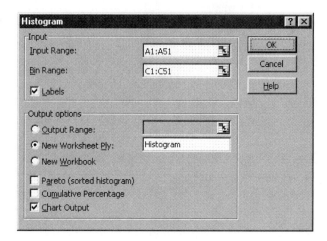

128

The add-in produces on a new sheet named Histogram both a frequency table and frequency histogram, similar to the one shown in Figure 2.E.9.

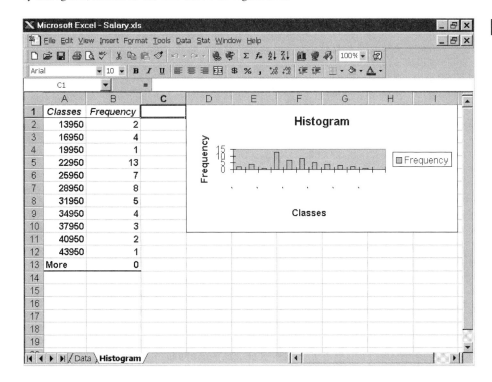

E **Figure 2.E.9**

Part C Solution: Correcting the Data Analysis Histogram output

Note that the Data Analysis Histogram output shown in Figure 2.E.9 contains two errors: there is an additional class, labeled More by Excel, in both the frequency distribution and the histogram and there are gaps between the bars that correspond to the class intervals in the histogram. To correct these errors, do the following:

1 Click the Histogram sheet tab.

2 Select the cell range A13:B13 (containing the More row of the frequency distribution table).

3 Select Edit | Delete.

4 In the Delete dialog box, select the Shift cells up option button and then click OK.

The "More" class is deleted from both the frequency distribution table and the histogram. Continue by removing the gaps between the bars.

5 Right-click on one of the histogram bars. (The mouse pointer is over a bar when a tool tip that includes the words "Series 'Frequency'" is displayed.)

6 Select Format Data Series from the shortcut menu.

7 In the Format Data Series dialog box, click the Options tab and change the value in the Gap width: edit box to 0 (see Figure 2.E.10). Click OK.

The errors are now eliminated and the resulting histogram will be similar to the one shown in Figure 2.6 on page 77.

E **Figure 2.E.10**

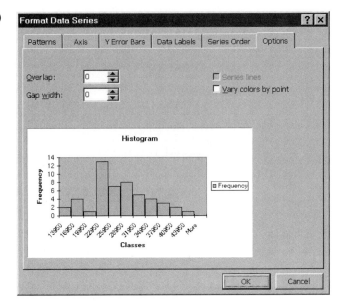

Part D Solution: Using the Chart Wizard to construct a relative frequency histogram

Constructing a relative frequency histogram requires a two-step approach that begins by adding relative frequencies to the frequency table produced by the Data Analysis Histogram tool and then using the Chart Wizard to construct the relative frequency histogram.

To modify the frequency table of starting salaries data, do the following:

1. If not previously done, follow the procedure given in Part B Solution to generate a Histogram sheet.

2. Click the Histogram sheet tab.

3. Enter the label Total: in cell A13

4. Enter the formula =SUM(B2:B12) in cell B13. Verify that the result of this formula is 50.

5. Enter the column heading Rel. Freq. in cell C1.

6. Enter the formula =B2/$B$13 in cell C2 and copy this formula down through cell C12.

The frequency table now includes the relative frequencies (see Figure 2.E.11).

E **Figure 2.E.11**

	A	B	C
1	Classes	Frequency	Rel. Freq.
2	13950	2	0.04
3	16950	4	0.08
4	19950	1	0.02
5	22950	13	0.26
6	25950	7	0.14
7	28950	8	0.16
8	31950	5	0.10
9	34950	4	0.08
10	37950	3	0.06
11	40950	2	0.04
12	43950	1	0.02
13	Total:	50	

To generate a relative frequency histogram from this table, we can use the Microsoft Excel Chart Wizard, a series of four linked dialog boxes. To use the Chart Wizard, follow this general procedure:

1. Click on the tab of the worksheet containing the source data for the charts.

2. Select Insert | Chart to begin the wizard.

3. Make selections and enter information, as necessary, in the Steps 1, 2, and 3 dialog boxes, clicking the Next button to advance to the next dialog box.

4. In the Step 3 dialog box, add titling information and use the settings of Table 2.E.2 (unless directed otherwise).

Table 2.E.2 Common Settings for the Axes, Gridlines, Legend, Data Labels, and Data Table Tabs of the Chart Wizard Step 3 Dialog Box

Axes tab	Select both the (X) axis and (Y) axis check boxes.
	Select the Automatic option button under the (X) axis check box.
Gridlines tab	Deselect all the choices under the (Y) axis heading and the (X) axis heading.
Legend tab	Deselect (uncheck) the Show legend check box.
Data Labels tab	Select the None option button under the Data labels heading.
Data Table tab	Deselect (uncheck) the Show data table check box.

Note: Not all settings will be enabled or displayed for all chart types. Depending on the chart type, (X) and (Y) axis labels will be preceded by either the word Category or Value, as in Category (X) axis or Value (Y) axis.

5. In the Step 4 dialog box, select the As new sheet: option button and type a self-descriptive name for the new chart sheet. (The authors recommend always placing new charts on their own sheets.) Click the Finish button.

To generate a relative frequency histogram from the data on the Histogram sheet, do the following:

1. Click on the Histogram sheet tab. If relative frequencies have not been added to the frequency table, follow the procedure at the beginning of this section before continuing.

2. Select Insert | Chart.

3. Enter the information and make the selections shown in Figure 2.E.12 in the Step 1 dialog box.

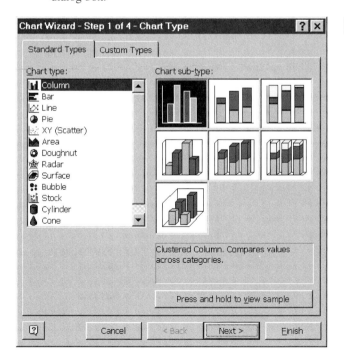

E Figure 2.E.12

4   Enter the information and make the selections shown in Figures 2.E.13 and 2.E.14 in
     the Step 2 dialog box. (The use of sheet names in the entries of the Series tab of this
     dialog box is mandatory.)

E Figure 2.E.13

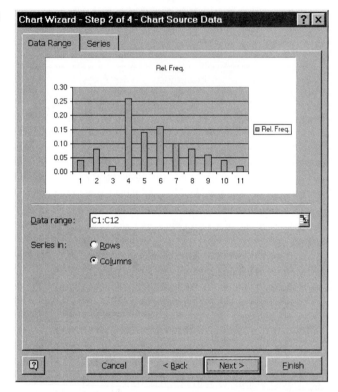

E Figure 2.E.14

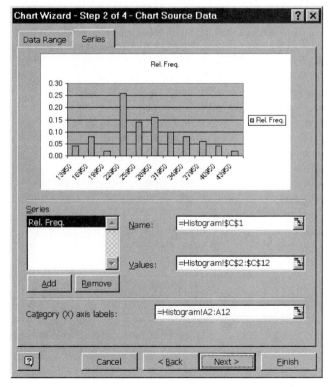

5 In the Step 3 dialog box, select the Titles tab and enter the values shown in Figure 2.E.15. For other selections, use the settings of Table 2.E.2 on page 131.

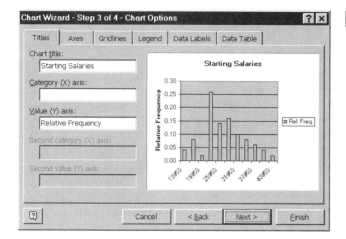

E Figure 2.E.15

6 In the Step 4 dialog box, select the As new sheet: option button and enter Rel. Frequency Histogram as the name for the new chart sheet (see Figure 2.E.16). Click the Finish button.

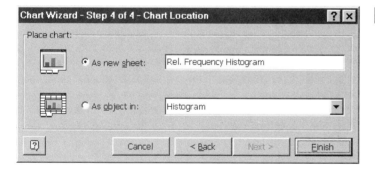

E Figure 2.E.16

The Chart Wizard produces on a new sheet named Rel. Frequency Histogram a relative frequency histogram similar to one shown in Figure 2.E.17. The gaps between the histogram bars can be adjusted using steps 5–7 of the "Correcting the Data Analysis Histogram output" procedure of Part C Solution. (The floating Chart toolbar that may appear after the chart is created can be closed or dragged to the side of the application window.)

E **Figure 2.E.17**

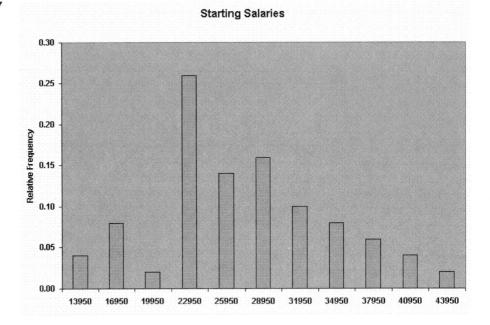

2.E.3 Using Microsoft Excel to Explore the Relationship between Two Qualitative Variables

Solution Summary:

A.   Use the PivotTable Wizard to construct a cross-classification table.

B.   Append the relative frequencies for each cell to the cross-classification table.

C.   Use the Chart Wizard to construct a side-by-side bar chart.

Alternatively, we can use the PHStat Add-In by selecting PHStat | Two-Way Tables & Charts to tabulate and graph bivariate categorical data.

Using the PivotTable Wizard to construct a cross-classification table

To construct a cross-classification table, we can use the Microsoft Excel PivotTable Wizard. To use this wizard, follow this general procedure:

1   Click the tab of the worksheet containing the source data for the PivotTable.

2   Select Data | PivotTable Report.

3   In the Step 1 dialog box, select the Microsoft Excel list or database option button. (see Figure 2.E.18)

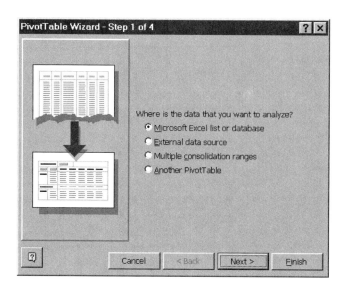

E Figure 2.E.18

4   In the Step 2 dialog box, enter the cell range containing the data to be tabulated.

5   In the Step 3 dialog box, specify the design of the summary table.

6   In the Step 4 dialog box, select the New worksheet option button and click the Options button to display the PivotTable Options dialog box.

7   In the PivotTable Options dialog box, make selections and enter information, including a worksheet name in the Name: edit box, as necessary. Click the OK button to return to the Step 4 dialog box. Click the Finish button.

Example: Speaker and Interrupter Gender Data of Table 2.7

Part A Solution:

To generate a cross-classification table for the data of Table 2.7 on page 86, do the following:

1   Open the Speaker workbook (SPEAKER.XLS) and click the Data sheet tab.

2   Select Data | PivotTable Report.

3   In the Step 1 dialog box, select the Microsoft Excel list or database option button. Click the Next button.

4   In the Step 2 dialog box, enter A1:C41 in the Range: edit box (see Figure 2.E.19). Click the next button.

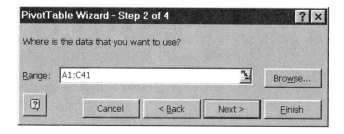

E Figure 2.E.19

5   In the Step 3 dialog box, drag the column label boxes at the right and drop them into the ROW, COLUMN, and DATA areas as shown in Figure 2.E.20. If the label in the DATA area reads Sum of Interruption instead of Count of Interruption, double-click the label to display the PivotTable Field dialog box (see Figure 2.E.21). In this dialog box:

a   Select Count from the Summarize by: list box

b   Click the Options button.

c   Select Normal in the Show data as: drop-down list box.

d   Click the OK button to redisplay the Step 3 dialog box.

E Figure 2.E.20

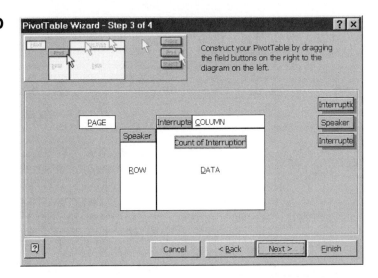

E Figure 2.E.21

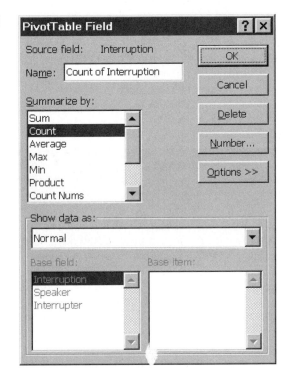

6 In the Step 4 dialog box, select the New worksheet option button and click the Options button. In the PivotTable Options dialog box, enter the information and make the selections shown in Figure 2.E.22 of the Step 4 dialog box. Click the OK button to close this dialog box. Then click the Finish button.

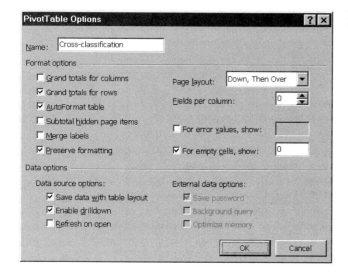

E Figure 2.E.22

The wizard constructs the PivotTable on a new worksheet. Continue by modifying the worksheet as follows:

7 Rename the new worksheet CrossClass. (The floating Query and Pivot toolbar that may appear after the table is created can be closed or dragged to the side of the application window.)

8 Select row 1 by clicking on the row 1 legend on the left side of the worksheet. Select Insert | Rows two times to insert two empty rows above the PivotTable. Enter a title for the cross-classification table in the new row 1.

The modified worksheet containing the new PivotTable will be similar to the one shown in Figure 2.E.23.

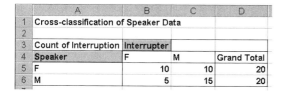

E Figure 2.E.23

Part B Solution: Appending relative frequencies

To append relative frequencies to the PivotTable constructed in Part A Solution, do the following:

1 Click the CrossClass sheet tab. (If the CrossClass sheet does not exist, follow the procedure in Part A Solution to generate such a sheet before continuing.)

2 Enter the column heading and formulas shown in Table 2.E.3 into their appropriate cells. Use the Increase Decimal or Decrease Decimal button on the formatting toolbar to adjust the decimal place.

Table 2.E.3 Additions to the CrossClass Sheet

	Column E	Column F
1		
2		
3	Relative Frequencies	
4	=B4	=C4
5	=B5/$D$5	=C5/$D$5
6	=B6/$D$6	=C6/$D$6

The relative frequencies now appear in the cell range E5:F6 of the CrossClass sheet as shown in Figure 2.E.24.

E Figure 2.E.24

	A	B	C	D	E	F
1	Cross-classification of Speaker Data					
2						
3	Count of Interruption	Interrupter			Relative Frequencies	
4	Speaker	F	M	Grand Total	F	M
5	F	10	10	20	0.50	0.50
6	M	5	15	20	0.25	0.75

Part C Solution: Using the Chart Wizard to construct a side-by-side bar chart

To construct a side-by-side bar chart based on the relative frequencies of the Table 2.9 data, do the following:

1. Click the CrossClass sheet tab. If relative frequencies have not been appended to the cross-classification table, follow the procedure given in Part B Solution before continuing.
2. Select Insert | Chart.
3. Enter the information and make the selections shown in Figures 2.E.25 and 2.E.26 in the Step 1 dialog box and in the Data Range tab of the Step 2 dialog box.

E Figure 2.E.25

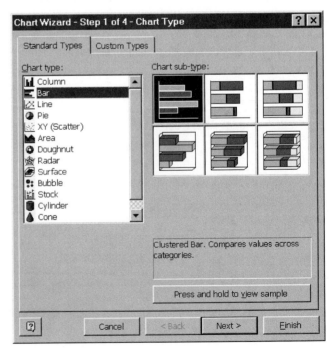

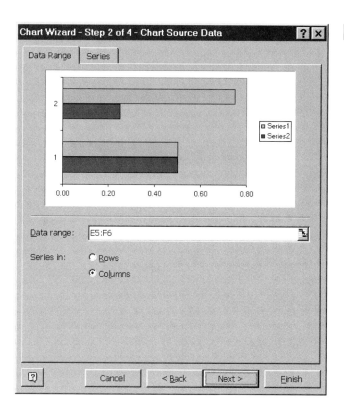

E Figure 2.E.26

4 In the Series tab of the Step 2 dialog box, enter =CrossClass!A5:A6 in the Category (X) axis labels: edit box as shown in Figure 2.E.27a. Then, in turn, select Series 1 and

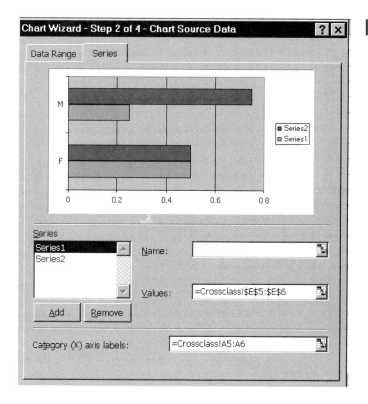

E Figure 2.E.27a

Series 2 from the Series list box, and enter the values Female Interrupter and Male Interrupter in the Name: edit box. After these values are entered the Series tab should resemble the one shown in Figure 2.E.27b.

E **Figure 2.E.27b**

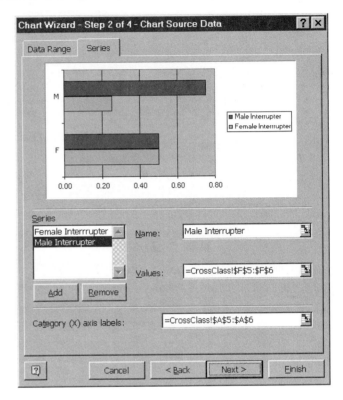

5 In the Step 3 dialog box, select the Titles tab and enter the values shown in Figure 2.E.28. For other selections, use the settings of Table 2.E.2.

E **Figure 2.E.28**

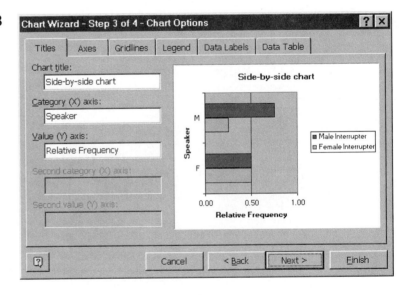

6 In the Step 4 dialog box, select the As new sheet: option button and enter SideBySide as the name for the new chart sheet. Click the Finish button.

The wizard constructs a side-by-side chart on a new sheet named SideBySide that is similar to the one shown in Figure 2.10 on page 88.

To use the PHStat Add-In to tabulate and graph the data of Table 2.7, do the following:

1 If the PHStat add-in has not been previously loaded, load the add-in using the instructions of Section EP.3.2. (You will also need to run the disk setup program using the procedure in Appendix D if you have never previously installed the files on the CD-ROM that accompanies this text.)

2 Open the Speaker workbook (SPEAKER.XLS) and click the Data sheet tab.

3 Select PHStat | Two-Way Tables & Charts

4 In the Two-Way Tables & Charts dialog box:

a Enter B1:B41 in the Row Variable Cell Range: edit box.

b Enter C1:C41 in the Column Variable Cell Range: edit box.

c Select the First cells in both ranges contain label check box.

d Enter Speaker Cross Tab in the Output Title: edit box.

e Select the Side-by-Side Bar Chart check box.

f Click the OK button.

2.E.4 Using Microsoft Excel to Explore the Relationship between Two Quantitative Variables

Solution Summary:

Use the Chart Wizard to construct a scatterplot.

  Carbon Monoxide (CO) and Nicotine FTC Data of Table 2.13

Solution:

1 Open the Smoking workbook (SMOKING.XLS) and click the Data sheet tab.

2 Select Insert | Chart.

3 Enter the information and make the selections shown in Figures 2.E.29–2.E.31 in the Steps 1 and 2 dialog boxes. When generating scatter diagrams, the Chart Wizard always assumes that the first column (or row) of data of the data range entered in the Step 2 dialog box contains values for the X variable. Since in the case of Figure 2.12, nicotine (the second column of data) is the X variable, it is necessary to select the Series Tab of the Step 2 dialog box and change the cell ranges in the X values: edit box to Data!C2:C21 and the Y values: edit box to Data!B2:B21. Note the revised cell ranges *must* be entered as formulas that include the sheet name.

Figure 2.E.29

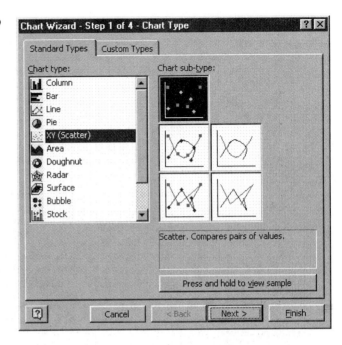

Figure 2.E.30

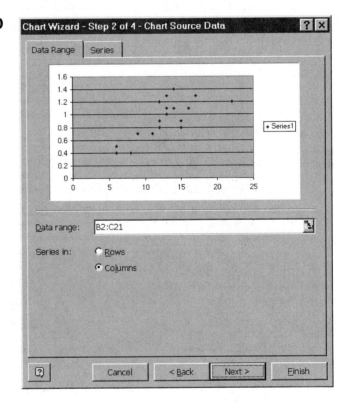

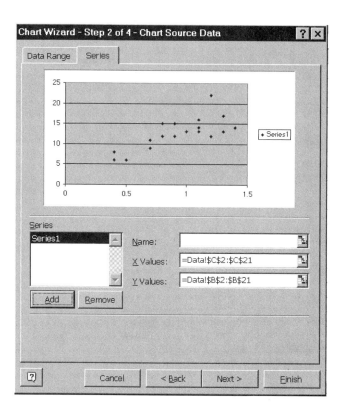

E **Figure 2.E.31**

4 In the Step 3 dialog box, select the Titles tab and enter the values shown in Figure 2.E.32. Select the Gridlines tab. Deselect (uncheck) all checkboxes in this tab. For the other tabs of this dialog box, use the settings of Table 2.E.2 on page 131.

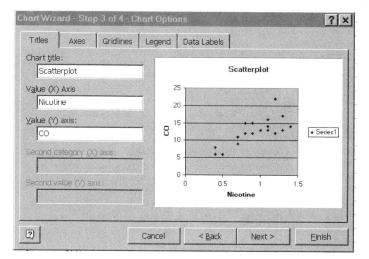

E **Figure 2.E.32**

5 In the Step 4 dialog box, select the As new sheet: option button and enter Scatter as the name for the new chart sheet. Click the Finish button.

The wizard constructs a scatterplot on a new sheet named Scatter that is similar to the one shown in Figure 2.12 on page 96.

Chapter 3

Exploring Data with Numerical Descriptive Measures

CONTENTS

3.1 Types of Numerical Descriptive Measures

3.2 Summation Notation

3.3 Measures of Central Tendency: Mean, Median, and Mode

3.4 Measures of Data Variation: Range, Variance, and Standard Deviation

3.5 Interpreting the Standard Deviation

3.6 Measures of Relative Standing: Percentiles and z Scores

3.7 Box-and-Whisker Plots

3.8 Methods for Detecting Outliers

3.9 A Measure of Association: Correlation

3.10 Numerical Descriptive Measures for Populations

EXCEL TUTORIAL

3.E.1 Using Microsoft Excel to Explore Data with Numerical Descriptive Measures

REAL-WORLD APPLICATION
Bid-Collusion in the Highway Contracting Industry

The Florida Attorney General's office maintains a database of bid prices for all sealed bids for road construction contracts. Due to collusion on the part of road contractors, some of those bids were determined to be noncompetitively bid. The Attorney General wants to describe the noncompetitive bids with a single number that characterizes the "fixed" prices. In this chapter we will show you how numbers called *numerical descriptive measures* can be used to describe the characteristics of a set of measurements. You will have an opportunity to apply this knowledge to the Florida Attorney General's database in Real-World Revisited at the end of this chapter.

3.1 TYPES OF NUMERICAL DESCRIPTIVE MEASURES

It is probably true that a picture is worth a thousand words, and it is certainly true when the goal is to describe a quantitative data set. But sometimes you need to discuss the major features of a data set and it may not be convenient to produce a stem-and-leaf display or histogram for the data. When this situation occurs, we seek a few summarizing numbers, called **numerical descriptive measures** (or *descriptive statistics*), that create in our minds a picture of the data's distribution.

Consider a data set containing the scores of all students who have taken a statistics exam. Two numbers that will help you construct a mental image of the distribution of exam scores are:

1. A number that is located near the *center* of the distribution (see Figure 3.1a)
2. A number that measures the *spread* of the distribution (see Figure 3.1b)

A number that describes the center of the distribution is located near the spot where most of the exam scores are concentrated. Consequently, numbers that fulfill this role are called **measures of central tendency**. We will define and describe several measures of central tendency for data sets in Section 3.3.

The amount of spread in the exam scores is a measure of the variation in the data. Consequently, numerical descriptive measures that perform this function are called **measures of variation** or **measures of dispersion**. As you will subsequently see (Section 3.4), there are several ways to measure the variation in a data set.

Measures of central tendency and data variation are not the only types of numerical measures for describing data sets. Some are constructed to measure the **skewness** of a distribution. Recall from Section 2.3 that skewness describes the tendency of the distribution to have long tails out to the right or left. (For example, the distributions in Figure 3.1 are skewed to the left—or negatively skewed.) Although numerical descriptive measures of skewness are beyond the scope of this text, knowing whether a distribution of data is skewed or symmetric is important when describing the data with measures of central tendency and variation (Section 3.5).

Two other important numerical descriptive measures covered in this chapter are **measures of relative standing** (Section 3.6) and **measures of association** (Section 3.9). A measure of relative standing is used to determine where a particular measurement lies *relative* to all other measurements in the data set. A measure of association is used to help determine the nature of the relationship between two different variables in a data set.

As you read this chapter, keep in mind our goal of using a few summarizing numbers to create a mental image of a distribution. Relate each numerical descriptive measure to this objective, and verify that it fulfills the role it is intended to play.

Figure 3.1 Numerical Descriptive Measures

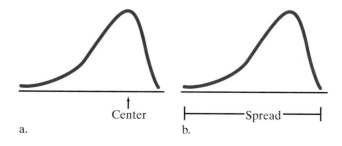

a.   Center   b.   Spread

Types of Numerical Descriptive Measures

1. Measures of *central tendency* locate the center of the distribution.
2. Measures of *variation* measure the spread of the distribution.
3. Measures of *skewness* measure the tendency of the distribution to have elongated tails.
4. Measures of *relative standing* represent the location of any one observation relative to the others in the distribution.
5. Measures of *association* measure the nature of the relationship between two variables.

3.2 SUMMATION NOTATION

Suppose a data set was obtained by observing a quantitative variable, x. For example, x may represent the GPA of a college student. If we want to represent a particular observation in a data set—say, the 35th—we represent it by the symbol x with a subscript of 35. If Linda Marcus, the 35th student on the list, has a GPA of 3.27, then we write $x_{35} = 3.27$. If we are interested in the GPAs of a group of 180 students, then the complete data set would be represented by the symbols $x_1, x_2, x_3, \ldots, x_{180}$.

Most of the formulas that we use require the summation of numbers. For example, we may want to sum the observations in a data set, or we may want to square each observation and then sum the squares of all the observations. The sum of the observations in a data set will be represented by the symbol

$$\sum x$$

This symbol is read "summation x." The symbol $\sum$ (sigma) gives an instruction: it is telling you to sum a set of numbers. The variable to be summed, x, is shown to the right of the $\sum$ symbol.

Example 3.1   Finding Sums

Suppose that the variable x is used to represent the length of time (in years) for a college student to obtain his or her bachelor's degree. Five graduates are selected and the value of x is recorded for each. The observations are 2, 4, 3, 5, 4.

a. Find $\sum x$.   **b.** Find $\sum x^2$

Solution

a. The symbol $\sum x$ tells you to sum the x values in the data set. Therefore,

$$\sum x = 2 + 4 + 3 + 5 + 4 = 18$$

b. The symbol $\sum x^2$ tells you to sum the squares of the x values in the data set. Therefore,

$$\sum x^2 = (2)^2 + (4)^2 + (3)^2 + (5)^2 + (4)^2$$
$$= 4 + 16 + 9 + 25 + 16 = 70$$

Example 3.2   Finding Sums

Refer to Example 3.1.

a. Find $\sum(x - 4)$.   **b.** Find $\sum(x - 4)^2$.   **c.** Find $\sum x^2 - 4$.

Solution

a. The symbol $\sum(x - 4)$ tells you to subtract 4 from each x value and then sum. Therefore,

$$\sum(x - 4) = (2 - 4) + (4 - 4) + (3 - 4) + (5 - 4) + (4 - 4)$$
$$= (-2) + 0 + (-1) + 1 + 0 = -2$$

b. The symbol $\sum(x - 4)^2$ tells you to subtract 4 from each x value in the data set, square these differences, and then sum them as follows:

$$\sum(x - 4)^2 = (2 - 4)^2 + (4 - 4)^2 + (3 - 4)^2 + (5 - 4)^2 + (4 - 4)^2$$
$$= (-2)^2 + (0)^2 + (-1)^2 + (1)^2 + (0)^2$$
$$= 4 + 0 + 1 + 1 + 0 = 6$$

> **The Meaning of Summation Notation $\sum x$**
>
> Sum observations on the variable that appears to the right of the summation symbol.

c. The symbol $\sum x^2 - 4$ tells you to first sum the squares of the x values, and then subtract 4 from this sum:

$$\sum x^2 - 4 = (2)^2 + (4)^2 + (3)^2 + (5)^2 + (4)^2 - 4$$
$$= 4 + 16 + 9 + 25 + 16 - 4 = 66 \qquad ❑$$

Self-Test 3.1

A data set contains the observations 5, 1, 3, 2, 1. Find:

a. $\sum x$ **b.** $\sum x^2$
c. $\sum(x - 1)$ **d.** $\sum(x - 1)^2$

PROBLEMS 3.1–3.6

Using the Tools

3.1 A data set contains the observations 10, 15, 10, 11, 12. Find:

a. $\sum x$ **b.** $\sum x^2$ **c.** $(\sum x)^2$
d. $\sum(x - 13)$ **e.** $\sum(x - 13)^2$

3.2 A data set contains the observations 3, 8, 4, 5, 3, 4, 6. Find:

a. $\sum x$ **b.** $\sum x^2$ **c.** $\sum(x - 5)^2$
d. $\sum(x - 2)^2$ **e.** $(\sum x)^2$

3.3   Refer to Problem 3.1. Find:

a. $\sum x^2 - \dfrac{(\sum x)^2}{5}$ b. $\sum(x - 10)^2$ c. $\sum x^2 - 10$

3.4   Refer to Problem 3.2. Find:

a. $\sum x^2 - \dfrac{(\sum x)^2}{7}$ b. $\sum(x - 4)^2$ c. $\sum x^2 - 15$

3.5   A data set contains the observations $6, 0, -2, -1, 3$. Find:

a. $\sum x$ b. $\sum x^2$ c. $\sum x^2 - \dfrac{(\sum x)^2}{5}$

3.6   A data set contains the observations $.7, 1.1, .5, .3$. Find:

a. $\sum x$ b. $\sum x^2$ c. $\sum x^2 - \dfrac{(\sum x)^2}{4}$

3.3   MEASURES OF CENTRAL TENDENCY: MEAN, MEDIAN, AND MODE

The word *center,* as applied to a distribution of data, is not a well-defined term. In our minds, we know vaguely what we mean: a number somewhere near the "middle" of the distribution, a single number that tends to "typify" the data set. The measures of central tendency that we define often generate different numbers for the same data set but all satisfy our general objective. If we imagine a hump-shaped distribution, all measures of central tendency will fall near the middle of the hump.

The most common measure of the central tendency of a data set is called the *arithmetic mean* of the data. The arithmetic mean, or *average,* is defined as indicated in the box.

Definition 3.1

The **arithmetic mean** of a sample of n observations, $x_1, x_2, \ldots, x_n$, is denoted by the symbol $\bar{x}$ (read "x-bar"), and is computed as

$$\bar{x} = \frac{\text{Sum of the } x \text{ values}}{\text{Number of observations}} = \frac{\sum x}{n} \qquad (3.1)$$

[*Note:* From now on, we will refer to an arithmetic mean simply as a **mean**.]

Example 3.3   Computing a Mean

Find the mean for the data set consisting of the observations $5, 1, 6, 2, 4$.

Solution

The data set contains $n = 5$ observations. Using equation 3.1, we have

$$\bar{x} = \frac{\sum x}{n} = \frac{5 + 1 + 6 + 2 + 4}{5} = \frac{18}{5} = 3.6$$ ❑

Example 3.4   Interpreting a Mean

FISH.XLS

Refer to the data on 144 contaminated fish captured in the Tennessee River (Alabama). Find the mean for the 144 fish weights (recorded in grams). Does the mean fall near the center of the distribution?

Solution

With such a large data set, it is impractical to calculate numerical descriptive measures by hand or calculator. In practice, we rely on an available software package. Figure 3.2 is an Excel printout giving descriptive statistics for the fish weight data. The mean of the 144 weights, highlighted on the printout, is (rounded) $\bar{x} = 1,049.7$ grams. This mean, or average, weight should be located near the center of the histogram for the 144 fish weights. Figure 3.3, also produced with Excel, is a frequency histogram for the data. Note that the mean $\bar{x}$ does indeed fall near the center of the mound-shaped portion of the distribution. If we did not have Figure 3.3 available, we could reconstruct the distribution in our minds as a mound-shaped figure centered in the vicinity of $\bar{x} = 1,049.7$ grams. ❏

A second measure of central tendency for a data set is the *median*. For large data sets, the median M is a number chosen so that half the observations are less than the median and half are larger. Since the areas of the bars used to construct

E Figure 3.2 Excel Descriptive Measures for the Fish Weight Data

	A	B
1	*Weight*	
2		
3	Mean	1049.715278
4	Standard Error	31.37884265
5	Median	1000
6	Mode	1186
7	Standard Deviation	376.5461118
8	Sample Variance	141786.9743
9	Kurtosis	0.368447045
10	Skewness	0.500551656
11	Range	2129
12	Minimum	173
13	Maximum	2302
14	Sum	151159
15	Count	144
16	Largest(1)	2302
17	Smallest(1)	173

E Figure 3.3 Excel Histogram for the Fish Weight Data

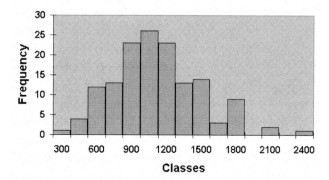

Figure 3.4 The Median Divides the Area of a Relative Frequency Distribution into Two Equal Portions

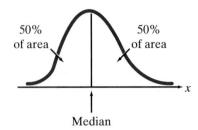

50% of area

50% of area

Median

x

the relative frequency histogram are proportional to the numbers of observations falling within the classes, it follows that the median is a value of x that divides the area of the histogram into two equal portions. Half the area will lie to the left of the median (see Figure 3.4) and half will lie to the right. For example, the median for the 144 fish weights (highlighted in Figure 3.2) is 1,000 grams. You can see from Figure 3.3 that this weight divides the data into two sets of equal size. Half the 144 weights are less than 1,000 grams; half are larger. The next two examples illustrate how to find the median for small data sets.

Example 3.5   **Computing a Median: *n* Odd**

Find the median for the data set consisting of the observations 7, 4, 3, 5, 3.

Solution
We first arrange the data in increasing (or decreasing) order:

3   3   4   5   7

With an *odd number of measurements,* the median is the middle value. Thus, the median is 4; half the remaining measurements are less than 4 and half are greater than 4. ❏

Example 3.6   **Computing a Median: *n* Even**

Find the median for the data set consisting of the observations: 5, 7, 3, 1, 4, 6.

Solution
If we arrange the data in increasing order, we obtain

1   3   4   5   6   7

You can see that there are now many choices for the median. Any number between 4 and 5 will divide the data set into two groups of three each. There are many ways to choose this number, but the simplest is to choose the median as the point halfway between the two middle numbers when the data are arranged in order. Thus, the median is

$$M = \frac{4 + 5}{2} = 4.5$$

❏

Definition 3.2

The **median** M of a sample of n observations $x_1, x_2, \ldots, x_n$, is defined as follows:

If n is odd: The middle observation when the data are arranged in order. [The number in the $(n + 1)/2$ position is the median.]

If n is even: The number halfway between the two middle observations—that is, the mean of the two middle observations—when the data are arranged in order. [The two middle observations are those in the $n/2$ and $(n/2 + 1)$ positions.]

A third measure of central tendency for a data set is the *mode*. The mode is the value of x that occurs with greatest frequency (see Figure 3.5). If the data have been grouped into classes, we can define the modal class as the class with the largest class frequency (or relative frequency). For example, you can see from the Excel printout, Figure 3.2 (page 149), that the modal fish weight is 1,186 grams. The modal class, however, is 900–1,050 grams (see Figure 3.3).

Definition 3.3

The **mode** of a data set is the value of x that occurs with greatest frequency. The **modal class** is the class with the largest class frequency.

Self-Test		Find the mean, median, and mode for the following sample of ten measurements: 6, 10, 3, 4, 4, 5, 8, 11, 4, 2.
3.2		

The mean, median, and mode are shown (Figure 3.6) on the histogram for the 144 fish weights. Which is the best measure of central tendency for this data set? The answer is that it depends on the type of descriptive information you want. If you want the "average" weight or "center of gravity" for the weight distribution, then the mean is the desired measure. If your notion of a typical or "central" weight is one that is larger than half the fish weights and less than the remainder, then you will prefer the median. The mode is rarely the choice measure of central tendency because the measurement that occurs most often does not necessarily lie in the "center" of the distribution. There are situations, however, where the mode is preferred. For example, a retailer of women's shoes would be interested in the modal shoe size of potential customers.

Figure 3.5 The Mode Is the Value of x that Occurs with Greatest Frequency

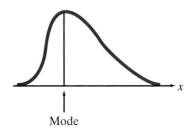

Figure 3.6 Location of the Mean, Median, Modal Class, and Mode for the Data on Weights of Contaminated Fish

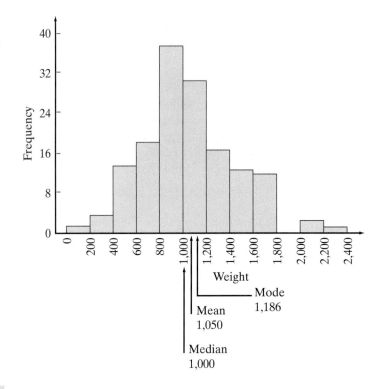

In making your decision, you should know *the mean is sensitive to very large or very small measurements.* Consequently, the mean will shift toward the direction of skewness and may be a misleading measure of central tendency in some situations. You can see from Figure 3.6 that the mean is slightly larger than the median and that the fish weights are slightly skewed to the right. The relatively few captured fish that have a very high weight will influence the mean more than the median. For this reason, the median is sometimes called a *resistant* measure of central tendency since it, unlike the mean, is resistant to the influence of extreme measurements. For data sets that are extremely skewed, the median would better represent the "center" of the distribution.

Most of the inferential statistical methods discussed in this text are based, theoretically, on mound-shaped distributions of data with little or no skewness. For these situations, the mean and the median will be the same, for all practical purposes. Since the mean has better mathematical properties than the median, it is the preferred measure of central tendency for these inferential techniques.

Caution

For data sets that are extremely skewed, be wary of using the mean as a measure of the "center" of the distribution. In this situation, a more meaningful measure of central tendency may be the median, which is more resistant to the influence of extreme measurements.

PROBLEMS 3.7–3.23

Using the Tools

3.7 Calculate the mean for samples with the following characteristics:

   a. $n = 5$ and $\Sigma x = 30$

   b. $n = 100$ and $\Sigma x = 1,220$

   c. $n = 800$ and $\Sigma x = 200$

3.8   Calculate the mean for samples with the following characteristics:

  a. $n = 10$ and $\Sigma x = 500$

  b. $n = 20$ and $\Sigma x = 400$

  c. $n = 500$ and $\Sigma x = 100$

3.9   Find the mean and median for a sample consisting of these five measurements: 3, 9, 0, 7, 4.

3.10   Find the mean and median for the following sample of $n = 6$ measurements: 7, 3, 4, 1, 5, 6.

3.11   Calculate the mean, median, and mode for each of the following samples:

  a.   3, 4, 4, 5, 5, 5, 6, 6, 7

  b.   3, 4, 4, 5, 5, 5, 6, 6, 70

  c.   −50, −49, 0, 0, 49, 50

  d.   −50, −49, 0, 9, 9, 81

3.12   Calculate the mean, median, and mode for each of the following samples:

  a.   1.3, 1.2, 1.2, 1.1, 1.3, 1.7, 1.5, 1.2, 1.8, 1.0

  b.   1.3, 1.2, 1.2, 1.1, 1.3, 1.7, 1.5, 1.2, 1.8, 10.0

  c.   .3, 1.2, 1.2, 1.1, 1.3, 1.7, 1.5, 1.2, 1.8, 1.0

Applying the Concepts

3.13   The accompanying data set contains quiz scores for 15 students in an introductory statistics class: 8, 7, 9, 6, 8, 10, 9, 9, 5, 7, 9, 1, 7, 9, 10.

  a.   Find a measure of central tendency that represents the quiz score that occurs most often.

  b.   Find a measure of central tendency that represents the average of the 15 quiz scores.

  c.   Find a measure of central tendency that is resistant to the influence of extremely small quiz scores.

3.14   The following data are the amount of calories in a 30-gram serving for a random sample of 10 types of fresh-baked chocolate-chip cookies:

COOKIES.XLS

Product	Calories	Product	Calories
Hillary Rodham Clinton's	153	David's	146
Original Nestlé Toll House	152	David's Chocolate Chunk	149
Mrs. Fields	146	Great American Cookie Company	138
Stop & Shop	138	Pillsbury Oven Lovin'	168
Duncan Hines	130	Pillsbury	147

Data Source: Adapted from "Chocolate-chip Cookies," *Consumer Reports,* October 1993, pp. 646–647. Although this data set originally appeared in *Consumer Reports,* the selective adaptation and resulting conclusions presented are those of the authors and are not sanctioned or endorsed in any way by Consumers Union, the publisher of *Consumer Reports.*

  a.   What is the mean amount of calories? The median amount of calories?

  b.   How would this information be of use to a company about to market a new fresh-baked chocolate-chip cookie? Discuss.

3.15 For the last 10 days in June, the "Shore Special" train was late arriving at its destination by the following times (in minutes):

| −3 | 6 | 4 | 10 | −4 | 124 | 2 | −1 | 4 | 1 |

TRAIN.XLS

(*Note:* A negative number means that the train was early by that number of minutes.)

a. If you were hired by the railroad to show that the railroad is providing good service, what are some of the numerical descriptive measures you would use to accomplish that?

b. If you were hired by a TV station that was producing a documentary to show that the railroad is providing bad service, what numerical descriptive measures would you use?

c. If you were trying to be objective and unbiased in assessing the railroad's performance, which numerical descriptive measures would you use? (This is the hardest part, because you cannot answer without making additional assumptions about the relative costs of being late by various amounts of time.)

d. What would be the effect on your conclusions if the value of 124 had been incorrectly recorded and should have been 12?

3.16 The *Journal of Applied Behavior Analysis* (Summer 1997) published a study of a treatment designed to reduce destructive behavior. A 16-year-old boy diagnosed with severe mental retardation was observed playing with toys for 10 sessions, each ten minutes in length. The number of destructive responses (e.g., head banging, pulling hair, pinching, etc.) per minute was recorded for each session in order to establish a baseline behavior pattern. The data for the 10 sessions are listed here:

| 3 | 2 | 4.5 | 5.5 | 4.5 | 3 | 2.5 | 1 | 3 | 3 |

BOY16.XLS

a. Find and interpret the mean for the data set.

b. Find and interpret the median for the data set.

c. Find and interpret the mode for the data set.

3.17 Refer to the Centers for Disease Control study of sanitation levels for 91 international cruise ships, Problem 2.24, page 82. An Excel printout of the descriptive statistics for the data is shown here. (Recall that sanitation scores range from 0 to 100.) Interpret the numerical descriptive measures of central tendency displayed on the printout.

	A	B
1	*Sanitation*	
2		
3	Mean	91.04395604
4	Standard Error	0.583432663
5	Median	92
6	Mode	89
7	Standard Deviation	5.565592886
8	Sample Variance	30.97582418
9	Kurtosis	5.867375284
10	Skewness	-1.873202434
11	Range	33
12	Minimum	66
13	Maximum	99
14	Sum	8285
15	Count	91
16	Largest(1)	99
17	Smallest(1)	66

E Problem 3.17

3.18 Computer scientists at the University of Virginia developed a computer algorithm for quickly solving problems that involve multiple input terminals on a workstation (*Networks*, Mar. 1995). The running time (in seconds) for each of 100 runs of the algorithm was recorded for different workstation configurations. The mean and median running times for workstations with 10, 20, and 30 terminals are given in the table. For each of the three workstation configurations, comment on the type of skewness present in the distribution of algorithm running times.

Number of Workstations	Mean	Median
10	0.3	0.3
20	4.0	2.5
30	2,210	171

3.19 More than 4,000 members of the American Institute of Certified Public Accountants (CPAs) responded to a survey on their use of software tools (*Journal of Accountancy,* Nov. 1997). Responses to the question, "How satisfied are you with the tax software program you use?" were recorded on a scale of 1 (worst) to 5 (best). The satisfaction scores for 13 tax software programs are listed below. Compute the mean, median, and mode of the 13 satisfaction scores. Interpret their values.

TAXRATE.XLS

Software Program	Satisfaction Score
AM Tax Pro	5.0
Tax $imple	4.5
Ultra Tax	4.5
TaxWorks	4.4
ProSystem fx	4.4
Lacerte	4.4
Tax Machine	4.2
Professional Tax System	4.2
Turbo Tax Pro	4.1
GoSystem	4.0
A-Plus-Tax	3.9
Tax Relief	3.8
Pencil Pushers	3.8

3.20 Spinifex pigeons are one of the few bird species that inhabit the desert of Western Australia. The pigeons rely almost entirely on seeds for food. The *Australian Journal of Zoology* (Vol. 43, 1995) reported on a study of the diets and water requirements of these spinifex pigeons. Sixteen pigeons were captured in the desert and the crop (i.e., stomach) contents of each examined. The accompanying table reports the weight (in grams) of dry seed in the crop of each pigeon. Calculate three different measures of central tendency for the 16 crop weights. Interpret the value of each.

CROPWT.XLS

.457	3.751	.238	2.967	2.509	1.384	1.454	.818
.335	1.436	1.603	1.309	.201	.530	2.144	.834

Source: Excerpted from Williams, J. B., Bradshaw, D., and Schmidt, L. "Field metabolism and water requirements of spinifex pigeons (*Geophaps plumifera*) in Western Australia." *Australian Journal of Zoology,* Vol. 43, No. 1, 1995, p. 7 (Table 2).

3.21 Patients with normal hearing in one ear and unaidable sensorineural hearing loss in the other are characterized as suffering from unilateral hearing loss. In a study reported in the *American Journal of Audiology* (Mar. 1995), eight patients with unilateral hearing loss were fitted with a special wireless hearing aid in the "bad" ear. The absolute sound pressure level (SPL) was then measured near the eardrum of the ear when noise was produced at a frequency of 500 hertz. The SPLs of the eight patients, recorded in decibels, are listed below.

73.0	80.1	82.8	76.8	73.5	74.3	76.0	68.1

SPL.XLS

Calculate three different measures of central tendency for the eight SPLs. Interpret the value of each.

3.22 Social psychologists label the process by which decision makers escalate their commitment to an ineffective course of action as "entrapment." Fifty-two introductory psychology students took part in a laboratory experiment designed to explore whether individuals' tendencies to view prior outcomes as revealing of their self-identity would heighten entrapment (*Administrative Science Quarterly*, Mar. 1986). The experiment consisted of 30 trials in which points were "awarded" based on the accuracy of students' judgments of geometric patterns of various shapes. The total points awarded on each trial are listed in the table.

ENTRAP.XLS

5	5	4	7	24	6
10	12	11	15	11	10
3	23	4	20	5	4
7	5	6	6	15	5
15	10	13	9	4	6

Source: Brockner, J., et al. "Escalation of commitment to an ineffective course of action: The effect of feedback having negative implications for self-identity." *Administrative Science Quarterly,* Vol. 31, No. 1, Mar. 1986, p. 115.

    a. Construct a stem-and-leaf display for the data.

    b. Compute the mean, median, and mode for the data set and locate them on the graph, part **a**.

    c. Which of these measures of central tendency appears to best locate the center of the distribution of data?

3.23 Refer to the *Journal of Geography* (May/June 1997) study of the effectiveness of computer software in the geography classroom, Problem 2.27 (page 84). Recall that two groups of seventh graders participated in the study. Group 1 (43 students) used computer resources to conduct research on Africa. Group 2 (22 students) used traditional methods to conduct the same research. All students were then given a test on the African regions studied. Test scores for the two groups of seventh graders are reproduced in the table below.

GEO.XLS

Group 1 (Computer Resources) Student Scores											
41	53	44	41	66	91	69	44	31	75	66	69
75	53	44	78	91	28	69	78	72	53	72	63
66	75	97	84	63	91	59	84	75	78	88	59
97	69	75	69	91	91	84					

Group 2 (No Computer) Student Scores											
56	59	9	59	25	66	31	44	47	50	63	19
66	53	44	66	50	66	19	53	78	78		

Source: Linn, S. E. "The effectiveness of interactive maps in the classroom: A selected example in studying Africa." *Journal of Geography,* Vol. 96, No. 3, May/June 1997, p. 167 (Table 1).

    a. Find the mean and median test scores for the Group 1 students.

    b. Find the mean and median test scores for the Group 2 students.

    c. Based on the results, parts **a** and **b**, determine the type of skewness (if any) present in the two test score distributions.

    d. Do you think using computer resources had a positive impact on student test scores? Explain.

3.4 MEASURES OF DATA VARIATION: RANGE, VARIANCE, AND STANDARD DEVIATION

Just as measures of central tendency locate the center of a relative frequency distribution, *measures of variation* measure its spread. The simplest measure of spread for a data set is the range.

Definition 3.4

The **range** of a quantitative data set is equal to the difference between the largest and the smallest measurements in the set.

$$\text{Range} = (\text{Largest measurement} - \text{Smallest measurement}) \quad \textbf{(3.2)}$$

Example 3.7   **Computing a Range**

Find the range for the data set consisting of the observations 3, 7, 2, 1, 8.

Solution

The smallest and largest members of the data set are 1 and 8, respectively. Using equation 3.2, we find

$$\text{Range} = \text{Largest measurement} - \text{Smallest measurement} = 8 - 1 = 7 \quad \square$$

The range of a data set is easy to find, but it is an insensitive measure of variation and is not very informative. For an example of its insensitivity, consider Figure 3.7. Both distributions have the *same range,* but it is clear that the distribution in Figure 3.7b indicates much less data variation than the distribution in Figure 3.7a. Most of the observations in Figure 3.7b lie close to the mean. In contrast, most of the observations in Figure 3.7a deviate substantially from the center of the distribution. Since the ranges for the two distributions are equal, it is clear that the range is a fairly insensitive measure of data variation. It was unable to detect the differences in data variation for the data sets represented in Figure 3.7.

A more useful measure of spread is the *variance.* The variance of a data set is based on how much the measurements "deviate" from their mean. The *deviation* between an observation x and the mean $\bar{x}$ of a sample is the difference

$$x - \bar{x}$$

If a sample data set contains n observations, the **sample variance** is equal to the "average" of the squared deviations of all n observations. The formula for computing the sample variance, denoted by the symbol s^2, is given in the box.

Figure 3.7 Two Distributions that Have Equal Ranges, but Show Differing Amounts of Data Variation

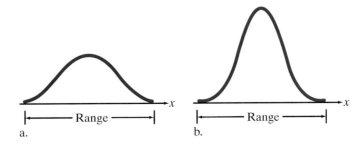

We use $(n - 1)$ rather than n in the denominator of s^2 in order to obtain an estimator of the true population variance with desirable mathematical properties. When n is used in the denominator, the value of s^2 tends to underestimate the population variance. Dividing by $(n - 1)$ adjusts for the underestimation problem.

Definition 3.5

The **variance**, s^2, of a set of n sample measurements is equal to the sum of squares of deviations of the measurements about their mean, divided by $(n - 1)$:

$$s^2 = \frac{\Sigma(x - \bar{x})^2}{n - 1} \qquad (3.3)$$

Note: An alternative (or *shortcut*) formula for calculating s^2, which may be easier to apply in practice, is:

$$s^2 = \frac{\Sigma x^2 - \dfrac{(\Sigma x)^2}{n}}{n - 1} = \frac{\Sigma x^2 - n(\bar{x})^2}{n - 1} \qquad (3.4)$$

The larger the value of s^2, the more spread out (i.e., the more variable) the sample data.

Example 3.8   Computing a Variance

Find the variance for the sample measurements: 3, 7, 2, 1, 8.

Solution

The five observations are listed in the first column of Table 3.1. You can see that $\Sigma x = 21$. Applying equation 3.1, the sample mean is

$$\bar{x} = \frac{\Sigma x}{n} = \frac{21}{5} = 4.2$$

This value of $\bar{x}$, 4.2, is subtracted from each observation to determine how much each observation deviates from the mean. These deviations are shown in the second column of Table 3.1. A *negative deviation* means that the observation fell *below* the mean; a *positive deviation* indicates that the observation fell *above* the mean. **Notice that the sum of the deviations equals 0. This will be true for all data sets**.

The squares of the deviations are shown in the third column of Table 3.1. The total at the bottom of the column gives the sum of squares of the deviations,

$$\Sigma(x - \bar{x})^2 = 38.8$$

Applying equation 3.3, the sample variance is

$$s^2 = \frac{\Sigma(x - \bar{x})^2}{n - 1} = \frac{38.8}{4} = 9.7 \qquad \square$$

How can we interpret the value of the sample variance calculated in Example 3.8? We know that data sets with large variances are more variable (i.e., more spread out) than data sets with smaller variances. But what information can we obtain from the number $s^2 = 9.7$? One interpretation is that the average squared deviation of the sample measurements from their mean is 9.7. However, a more practical interpretation can be obtained by calculating the square root of this number.

Table 3.1 Data and Computation Table

Observation x	$x - \bar{x}$	$(x - \bar{x})^2$
3	−1.2	1.44
7	2.8	7.84
2	−2.2	4.84
1	−3.2	10.24
8	3.8	14.44
TOTALS 21	0	38.8

A third measure of data variation, the *standard deviation,* is obtained by taking the square root of the variance. This results in a number with units of measurement equal to the units of the original data. That is, if the units of measurement for the sample observations are feet, dollars, or hours, the standard deviation of the sample is measured in feet, dollars, or hours (instead of feet^2, dollars^2, or hours^2). Like the variance, the standard deviation measures the amount of spread in a quantitative data set.

Definition 3.6

The **standard deviation** of a set of n sample measurements is equal to the square root of the variance:

$$s = \sqrt{s^2} = \sqrt{\frac{\Sigma(x - \bar{x})^2}{n - 1}} \tag{3.5}$$

The standard deviation of the five sample measurements in Example 3.8 is

$$s = \sqrt{s^2} = \sqrt{9.7} = 3.1$$

Now that you know how to calculate a standard deviation, we will demonstrate in the next section how it can be used to measure the spread or variation of a distribution of data.

Self-Test

3.3

Find the range, variance, and standard deviation of the following sample of ten measurements: 6, 10, 3, 4, 4, 5, 8, 11, 4, 5.

3.5 INTERPRETING THE STANDARD DEVIATION

In this section, we give two rules for interpreting the standard deviation. Both rules use the mean and standard deviation of a data set to determine an interval of values within which most of the measurements fall. For samples, the intervals take the form

$$\bar{x} \pm (k)s$$

where k is any positive constant (usually 1, 2, or 3). The particular rule applied will depend on the shape of the relative frequency histogram for the data set, as the following examples illustrate.

Example 3.9   Forming Intervals around the Mean

Periodically, the Federal Trade Commission (FTC) ranks domestic cigarette brands according to hazardous substances such as tar, nicotine, and the amount of carbon monoxide (CO) in smoke residue. The CO amounts (measured in milligrams) of a sample of 500 cigarette brands for a recent year are available in the Excel worksheet FTC.XLS. Suppose we want to describe the distribution of CO measurements for this sample. To do so, we require the mean and standard deviation of the measurements.

a. Calculate $\bar{x}$ and s for the data set.

b. Form an interval by measuring 1 standard deviation on each side of the mean, i.e., $\bar{x} \pm s$. Also, form the intervals $\bar{x} \pm 2s$ and $\bar{x} \pm 3s$.

c. Find the proportions of the total number (500) of CO measurements falling within these intervals.

Solution

a. An Excel printout giving numerical descriptive measures for the data set is shown in Figure 3.8. The mean and standard deviation, highlighted on the printout, are (rounded)

$$\bar{x} = 11 \quad \text{and} \quad s = 4$$

b. The intervals $\bar{x} \pm s, \bar{x} \pm 2s$, and $\bar{x} \pm 3s$, are formed as follows:

$$\bar{x} \pm s = 11 \pm 4 = (11 - 4, 11 + 4) = (7, 15)$$
$$\bar{x} \pm s = 11 \pm 2(4) = (11 - 8, 11 + 8) = (3, 19)$$
$$\bar{x} \pm s = 11 \pm 3(4) = (11 - 12, 11 + 12) = (-1, 23)$$

c. We used Excel to determine the total number of CO measurements falling within the three intervals. The proportions of the total number of CO measurements falling within the three intervals are shown in Table 3.2. The three intervals—$\bar{x} \pm s, \bar{x} \pm 2s$, and $\bar{x} \pm 3s$—are also shown on an Excel frequency histogram for the CO data displayed in Figure 3.9. If you estimate the proportions of the total area under the histogram that lie over the three intervals, you will obtain proportions approximately equal to those given in Table 3.2. ❏

E Figure 3.8 Excel Printout: Descriptive Statistics for 500 CO Measurements

	A	B
1	CO	
2		
3	Mean	11.015
4	Standard Error	0.179373632
5	Median	12
6	Mode	12
7	Standard Deviation	4.010916342
8	Sample Variance	16.0874499
9	Kurtosis	-0.360317707
10	Skewness	-0.425175549
11	Range	18.5
12	Minimum	0.5
13	Maximum	19
14	Sum	5507.5
15	Count	500
16	Largest(1)	19
17	Smallest(1)	0.5

Will the proportions of the total number of observations falling within the intervals $\bar{x} \pm s, \bar{x} \pm 2s$, and $\bar{x} \pm 3s$ remain fairly stable for most distributions of data? To examine this possibility, consider the next example.

Table 3.2 Proportions of the Total Number of CO Measurements in Intervals $\bar{x} \pm s, \bar{x} \pm 2s$, and $\bar{x} \pm 3s$

Interval	Number of Observations in Interval	Proportion in Interval
$\bar{x} \pm s$ or (7, 15)	324	.648
$\bar{x} \pm 2s$ or (3, 19)	485	.970
$\bar{x} \pm 3s$ or (-1, 23)	500	1.000

E Figure 3.9 Excel Histogram for Carbon Monoxide Measurements

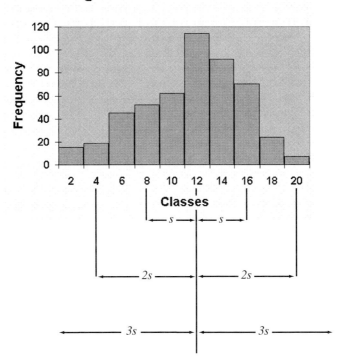

Histogram for CO Measurements

Example 3.10   Descriptive Statistics

Calculate the mean and standard deviation of each of the following data sets.

   a. The 98 driver head-injury ratings for crash-tested cars in the CRASH.XLS Excel workbook

   b. The 144 weights of fish specimens in the FISH.XLS Excel workbook

   c. The 500 tar contents of cigarettes in the FTC.XLS Excel workbook

   d. The 500 checkout times of supermarket customers in the CHKOUT.XLS Excel workbook

Solution

We used Excel to compute the means and standard deviations of each of the four data sets. They are shown in Table 3.3. The accompanying Excel frequency histograms are shown in Figures 3.10–3.13. ❑

The means and standard deviations of Table 3.3 were used to calculate the intervals $\bar{x} \pm s, \bar{x} \pm 2s$, and $\bar{x} \pm 3s$ for each data set. From Excel, we obtained

Table 3.3 Means and Standard Deviations for Four Data Sets

Data Set	Mean	Standard Deviation
a. Driver head-injury ratings (impact points)	604	185
b. Fish weights (grams)	1,050	377
c. Cigarette tar contents (milligrams)	11.1	4.9
d. Supermarket checkout times (seconds)	50	49

E **Figure 3.10** Excel Frequency Histogram for the 98 Driver Head-Injury Ratings

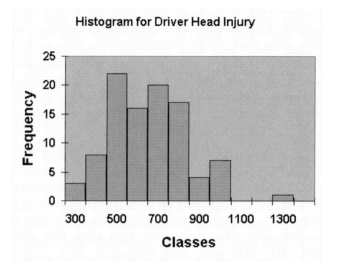

E **Figure 3.11** Excel Frequency Histogram for the 144 Fish Weights

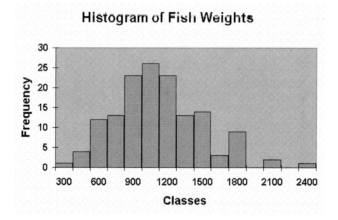

E **Figure 3.12** Excel Frequency Histogram for the 500 Tar Contents of Cigarettes

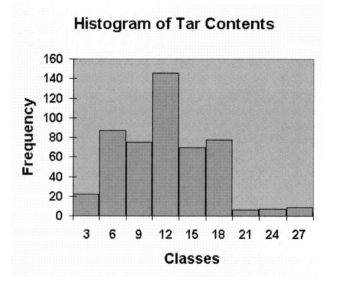

E **Figure 3.13** Excel Frequency Histogram for the 500 Supermarket Checkout Times

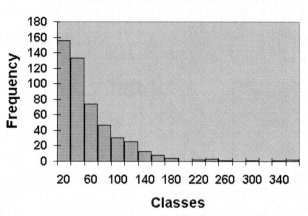

Table 3.4 Proportions of the Total Number of Observations Falling within $\bar{x} \pm s$, $\bar{x} \pm 2s$, and $\bar{x} \pm 3s$

A. DRIVER HEAD-INJURY RATINGS

Interval (impact points)	Proportion in Interval
$\bar{x} \pm s$ or $(419, 789)$	.694
$\bar{x} \pm 2s$ or $(234, 974)$	.980
$\bar{x} \pm 3s$ or $(49, 1{,}159)$	.990

B. FISH WEIGHTS

Interval (grams)	Proportion in Interval
$\bar{x} \pm s$ or $(673, 1{,}427)$	.681
$\bar{x} \pm 2s$ or $(296, 1{,}804)$	.972
$\bar{x} \pm 3s$ or $(-81, 2{,}181)$	.986

C. TAR CONTENTS OF CIGARETTES

Interval (milligrams)	Proportion in Interval
$\bar{x} \pm s$ or $(6.2, 16.0)$	.582
$\bar{x} \pm 2s$ or $(1.3, 20.9)$	.940
$\bar{x} \pm 3s$ or $(-3.6, 25.8)$	.999

D. SUPERMARKET CHECKOUT TIMES

Interval (seconds)	Proportion in Interval
$\bar{x} \pm s$ or $(1, 99)$	.860
$\bar{x} \pm 2s$ or $(-48, 148)$	.962
$\bar{x} \pm 3s$ or $(-97, 197)$	.980

the number and proportion of the total number of observations falling within each interval. These proportions are presented in Table 3.4a–d.

Tables 3.2 and 3.4 demonstrate a property that is common to many data sets. The percentage of observations that lie within one standard deviation of the mean $\bar{x}$, i.e., in the interval $\bar{x} \pm s$, is fairly large and variable, usually from 60% to 80% of the total number, but the percentage can reach 90% or more for highly skewed distributions of data. The percentage within two standard deviations of $\bar{x}$, i.e., in

Figure 3.14 Illustration of the Empirical Rule

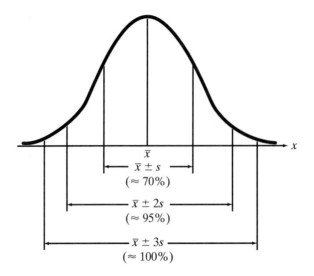

the interval $\bar{x} \pm 2s$, is close to 95% but, again, the percentage will be larger for highly skewed sets of data. Finally, the percentage of observations within three standard deviations of $\bar{x}$, i.e., in the interval $\bar{x} \pm 3s$, is almost 100%, meaning that almost all of the observations in a data set will fall within this interval. This property, which seems to hold for most data sets that contain at least 20 observations and are *mound-shaped,* is called the **Empirical Rule**. The Empirical Rule provides a very good general rule for forming a mental image of a distribution of data when you know the mean and standard deviation of the data set. Calculate the intervals $\bar{x} \pm s$, $\bar{x} \pm 2s$, and $\bar{x} \pm 3s$ and then picture the observations grouped as described in the box below and shown in Figure 3.14.

The Empirical Rule

If a distribution of sample data is mound-shaped with mean $\bar{x}$ and standard deviation s, then the proportions of the total number of observations falling within the intervals $\bar{x} \pm s$, $\bar{x} \pm 2s$, and $\bar{x} \pm 3s$ are as follows:

$\bar{x} \pm s$: Usually between 60% and 80%. The percentage will be approximately 70% for distributions that are nearly symmetric, but larger (near 90%) for highly skewed distributions.

$\bar{x} \pm 2s$: Close to 95% for symmetric distributions. The percentage will be larger (near 100%) for highly skewed distributions.

$\bar{x} \pm 3s$: Near 100%.

Note that the frequency histogram of supermarket checkout times (Figure 3.13) is highly skewed. Consequently, actual percentages of observations falling within the intervals $\bar{x} \pm s$, $\bar{x} \pm 2s$, and $\bar{x} \pm 3s$ for this data set will tend to be on the high side of the range of values given by the Empirical Rule. On the other hand, the mound-shaped distributions of head-injury ratings (Figure 3.10), fish weights (Figure 3.11), and tar contents (Figure 3.12) are nearly symmetric; consequently, the percentages falling within the three intervals are very close to the values given by the Empirical Rule.

Self-Test 3.4

Suppose a set of sample data has a mound-shaped, symmetric distribution. Make a statement about the percentage of measurements contained in each of the following intervals:

a. $\bar{x} \pm s$ b. $\bar{x} \pm 2s$ c. $\bar{x} \pm 3s$

For $k = 1$,

$$1 - \frac{1}{k^2} = 1 - \frac{1}{(1)^2} = 0.$$

Thus, Tchebysheff's Theorem states that at least 0% of the observations fall within $\bar{x} \pm s$. Consequently, no useful information is provided about the interval.

Can the Empirical Rule be applied to data sets with non-mound-shaped histograms or histograms of unknown shape? The answer, unfortunately, is no. However, in these situations we can apply a more conservative rule, called **Tchebysheff's Theorem**.

Tchebysheff's Theorem

For any set of sample measurements with mean $\bar{x}$ and standard deviation s, at least $1 - 1/k^2$ of the total number of observations in a sample data set will fall within the interval $\bar{x} \pm ks$, where k is a constant. For two useful values of k, $k = 2$ and $k = 3$, the theorem states that the proportions of the total number of observations in the sample falling within the intervals $\bar{x} \pm 2s$ and $\bar{x} \pm 3s$ are:

$\bar{x} \pm 2s$: At least 75% $\bar{x} \pm 3s$: At least 89%

Note that Tchebysheff's Theorem applies to any set of sample measurements, regardless of the shape of the relative frequency histogram. The rule is conservative in the sense that the specified percentage for any interval is a lower bound on the actual percentage of measurements falling in that interval. For example, Tchebysheff's Theorem states that at least 75% of the 500 CO measurements discussed in Example 3.9 will fall in the interval $\bar{x} \pm 2s$. We know (from Table 3.2) that the actual percentage (97%) is closer to the Empirical Rule's value of 95%. Consequently, whenever you know that a relative frequency histogram for a data set is mound-shaped, the Empirical Rule will give more precise estimates of the true percentages falling within the intervals $\bar{x} \pm s$, $\bar{x} \pm 2s$, and $\bar{x} \pm 3s$.

The last example in this section demonstrates the use of these rules in statistical inference.

Example 3.11   Application of the Empirical Rule

Travelers who have no intention of showing up often fail to cancel their hotel reservations in a timely manner. These travelers are known, in the parlance of the hospitality trade, as "no-shows." To protect against no-shows and late cancellations, hotels invariably overbook rooms. A study reported in the *Journal of Travel Research* examined the problems of overbooking rooms in the hotel industry. The data in Table 3.5, extracted from the study, represent daily numbers of late cancellations and no-shows for a random sample of 30 days at a large (500-room) hotel. Based on this sample, how many rooms, at minimum, should the large hotel overbook each day?

NOSHOW.XLS

Table 3.5 Hotel No-Shows for a Sample of 30 Days

18	16	16	16	14	18	16	18	14	19
15	19	9	20	10	10	12	14	18	12
14	14	17	12	18	13	15	13	15	19

Source: Toh, R. S. "An inventory depletion overbooking model for the hotel industry." *Journal of Travel Research,* Vol. 23, No. 4, Spring 1985, p. 27. The *Journal of Travel Research* is published by the Travel and Tourism Research Association (TTRA) and the Business Research Division, University of Colorado at Boulder.

Solution

To answer this question, we need to know a range of values where most of the daily numbers of no-shows fall. This requires that we compute $\bar{x}$ and s, and examine the shape of the relative frequency distribution for the data.

Figures 3.15a and b are Excel printouts that show both a histogram and descriptive statistics for the sample data. Notice from the histogram that the distribution of daily no-shows is mound-shaped, and only slightly skewed on the low (left) side of Figure 3.15a. Thus, the Empirical Rule should give a good estimate of the percentage of days that fall within one, two, and three standard deviations of the mean.

The mean and standard deviation of the sample data, highlighted on Figure 3.15b, are $\bar{x} = 15.133$ and $s = 2.945$. From the Empirical Rule, we know that

E **Figure 3.15** Excel Printout Describing the No-Show Data, Example 3.11

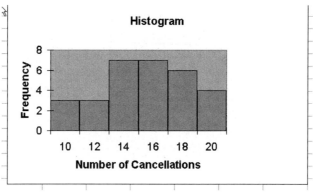

a.

	A	B
1	*Noshows*	
2		
3	Mean	15.13333333
4	Standard Error	0.5376264
5	Median	15
6	Mode	18
7	Standard Deviation	2.944701066
8	Sample Variance	8.671264368
9	Kurtosis	-0.670926394
10	Skewness	-0.296121431
11	Range	11
12	Minimum	9
13	Maximum	20
14	Sum	454
15	Count	30
16	Largest(1)	20
17	Smallest(1)	9

b.

about 95% of the daily number of no-shows fall within two standard deviations of the mean, i.e., within the interval

$$\bar{x} \pm 2s = 15.133 \pm 2(2.945)$$
$$= 15.133 \pm 5.890$$

or between 9.243 no-shows and 21.023 no-shows. (If we count the number of measurements in this data set, we find that actually 29 out of 30, or 96.7%, fall in this interval).

From this result, the large hotel can infer that there will be at least 9.243 (or, rounding up, 10) no-shows per day. Consequently, the hotel can overbook at least 10 rooms per day and still be highly confident that all reservations can be honored. ❑

PROBLEMS 3.24–3.41

Using the Tools

3.24 Calculate the variance and standard deviation of samples for which
   a. $n = 10$ $\sum x^2 = 331$ $\sum x = 50$
   b. $n = 25$ $\sum x^2 = 163{,}456$ $\sum x = 2{,}000$
   c. $n = 5$ $\sum x^2 = 26.46$ $\sum x = 11.5$

3.25 Calculate the range, variance, and standard deviation of each of the following examples:
   a. $0, 2, 4, 6, 8, 10$ **b.** $0, 4, 5, 5, 6, 10$ **c.** $4, 4, 4, 4, 4, 4$

3.26 Find the range, variance, and standard deviation for the following data set: $3, 9, 0, 7, 4$.

3.27 Find the mean and standard deviation for the following $n = 25$ measurements: $2, 1, 7, 6, 5, 3, 8, 5, 2, 4, 5, 6, 3, 4, 4, 6, 9, 4, 3, 4, 5, 5, 7, 3, 5$.

3.28 Refer to the data given in Problem 3.27.
   a. Construct the intervals $\bar{x} \pm s, \bar{x} \pm 2s$, and $\bar{x} \pm 3s$.
   b. Count the number of observations falling within each interval and find the corresponding proportions. Compare your results to the Empirical Rule and Tchebysheff's Theorem.

3.29 Find the mean and standard deviation for the following $n = 20$ measurements: $11, 16, 12, 12, 21, 4, 13, 10, 17, 12, 15, 12, 9, 11, 14, 13, 12, 15, 10, 16$.

3.30 Refer to the data given in Problem 3.29.
   a. Construct the intervals $\bar{x} \pm s, \bar{x} \pm 2s$, and $\bar{x} \pm 3s$.
   b. Count the number of observations falling within each interval and find the corresponding proportions. Compare your results to the Empirical Rule and Tchebysheff's Theorem.

Applying the Concepts

3.31   The data on retail prices for a random sample of 22 VCR models, Problem 2.21 (page 80), are repeated below.

VCRPRIC.XLS

350	300	340	220	320	450	270	265	210	250	180	300
190	170	190	170	170	200	180	220	200	250		

Data Source: Adapted from "VCRs," *Consumer Reports,* Nov. 1996, pp. 36–38.

  a. Find the range for the data set.

  b. Find the variance for the data set.

  c. Find the standard deviation for the data set.

  d. Find the mean for the data set.

  e. Construct an interval that will include about 95% of the 22 VCR retail prices.

3.32   The data on calories per 12-ounce serving for 22 beer brands, Problem 2.22 (page 81), are repeated in the table.

BEERCAL.XLS

Beer	Calories	Beer	Calories
Old Milwaukee	145	Miller High Life	143
Stroh's	142	Pabst Blue Ribbon	144
Red Dog	147	Milwaukee's Best	133
Budweiser	148	Miller Genuine Draft	143
Icehouse	149	Rolling Rock	143
Molson Ice	155	Michelob Light	134
Michelob	159	Bud Light	110
Bud Ice	148	Natural Light	110
Busch	143	Coors Light	105
Coors Original	137	Miller Light	96
Gennessee Cream Ale	153	Amstel Light	95

Source: Consumer Reports, June 1996, Vol. 61, No. 6, p. 16.

  a. Find the range for the data set.

  b. Find the variance for the data set.

  c. Find the standard deviation for the data set.

  d. Find the mean for the data set.

  e. Construct an interval that will include about 95% of the 22 caloric measurements.

3.33   Refer to the *Journal of Applied Behavior Analysis* (Summer 1997) study of a 16-year-old boy's destructive behavior, Problem 3.16, page 154. The data on the number of destructive responses per minute observed during each of 10 sessions is reproduced below.

  3   2   4.5   5.5   4.5   3   2.5   1   3   3

BOY16.XLS

  a. Find the range for the data set.

  b. Find the standard deviation for the data set.

  c. Use the mean from Problem 3.16 and the standard deviation from part **b** to construct an interval that will be likely to include the number of destructive responses per minute in any given 10-minute session.

3.34 Refer to the *American Journal of Audiology* (Mar. 1995) study of patients with unilateral hearing loss, Problem 3.21, page 155. The absolute sound pressure levels (SPLs) for the eight patients (recorded in decibels) is reproduced below.

SPL.XLS

| 73.0 | 80.1 | 82.8 | 76.8 | 73.5 | 74.3 | 76.0 | 68.1 |

a. Find the range for the data set.

b. Find the standard deviation for the data set.

c. Use the mean from Problem 3.21 and the standard deviation from part **b** to construct an interval that will be likely to include the SPL of any given patient.

3.35 The *Journal of Performance of Constructed Facilities* (Feb. 1990) published a study of water distribution networks. The internal pressure readings (measured in pounds per square inch, psi) for a sample of pipe sections had a mean of 7.99 psi and a standard deviation of 2.02 psi.

a. Use this information to construct an interval that captures about 95% of the pressure readings sampled.

b. Would you expect to observe an internal pressure reading of 20 psi? Explain.

3.36 The Trail Making Test (TMT) is frequently used in neuropsychological assessment to provide a quick estimate of brain damage in humans. To investigate the neuropsychological deficits in alcoholics, 50 problem drinkers (25 drinkers under the age of 40 and 25 drinkers 40 years or older) were given the TMT and their performance scores observed (the higher the score, the more extensive the brain damage). The results are reported in the accompanying table.

	Alcoholics under Age 40	**Alcoholics 40 or Older**
Mean performance score	39.6	49.7
Standard deviation	19.7	19.1

a. Use the information in the table, in conjunction with either the Empirical Rule or Tchebysheff's Theorem, to sketch your mental images of the relative frequency histograms of TMT performance scores for the two groups of alcoholics.

b. Estimate the fraction of alcoholics under age 40 who score between 19.9 and 59.3 on the TMT.

c. Approximately what percentage of alcoholics aged 40 or older score between 11.5 and 87.9 on the TMT?

3.37 Based on an analysis of automobile insurance claims, the Highway Loss Data Institute (HLDI) compiles injury and collision-loss data for popular cars, station wagons, and vans. The data on page 170, reported in *Consumer's Research* (Nov. 1993), are the HLDI collision-damage ratings of large station wagons and minivans. The collision-damage rating reflects how much, compared to other cars, is paid out by insurance companies to the model's owners for collision damage repairs. The higher the rating, the greater the amounts paid for collision-damage repairs.

HLDI.XLS

Vehicle Model	Collision-Damage Rating
Chevrolet Astro 4-wheel drive	50
Plymouth Voyager	59
Chevrolet Caprice	77
Oldsmobile Silhouette	72
Dodge Caravan	60
GMC Safari	60
Mazda MPV 4-wheel drive	121
Toyota Previa	77
Chevrolet Lumina APV	71
Ford Aerostar	74
Chevrolet Astro	59
Mazda MPV	114
Pontiac Trans Sport	72

Problem 3.37

a. Compute the mean collision-damage ratings of the cars listed in the table.

b. Compute the standard deviation of the collision-damage ratings.

c. You have recently purchased a new minivan. Give an interval that is highly likely to contain the collision-damage rating of your new minivan.

3.38 A Harris Corporation/University of Florida study was undertaken to compare the voltage readings of a manufacturing process performed at two locations—an old, remote location and a new location closer to the plant. Test devices were set up at both locations and voltage readings on the process were obtained. [*Note:* A "good process" was considered to be one with voltage readings of at least 9.2 volts.] The table contains voltage readings for 30 production runs at each location. Descriptive statistics for both sample data sets are provided in the accompanying Excel printout. Use the Empirical Rule to compare the voltage reading distributions for the two locations.

VOLTAGE.XLS

Old Location			New Location		
9.98	10.12	9.84	9.19	10.01	8.82
10.26	10.05	10.15	9.63	8.82	8.65
10.05	9.80	10.02	10.10	9.43	8.51
10.29	10.15	9.80	9.70	10.03	9.14
10.03	10.00	9.73	10.09	9.85	9.75
8.08	9.87	10.01	9.60	9.27	8.78
10.55	9.55	9.98	10.05	8.83	9.35
10.26	9.95	8.72	10.12	9.39	9.54
9.97	9.70	8.80	9.49	9.48	9.36
9.87	8.72	9.84	9.37	9.64	8.68

Source: Harris Corporation, Melbourne, Fla.

E

	A	B	C
1		Old	New
2			
3	Mean	9.803666667	9.422333333
4	Standard Error	0.098757201	0.087430344
5	Median	9.975	9.455
6	Mode	9.98	8.82
7	Standard Deviation	0.540915465	0.478875719
8	Sample Variance	0.29258954	0.229321954
9	Kurtosis	3.473527903	-0.891340632
10	Skewness	-1.877868645	-0.26698597
11	Range	2.5	1.61
12	Minimum	8.05	8.51
13	Maximum	10.55	10.12
14	Sum	294.11	282.67
15	Count	30	30
16	Largest(1)	10.55	10.12
17	Smallest(1)	8.05	8.51

3.39 Marine scientists at the University of South Florida investigated the feeding habits of midwater fish inhabiting the eastern Gulf of Mexico (*Prog. Oceanog.*, Vol. 38, 1996). Most of the fish captured fed heavily on copepods, a type of zooplankton. The table below gives the percent of shallow-living copepods in the diets of each of 35 species of myctophid fish.

COPEPOD.XLS

79	100	100	98	95	96	56	51	90	95	94	93
92	71	100	100	99	100	100	100	99	80	54	56
59	92	88	81	88	59	100	85	82	74	66	

Source: Hopkins, T. L., Sutton, T. T,. and Lancraft, T. M. "The tropic structure and predation impact of low latitude midwater fish assemblage." *Prog. Oceanog.*, Vol. 38, No. 3, 1996, p. 223 (Table 6).

a. Use Excel to find the mean, variance, and standard deviation of the data set.

b. Form the interval $\bar{x} \pm 2s$. Make a statement about the proportion of measurements that will fall within this interval.

c. Count the number of measurements that fall within the interval $\bar{x} \pm 2s$. Compare the result to your answer in part **b**.

3.40 It is customary practice in the U.S. to base roadway design on the 30th highest hourly volume in a year. Thus, all roadway facilities are expected to operate at acceptable levels of service for all but 29 hours of the year. The Florida Department of Transportation (DOT), however, has shifted from the 30th highest hour to the 100th highest hour as the basis for level-of-service determinators. Florida Atlantic University researcher Reid Ewing investigated whether this shift was warranted in the *Journal of STAR Research* (July 1994). The table below gives the traffic counts at the 30th highest hour and the 100th highest hour of a recent year for 20 randomly selected DOT permanent count stations. Excel histograms for the two variables are provided on page 172 as well as a summary statistics printout.

DOTHOUR.XLS

Station	30th Highest Hour	100th Highest Hour
0117	1,890	1,736
0087	2,217	2,069
0166	1,444	1,345
0013	2,105	2,049
0161	4,905	4,815
0096	2,022	1,958
0145	594	548
0149	252	229
0038	2,162	2,048
0118	1,938	1,748
0047	879	811
0066	1,913	1,772
0094	3,494	3,403
0105	1,424	1,309
0113	4,571	4,425
0151	3,494	3,359
0159	2,222	2,137
0160	1,076	989
0164	2,167	2,039
0165	3,350	3,123

Source: Ewing, R. "Roadway levels of service in an era of growth management." *Journal of STAR Research*, Vol. 3, July 1994, p. 103 (Table 2).

E

	A	B	C
		30th Highest	*100th Highest*
1			
2			
3	Mean	2205.95	2095.6
4	Standard Error	273.6523291	269.0249392
5	Median	2063.5	1998.5
6	Mode	3494	#N/A
7	Standard Deviation	1223.81042	1203.116103
8	Sample Variance	1497711.945	1447488.358
9	Kurtosis	0.272765229	0.375983274
10	Skewness	0.709281198	0.784619654
11	Range	4653	4586
12	Minimum	252	229
13	Maximum	4905	4815
14	Sum	44119	41912
15	Count	20	20
16	Largest(1)	4905	4815
17	Smallest(1)	252	229

Problem 3.40

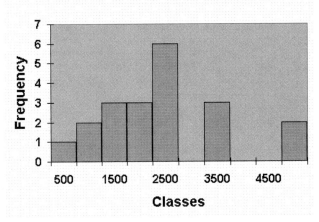

Problem 3.40

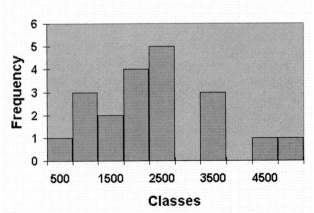

Problem 3.40

a. Determine the mean and standard deviation for the 30th highest hour and form the interval $\bar{x} \pm 2s$.

b. Make a statement about the percentage of stations that have 30th highest hour traffic counts in the interval $\bar{x} \pm 2s$.

c. Locate the mean and standard deviation for the 100th highest hour and form the interval $\bar{x} \pm 2s$.

d. Make a statement about the percentages of stations that have 100th highest hour traffic counts in the interval $\bar{x} \pm 2s$.

3.41 Refer to the *Chemosphere* study of Vietnam veterans possibly exposed to Agent Orange, Problem 2.28 (page 84). The data on TCDD levels (in parts per trillion) in the plasma of 20 veterans is reproduced in the table.

TCDD.XLS

TCDD Levels in Plasma									
2.5	1.8	6.9	1.8	3.5	2.5	1.6	36.0	6.8	3.1
20.0	3.3	4.7	3.1	4.1	7.2	4.6	3.0	2.1	2.0

Source: Schecter, A., et al. "Partitioning of 2,3,7,8-chlorinated dibenzo-p-dioxins and dibenzofurans between adipose tissue and plasma lipid of 20 Massachusetts Vietnam veterans." *Chemosphere,* Vol. 20, Nos. 7–9, 1990, pp. 954–955 (Table I).

a. Compute $\bar{x}$, s^2, and s for this data set.

b. What percentage of measurements would you expect to find in the interval $\bar{x} \pm 2s$?

c. Count the number of measurements that actually fall within the interval of part **b** and express the interval count as a percentage of the total number of measurements. Compare this result with the answer to part **b**.

d. Suppose the veteran that had the highest TCDD level (36) was omitted from the analysis. Would you expect $\bar{x}$ to increase or decrease? Would you expect s to increase or decrease? Explain.

3.6 MEASURES OF RELATIVE STANDING: PERCENTILES AND z SCORES

In some situations, you may want to describe the relative position of a particular measurement in a data set. For example, suppose a recent graduate of your college secured a job with a starting salary of $46,000. You might want to know whether this is a relatively low or high starting salary, etc. What percentage of the starting salaries of graduates of your college were less than $46,000; what percentage were larger? Descriptive measures that locate the relative position of a measurement—in relation to the other measurements—are called *measures of relative standing.* One measure that expresses this position in terms of a percentage is called a *percentile* for the data set.

Definition 3.7

Let $x_1, x_2, \ldots, x_n$ be a set of n measurements arranged in increasing (or decreasing) order. The pth **percentile** is a number x such that $p\%$ of the measurements fall below the pth percentile and $(100 - p)\%$ fall above it.

Example 3.12   Interpreting a Percentile

Suppose a starting salary of $46,000 falls at the 95th percentile of the distribution of starting salaries for recent graduates of your college. What does this imply?

Solution

Using $p = 95$ in Definition 3.7, 95% of the numbers in a data set fall below the 95th percentile and $(100 - 95)\% = 5\%$ fall above it. Thus, 95% of the starting salaries are less than $46,000 and 5% are greater. ❏

The median, by definition, is the 50th percentile. The 25th percentile, the median, and the 75th percentiles are often used to describe a data set because they divide the data set into four groups, with each group containing one-fourth (25%) of the observations. They also divide the relative frequency histogram for a data set into four parts, each containing the same area (.25), as shown in Figure 3.16. Consequently, the 25th percentile, the median, and the 75th percentile are called the *1st quartile*, the *mid quartile*, and the *3rd quartile*, respectively, for a data set.

Definition 3.8

The **1st quartile**, Q_1, for a data set is the 25th percentile.

Definition 3.9

The **mid-quartile** (median), M, for a data set is the 50th percentile.

Definition 3.10

The **3rd quartile**, Q_3, for a data set is the 75th percentile.

For large data sets, percentiles can be found by locating the corresponding areas under the relative frequency distribution. They also can be found by ranking the data and determining the percentiles of interest. Such computations can be performed using Excel.

Figure 3.16 Locations of the Lower and Upper Quartiles

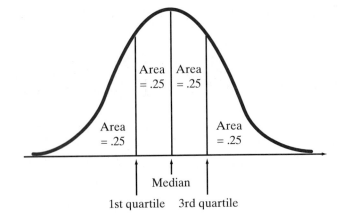

E **Figure 3.17** Excel Descriptive Statistics for Driver Head-Injury Ratings

	A	B
1	Five-Number Summary	
2	for Drivhead	
3		
4	Minimum	216
5	First Quartile	476.25
6	Median	605
7	Third Quartile	724
8	Maximum	1240

a.

	A	B
1	Drivhead	
2		
3	Mean	603.744898
4	Standard Error	18.72427929
5	Median	605
6	Mode	528
7	Standard Deviation	185.360908
8	Sample Variance	34358.66621
9	Kurtosis	0.473236183
10	Skewness	0.473483012
11	Range	1024
12	Minimum	216
13	Maximum	1240
14	Sum	59167
15	Count	98
16	Largest(1)	1240
17	Smallest(88)	874

b.

Example 3.13   Finding Percentiles

Refer to the NCAP crash test data in the Excel workbook, CRASH.XLS. Suppose we are interested in the driver head-injury ratings for the 98 crash-test cars.

  a. Find and interpret the values of Q_1, M, and Q_3.

  b. Find and interpret the 90th percentile for the data set.

Solution

  a. Descriptive statistics for the driver head-injury ratings are displayed in the Excel printouts, Figure 3.17. The values of Q_1, M, and Q_3, highlighted on Figure 3.17a, are 476.25, 605, and 724, respectively. From these values, we know that 25% of the 98 head-injury ratings fall below the 1st quartile, 476.25; 50% fall below the median, 605; and 75% fall below the 3rd quartile, 724.

  b. By definition, we know that 90% of the head-injury ratings in the data set fall below the 90th percentile. If we rank the 98 head-injury ratings in increasing order, where the rating with rank 1 is the smallest and the rating with rank 98 is the largest, then the 90th percentile is the value with rank $(.90)(98) = 88.2$, or 88. This value, labeled **Smallest (88)** on the Excel printout, Figure 3.17b, is 874. Thus, 90% of the head-injury ratings in the data set fall below 874. ❑

Self-Test

3.5

State the percentage of measurements that are above and below each of the following percentiles:

  a. 20th percentile   **b.** Median

  c. 86th percentile   **d.** Lower quartile

  e. Upper quartile   **f.** 10th percentile

When the sample data set is small, it may be impossible to find a measurement in the data set that exceeds, for example, *exactly* 25% of the remaining measurements. Consequently, percentiles for small data sets are not well defined. The box describes a procedure for finding quartiles and other percentiles with small data sets.

FINDING QUARTILES (AND PERCENTILES) WITH SMALL DATA SETS

Step 1   Rank the n measurements in the data set in increasing order of magnitude.

Step 2   Calculate the quantity $\frac{1}{4}(n + 1)$ and round to the nearest integer. The measurement with this rank represents the lower quartile or 25th percentile.

Step 3   Calculate the quantity $\frac{3}{4}(n + 1)$ and round to the nearest integer. The measurement with this rank represents the upper quartile or 75th percentile.

General   To find the pth percentile, calculate the quantity

$$\frac{p(n + 1)}{100} \tag{3.6}$$

and round to the nearest integer. The measurement with this rank is the pth percentile.

Another measure of relative standing is the **z score** for a measurement. For example, suppose you were told that $42,000 lies 1.44 standard deviations above the mean of the distribution of starting salaries of recent graduates of your college. Knowing that most of the starting salaries will be less than two standard deviations from the mean and almost all will be within three, you would have a good idea of the relative standing of the $42,000 starting salary. The distance that a measurement x lies above or below the mean $\bar{x}$ of a data set, measured in units of the standard deviation s, is called the *z score* for the measurement. A negative z score indicates that the observation lies to the left of the mean; a positive z score indicates that the observation lies to the right of the mean.

Definition 3.11

The sample **z score** for the measurement x is

$$z = \frac{x - \bar{x}}{s} \tag{3.7}$$

A negative z score indicates that the observation lies to the left of the mean; a positive z score indicates that the observation lies to the right of the mean.

Example 3.14   Computing a z Score

In Example 3.10, we noted that the mean and standard deviation for the 98 driver head-injury ratings in CRASH.XLS are $\bar{x} = 604$ and $s = 185$. Use these values to find the z score for a driver head-injury rating of 408. Interpret the result.

Solution

Substituting the values of x, $\bar{x}$, and s into the formula for z (equation 3.7), we obtain

$$z = \frac{x - \bar{x}}{s} = \frac{408 - 604}{185} = -1.06$$

Since the z score is negative, we conclude that the head-injury rating of 408 lies a distance of 1.06 standard deviations below (to the left of) the mean of 604. ❑

In the following sections, we discuss how percentiles and z scores can be used to detect skewness and unusual observations in a data set.

**PROBLEMS
3.42–3.56**

Using the Tools

3.42   Compute the z score corresponding to each x value, assuming $\bar{x} = .7$ and $s = .2$.

    a. $x = .9$ **b.** $x = .4$ **c.** $x = 1.3$

3.43   Compute the z score corresponding to each x value, assuming $\bar{x} = 20$ and $s = 5$.

    a. $x = 12$ **b.** $x = 23$ **c.** $x = 28$

3.44   The 24 sample measurements of Problem 2.17 (page 80) are reproduced here.

| 213 | 228 | 241 | 268 | 234 | 303 | 274 | 316 | 319 | 320 | 227 | 226 |
| 224 | 267 | 303 | 266 | 265 | 237 | 288 | 291 | 285 | 270 | 254 | 215 |

    a.   Use the stem-and-leaf display you constructed in Problem 2.17 to find Q_1, M, and Q_3.

    b.   Find the 90th percentile of the data set.

3.45   The 28 sample measurements of Problem 2.18 (page 80) are reproduced here.

5.9	5.3	1.6	7.4	8.6	1.2	2.1	4.0	7.3	8.4	8.9	6.7
4.5	6.3	7.6	9.7	3.5	1.1	4.3	3.3	8.4	1.6	8.2	6.5
1.1	5.0	9.4	6.4								

    a.   Use the stem-and-leaf display you constructed in Problem 2.18 to find Q_1, M, and Q_3.

    b.   Find the 10th percentile of the data set.

3.46   Compute z scores for each of the following situations. Then determine which x value lies the greatest distance above the mean; the greatest distance below the mean.

    a. $x = 77, \bar{x} = 58, s = 8$ **b.** $x = 8.8, \bar{x} = 11, s = 2$

    c. $x = 0, \bar{x} = -5, s = 1.5$ **d.** $x = 2.9, \bar{x} = 3, s = .1$

3.47   State the percentage of measurements that are larger than each of the following percentiles:

    a.   60th percentile   **b.**   95th percentile

    c.   33rd percentile   **d.**   15th percentile

Applying the Concepts

3.48   The mean and standard deviation of the 50 starting salaries listed in Table 2.5 are $\bar{x} = \$25{,}708$ and $s = \$9{,}338$. Find and interpret the z score for a college graduate with a starting salary of:

    a. $15,300   **b.** $30,900

3.49   The 3rd quartile for the 500 supermarket customer checkout times in the Excel workbook, CHKOUT.XLS, is 65 seconds. Interpret this value.

3.50   The hotel no-show data presented in Example 3.11 are reproduced here.

NOSHOW.XLS

18	16	16	16	14	18	16	18	14	19
15	19	9	20	10	10	12	14	18	12
14	14	17	12	18	13	15	13	15	19

Source: Toh, R. S. "An inventory depletion overbooking model for the hotel industry." *Journal of Travel Research,* Vol. 23, No. 4, Spring 1985, p. 27. The *Journal of Travel Research* is published by the Travel and Tourism Research Association (TTRA) and the Business Research Division, University of Colorado at Boulder.

    a.   Construct a stem-and-leaf display for the data.

    b.   Compute and interpret the values of Q_1, M, and Q_3.

    c.   Compute and interpret the 90th percentile for the data.

3.51   Refer to the CDC study of sanitation levels for 72 international cruise ships, Problem 3.17 (page 154). An Excel printout of descriptive statistics for the data is reproduced here. Locate Q_1, M, and Q_3 on the printout and interpret these values.

	A	B
1	Five-Number Summary	
2	for Sanitation Inspection Scores	
3		
4	Minimum	66
5	First Quartile	89
6	Median	92
7	Third Quartile	95
8	Maximum	99

E Problem 3.51

3.52   A biologist investigating the parent-young conflict in herring gulls recorded the feeding rates (the number of feedings per hour) for two groups of gulls: those parents with only one chick to feed and those with two or three chicks to feed. The results are summarized in the table.

    a.   A particular pair of parent gulls fed the three chicks in their brood at a rate of .44 times per hour. Find the z score for this feeding rate and interpret its value.

    b.   Would you expect to observe a feeding rate of .44 for parent gulls with a brood size of only one chick? Explain.

	BROOD SIZE (NUMBER OF CHICKS)	
	1	2 or 3
Mean feeding rate	.18	.31
Standard deviation	.15	.13

Problem 3.52

3.53   Refer to the *Administrative Science Quarterly* entrapment experiment, Problem 3.22 (page 155). The data (i.e., total points awarded) for the 30 trials are reproduced on page 179.

    a.   Calculate and interpret a measure of relative standing for a trial in which five points were awarded.

    b.   Calculate and interpret a measure of relative standing for a trial in which 20 points were awarded.

5	5	4	7	24	6
10	12	11	15	11	10
3	23	4	20	5	4
7	5	6	6	15	5
15	10	13	9	4	6

ENTRAP.XLS

Source: Brockner, J., et al. "Escalation of commitment to an ineffective course of action: The effect of feedback having negative implications for self-identity." *Administrative Science Quarterly,* Vol. 31, No. 1, Mar. 1986, p. 115.

Problem 3.53

3.54 Refer to the *Journal of Geography* study comparing test scores of geography students who used computer resources and those who did not, Problem 3.23 (page 156). Excel descriptive statistics for the test scores of the two groups of students are shown below.

GEO.XLS

	A	B	C
1		**Computer Resources**	**No Computer Resources**
2			
3	Mean	68.86046512	50.04545455
4	Standard Error	2.693164596	4.068396449
5	Median	69	53
6	Mode	91	66
7	Standard Deviation	17.66026128	19.08247082
8	Sample Variance	311.8848283	364.1406926
9	Kurtosis	-0.352248464	-0.255178597
10	Skewness	-0.482559525	-0.665443331
11	Range	69	69
12	Minimum	28	9
13	Maximum	97	78
14	Sum	2961	1101
15	Count	43	22
16	Largest(1)	97	78
17	Smallest(1)	28	9

a. A student in group 1 (computer resources group) scored a 72 on the test. Calculate the *z* score for this student.

b. A student in group 2 (no computer) scored a 66 on the test. Calculate the *z* score for this student.

c. One student scored a 30 on the test. (This test score was previously unrecorded.) What group is this student more likely to come from? Explain.

3.55 Refer to the *Prog. Oceanog.* study of the diets of myctophid fish living in the Gulf of Mexico, Problem 3.39 (page 170). The data on percent of shallow-living copepods in the diets of the 35 fish species are reproduced below.

COPEPOD.XLS

79	100	100	98	95	96	56	51	90	95	94	93
92	71	100	100	99	100	100	100	99	80	54	56
59	92	88	81	88	59	100	85	82	74	66	

Source: Hopkins, T. L., Sutton, T. T. and Lancraft, T. M. "The tropic structure and predation impact of low latitude midwater fish assemblage." *Prog. Oceanog.,* Vol. 38, No. 3, 1996, p. 223 (Table 6).

    a. Find Q_1, M, and Q_3 for this data set. Interpret these values.

    b. Find the 90th percentile for the data set. Interpret this value.

    c. Find the z score for the fish species with 81% of copepods in its diet.

3.56 The data on TCDD levels in the plasma of 20 Vietnam veterans from Problem 3.41 is reproduced here. In addition, the TCDD levels in fat tissue are also recorded in the table.

TCDDFAT.XLS

TCDD Levels in Plasma				TCDD Levels in Fat Tissue			
2.5	1.8	6.9	1.8	4.9	1.1	7.0	4.2
3.5	2.5	1.6	36.0	6.9	2.3	1.4	41.0
6.8	3.1	20.0	3.3	10.0	5.9	11.0	2.9
4.7	3.1	4.1	7.2	4.4	7.0	2.5	7.7
4.6	3.0	2.1	2.0	4.6	5.5	4.4	2.5

Source: Schecter, A., et al. "Partitioning of 2,3,7,8-chlorinated dibenzo-p-dioxins and dibenzofurans between adipose tissue and plasma lipid of 20 Massachusetts Vietnam veterans." *Chemosphere,* Vol. 20, Nos. 7–9, 1990, pp. 954–955 (Tables I and II).

    a. Calculate $\bar{x}$ and s for the TCDD levels in plasma.

    b. Calculate $\bar{x}$ and s for the TCDD levels in fat tissue.

    c. Calculate the z score for a TCDD level in plasma of 20.

    d. Calculate the z score for a TCDD level in fat tissue of 20.

    e. Based on the results of parts **c** and **d**, is a TCDD level of 20 or more likely to occur in plasma or fat tissue?

3.7 BOX-AND-WHISKER PLOTS

To this point, we have presented numerical descriptive measures of central tendency, variation, and relative standing for a quantitative data set. When conducting a preliminary analysis of the sample data—often called an *exploratory data analysis*—it is helpful to present these descriptive measures in a summarized format.

One obvious approach is to give a **two-number summary** consisting of the mean, $\bar{x}$, and standard deviation, s. As demonstrated in Section 3.5, these two numbers and the Empirical Rule provide an interval that captures most of the values in the data set. Two other useful summary techniques are a *five-number summary* and a *box-and-whisker plot*.

A **five-number summary** for a set of n sample observations, $x_1, x_2, \ldots, x_n$, consists of the smallest measurement, the 1st quartile, the median, the 3rd quartile, and the largest measurement:

$$x_{\text{smallest}} \qquad Q_1 \qquad M \qquad Q_3 \qquad x_{\text{largest}}$$

Note that this summary consists of one measure of central tendency (the median), two measures of relative standing (Q_1 and Q_3), and a measure of variation (the range).

From the five-number summary, several other measures of variation can be computed. An important one is the *interquartile range*.

Figure 3.18 The
Interquartile Range

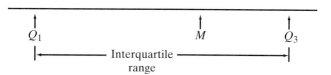

The **interquartile range**, **IQR**, is the distance between the 1st and 3rd quartiles:

$$IQR = Q_3 - Q_1 \tag{3.8}$$

You can see from Figure 3.18 that the interquartile range is a measure of data variation. The larger the interquartile range, the more variable the data tend to be. Also, since Q_3 is the 75th percentile and Q_1 is the 25th percentile, we know that $(75\% - 25\%) = 50\%$ of the data lie within the interquartile range.

Example 3.15   Five-Number Summary

The monthly rental prices for a random sample of 25 two-bedroom apartments are listed in Table 3.6.

a. Give a five-number summary for the data.

b. Find IQR.

Solution

a. We used Excel to produce descriptive statistics for the 25 monthly rental prices. These values are displayed in Figure 3.19. From Figure 3.19, we find $x_{\text{smallest}} = 367$, $Q_1 = 650$, $M = 760$, $Q_3 = 950$, and $x_{\text{largest}} = 1,480$. Thus, the five-number summary is:

| 367 | 650 | 760 | 950 | 1,480 |

b. Using equation 3.8, the interquartile range is

$$IQR = Q_3 - Q_1 = 950 - 650 = 300 \qquad \square$$

E **Figure 3.19**
Descriptive Statistics
for the Rental Price
Data of Table 3.6

	A	B
1	Five-Number Summary	
2	for Monthly Rents	
3		
4	Minimum	367
5	First Quartile	650
6	Median	760
7	Third Quartile	950
8	Maximum	1480

A **box-and-whisker plot** is a graphical representation of the five-number summary. The next example illustrates how to construct such a plot.

MONRENTS.XLS

Table 3.6 Monthly Rents for 25 Two-Bedroom Apartments

Monthly Rents				
$ 660	$ 595	$1,060	$ 500	$ 630
899	1,295	749	820	843
710	950	720	575	760
1,090	770	682	1,016	650
425	367	1,480	945	1,120

E **Figure 3.20** PHStat Add-In for Excel Box-and-Whisker Plot for the Rent Data of Table 3.6

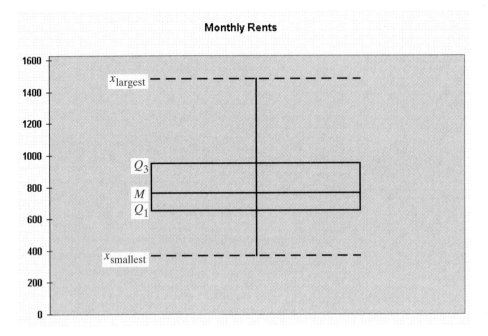

Monthly Rents

Example 3.16   Constructing a Box-and-Whisker Plot

Refer to the monthly rental prices for a random sample of 25 two-bedroom apartments as listed in Table 3.6. Construct a box-and-whisker plot for the data.

Solution

A vertical box-and-whisker plot for the 25 monthly rental prices, generated by the PHStat Add-in for Excel, is shown in Figure 3.20. Recall from Example 3.15 that $Q_1 = 650$, $M = 760$, and $Q_3 = 950$. You can see that the box-and-whisker plot is a box with Q_1 and Q_3 located at the lower and upper corners, respectively. The height of the box is equal to the interquartile range, IQR; consequently, the box contains the middle 50% of the rental prices in the data set. The horizontal line within the box locates the median, M.

The lower 25% of the data are represented by a line (i.e., a *whisker*) connecting the bottom of the box to the location of the smallest rental price, $x_{smallest}$. Similarly, the upper 25% of the data are represented by a line connecting the top of the box to $x_{largest}$. ☐

STEPS TO FOLLOW IN CONSTRUCTING A VERTICAL BOX-AND-WHISKER PLOT

Step 1   Calculate the median, M, 1st and 3rd quartiles, Q_1 and Q_3, and the interquartile range, IQR, for the measurements in a data set.

Step 2   Construct a box with Q_1 and Q_3 located at the lower and upper corners, respectively. The height will then be equal to IQR. Draw a horizontal line inside the box to locate the median M.

Step 3   Use a whisker (vertical line) to connect the smallest observation in the data set, $x_{smallest}$, to the bottom of the box.

Step 4   Use a whisker to connect the largest observation in the data set, x_{largest}, to the top of the box.

A box-and-whisker plot helps us detect possible skewness in the distribution for a data set. For example, if Q_1 is farther away from the median than Q_3, then the distribution is likely to be skewed to the left, as shown in Figure 3.21b. If Q_3 is farther away from the median than Q_1, then the distribution is likely to be skewed to the right, as in Figure 3.21c. Symmetry or lack of skewness is suggested when Q_1 and Q_3 are approximately equidistant from the median and when the whiskers are of approximately equal length, as depicted in Figure 3.21a.

The key ideas of this section are: (1) A two-number summary (i.e., $\bar{x}$ and s) in combination with the Empirical Rule provides us with an interval that contains most of the measurements in a data set; and (2) A five-number summary (i.e., x_{smallest}, Q_1, M, Q_3, and x_{largest}) illustrated with a box-and-whisker plot allows us to study the shape of the distribution of data.

In the next section, we demonstrate how to use both approaches to detect highly unusual observations in the data set.

Figure 3.21 Horizontal Box-and-Whisker Plots for Three Types of Distributions

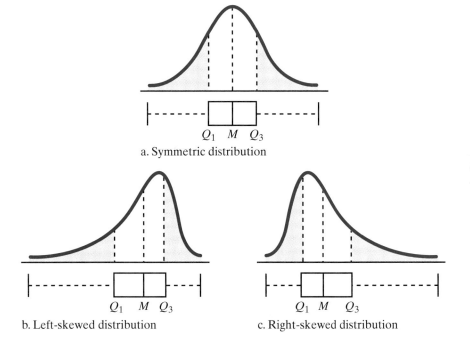

a. Symmetric distribution

b. Left-skewed distribution

c. Right-skewed distribution

3.8 METHODS FOR DETECTING OUTLIERS

Sometimes inconsistent observations are included in a data set. For example, when we discuss starting salaries for college graduates with bachelor's degrees, we generally think of traditional college graduates—those near 22 years of age with four years of college education. But suppose one of the graduates is a 34-year-old Ph.D. chemist who has returned to the university to obtain a bachelor's degree in metallurgy. Clearly, the starting salary for this graduate could be much larger than the other starting salaries because of the graduate's additional education and experience, and we probably would not want to include it in the data set. An unusual observation that lies outside the range of the data values that we want to describe is called an *outlier*.

Outliers are often attributable to one of several causes. First, the measurement associated with the outlier may be invalid. For example, the experimental procedure used to generate the measurement may have malfunctioned, the experimenter may have misrecorded the measurement, or the data might have been coded incorrectly when entered in the computer. Second, the outlier may be the result of a misclassified measurement. That is, the measurement belongs to a population different from that from which the rest of the sample was drawn, as in the case of the chemist's salary described in the preceding paragraph. Finally, the measurement associated with the outlier may be recorded correctly and from the same population as the rest of the sample, but represents a rare (chance) event. Such outliers occur most often when the relative frequency histogram of the sample data is extremely skewed, because such a distribution has a tendency to include extremely large or small observations relative to the others in the data set.

E **Figure 3.22** Excel Descriptive Statistics for the 144 Fish Specimen DDT Measurements

	A	B
1	DDT	
2		
3	Mean	24.355
4	Standard Error	8.198215451
5	Median	7.15
6	Mode	12
7	Standard Deviation	98.37858542
8	Sample Variance	9678.346069
9	Kurtosis	102.2387637
10	Skewness	9.61165982
11	Range	1099.89
12	Minimum	0.11
13	Maximum	1100
14	Sum	3507.12
15	Count	144
16	Largest(1)	1100
17	Smallest(1)	0.11

Definition 3.13

An observation (or measurement) that is unusually large or small relative to the other values in a data set is called an **outlier**. Outliers typically are attributable to one of the following causes:

1. The measurement is observed, recorded, or entered into the computer incorrectly.

2. The measurement comes from a different population.

3. The measurement is correct, but represents a rare (chance) event.

The most obvious method for determining whether an observation is an outlier is to calculate its z score (Section 3.6), as the following example illustrates.

Example 3.17   Outliers Using z Scores

Consider the DDT measurements of 144 fish specimens in the Excel workbook FISH.XLS. The fish specimen identified by observation 115 has a DDT measurement of 1,100 ppm. Is this observation an outlier?

Solution

To obtain a z score, we need a two-number summary ($\bar{x}$ and s) for the data set. An Excel descriptive statistics printout for the 144 DDT measurements is displayed in Figure 3.22.

Note that $\bar{x} = 24.36$ ppm and $s = 98.38$ ppm. Therefore, the z score for the DDT measurement of fish 115 is

$$z = \frac{x - \bar{x}}{s} = \frac{1,100 - 24.36}{98.38} = 10.93$$

Both the Empirical Rule and Tchebysheff's Theorem (Section 3.5) tell us that almost all the observations in a data set will have z scores less than 3 in absolute value. Since a z score as large as 10.93 is highly improbable, the DDT measurement of 1,100 ppm is called an outlier. Some research by the Army Corps of Engineers revealed that the DDT value for this fish specimen was correctly recorded, but that the fish was one of the few found at the exact location where the manufacturing plant was discharging its toxic waste materials into the river. ❏

Self-Test 3.7

A sample of measurements has a mean of $\bar{x} = 100$ and a standard deviation of $s = 15$. For each of the following values of x, calculate the z score and determine if the measurement is an outlier.

    a. $x = 90$ **b.** $x = 190$

Another procedure for detecting outliers utilizes the information in a five-number summary (Section 3.7). With this method, we construct intervals similar to the intervals of the Empirical Rule; however, the intervals are based on the 1st and 3rd quartiles and the interquartile range.

Example 3.18   Outliers Using Quartiles

Refer to the monthly rental prices for 25 two-bedroom apartments, Examples 3.15 and 3.16 (pages 181–182). The five-number summary for the data set is:

  367   650   760   950   1,480

Use this information to check for outliers in the rental price data set.

Solution
Recall that 50% of the data lie between Q_1 and Q_3. Consequently, observations that fall too far to the left of Q_1 or too far to the right of Q_3 are deemed to be *outliers*. Specifically, observations that fall a distance of 1.5(IQR) below Q_1 or a distance of 1.5(IQR) above Q_3 are suspect. Thus, any observation in a data set that falls outside the interval $(Q_1 - 1.5 \cdot \text{IQR}, Q_3 + 1.5 \cdot \text{IQR})$ is considered an outlier. For the rental price data, $Q_1 = 650, Q_3 = 950$, and IQR $= Q_3 - Q_1 = 950 - 650 = 300$. Thus,

$$Q_1 - 1.5 \cdot \text{IQR} = 650 - (1.5)(300) = 200$$

$$Q_3 + 1.5 \cdot \text{IQR} = 950 + (1.5)(300) = 1,400$$

Checking the data set in Table 3.6, you can see that only the rental price of \$1,480 falls outside the interval (200, 1,400); consequently, it is judged to be an outlier. ❏

The z score and quartiles methods both establish limits outside of which a measurement is deemed to be an outlier (see accompanying box). Usually, the two methods produce similar results. However, the presence of one or more outliers in a data set can inflate the computed value of s. Consequently, it will be less likely that an errant observation would have a z score larger than 3 in absolute value. In contrast, the values of the quartiles used to calculate the intervals are not as affected by the presence of outliers.

General Rule for Detecting Outliers

1. *z scores:*   Observations with z scores greater than 3 in absolute value are considered outliers. (For some highly skewed data sets, observations with z scores greater than 2 in absolute value may be outliers.)

2. *Quartiles:*   Observations falling outside the interval $(Q_1 - 1.5 \cdot IQR, Q_3 + 1.5 \cdot IQR)$ are deemed outliers.

Example 3.19

Identifying Outliers

Each county in the state of Florida negotiates an annual contract for bread to supply the county's public schools. Sealed bids are submitted by vendors, and the lowest bid (price per pound of bread) is selected as the bid winner. The winning (or low) bid prices (in dollars) for a random sample of 303 white-bread contracts awarded in eight geographic markets over a 6-year period in Florida are available in the Excel workbook BREAD.XLS. For confidentiality, the specific years and markets are not identified.

State officials are concerned that some vendors have conspired to illegally set the winning bid price above the fair (or competitive) price. Suppose our goal is to identify any contracts with unusually high prices in the data set. These contracts and the vendors that bid on them will be investigated for bid collusion.

Descriptive statistics for the data are shown in the Excel printout, Figure 3.23.

a.   Use the z score method to identify any outliers in the bid-price data.

b.   Use the quartiles method to identify any outliers in the bid-price data.

E **Figure 3.23** Excel Descriptive Statistics for Bid-Price Data

	A	B
1	Z Scores For Bid-Price Data	
2		
3	Mean	0.24278
4	Standard Deviation	0.05170
5		
6	Ten Highest Prices	Z Score
7	0.44000	3.81468
8	0.41000	3.23442
9	0.40533	3.14416
10	0.37500	2.55746
11	0.36400	2.34469
12	0.36000	2.26733
13	0.35344	2.14044
14	0.35200	2.11259
15	0.34667	2.00943
16	0.34000	1.88049

a.

Five-number Summary	
Minimum	0.145
First Quartile	0.2055
Median	0.23
Third Quartile	0.277983
Maximum	0.44

b.

Solution

a.   The mean and standard deviation of the sample bid prices are highlighted on the Excel printout, Figure 3.23a. To check for outliers we would use these values, $\bar{x} = .243$ and $s = .052$, to calculate z scores for all $n = 303$ bid prices in the data set. For the purposes of this example, we will focus on only the smallest and largest bid prices in the sample. The three highest prices, highlighted on the printout, Figure 3.23a, are .40533, .410, and .440 dollars. The z scores for these three prices, also shown on the printout, are 3.14, 3.23, and 3.81, respectively. All three prices have z scores that exceed 3 in absolute value; consequently, these bid prices are outliers. Further investigation of the bread contracts associated with these prices may reveal that they were not from the population of competitively bid prices, but were "fixed" during collusion.

b.   The 1st and 3rd quartiles, $Q_1 = .205$ and $Q_3 = .278$, are highlighted on the Excel printout, Figure 3.23b. Then,

$$\text{IQR} = .278 - .205 = .074$$

$$Q_1 - 1.5 \cdot \text{IQR} = .205 - (1.5)(.074) = .094$$

$$Q_3 + 1.5 \cdot \text{IQR} = .278 + (1.5)(.074) = .389$$

The three highest bid prices fall outside the interval (.094, .390). Thus, we recommend that state officials investigate the vendors who bid on these three contracts for collusion. ❑

PROBLEMS
3.57–3.68

Using the Tools

3.57   Consider a sample data set with $n = 50$ observations and the following five-number summary: 75   98   110   122   143.

    a.   Find Q_1.   **b.**   Find Q_3.   **c.**   Find IQR.

    d.   Find x_{smallest}.   **e.**   Find x_{largest}.

    f.   Construct a box-and-whisker plot for the data.

    g.   What type of skewness, if any, is present in the data?

3.58   Consider a sample data set with $n = 30$ observations and the following five-number summary: 1.7   3.5   5.2   7.8   8.5.

    a.   Find Q_1.   **b.**   Find Q_3.   **c.**   Find IQR.

    d.   Find x_{smallest}.   **e.**   Find x_{largest}.

    f.   Construct a box-and-whisker plot for the data.

    g.   What type of skewness, if any, is present in the data?

3.59   Construct a box-and-whisker plot for the 24 sample measurements reproduced in Problem 3.44, page 177.

3.60   Construct a box-and-whisker plot for the 28 sample measurements reproduced in Problem 3.45, page 177.

3.61   Compute the z score for the value of x in each of the following:

    a. $x = 75, \bar{x} = 70, s = 10$ **b.** $x = .28, \bar{x} = .13, s = .02$

    c. $x = 158, \bar{x} = 275, s = 30$ **d.** $x = 3, \bar{x} = 4, s = .5$

Applying the Concepts

3.62   *Medical Interface* (Oct. 1992) published a study on the cost effectiveness of a sample of 186 physicians in a managed-care HMO. One variable of interest was the monthly cost accrued by each physician divided by the number of patients treated—called the per-patient, per-month cost. The mean and standard deviation of the per-patient, per-month costs were $\bar{x}$ = \$102.96 and s = \$366.31, respectively. One of the physicians in the data set had a per-patient, per-month cost of \$4,725.10.

a. Calculate and interpret the z score for this physician's cost. Is the observation an outlier?

b. A careful examination of this physician's data revealed that this physician treated only a single patient during a single month of the year at a cost of $4,725.10. Based on this information, how would you classify this outlier?

3.63 Refer to the hotel no-show data of Example 3.11 and Problem 3.50 (page 178).

a. Construct a box-and-whisker plot for the data.

b. Identify the type of skewness present in the data.

c. Use the quartiles method to identify outliers in the data.

d. Use z scores to detect outliers in the data. Do your results agree with part **c**? Explain.

3.64 The accompanying table lists the lymphocyte (LYMPHO) and white blood cell (WBC) count results from hematology tests administered to a sample of 50 black (West Indian or African) workers.

HEMATO.XLS

Case Number	WBC	LYMPHO	Case Number	WBC	LYMPHO
1	4,100	14	26	4,300	9
2	5,000	15	27	5,200	16
3	4,500	19	28	3,900	18
4	4,600	23	29	6,000	17
5	5,100	17	30	4,700	23
6	4,900	20	31	7,900	43
7	4,300	21	32	3,400	17
8	4,400	16	33	6,000	23
9	4,100	27	34	7,700	31
10	8,400	34	35	3,700	11
11	5,600	26	36	5,200	25
12	5,100	28	37	6,000	30
13	4,700	24	38	8,100	32
14	5,600	26	39	4,900	17
15	4,000	23	40	6,000	22
16	3,400	9	41	4,600	20
17	5,400	18	42	5,500	20
18	6,900	28	43	6,200	20
19	4,600	17	44	4,900	26
20	4,200	14	45	7,200	40
21	5,200	8	46	5,800	22
22	4,700	25	47	8,400	61
23	8,600	37	48	3,100	12
24	5,500	20	49	4,000	20
25	4,200	15	50	6,900	35

Source: Royston, J. P. "Some techniques for assessing multivariate normality based on the Shapiro-Wilk W." *Applied Statistics,* Vol. 32, No. 2, 1983, pp. 121–133.

a. Construct a box-and-whisker plot for the white blood cell count data. Do you detect any skewness?

b. Do you detect any outliers in the white blood cell count data?

c. Repeat part **a** for lymphocyte cell counts.

d. Repeat part **b** for lymphocyte cell counts.

FTC.XLS

3.65 Refer to the Excel workbook FTC.XLS described in Appendix C. An Excel box-and-whisker plot and descriptive statistics for the nicotine contents of cigarettes data are shown here.

   a. Identify the type of skewness in the data.

   b. Do you detect any outliers? If so, identify them.

	A	B
1	*Nicotine*	
2		
3	Mean	0.8425
4	Standard Error	0.015452348
5	Median	0.9
6	Mode	0.8
7	Standard Deviation	0.345524999
8	Sample Variance	0.119387525
9	Kurtosis	0.162504345
10	Skewness	-0.170360155
11	Range	1.85
12	Minimum	0.05
13	Maximum	1.9
14	Sum	421.25
15	Count	500
16	Largest(1)	1.9
17	Smallest(1)	0.05

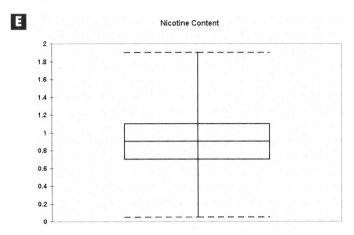

CRASH.XLS

3.66 Refer to the Excel workbook CRASH.XLS, described in Appendix C. An Excel box-and-whisker plot and descriptive statistics for the passenger head-injury ratings data are shown here.

   a. Identify the type of skewness in the data.

   b. Do you detect any outliers? If so, identify them.

	A	B
1	*Passhead*	
2		
3	Mean	549.4591837
4	Standard Error	21.55076192
5	Median	531.5
6	Mode	419
7	Standard Deviation	213.3416585
8	Sample Variance	45514.66327
9	Kurtosis	0.60266587
10	Skewness	0.568088223
11	Range	1077
12	Minimum	128
13	Maximum	1205
14	Sum	53847
15	Count	98
16	Largest(1)	1205
17	Smallest(1)	128

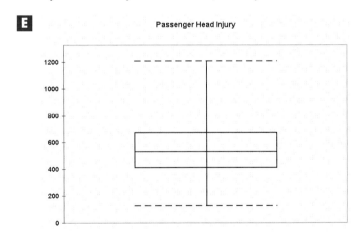

3.67 The data in the accompanying table represent sales, in hundreds of thousands of dollars per week, for a random sample of 24 fast-food outlets located in four cities.

FASTFOOD.XLS

City	Weekly Sales (hundred thousand $)
A	6.3, 6.6, 7.6, 3.0, 9.5, 5.9, 6.1, 5.0, 3.6
B	2.8, 6.7, 5.2
C	82.0, 5.0, 3.9, 5.4, 4.1, 3.1, 5.4
D	8.4, 9.5, 8.7, 10.6, 3.3

a. Generate a five-number summary for the data.

b. Using quartiles, do you detect any outliers in the data? If so, identify the city and the weekly sales measurement associated with an outlier.

c. Calculate $\bar{x}$ and s for the sample data. Use this information to compute the z score for the outlier(s) identified in part **b**.

d. A careful check of the sales records revealed that the weekly sales value for the first fast-food outlet in city C was actually 8.2, but was incorrectly coded as 82.0. Repeat parts **a–c** for the corrected sales data set.

BOND.XLS

3.68 Refer to the *Dalton Transactions* (Dec. 1997) study of the bond strength of a new copper-selenium metal compound, Problem 2.67 (page 118). The bond lengths (pm) of 36 copper-selenium metal clusters are reproduced in the table.

256.1	259.0	256.0	264.7	262.5	239.0
238.9	237.4	265.9	240.2	239.7	238.3
273.8	246.4	246.0	247.4	290.6	260.9
255.6	253.1	283.7	261.4	271.9	267.9
238.7	237.6	240.0	241.1	240.3	240.8
246.5	248.3	246.0	250.3	250.1	254.6

Source: Deveson, A., et al. "Syntheses and structures of four new copper (I)-selenium clusters: Size dependence of the cluster on the reaction conditions." *Dalton Transactions,* No. 23, Dec. 1997, p. 4492 (Table 1).

A	B
Length	
Mean	252.5194
Standard Error	2.264389
Median	249.2
Mode	246
Standard Deviation	13.58633
Sample Variance	184.5885
Kurtosis	0.621601
Skewness	1.008324
Range	53.2
Minimum	237.4
Maximum	290.6
Sum	9090.7
Count	36
Largest(9)	261.4
Smallest(9)	240.2

E Problem 3.68

Descriptive statistics for the 36 bond lengths are shown in the accompanying Excel printout.

a. Determine the measures of central tendency on the printout and interpret their values.

b. Determine the measures of variation on the printout and interpret their values.

c. The bond lengths for 7 copper-phosphorus metal clusters are listed below. Comment on whether or not these bond lengths are from the same population as the 36 copper-selenium bond lengths. Explain.

223.8   222.7   223.3   226.1   226.3   226.0   225.0

3.9 A MEASURE OF ASSOCIATION: CORRELATION

In Section 2.5 we used a scatterplot to investigate the relationship between two quantitative variables, x and y. Depending on the pattern of the points plotted, the two variables may be positively associated, negatively associated, or have little or no association. The strength of the association can be expressed numerically with the sample coefficient of correlation, denoted by the symbol r.

Definition 3.14

The **coefficient of correlation,** r, is computed as follows for a sample of n measurements x and y:

$$r = \frac{SS_{xy}}{\sqrt{SS_{xx}SS_{yy}}}$$

(3.9)

where

$$SSxy = \sum xy - \frac{(\sum x)(\sum y)}{n}$$

$$SSxx = \sum x^2 - \frac{(\sum x)^2}{n}$$

$$SSyy = \sum y^2 - \frac{(\sum y)^2}{n}$$

It is a measure of the strength of the *linear* relationship between two quantitative variables x and y. Values of r near $+1$ imply a strong positive linear association between the variables; values of r near -1 imply a strong negative linear association; and values of r near 0 imply little or no linear association.

We illustrate how to calculate r in the following example.

Example 3.20   Computing r

Suppose a psychologist wants to investigate the relationship between the creativity score y and the flexibility score x of a mentally retarded child. The creativity and flexibility scores (both measured on a scale of 1 to 20) for each in a random sample of five mentally retarded children are listed in Table 3.7. Calculate the coefficient of correlation r between flexibility score and creativity score.

Solution

Step 1   To calculate r, we need to find the following sums: $\sum x$, $\sum y$ $\sum x^2$, $\sum y^2$, and $\sum xy$. As an aid in finding these quantities, construct a *sum of squares table* of the type shown in Table 3.8. Notice that the quantities needed to compute r appear in the bottom row of the table.

Table 3.7 Creativity–Flexibility Scores for Example 3.20

Child	Flexibility Score, x	Creativity Score, y
1	2	2
2	3	8
3	4	7
4	5	10
5	6	8

Table 3.8 Sum of Squares for Data of Table 3.7

	x	y	x^2	y^2	xy
	2	2	4	4	4
	3	8	9	64	24
	4	7	16	49	28
	5	10	25	100	50
	6	8	36	64	48
TOTALS	$\sum x = 20$	$\sum y = 35$	$\sum x^2 = 90$	$\sum y^2 = 281$	$\sum xy = 154$

Step 2   Now we compute the quantities SS*xy*, SS*xx*, and SS*yy*, as shown below.

$$SSxy = \sum xy - \frac{(\sum x)(\sum y)}{n} = 154 - \frac{(20)(35)}{5} = 154 - 140 = 14$$

$$SSxx = \sum x^2 - \frac{(\sum x)^2}{n} = 90 - \frac{(20)^2}{5} = 90 - 80 = 10$$

$$SSyy = \sum y^2 - \frac{(\sum y)^2}{n} = 281 - \frac{(35)^2}{5} = 281 - 245 = 36$$

Step 3   Finally, compute the *coefficient of correlation r* as follows:

$$r = \frac{SSxy}{\sqrt{SSxx\,SSyy}} = \frac{14}{\sqrt{(10)(36)}} = .74 \qquad \square$$

The formal name given to *r* is the sample *Pearson product moment coefficient of correlation,* named for Karl Pearson (1857–1936).

Caution

High correlation does not imply causality. If a large positive or negative value of the sample correlation coefficient is observed, it is incorrect to conclude that a change in *x* causes a change in *y*. The only valid conclusion is that a linear trend *may* exist between *x* and *y*.

The coefficient of correlation *r* is *scaleless* (it is not measured in dollars, pounds, etc.); values of *r* will always be between −1 and +1, regardless of the units of measurement of the variables *x* and *y*. More importantly, *r* is a measure of the strength of the *linear* (i.e., straight-line) *association* between *x* and *y* in the sample. We can gain insight into this interpretation by examining the scatterplots presented in Figure 3.24.

Consider first the scatterplot in Figure 3.24b. The correlation coefficient *r* for this set of points is near 0, implying little or no linear relationship between *x* and *y*. In contrast, positive values of *r* imply a positive linear association between *y* and *x*. Consequently, the value of *r* for the data illustrated in Figure 3.24a is positive. Similarly, a negative value of *r* implies a negative linear association between *y* and *x* (see Figure 3.24c). A perfect linear relationship exists when all the (*x, y*) points fall exactly along a straight line. A value of *r* = +1 implies a perfect positive linear relationship between *y* and *x* (see Figure 3.24d), and a value of *r* = −1 implies a perfect negative linear relationship between *y* and *x* (see Figure 3.24e).

Example 3.21

Interpreting r

Interpret the value of *r* calculated in Example 3.20.

Solution

The value *r* = .74 implies that creativity score and flexibility score are positively correlated—at least for this sample of five mentally retarded children. The implication is that a positive linear association exists between these variables. We must be careful, however, not to jump to any unwarranted conclusions. For instance, the psychologist may be tempted to conclude that a high flexibility score will *always* result in a higher creativity score. The implication of such a conclusion is that there is a *causal* relationship between the two variables. However, *high correlation does not imply causality.* Many other factors, such as severity of illness, parental neglect, and IQ, may contribute to the change in creativity score.   ❏

Figure 3.24 Values of *r* and Their Implications

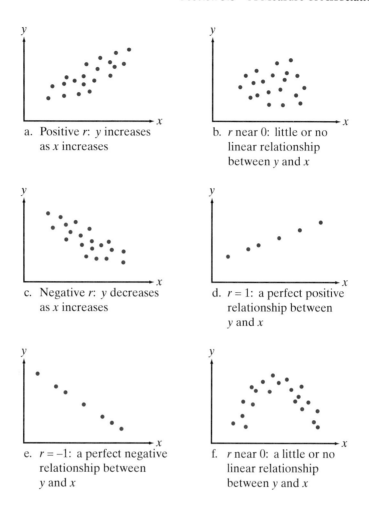

a. Positive *r*: *y* increases as *x* increases

b. *r* near 0: little or no linear relationship between *y* and *x*

c. Negative *r*: *y* decreases as *x* increases

d. *r* = 1: a perfect positive relationship between *y* and *x*

e. *r* = −1: a perfect negative relationship between *y* and *x*

f. *r* near 0: a little or no linear relationship between *y* and *x*

Self-Test 3.8

Find and interpret the coefficient of correlation *r* for the values of *x* and *y* listed in the table.

x	−1	2	1	0	3
y	8	4	6	5	1

Example 3.22   Interpreting Correlations

Refer to the NCAP test data in the Excel workbook CRASH.XLS. In Example 2.11 we used scatterplots to examine the relationship between three pairs of quantitative variables: (a) car weight (WEIGHT) vs. driver's head-injury rating (DRIVHEAD), (b) DRIVHEAD vs. driver's rate of chest deceleration (DRIVCHST), and (c) DRIVHEAD vs. overall driver injury rating (DRIVSTAR). Find the correlation coefficient *r* relating these three pairs of variables. Interpret the results.

E Figure 3.25
Excel Printout of
Correlations for NCAP
Data of Example 3.22

	A	B
1	Coefficients of Correlation	
2	for NCAP Crash Data	
3		
4	Drivhead to Weight	0.04563
5	Drivhead to Drivchst	0.53514
6	Drivhead to Drivstar	-0.74174

Solution

We used Excel to calculate the correlation coefficients. They are displayed in Figure 3.25.

The correlation between car weight and driver's head-injury rating, $r = .046$, is listed first on the printout. Since this value of r is very close to 0, there is little or no linear association between car weight and driver's head-injury rating in this sample.

The correlation between driver head-injury rating and driver chest deceleration, $r = .535$, implies that a moderate positive linear association exists between these variables. The correlation between driver head-injury rating and overall injury rating, $r = -.742$, implies that these two variables are fairly strongly negatively associated. If you refer back to the solution to Example 2.10 (page 98), you will see that these conclusions agree with those derived from the scatterplots. ❏

PROBLEMS 3.69–382

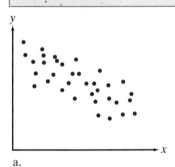

a.

b.

c.

Problem 3.70

Using the Tools

3.69 Calculate the coefficient of correlation r for a sample data set with the following characteristics:

   a. $SSxy = 10, SSxx = 100, SSyy = 25$

   b. $SSxy = -35, SSxx = 100, SSyy = 25$

   c. $SSxy = 125, SSxx = 400, SSyy = 100$

   d. $SSxy = -200, SSxx = 400, SSyy = 100$

3.70 For each of the three scatterplots at left, decide whether the coefficient of correlation r is positive or negative.

3.71 Calculate the coefficient of correlation r for each of the following sample data sets:

x	y		x	y		x	y
−1	−1		−1	4		−1	11
0	1		0	1		0	8
1	2		1	5		1	7
2	4		2	2		2	3
3	5		3	1		3	1
4	8		4	7		4	0
5	9		5	3		5	−2

  a.   b.   c.

3.72 Calculate the coefficient of correlation r for the following sample values of x and y:

y	7	5	8	3	6	10	12	4	9	15	18
x	21	15	24	9	18	30	36	12	27	45	54

3.73 Calculate the coefficient of correlation *r* for the following sample values of *x* and *y*:

y	100	200	250	400	350	300	450	500	100	250
x	6	20	10	4	12	30	35	15	18	5

Applying the Concepts

3.74 Do you believe the grade point average of a college student is correlated with the student's intelligence quotient (IQ)? If so, will the correlation be positive or negative? Explain.

3.75 Research by law enforcement agencies has shown that the crime rate is correlated with the U.S. population. Would you expect the correlation to be positive or negative? Explain.

3.76 The data on retail price and energy cost for 9 large side-by-side refrigerators, Problem 2.45 (page 101), are reproduced below. Find the coefficient of correlation *r* and interpret the result.

REFRIDGE.XLS

Brand	Price ($)	Energy Cost Per Year ($)
KitchenAid Superba KSRS25QF	1,600	73
Kenmore (Sears) 5757	1,200	73
Whirlpool ED25DQXD	1,550	78
Amana SRD25S3	1,350	85
Kenmore (Sears) 5647	1,700	93
GE Profile TPX24PRY	1,700	93
Frigidaire Gallery FRS26ZGE	1,500	95
Maytag RSW2400EA	1,400	96
GE TFX25ZRY	1,200	94

Data Source: Adapted from "The Kings of Cool," *Consumer Reports,* Jan. 1998, p. 52.

3.77 Refer to the *American Statistician* (Feb. 1991) study of rock drilling, Problem 2.46 (page 101). The data on depth at which drilling begins and length of time to drill 5 feet are reproduced on page 196. Find the coefficient of correlation *r* and interpret the result.

3.78 A math and computer science researcher at Duquesne University investigated the relationship between a student's confidence in math and interest in computers (*Educational Technology,* May–June 1995). A sample of 1,730 high school students—902 boys and 828 girls—from public schools in Pittsburgh, Pennsylvania, participated in the study. Using 5-point Likert scales, where 1 = "strongly disagree" and 5 = "strongly agree," the researcher measured a student's confidence in mathematics (*x*) and interest in computers (*y*).

a. For boys, math confidence (*x*) and computer interest (*y*) were correlated at *r* = .14. Fully interpret this result.

b. For girls, math confidence (*x*) and computer interest (*y*) were correlated at *r* = .33. Fully interpret this result.

3.79 An investigation sought to determine whether certain collective behaviors, affective reactions, or performance outcomes are associated with the maturity level of small groups (*Small Group Behavior,* May 1988). Fifty-eight undergraduate students enrolled in MIS or communications

DRILL.XLS

Depth at which drilling begins (feet)	Time to drill 5 feet (minutes)
0	4.90
25	7.41
50	6.19
75	5.57
100	5.17
125	6.89
150	7.05
175	7.11
200	6.19
225	8.28
250	4.84
275	8.29
300	8.91
325	8.54
350	11.79
375	12.12
395	11.02

Source: Penner, R., and Watts, D. G. "Mining information." *The American Statistician,* Vol. 45, No. 1, Feb. 1991, p. 6 (Table 1).

Problem 3.77

courses at a medium-size university participated in the experiment. A 10-item questionnaire was used to measure the maturity level, y, of the students on a scale of 0–100, where more mature students received higher scores. One of several other variables measured was the number, x, of meetings held with their groups outside of regular class sessions. The correlation coefficient relating y to x was found to be $r = .46$. Fully interpret this result.

3.80   Refer to the *American Journal of Psychiatry* (July 1995) study relating verbal memory retention (y) and right hippocampus volume (x) of 21 Vietnam veterans, Problem 2.48 (page 102). The data for the study are provided in the table. Find and interpret the coefficient of correlation r for the two variables.

BRAIN.XLS

Veteran	Verbal Memory Retention, y	Right Hippocampus Volume, x	Veteran	Verbal Memory Retention, y	Right Hippocampus Volume, x
1	26	960	12	60	1,210
2	22	1,090	13	62	1,220
3	30	1,180	14	66	1,215
4	65	1,000	15	76	1,220
5	70	1,010	16	65	1,300
6	60	1,070	17	66	1,350
7	60	1,080	18	84	1,350
8	58	1,040	19	90	1,370
9	65	1,040	20	102	1,400
10	83	1,030	21	102	1,460
11	69	1,045			

3.81   The fertility rate of a country is defined as the number of children a woman citizen bears, on average, in her lifetime. *Scientific American* (Dec. 1993) reported on the declining fertility rate in developing countries. The researchers found that family planning can have a great effect on fertility rate. The accompanying table gives the fertility rate, *y*, and contraceptive prevalence, *x* (measured as the percentage of married women who use contraception), for each of 27 developing countries. Find the coefficient of correlation between fertility rate and contraceptive prevalence. Interpret the result.

FERTIL.XLS

Country	Contraceptive Prevalence, *x*	Fertility Rate, *y*	Country	Contraceptive Prevalence, *x*	Fertility Rate, *y*
Mauritius	76	2.2	Egypt	40	4.5
Thailand	69	2.3	Bangladesh	40	5.5
Colombia	66	2.9	Botswana	35	4.8
Costa Rica	71	3.5	Jordan	35	5.5
Sri Lanka	63	2.7	Kenya	28	6.5
Turkey	62	3.4	Guatemala	24	5.5
Peru	60	3.5	Cameroon	16	5.8
Mexico	55	4.0	Ghana	14	6.0
Jamaica	55	2.9	Pakistan	13	5.0
Indonesia	50	3.1	Senegal	13	6.5
Tunisia	51	4.3	Sudan	10	4.8
El Salvador	48	4.5	Yemen	9	7.0
Morocco	42	4.0	Nigeria	7	5.7
Zimbabwe	46	5.4			

Source: Robey, B., et al. "The fertility decline in developing countries." *Scientific American,* Dec. 1993, p. 62. [*Note:* The data values are estimated from a scatterplot.]

3.82 The data on temperature at flight time and O-ring damage index for 23 launches of the space shuttle *Challenger,* first presented in Problem 2.50 (page 103), are reproduced below. Find the coefficient of correlation between temperature and damage index. Interpret the result.

ORING.XLS

Flight Number	Temperature (°F)	O-Ring Damage Index	Flight Number	Temperature (°F)	O-Ring Damage Index
1	66	0	51-A	67	0
2	70	4	51-B	75	0
3	69	0	51-C	53	11
5	68	0	51-D	67	0
6	67	0	51-F	81	0
7	72	0	51-G	70	0
8	73	0	51-I	67	0
9	70	0	51-J	79	0
41-B	57	4	61-A	75	4
41-C	63	2	61-B	76	0
41-D	70	4	61-C	58	4
41-G	78	0			

Note: Data for flight number 4 are omitted due to an unknown O-ring condition.

Primary Sources: Report of the Presidential Commission on the Space Shuttle Challenger Accident, Washington, D.C., 1986, Vol. II (pp. H1–H3) and Volume IV (p. 664). *Post-Challenger Evaluation of Space Shuttle Risk Assessment and Management,* Washington, D.C., 1988, pp. 135–136.

Secondary Source: Tufte, E. R., *Visual and Statistical Thinking: Displays of Evidence for Making Decisions.* Graphics Press, 1997.

3.10 NUMERICAL DESCRIPTIVE MEASURES FOR POPULATIONS

When you analyze a sample, you are doing so for a reason. Most likely, you want to use the information in the sample to infer the nature of some larger set of data—the population. Numerical descriptive measures that characterize the distribution for a population are called *parameters*. Since we will often use numerical descriptive measures of a sample to estimate the corresponding unknown descriptive measures of the population, we need to make a distinction between the numerical descriptive measure symbols for the population and for the sample.

Definition 3.15

Numerical descriptive measures (e.g., mean, median, standard deviation) of a population are called **parameters**.

In our previous discussion, we used the symbols $\bar{x}$ and s to denote the mean and standard deviation, respectively, of a sample of n observations. Similarly, we will use the symbol μ (mu) to denote the mean of a population and the symbol σ (sigma) to denote the standard deviation of a population. As you will subsequently see, we will use the sample mean $\bar{x}$ to estimate the population mean μ, and the sample standard deviation s to estimate the population standard deviation σ. In doing so, we will be using the sample to help us infer the nature of the population relative frequency distribution.

Sample and Population Numerical Descriptive Measures

Sample	Population
mean: $\bar{x}$	mean: μ
variance: s^2	variance: σ^2
standard deviation: s	standard deviation: σ
z score: $z = \dfrac{x - \bar{x}}{s}$	z score: $z = \dfrac{x - \mu}{\sigma}$
correlation coefficient: r	correlation coefficient: ρ

REAL-WORLD REVISITED
Bid-Collusion in the Highway Contracting Industry

Many products and services are purchased by governments, cities, states, and businesses on the basis of competitive bids, and frequently contracts are awarded to the lowest bidders. As we learned in Example 3.19, such a process is prone to price-fixing or bid-collusion by the bidders. Recall that price-fixing involves setting the bid price above the fair, or competitive, price in order to increase profit margin.

Numerous methods exist for detecting the possibility of collusive practices among bidders. According to Rothrock and McClave (1979), these procedures involve the detection of significant departures from normal market conditions such as (1) systematic rotation of the winning bid, (2) stable market shares over time, (3) geographic market divisions, (4) lack of relationship between delivery costs and bid levels, (5) high degree of uniformity and stability in bid levels over time, and (6) presence of a baseline point pricing scheme.*

*Rothrock, T. P., and McClave, J. T. "An analysis of bidding competition in the Florida school bidding competition using a statistical model." Paper presented at the TIMS/ORSA Joint National Meeting, Chicago, 1979.

In this application, we examine a data set collected during the early 1980s involving road construction contracts in the state of Florida. During this time period, the Office of the Florida Attorney General suspected numerous contractors of practicing bid-collusion. The investigations led to an admission of guilt by several of the contractors. Although these contractors were heavily fined, they avoided harsher punishment by identifying which road construction contracts were competitively bid and which involved fixed bids. By comparing the bid prices (and other important bid variables) of the fixed contracts to the competitive contracts, the Florida Attorney General was able to establish invaluable benchmarks for detecting bid-rigging in the future. (In fact, the benchmarks led to a virtual elimination of bid-rigging in road construction in the state.)

Table 3.9 shows two of the many variables measured for each contract in the data set. STATUS is a qualitative variable representing the bid status (fixed or competitive) of the contract. LBERATIO is a quantitative variable representing the ratio of the winning (low) bid to the Department of Transportation engineer's estimate of the "fair" bid price. Theoretically, the ratio will be near 1 for competitive bids. The larger the ratio, the more evidence that an unusually high bid (and possibly price-fixing) has occurred.

BIDRIG.XLS

Table 3.9 Low-Bid–Estimate Ratios for Competitive and Fixed Contracts

Competitive Contracts					
0.84706	0.81294	0.84947	0.82045	0.74058	0.95659
0.79604	0.94029	0.83473	1.01686	0.86738	0.83417
0.99504	0.96421	0.56458	1.20942	0.98283	0.85788
0.75519	0.55926	0.82667	0.79416	0.80052	0.53256
0.80993	0.61726	0.87136	0.70124	0.84679	0.58660
0.85818	1.02779	1.02066	0.82872	0.56154	0.80604
0.83678	0.88429	0.67427	0.85418	0.80769	0.82356
0.90790	0.84337	0.97863	0.85910	0.95524	0.99704
0.93263	0.98418	1.01218	0.91837	0.75132	0.56958
0.76891	0.98350	0.90747	0.86848	0.87175	1.06275
1.03847	0.81453	0.97092	0.84639	0.91569	0.80036
0.74178	1.15942	0.92574	1.05129	0.93190	0.94623
1.02971	1.00238	0.99094	1.08052	0.82088	0.86044
0.91729	0.99481	0.90621	1.04081	0.96586	0.93358
0.75446	0.92454	0.90958	0.81181	1.04067	0.85952
1.02410	0.91999	0.66151	0.77619	0.91446	0.99429
0.75085	0.94714	0.79414	1.23987	0.91706	1.17211
0.94463	0.95872	0.86617	0.70829	0.63516	0.88898
0.94505	0.86283	0.89214	1.03969	0.81636	0.76920
1.28562	0.89310	0.78920	0.86720	0.72732	1.00800
0.76011	0.74089	1.12640	0.84523	0.90192	0.76520
1.06395	0.96832	1.05070	0.86668	0.81801	0.95358

continued

Table 3.9 Continued

Competitive Contracts					
0.79852	1.17036	1.05659	0.96512	1.04657	0.78969
0.91536	0.87965	0.99660	0.93633	0.79439	0.98118
0.98819	0.99483	1.10669	1.03936	1.08561	1.11340
0.92861	1.07797	1.03524	1.12084	1.00052	1.02635
1.35009	0.96836	1.10043	0.91859	1.02384	0.86765
0.91293	1.30215	0.92623	0.88246	0.96999	0.97556
0.87291	0.91385	0.87678	0.91367	0.89729	1.05454
0.87506	1.02810	0.79883	0.98221	0.99722	0.85552
0.80171	0.83699	0.79261	0.77651	0.66907	0.82356
1.01985	1.04099	0.93329	0.99935	0.98071	0.94421
0.86350	0.80633				

Fixed Contracts					
1.05701	1.18705	0.99505	1.56376	1.19798	1.34676
1.11814	1.07163	0.84640	0.93359	1.14827	1.09177
0.86693	1.52402	1.10792	1.23126	1.23735	0.88570
1.21388	0.99578	1.20061	1.16075	0.83974	1.03118
1.34423	1.17919	1.25092	1.02069	1.05933	0.96487
1.01136	1.16030	0.89721	0.94805	1.33058	1.33976
1.19888	1.17608	1.17053	1.10921	1.29868	1.26369
1.10506	1.12001	0.97225	1.03805	1.34698	1.18351
1.33447	1.04488	1.00582	0.87470	1.21535	0.98325
1.05964	1.41907	0.98426	1.07672	1.03658	1.10761
1.01960	1.13592	1.29874	1.10312	1.15079	1.09186
1.17556	1.28360	1.07151	1.02573	0.99354	1.22666
1.08694	0.99070	1.08237	0.81352	1.07066	1.09208
0.96967	1.05283	1.21457	0.91678	1.12155	1.04438
1.03024					

The data set of Table 3.9 contains 194 competitively bid and 85 fixed contracts—a total of 279 contracts. These data are stored in the Excel workbook BIDRIG.XLS. Use the methods of Chapters 2 and 3 to analyze the road construction contract data. In particular, comment on the belief that fixed bids result in higher low-bid–estimate ratios than competitively bid contracts. Support your conclusions with graphs, whenever possible.

Key Terms

Arithmetic mean   148
Box-and-whisker plot   181
Coefficient of correlation   190
Empirical Rule   164
Five-number summary   180
Interquartile range   181
Mean   148
Measures of central tendency   145
Measures of relative standing   145

Measures of variation   145
Median   151
Modal class   151
Mode   151
Numerical descriptive
   measures   145
Outlier   184
Parameters   198
Percentile   173

Quartiles   174
Range   157
Sample variance   157
Skewness   145
Standard deviation   159
Tchebysheff's Theorem   165
Two-number summary   180
Variance   158
z score   176

Key Formulas and Rules

Sample mean:	$\bar{x} = \dfrac{\Sigma x}{n}$	**(3.1)**, 148
Sample variance:	$s^2 = \dfrac{\Sigma(x - \bar{x})^2}{n-1} = \dfrac{\Sigma x^2 - \dfrac{(\Sigma x)^2}{n}}{n-1}$	**(3.3)** and **(3.4)**, 158
Sample standard deviation:	$s = \sqrt{s^2}$	**(3.5)**, 159
Rank of pth percentile:	$\dfrac{p(n-1)}{100}$	**(3.6)**, 176
Sample z score:	$z = \dfrac{x - \bar{x}}{s}$	**(3.7)**, 176
Interquartile range:	$\text{IQR} = Q_3 - Q_1$	**(3.8)**, 181
Sample coefficient of correlation:	$r = \dfrac{\text{SS}xy}{\sqrt{\text{SS}xx\,\text{SS}yy}}$	**(3.9)**, 190

(see Definition 3.14, page 190, for SSxy, SSxx, and SSyy formulas)

Population z score:	$z = \dfrac{x - \mu}{\sigma}$	198

Key Symbols

Symbol	Definition
$\bar{x}$	Sample mean
s^2	Sample variance
s	Sample standard deviation
r	Sample coefficient of correlation
μ	Population mean
σ^2	Population variance
σ	Population standard deviation
ρ	Population coefficient of correlation
M	Median
Q_1	1st quartile
Q_3	3rd quartile

Supplementary Problems 3.83–3.96

Applying the Concepts

3.83   The table at the top of page 202 describes the sale prices for properties sold in six residential neighborhoods located in Tampa, Florida. Use this information to comment on the skewness of the relative frequency distributions for the six data sets.

3.84   According to one study, "The majority of people who die from fire and smoke in compartmented fire-resistive buildings—the type used for hotels, motels, apartments, and other health care facilities—die in the attempt to evacuate" (*Risk Management,* Feb. 1986). The accompanying

Neighborhood	Mean Sale Price	Median Sale Price
Avila	$297,004	$192,000
Carrollwood Village	137,492	125,000
Northdale	94,391	91,000
Tampa Palms	198,428	170,000
Town & Country	70,477	69,000
Ybor City	25,764	23,400

Source: Hillsborough County (Florida) property appraiser's office.

Problem 3.83

data represent the numbers of victims who attempted to evacuate for a sample of 14 fires at compartmented fire-resistive buildings reported in the study.

FIRE.XLS

Fire	Number of Victims
Las Vegas Hilton (Las Vegas)	5
Inn on the Park (Toronto)	5
Westchase Hilton (Houston)	8
Holiday Inn (Cambridge, Ohio)	10
Conrad Hilton (Chicago)	4
Providence College (Providence)	8
Baptist Towers (Atlanta)	7
Howard Johnson (New Orleans)	5
Cornell University (Ithaca, New York)	9
Westport Central Apartments (Kansas City, Missouri)	4
Orrington Hotel (Evanston, Illinois)	0
Hartford Hospital (Hartford, Connecticut)	16
Milford Plaza (New York)	0
MGM Grand (Las Vegas)	36

Source: Macdonald, J. N. "Is evacuation a fatal flaw in fire fighting philosophy?" *Risk Management,* Vol. 33, No. 2, Feb. 1986, p. 37.

a. Construct a stem-and-leaf display for the data.

b. Compute the mean, median, and mode for the data set. Which measure of central tendency appears to best describe the center of the distribution of data?

c. The MGM Grand fire in Las Vegas was treated separately in the *Risk Management* analysis because of the size of the high-rise hotel and other unique factors. Do the data support treating the MGM Grand fire deaths differently than the other measurements in the sample? Explain.

3.85 In a follow-up study of alcoholics treated at Maudsley Hospital in London, the data in the table on page 203 were collected at the time of admission for each of seven subjects suffering from alcohol addiction.

a. Find the mean and median of the age of the seven patients.

b. Find the mean and median of the number of years of excessive drinking reported by the seven patients.

c. Which measure of central tendency, the mean or the median, better describes the age distribution of the seven subjects?

d. Which measure of central tendency, the mean or the median, better describes the distribution of number of years of excessive drinking for the seven subjects?

ALCOHOL.XLS

Case Number	Age	Years of Excessive Drinking
1	41	10
2	42	8
3	47	13
4	26	3
5	28	2
6	40	5
7	35	6

Data Source: Edwards, G. "A later follow-up of a classic case series: D. L. Davies's 1962 report and its significance for the present." *Journal of Studies on Alcohol,* Vol. 46, No. 3, 1985, pp. 181–190, Alcohol Research Documentation, Inc., Rutgers Center of Alcohol Studies, New Brunswick, N.J. 08903.

Problem 3.85

3.86   The $n = 50$ starting salaries of Table 2.5 are reproduced here.

SAL50.XLS

$30,000	$20,600	$32,400	$14,200	$15,700
25,200	26,100	30,300	20,400	38,900
22,800	20,400	17,400	30,900	29,400
15,300	24,400	32,700	23,200	23,100
34,600	31,400	22,100	26,700	33,100
26,600	36,700	22,800	21,400	16,800
20,700	43,900	25,200	26,700	40,500
11,000	36,600	28,200	26,300	20,100
21,500	20,500	25,600	24,900	20,000
20,400	37,500	28,700	13,200	28,300

a.   Compute the mean and median, and locate these values on the relative frequency histogram for the data set (see Figure 2.6, page 77). Notice that they fall near the center of the distribution.

b.   Find the modal class (the class with the greatest relative frequency) for the relative frequency histogram shown in Figure 2.6. Compare your answer with the mean and median obtained in part **a**.

c.   Suppose that a distribution of data is skewed to the right. Would you expect the mean of this data set to be larger or smaller than the median? Does your answer agree with the results of part **a**?

3.87   The sample data on fat and cholesterol levels for popular protein foods, first presented in Problem 2.58 (page 113), is reproduced on page 204. For each food type, five quantitative variables are measured: amount of calories, amount of protein, percentage of calories from fat, percentage of calories from saturated fat, and amount of cholesterol.

a.   For each of these five variables, find the following descriptive statistics: mean, median, mode, range, variance, standard deviation, lower quartile, upper quartile, interquartile range.

b.   Construct a box-and-whisker plot for each variable.

c.   Identify the type of skewness in the distribution of each variable.

d.   For each variable, compute the intervals $\bar{x} \pm s$, $\bar{x} \pm 2s$, and $\bar{x} \pm 3s$. Determine the percentage of measurements that fall within each interval. Compare your results to the Empirical Rule.

e.   Identify the outliers for each variable.

PROTEIN.XLS

Food	Calories	Protein (g)	Fat	Saturated Fat	Cholesterol (mg)
			PERCENT CALORIES FROM		
Beef, ground, extra lean	250	25	58	23	82
Beef, ground, regular	287	23	66	26	87
Beef, round	184	28	24	12	82
Brisket	263	28	54	21	91
Flank steak	244	28	51	22	71
Lamb leg roast	191	28	38	16	89
Lamb loin chop, broiled	215	30	42	17	94
Liver, fried	217	27	36	12	482
Pork loin roast	240	27	52	18	90
Sirloin	208	30	37	15	89
Spareribs	397	29	67	27	121
Veal cutlet, fried	183	33	42	20	127
Veal rib roast	175	26	37	15	131
Chicken, with skin, roasted	239	27	51	14	88
Chicken, no skin, roast	190	29	37	10	89
Turkey, light meat, no skin	157	30	18	6	69
Clams	98	16	6	0	39
Cod	98	22	8	1	74
Flounder	99	21	12	2	54
Mackerel	199	27	77	20	100
Ocean perch	110	23	13	3	53
Salmon	182	27	24	5	93
Scallops	112	23	8	1	56
Shrimp	116	24	15	2	156
Tuna	181	32	41	10	48

Source: United States Department of Agriculture.

Problem 3.87

f.   For each pair of variables, find and interpret the coefficient of correlation, r.

3.88   Passive exposure to environmental tobacco has been associated with growth suppression and an increased frequency of respiratory tract infections in normal children. Is this association more pronounced in children with cystic fibrosis? To answer this question, 43 children (18 girls and 25 boys) attending a 2-week summer camp for cystic fibrosis patients were studied (*New England Journal of Medicine,* Sept. 20, 1990). Among several variables measured were the child's weight percentile (y) and the number of cigarettes smoked per day in the child's home (x).

a.   For the 18 girls, the coefficient of correlation between y and x was reported as $r = -.50$. Interpret this result.

b.   For the 25 boys, the coefficient of correlation between y and x was reported as $r = -.12$. Interpret this result.

DEFAULT.XLS

3.89   Refer to the data on student-loan default rates for 66 Florida colleges, Problem 2.61 (page 115). An Excel printout giving descriptive statistics for the data set is displayed on page 205.

a.   What is the mean default rate?

Student Loan Default Rate

Mean	14.68182
Standard Error	1.740663
Median	9.6
Mode	11.8
Standard Deviation	14.14121
Sample Variance	199.9738
Kurtosis	5.42669
Skewness	2.204234
Range	74.7
Minimum	1.5
Maximum	76.2
Sum	969
Count	66
Largest(17)	16.9
Smallest(17)	5.6

 Problem 3.89

b. What are the variance and standard deviation of the default rates?

c. What proportion of measurements would you expect to find within two standard deviations of the mean?

d. Determine the proportion of measurements (default rates) that actually fall within the interval of part **c.** Compare this result with your answer to part **c.**

e. Suppose the college with the highest default rate (Florida College of Business, 76.2%) was omitted from the analysis. Would you expect the mean to increase or decrease? Would you expect the standard deviation to increase or decrease?

f. Calculate the mean and standard deviation for the data set with Florida College of Business excluded. Compare these results with your answer to part **e.**

g. Answer parts **c** and **d** using the recalculated mean and standard deviation. Compare these results to the results of parts **c** and **d.** What does this tell you about the effect a single observation can have on the analysis?

3.90 Nevada continues to be the leading gold producer in the U.S., and according to the U.S. Bureau of Mines, it ranks among the top four regional producers worldwide (trailing South Africa, Russia, and Australia). The data in the table represent the production (in thousands of ounces) for the top 30 gold mines in the state.

GOLD.XLS

1,467.8	228.0	111.3	76.0	55.1	40.0
318.0	222.6	89.1	72.5	54.1	32.4
296.9	214.6	82.0	66.0	50.0	30.9
256.0	207.3	81.5	60.4	50.0	30.3
254.5	120.7	78.8	60.0	44.5	30.0

Source: Engineering & Mining Journal, June 1990, p. 38.

a. Summarize the data with a graphical technique.

b. Calculate the mean, median, and standard deviation of the data.

c. What proportion of Nevada mines have production values that lie within two standard deviations of the mean?

d. Note the extremely large production value, 1,467.8, for the first mine listed in the table. Recalculate the mean, median, and standard deviation with the production measurement for this mine deleted.

e. Explain how the three numerical descriptive measures (mean, median, and standard deviation) are affected by the deletion of the measurement 1,467.8.

3.91 To investigate the phenomenon of ants protecting plants, a naturalist sampled 50 flower heads of a certain sunflower plant that attracts ants. (Each flower head was on a different sunflower plant.) The flower heads were divided into two groups of 25 each. Ants were prevented from reaching the flower heads of one group (this was accomplished by painting ant repellent around the flower stalks of 25 plants). After a specified period of time, the number of insect seed predators on each flower head was counted. The results are summarized in the table on page 206.

	Plants with Ants	Plants without Ants
Mean number of predators per flower head	2.9	7.6
Standard deviation	2.4	4.4

    a. For each flower head group, compute the interval $\bar{x} \pm 2s$.

    b. Estimate the percentage of measurements that fall within each of the intervals, part **a**.

    c. Does it appear that ants protect sunflower plants from insect seed predators?

3.92 A study was conducted to determine whether a student's final grade in an introductory sociology course was linearly related to his or her performance on the verbal ability test administered before entrance to college. The verbal test scores and final grades for a random sample of 10 students are shown in the following table. Calculate and interpret a measure of association between verbal ability test score and final sociology grade.

SOCIOL.XLS

Student	Verbal Ability Test Score	Final Sociology Grade
1	39	65
2	43	78
3	21	52
4	64	82
5	57	92
6	47	89
7	28	73
8	75	98
9	34	56
10	52	75

3.93 Industrial engineers periodically conduct "work measurement" analyses to determine the time used to produce a single unit of output. At a large processing plant, the total number of man-hours required per day to perform a certain task was recorded for 50 days. This information will be used in a work measurement analysis. The total number of man-hours required for each of the 50 days is listed here, accompanied by an Excel printout on page 207 summarizing and describing the data.

MANHRS.XLS

128	119	95	97	124	113	109	124	132	97
146	128	103	135	114	124	131	133	131	88
100	112	111	150	117	128	142	98	108	120
138	133	136	120	112	109	100	111	131	113
118	116	98	112	138	122	97	116	92	122

    a. Find the mean, median, and mode of the data set and interpret their values.

    b. Find the range, variance, and standard deviation of the data set and interpret their values.

    c. Construct the intervals $\bar{x} \pm s, \bar{x} \pm 2s$, and $\bar{x} \pm 3s$. Count the number of observations that fall within each interval and find the corresponding proportions. Compare the results to the Empirical Rule. Do you detect any outliers?

	A	B
1	Hours	
2		
3	Mean	117.82
4	Standard Error	2.122896
5	Median	117.5
6	Mode	128
7	Standard Deviation	15.01114
8	Sample Variance	225.3343
9	Kurtosis	-0.69123
10	Skewness	0.00906
11	Range	62
12	Minimum	88
13	Maximum	150
14	Sum	5891
15	Count	50
16	Largest(1)	150
17	Smallest(1)	88

Problem 3.93

3.94 Neuropsychologists often measure the impact of brain damage on a patient's verbal skills by means of a verbal fluency test. One such test requires subjects to produce as many words as they can that begin with a particular letter. The production figures for one such study are summarized in the accompanying table.

	Mean Productivity	Standard Deviation
Normal patients	19.30	5.73
Brain-damaged patients	9.20	2.50

   a. A particular brain-damaged patient produced eight words. Find the z score for this observation and interpret the value.

   b. Would you expect a normal patient to produce a total of eight words? Explain.

3.95 *Muck* is a rich, highly organic type of soil that serves as a growth medium for most vegetation in the Florida Everglades. Because of its high concentration of organic material, muck can be destroyed by drought, fire, and windstorms. During a recent drought in south Florida, the Everglades lost a considerable amount of muck. To assess the loss, members of the Florida Fish and Game Commission marked 40 plots with stakes at various locations in the Everglades, and measured the depth of the muck (in inches) at each stake. The data are given in the accompanying table.

MUCK.XLS

27	30	40	21	41	23	24	30	26	40
35	32	35	7	37	19	30	28	39	33
33	23	35	22	15	32	26	38	26	30
45	15	29	57	27	27	18	16	31	36

   a. Give a two-number summary of the data set. Interpret these numbers.

   b. Give a five-number summary of the data set. Interpret these numbers.

c. Construct the intervals $\bar{x} \pm s, \bar{x} \pm 2s$, and $\bar{x} \pm 3s$. Count the number of observations that fall within each interval and find the corresponding proportions. Compare the results to the Empirical Rule.

d. Find the 10th percentile for the 40 muck depths.

e. Construct a box-and-whisker plot for the data. Use the plot to detect any skewness in the data.

f. Identify any outliers in the data.

3.96 Behavior geneticists are scientists who study the ways in which genetic factors influence behavior. One characteristic that has been studied is the "emotional" behavior of different strains of rats. Emotional behavior is sometimes defined as the tendency to "freeze"—i.e., not move—when presented with a new situation. For rats from different strains that are raised under identical circumstances, differences in emotional behavior may suggest a genetic base. Summary statistics on the number of meters traversed by 500 rats of a particular strain when put in a box with a bright light and white noise are listed here.

$$\bar{x} = 7.65 \qquad s = .88 \qquad M = 7.3$$

a. Find the 50th percentile of the distribution of the number of meters traversed by this sample of rats.

b. Compute the interval $\bar{x} \pm 2s$. Use the Empirical Rule to approximate the percentage of rats that had distance measurements within this interval.

c. Would you expect to observe a rat of this particular strain traverse a distance of 9.50 meters?

3.E.1   Using Microsoft Excel to Explore Data with Numerical Descriptive Measures

Solution Summary:

A.   Use the Data Analysis Descriptive Statistics tool to generate a table of numerical descriptive measures.

B.   Use the PHStat add-in to construct a box-and-whisker plot.

C.   Use the Excel worksheet CORREL function to compute the coefficient of correlation.

Example: Weights of 144 contaminated fish captured in Tennessee River from Example 3.4

Part A Solution: Using the Data Analysis Descriptive Statistics tool to generate a table of numerical descriptive measures

To generate a table of descriptive statistics from the weights of 144 contaminated fish, do the following:

1   Open the fish weights workbook (FISH.XLS) and click the Data sheet tab.

2   Select Tools | Data Analysis. Select Descriptive Statistics from the Analysis Tools list box in the Data Analysis dialog box. Click the OK button.

3   In the Descriptive Statistics dialog box, enter the information and make the selections shown in Figure 3.E.1. Click the OK button.

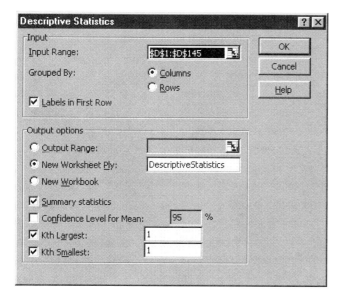

E Figure 3.E.1

The add-in produces a table of descriptive statistics similar to the one shown in Figure 3.2 (page 149).

Part B Solution: Using the PHStat add-in to construct a box-and-whisker plot

To generate a box-and-whisker plot from the weights of 144 contaminated fish, do the following:

1   If the PHStat add-in has not been previously loaded, load the add-in using the instructions of Section EP.3.2.

2   Open the fish weights workbook (FISH.XLS) and click the Data sheet tab.

3 Select PHStat I Box-and-Whisker Plot.

4 In the Box-and-Whisker Plot dialog box, enter the information and make the selections shown in Figure 3.E.2. Click the OK button.

E Figure 3.E.2

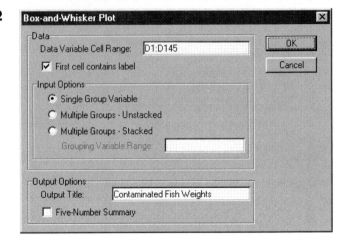

The add-in produces a box-and-whisker plot similar to the one shown in Figure 3.E.3.

E Figure 3.E.3

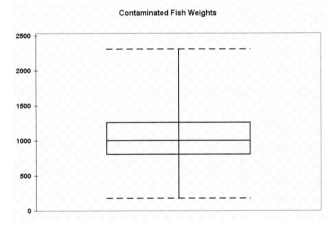

Example: NCAP crash test data of Example 3.22

Part C Solution: Using the Excel worksheet CORREL function to compute the coefficient of correlation

To compute a coefficient of correlation, use the Excel worksheet CORREL function, the format of which is:

$$CORREL(cell\ range\ of\ Y\ variable,\ cell\ range\ of\ X\ variable)$$

To compute the coefficients of correlation that use the NCAP test data to relate the driver's head-injury rating (Drivhead) to each of car weight (Weight), driver's rate of chest deceleration (Drivchst), and driver's overall injury rating (Drivstar), do the following:

1 Open the NCAP crash test data workbook (CRASH.XLS) and click the Data sheet tab.

2 Verify that this sheet contains the Drivhead variable data in the cell range J1:J99, the Weight variable in cell range E1:E99, the Drivchst variable in cell range L1:L99, and the Drivstar variable in cell range F1:F99.

3 To compute the coefficient of correlation that relates Drivhead to Weight, enter the formula =CORREL(Data!J1:J99, Data!E1:E99) in any blank cell of any worksheet in the workbook.

4 To compute the coefficient of correlation that relates Drivhead to Drivchst, enter the formula =CORREL(Data!J1:J99, Data!L1:L99) in any blank cell of any worksheet in the workbook.

5 To compute the coefficient of correlation that relates Drivhead to Drivstar, enter the formula =CORREL(Data!J1:J99, Data!F1:F99) in any blank cell of any worksheet in the workbook.

Figure 3.25 on page 194 illustrates a worksheet in which all three of these formulas have been entered, along with suitable labels.

Chapter 4

Probability: Basic Concepts

CONTENTS

4.1 The Role of Probability in Statistics

4.2 Experiments, Events, and the Probability of an Event

4.3 Probability Rules for Mutually Exclusive Events

4.4 The Combinatorial Rule for Counting Simple Events (Optional)

4.5 Conditional Probability and Independence

4.6 The Additive and Multiplicative Laws of Probability (Optional)

REAL-WORLD APPLICATION
Lottery Buster!

How would you like to win a state lottery, choose the right door in order to win the grand prize in "Let's Make a Deal," or obtain a winning edge in casino blackjack? What does each of these ventures have to do with statistics? The answer is uncertainty. The return in real dollars on most investments cannot be predicted with certainty. Neither can we be certain that an inference about a population, based on the partial information contained in a sample, will be correct. In this chapter we learn how probability can be used to measure uncertainty, and we take a brief glimpse at its role in assessing the reliability of statistical inferences. The probability concepts developed are used to assess your chances of winning a state lottery in Real-World Revisited at the end of the chapter.

4.1 THE ROLE OF PROBABILITY IN STATISTICS

If you play blackjack, a popular casino table game, you know that whether you win in any one game is an outcome that is very uncertain. Similarly, investing in bonds, stock, or a new business is a venture whose success is subject to uncertainty. (In fact, some would argue that investing is a form of educated gambling—one in which knowledge, experience, and good judgment can improve the odds of winning.)

Much like playing blackjack and investing, making inferences based on sample data is also subject to uncertainty. A sample rarely tells a perfectly accurate story about the population from which it was selected. There is always a margin of error (as the pollsters tell us) when sample data are used to estimate the proportion of people in favor of a particular political candidate, some consumer product, or some political or social issue. There is always uncertainty about how far the sample estimate will depart from the true population proportion of affirmative answers that you are attempting to estimate. Consequently, a measure of the amount of uncertainty associated with an estimate (which we called the *reliability of an inference* in Chapter 1) plays a major role in statistical inference.

How do we measure the uncertainty associated with events? Anyone who has observed a daily newscast can answer that question. The answer is *probability*. For example, it may be reported that the probability of rain on a given day is 20%. Such a statement acknowledges that it is uncertain whether it will rain on the given day and indicates that the forecaster measures the likelihood of its occurrence as 20%.

Probability also plays an important role in decision making. To illustrate, suppose you have an opportunity to invest in an oil exploration company. Past records show that for 10 previous oil drillings (a sample of the company's experiences), all 10 resulted in dry wells. What do you conclude? Do you think the chances are better than 50-50 that the company will hit a producing well? Should you invest in this company? We think your answer to these questions will be an emphatic "no." If the company's exploratory prowess is sufficient to hit a producing well 50% of the time, a record of 10 dry wells out of 10 drilled is an event that is just too improbable. Do you agree?

In this chapter we will examine the meaning of probability and develop some properties of probability that will be useful in our study of statistics.

4.2 EXPERIMENTS, EVENTS, AND THE PROBABILITY OF AN EVENT

In the language employed in a study of probability, the word *experiment* has a very broad meaning. In this language, an experiment is a process of making an observation or taking a measurement on the experimental unit. For example, suppose you are dealt a single card from a standard 52-card deck. Observing the outcome (i.e., the number and suit of the card) could be viewed as an experiment. Counting the number of U.S. citizens who live in a particular state or county is an experiment. Similarly, recording a voter's opinion on an important political issue is an experiment. Observing the fraction of insects killed by a new insecticide is an experiment. Note that most experiments result in outcomes (or measurements) that cannot be predicted with certainty in advance.

> **Definition 4.1**
> The process of making an observation or taking a measurement on one or more experimental units is called an **experiment**.

Example 4.1   Listing Outcomes

Consider the following experiment. You are dealt one card from a standard 52-card deck. List some possible outcomes of this experiment that cannot be predicted with certainty in advance.

Solution

Some possible outcomes of this experiment that cannot be predicted with certainty in advance are as follows (see Figure 4.1):

a. You draw an ace of hearts.   **b.** You draw an eight of diamonds.

c. You draw a spade.   **d.** You do not draw a spade. ❑

Example 4.2   Listing Outcomes

Consider the following experiment. Five hundred urban residents with children are selected from a large number of urban residents with children to determine the proportion who favor a busing plan that would, in theory, racially balance public schools in the city. The response of each urban resident is recorded. List some possible outcomes of this experiment that cannot be predicted with certainty in advance.

Solution

Since we are observing the responses of 500 urban residents with children (where urban residents are the experimental units), this experiment can result in a very large number of outcomes. Three of the many possible outcomes are listed below.

a. Exactly 387 of the 500 urban residents favor the busing plan.

b. Exactly 388 favor the busing plan.

c. A particular resident, the Jones family, favors the busing plan.

Clearly, we could define many other outcomes of this experiment that cannot be predicted in advance. ❑

In the language of probability theory, outcomes of experiments are called *events*.

Figure 4.1 Possible Outcomes of Card-Drawing Experiment

a.   b.   c.   d.

Definition 4.2

Outcomes of experiments are called **events**. [*Note:* To simplify our discussion, we will use italic capital letters *A, B, C,* ..., to denote specific events.]

The outcome for each of the experiments described in Examples 4.1 and 4.2 is shrouded in uncertainty; that is, prior to conducting the experiment, we could not be certain whether a particular event would occur. This uncertainty is measured by the *probability* of the event.

Example 4.3 **Interpreting a Probability**

Suppose we perform the following experiment: Toss a coin and observe whether it results in a head or a tail. Define the event *H* by *H:* Observe a head.

What do we mean when we say that the probability of *H,* denoted by $P(H)$, is equal to $\frac{1}{2}$?

Solution

Stating that the probability of observing a head is $P(H) = \frac{1}{2}$ does *not* mean that exactly half of a number of tosses will result in heads. (For example, we do not expect to observe exactly 1 head in 2 tosses of a coin or exactly 5 heads in 10 tosses of a coin.) Rather, it means that, in a very long series of tosses, we believe that approximately half would result in a head. Therefore, the number $\frac{1}{2}$ measures the likelihood of observing a head on a single toss of the coin. ❏

The "relative frequency" concept of probability discussed in Example 4.3 is illustrated in Figure 4.2. The graph shows the proportion of heads observed after $n = 25, 50, 75, 100, 125, ..., 1,450, 1,475,$ and $1,500$ computer-simulated repetitions of a coin-tossing experiment. The number of tosses is marked along the horizontal

Figure 4.2 The Proportion of Heads in *n* Tosses of a Coin

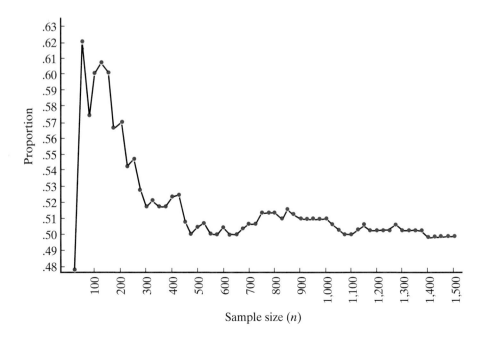

Sample size (*n*)

axis of the graph, and the corresponding proportions of heads are plotted on the vertical axis above the values of n. We have connected the points by line segments to emphasize the fact that the proportion of heads moves closer and closer to .5 as n gets larger (as you move to the right on the graph).

Although most people think of the probability of an event as the proportion of times the event occurs in a very long series of trials, some experiments can never be repeated. For example, if you invest $50,000 in starting a new business, the probability that your business will survive 5 years has some unknown value that you will never be able to evaluate by repetitive experiments. The probability of this event occurring is a number that has some value, but it is unknown to us. The best that we could do, in estimating its value, would be to attempt to determine the proportion of similar businesses that survived 5 years and take this as an approximation to the desired probability. In spite of the fact that we may not be able to conduct repetitive experiments, the relative frequency definition for probability appeals to our intuition.

Definition 4.3

The **probability of event** A, denoted by $P(A)$, is a number between 0 and 1 that measures the likelihood that A will occur when the experiment is performed. $P(A)$ can be approximated by the proportion of times that A is observed when the experiment is repeated a very large number of times.

Self-Test 4.1

Consider the following experiment: A family with three children whose gender is unknown is selected and the objective is to count the number of boys in the family.

a. List some possible outcomes of this experiment that cannot be predicted in advance.

b. We will learn that the probability of observing three boys is $\frac{1}{8}$. Explain what this probability represents.

In the next section we give some rules for calculating the exact probability of an event.

PROBLEMS 4.1–4.8

Using the Tools

4.1 Consider the following experiment: Toss a fair six-sided die and observe the number of dots showing on the upper face.

a. If this experiment were to be repeated over and over again in a very long series of trials, what proportion of the experimental outcomes do you think would result in a 5?

b. What does it mean to say, "The probability that the outcome is a 5 is $\frac{1}{6}$"?

c. Perform the experiment a large number of times and calculate the proportion of outcomes that result in a 5. Note that as the number of repetitions becomes larger and larger, this proportion moves closer and closer to $\frac{1}{6}$.

Applying the Concepts

4.2   The following facts are extracted from *True Odds: How Risk Affects Your Everyday Life* (Walsh, 1997). Convert each to a probability statement.

a.   The prospect of dying from flesh-eating bacteria are one in a million.

b.   The chance of being the victim of a violent crime is one in 135.

c.   Nine of ten premature deaths are linked to smoking, overeating, alcohol abuse, high blood pressure, not exercising, or not wearing seat belts.

4.3   In his book *100% American* (Poseidon Press, 1988), Daniel Weiss presents "fascinating facts about who we are and what some of us—or even most of us—are thinking, doing, believing today." Some excerpts are given below.* Form a probability statement about each of these events.

a.   8% of American 17-year-olds think Abraham Lincoln wrote *Uncle Tom's Cabin.*

b.   43% of Americans think it is likely that some UFOs are really space vehicles from other civilizations.

c.   71% of American men think women should call men for a date.

d.   95% of American peanut eaters eat at least nine at a sitting.

4.4   Each year the Florida Game and Fresh Water Fish Commission issues permits for trappers to hunt and kill alligators. In 1996, there were 12,685 applications for permits. The Commission issued 678 permits using a computerized random selection system (*Tampa Tribune*, Sept. 3, 1996). If you applied for a permit to hunt alligators in 1996 in Florida, what was the probability that you were issued a permit?

4.5   What is the probability that you understand basic probability? As a partial answer to this question, a random sample of 287 female military veterans were asked: "Imagine that we flip a fair coin 1,000 times. What is your best guess about how many times the coin would come up heads in 1,000 flips?" Surprisingly, only 155 of the women correctly answered "about 500 heads" with most of the incorrect answers ranging between 60 and 300 heads (*Annals of Internal Medicine,* Dec. 1, 1997). Estimate the probability that a female veteran will correctly answer "about 500 heads" to the coin flip question.

4.6   During the 1989 U.S. Open, four professional golfers (Doug Weaver, Jerry Pate, Nick Price, and Mark Wiebe) made holes in one (aces) on the sixth hole at Oak Hill Country Club—all on the same day! How unlikely is such a feat? According to *Golf Digest* (Mar. 1990), the probability of a Professional Golf Association (PGA) tour pro making an ace on a given hole is approximately $\frac{1}{3,000}$. The estimate is based on the ratio of the number of aces made on the PGA tour to the total number of rounds played.

a.   Interpret the probability of $\frac{1}{3,000}$.

b.   *Golf Digest* also estimates that the probability of any four players getting aces on the same hole on the same day during the next U.S. Open as $\frac{1}{150,000}$. Interpret this probability.

4.7   A survey of 1,035 family doctors found that 70% have increased tests in an effort to detect cancer early in their patients (*Ca—A Cancer Journal for Clinicians,* July/Aug. 1985). Suppose that one of the 1,035 doctors is selected at random. Consider the following statement: "The probability that this doctor has increased tests to detect cancer early in patients is only $\frac{1}{1,035} = .00097$." Do you agree with the statement? If not, give the correct probability.

4.8   The *San Francisco Examiner* (Oct. 15, 1992) reported on a large-scale study of expectant parents in order to assess the risk of their children being born with cystic fibrosis. The study used genetic DNA analysis of the parents to determine whether the unborn child is likely to have the disease. Based on the study, the following risk assessment table was compiled.

		FATHER		
		Untested	**Test Positive**	**Test Negative**
Mother	Untested	Unknown	1 in 100	1 in 16,500
	Test Positive	1 in 100	1 in 4	1 in 660
	Test Negative	1 in 16,500	1 in 660	1 in 106,000

a.   If both parents test positive, what is the probability of the child developing cystic fibrosis?

b.   If both parents test negative, what is the probability of the child developing cystic fibrosis?

4.3 PROBABILITY RULES FOR MUTUALLY EXCLUSIVE EVENTS

One special property of events can be seen in the examples of the preceding section. Two events are said to be mutually exclusive if, when one occurs, the other cannot occur. Consider the experiment of Example 4.1, where we draw one card at random from a 52-card deck. The events listed under parts **c** and **d** of Example 4.1 are mutually exclusive. You cannot conduct an experiment and "draw a spade" (the event listed under part **c**) and at the same time "not draw a spade" (the event listed under part **d**). If one of these two events occurs when an experiment is conducted, the other event cannot have occurred. Therefore, we say that they are mutually exclusive events.

Definition 4.4

Two events are said to be **mutually exclusive** if, when one of the two events occurs in an experiment, the other cannot occur.

Example 4.4 Mutually Exclusive Events

Refer to Example 4.2 (page 214) where 500 urban residents are surveyed about a proposed busing plan. Consider the following events:

A: Exactly 387 of the 500 urban residents favor the busing plan.

B: Exactly 388 favor the busing plan.

C: A particular resident, the Jones family, favors the busing plan.

State whether the following pairs of events are mutually exclusive:

    a. *A* and *B*   **b.** *A* and *C*   **c.** *B* and *C*

Solution

    a. Events *A* and *B* are mutually exclusive. If you have observed exactly 387 residents who favor the busing plan, then you could not, at the same time, have observed exactly 388.

    b. Events *A* and *C* are *not* mutually exclusive. The Jones family may be one of the residents among the 387 residents in event *A* who favor busing. Therefore, it is possible for both events *A* and *C* to occur simultaneously.

    c. Events *B* and *C* are not mutually exclusive for the same reason given in part **b.** ❑

Example 4.5 Mutually Exclusive Events

Suppose an experiment consists of selecting two electric light switches from an assembly line for inspection. Define the following events:

A: The first switch is defective.

B: The second switch is defective.

Are *A* and *B* mutually exclusive events?

Solution

Events *A* and *B* are not mutually exclusive because both the first and the second switch could be defective when the inspection is made. That is, both *A* and *B* can occur together. ❑

Self-Test

4.2

An experiment consists of observing whether or not each of the next two people who check out at a supermarket use coupons. List two events that are mutually exclusive.

If two events *A* and *B* are mutually exclusive, then the probability that *either A or B* occurs is equal to the sum of their probabilities. We will illustrate with an example.

Example 4.6 Summing Probabilities

Consider the following experiment: You roll a single six-sided die and observe the number of dots on the upper face of the die. What is the probability that you observe an even number?

Figure 4.3 Six Mutually Exclusive Outcomes (Simple Events) and Associated Probabilities when Rolling a Die

$$1 \quad (1/6) \qquad 2 \quad (1/6) \qquad 3 \quad (1/6)$$

$$4 \quad (1/6) \qquad 5 \quad (1/6) \qquad 6 \quad (1/6)$$

Solution

The experiment can result in one of six mutually exclusive events: 1, 2, 3, 4, 5, or 6 dots showing on the die. These six events represent the most basic outcomes of the experiment and are called *simple events*. The collection of all possible simple events in an experiment is called the *sample space*. The sample space for this die-rolling experiment is shown diagrammatically in Figure 4.3.

With a "fair" die, we expect each of the six simple events of Figure 4.3 to occur with approximately equal relative frequency ($\frac{1}{6}$) if the die-rolling experiment were repeated a large number of times. Since you will observe an even number only if a 2 or a 4 or a 6 occurs, and these simple events are mutually exclusive, either one or the other of these events will occur ($\frac{1}{6} + \frac{1}{6} + \frac{1}{6}$) = $\frac{1}{2}$ of the time. Therefore, the probability of observing an even number in the roll of a die is equal to the probability of observing *either a 2 or a 4 or a 6,* which is $\frac{1}{2}$. You can verify this result experimentally, using the procedure employed in Section 4.2. ❑

Definition 4.5

Simple events are mutually exclusive events that represent the most basic outcomes of an experiment.

Definition 4.6

The collection of all possible simple events in an experiment is called the **sample space**.

Although Example 4.6 utilized the concept of simple events, the additive probability rule shown in the next box applies to any two mutually exclusive events. You can now see why the concept of mutually exclusive events is important. We will illustrate with several more examples.

Probability Rule #1

The Additive Rule for Mutually Exclusive Events   If two events A and B are mutually exclusive, then *the probability that either A or B occurs* is equal to the sum of their respective probabilities:

$$P(A \text{ or } B) = P(A) + P(B) \qquad \textbf{(4.1)}$$

Example 4.7   Applying Probability Rule #1

A local weather reporter forecasts the following three mutually exclusive weather conditions, with their associated probabilities:

Overcast with some rain: 40%

No rain, but partially or completely overcast: 30%

No rain and a clear day: 30%

What is the probability that the day will be overcast?

Solution

The forecaster suggests that the experiment of observing the day's weather can result in one (and only one) of three mutually exclusive events:

    A: Overcast with some rain

    B: No rain, but partially or completely overcast

    C: No rain and a clear day

where

$$P(A) = .4 \qquad P(B) = .3 \qquad P(C) = .3$$

The event of interest, that the day will be overcast, will occur if either event *A* or event *B* occurs. Therefore, using Probability Rule #1, we have:

$$P(A \text{ or } B) = P(A) + P(B) = .4 + .3 = .7$$

The implication is that if we were to forecast the weather for a large number of days with similar weather conditions, 70% of the days would be overcast. ❑

Example 4.8 Applying Probability Rule #1

Consider the following experiment: Roll two dice and observe the sum of the dots showing on the two dice. Find the probability that the sum is equal to 7 (an important number in the casino game of craps).

Solution

Mark the dice so that they are identified as die #1 and die #2. Then there are $6 \times 6 = 36$ distinctly different ways that the dice could fall. You could observe a 1 on die #1 and a 1 on die #2; a 1 on die #1 and a 2 on die #2; a 1 on die #1 and a 3 on die #2, etc. In other words, you can pair the six values (shown on the six sides) of die #1 with the six values of die #2 in $6 \times 6 = 36$ mutually exclusive ways. These 36 possibilities represent the simple events for the experiment. The sums associated with the simple events are shown in Figure 4.4 (page 222).

Since there are 36 possible ways the dice could fall and since these ways should occur with equal frequency, the probability of observing any one of the 36 events shown in the figure is $\frac{1}{36}$. Then to find the probability of rolling a 7, we need only to add the probabilities of those events corresponding to a sum on the dice equal to 7.

If we denote the simple event that you observe a 6 on die #1 and a 1 on die #2 as $(6, 1)$, etc., then as shown in Figure 4.4, you will toss a 7 if you observe a $(6, 1)$, $(5, 2), (4, 3), (3, 4), (2, 5),$ or $(1, 6)$. Therefore, the probability of tossing a 7 is

$$\begin{aligned} P(7) &= P[(6, 1) \text{ or } (5, 2) \text{ or } (4, 3) \text{ or } (3, 4) \text{ or } (2, 5) \text{ or } (1, 6)] \\ &= P(6, 1) + P(5, 2) + P(4, 3) + P(3, 4) + P(2, 5) + P(1, 6) \\ &= \tfrac{1}{36} + \tfrac{1}{36} + \tfrac{1}{36} + \tfrac{1}{36} + \tfrac{1}{36} + \tfrac{1}{36} = \tfrac{6}{36} = \tfrac{1}{6} \end{aligned}$$

    ❑

Figure 4.4 The Sum of the Dots for the 36 Mutually Exclusive Outcomes in the Tossing of a Pair of Dice

Die #1

	2	3	4	5	6	7
	3	4	5	6	7	8
	4	5	6	7	8	9
	5	6	7	8	9	10
	6	7	8	9	10	11
	7	8	9	10	11	12

Die #2

Self-Test 4.3

Two marbles are selected at random and without replacement from a box containing two blue marbles and three red marbles. Determine the probability of observing each of the following events:

A: Two blue marbles are selected.

B: A red and a blue marble are selected.

C: Two red marbles are selected.

Example 4.8 suggests a modification of Probability Rule #1 when an experiment can result in one and only one of a number of equally likely (equiprobable) mutually exclusive events.

Probability Rule #2

The Probability Rule for an Experiment That Results in One of a Number of Equally Likely Mutually Exclusive Events   Suppose an experiment can result in one and only one of M equally likely mutually exclusive events and that m of these events result in event A. Then the probability of event A is

$$P(A) = \frac{m}{M} \tag{4.2}$$

Example 4.9   **Applying Probability Rule #2**

As a contestant on the popular television game show "Let's Make A Deal," you have selected, by chance, two doors from among four doors offered. Unknown to you, there are prizes behind only two of the four doors. What is the probability that you win at least one of the two prizes?

Solution

Identify the four doors as D_1, D_2, D_3, and D_4, and let D_3 and D_4 be the two doors that hide the two prizes. Then the six distinctly different and mutually exclusive ways that the two doors may be selected from the four are:

(D_1, D_2) (D_1, D_3) (D_1, D_4) (D_2, D_3) (D_2, D_4) (D_3, D_4)

Step 1   Since the doors were selected by chance, we would expect the likelihood that any one pair would be chosen to be the same as for any other pair. Also, since the experiment can result in only one of these pairs, $M = 6$.

Step 2   You can see from the listed outcomes that the number of pairs that result in a choice of D_3 or D_4 is $m = 5$. (These are the last five pairs listed.)

Step 3   Using Probability Rule #2, the probability of selecting at least one of the two doors with prizes is:

$$P(\text{at least one of } D_3 \text{ or } D_4) = \frac{m}{M} = \frac{5}{6}$$ ❑

Examples 4.6–4.9 identify the properties of the probabilities of all events, as summarized in the box.

Properties of Probabilities

1. The probability of an event always assumes a value between 0 and 1.
2. If two events A and B are mutually exclusive, then the probability that either A or B occurs is equal to $P(A) + P(B)$.
3. If we list all possible simple events associated with an experiment, then the sum of their probabilities will always equal 1.

Before concluding this section, we will comment on two important mutually exclusive events and their probabilities. Consider again the dice-tossing experiment (Example 4.8, page 221) and define the following two events:

A: The sum of the dots on the two dice is 7.

A': The sum of the dots on the two dice is not 7.

Thus, A' is the event that A *does not occur.* You can see that A and A' are mutually exclusive events and, further, that if A occurs $\frac{1}{6}$ of the time in a long series of trials (see Example 4.8), then A' will occur $\frac{5}{6}$ of the time. In other words,

$$P(A) + P(A') = 1$$

Definition 4.7

The Rule of Complements   The **complement** of an event A, denoted by the symbol A', is the event that A does not occur (see Figure 4.5).

Probability Rule #3

Probability Relationship for Complementary Events

$P(A) = 1 - P(A')$ **(4.3)**

Complementary events are important because sometimes it is difficult to find the probability of an event A, but easy to find the probability of its complement A'. In this case, we can find $P(A)$ using the relationship stated in Probability Rule #3.

Figure 4.5 Complementary Events A and A' in a Sample Space

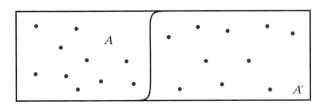

**Self-Test
4.4**

Refer to Self-Test 4.3 (page 222). Recall that two marbles are selected from a box containing two blue and three red marbles. Use the Rule of Complements to find the probability that at least one red marble is selected.

**PROBLEMS
4.9–4.25**

Using the Tools

Simple Event	Probability
S_1	.15
S_2	.20
S_3	.20
S_4	.25
S_5	.20

Problem 4.9

4.9 An experiment has five possible outcomes (simple events) with the probabilities at left:

a. Find the probability of each of the following events:

> A: Outcome S_1, S_2, or S_4 occurs.
> B: Outcome S_2, S_3, or S_5 occurs.
> C: Outcome S_4 does not occur.

b. List the simple events in the complements of events A, B, and C.

c. Find the probabilities of A', B', and C'.

Simple Event	Probability
1	$1/12$
2	$1/6$
3	$1/6$
4	$1/6$
5	$1/6$
6	$3/12$

Problem 4.10

4.10 In the roll of an unbalanced ("loaded") die, the outcomes (simple events) occur with the probabilities in the table. Define the following events:

> A: Observe a number less than 4.
> B: Observe an odd number.
> C: Observe an even number.

a. Find the probabilities of events A, B, and C.

b. List the outcomes (simple events) in the complements of events A, B, and C.

c. Find the probabilities of A', B', and C'.

d. Are any of the pairs of events, A and B, A and C, or B and C, complementary? Explain.

4.11 Suppose an experiment involves tossing two coins and observing the faces of the coins. Find the probabilities of:

a. A: Observing exactly two heads.

b. B: Observing exactly two tails.

c. C: Observing at least one head.

d. Describe the complement of event A and find its probability.

4.12 Two dice are tossed. Use Figure 4.4 on page 222 to find the probability that the sum of the dots showing on the two dice is equal to:

a. 12   **b.** 5   **c.** 11

4.13 Refer to the dice-tossing experiment in Problem 4.12. Find the approximate probability of tossing a sum of 7 by conducting an experiment similar to the experiments illustrated in Figure 4.2. Toss a pair of dice a large number of times and record the proportion of times a sum of 7 is observed. Compare your value with the exact probability, $\frac{1}{6}$.

4.14   Refer to Examples 4.6 and 4.7 (pages 219–220). Verify that property 3 (shown in the box titled Properties of Probabilities, page 223) holds for each of these examples.

Applying the Concepts

4.15   The bar graph below summarizes the market shares of diaper brands in the United States. Suppose we randomly select a recently purchased package of diapers and observe its brand.

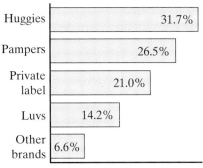

U.S. diaper market shares

Data Source: A. C. Nielsen.

   a.   List the simple events for this experiment.

   b.   Assign probabilities to the simple events.

   c.   What is the probability the brand selected is Pampers or Luvs?

   d.   What is the probability the brand selected is not Huggies?

4.16   The National Highway Traffic Safety Administration maintains the Fatal Accident Report System (FARS) file. The FARS file documents every motor vehicle crash (since Jan. 1, 1975) on a U.S. public road in which a fatality occurred. Based on over 800,000 deaths recorded in the FARS file, the types of crashes that account for all car occupant fatalities is distributed as follows:

Type of Crash	Proportion
Single-car crash	.45
Two-car crash	.22
Car crashing with another noncar vehicle (e.g., a truck)	.25
Crash involving 3 or more vehicles	.08
TOTAL	1.00

Source: Evans, L. "Small cars, big cars: What is the safety difference?" *Chance,* Vol. 7, No. 3, Summer 1994, p. 9.

Consider a recent car crash that resulted in a fatality.

   a.   What is the probability that a single-car crash occurred?

   b.   What is the probability that a two-car crash occurred?

   c.   What is the probability that two or more vehicles were involved in the crash?

4.17   A national study was commissioned to gauge the performance of property managers who are employed by nonprofit organizations. The study sample was comprised of 23 rental properties located in six cities—Boston, Chicago, Miami, Minneapolis/St. Paul, New York, and Oakland—all managed by nonprofit organizations (*Journal of the American Planning Association,* Winter 1998). The percentage of total rent collected by each property manager was reported and is summarized in the table. Suppose we randomly select one of the 23 rental properties and observe the percentage of total rent collected by the property manager.

Percentage of Rent Collected	Number of Properties
Less than 80	4
80–84	2
85–89	0
90–94	2
95–99	2
100	13
TOTAL	23

Source: Bratt, R. G., et al. "The status of nonprofit-owned affordable housing." *Journal of the American Planning Association,* Vol. 64, No. 1, Winter 1998, p. 41 (Table 1).

a.   Let A be the event that the percentage of total rent collected is less than 80%. Find $P(A)$.

b.   Let B be the event that between 85% and 89% of the total rent is collected by the property manager. Find $P(B)$.

c.   Are events A and B mutually exclusive? Explain.

d.   A widely accepted standard in the rental industry is that management should collect at least 95% of the total rent due. Find the probability that the property manager is meeting this standard.

4.18   Local area merchants often use "scratch off" tickets with promises of grand prize giveaways to entice customers to visit their showrooms. One such game, called Jackpot, was recently used in Tampa, Florida. Residents were mailed game tickets with ten "play squares" and the instructions: "Scratch off your choice of ONLY ONE Jackpot Play Square. If you reveal any 3-OF-A-KIND combination, you WIN. Call NOW to make an appointment to claim your prize." Unknown to the customer, all ten play squares have *winning* 3-of-a-kind combinations. On one such ticket, three cherries (worth up to $1,000 in cash or prizes) appeared in play squares numbered 1, 2, 3, 4, 5, 6, 7, and 9; three lemons (worth a 10-piece microwave cookware set) appeared in play square number 8; and three sevens (worth a whirlpool spa) appeared in play square 10.

a.   If you play Jackpot, what is the probability that you win a whirlpool spa?

b.   If you play Jackpot, what is the probability that you win $1,000 in cash or prizes?

4.19   In a study conducted by a marketing honors class at the University of South Florida, 257 randomly selected Tampa Bay area residents were asked to select their favorite local attraction. The results are shown in the accompanying table. Suppose one of the Tampa Bay area residents surveyed is selected at random. Find the probability of each of the following events.

Favorite Attraction	Number Surveyed
Adventure Island Water Park	10
Beaches	21
Busch Gardens	69
Disney Theme Parks (Orlando)	31
Museum of Science and Industry	3
Parks	8
Sea World (Orlando)	8
Sports	29
Other	78
TOTAL	257

Source: Department of Marketing, University of South Florida.

a. The resident's favorite local attraction is Busch Gardens.

b. The resident's favorite local attraction is Sports.

c. The resident's favorite local attraction is not any of the Disney Theme Parks.

4.20 The American Association for Marriage and Family Therapy (AAMFT) is a group of professional therapists and family practitioners who treat many of the nation's couples and families. The AAMFT released the findings of a study that tracked the postdivorce history of 98 pairs of former spouses with children. Each divorced couple was classified into one of four groups, nicknamed "perfect pals," "cooperative colleagues," "angry associates," and "fiery foes." The proportions classified into each group are given here.

> "Perfect pals": .12
>   Joint-custody parents who get along well, do not remarry.
> "Cooperative colleagues": .38
>   Occasional conflict, likely to be remarried.
> "Angry associates": .25
>   Cooperate on issues related to children only, conflicting otherwise.
> "Fiery foes": .25
>   Communicate only through children, hostile toward each other.

Suppose one of the 98 couples is selected at random.

a. What is the probability that the former spouses are "cooperative colleagues"?

b. What is the probability that the former spouses are "angry associates"?

c. What is the probability that the former spouses are not "perfect pals"?

4.21 Refer to the *Archaeol. Oceania* (Apr. 1997) study of archaeological digs in Papua, New Guinea, Problem 2.9 (page 69). The table on page 228 describes the distribution of artifacts (stone drillpoints, shell beads, and grinding slabs) discovered on Motupore Island. If an artifact is found at the site, what is the probability that it is:

Type of Artifact	Number Discovered
Drillpoint	566
Shell Bead	385
Grinding slab	43

Source: Allen, J., et al. "Identifying specialization, production, and exchange in the archaeological record: The case of shell bead manufacture on Motupore Island, Papua." *Archaeol. Oceania,* Vol. 32, No. 1, Apr. 1997, p. 27 (Table 1).

a. a stone drillpoint?   **b.** a shell bead?   **c.** a grinding slab?
d. a stone drillpoint or grinding slab?   **e.** not a shell bead?

4.22 In a study of the effect of traffic noise on sleep, acoustical engineers monitored the sleep of 10 married couples for twelve consecutive nights (*Journal of Sound and Vibration,* Mar. 1986). The researchers observed whether each of the couples slept with or without earplugs and with or without the windows open. Suppose we randomly select one couple from the experiment on one randomly selected night and note whether the couple is wearing earplugs and whether the windows are open or closed.

a. List the simple events for the experiment.

b. During the experimental phase of the study, all 10 couples were required to sleep with the windows open. However, half slept with earplugs and half slept without earplugs. Assign probabilities to the simple events for this phase of the study.

c. What is the probability that a randomly selected couple is wearing earplugs during the experimental phase?

d. What is the probability that the windows are closed during the experimental phase?

4.23 The *Journal of Engineering for Industry* (Aug. 1993) reported on an automated system designed to replace the cutting tool of a drilling machine at optimum times. To test the system, data were collected over a broad range of materials, drill sizes, drill speeds, and feed rates—called machining conditions. Although a total of 168 different machining conditions were possible, only eight were employed in the study. These are described below:

Experiment	Workpiece Material	Drill Size (in.)	Drill Speed (rpm)	Feed Rate (ipr)
1	Cast iron	.25	1,250	.011
2	Cast iron	.25	1,800	.005
3	Steel	.25	3,750	.003
4	Steel	.25	2,500	.003
5	Steel	.25	2,500	.008
6	Steel	.125	4,000	.0065
7	Steel	.125	4,000	.009
8	Steel	.125	3,000	.010

a. Suppose one (and only one) of the 168 possible machining conditions will detect a flaw in the system. What is the probability that the experiment conducted in the study will detect the system flaw?

b. Suppose the system flaw occurs when drilling steel material with a .25-inch drill size at a speed of 2,500 rpm. Of the 168 possible machining conditions, only 7 of these have such a configuration. Find the probability that the experiment conducted in the actual study will detect the system flaw.

4.24 Ohio's Learning, Earning, and Parenting (LEAP) program is designed to encourage school attendance among pregnant and parenting teens on welfare. Eligible teens who provide evidence of school enrollment receive a bonus payment, while those who do not attend school have money deducted from their grant (i.e., the teens are "sanctioned"). In a survey of LEAP teens, published in *Children and Youth Services Review* (Vol. 17, 1995), the bonus and/or sanction requests were recorded for each teen. The results are summarized in the accompanying table.

Request	Percentage
No bonuses or sanction (N)	7
Only bonuses (OB)	37
Only sanction (OS)	18
Both, more bonuses than sanctions (BB)	14
Both, more sanctions than bonuses (BS)	18
Both, equal number of bonuses and sanctions (BE)	6
TOTAL	100%

Source: Reprinted from *Children and Youth Services Review,* Vol. 17, Nos. 1/2, Wood, R. G., et al. "Encouraging school enrollment and attendance among teenage parents on welfare: Early impacts of Ohio's LEAP program," p. 302, Figure 1, Elsevier Science Ltd., The Boulevard, Langford Lane, Kidlington OX5 1GB, UK.

a. Define the experiment that generated the data in the table, and list the simple events.

b. Assign probabilities to the simple events.

c. What is the probability that both bonuses and sanctions are requested for a LEAP teen?

d. What is the probability that only one type of request (either a bonus or a sanction, but not both) is made?

4.25 Entomologists are often interested in studying the effect of chemical attractants (*pheromones*) on insects. One common technique is to release several insects equidistant from the pheromone being studied and from a control substance. If the pheromone has an effect, more insects will travel toward it than toward the control. Otherwise, the insects are equally likely to travel in either direction. Suppose five insects are released.

a. If we are interested in which insects travel toward the pheromone, how many outcomes (simple events) are possible? [*Hint:* If the five insects are denoted as *A, B, C, D,* and *E,* one possible outcome is that insects *A* and *B* travel toward the pheromone. Another possible outcome is that insects *A, B, C,* and *D* travel toward the pheromone.]

b. Suppose the pheromone under study has no effect and, therefore, it is equally likely that an insect will move toward the pheromone or toward the control—i.e., the possible outcomes are equiprobable. Find the probability that both insects *A* and *E* travel toward the pheromone.

4.4 THE COMBINATORIAL RULE FOR COUNTING SIMPLE EVENTS (OPTIONAL)

Consider an experiment that results in M equally likely mutually exclusive outcomes (i.e., M simple events). To utilize Probability Rule #2, we need to determine M. When M is small (in Example 4.9 on page 222), we can list the simple events. However, when M is large, it may be inconvenient or practically impossible to list all the different outcomes of the experiment. In this situation, we can rely on a rule for counting the number M of different outcomes in the experiment. The next two examples demonstrate how to use this counting rule.

Example 4.10   Counting Simple Events

Refer to Example 4.9 and the "Let's Make a Deal" problem of determining the number of ways in which you can select two doors from among four doors offered. Use a counting rule to determine M, the number of mutually exclusive outcomes.

Solution

In this problem, we want to know how many different ways we can select two of the four doors. In other words, we want to find M, the number of distinctly different combinations of two doors that can be selected from the four doors offered. The box below gives a formula—called the *Combinatorial Rule*—for counting the number of ways of selecting n objects from a total of N.

Combinatorial Rule for Determining the Number of Different Samples That Can Be Selected from a Population

---

The number of different samples of n objects that can be selected from among a total of N is denoted $\binom{N}{n}$ and is equal to

$$\binom{N}{n} = \frac{N!}{n!(N-n)!} \tag{4.4}$$

where
$$N! = N(N-1)(N-2)(N-3)\cdots(1) \text{ and is read } N \text{ factorial}$$
$$n! = n(n-1)(n-2)(n-3)\cdots(1)$$
$$0! = 1$$

In this example, $n = 2$ and $N = 4$. Applying the formula, we have

$$M = \binom{N}{n} = \frac{N!}{n!(N-n)!} = \frac{4!}{2!(4-2)!} = \frac{4!}{2!2!} = \frac{(4)(3)(2)(1)}{[(2)(1)][(2)(1)]}$$

or

$$M = \frac{(4)(3)}{(2)(1)} = 6$$

Our answer, 6, agrees with the value of M determined by listing the outcomes in Example 4.9. ❑

Example 4.11   Counting Simple Events

In a standard 52-card deck, find the number of different ways in which you can select 2 cards from the deck.

Solution

We want to find M, the number of distinctly different combinations of $n = 2$ cards that can be selected from a total of $N = 52$ cards. Applying the Combinatorial Rule, we have

$$M = \binom{N}{n} = \frac{N!}{n!(N-n)!}$$

$$= \frac{52!}{2!50!} = \frac{(52)(51)(50)\cdots(3)(2)(1)}{[(2)(1)][(50)(49)(48)\cdots(3)(2)(1)]}$$

or

$$M = \frac{(52)(51)}{2} = 1{,}326$$

Therefore, there are 1,326 different ways in which we can select 2 cards from a standard 52-card deck. ❑

Self-Test
4.5

Find the number of different ways that you can randomly select three fighter pilots for a dangerous mission from a squadron of ten pilots.

PROBLEMS
4.26–4.35

Using the Tools

4.26   Find numerical values for each of the following:

    a. $\binom{7}{2}$ **b.** $\binom{5}{3}$ **c.** $\binom{6}{4}$ **d.** $\binom{6}{2}$

4.27   Find numerical values for each of the following:

    a. $\binom{50}{50}$ **b.** $\binom{50}{49}$ **c.** $\binom{50}{1}$ **d.** $\binom{9}{0}$

4.28a.   List the different combinations of two letters that can be formed using the letters A, B, C, D, and E.

    b.   Suppose you were to select a sample of two letters from among the five letters given in part **a**. Use the Combinatorial Rule for finding the number of different samples of two letters that could be selected. Compare your answer with the number of combinations listed in part **a**.

4.29 How many different samples of size $n = 3$ can be selected from a population containing $N = 5$ elements?

   a. Use the Combinatorial Rule to find the answer.

   b. Suppose that the $N = 5$ elements are denoted as *A, B, C, D*, and *E*. List the different samples (e.g., one sample would be *ABC*, another would be *ABE*, etc.).

Applying the Concepts

4.30 A reading list of articles for a sociology course contains 20 articles. How many ways are there to choose three articles from the list for an exam?

4.31 The Quinella bet at jai-alai consists of picking the jai-alai players that will place first and second in a game *irrespective* of order. In jai-alai, eight players (numbered $1, 2, 3, ..., 8$) compete in every game. How many different Quinella bets are possible?

4.32 A National Merit Scholarship award committee recently claimed that each of five grant applications received equal consideration for two grants and that, in fact, the recipients were randomly selected from among the five. Three of the applicants were from a majority group and two were from a minority group. Suppose that both grants were awarded to members of a majority group.

   a. What is the probability of this event occurring if, in fact, the committee's claim is true?

   b. Is the probability computed in part **a** inconsistent with the committee's claim that the selection was at random?

4.33 To evaluate the air traffic control systems of four facilities relying on computer-based equipment, the Federal Aviation Administration (FAA) formed a 16-member task force. If the FAA wants to assign four task-force members to a specific facility, how many different assignments are possible?

4.34 Researchers at the University of Pennsylvania conducted a study of consumer preferences for shampoo benefits (*Journal of the Market Research Society*, Jan. 1984). Part of the study involved a survey of 186 undergraduate business students. Each respondent was shown a list of 13 benefits and asked to select up to four benefits that he or she most strongly desires in a shampoo brand. The group of benefits selected by a respondent is termed a *benefit bundle*. The list of 13 benefits is shown in the accompanying table.

Shampoo Benefits		
Body	Thickness	Natural ingredients
Bounciness	Softness	Repairs split ends
Control	Manageability	Conditions hair
Luster	Gentle action	
Protection against dandruff	Contains protein	

   a. How many bundles of four shampoo benefits can be specified?

b.   Assuming that the benefit bundles of part **a** are equally likely, what is the probability that a respondent will select the "Body, Bounciness, Luster, Protein" bundle as most desirable?

c.   Assuming that the benefit bundles of part **a** are equally likely, what is the probability that a respondent will select the "Softness" benefit as one of the four benefits? [*Hint:* The number of bundles that include "Softness" is $\binom{12}{3}$.]

4.35   Consider five-card stud poker hands dealt from a standard 52-card deck. A *royal flush*—the highest poker hand possible—consists of the cards ace, king, queen, jack, and ten, all of the same suit.

a.   How many different 5-card poker hands can be dealt?

b.   How many different royal flushes are possible?

c.   Use the results, parts **a** and **b**, to find the probability of being dealt a royal flush.

4.5   CONDITIONAL PROBABILITY AND INDEPENDENCE

The event probabilities we have discussed thus far give the relative frequencies of occurrence of the events when an experiment is repeated a very large number of times. They are called *unconditional probabilities* because no special conditions are assumed other than those that define the experiment.

Sometimes we may want to revise the probability of an event when we have additional knowledge that might affect its outcome. To give a simple example, we found that the probability of observing a 7 when two dice are tossed is $\frac{1}{6}$ (see Example 4.8 on page 221). But suppose you are given the information that the sum of the two numbers showing on the dice is even. Would you still believe that the probability of observing a 7 on that particular toss is $\frac{1}{6}$? Intuitively, you will realize that the probability of observing a 7 is now 0. Since you know that an even number occurred, the outcome 7 cannot have occurred (because 7 is an odd number). The probability of observing a 7, *given that you know some other event has already occurred*, is called the *conditional probability* of the event.

> ### Definition 4.8
>
> The probability of an event *A*, given that an event *B* has occurred, is called the **conditional probability of *A* given *B*** and is denoted by the symbol
>
> $$P(A \mid B)$$
>
> [*Note:* The vertical bar between *A* and *B* is read "given."]

Example 4.12   Conditional Probability

A box contains three fuses, one good and two defective. Two fuses are drawn in sequence, first one and then the other.

a.   What is the probability that the second fuse drawn is defective?

b.   What is the probability that the second fuse drawn is defective if you know, for certain, that the first fuse drawn is defective?

Solution

a. We will denote the good fuse by G and the two defective fuses as D_1 and D_2. If the fuses are drawn at random from the box, the six possible orders of selection, i.e., the six simple events, are

$$(G, D_1) \quad (G, D_2) \quad (D_1, G) \quad (D_1, D_2) \quad (D_2, G) \quad (D_2, D_1)$$

Step 1 Since these six mutually exclusive events are equally likely and comprise all possible outcomes of the draw, we have $M = 6$.

Step 2 Next, we must find the number of selections in which a defective fuse is selected in the second draw. You can see from the listed draws that $m = 4$.

Step 3 Using Probability Rule #2, we conclude that the unconditional probability of obtaining a defective fuse on the second draw is

$$P(\text{defective fuse on the second draw}) = \frac{m}{M} = \frac{4}{6} = \frac{2}{3}$$

b. The probability of observing a defective fuse on the second draw, given that you have observed a defective fuse on the first draw, is the conditional probability $P(A \mid B)$, where

    A: Observe a defective fuse on the second draw.

    B: Observe a defective fuse on the first draw.

If the first fuse drawn from the box is defective, then the box now contains only two fuses, one defective and one nondefective. This means that there is a 50% chance of drawing a defective fuse on the second draw, given that a defective fuse has already been drawn. That is,

$$P(A \mid B) = \tfrac{1}{2}$$

The probability obtained in part **a** (the unconditional probability of event A) was equal to $\tfrac{2}{3}$. Clearly, the probability has changed when we know that event B has occurred. ❏

Example 4.13   Conditional Probability

A fair coin is tossed 10 times, resulting in 10 tails. If the coin is tossed one more time, what is the probability of observing a head?

Solution

We are asked to find the conditional probability of event A, given that event B has occurred, where

    A: The 11th toss results in a head.

    B: The first 10 tosses resulted in 10 heads.

Intuitively, it may seem reasonable to expect the probability of observing a head on the 11th toss (given that the 10 previous tosses resulted in heads) to be greater than $\tfrac{1}{2}$, but such is not the case. If the coin is truly balanced and is tossed in an unbiased manner, then the probability of observing a head on the 11th toss is still $\tfrac{1}{2}$. (This has been verified both theoretically and experimentally.) Therefore, this is a case where the conditional probability of an event A is equal to the unconditional probability of A. ❑

Self-Test
4.6

Consider the following experiment: Toss a single, fair die and observe the number of dots on the upper face. Given that the number is even, what is the probability of observing a 6?

Example 4.13 illustrates an important relationship that exists between some pairs of events. If the probability of one event does not depend on whether a second event has occurred, then the events are said to be *independent*.

The notion of independence is particularly important when we want to find the probability that *both* of two events will occur. When the events are independent, the probability that both events will occur is equal to the product of their unconditional probabilities.

Definition 4.9

Two events A and B are said to be **independent** if

$$P(A \mid B) = P(A) \tag{4.5}$$

or if

$$P(B \mid A) = P(B) \tag{4.6}$$

[*Note:* If one of these equalities is true, then the other will also be true.]

Probability Rule #4

The Probability that Both of Two Independent Events A and B Occur   If two events A and B are independent, then the *probability that both A and B occur* is equal to the product of their respective unconditional probabilities:

$$P(A \text{ and } B) = P(A)P(B) \tag{4.7}$$

Probability Rule #4 can be extended to apply to any number of independent events. For example, if A, B, and C are independent events, then

$$P(\text{all of the events } A, B, \text{ and } C \text{ occur}) = P(A \text{ and } B)P(C) = P(A)P(B)P(C)$$

Example 4.14   Applying Probability Rule #4

Find the probability of observing two heads in two tosses of a fair coin.

Solution

Define the following events:

    A: Observe a head on the first toss.

    B: Observe a head on the second toss.

Since we know that events A and B are independent and that $P(A) = P(B) = \frac{1}{2}$, the probability that we observe two heads, i.e., both events A and B, is

$$P(\text{observe two heads}) = P(A)P(B)$$

$$= \left(\tfrac{1}{2}\right)\left(\tfrac{1}{2}\right) = \tfrac{1}{4}$$ ❑

Self-Test

4.7

Consider the following experiment: Roll two dice and observe the number of dots showing on each. Find the probability of observing two 1's (called "snake-eyes" in the game of craps).

We now consider a problem in statistical inference.

Example 4.15   Using Probability to Make an Inference

Experience has shown that a manufacturing operation produces, on the average, only one defective unit in 10. These are removed from the production line, repaired, and returned to the warehouse. Suppose that during a given period of time you observe five defective units emerging in sequence from the production line.

a. If experience has shown that the defective units usually emerge randomly from the production line, what is the probability of observing a sequence of five consecutive defective units?

b. If the event in part **a** really occurred, what would you conclude about the process?

Solution

a. If the defectives really occur randomly, then whether any one unit is defective should be independent of whether the others are defective. Second, the unconditional probability that any one unit is defective is known to be $\frac{1}{10}$. We will define the following events:

D_1: The first unit is defective.

D_2: The second unit is defective.

$\vdots$

D_5: The fifth unit is defective.

Then

$$P(D_1) = P(D_2) = P(D_3) = P(D_4) = P(D_5) = \tfrac{1}{10}$$

and the probability that all five are defective is

$$P(\text{all five are defective}) = P(D_1)P(D_2)P(D_3)P(D_4)P(D_5)$$

$$= \left(\tfrac{1}{10}\right)\left(\tfrac{1}{10}\right)\left(\tfrac{1}{10}\right)\left(\tfrac{1}{10}\right)\left(\tfrac{1}{10}\right)$$

$$= \left(\tfrac{1}{100,000}\right)$$

b. We do not need a knowledge of probability to know that something must be wrong with the production line. Intuition would tell us that observing five defectives in sequence is highly improbable (given the past), and we would immediately infer that past experience no longer describes the condition of the process. In fact, we would infer that something is disturbing the stability of the process. ❏

Example 4.15 illustrates how you can use your knowledge of probability and the probability of a sample event to make an inference about some population. The technique, called the **rare event approach**, is summarized in the box.

Rare Event Approach to Making Statistical Inferences

Suppose we calculate the probability of event *A based on certain assumptions about the sampled population.* If $P(A) = p$ is small (say, p less than .05) and we observe that A occurs, then we can reach one of two conclusions:

1. Our original assumption about the sampled population is correct, and we have observed a rare event, i.e., an event that is highly improbable. (For example, we would conclude that the production line of Example 4.15, in fact, produces only 10% defectives. The fact that we observed five defectives in a sample of five was an unlucky and rare event.)

2. Our original assumption about the sampled population is incorrect. (In Example 4.15, we would conclude that the line is producing more than 10% defectives, a situation that makes the observed sample—five defectives in a sample of five—more probable.)

Using the rare event approach, we prefer conclusion 2. The fact that event A did occur in Example 4.15 leads us to believe that $P(A)$ is much higher than p, and that our original assumption about the population is incorrect.

PROBLEMS
4.36–4.50

Using the Tools

4.36 Assume that $P(A) = .6$ and $P(B) = .3$. If A and B are independent, find $P(A \text{ and } B)$.

4.37 Assume that $P(A) = .2$ and $P(B) = .9$. If A and B are independent, find $P(A \text{ and } B)$.

4.38 If $P(A) = .5$, $P(B) = .7$, and $P(B \mid A) = .5$, are A and B independent?

4.39 Assume that $P(A) = .6$, $P(B) = .4$, $P(C) = .5$, $P(A \mid B) = .15$, $P(A \mid C) = .5$, and $P(B \mid C) = .3$.

   a. Are events A and B independent?

   b. Are events A and C independent?

   c. Are events B and C independent?

4.40 Consider an experiment that consists of two trials and has the nine possible outcomes (simple events) listed here:

$$AA \quad AB \quad AC \quad BA \quad BB \quad BC \quad CA \quad CB \quad CC$$

where AC indicates that A occurs on the first trial and C occurs on the second trial. Suppose the following events are defined:

D: Observe an A on the first trial.
E: Observe a B on the second trial.

a. List the simple events associated with event D and find $P(D)$.

b. List the simple events associated with event E and find $P(E)$.

c. Find $P(E \mid D)$.

d. Are D and E independent?

e. Find $P(\text{both } D \text{ and } E \text{ occur})$.

Applying the Concepts

4.41 According to the Newspaper Advertising Bureau, 40% of all those primarily responsible for car maintenance are women. Consider the population consisting of all primary car maintainers.

a. What is the probability that a primary car maintainer, selected from the population, is a woman? A man?

b. What is the probability that both primary car maintainers in a sample of two selected from the population are women?

4.42 To develop programs for business travelers staying at convention hotels, Hyatt Hotels Corp. commissioned a study of executives who play golf (*Tampa Tribune*, July 10, 1993). The research revealed two surprising results: (1) 55% of the respondents admitted they had cheated at golf; (2) more than one-third of the respondents that admitted cheating at golf said they have also lied in business. Let A represent the event that an executive has cheated at golf and let B represent the event that the executive has lied in business. Convert the research results into probability statements involving events A and B.

4.43 Recall that the Excel workbook, FISH.XLS, contains data on the DDT contamination of fish in the Tennessee River in Alabama. Part of the investigation by the U.S. Army Corps of Engineers focused on how far upstream the contaminated fish have migrated. (A fish is considered to be contaminated if its measured DDT concentration is greater than 5.0 parts per million.) In addition to DDT concentration, the species and capture location (in miles from the river's mouth) for each contaminated fish were recorded. The accompanying cross-classification table gives the number of contaminated fish found for each species–location combination. Suppose a contaminated fish is captured from the river.

FISH.XLS

		CAPTURE LOCATION (MILES)	
		275–300	**305–350**
Species	Bass/Buffalo	9	8
	Channel Catfish	31	29

a. Given that the fish is a channel catfish, what is the probability that it is captured 305–350 miles upstream?

b. Given that the fish is captured 275–300 miles upstream, what is the probability that it is of the bass/buffalo species?

4.44 An article in *IEEE Computer Applications in Power* (Apr. 1990) describes "an unmanned watching system to detect intruders in real time without spurious detections, both indoors and outdoors, using video cameras and microprocessors." The system was tested outdoors under various weather conditions in Tokyo, Japan. The numbers of intruders detected and missed under each condition are provided in the table.

| | WEATHER CONDITION | | | | |
	Clear	Cloudy	Rainy	Snowy	Windy
Intruders detected	21	228	226	7	185
Intruders missed	0	6	6	3	10
TOTALS	21	234	232	10	195

Source: Kaneda, K., et al. "An unmanned watching system using video cameras." *IEEE Computer Applications in Power,* Apr. 1990, p. 24.

a. Under cloudy conditions, what is the probability that the unmanned system detects an intruder?

b. Given that the unmanned system missed detecting an intruder, what is the probability that the weather condition was snowy?

4.45 A recent survey of 12th-graders revealed the following information on youth smoking: 34% of the 12th-graders were classified as smokers; 32% own merchandise promoting cigarette brand names and trademarks; of those 12th-graders who own a promotional item, 58% smoke; and of those 12th-graders who do not own a promotional item, 23% are smokers (*Tampa Tribune,* Dec. 15, 1997).

a. Convert each of these percentages into a probability.

b. Consider the events $S = \{$a 12th-grader smokes$\}$ and $I = \{$a 12th-grader owns merchandise promoting cigarettes$\}$. Are S and I independent? Explain.

4.46 Nightmares about college exams appear to be common among college graduates. In a recent survey of 30- to 45-year-old graduates from Transylvania University (Kentucky), 50 of 188 respondents admitted they had recurring dreams about college exams (*Tampa Tribune,* Dec. 12, 1988). Of these 50, 47 felt distress, anguish, fear, or terror in their dreams. (For example, some dreamers "couldn't find the building or they walked in and all the students were different." Other dreamers either overslept or didn't realize they were enrolled in the class.)

a. Calculate the approximate probability that a 30- to 45-year-old graduate of Transylvania University has recurring dreams about college exams. Why is this probability approximate?

b. Refer to part **a**. Given that the graduate has recurring dreams, what is the approximate probability that the dreams are unpleasant (i.e., that the graduate feels distress, anguish, fear, or terror in the dreams)?

c. Are the events $\{$graduate has recurring dreams$\}$ and $\{$dreams are unpleasant$\}$ independent?

4.47 Certain magazine publishers promote their products by mailing sweepstakes packets to consumers. These packets offer the chance to win a grand prize of $1 million or more, with no obligation to purchase any of the advertised products. Three popular sweepstakes are conducted by Publishers Clearing House, American Family Publishers, and Reader's Digest. On a nationwide basis, the odds of winning the grand prize are 1 in 181,795,000 for the Publishers Clearing House sweepstakes, 1 in 200,000,000 for the American Family Publishers sweepstakes, and 1 in 84,000,000 for the Reader's Digest sweepstakes.

a. Calculate the probability of winning the grand prize for each of the three sweepstakes.

b. Suppose you enter the sweepstakes contests of all three companies. What is the probability that you win the grand prize in all three contests?

c. What is the probability that you do not win any of the three grand prizes?

d. Use the probability computed in part **c** to calculate the probability of winning at least one of the three grand prizes. [*Hint:* The complement of "at least one" is "none."]

4.48 According to NASA, each space shuttle in the U.S. fleet has 1,500 "critical items" that could lead to catastrophic failure if rendered inoperable during flight. NASA estimates that the chance of at least one critical-item failure within the shuttle's main engines is about 1 in 63 for each mission (*Tampa Tribune*, Dec. 3, 1993). To build the space station *Freedom*, NASA plans to fly eight shuttle missions a year.

a. Find the probability that none of the eight shuttle flights scheduled next year results in a critical-item failure.

b. Use the Rule of Complements and the result, part **a**, to find the probability that at least one of the eight shuttle flights scheduled next year results in a critical-item failure.

	English	**Chinese**
$P(I|I)$	.60	.78
$P(I|G)$	.30	.38
$P(G|I)$	.11	.12
$P(G|G)$	.50	.56

Source: Trafimow, D., et al. "The effects of language and priming on the relative accessibility of the private self and the collective self." *Journal of Cross-Cultural Psychology,* Vol. 28, No. 1, Jan. 1997, p. 116 (Table 2).

Problem 4.49

4.49 An experiment was performed to test whether language influences self cognition (*Journal of Cross-Cultural Psychology,* Jan. 1997). English-speaking Chinese students completed a self-attitudes questionnaire in which responses were categorized as idiocentric (*I*) or group (*G*). The conditional probability table at left gives the probability of a particular type of response given the previous response was either *I* or *G* when the questions concerned family and friends. The conditional probabilities in the table were computed for students answering questions written in English and for those answering questions written in Chinese. Interpret, in the words of the problem, each of the conditional probabilities in the table.

4.50 The *Journal of Teaching in Physical Education* (Apr. 1997) published a research note on sequential analysis of behavioral data collected in physical education classes. An example was given on teacher instruction (event *A*) and student behavior response (event *B*). During one 4-minute session with an expert teacher, the following chain of events occurred.

Expert: *ABABAABABABABAABABBABABABABABA*

During another 4-minute session with a novice teacher, the following chain of events occurred.

Novice: *ABAABABBABBAAABABBABABAAABBAAABB*

a. For the session with an expert teacher, find the probability of the student behavior response (event *B*) occurring.

b. Repeat part **a** for the session with a novice teacher.

c. Given that the expert teacher provides an instruction (event *A*), find the probability that the student behavior (event *B*) occurs immediately after.

d. Repeat part **c** for the novice teacher.

4.6 THE ADDITIVE AND MULTIPLICATIVE LAWS OF PROBABILITY (OPTIONAL)

In this optional section, we define some standard probability notation and give two laws for finding probabilities. Although these laws are not required for a study of the remaining material in the text, they are needed to complete an introductory coverage of probability.

When both of two events *A* and *B* occur (Section 4.4), this is called the *intersection of A and B* and is denoted $A \cap B$. When either *A* or *B* occurs (Section 4.3), this is called the *union of A and B* and is denoted $A \cup B$.

> **Definition 4.10**
>
> The **intersection of A and B**, denoted by $A \cap B$, is the event that both *A* and *B* occur. (See Figure 4.6.)

> **Definition 4.11**
>
> The **union of A and B**, denoted by $A \cup B$, is the event that either *A* or *B* occurs. (See Figure 4.7.)

The graphical representation that shows the relationships among events in an experiment (as in Figures 4.6 and 4.7) is called a **Venn diagram**.

Probability Rule #4 (Section 4.5) gave a formula for finding the probability that both events *A* and *B* occur (i.e., $A \cap B$) for the special case where *A* and *B* are independent events. We now give a formula, called the *Multiplicative Law of Probability*, that applies in general—that is, regardless of whether *A* and *B* are independent events.

Figure 4.6 Illustration of the Intersection of Two Events, $A \cap B$

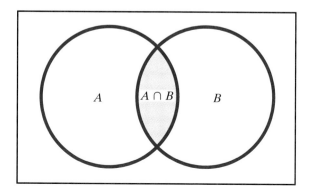

Figure 4.7 Illustration of the Union of Two Events, $A \cup B$

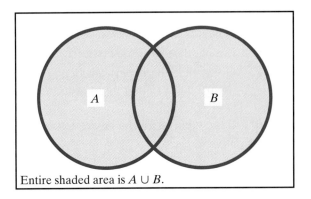

Entire shaded area is $A \cup B$.

Probability Rule #5

The Multiplicative Law of Probability   The probability that *both* of two events A and B occur is

$$P(A \text{ and } B) = P(A \cap B) = P(A)(P(B \mid A)$$
$$= P(B)(P(A \mid B) \tag{4.8}$$

Example 4.16   **Applying Probability Rule #5**

Refer to Example 4.12 (pages 233–234), where we selected two fuses from a box that contained three, two of which were defective. Use the Multiplicative Law of Probability to find the probability that you first select defective fuse D_1 and then select D_2.

Solution

Define the following events:

  A: The second fuse selected is D_2.

  B: The first fuse selected is D_1.

The probability of event B is $P(B) = \frac{1}{3}$. Also, from Example 4.12, the conditional probability of A given B is $P(A \mid B) = \frac{1}{2}$. Then the probability that both events A and B occur is

$$P(A \text{ and } B) = P(B)P(A \mid B) = (\tfrac{1}{3})(\tfrac{1}{2}) = \tfrac{1}{6}$$

You can verify this result by rereading Example 4.12. ❑

   An additive probability rule, Probability Rule #1, was given for the event that either A or B occurs (i.e., $A \cup B$), but it applies only to the case where A and B are mutually exclusive events. A rule that applies in general is given by the *Additive Law of Probability.*

Probability Rule #6

The Additive Law of Probability   The probability that *either* an event A *or* an event B *or both* occur is

$$P(A \text{ or } B) = P(A \cup B) = P(A) + P(B) - P(A \cap B) \tag{4.9}$$

Example 4.17   Applying Probability Rule #6

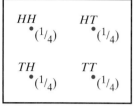

An experiment consists of tossing two coins with the sample space shown in Figure 4.8. Define the following events:

> *A:* Observe at least one head.
>
> *B:* Observe at least one tail.

Use the Additive Law of Probability to find the probability of observing either *A* or *B* or both.

Figure 4.8 Sample Space for Example 4.17

HH •($^1/_4$)	*HT* •($^1/_4$)
TH •($^1/_4$)	*TT* •($^1/_4$)

Solution

We know the answer to this question before we start because the probability of observing at least one head or at least one tail is 1—i.e., the event is a certainty. To obtain this answer using the Additive Law of Probability, we first find

$$P(A) = P(\text{at least one head}) = P(HH) + P(HT) + P(TH) = \tfrac{3}{4}$$

$$P(B) = P(\text{at least one tail}) = P(TT) + P(HT) + P(TH) = \tfrac{3}{4}$$

The event that both *A* and *B* occur—observing at least one head and at least one tail—is the event that you observe exactly one head and exactly one tail. The probability of this event is

$$P(\text{both } A \text{ and } B \text{ occur}) = P(HT) + P(TH) = \tfrac{1}{4} + \tfrac{1}{4} = \tfrac{1}{2}$$

Now, we apply equation 4.6 to obtain

$$P(\text{either } A \text{ or } B \text{ or both occur}) = P(A) + P(B) - P(\text{both } A \text{ and } B \text{ occur})$$

$$= \tfrac{3}{4} + \tfrac{3}{4} - \tfrac{1}{2} = 1 \qquad ❏$$

Self-Test

4.8

For two events *A* and *B*, suppose $P(A) = .8$, $P(B) = .5$, and $P(A \cap B) = .6$.

   a. Find $P(A \cup B)$.   **b.** Find $P(B \mid A)$.

In Examples 4.16 and 4.17, two key words helped us identify which probability law to employ. In Example 4.16, the key word was *and*, as in "find the probability that both *A and B* occur." The word *and* implies intersection; therefore, we use the Multiplicative Law of Probability. Alternatively, the key word was *or* in Example 4.17, as in "find the probability that either *A or B or both* occur." The word *or* implies union; therefore, we use the Additive Law of Probability.

Example 4.18   Applying Several Probability Rules

Psychologists believe that there is a relationship between aggressiveness and order of birth. To test this belief, a psychologist randomly chose 1,000 elementary school children and administered each a test designed to measure the student's aggressiveness. Each student was then classified according to aggressiveness (aggressive or unaggressive) and order of birth (firstborn, secondborn, or other).

Table 4.1 Results of Aggressiveness Test

		ORDER OF BIRTH			
		Firstborn	**Secondborn**	**Other**	**TOTALS**
Aggressiveness	Aggressive	60	100	80	240
	Unaggressive	120	330	310	760
TOTALS		180	430	390	1000

The number of students falling in the six categories are shown in the cross-classi-fication table, Table 4.1.

Suppose a single elementary school student is randomly selected from the 1,000 in the study. Define the following events:

A: The student is aggressive.

B: The student was firstborn.

a. Find the probability that both *A* and *B* occur.

b. Find the conditional probability that *A* will occur given that *B* has occurred.

c. Find the probability that *A* will not occur.

d. Find the probability that either *A* or *B* or both occur.

Solution

a. We can see from the table that 60 of the 1,000 students were classified as both aggressive (*A*) and firstborn (*B*). Therefore,

$$P(A \text{ and } B) = P(A \cap B) = \frac{60}{1,000} = .06$$

b. Intuitively, this conditional probability can be found by focusing on the "firstborn" column of Table 4.1, i.e., the column corresponding to the conditional event *B*. Now, of the 180 firstborn children, 60 are "aggres-sive." Hence,

$$P(A \mid B) = \frac{60}{180} = .333$$

Alternatively, we can use the Multiplicative Law of Probability to find $P(A \mid B)$. Examining Table 4.1, we find that 240 of all the students were classified as aggressive and 180 were the firstborn in their family. Therefore,

$$P(A) = \tfrac{240}{1,000} = .24 \quad \text{and} \quad P(B) = \tfrac{180}{1,000} = .18$$

To find $P(A \mid B)$, we substitute the answer to part **a** and the value of $P(B)$ into the formula for the Multiplicative Law of Probability. Thus,

$$P(A \text{ and } B) = P(A \cap B) = P(B)P(A \mid B)$$

or

$$.06 = (.18)P(A \mid B)$$

Solving for $P(A \mid B)$ yields

$$P(A \mid B) = \frac{.06}{.18} = .333$$

This answer agrees with our intuitive approach.

c. The event that A does not occur is the complement of A, denoted by the symbol A'. Since A is the event that a student is aggressive, A' is the event that a student is unaggressive. Recall that $P(A)$ and $P(A')$ bear a special relationship to each other:

$$P(A) + P(A') = 1$$

From part **b**, we have $P(A) = .24$. Therefore,

$$P(A') = 1 - .24 = .76$$

which can be verified by examining Table 4.1.

d. The probability that either A or B or both occur is given by the Additive Law of Probability. From parts **a** and **b** we know that

$$P(A) = .24 \qquad P(B) = .18 \qquad P(A \cap B) = .06$$

Then,

$$P(A \text{ or } B) = P(A) + P(B) - P(A \cap B)$$
$$= .24 + .18 - .06 = .36 \qquad \square$$

PROBLEMS 4.51–4.68

Using the Tools

4.51   For two events A and B, $P(A) = .3, P(B) = .5$, and $P(A \cap B) = .2$. Find $P(A \cup B)$.

4.52   For two events A and B, $P(A) = .5, P(B) = .6$, and $P(A \cap B) = .4$. Find $P(A \cup B)$.

4.53   For two events A and B, $P(A) = .7, P(B) = .2$, and $P(A \mid B) = .3$. Find $P(A \cap B)$.

4.54   For two events A and B, $P(A) = .1, P(B) = .3$, and $P(B \mid A) = .8$. Find $P(A \cap B)$.

4.55   Consider the experiment of rolling a pair of dice. Define events A and B as follows:

> A: Observe a sum of 7.
> B: Observe a 4 on at least one die.

a.   List the possible outcomes in event A and find $P(A)$.

b. List the possible outcomes in event B and find $P(B)$.

c. Find $P(A \cap B)$ using the Multiplicative Law of Probability. Then list the possible outcomes associated with this event, and find $P(A \cap B)$ by summing the probabilities of these outcomes.

d. Find $P(A \cup B)$ using the Additive Law of Probability. Then list the possible outcomes associated with this event and find $P(A \cup B)$ by summing the probabilities of these outcomes.

Applying the Concepts

4.56 The National Acid Precipitation Assessment Program (NAPAP) conducted a 10-year study of acid rain. In its report, NAPAP estimates the probability of an Adirondack lake being acidic at .14. Given that the Adirondack lake is acidic, the probability that the lake comes by its acidity naturally is .25 (*Science News,* Sept. 15, 1990). Use this information to find the probability that an Adirondack lake is naturally acidic.

4.57 According to the National Association of State Boards of Accountancy, the probability is .20 that a first-time candidate passes all subjects in the Uniform CPA Examination (*New Accountant,* Sept. 1993). Given that a first-time candidate fails, the probability that such a candidate passes all subjects increases to .30 on repeat examinations. Use this information to find the probability that a candidate passes the Uniform CPA Examination on his or her second attempt.

4.58 The National Fire Incident Reporting System (NFIRS) is an accumulation of fire reports submitted by fire departments across the country. According to the NFIRS, 5% of all fires in the U.S. are started by children. Of the fires started by children, 25% occur inside the home. (*Tampa Tribune,* Nov. 9, 1996.)

a. Based on this information, a newspaper reporter wrote that "25 percent of all fires in the U.S. are started by children inside the home." Do you agree with this statement?

b. Find the probability that a fire is started by a child inside the home.

4.59 Refer to the Jackpot scratch-off game described in Problem 4.18 (page 226). Recall that eight of the 10 play squares reveal three cherries, "worth up to $1,000 in cash or prizes." Given that you scratch off three cherries, you will receive one of the following prizes (with associated probabilities):

$1,000 cash (.0000125)
$500 cash (.0000125)
VCR (.0000125)
Grandfather clock (.0000625)
Designer watch (.0000625)
Assorted prizes (.9998375)

a. Given that you scratch off three cherries, find the probability that you win cash.

b. What is the unconditional probability that you win cash in Jackpot?

4.60 As part of a study of phonological dyslexia in a highly literate individual, researchers asked 22 undergraduate students to spell a list of familiar but commonly misspelled words (*Quarterly Journal of Experimental*

Word	Number Misspelled
sacrilegious	21
Gandhi	19
professor	8
khaki	5

Problem 4.60

Psychology, Aug. 1985). The purpose of the study was to compare the misspellings of the control group (i.e., the 22 undergraduates) with the misspellings of the dyslexic subject. Several words on the list, along with the number of students in the control group who misspelled the words, are given in the accompanying table. Suppose one of the 22 students in the control group is selected at random.

a. What is the probability that the student misspelled the word *sacrilegious*?

b. What is the probability that the student spelled the word *professor* correctly?

c. Suppose that 14% of the students in the control group misspelled both the words *Gandhi* and *khaki.* What is the probability that the selected student misspelled either *Gandhi* or *khaki* or both?

4.61 Refer to the study of a drug designed to reduce blood loss in coronary artery bypass patients, Problem 2.37 (page 93). Recall that half of the 114 patients were administered the drug prior to the operation and half received a placebo (no drug). The type of complication (if any) experienced by the patient was recorded. The results are reproduced in the accompanying cross-classification table. Suppose we randomly select one of the 114 coronary artery bypass patients in the study.

		DRUG		
		Yes	**No**	**TOTALS**
Complication	Yes	17	10	27
	No	40	47	87
	TOTALS	57	57	114

a. Find the probability that the patient received the drug.

b. Find the probability that the patient had no complications.

c. Find the probability that the patient received the drug and had no complications.

d. Given that the patient received the drug, what is the probability that no complications occurred?

e. Given that the patient did not receive the drug, what is the probability that no complications occured?

4.62 Refer to the *American Journal on Mental Retardation* study of social interactions of children, Problem 2.38 (page 94). The 30 children in the study were classified according to type of behavior (disruptive or not) and whether or not they displayed developmental delays. The data are summarized in the cross-classification table at the top of page 248. Consider a randomly selected child from the 30 studied.

a. What is the probability that the child exhibits disruptive behavior and developmental delays?

b. What is the probability that the child exhibits disruptive behavior, given that the child does not display developmental delays?

4.63 A study conducted by the National Opinion Research Center found that Hispanics who are proficient in English do better in school than those who communicate primarily in Spanish. Of those students who speak

	Disruptive Behavior	Nondisruptive Behavior	TOTALS
With developmental delays	12	3	15
Without developmental delays	5	10	15
TOTALS	17	13	30

Source: Kopp, C. B., Baker, B., and Brown, K. W. "Social skills and their correlates: Preschoolers with developmental delays." *American Journal on Mental Retardation,* Vol. 96, No. 4, Jan. 1992.

Problem 4.62

only Spanish at home, 41% earn a high school grade average of B or better. In contrast, 53% of all students earn a grade of B or better. It is also known that 5% of all students speak only Spanish at home, while 89% speak only English at home. Consider the following events:

 A: A student earns a grade of B or better.
 B: A student speaks only Spanish at home.
 C: A student speaks only English at home.

a. Find $P(A)$.   **b.** Find $P(B)$.   **c.** Find $P(C)$.

d. Find $P(A \mid B)$.   **e.** Find $P(A \text{ and } B)$.

f. Which (if any) of the pairs of events, A and B, A and C, or B and C, are mutually exclusive?

4.64 A survey of 300 children in the San Francisco Bay area showed that 90% owned pets. Of these, 30% owned dogs only, 18% owned dogs and other pets, 22% owned cats only, 22% owned cats and other pets, and 8% owned only pets other than dogs and cats (*Psychological Reports,* 1985, Vol. 57). For a randomly selected child, find the following probabilities:

a. The child owns a pet.   **b.** The child owns a dog only.

c. The child owns a cat only.   **d.** The child owns a dog.

4.65 In 1941, Ted Williams of the Boston Red Sox became the last major league baseball player to record a batting average over .400, and Joe DiMaggio of the New York Yankees had the longest hitting streak in major league baseball history—56 games. *Chance* (Fall 1991) assessed the probabilities of the events occurring within the next 50 years as $P(\text{bat over } .400) = .035$ and $P(\text{56-game hitting streak}) = .016$ under specific conditions (i.e., a league batting average of .270).

a. Given a current major league baseball star (e.g., Ken Griffey, Jr., of the Seattle Mariners) has just completed a 56-game hitting streak, explain why it is not possible to calculate the probability that he will end the season with a batting average of over .400. Assume the two events, {bat over .400} and {56-game hitting streak}, are dependent.

b. Calculate the conditional probability of part **a**, assuming the two events are independent. (If you have knowledge of baseball, batting averages, and hitting streaks, do you believe the two events are independent?)

4.66 Refer to the *Psychological Science* experiment in which 120 subjects participated in the "water-level task," Problem 2.39 (page 94). Recall that the subjects were shown a drawing of a water glass tilted at a 45°

angle and asked to draw a line representing the surface of the water. A summary of the results is reproduced in the table.

| | GROUP | | | | | | |
| | FEMALES | | | MALES | | | |
Judged Line	Students	Waitresses	Housewives	Students	Bartenders	Bus Drivers	TOTALS
More than 5° below surface	0	0	1	1	1	1	4
More than 5° above surface	7	15	13	3	11	4	53
Within 5° of surface	13	5	6	16	8	15	63
TOTALS	20	20	20	20	20	20	120

Source: Hecht, H., and Proffitt, D. R. "The price of experience: Effects of experience on the water-level task." *Psychological Science,* Vol. 6, No. 2, Mar. 1995, p. 93 (Table 1).

Suppose that we select one subject at random from the 120 subjects in the study.

  a. What is the probability that the subject is a female?

  b. What is the probability that the subject's line is within 5° of the water surface?

  c. What is the probability that the subject is a waitress or bartender?

  d. What is the probability that the subject is a student and judges the line to be within 5° of the water surface?

  e. Given the subject is a waitress or bartender, what is the probability that the subject judges the line to be within 5° of the water surface?

  f. Given the subject is a male, what is the probability that the subject judges the line to be within 5° of the water surface?

  g. Given the subject judges the line to be within 5° of the water surface, is the subject more likely to be a bus driver or a housewife?

4.67 The merging process from an acceleration lane to the through lane of a freeway constitutes an important aspect of traffic operation at interchanges. From a study of parallel and tapered interchange ramps in Israel, the accompanying table provides information on traffic lags (where a lag is defined as an interval of time between arrivals of major streams of vehicles) accepted and rejected by drivers in the merging lane.

Type of Interchange Lane	Traffic Condition on Freeway	Number of Merging Drivers Accepting the First Available Lag	Number of Merging Drivers Rejecting the First Available Lag
Tapered	Heavy traffic	16	115
	Little traffic	67	121
Parallel	Heavy traffic	40	139
	Little traffic	144	331

Source: Polus, A., and Livneh, M. "Vehicle flow characteristics on acceleration lanes." *Journal of Transportation Engineering,* Vol. III, No. 6, Nov. 1985, pp. 600–601 (Table 4).

  a. What is the probability that a driver in a tapered merging lane with heavy traffic will accept the first available lag?

  b. What is the probability that a driver in a parallel merging lane will reject the first available lag in traffic?

c. Given that a driver accepts the first available lag in little traffic, what is the probability that the driver is in a parallel merging lane?

4.68 Children who develop unexpected difficulties with acquisition of spoken language are often diagnosed with specific language impairment (SLI). A study published in the *Journal of Speech, Language, and Hearing Research* (Dec. 1997) investigated the incidence of SLI in kindergarten children. As an initial screen, each in a national sample of over 7,000 children was given a test for language performance. The percentages of children who passed and failed the screen were 73.8% and 26.2%, respectively. All children who failed the screen were tested clinically for SLI. About one-third of those who passed the screen were randomly selected and also tested for SLI. The percentage of children diagnosed with SLI in the "failed screen" group was 20.5%; the percentage diagnosed with SLI in the "pass screen" group was 2.8%.

a. For this problem, let "pass" represent a child who passed the language performance screen, "fail" represent a child who failed the screen, and "SLI" represent a child diagnosed with SLI. Now find each of the following probabilities: $P(\text{Pass})$, $P(\text{Fail})$, $P(\text{SLI} \mid \text{Pass})$, and $P(\text{SLI} \mid \text{Fail})$.

b. Use the probabilities, part **a**, to find $P(\text{Pass} \cap \text{SLI})$ and $P(\text{Fail} \cap \text{SLI})$. What probability law did you use to calculate these probabilities?

c. Use the probabilities, part **b**, to find $P(\text{SLI})$. What probability law did you use to calculate this probability?

REAL-WORLD REVISITED
Lottery Buster!

"Welcome to the Wonderful World of Lottery Bu$ters." So begins the premier issue of *Lottery Buster,* a monthly publication for players of the state lottery games. *Lottery Buster* provides interesting facts and figures on the 36 state lotteries currently operating in the United States and, more importantly, tips on how to increase a player's odds of winning the lottery.

New Hampshire, in 1963, was the first state in modern times to authorize a state lottery as an alternative to increasing taxes. (Prior to this time, beginning in 1895, lotteries were banned in America because of corruption.) Since then, lotteries have become immensely popular for two reasons. First, they lure you with the opportunity to win millions of dollars with a $1 investment, and second, when you lose, at least you believe your money is going to a good cause.

The popularity of the state lottery has brought with it an avalanche of "experts" and "mathematical wizards" (such as the editors of *Lottery Buster*) who provide advice on how to win the lottery—for a fee, of course! These experts—the legitimate ones, anyway—base their "systems" of winning on their knowledge of probability and statistics.

For example, most experts would agree that the "golden rule" or "first rule" in winning lotteries is *game selection*. State lotteries generally offer three types of games: Instant (scratch-off) tickets, Daily Numbers (Pick-3 or Pick-4), and the weekly Pick-6 Lotto game.

Figure 4.9 Reproduction of Florida's 6/49 Lotto Ticket

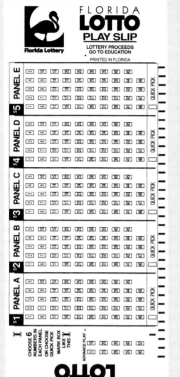

The Instant game involves scratching off the thin opaque covering on a ticket with the edge of a coin to determine whether you have won or lost. The cost of a ticket ranges from 50¢ to $1, and the amount won ranges from $1 to $100,000 in most states, and to as much as $1 million in others. *Lottery Buster* advises against playing the Instant game because it is "a pure chance play, and you can win only by dumb luck. No skill can be applied to this game."

The Daily Numbers game permits you to choose either a three-digit (Pick-3) or four-digit (Pick-4) number at a cost of $1 per ticket. Each night, the winning number is drawn. If your number matches the winning number, you win a large sum of money, usually $100,000. You do have some control over the Daily Numbers game (since you pick the numbers that you play) and, consequently, there are strategies available to increase your chances of winning. However, the Daily Numbers game, like the Instant game, is not available for out-of-state play.

To play Pick-6 Lotto, you select six numbers of your choice from a field of numbers ranging from 1 to N, where N depends on which state's game you are playing. For example, Florida's Lotto game involves picking six numbers ranging from 1 to 49 (denoted 6/49) as shown on the Florida Lotto ticket, Figure 4.9; Delaware's Lotto is a 6/36 game, and New York's is a 6/54 game. The cost of a ticket is $1 and the payoff, if your six numbers match the winning numbers drawn at the end of each week, is $7 million or more, depending on the number of tickets purchased. (To date, PowerBall has had the largest weekly payoff of over $200 million.) In addition to the grand prize, you can win second-, third-, and fourth-prize payoffs by matching five, four, and three of the six numbers drawn, respectively. And you don't have to be a resident of the state to play the state's Lotto game.

Study Questions

a. Consider Florida's 6/49 Lotto game. Calculate the number of possible ways in which you can choose the six numbers from the 49 available. If you purchase a single $1 ticket, what is the probability that you win the grand prize (i.e., match all six numbers)?

b. Repeat part **a** for Delaware's 6/36 game.

c. Repeat part **a** for New York's 6/54 game.

d. Since you can play any state's Lotto game, which of the three, Florida, Delaware, or New York, would you choose to play? Why?

e. Prior to 1987, New York operated a 6/48 Lotto game. Compute the probability of winning a 6/48 Lotto. Compare it to your answer to part **c**.

f. One strategy used to increase your odds of winning a Lotto is to employ a *wheeling system.* In a complete wheeling system, you select more than six numbers, say, seven, and play every combination of six of those seven numbers. Suppose you choose to "wheel" the following seven numbers in a 6/40 game: 2, 7, 18, 23, 30, 32, 39. How many tickets would you need to purchase to have every possible combination of the seven numbers? List the six numbers on each of these tickets.

g. Refer to part **e**. What is the probability of winning the 6/40 Lotto when you wheel seven numbers? Does the strategy, in fact, increase your odds of winning?

Key Terms

Starred (*) terms are from the optional sections of this chapter.

Complementary events 223	Intersection* 241	Simple events 220
Conditional probability 233	Mutually exclusive events 218	Union* 241
Event 215	Probability 216	Venn diagram 241
Experiment 214	Rare event approach 237	
Independent events 235	Sample space 220	

Key Formulas and Rules

Starred (*) problems refer to the optional sections of this chapter.

1. *Additive Rule:* If two events A and B are mutually exclusive, then

$$P(A \text{ or } B) = P(A \cup B) = P(A) + P(B) \qquad \textbf{(4.1)}, 220$$

   In general,*

$$P(A \text{ or } B) = P(A \cup B) = P(A) + P(B) - P(A \text{ and } B) \qquad \textbf{(4.9)}, 242$$

2. *Modified Additive Rule for Equally Likely Mutually Exclusive Events:* If an experiment results in one and only one of M equally likely mutually exclusive (simple) events, of which m of these result in an event A, then

$$P(A) = \frac{m}{M} \qquad \textbf{(4.2)}, 222$$

3. *Multiplicative Rule:* If two events A and B are independent, then

$$P(A \text{ and } B) = P(A \cap B) = P(A)P(B) \qquad \textbf{(4.7)}, 235$$

   In general,*

$$P(A \text{ and } B) = P(A \cap B) = P(A)P(B \mid A) = P(B)P(A \mid B) \qquad \textbf{(4.8)}, 242$$

4. *Rule of Complements:*

$$P(A) = 1 - P(A') \qquad \textbf{(4.3)}, 223$$

5. *Combinatorial Rule*:* The number of different ways to select n objects from a total of N is

$$\binom{N}{n} = \frac{N!}{n!(N-n)!} \quad \text{where} \quad N! = N(N-1)(N-2)\cdots(2)(1) \qquad \textbf{(4.4)}, 230$$

Key Symbols

Symbol	Definition
$P(A)$	Probability of an event A
A'	Complement of an event A
$\binom{N}{n}$	Number of ways to choose n objects from N objects
$n!$	n factorial
$P(A \mid B)$	Probability of A given B has occurred
$A \cup B$	A union (or) B
$A \cap B$	A intersect (and) B

Supplementary Problems 4.69–4.88

Starred (*) problems refer to the optional sections of this chapter.

4.69 Environmental engineers classify U.S. consumers into five mutually exclusive groups based on consumers' feelings about environmentalism:

1. "Basic browns" claim they don't have the knowledge to understand environmental problems.
2. "True-blue greens" use biodegradable products.
3. "Greenback greens" support requiring new cars to run on alternative fuel.
4. "Sprouts" recycle newspapers regularly.
5. "Grousers" believe industries, not individuals, should solve environmental problems.

The proportion of consumers in each group is shown in the table. Suppose a U.S. consumer is selected at random and his or her feelings about environmentalism are determined.

Basic browns	.28
True-blue greens	.11
Greenback greens	.11
Sprouts	.26
Grousers	.24

Source: The Orange County (Calif.) Register, Aug. 7, 1990.

Problem 4.69

a. List the simple events for the experiment.

b. Assign probabilities to the simple events.

c. Find the probability that the consumer is either a basic brown or a grouser.

d. Find the probability that the consumer supports environmentalism in some fashion (i.e., is a true-blue green, greenback green, or sprout).

4.70 Video card games (e.g., video poker), once only legal in casinos, are now available for play in laundromats and grocery stores of states that employ a lottery. Because these video games are less intimidating than the traditional casino table games (e.g., blackjack, roulette, and craps), experts speculate that more women are becoming addicted to them. *The Wall Street Journal* (July 14, 1992) reported on the gambling habits of 52 women in Gamblers Anonymous.

a. Of the 52 women, 47 were video poker players. Use this information to assess the likelihood that a female member of Gamblers Anonymous (GA) is or was addicted to video poker.

b. Refer to part **a**. Of the 47 video poker players, 35 gambled until they exhausted their family savings. Express this information in a probability statement.

*c. Use the probabilities, parts **a** and **b**, to find the probability that a female member of GA was addicted to video poker and gambled until she exhausted her family savings.

4.71 A child psychologist is interested in the ability of 5-year-old children to distinguish between imaginative and nonfiction stories. Two 5-year-old children are selected from a kindergarten class and are given a test to determine if they can distinguish imagination from reality. Consider the following events:

A: The first child can distinguish imagination from reality.
B: Neither child can distinguish imagination from reality.
C: The second child can distinguish imagination from reality, but the first child cannot.

Explain whether the pairs of events, *A* and *B, A* and *C,* and *B* and *C,* are mutually exclusive.

4.72   There are very few (if any) tests for pregnancy that are 100% accurate. Sometimes a test may indicate that a woman is pregnant even though she really is not. This is known as a *false positive* test. Similarly, a *false negative* result occurs when the test indicates that the woman is not pregnant even though she really is. Suppose a woman takes a certain pregnancy test. Define the events *A, B,* and *C* as follows:

> *A:* The woman is really pregnant.
> *B:* The test gives a false positive result.
> *C:* The test gives a false negative result.

Which pair(s) of events are mutually exclusive?

4.73   An improved method for measuring the electrical resistivity of concrete has been developed. The method was tested on concrete specimens with different water-cement mixes. Three different water-weight ratios (40%, 45%, and 50%) and three different amounts of cement (300, 350, and 400 kilograms per cubic meter) were examined.

a.   List all possible water-cement mixes for this experiment.

b.   Suppose we determine the water-cement mix that yields the highest electrical resistivity. Before the experiment is performed, should equal probabilities be assigned to the simple events? Why or why not?

4.74   Californians living along the San Andreas fault constantly fear the "Big One," i.e., a devastating major earthquake. According to a U.S. Geological Survey, "the question is not whether a [major] earthquake is coming. The question is when." The table below gives survey estimates of the probability of a major earthquake striking a particular area along the fault line before the year 2018.

Location	Probability of Earthquake
North Coast	Less than .10
Northern East Bay	.20
San Francisco Peninsula	.20
Southern East Bay	.20
South Santa Cruz Mountains	.30
Parkfield	.90
Cholame	.30
Carrizo	.10
Mojave	.30
San Bernardino Mountain	.20
San Bernardino Valley	.20
San Jacinto Valley	.10
Anza	.30
Coachella Valley	.40
Borrego Mountain	Less than .10
Imperial	.50

Source: U.S. Geological Survey; *Time,* Oct. 30, 1989.

a.   Are the events listed in the table mutually exclusive? Why?

b.   Do you believe the events listed in the table are independent? Why?

c. Explain why the probabilities listed in the table do not sum to 1.

d. Northern East Bay, San Francisco Peninsula, Southern East Bay, and South Santa Cruz Mountains are fault line segments in the San Francisco area. What is the probability that a major earthquake will strike the San Francisco area before the year 2018? What did you assume about the four locations to answer this question?

4.75 A nationwide telephone survey of 1,200 people, aged 22 to 61, was conducted to gauge Americans' plans for retirement (*Tampa Tribune,* May 20, 1997). Some of the results are listed below.

- 46% have saved less than $10,000 for their retirement.
- 68% acknowledge they could save more if they made the effort.
- Of those who acknowledge they could save more, 20% say they will make the effort to do so.

a. What is the probability that an American aged 22 to 61 has saved less than $10,000 for retirement?

b. What is the probability that an American aged 22 to 61 can save more for retirement with more effort?

c. Given an American aged 22 to 61 can save more for retirement, what is the probability that he or she will do so?

4.76 To find the probability that a randomly selected United States citizen was born in a given state—Virginia, for example—an introductory statistics student divides the number of favorable outcomes, one, by the total number of states, 50. Thus, the student reports that the probability that John Q. Citizen was born in Virginia is $\frac{1}{50}$. Explain why this probability is incorrect. If you had access to the complete 1990 census, how could the correct probability be obtained?

*4.77 In Florida's daily "Cash 5" lottery game, players select five different whole numbers ranging from 1 to 30. To win, all five numbers must match the five numbers randomly selected each day.

a. How many different combinations of five numbers from 30 are possible?

b. Based on your answer to part **a**, what is the probability of winning the "Cash 5" lottery?

4.78 A Northwestern University Department of Pediatrics study of bicycle-related injuries over a 7-year period found that more than 2,500 babies and toddlers were hurt while riding in seats mounted on an adult's bike. The accompanying table gives a breakdown of the causes of the injuries.

Cause of Bike-Related Injuries to Children	Percent
Fell out of seat	39
Accident with car	10
Stationary bike fell over	24
Seat fell off bike	6
Extremity caught in spoke	21
TOTAL	100

Suppose a child is injured in a bicycle accident.

a. What is the probability that the injury results from an extremity getting caught in the bicycle spoke?

b. What is the probability that the injury results from either the child falling out of the seat or the seat falling off the bike?

c. What is the probability that the injury did not occur as a result of an accident with a car?

4.79 Refer to the *Journal of the American Geriatric Society* (Jan. 1998) study of age-related macular degeneration (AMD), Problem 2.40 (page 94). To investigate the link between AMD incidence and alcohol consumption, 3,072 adults were tested for AMD and classified according to type of alcohol most frequently consumed. The cross-tabulation table for the study is given below.

		CLINICAL DIAGNOSIS		
		AMD	No AMD	TOTALS
Most Frequent Alcohol Type Consumed	None	87	878	965
	Beer	30	400	430
	Wine	51	1,233	1,284
	Liquor	20	373	393
	TOTALS	188	2,884	3,072

Consider an adult randomly selected from the 3,072 adults in the study.

a. What is the probability that the adult is diagnosed with AMD?

b. What is the probability that the adult consumes beer most frequently?

c. What is the probability that the adult is diagnosed with AMD and consumes liquor most frequently?

d. Given that the adult does not consume alcohol, what is the probability that he or she is diagnosed with AMD?

e. Given that the adult consumes wine most frequently, what is the probability that he or she is diagnosed with AMD?

f. What is the probability that the adult is not diagnosed with AMD or does not consume alcohol?

4.80 Refer to the *Family Planning Perspectives* study of New Jersey births, Problem 2.76 (page 124). The number of births over a 2-year period, by race of mother and maternal age group, are reproduced in the table on page 257. Consider a randomly selected New Jersey birth mother during this 2-year period.

a. Find the probability that the birth mother is white.

b. Find the probability that the birth mother is a teenager.

c. Find the probability that the birth mother is black and 40 or more years old.

d. Given the birth mother is white, what is the probability that she is a teenager?

e. Given the birth mother is a teenager, what is the probability that she is black?

| | | NUMBER OF NEW JERSEY BIRTHS | |
		White	Black
	<15	162	292
	15–17	3,084	3,551
	18–19	6,655	5,180
Maternal	20–24	29,057	12,699
Age Group	25–29	55,517	10,944
	30–34	49,023	6,588
	35–39	16,558	2,172
	≥40	2,273	376
	TOTALS	162,329	41,802

Source: Reichman, N. E., and Pagnini, D. L. "Maternal age and birth outcomes: Data from New Jersey." *Family Planning Perspectives,* Vol. 29, No. 6, Nov./Dec. 1997, p. 269 (Table 1).

Problem 4.80

    f. What is the probability that the birth mother is white or a teenager?

4.81 The transport of neutral particles in an evacuated duct is an important aspect of nuclear fusion reactor design. In one experiment, particles entering through the duct ends streamed unimpeded until they collided with the inner duct wall. Upon colliding, they are either scattered (reflected) or absorbed by the wall (*Nuclear Science and Engineering,* May 1986). The reflection probability (i.e., the probability a particle is reflected off the wall) for one type of duct was found to be .16.

    a. If two particles are released into the duct, find the probability that both will be reflected.

    b. If five particles are released into the duct, find the probability that all five will be absorbed.

    c. What assumption about the simple events in parts **a** and **b** is required to calculate the probabilities?

*4.82 Despite penicillin and other antibiotics, bacterial pneumonia still kills thousands of Americans every year. An antipneumonia vaccine, called Pneumovax, is designed especially for elderly or debilitated patients, who are usually the most vulnerable to bacterial pneumonia. Suppose the probability that an elderly or debilitated person will be exposed to these bacteria is .45. If exposed, the probability that an elderly or debilitated person inoculated with the vaccine acquires pneumonia is only .10. What is the probability that an elderly or debilitated person inoculated with the vaccine does acquire bacterial pneumonia?

*4.83 One of the most popular card games in the world is the game of bridge. Each player in the game is dealt 13 cards from a standard 52-card deck. Determine the number of different 13-card bridge hands that can be dealt from a 52-card deck.

4.84 A common data-collection method in consumer and market research is the telephone survey. However, a major problem with consumer telephone surveys is nonresponse. How likely are consumers to be at home to take the call and, if at home, how likely are they to take part in the survey? In a study of over 250,000 random-digit dialings of both listed and unlisted telephone numbers across the United States, the probabilities of various outcomes on the first dialing attempt were assessed and are shown in the table.

Result of Dialing Attempt	Probability
No answer	.347
Busy signal	.020
Out-of-service	.203
No eligible person at home	.291
Business number	.041
Eligible person at home—refusal	.014
Eligible person at home—completed interview	.084

Source: Kerin, R. A., and Peterson, R. A. *Journal of Advertising Research,* Apr./May 1983.

a. What is the probability that a single call will result in no answer or a busy signal or an out-of-service number?

b. What is the probability that an eligible person will be at home to take the call?

c. Given that an eligible person is at home to take the call, what is the probability that he or she will refuse to participate in the interview? (This probability is known as the *refusal rate.*)

4.85 *Science Digest* magazine features an interesting and informative columnist, known only by the eerie name of Dr. Crypton. Each month Dr. Crypton presents mind-twisters, riddles, puzzles, and enigmas that very often can be solved using the laws of probability. In the January 1982 issue, Dr. Crypton offered this view of what is commonly called the "birthday problem":

> Surely you have been in a situation in which a small group of people compared birthdays and found, to their surprise, that at least two of them were born on the same day of the same month. Suppose there are 10 people in the group. Intuition may suggest that the odds of two sharing a birthday are quite poor. Probability theory, however, shows the odds are better than one in nine.
>
> If you like to win bets, you should keep in mind that for a group of 23 people the odds are in favor of at least two of them sharing a birthday. (For 22 people, the odds are slightly against this.)*

a. Find the probability that no two of a group of 23 people share the same birthday. [*Hint:* Since there are 365 days in a year, any of which may be the birthday of one of the 23 people, the probability that none of the people in the group share a birthday is

$$\left(\frac{365}{365}\right) \cdot \left(\frac{364}{365}\right) \cdot \left(\frac{363}{365}\right) \cdots \left(\frac{343}{365}\right)$$

$$\begin{array}{cccc} \text{Person} & \text{Person} & \text{Person} & \text{Person} \\ \#1 & \#2 & \#3 & \#23 \end{array}$$

Compute this probability.]

b. Using the rule of complements (Probability Rule #3), find the probability that at least two of 23 people will share the same birthday.

c. Use the steps outlined in parts **a** and **b** to find the probability that at least two of 10 people will share the same birthday.

Parade Magazine, Feb. 17, 1991.

4.86 Marilyn vos Savant, who is listed in the *Guinness Book of World Records Hall of Fame* for "Highest IQ," writes a weekly column in the Sunday newspaper supplement *Parade Magazine.* Her column, "Ask Marilyn," is devoted to games of skill, puzzles, and mind-bending riddles. In one issue, vos Savant posed the following question:* Suppose you're on a game show, and you're given a choice of three doors. Behind one door is a car; behind the others, goats. You pick a door—say, #1—and the host, who knows what's behind the doors, opens another door—say, #3—which has a goat. He then says to you, "Do you want to pick door #2?" Is it to your advantage to switch your choice? By answering the following series of questions, you will arrive at the correct solution.

a. Before taping of the game show, the host randomly decides behind which of the three doors to put the car; the goats will go behind the remaining two doors. List the simple events for this experiment. [*Hint:* One simple event is $C_1 G_2 G_3$, i.e., car behind door #1 and goats behind door #2 and door #3.]

b. Randomly choose one of the three doors, as the contestant does in the game show. Now, for each simple event in part **a**, circle the selected door and put an X through one of the remaining two doors that hides a goat. (This is the door that the host shows the contestant—always a goat.)

c. Refer to the altered simple events in part **b**. Assume your strategy is to keep the door originally selected. Count the number of simple events for which this is a "winning" strategy (i.e., you win the car). Assuming equally likely simple events, what is the probability that you win the car?

d. Repeat part **c**, but assume your strategy is to always switch doors.

e. Based on the probabilities of parts **c** and **d**, "Is it to your advantage to switch your choice?"

4.87 Consider another problem posed by Marilyn vos Savant in her weekly column "Ask Marilyn": "A woman and a man (unrelated) each have two children. At least one of the woman's children is a boy, and the man's older child is a boy. Do the chances that the woman has two boys equal the chances that the man has two boys?" (*Parade Magazine,* Oct. 17, 1997.)

a. Answer the question based on your intuition.

b. Use simple events and the rules of probability to solve the problem. Does your answer agree with part **a**?

4.88 Refer to problem 4.87. A letter from one of vos Savant's readers reported the following statistics from the U.S. Census Bureau's National Health Interview Survey: Of all the households interviewed from 1987 to 1993, 42,888 families had exactly two children. Of these, 9,523 had two girls and 33,365 had at least one boy. Use these figures to support your answer to Problem 4.87.

Chapter 5

Discrete Probability Distributions

CONTENTS

5.1 Random Variables

5.2 Probability Models for Discrete Random Variables

5.3 The Binomial Probability Distribution

5.4 The Poisson Probability Distribution

5.5 The Hypergeometric Probability Distribution (Optional)

EXCEL TUTORIAL

5.E.1 Using Microsoft Excel to Obtain the Expected Value and Variance of a Probability Distribution

5.E.2 Using Microsoft Excel to Obtain Binomial Probabilities

5.E.3 Using Microsoft Excel to Obtain Poisson Probabilities

5.E.4 Using Microsoft Excel to Obtain Hypergeometric Probabilities

REAL-WORLD APPLICATION
Commitment to the Firm—Stayers vs. Leavers

Organizational behavior researchers have found a causal link between employees' commitment to a firm and employee turnover. Consider the number x of employees who leave the firm for a job at another company. What values of x signal a general lack of employee commitment to the firm? To answer this question, we must know something about the proba- bility distribution of x. In this chapter, we learn about several different types of discrete probability distributions that provide good models for many types of data. We apply one of these probability distributions to answer questions concerning stayers and leavers at an organization in Real-World Revisited at the end of this chapter.

5.1 RANDOM VARIABLES

In a practical setting, an experiment (as defined in Chapter 4) involves selecting a sample of data consisting of one or more observations on some variable. For example, we might survey 1,000 physicians concerning their preferences for aspirin and record x, the number who prefer a particular brand. Or, we might randomly select a single supermarket customer from the Excel workbook CHKOUT.XLS and observe his or her total checkout time, x. Since we can never know with certainty the exact value that we will observe when we record x for a single performance of the experiment, we call x a *random variable*.

> ### Definition 5.1
> A **random variable** is a variable that assumes numerical values associated with events of an experiment.

The random variables described above are examples of two different types of random variables—*discrete* and *continuous*. The number x of physicians in a sample of 1,000 who prefer a particular brand of aspirin is a discrete random variable because it can assume only a *countable* number of values—namely the whole numbers 0, 1, 2, 3, ..., 999, 1,000. In contrast, the checkout time of a supermarket customer is a continuous random variable because it could theoretically assume any one of an *infinite* number of values—namely, any value from 0 seconds upward. Of course, in practice we record checkout time to the nearest second but, in *theory*, the checkout time of a supermarket customer could assume any value—say 137.21471 seconds—making it impossible to count or list its values.

A good way to distinguish between discrete and continuous random variables is to imagine the values that they may assume as points on a line. Discrete random variables may assume any one of a countable number (say, 10, 21, or 100) of values corresponding to points on a line. In contrast, a continuous random variable can theoretically assume *any* value corresponding to the points in one or more intervals on a line.

> ### Definition 5.2
> A **discrete random variable** is one that can assume only a countable number of values.

> ### Definition 5.3
> A **continuous random variable** can assume any of the infinite number of values in one or more intervals on a line.

Example 5.1 Types of Random Variables

Suppose you randomly select a student attending your college or university. Classify each of the following random variables as discrete or continuous:

a. Number of credit hours taken by the student this semester

b. Current grade point average of the student

Solution

a. The number of credit hours taken by the student this semester is a discrete random variable because it can assume only a countable number of values (for example, 15, 16, 17, and so on). It is not continuous since the number of credit hours must be an integer; the variable can never assume values such as 15.2062, 16.1134, or 17.0398 hours.

b. The grade point average for the student is a continuous random variable since it could theoretically assume any value (for example, 2.87355) corresponding to the points on the line interval from 0 to 4. ❑

Self-Test

5.1

Determine whether the following random variables are discrete or continuous:

a. The time it takes a student to complete an examination
b. The number of registered voters in a national election
c. The number of winners each week in a state lottery
d. The weight of a professional wrestler

The focus of this chapter is on discrete random variables. Continuous random variables are the topic of Chapter 6.

5.2 PROBABILITY MODELS FOR DISCRETE RANDOM VARIABLES

We learned in Chapter 4 that we make inferences based on the probability of observing a particular sample outcome. Since we never know the *exact* probability of some events, we must construct probability models for the values assumed by random variables.

For example, if we toss a die, we assume that the values 1, 2, 3, 4, 5, and 6 represent equally likely events, i.e., $P(x = 1) = P(x = 2) = \ldots = P(x = 6) = \frac{1}{6}$. In doing so, we have constructed a probabilistic model for a theoretical population of x values, where x is the number of dots showing on the upper face of the die. The population is "theoretical" in the sense that we would observe an x for an infinite number of die tosses. A bar graph for the theoretical population would appear as shown in Figure 5.1.

Figure 5.1, which gives the relative frequency for each value of x in a very large number of tosses of a die, is called the *probability distribution for the discrete random variable, x.*

Figure 5.1 The Probability Distribution for x, the Number of Dots Observed on a Balanced Die

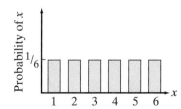

> **Definition 5.4**
>
> The **probability distribution for the discrete random variable** x is a table, graph, or formula that gives the probability of observing each value of x. If we denote the probability of x by the symbol $p(x)$, the probability distribution has the following properties:
>
> **1.** $0 \le p(x) \le 1$ for all values of x **2.** $\sum_{\text{all } x} p(x) = 1$

Example 5.2   Properties of a Probability Distribution

Consider the following sampling situation: Suppose we select a random sample of $n = 5$ physicians from a very large number—say, 10,000—and record the number x of physicians who recommend aspirin brand A. Suppose also that 2,000 of the physicians (i.e., 20%) actually prefer brand A. Now suppose we replace the five physicians in the population, randomly select a new sample of $n = 5$ physicians, and record the value of x again. If we repeat this process over and over again 100,000 times, a hypothetical data set containing the 100,000 values of x can be obtained. Table 5.1 is a summary of one such data set.

   a. Construct a relative frequency histogram for the 100,000 values of x summarized in Table 5.1.

   b. Assuming that the relative frequencies of Table 5.1 are good approximations to the probabilities of x, show that the properties of a probability distribution are satisfied.

Solution

   a. The relative frequency histogram for the 100,000 values of x is shown in Figure 5.2 on page 264. This figure provides a very good approximation to the probability distribution for x, the number of physicians in a sample of $n = 5$ who prefer brand A (assuming that 20% of the physicians in the population prefer brand A).

   b. We use the relative frequencies of Table 5.1 as approximations to the probabilities of $x = 0, x = 1, \ldots, x = 5$. For example, $P(x = 1) = p(1) \approx .32807$. Similarly, $P(x = 2) = p(2) \approx .40949$. Note that each probability (relative frequency) is between 0 and 1. Summing these probabilities, we obtain

Table 5.1 Relative Frequencies for 100,000 Hypothetical Observations on x, the Number of Physicians in a Sample of $n = 5$ Who Prefer Aspirin Brand A

x	Frequency	Relative Frequency
0	32,807	.32807
1	40,949	.40949
2	20,473	.20473
3	5,122	.05122
4	645	.00645
5	4	.00004
TOTALS	100,000	1.00000

Figure 5.2 Relative Frequency Histogram for Example 5.2

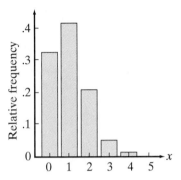

$$.32807 + .40949 + .20473 + .05122 + .00645 + .00004 = 1$$

Thus, the properties of a discrete probability distribution are satisfied. ❏

Often, the sample collected in an experiment (such as the physician survey of Example 5.2) is quite large. For example, the Gallup and Harris survey results reported in the news media are usually based on sample sizes from $n = 1,000$ to $n = 2,000$ people. Since we would not want to calculate $p(x)$ for values of n this large, we need an easy way to describe the probability distribution for x. To do this, we need to know the mean and standard deviation for the distribution. That is, we must know μ and σ for the theoretical population of x values modeled by the probability distribution. Then we can describe it using either the Empirical Rule or Tchebysheff's Theorem from Chapter 3.

Definition 5.5

The **mean μ** (or **expected value**) of a discrete random variable x is equal to the sum of the products of each value of x and the corresponding value of $p(x)$:

$$\mu = \Sigma x p(x) \qquad \text{(5.1)}$$

Definition 5.6

The **variance σ^2** of a discrete random variable x is equal to the sum of the products of $(x - \mu)^2$ and the corresponding value of $p(x)$:

$$\sigma^2 = \Sigma(x - \mu)^2 p(x) \qquad \text{(5.2)}$$

Definition 5.7

The **standard deviation σ** of a random variable x is equal to the positive square root of the variance:

$$\sigma = \sqrt{\sigma^2} \qquad \text{(5.3)}$$

Example 5.3   Finding μ and σ

Consider the sample survey of $n = 5$ physicians, described in Example 5.2. The graph of the probability distribution for x, the number of physicians in the sample who favor aspirin brand A, is reproduced in Figure 5.3.

a. Find the mean μ for this distribution. That is, find the expected value of x.

Figure 5.3 Probability Distribution for x, the Number of Physicians in a Sample of $n = 5$ Who Prefer Aspirin Brand A

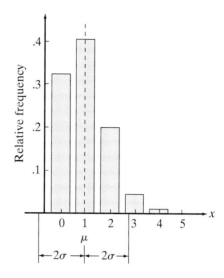

b. Interpret the value of μ.

c. Find the standard deviation of x.

Solution

a. According to Definition 5.5 (and equation 5.1), the mean μ is given by

$$\mu = \Sigma x p(x)$$

where $p(x)$ is given in Table 5.1. Since x can take values $x = 0, 1, 2, ..., 5$, we have

$$\mu = 0p(0) + 1p(1) + 2p(2) + 3p(3) + 4p(4) + 5p(5)$$
$$= 0(.32807) + 1(.40949) + 2(.20473) + 3(.05122) + 4(.00645) + 5(.00004)$$
$$= 1.0$$

b. As we learned in Chapter 3, the value of μ can also be obtained by adding the 100,000 values of x shown in Table 5.1 and dividing the sum by $n =$ 100,000. Consequently, the value $\mu = 1.0$ implies that *over a long series of surveys* similar to the one described in Example 5.2, the average number of physicians in the sample of 5 who favor aspirin brand A will equal 1. The key to the interpretation of μ is to think in terms of repeating the experiment over a long series of trials (i.e., in the *long run*). Then μ represents the average x value of this large number of trials.

c. By Definition 5.6 (and equation 5.2), we obtain

$$\sigma^2 = \sum (x - \mu)^2 p(x)$$
$$= (0 - 1)^2 p(0) + (1 - 1)^2 p(1) + (2 - 1)^2 p(2) + (3 - 1)^2 p(3)$$
$$+ (4 - 1)^2 p(4) + (5 - 1)^2 p(5)$$
$$= (1)(.32807) + (0)(.40949) + (1)(.20473) + (4)(.05122)$$
$$+ (9)(.00645) + (16)(.00004)$$
$$= .79$$

Then by Definition 5.7 (and equation 5.3), the standard deviation σ is given by

$$\sigma = \sqrt{\sigma^2} = \sqrt{.79} = .89 \qquad \qquad \square$$

Self-Test

5.2

Consider the probability distribution for a discrete random variable x, shown below:

x	0	1	2	3	4
$p(x)$	.4	.2	.2	—	.1

   a. Find $p(3)$.   **b.** Find μ.   **c.** Find σ.

Example 5.4   Empirical Rule Application

Refer to Example 5.3. Locate the interval $\mu \pm 2\sigma$ on the graph of the probability distribution for x. Confirm that most of the (theoretical) population falls within this interval.

Solution
Recall that $\mu = 1.0$ and $\sigma = .89$ for this distribution. Then

$$\mu - 2\sigma = 1.0 - 2(.89) = -.78$$

$$\mu + 2\sigma = 1.0 + 2(.89) = 2.78$$

The interval from $-.78$ to 2.78, shown in Figure 5.3, includes the values of $x = 0, x = 1$, and $x = 2$. Thus, the probability (relative frequency) that a population value falls within this interval is

$$p(0) + p(1) + p(2) = .32807 + .40949 + .20473 = .94229$$

This certainly agrees with the Empirical Rule, which states that approximately 95% of the data will lie within 2σ of the mean μ. ❑

   In the next section, we present a special discrete probability distribution that provides a useful model for many types of data encountered in the real world.

PROBLEMS 5.1–5.12

Using the Tools

x	-5	0	2	5
$p(x)$	.2	.3	.4	.1

Problem 5.1

5.1 Consider the probability distribution shown in the table at left.

   a. List the values that x may assume.

   b. What value of x is most probable?

   c. Find the probability that x is greater than 0.

   d. What is the probability that $x = -5$?

   e. Verify that the sum of the probabilities equals 1.

   f. Find μ and σ.

5.2 Suppose a discrete random variable can assume five possible values with the probability distribution shown on page 267.

x	1	2	3	4	5
$p(x)$	.20	.25	—	.30	.10

a.   Find the missing value for $p(3)$.

b.   Find the probability that $x = 2$ or $x = 4$.

c.   Find the probability that x is less than or equal to 4.

d.   Find μ and σ.

5.3   The probability distribution for a discrete random variable x is given by the formula,

$$p(x) = (.8)(.2)^{x-1}, x = 1, 2, 3, \dots$$

a.   Calculate $p(x)$ for $x = 1, x = 2, x = 3, x = 4$, and $x = 5$.

b.   Sum the probabilities of part **a**. Is it likely to observe a value of x greater than 5?

c.   Find $P(x = 1 \text{ or } x = 2)$.

5.4   Given the following probability distributions:

DISTRIBUTION A		**DISTRIBUTION B**	
x	$p(x)$	x	$p(x)$
0	.50	0	.05
1	.20	1	.10
2	.15	2	.15
3	.10	3	.20
4	.05	4	.50

a.   Compute the mean for each distribution.

b.   Compute the standard deviation for each distribution.

c.   For each distribution, compute an interval that captures most of the values of x.

Applying the Concepts

5.5   Researchers at Oregon State University attempted to develop a model for predicting the risk of early police arrest (*Journal of Quantitative Criminology*, Vol. 8, 1992). The model was based, in part, on data collected for 80 boys arrested for crimes prior to the age of 16. The accompanying table gives the probability distribution for x, the age at first arrest, for the 80 sampled boys.

Age, x, at First Arrest	$p(x)$
6	.0250
7	.0250
8	.0125
9	.0625
10	.0625
11	.0625
12	.1625
13	.2375
14	.1625
15	.1875

Source: Patterson, G. R., Crosby, L., and Vuchinich, S. "Predicting risk for early police arrest." *Journal of Quantitative Criminology*, Vol. 8, No. 4, 1992, p. 345 (adapted from Table II).

Problem 5.5

a.   Verify that the probabilities in the table sum to 1.

b.   Find $P(x \leq 10)$.

c.   Find $P(x > 12)$.

d.   Find and interpret μ.

e.   Find σ.

f.   Compute the interval $\mu \pm 2\sigma$.

g.   Find the probability that x falls in the interval of part **f**. Compare your results with the Empirical Rule.

5.6   *Developmental Psychology* (Mar. 1986) published a study on the sexual maturation of college students. The study included a probability distribution for the age x (in years) when males begin to shave regularly, shown in the accompanying figure.

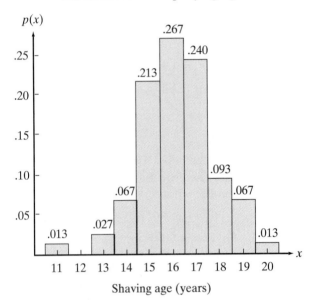

Shaving age (years)

a. Is this a valid probability distribution? Explain.

b. Display the probability distribution in tabular form.

c. What is the probability that a randomly selected male college student began shaving at age 16?

d. What is the probability that a randomly selected male college student began shaving before age 15?

e. Calculate μ and σ.

f. Interpret the value of μ.

g. Compute the interval $(\mu - 2\sigma, \mu + 2\sigma)$.

h. Find the probability that x falls within the interval of part **g**. Compare your results to the Empirical Rule.

5.7   A panel of meteorological and civil engineers studying emergency evacuation plans for Florida's Gulf Coast in the event of a hurricane has estimated that it would take between 13 and 18 hours to evacuate people living in low-lying land, with the probabilities shown in the table.

Time to Evacuate (nearest hour)	Probability
13	.04
14	.25
15	.40
16	.18
17	.10
18	.03

a. Calculate the mean and standard deviation of the probability distribution of the evacuation times.

b. Within what range would you expect the time to evacuate to fall?

c. Weather forecasters say they cannot accurately predict a hurricane landfall more than 14 hours in advance. If the Gulf Coast Civil Engineering Department waits until the 14-hour warning before beginning evacuation, what is the probability that all residents of low-lying areas are evacuated safely (i.e., before the hurricane hits the Gulf Coast)?

TRAINED LISTENERS	
Note	**Frequency**
D_3	1
G_3	1
$A_3^\#$	1
C_4	12
$C_4^\#$	1
D_4	1
E_4	15
F_4	7
G_4	8
A_4	3
TOTALS	50

Source: Thompson, W. F., et al. "Expectancies generated by melodic intervals: Evaluation of principles of melodic implication in a melody-completion task." *Perception & Psychophysics,* Vol. 59, No. 7, Oct. 1997, p. 1076.

Problem 5.8

5.8 Refer to the *Perception & Psychophysics* (Oct. 1997) melody-completion task, Problem 2.10 (page 70). After listening to a short (2-note) melody, each of 50 people trained in music completed the melody on a piano keyboard. The frequency distribution of the initial note selected by the subjects is reproduced at left. Consider a sample of two subjects randomly selected from the 50. Let x be the number of subjects who select the note E_4.

a. Find the probability distribution for x. **b.** Compute μ and σ.

c. Interpret the value of μ.

5.9 According to a survey of businesses that use the Internet and its World Wide Web, 70% use Netscape Navigator as a Web browser. If two surveyed businesses are selected, find the probability distribution of x, the number of businesses that use Netscape Navigator as a Web browser. [*Hint:* List the possible outcomes of the experiment. For each outcome, determine the value of x and then calculate its probability using Probability Rule #4 (page 235).]

5.10 In a carnival game called "Under-or-over 7," a pair of fair dice are rolled once. A player bets $1 that the sum showing on the dice is either under 7, over 7, or equal to 7.

a. If the player bets that the sum is *under* 7 and wins, he wins $1. Let x be the outcome (win $1 or lose $1) for this bet. Find the probability distribution for x.

b. If the player bets that the sum is *over* 7 and wins, he wins $1. Let y be the outcome (win $1 or lose $1) for this bet. Find the probability distribution for y.

c. If the player bets that the sum *equals* 7 and wins, he wins $4. Let z be the outcome (win $4 or lose $1) for this bet. Find the probability distribution for z.

d. For each of the three bets described in parts **a**, **b**, and **c**, find the expected long-run profit (or loss).

e. For each of the three bets described in parts **a**, **b**, and **c**, find the variance of the profit (or loss).

f. Based on the results, parts **d** and **e**, which bet would you prefer to make in "Under-or-over 7"? Explain.

5.11 The cornrake is a European species of bird in danger of worldwide extinction. A census of singing cornrakes on agricultural land in Great Britain and Ireland was recently conducted (*Journal of Applied Ecology,* Vol. 30, 1993). The census revealed that 12 cornrakes inhabit Mainland Scotland. Suppose that two of these Scottish cornrakes are captured for mating purposes. Let x be the number of these captured cornrakes that are capable of mating. Suppose exactly four of the original 12 cornrakes inhabiting Scotland are infertile.

a. Find the probability distribution for x.   **b.** Find μ and σ.

c. Interpret the value of μ.

5.12 A survey of American workers commissioned by Northwestern National Life Insurance Company found that only 1% of all workers do not experience stress on the job. Let x be the number of workers that must be sampled until the first worker with no stress is found.

a. List the possible values of x.

b. Refer to part **a**. If you sum the probabilities for these x values, what result will you obtain?

c. Find $p(1)$.

d. Find $p(2)$. [*Hint:* Use Probability Rule #4 for two independent events (Chapter 4).]

e. It can be shown (proof omitted) that the probability distribution of x is given by the formula $p(x) = \pi(1 - \pi)^{x-1}$, where π is the probability that a single worker has no on-the-job stress. Use this formula to find $p(12)$.

5.3 THE BINOMIAL PROBABILITY DISTRIBUTION

Opinion polls (i.e., sample surveys) are conducted frequently in politics, psychology, sociology, medicine, and business. Some recent examples include polls to determine (1) the number of voters in favor of legalizing casino gambling in a particular state, (2) the number of Americans who feel our president is a moral leader, and (3) the number of baseball fans who support league realignment of Major League Baseball franchises based on geography. Consequently, it is useful to know the probability distribution of the number x in a random sample of n experimental units (people) who exhibit some characteristic or prefer some specific proposition. This probability distribution can be approximated by a **binomial probability distribution** when the sample size n is small relative to the total number N of experimental units in the population.

Strictly speaking, the binomial probability distribution applies only to sampling that satisfies the conditions of a **binomial experiment**, as listed in the box.

Conditions Required for a Binomial Experiment

1. A sample of n experimental units is selected from the population *with replacement* (called **sampling with replacement**).

2. Each experimental unit possesses one of two mutually exclusive characteristics. By convention, we call the characteristic of interest a *success* and the other a *failure*.

3. The probability that a single experimental unit possesses the characteristic is equal to π. This probability is the same for all experimental units.

4. The outcome for any one experimental unit is independent of the outcome for any other experimental unit (i.e., the draws are independent).

5. The **binomial random variable** x is the number of "successes" in n trials.

In real life, there are probably few experiments that satisfy exactly the conditions for a binomial experiment. However, there are many that satisfy approximately—at least for all practical purposes—these conditions. Consider, for example, a sample survey. When the number N of elements in the population is large and the sample size n is small relative to N, the sampling satisfies, approximately, the conditions of a binomial experiment. The next two examples illustrate the point.

Example 5.5   Checking Binomial Conditions

Suppose that a sample of $n = 2$ elements is randomly selected from a population containing $N = 10$ elements, three of which are designated as successes and seven as failures. Explain why this sampling procedure violates the conditions of a binomial experiment.

Solution

The probability of selecting a success on the first draw is equal to $\frac{3}{10}$—that is, the number of successes in the population (3) divided by the total number of elements in the population (10). In contrast, the probability of a success on the second draw is either $\frac{2}{9}$ or $\frac{3}{9}$, depending on whether a success was or was not selected on the first draw. In other words, selecting a success on the second draw is dependent on the outcome of the first draw and this is a violation of condition 4 required for a binomial experiment. ❏

Example 5.6   Checking Binomial Conditions

Suppose that a sample of size $n = 2$ is randomly selected from a population containing $N = 1,000$ elements, 300 of which are successes and 700 of which are failures. Explain why this sampling procedure satisfies, approximately, the conditions required for a binomial experiment.

Solution

The probability of success on the first draw is the same as in Example 5.5—namely, $\frac{300}{1,000} = \frac{3}{10}$. The probability of a success on the second draw is either $\frac{299}{999} = .2993$ or $\frac{300}{999} = .3003$, depending on whether the first draw resulted in a success or a failure. But since these conditional probabilities are approximately equal to the unconditional probability $\left(\frac{3}{10}\right)$ of drawing a success on the second draw, we can say that, *for all practical purposes*, the sampling satisfies the conditions of a binomial experiment. Thus, when N is large and n is small relative to N (say, n/N less than .05), we can use the binomial probability distribution to calculate the probability of observing x successes in a survey sample. ❏

The binomial probability distributions for a sample of $n = 10$ and $\pi = .1$, $\pi = .3$, $\pi = .5$, $\pi = .7$, and $\pi = .9$ are shown in Figure 5.4. Note that the probability distribution is skewed to the right for small values of π, skewed to the left for large values of π, and symmetric for $\pi = .5$.

The formula used for calculating probabilities of the binomial probability distribution is shown in the box.

Figure 5.4 Binomial
Probability Distributions for
$n = 10, \pi = .1, .3, .5, .7, .9$

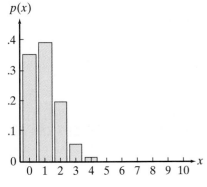

a. $\pi = .1$

b. $\pi = .3$

c. $\pi = .5$

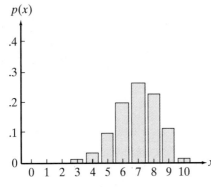

d. $\pi = .7$

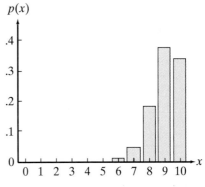

e. $\pi = .9$

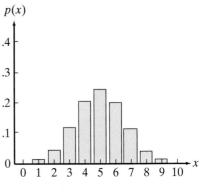

The Binomial Probability Distribution

$$p(x) = \binom{n}{x} \pi^x (1 - \pi)^{n-x} \qquad x = 0, 1, 2, \ldots, n \qquad \text{(5.4)}$$

where

n = sample size (number of trials)
x = number of successes in n trials
π = probability of success on a single trial

$$\binom{n}{x} = \frac{n!}{x!(n - x)!}$$

Assumption: The sample size n is small relative to the number N of elements in the population (say, n/N smaller than 1/20).

[*Note:* $x! = x(x - 1)(x - 2)(x - 3) \cdots (1)$, and $0! = 1$.]

Example 5.7 Computing Binomial Probabilities

Refer to the physician survey, Example 5.2. Recall that a random sample of 5 physicians is selected from a population of 10,000 physicians. The number x of physicians in the sample who prefer aspirin brand A is recorded. It is known that 20% of the physicians in the population prefer brand A.

a. Verify that the physician survey is a binomial experiment.

b. Use the binomial probability distribution to calculate the probabilities of $x = 0, x = 1, ..., x = 5$.

Solution

a. The sample survey satisfies the five requirements for a binomial experiment given in the box:

1. A sample of $n = 5$ physicians (experimental units) is selected from a population.

2. Each physician surveyed possesses one of two mutually exclusive characteristics: favors aspirin brand A (a *success*) or does not favor aspirin brand A (a *failure*).

3. The proportion of physicians in the population who prefer aspirin brand A is .2; thus, the probability of a success is $\pi = .2$. Since the sample size $n = 5$ is small relative to the population size $N = 10,000$, this probability remains approximately the same for all trials.

4. The response for any one physician is independent of the response for any other physician.

5. We are counting $x =$ the number of physicians in the sample who favor aspirin brand A.

b. To calculate the binomial probabilities, we will substitute the values of $n = 5$ and $\pi = .2$ and each value of x into the formula for $p(x)$ given in equation 5.4:

$$p(x) = \binom{n}{x}\pi^x(1 - \pi)^{n-x} = \frac{n!}{x!(n - x)!}\pi^x(1 - \pi)^{n-x}$$

Thus, remembering that $0! = 1$, we have

$$P(x = 0) = p(0) = \binom{5}{0}(.2)^0(.8)^5$$

$$= \frac{5!}{0!5!}(.2)^0(.8)^5 = (1)(1)(.32768)$$

$$= .32768$$

Similarly,

$$P(x = 1) = p(1) = \binom{5}{1}(.2)^1(.8)^4 = \frac{5!}{1!4!}(.2)^1(.8)^4 = .40960$$

$$P(x = 2) = p(2) = \binom{5}{2}(.2)^2(.8)^3 = \frac{5!}{2!3!}(.2)^2(.8)^3 = .20480$$

$$P(x = 3) = p(3) = \binom{5}{3}(.2)^3(.8)^2 = \frac{5!}{3!2!}(.2)^3(.8)^2 = .05120$$

E Figure 5.5
Binomial Probabilities
for Example 5.7
Generated Using the
PHStat Add-in
for Excel

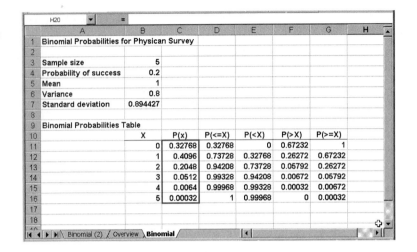

$$P(x = 4) = p(4) = \binom{5}{4}(.2)^4(.8)^1 = \frac{5!}{4!1!}(.2)^4(.8)^1 = .00640$$

$$P(x = 5) = p(5) = \binom{5}{5}(.2)^5(.8)^0 = \frac{5!}{5!0!}(.2)^5(.8)^0 = .00032$$

(Note that these probabilities are approximately equal to the relative frequencies reported in Table 5.1.) Binomial probabilities can be easily obtained using the PHStat Add-in for Excel. Figure 5.5 is an Excel printout that lists the probabilities for the physician survey. ❑

Self-Test

5.3

Consider a binomial distribution with $n = 10$.

a. Find $p(3)$ when $\pi = .7$.
b. Find $p(5)$ when $\pi = .1$.
c. Find $p(8)$ when $\pi = .3$.

Example 5.8 Summing Binomial Probabilities

Refer to Example 5.7. Find the probability that three or more physicians in the sample prefer brand A.

Solution
The values that a random variable x can assume are always mutually exclusive events—i.e., you could not observe $x = 2$ and, at the same time, observe $x = 3$. Therefore, the event "x is 3 or more" (the event that $x = 3$ or $x = 4$ or $x = 5$) can be found using Probability Rule #1 (Chapter 4). Thus,

$$P(x = 3 \text{ or } x = 4 \text{ or } x = 5) = P(x = 3) + P(x = 4) + P(x = 5)$$
$$= p(3) + p(4) + p(5)$$

Substituting the probabilities found in Example 5.7, we obtain

$$P(x = 3 \text{ or } x = 4 \text{ or } x = 5) = .05120 + .00640 + .00032 = .05792$$

In some situations, we will want to compare an observed value of x obtained from a binomial experiment with some theory or claim associated with the sampled population. In particular, we want to see if the observed value of x represents a **rare event**, assuming that the claim is true.

Example 5.9   Statistical Inference

A manufacturer of O-ring seals used to prevent hot gases from leaking through the joints of rocket boosters claims that 95% of all seals that it produces will function properly. Suppose you randomly select 10 of these O-ring seals, test them, and find that only six prevent gas from leaking. Is this sample outcome highly improbable (that is, does it represent a *rare event*) if in fact the manufacturer's claim is true?

Solution

If π is in fact equal to .95 (or some larger value), then observing a small number x of O-ring seals that function properly would represent a rare event. Since we observed $x = 6$, we want to know the probability of observing a value of $x = 6$ or some other value of x even more contradictory to the manufacturer's claim, i.e., we want to find the probability that $x = 0$ or $x = 1$ or $x = 2 \ldots$ or $x = 6$. Using the additive rule for values of $p(x)$, we have

$$P(x = 0 \text{ or } x = 1 \text{ or } x = 2 \ldots \text{ or } x = 6)$$
$$= p(0) + p(1) + p(2) + p(3) + p(4) + p(5) + p(6)$$
$$= P(x \le 6)$$

As you saw in Example 5.7, the calculation of binomial probabilities in the sum can be very tedious. In practice, we can refer to one of many tables that give partial sums of the values of $p(x)$, called **cumulative probabilities**, or to computer software. Figure 5.6 is a printout of cumulative probabilities for $\pi = .95$ and $n = 10$

Figure 5.6
Binomial Probabilities for $\pi = .95$, $n = 10$ Obtained from the PHStat Add-in for Excel

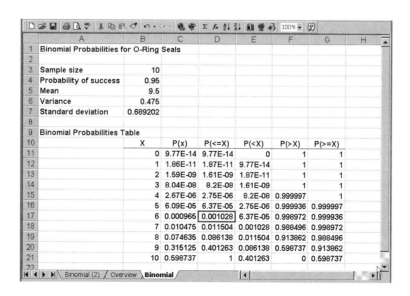

generated by the PHStat Add-in for Excel. To find $P(x \leq 6)$ on Figure 5.6, locate the probability in the row corresponding to $x = 6$ and the column labeled **P(< = x)**. This value, highlighted on Figure 5.6, is .001028.

This small probability tells us that observing as few as six good O-ring seals out of 10 is indeed a rare event, if in fact the manufacturer's claim is true. Such a sample result suggests either that the manufacturer's claim is false or that the 10 O-ring seals tested do not represent a random sample from the manufacturer's total production. Perhaps they came from a particular production line that was temporarily malfunctioning. ❏

**Self-Test
5.4**

A recent survey of American hotels found that 90% offer complimentary shampoo in their guest rooms. In a random sample of 20 hotels, less than 14 offered complimentary shampoo. Does this sample result contradict the survey percentage of 90%? Explain.

Using the formulas in Definitions 5.5, 5.6, and 5.7, you can show (proof omitted) that the mean, variance, and standard deviation for a binomial probability distribution are as listed in the following box.

Mean, Variance, and Standard Deviation for a Binomial Probability Distribution

$$\mu = n\pi \tag{5.5}$$

$$\sigma^2 = n\pi(1 - \pi) \tag{5.6}$$

$$\sigma = \sqrt{n\pi(1 - \pi)} \tag{5.7}$$

where
$\quad n$ = sample size
$\quad \pi$ = probability of success on a single trial or the proportion of experimental units in a large population that are successes

---

For *large* samples, the mean and standard deviation of the binomial probability distribution (in conjunction with the Empirical Rule or Tchebysheff's Theorem) can be used to describe and make inferences about the sampled population.

Example 5.10   Computing μ and σ

People who work at high-stress jobs frequently develop stress-related physical problems (for example, high blood pressure, ulcers, and irritability). In a recent study it was found that 40% of a large number of business executives surveyed have symptoms of stress-induced problems. Consider a group of 1,500 randomly selected business executives and assume that the probability of an executive with stress-induced problems is $\pi = .40$. Let x be the number of business executives in the sample of 1,500 who develop stress-related problems.

a. What are the mean and standard deviation of x?

b. Based on the Empirical Rule, within what limits would you expect x to fall?

c. Suppose you observe $x = 800$ executives with symptoms of stress-induced problems. What can you infer about the value of π?

Solution

a. Since $n = 1,500$ executives in the sample is small relative to the large number of business executives in the population, the number x of executives with stress-induced problems is a binomial random variable with $\pi = .4$. We use the formulas for μ and σ given in the box (i.e., equations 5.5 and 5.7) to obtain the mean and standard deviation of this binomial distribution:

$$\mu = n\pi = (1,500)(.4) = 600$$

$$\sigma = \sqrt{n\pi(1 - \pi)} = \sqrt{(1,500)(.4)(.6)} = \sqrt{360} = 18.97$$

b. According to the Empirical Rule, most (about 95%) of the x values will fall within two standard deviations of the mean. Thus, we would expect the number of sampled executives with stress-induced problems to fall in the interval from

$$\mu - 2\sigma = 600 - 2(18.97) = 562.06$$

to

$$\mu + 2\sigma = 600 + 2(18.97) = 637.94$$

c. Using the rare event approach of Chapter 4, we want to determine whether observing $x = 800$ executives with stress-induced problems is unusual, assuming $\pi = .40$. You can see that this value of x is highly improbable when $\pi = .40$, since it lies a long way outside the interval $\pi \pm 2\sigma$. The z score for this value of x is

$$z = \frac{x - \mu}{\sigma} = \frac{800 - 600}{18.97} = 10.54$$

Clearly, if $\pi = .40$, the probability that the number x of executives with stress-induced problems in the sample is 800 or larger is almost 0. Therefore, we are inclined to believe (based on the sample value of $x = 800$) that the proportion of executives with stress-induced problems is much higher than $\pi = .40$. ❑

Self-Test

5.5

Suppose that 80% of all American hotels provide complimentary shampoo in their guest rooms. In a random sample of 250 hotels, find the mean and standard deviation of x, the number of hotels that actually offer complimentary shampoo.

PROBLEMS
5.13–5.35

Using the Tools

5.13   Compute each of the following:

   a. $4!$ **b.** $\dfrac{4!}{1!3!}$ **c.** $\dbinom{5}{3}$ **d.** $(.4)^3$ **e.** $\dbinom{5}{3}(.4)^3(.6)^2$

5.14   A coin is tossed ten times and the number of heads is recorded. To a reasonable degree of approximation, is this a binomial experiment? Check to determine whether each of the five conditions required for a binomial experiment is satisfied.

5.15   Four coins are selected without replacement from a group of five pennies and five dimes. Let x equal the number of pennies in the sample of four coins. To a reasonable degree of approximation, is this a binomial experiment? Determine whether each of the five conditions required for a binomial experiment is satisfied.

5.16   Consider a binomial experiment with $n = 4$ trials and probability of success $\pi = .5$.

   a.   Find the probabilities for $x = 0, 1, 2, 3,$ and 4.
   b.   Find the probability that x is less than 2.
   c.   Find the probability that x is less than or equal to 2.
   d.   Verify that (except for rounding) the sum of the probabilities for $x = 0, 1, 2, 3,$ and 4 equals 1.
   e.   Assume $\pi = .3$ and repeat parts **a–d**.
   f.   Assume $\pi = .8$ and repeat parts **a–d**.

5.17   Let x be a binomial random variable with parameters $n = 4$ and $\pi = .2$.

   a.   Find the probability that x is less than 2.
   b.   Find the probability that x is equal to 2 or more.
   c.   How are the events in parts **a** and **b** related?
   d.   What relationship must the probabilities of the events in parts **a** and **b** satisfy?

5.18   For a binomial probability distribution with $n = 5$, use the PHStat Add-in for Excel to find:

   a. $\displaystyle\sum_{x=0}^{2} p(x) = p(0) + p(1) + p(2)$ when $\pi = .3$
   b. $\displaystyle\sum_{x=0}^{4} p(x)$ when $\pi = .3$
   c. $\displaystyle\sum_{x=4}^{5} p(x)$ when $\pi = .3$
   d. $p(2)$ when $\pi = .3$

5.19   For a binomial probability distribution with $n = 10$ and $\pi = .4$, find:

    a. the probability that x is less than or equal to 8

    b. the probability that x is less than 8

    c. the probability that x is larger than 8

5.20 Calculate μ and σ for a binomial probability distribution with

    a. $n = 15$ and $\pi = .1$ **b.** $n = 15$ and $\pi = .5$

5.21 Find the mean and standard deviation for each of the following binomial probability distributions:

    a. $n = 300$ and $\pi = .99$ **b.** $n = 200$ and $\pi = .8$

    c. $n = 100$ and $\pi = .5$ **d.** $n = 100$ and $\pi = .2$

    e. $n = 1,000$ and $\pi = .01$

Applying the Concepts

5.22 Suppose that warranty records show that the probability that a new car needs a warranty repair in the first 90 days is .05. If a sample of three new cars is selected,

    a. What is the probability that

        (1) none needs a warranty repair?

        (2) at least one needs a warranty repair?

        (3) more than one need a warranty repair?

    b. What assumptions are necessary for part **a**?

    c. What are the mean and standard deviation of the probability distribution in part **a**?

    d. Answer part **a–c** if the probability of needing a warranty repair was .10.

5.23 Based on past experience, printers in a university computer lab are available 90% of the time. If a random sample of 10 time periods are selected

    a. What is the probability that printers are available

        (1) exactly nine times?   **(2)** at least nine times?

        (3) at most nine times?   **(4)** more than nine times?

        (5) fewer than nine times?

    b. How many times can the printers be expected to be available?

    c. What would be your answers to parts **a** and **b** if the printers were available 95% of the time?

5.24 The warning signs of a stroke are a sudden severe headache, unexplained dizziness, unsteadiness, loss of vision, difficulty speaking, and numbness on one side of the body. According to the *Journal of the American Medical Association* (Apr. 22, 1998), only 57% of Americans who are susceptible to stroke recognize even one warning sign. Consider a random sample of five Americans susceptible to stroke and similar in age, gender, and race. Let x be the number who know at least one warning sign of a stroke.

    a. To a reasonable degree of approximation, is this a binomial experiment?

    b. What is a "success" in the context of this experiment?

    c. What is the value of π?

d.   Find the probability that $x = 4$.

e.   Find the probability that $x \geq 4$.

5.25   Would most wives marry the same man again, if given the chance? According to a poll of 608 married women conducted by *Ladies Home Journal* (June 1988), 80% would, in fact, marry their current husbands. Assume the women in the sample were randomly selected from among all married women in the United States. Does the number x in the sample who would marry their husbands again possess (approximately) a binomial probability distribution? Explain.

5.26   In a recent study, *Consumer Reports* (Feb. 1992) found widespread contamination and mislabeling of seafood in supermarkets in New York City and Chicago. One alarming statistic: 40% of the swordfish pieces available for sale had a level of mercury above the Food and Drug Administration (FDA) minimum amount. For a random sample of three swordfish pieces, find the probability that:

a.   All three swordfish pieces have mercury levels above the FDA minimum.

b.   Exactly one swordfish piece has a mercury level above the FDA minimum.

c.   At most one swordfish piece has a mercury level above the FDA minimum.

5.27   USAir (now known as USAirways) once operated 20% of all domestic flights in the U.S. However, the airline had been involved in four of the last seven major American air disasters and had five fatal crashes in five years of operation (*New York Times*, Sept. 11, 1994). Do these statistics imply that USAir was more negligent than its competitors during this time period or can USAir's tragic crash streak be attributed to the vagaries of chance? If a major airline crash occurs, assume that the probability that the plane involved is one of USAir's is $\pi = .2$. (This probability assumes that USAir is no more or less likely to be involved in a crash than any other airline.)

a.   Find the probability that exactly four of the next seven airline crashes involve USAir.

b.   Find the probability that at least four of the next seven airline crashes involve USAir.

c.   Use the probability, part **b**, to make an inference about USAir's tragic crash streak.

5.28   Do you have the basic skills necessary to succeed in college? Most likely, your college professor does not think so. According to a survey conducted by the Carnegie Foundation for the Advancement of Teaching, over 70% of college professors consider their students "seriously unprepared in basic skills" that should have been learned in high school (*Tampa Tribune*, Nov. 6, 1989). In a sample of 25 college faculty at your institution, let x represent the number who agree that students lack basic skills. Assume that $\pi = .70$ at your institution.

a.   What is the probability that x is less than 20?

b.   What is the probability that x is less than nine?

c.   What is the probability that x is more than nine?

d.   If, in fact, x is less than nine, make an inference about the value of π at your institution.

5.29   Two University of Illinois researchers applied the binomial distribution to the prescription fitting of hearing aids (*American Journal of Audiology,* Mar. 1995). The experiment was designed as follows. Each subject (a hearing aid wearer) listened alternately to speech samples recorded on tape at two different frequency-gain settings—call these two frequencies F1 and F2. The listener chose the frequency-gain setting that best amplified the recorded sounds. As an illustration, the researchers considered a trial with $n = 3$ subjects and counted x, the number of subjects who chose frequency-gain setting F1. The researchers also assumed that there was no real preference for the two settings, i.e., $\pi = .5$.

  a.   Find the probability distribution for x. Give the results in table form.

  b.   Find $P(x \le 2)$.

  c.   Suppose two of the three subjects chose frequency-gain setting F1 over F2. Would the result lead you to believe that π exceeds .5, i.e., that hearing aid wearers, in general, prefer F1 over F2? Explain.

5.30   Refer to the *IEEE Computer Applications in Power* study of an outdoor unmanned watching system designed to detect trespassers, Problem 4.44. In snowy weather conditions, the system detected seven out of 10 intruders; thus, the researchers estimated the system's probability of intruder detection in snowy conditions at .70.

  a.   Assuming the probability of intruder detection in snowy conditions is only .50, find the probability that the unmanned system detects at least seven of the 10 intruders.

  b.   Based on the result, part **a**, comment on the reliability of the researcher's estimate of the system's detection probability in snowy conditions.

5.31   *Chance* (Summer 1993) reported on a study of the success rate of National Football League (NFL) field goal kickers. Data collected over three NFL seasons (1989–1991) revealed that NFL kickers, in aggregate, made 75% of all field goals attempted. However, when the kicks were segmented by distance (yards), the rate of successful kicks varied widely. For example, 95% of all short field goals (under 30 yards) were made, while only 58% of all long field goals (over 40 yards) were successful. Consider x, the number of field goals made in a random sample of 20 kicks attempted last year in an NFL game. Is x (approximately) a binomial random variable? Explain.

5.32   Zoologists have discovered that animals spend a great deal of time resting, although this rest time can have functional importance (e.g., predators lying in wait for their prey). Discounting time spent in deep sleep, a University of Vermont researcher estimated the percentage of time various species spend at rest (*National Wildlife,* Aug.–Sept. 1993). For example, the probability that a female fence lizard will be resting at any given point in time is .95.

  a.   In a random sample of 20 female fence lizards, what is the probability that at least 15 will be resting at any given point in time?

  b.   In a random sample of 20 female fence lizards, what is the probability that fewer than 10 will be resting at any given point in time?

   c. In a random sample of 200 female fence lizards, would you expect to observe fewer than 190 at rest at any given point in time? Explain.

5.33 Patients on dialysis are susceptible to heart problems. Doctors can often attribute these problems to a condition called mitral annular calcification (MAC). In one study, medical researchers have found that 30% of patients on dialysis suffer moderate or severe cases of MAC (*Collegium Antropologicum,* Vol. 21, 1997). Consider a random sample of 80 dialysis patients.

   a. On average, how many of the 80 sampled patients would you expect to suffer a moderate or severe case of MAC? Assume the percentage reported in the study applies to all dialysis patients.

   b. Suppose 20 of the 80 dialysis patients are diagnosed with a moderate or severe case of MAC. Does this result tend to support or refute the percentage reported in the study? Explain.

5.34 *Psychological Reports* (Vol. 57, 1985) examined sex differences in unique and common first names of infants. A name was judged to be unique if it was given only once a year at the participating study hospital. Results showed that 45% of the females and 27% of the males received unique first names. Assuming these percentages can be generalized to the entire U.S. population, consider a nationwide random sample of 500 newborn males and 450 newborn females.

   a. What is the mean number of males in the sample to receive a unique first name? Within what limits would you expect x, the number of males receiving a unique first name, to fall?

   b. Repeat part **a** for females.

5.35 An exchange rate specifies the price of one nation's currency in terms of another nation's currency. Periodically, the United States enters the foreign exchange market in an attempt to stabilize exchange rates. Are these attempts—called *interventions*—successful? Between 1987 and 1990, economic advisor O. W. Humpage found that exchange rates showed no movement on 65% of the days where U.S. monetary authorities *did not* intervene. Then, if we assume that U.S. interventions are not effective, the probability that an intervention will result in a stable exchange rate by chance is .65 (*Economic Commentary,* Federal Reserve Bank of Cleveland, Mar. 1, 1996).

Consider that between 1987 and 1990 the U.S. intervened 64 times in an attempt to stabilize the exchange rate of the Japanese yen. Let x be the number of successful interventions of the 64. Humpage showed that x is approximately a binomial random variable with $\pi = .65$, if, in fact, U.S. interventions are not successful (i.e., no better than chance) in stabilizing exchange rates.

   a. Find $P(x = 30)$.   **b.** Find $P(x \geq 50)$.

   c. Find μ and interpret its value.   **d.** Find σ and interpret its value.

   e. Humpage determined that 50 of the interventions were successful in stabilizing the exchange rate of the Japanese yen. Use this result and your answer to part **b** to make an inference about the true value of π. Do you believe the U.S. intervention policy was effective or that the stabilized exchange rates can be attributed to chance?

5.4 THE POISSON PROBABILITY DISTRIBUTION

Many real-world discrete random variables can be modeled using the **Poisson probability distribution**. Named for the French mathematician Siméon D. Poisson (1781–1840), the Poisson distribution provides a good model for the probability distribution of the number of "rare events" that occur randomly in time, distance, or space. Some random variables that may possess (approximately) a Poisson probability distribution are the following:

1. The number of car accidents that occur per month (or week, day, etc.) at a busy intersection

2. The number of arrivals per minute (or hour, etc.) at a medical clinic or other servicing facility

3. The number of fleas found on a certain breed of dog

4. The parts per million of a toxic substance found in polluted water

5. The number of diseased Dutch elm trees per acre of a certain woodland

The characteristics of a Poisson random variable are listed in the box, followed by a formula for its probability distribution.

Characteristics of a Poisson Random Variable

1. The experiment consists of counting the number x of times a particular event occurs during a given unit of time, or a given area or volume (or weight, or distance, or any other unit of measurement).

2. The probability that an event occurs in a given unit of time, area, or volume is the same for all the units.

3. The number of events that occur in one unit of time, area, or volume is independent of the number that occur in other units.

The Poisson Probability Distribution

$$p(x) = \frac{\mu^x e^{-\mu}}{x!}, x = 0, 1, 2, \dots, \infty \qquad (5.8)$$

where

x = number of rare events per unit of time, distance, or space
μ = mean value of x
$e = 2.71828\dots$

Graphs of $p(x)$ for $\mu = 1, 2, 3,$ and 4 are illustrated in Figure 5.7. The figure shows how the shape of the Poisson distribution changes as its mean μ changes. Poisson probabilities can also be found using software such as Excel.

Example 5.11   Cumulative Poisson Probability

The quality control inspector in a diamond-cutting operation has found that the mean number of defects per diamond (defects discernible to the eye of a jeweler) is $\mu = 1.5$. What is the probability that a randomly selected diamond contains more than one such defect?

Figure 5.7 Bar Graphs of
the Poisson Distribution for
$\mu = 1, 2, 3,$ and 4

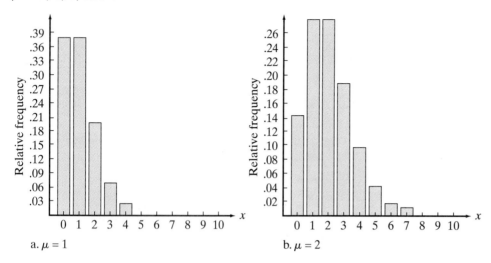

a. $\mu = 1$

b. $\mu = 2$

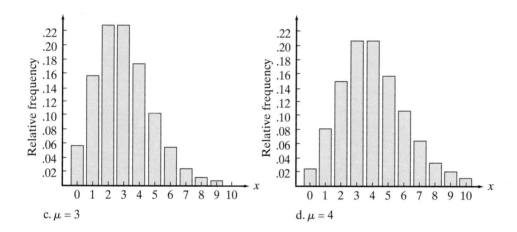

c. $\mu = 3$

d. $\mu = 4$

Solution

The probability distribution of x, the number of discernible defects per diamond, can be approximated by a Poisson probability distribution with $\mu = 1.2$. The probability that a diamond contains more than one defect is

$$P(x > 1) = p(2) + p(3) + p(4) + \cdots$$
$$= 1 - [p(0) + p(1)]$$
$$= 1 - P(x \leq 1)$$

To find $p(0)$ and $p(1)$, we can substitute into the formula for $p(x)$. Or we can find the probability using Excel. Figure 5.8 is a PHStat Add-in for Excel printout of Poisson probabilities for $\mu = 1.5$. The value of $P(x \leq 1)$, highlighted on the printout, is .557825. Therefore, the probability that the number x of defects per diamond is more than one is

E **Figure 5.8**
Poisson Probabilities
Generated Using the
PHStat Add-in for
Excel

File	Edit	View	Insert	Format	Tools	Data	Window	Help				
	A	B	C	D	E	F	G	H	I		
1	Poisson Probabilities for Defects per Diamond										
2											
3	Average/Expected number of successes			1.5							
4											
5	Poisson Probabilities Table										
6		X	P(X)	P(<=X)	P(<X)	P(>X)	P(>=X)				
7		0	0.223130	0.223130	0.000000	0.776870	1.000000				
8		1	0.334695	0.557825	0.223130	0.442175	0.776870				
9		2	0.251021	0.808847	0.557825	0.191153	0.442175				
10		3	0.125511	0.934358	0.808847	0.065642	0.191153				
11		4	0.047067	0.981424	0.934358	0.018576	0.065642				
12		5	0.014120	0.995544	0.981424	0.004456	0.018576				
13		6	0.003530	0.999074	0.995544	0.000926	0.004456				
14		7	0.000756	0.999830	0.999074	0.000170	0.000926				
15		8	0.000142	0.999972	0.999830	0.000028	0.000170				
16		9	0.000024	0.999996	0.999972	0.000004	0.000028				
17		10	0.000004	0.999999	0.999996	0.000001	0.000004				
18		11	0.000000	1.000000	0.999999	0.000000	0.000001				
19		12	0.000000	1.000000	1.000000	0.000000	0.000000				
20		13	0.000000	1.000000	1.000000	0.000000	0.000000				
21		14	0.000000	1.000000	1.000000	0.000000	0.000000				
22		15	0.000000	1.000000	1.000000	0.000000	0.000000				
23											

Overview \ **Poisson** / Poisson (2) / Poisson Chart /

$$P(x > 1) = 1 - P(x \le 1)$$
$$= 1 - .557825 = .442175$$

The mean μ of the Poisson probability distribution appears in the formula for $p(x)$. It is given, with the variance and standard deviation of x, in the accompanying box.

Mean, Variance, and Standard Deviation of a Poisson Probability Distribution

$$\text{Mean} = \mu \qquad \sigma^2 = \mu \qquad \sigma = \sqrt{\mu}$$

Example 5.12   **Computing Poisson μ and σ**

Suppose the mean number of employee accidents per month at a particular manufacturing plant is 3.4. Find the mean, variance, and standard deviation of x, the number of accidents in a randomly selected month. Is it likely that x will be as large as 12?

Solution
We are given the fact that $\mu = 3.4$. Therefore,

$$\sigma^2 = 3.4 \qquad \text{and} \qquad \sigma = \sqrt{3.4} = 1.84$$

Then the z score corresponding to $x = 12$ is

$$z = \frac{x - \mu}{\sigma} = \frac{12 - 3.4}{1.84} = 4.67$$

Since $x = 12$ lies 4.67 standard deviations away from its mean, we know (from the Empirical Rule) that this is a highly improbable event. ❑

Self-Test
5.6

A can company reports that the number of breakdowns per 8-hour shift on its machine-operated assembly line follows a Poisson distribution, with a mean of 1.5.

a. What is the probability of exactly two breakdowns during the midnight shift?

b. What is the probability of fewer than two breakdowns during the afternoon shift?

c. What is the variance of the number of breakdowns per 8-hour shift?

PROBLEMS
5.36–5.45

Using the Tools

5.36   Assume that x has a Poisson probability distribution. Find each of the following probabilities:

a. $P(x = 5)$ when $\mu = 3.0$

b. $P(x \geq 4)$ when $\mu = 7.5$

c. $P(x \leq 2)$ when $\mu = 5.5$

d. $P(x = 8)$ when $\mu = 4.0$

5.37   Suppose x has a Poisson probability distribution with $\mu = 2.5$.

a.   Find and graph $p(x)$ for $x = 0, 1, 2, ..., 10$.

b.   Find μ and σ.

c.   Locate $\mu \pm 2\sigma$ on the graph of part **a**. What is the probability that x falls within $\mu \pm 2\sigma$?

5.38   Suppose x has a Poisson probability distribution with $\mu = 6.0$.

a.   Find and graph $p(x)$ for $x = 0, 1, 2, ..., 12$.

b.   Find μ and σ.

c.   Locate $\mu \pm 2\sigma$ on the graph of part **a**. What is the probability that x falls within $\mu \pm 2\sigma$?

5.39   Assume x has a Poisson probability distribution. Find $P(x \leq 3)$ when

a. $\mu = 1$ b. $\mu = 2$ c. $\mu = 3.5$

d. $\mu = 5$ e. $\mu = 7.5$

Applying the Concepts

5.40   Ecologists often use the number of reported sightings of a rare species of animal to estimate the remaining population size. For example, suppose the number x of reported sightings per week of blue whales is recorded. Assume that x has (approximately) a Poisson probability distribution. Furthermore, assume that the average number of weekly sightings is 2.6.

a.   Find the mean and standard deviation of x, the number of blue whale sightings per week.

b. Find the probability that fewer than two sightings are made during a given week.

c. Find the probability that more than five sightings are made during a given week.

d. Find the probability that exactly five sightings are made during a given week.

5.41 Based on past records, the average number of two-car accidents in a New York City police precinct is 3.4 per day. What is the probability that there will be

a. At least six such accidents in this precinct on any given day?

b. Not more than two such accidents in this precinct on any given day?

c. Fewer than two such accidents in this precinct on any given day?

d. At least two but no more than six such accidents in this precinct on any given day?

e. What would be your answers to parts **a–d** if the average was five such accidents per day?

5.42 The quality-control manager of Marilyn's Cookies is inspecting a batch of chocolate-chip cookies that has just been baked. If the production process is in control, the average number of chip parts per cookie is 6.0. What is the probability that in any particular cookie being inspected

a. Fewer than five chip parts will be found?

b. Exactly five chip parts will be found?

c. Five or more chip parts will be found?

d. Four or five chip parts will be found?

e. What would be your answers to parts **a–d** if the average number of chip parts per cookie is 5.0?

5.43 Researchers at the Massachusetts Institute of Technology (MIT) studied the spectroscopic properties of main-belt asteroids having diameters smaller than 10 kilometers (*Science,* Apr. 3, 1993). Research revealed that, on average, 2.5 independent spectral image exposures are observed per asteroid.

a. Assuming a Poisson distribution, find the probability that exactly one independent spectral image exposure is observed during a main-belt asteroid sighting.

b. Assuming a Poisson distribution, find the probability that at most two independent spectral image exposures are observed during a main-belt asteroid sighting.

c. Would you expect to observe seven or more independent spectral image exposures during a main-belt asteroid sighting? Explain.

5.44 The number of cars that arrive at an intersection during a specified period of time often possesses (approximately) a Poisson probability distribution. When the mean arrival rate μ is known, the Poisson probability distribution can be used to aid a traffic engineer in the design of a traffic control system. Suppose you estimate that the mean number of arrivals per minute at the intersection is one car per minute.

a. What is the probability that in a given minute, the number of arrivals will equal three or more?

b. Can you assure the engineer that the number of arrivals will rarely exceed three per minute?

5.45 The Environmental Protection Agency (EPA) has established national ambient air quality standards in an effort to control air pollution. Currently, The EPA limit on ozone levels in air is set at 12 parts per hundred million (pphm). A 1990 study examined the long-term trend in daily ozone levels in Houston, Texas.* One of the variables of interest is *x*, the number of days in a year on which the ozone level exceeds the EPA threshold of 12 pphm. The mean number of exceedences in a year is estimated to be 18. Assume that the probability distribution for *x* can be modeled with the Poisson distribution.

a. Compute $P(x \geq 20)$.   b. Compute $P(5 \leq x \leq 10)$.

c. Estimate the standard deviation of *x*. Within what range would you expect *x* to fall in a given year?

d. The study revealed a decreasing trend in the number of exceedences of the EPA threshold over the past several years. The observed values of *x* for the past six years were 24, 22, 20, 15, 14, and 16. Explain why this trend casts doubt on the validity of the Poisson distribution as a model for *x*. [*Hint:* Consider characteristic #3 of the Poisson random variable.]

*Shively, Thomas S. "An analysis of the trend in ozone using nonhomogeneous Poisson processes." Paper presented at the annual meeting of the American Statistical Association, Anaheim, Calif., Aug. 1990.

5.5 THE HYPERGEOMETRIC PROBABILITY DISTRIBUTION (OPTIONAL)

Characteristics That Define a Hypergeometric Random Variable

1. The experiment consists of randomly drawing *n* elements without replacement from a set of *N* elements, *S* of which are "successes" and *N* − *S* of which are "failures."

2. The hypergeometric random variable *x* is the number of *S*'s in the draw of *n* elements.

We noted in Section 5.3 that one of the applications of the binomial probability distribution is its use as the probability distribution for the number *x* of favorable (or unfavorable) responses in a public opinion or market survey. Ideally, it is applicable when sampling is *with replacement*—that is, each item drawn from the population is observed and returned to the population before the next item is drawn. Practically speaking, sampling in surveys is rarely conducted with replacement. Nevertheless, the binomial probability distribution is still appropriate if the number *N* of elements in the population is large and the sample size *n* is small relative to *N*.

When **sampling is without replacement**, and the number of elements *N* in the population is small (or when the sample size *n* is large relative to *N*), the number of "successes" in a random sample of *n* items has a **hypergeometric probability distribution**. The defining characteristics and probability for a hypergeometric random variable are stated in the boxes.

The Hypergeometric Probability Distribution

$$p(x) = \frac{\binom{S}{x}\binom{N-S}{n-x}}{\binom{N}{n}}$$

(5.9)

where

$$N = \text{number of elements in the population}$$
$$S = \text{number of “successes” in the population}$$
$$n = \text{sample size}$$
$$x = \text{number of “successes” in the sample}$$
$$\binom{S}{x} = \frac{S!}{x!(S - x)!}$$
$$\binom{N - S}{n - x} = \frac{(N - S)!}{(n - x)!(N - S - n + x)!}$$
$$\binom{N}{n} = \frac{N!}{n!(N - n)!}$$

Assumptions:
1. The sample of n elements is randomly selected from the N elements of the population.
2. The value of x is restricted so that $x \leq n$ and $x \leq S$.

Example 5.13   Computing Hypergeometric Probabilities

During a drug "bust," the police seized 10 foil packets of a white, powdery substance, presumably cocaine. Unknown to police, seven of the 10 packets contained cocaine, but three were nonnarcotic (inert) powders. The police randomly selected three of the 10 packets for chemical testing.

a. What is the probability that all three of the selected packets test positive for cocaine?

b. What is the probability that at least one of the packets tests positive for cocaine? [*Note:* Since it is not a crime to buy or sell inert powders, at least one of the packets needs to contain cocaine for a conviction.]

Solution

For this problem, define a "success" as a packet that contains cocaine. Then, the number of elements (packets) in the population of interest is $N = 10$, the number of successes is $S = 7$, and the sample size (i.e., the number of packets tested) is $n = 3$.

a. Using equation 5.9, the probability of observing exactly $x = 3$ successes in the sample of $n = 3$ is:

$$p(3) = \frac{\binom{S}{3}\binom{N - S}{n - 3}}{\binom{N}{n}} = \frac{\binom{7}{3}\binom{3}{0}}{\binom{10}{3}} = \frac{\left(\frac{7!}{3!4!}\right)\left(\frac{3!}{0!3!}\right)}{\left(\frac{10!}{3!7!}\right)}$$

$$= \frac{\left(\frac{7 \cdot 6 \cdot 5}{3 \cdot 2}\right)(1)}{\left(\frac{10 \cdot 9 \cdot 8}{3 \cdot 2}\right)} = \frac{35}{120} = .292$$

Hypergeometric probabilities can also be obtained using the PHStat Add-in for Excel. The probabilities for $x = 0, 1, 2, 3$ are shown in the Excel printout, Figure 5.9.

E **Figure 5.9**
Hypergeometric
Probabilities for
Example 5.13
Obtained from the
PHStat Add-in for
Excel

	A	B	C
1	Hypergeometric Probabilities		
2			
3	Sample size	3	
4	No. of successes in population	7	
5	Population size	10	
6			
7	Hypergeometric Probabilities Table		
8		X	P(x)
9		0	0.008333
10		1	0.175
11		2	0.525
12		3	0.291667
13			

b. The probability of observing at least one success can be found by using the Rule of Complements (page 223).

$$P(x \geq 1) = 1 - P(x = 0)$$

Checking the Excel printout, Figure 5.9, we find $P(x = 0) = .008333$. Therefore,

$$P(x \geq 1) = 1 - P(x = 0) = 1 - .008333 = .991667$$

Consequently, the chance of at least one of the foil packets testing positive for cocaine is extremely high. ❑

Self-Test

5.7

Suppose you are purchasing small lots of hard disk drives for laptop computers. Assume that each lot contains seven drives. You decide to sample three disk drives per lot and to reject the lot if you observe one or more defectives in the sample.

a. If the lot contains one defective disk drive, what is the probability that you will accept the lot?

b. What is the probability that you will accept the lot if it contains three defective disk drives?

The mean, variance, and standard deviation for a hypergeometric probability distribution are shown in the box.

Mean, Variance, and Standard Deviation for a
Hypergeometric Probability Distribution

$$\mu = \frac{nS}{N} \tag{5.10}$$

$$\sigma^2 = n\left(\frac{S}{N}\right)\left(\frac{N-S}{N}\right)\left(\frac{N-n}{N-1}\right) \tag{5.11}$$

$$\sigma = \sqrt{n\left(\frac{S}{N}\right)\left(\frac{N-S}{N}\right)\left(\frac{N-n}{N-1}\right)} \tag{5.12}$$

Example 5.14   Finding μ and σ

Refer to Example 5.13 on page 289.

   a. Find μ and σ for the number x of packets that contain cocaine.
   b. Interpret μ.

Solution
   a. From Example 5.13, we have $N = 10$, $S = 7$, and $n = 3$. Therefore,

$$\mu = \frac{nS}{N} = \frac{(3)(7)}{10} = 2.1$$

$$\sigma = \sqrt{n\left(\frac{S}{N}\right)\left(\frac{N-S}{N}\right)\left(\frac{N-n}{N-1}\right)} = \sqrt{3\left(\frac{7}{10}\right)\left(\frac{3}{10}\right)\left(\frac{7}{9}\right)}$$

$$= 0.7$$

   b. The value $\mu = 2.1$ implies that, on average, we expect to observe 2.1 packets of the three sampled that actually contain cocaine.   ❏

PROBLEMS 5.46–5.55

Using the Tools

5.46 Suppose x has a hypergeometric probability distribution with $N = 6$, $n = 4$, and $S = 2$.
   a. Compute $p(x)$ for $x = 0, 1, 2$.
   b. Graph the probability distribution $p(x)$.
   c. Compute μ and σ.

5.47 Suppose x has a hypergcometric probability distribution with $N = 12$, $n = 7$, and $S = 5$. Compute each of the following:
   a. $P(x = 3)$ **b.** $P(x \leq 2)$ **c.** $P(x = 5)$
   d. $P(x > 3)$ **e.** μ **f.** σ

5.48 Suppose x has a hypergeometric probability distribution with $N = 8$, $n = 5$, and $S = 3$. Compute each of the following:
   a. $P(x = 1)$ **b.** $P(x \leq 1)$ **c.** $P(x \geq 2)$
   d. $P(x = 4)$ **e.** μ **f.** σ

5.49 Suppose that x is a hypergeometric random variable. Compute $p(x)$ for each of the following cases:
   a. $N = 5, n = 3, S = 4, x = 1$ **b.** $N = 10, n = 5, S = 3, x = 3$
   c. $N = 3, n = 2, S = 2, x = 2$ **d.** $N = 4, n = 2, S = 2, x = 0$

Applying the Concepts

5.50 The dean of a liberal arts school wishes to form an executive committee of five from among the 40 tenured faculty members at the school. The selection is to be random, and at the school there are eight tenured faculty members in sociology.

    a. What is the probability that the committee will contain

        (1) none of them?

        (2) at least one of them?

        (3) not more than one of them?

    b. What would be your answers to **a** if the committee consisted of seven members?

5.51 A state lottery is conducted in which six winning numbers are selected from a total of 54 numbers. What is the probability that if six numbers are randomly selected

    a. All six numbers will be winning numbers?

    b. Five numbers will be winning numbers?

    c. Four numbers will be winning numbers?

    d. Three numbers will be winning numbers?

    e. None of the numbers will be winning numbers?

    f. What would be your answers to parts **a–d** if the six winning numbers were selected from a total of 40 numbers?

5.52 Based on data provided by the U.S. Department of Health and Human Services, *U.S. News & World Report* (Sept. 28, 1992) estimates that one out of every five kidney transplants fails within a year. Suppose that exactly three of the next 15 kidney transplants will fail within a year. Consider a random sample of three of these 15 patients.

    a. Find the probability that all three sampled transplants fail within a year.

    b. Find the probability that at least one of the three sampled transplants fails within a year.

5.53 "Hotspots" are species-rich geographical areas. A *Nature* (Sept. 1993) study estimated the probability of a bird species in Great Britain inhabiting a butterfly hotspot at .70. Consider a random sample of four British bird species selected from a total of 10 tagged species. Assume that seven of the 10 tagged species inhabit a butterfly hotspot.

    a. What is the probability that exactly two of the bird species sampled inhabit a butterfly hotspot?

    b. What is the probability that at least one of the bird species sampled inhabits a butterfly hotspot?

5.54 According to the Centers for Disease Control (CDC), a recent increase in outbreaks of salmonella poisoning can be attributed to the consumption of raw or undercooked eggs. One such outbreak occurred at a cookout in Jacksonville, Florida. Among the people who attended the cookout, 12 cases of salmonella poisoning were identified. Eleven of the 12 ill persons had eaten homemade ice cream served at the cookout. The ice cream, made with raw eggs, tested positive for salmonella contamination (*Morbidity and Mortality Weekly Report,* Sept. 16, 1994). Suppose the CDC randomly selects four of the 12 persons with salmonella poisoning and interviews each.

a.   What is the probability that all four of the people ate homemade ice cream at the cookout?

b.   What is the probability that at least one person ate homemade ice cream at the cookout?

5.55   A commercial for a brand of sugarless chewing gum claims that "three out of four dentists who recommend sugarless gum to their patients recommend our brand." Suppose this claim was established following a survey of four dentists randomly selected from a group of 20 dentists who were known to recommend sugarless gum to their patients. What is the probability that at least three of the four dentists surveyed would recommend the advertised brand if in fact only half of the original group of 20 dentists favor that brand? Does your probability calculation strengthen or weaken the gum manufacturer's claim? Explain.

REAL-WORLD REVISITED
Commitment to the Firm—Stayers vs. Leavers

COMMIT.XLS

The idea that commitment to a business organization is in some way related to employee turnover has received considerable attention from organizational behaviorists. Numerous researchers have found a causal link between an employee's commitment to the firm and his or her tendency to leave the firm voluntarily. The *Academy of Management Journal* (Oct. 1993) published one such study involving employees of an aerospace firm located in a major metropolitan area in the southeastern United States.

At the time of the study, the firm operated at the leading edge of military aerospace technology in designing, manufacturing, and marketing aviation components such as electronic flight instrument systems, weapons-aiming computers, and radar display units. Top management, however, was concerned with rising discontent among operatives in the repair and overhaul group, the division most responsible for generating revenue growth and maintaining financial ratios. To investigate this problem, researchers at the University of South Florida designed and administered a questionnaire to a sample of 270 of the firm's employees.[*]

Several of the numerous variables measured in the study are listed and described in Table 5.2. The complete data set, described in Appendix C, is stored in the Excel workbook COMMIT.XLS. Note that variable 11 classifies employees as either "stayers" or "leavers." For this application, we are interested in the random variable x, where x is the number of the 270 employees who are "leavers." Assume that the $n = 270$ employees represent a random sample from the population of employees at all similar aerospace firms in the U.S.

Study Questions

a.   Explain why the binomial probability distribution is the best approximation for the distribution of x.

b.   A claim is made that the true proportion of "leavers" at all U.S. aerospace firms is $\pi = .10$. Find the expected value and variance of x. Interpret these values.

[*]Jaros, S. J., et al. "Effects of continuance, affective, and moral commitment on the withdrawal process: An evaluation of eight structural equation models." *Academy of Management Journal*, Vol. 36, No. 5, Oct. 1993, pp. 951–995.

Table 5.2 Variables Measured in *Academy of Management Journal* Study

1. **Age** (years)
2. **Gender** (1 = male, 0 = female)
3. **Organizational tenure** (months): length of time with organization
4. **Job tenure** (months): length of time in current job
5. **Continuance commitment** (3- to 21-point scale): measure of the degree to which past behaviors force the employee to be committed to the firm
6. **Affective commitment** (14- to 98-point scale): measure of the degree to which emotional ties commit the employee to the firm
7. **Moral commitment** (4- to 28-point scale): measure of the degree to which duty commits the employee to the firm
8. **Thinking of quitting** (1- to 5-point scale): measure of how often the employee thinks about quitting
9. **Search** (1- to 5-point scale): measure of how likely the employee is to search for a new job
10. **Intent to leave** (2- to 11-point scale): measure of the employee's intentions to leave the firm
11. **Leave** (1 = employee quit the firm prior to the end of the study, 0 = employee stayed with the firm until the end of the study)

---

    c. Use Excel to generate a relative frequency table for the variable LEAVE. Based on the relative frequency of "leavers" in the data, make an inference about the claim in part **b**.

    d. Use Excel to generate a cross-tabulation table for the variables LEAVE and GENDER. Make an inference about whether π, the probability of leaving the firm, depends on gender.

Key Terms
Starred (*) terms are from the optional section of this chapter.

Binomial experiment   270
Binomial probability
  distribution   270
Binomial random variable   270
Continuous random variable   261
Cumulative probabilities   275
Discrete random variable   261

Hypergeometric probability
  distribution*   288
Poisson probability
  distribution   283
Probability distribution for the
  discrete random variable x 263
Random variable   261

Rare event   275
Sampling with replacement   270
Sampling without
  replacement*   288

Key Formulas
Starred (*) formulas are from the optional section of this chapter.

Probability Distribution	Mean (μ)	Variance (σ^2)
General discrete random variable		
$p(x)$	$\sum\limits_{\text{all } x} x p(x)$ **(5.1)**, 264	$\sum\limits_{\text{all } x} (x - \mu)^2 p(x)$ **(5.2)**, 264
Binomial random variable		
$\binom{n}{x} \pi^x (1-\pi)^{n-x}$ **(5.4)**, 272	$n\pi$ **(5.5)**, 276	$n\pi(1-\pi)$ **(5.6)**, 276
Poisson random variable		
$\dfrac{\mu^x e^{-\mu}}{x!}$ **(5.8)**, 283	μ	μ

| Probability Distribution | Mean (μ) | Variance (σ^2) |

*Hypergeometric random variable

$$\frac{\binom{S}{x}\binom{N-S}{n-x}}{\binom{N}{n}} \quad \textbf{(5.9)}, 288 \qquad \frac{nS}{N} \quad \textbf{(5.10)}, 290 \qquad n\left(\frac{S}{N}\right)\left(\frac{N-S}{N}\right)\left(\frac{N-n}{N-1}\right) \quad \textbf{(5.11)}, 290$$

Key Symbols

Starred (*) symbols are from the optional section of this chapter.

Symbol	Definition
$p(x)$	Probability distribution of the discrete random variable x
μ	Mean of a discrete random variable x
σ	Standard deviation of a discrete random variable x
π	Probability of success in a binomial trial
e	Euler's constant (approximately 2.718) used with a Poisson probability distribution
S	*Number of successes in a hypergeometric probability distribution
N	*Total number of elements in a hypergeometric probability distribution

Supplementary Problems 5.56–5.70

Starred (*) problems refer to the optional section of this chapter.

x	$p(x)$
0	.10
1	.15
2	.50
3	.25

Problem 5.56

5.56 The probability distribution for a discrete random variable x is given in the table at left.

    a. Verify that the properties of a probability distribution are satisfied.

    b. Find $P(x \geq 2)$.

    c. Calculate μ and σ for the distribution.

5.57 A balanced coin is tossed eight times and the number x of heads is recorded. Find the probability that:

    a. $x = 4$ **b.** x is larger than 4   **c.** x is less than 2

5.58 A random sample of 100 persons was selected to take the "Pepsi Challenge" taste test. Each person tasted one cup of Pepsi and one cup of Coke in random order, then identified the cup of soda that "tastes best." (The cups were disguised so that the taster did not know, a priori, which cup was which.) Let x represent the number of persons preferring Pepsi.

    a. Is x a binomial random variable?

    b. Suppose that the majority of tasters in fact prefer Pepsi. What does this imply about the value of π?

    c. If $\pi = .5$, what is the implication?

5.59 According to a survey sponsored by the American Society for Microbiology, only 60% of people using the restrooms in New York City's Penn Station washed their hands after going to the bathroom (*Tampa Tribune*, Sept. 17, 1996). Suppose you randomly observe people using the Penn Station restrooms and record x, the number of people that must be sampled until the first person who does wash up is found. The distribution of x, where x is the number of trials until the first "success" is observed, is called a *geometric distribution*. It can be shown that $P(x = k) = \pi(1 - \pi)^{k-1}$, where π is the probability of a "success."

a. Find $P(x = 1)$.   **b.** Find $P(x = 2)$.

c. Find $P(x = 3)$.   **d.** Find $P(x = 4)$.

5.60 Electrical engineers recognize that high neutral current in computer power systems is a potential problem. A recent survey of computer power system load currents at U.S. sites found that 10% of the sites had high neutral current (*IEEE Transactions on Industry Applications,* July/Aug. 1990). In a sample of 20 computer power systems selected from the large number of sites in the country, let x be the number with high neutral current.

a. Find and interpret the mean of x.

b. Find and interpret the standard deviation of x.

5.61 As a college student, are you frequently depressed and overwhelmed by the pressure to succeed? According to the American Council on Education (ACE), the level of stress among college freshmen is rising rapidly. In a survey of 380,000 full-time college freshmen conducted for the ACE by UCLA's Higher Education Research Institute, over 10% reported frequently "feeling depressed" (*Tampa Tribune,* Jan. 9, 1989). Assume that 10% of all college freshmen frequently feel depressed.

a. On average, how many of the 380,000 college freshmen surveyed would you expect to report frequent feelings of depression if the true percentage in the population is 10%?

b. Find the standard deviation of the number of the 380,000 freshmen surveyed who frequently feel depressed if the true percentage is 10%.

c. What can you infer about the true percentage if 35,000 of the college freshmen surveyed reported frequent feelings of depression? Explain.

5.62 A study of vehicle flow characteristics in acceleration lanes (i.e., merging ramps) was conducted at a major freeway (*Journal of Transportation Engineering,* Nov. 1985). Suppose the number of vehicles using the acceleration lane per minute has a mean equal to 1.1.

a. What is the probability that more than two vehicles will use the acceleration lane in the next minute?

b. What is the probability that exactly three vehicles will use the acceleration lane in the next minute?

***5.63** One recent survey found that almost half of all business executives report symptoms of stress-induced problems. Suppose a project team consisting of four business executives is to be selected at random from a group of 10 executives at an international trade organization. Unknown to the firm, three of the 10 executives are currently experiencing stress-related physical problems.

a. What is the probability that the project team will include at least one executive with stress-related physical problems?

b. What is the probability that the project team will include exactly one executive with stress-related physical problems?

c. What is the probability that the project team will include all three executives with stress-related physical problems?

d. Within what limits would you expect the number of executives with stress-related physical problems on the project team to fall?

*5.64 Refer to Problem 4.44 (page 239). As reported in *IEEE Computer Applications in Power* (Apr. 1990), an outdoor, unmanned computerized video monitoring system detected seven out of 10 intruders in snowy conditions. Suppose that two of the intruders had criminal intentions. What is the probability that both of these intruders were detected by the system?

5.65 A group of 20 college graduates contains 10 highly motivated persons, as determined by a company psychologist. Suppose a personnel director selects 10 persons from this group of 20 for employment. Let x be the number of highly motivated persons included in the personnel director's selection. Is this a binomial experiment? Explain.

5.66 According to a Federal Bureau of Investigation (FBI) report, 40% of all assaults on federal workers are directed at IRS employees. For a random sample of 50 federal workers who were assaulted, suppose the FBI records the number x of workers employed by the IRS. Is x a binomial random variable? If so, what are the values of n and π?

5.67 A National Basketball Association (NBA) player makes 80% of his free throws. Near the conclusion of a playoff game, the player is fouled and is rewarded two free throws. If he makes at least one free throw, his team wins the game.

a. Do you think the probability of the NBA player making his first free throw is equal to .8?

b. Is the outcome of the second free throw likely to be independent of the outcome of the first free throw? Why or why not?

c. Considering your answers to parts **a** and **b**, is it likely that the two free throws constitute a binomial experiment? Explain.

5.68 On the average, only 3.5 out of every 1,000 childbirths result in identical twins. That is, the probability that a childbirth will result in identical twins is $\pi = .0035$. If a random sample of 20 childbirths is selected, what is the probability of observing at least one childbirth in the sample that results in identical twins?

5.69 An international survey involving thousands of past Olympic participants indicated that an overwhelming majority—70%—believe that athletes should be barred from Olympic competition if they have been treated with anabolic steroids during the month prior to competition. In a random sample of 10 former Olympic participants, let x be the number who believe that athletes should be barred from Olympic competition if they have been treated with anabolic steroids during the month prior to competition. What is the probability that:

a. x is equal to 10? **b.** x is at least 3? **c.** x is at most 1?

5.70 James S. Trefil wrote on the role of probabilities and expected values in our everyday lives (*Smithsonian,* Sept. 1984). Trefil illustrates expected value theory by applying it to junk-mail contests, ploys used by advertisers to interest readers in their products. He writes:

If you are on a junk-mail list, you probably get regular notices announcing that "You may have already won the $10,000 jackpot!" Is it really worth answering the ad? Well, suppose the mailing went to 100,000 people. Your chances of winning are then 1 in 100,000. Over many contests, therefore, you would expect to win an average of $10,000 $\times$ $(\frac{1}{100,000})$, or ten cents per game .... You will note that the expected [winnings] in this case is less than the price of the postage stamp you need to enter the contest. In fact, you can expect to lose ... cents every time you play, so the reaction "It's not worth answering this" is correct.

Suppose you randomly select $n = 10$ junk-mail contests to enter, each offering a $10,000 jackpot if you win. Assume that the probability of your winning any one contest is 1 in 100,000 or .00001. Let x be the number of contests that you win.

a. Find the expected value of x.

b. Now let y be the amount of money you win on any one contest. In Trefil's illustration, either $y = $10,000$ (i.e., you win the contest) or $y = 0$ (i.e., you lose the contest). In this particular case, it can be shown that $E(y) = $10,000E(x)$. Find your expected winnings, $E(y)$.

c. Find $E(y)$ if you enter only a single contest (i.e., $n = 1$). Does this value agree with Trefil's "ten cents" per game?

d. Now suppose you decide to enter a series of junk-mail contests until you win one contest. What is the probability that you will not win a junk-mail contest until after the fifth contest you enter? [Hint: See Problem 5.59 on page 295.]

Using Microsoft Excel

5.E.1 Using Microsoft Excel to Obtain the Expected Value and Variance of a Probability Distribution

Solution Summary:

Implement a worksheet that uses the SUMPRODUCT function to calculate the expected value and variance of a probability distribution.

Example: Physician sampling situation of Example 5.2

Solution:

To obtain the expected value and variance for the physician sampling situation of Example 5.2 on page 263, using the relative frequencies of Table 5.1 as approximations for the probabilities, implement a worksheet based on the design of Table 5.E.1. This design uses the SUMPRODUCT function, the format of which is

$$\text{SUMPRODUCT}(\textit{multiplier cell range, multiplicand cell range})$$

where
   multiplier cell range is the cell range containing the multipliers
   multiplicand cell range is the cell range containing the multiplicands

to sum the products of pairs of numbers.

Table 5.E.1 Calculations Sheet Design for Obtaining Expected Value and Variance for the Physician Sampling Problem of Example 5.2

	A	B	C	D	E
1	Expected Value and Variance				
2					Calculations
3	Probabilities:	X	P(X)		[(X-E(X)]^2
4		0	0.32807		=(B4-$B$12)^2
5		1	0.40949		=(B5-$B$12)^2
6		2	0.20473		=(B6-$B$12)^2
7		3	0.05122		=(B7-$B$12)^2
8		4	0.00645		=(B8-$B$12)^2
9		5	0.00004		=(B9-$B$12)^2
10		Total:	=SUM(C4:C9)		
11					
12	Expected Value	=SUMPRODUCT(B4:B9,C4:C9)			
13	Variance	=SUMPRODUCT(E4:E9,C4:C9)			
14	Standard Deviation	=SQRT(B13)			

To implement the Table 5.E.1 design, do the following:

1. Select File | New to open a new workbook (or open the existing workbook into which the new worksheet is to be placed).

299

2. Select an unused worksheet (or select Insert | Worksheet) and rename the worksheet Calculations.

3. Enter the title in row 1 and the headings for rows 2 and 3 as shown in Table 5.E.1.

4. Enter the X and P(X) values in cell range B4:C9 as shown in Table 5.E.1.

5. Enter the labels and formulas for rows 10, 12, 13, and 14 and enter the formulas for column E as shown in Table 5.E.1.

6. Adjust column widths as is necessary by selecting columns and then selecting Format | Column | AutoFit Selection.

The implemented worksheet will be similar to the one shown in Figure 5.E.1.

E **Figure 5.E.1**
Expected Value and Variance Worksheet for the Physician Sampling Situation of Example 5.2

	A	B	C	D	E
1	Expected Values and Variances				
2					Calculations
3	Probabilities:	X	P(X)		[(X-E(X)]^2
4		0	0.32807		0.99722193
5		1	0.40949		1.9321E-06
6		2	0.20473		1.00278193
7		3	0.05122		4.00556193
8		4	0.00645		9.00834193
9		5	0.00004		16.0111219
10		Total:	1		
11					
12	Expected Value	0.99861			
13	Variance	0.796368			
14	Standard Deviation	0.892395			

5.E.2 Using Microsoft Excel to Obtain Binomial Probabilities

Solution Summary:

Use the PHStat add-in to generate a worksheet that displays binomial probabilities.

Example: Physician survey problem of Example 5.7

Solution:

Use the Probability Distributions | Binomial choice of the PHStat add-in to obtain binomial probabilities. As an example, to obtain the binomial probabilities for the physician survey problem of Example 5.7 on page 273, do the following:

1. If the PHStat add-in has not been previously loaded, load the add-in using the instructions of Section EP.3.2.

2. Select File | New to open a new workbook (or open the existing workbook into which the binomial probabilities worksheet is to be inserted).

3. Select PHStat | Probability Distributions | Binomial.

4. In the Binomial Probability Distribution dialog box (see Figure 5.E.2):
   a. Enter 5 in the Sample Size: edit box.
   b. Enter 0.2 in the Probability of Success: edit box
   c. Enter 0 (zero) in the Outcomes From: edit box and enter 5 in the Outcomes To: edit box. To obtain cumulative probabilities, select the Cumulative Probabilities check box. To obtain a histogram, select the Histogram check box.
   d. Enter Binomial Probabilities in the Output Title: edit box.
   e. Click the OK button.

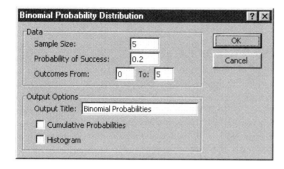

E **Figure 5.E.2** PHStat Binomial Probability Distribution Dialog Box

The add-in inserts a table of binomial probabilities on a new worksheet that is similar to the one shown in Figure 5.5 on page 274.

5.E.3 Using Microsoft Excel to Obtain Poisson Probabilities

Solution Summary:

Use the PHStat add-in to generate a worksheet that displays Poisson probabilities.

 Quality control problem of Example 5.11

Solution:

Use the Probability Distributions | Poisson choice of the PHStat add-in to obtain Poisson probabilities. As an example, to obtain the Poisson probabilities for the diamond-cutting quality control problem of Example 5.11 on page 283, do the following:

[1] If the PHStat add-in has not been previously loaded, load the add-in using the instructions of Section EP.3.2.

[2] Select File | New to open a new workbook (or open the existing workbook into which the Poisson probabilities worksheet is to be inserted).

[3] Select PHStat | Probability Distributions | Poisson.

[4] In the Poisson Probability Distribution dialog box (see Figure 5.E.3):

    [a] Enter 1.5 in the Average/Expected No. of Successes: edit box.

    [b] Enter Poisson Probabilities in the Output Title: edit box.

    [c] Optional: To obtain cumulative probabilities, select the Cumulative Probabilities check box. To obtain a histogram, select the Histogram check box.

    [d] Click the OK button.

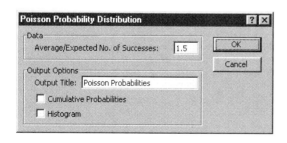

E **Figure 5.E.3** PHStat Poisson Probability Distribution Dialog Box

The add-in inserts a table of Poisson probabilities on a new worksheet similar to the one shown in Figure 5.8 on page 285.

5.E.4   Using Microsoft Excel to Obtain Hypergeometric Probabilities

Solution Summary:

Use the PHStat add-in to generate a worksheet that displays hypergeometric probabilities.

Example: Evidence testing problem of Example 5.13

Solution:

Use the Probability Distributions I Hypergeometric choice of the PHStat add-in to obtain hypergeometric probabilities. As an example, to obtain the hypergeometric probabilities for the drug evidence testing problem of Example 5.13 on page 289, do the following:

1.   If the PHStat add-in has not been previously loaded, load the add-in using the instructions of Section EP.3.2.

2.   Select File I New to open a new workbook (or open the existing workbook into which the hypergeometric probabilities worksheet is to be inserted).

3.   Select PHStat I Probability Distributions I Hypergeometric.

4.   In the Hypergeometric Probability Distribution dialog box (see Figure 5.E.4):

   a.   Enter 3 in the Sample Size: edit box.

   b.   Enter 7 in the No. of Successes in Population: edit box.

   c.   Enter 10 in the Population Size: edit box.

   d.   Enter Hypergeometric Probabilities in the Output Title: edit box.

   e.   Optional: To obtain a histogram, select the Histogram check box.

   f.   Click the OK button.

E   **Figure 5.E.4** PHStat Hypergeometric Probability Distribution Dialog Box

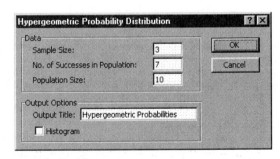

The add-in inserts a table of hypergeometric probabilities on a new worksheet similar to the one shown in Figure 5.9 on page 290.

Chapter 6

Normal Probability Distributions

CONTENTS

6.1 Probability Models for Continuous Random Variables

6.2 The Normal Probability Distribution

6.3 Descriptive Methods for Assessing Normality

6.4 Sampling Distributions

6.5 The Sampling Distribution of the Mean and the Central Limit Theorem

EXCEL TUTORIAL

6.E.1 Using Microsoft Excel to Obtain Normal Probabilities

6.E.2 Using Microsoft Excel to Construct Normal Probability Plots

6.E.3 Using Microsoft Excel to Simulate Sampling Distributions

REAL-WORLD APPLICATION
Detecting Fire Insurance Fraud

Insurance companies require an estimate of "lost" profit for retail items destroyed by fire before replacing them. Interestingly, the mean profit losses for certain items produce a relative frequency distribution that has a familiar bell-shaped curve known as a normal distribution. For reasons that you will subsequently learn, many random variables possess normal distributions. The characteristics of a normal distribution are presented in this chapter. Knowledge of the normal distribution helps us to identify improbable or rare events; this knowledge will help us solve a case of insurance fraud in Real-World Revisited at the end of this chapter.

6.1 PROBABILITY MODELS FOR CONTINUOUS RANDOM VARIABLES

Suppose you want to predict your waiting time in a dentist's office, the sale price of a home, the annual amount of rainfall in your city, or your weekly winnings playing the lottery. If you do, you will need to know something about continuous random variables.

Recall that continuous random variables are those that can assume (at least theoretically) any of the infinitely large number of values contained in an interval. Thus, we might envision a population of patient waiting times in a dentist's office, the sale prices of houses, the annual amounts of rainfall in your city since 1900, or the gains (or losses) of many weeks of playing the lottery. Since our ultimate goal is to make inferences about a population based on the measurements contained in a sample, we need to know the probability that the sample observations (or sample statistics) assume specific values.

For example, suppose we are interested in the intelligence quotient (IQ) of a college student. Then the target population is the set of IQs for all students who attend college. What is the probability that a student's IQ will exceed 115? This probability can be obtained only if we know the relative frequency distribution of the population of IQs. For example, let Figure 6.1(on page 305) represent the *hypothetical* relative frequency distribution for the population, called a **continuous probability distribution**. Then, the probability that a randomly selected college student has an IQ exceeding 115 is .3, the area shaded under the curve of Figure 6.1.

The problem, of course, is that we do not know the IQs of all college students. Hence, we do not know the exact shape of the population relative frequency distribution sketched in Figure 6.1. Then, as in the case of a coin-tossing experiment, we select a smooth curve (similar to the one shown in Figure 6.1) as a **probability model** for the population relative frequency distribution. To find the probability that a particular observation (say, an IQ) will fall in a particular interval, we use the model and find the area under the curve that falls over that interval. Of course, in order for this approximate probability to be realistic, we need to be fairly certain that the model and the population relative frequency distribution are very similar.

Areas and Probabilities

The probability that a continuous random variable x falls between two points a and b equals the area under the relative frequency distribution curve that falls over the interval (a, b).

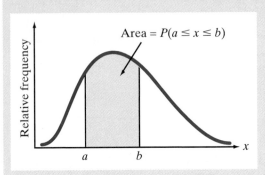

Figure 6.1 Hypothetical Relative Frequency Distribution of College Student IQs

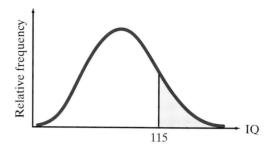

In the next section, we introduce one of the most important and useful probability models for a continuous random variable and show how it can be used to find probabilities associated with specific values of the random variable.

6.2 THE NORMAL PROBABILITY DISTRIBUTION

The normal distribution was proposed by C. F. Gauss (1777–1855) as a model for the relative frequency distribution of *errors*, such as errors of measurement. Amazingly, this curve provides an adequate model for the relative frequency of data collected from many different disciplines.

One of the most useful probability models for population relative frequency distributions is known as the **normal distribution**. A graph of the normal distribution (often called the **normal**, or **bell**, **curve**) is shown in Figure 6.2.

The mathematical model for the normal curve, denoted $f(x)$ and called the *normal probability density function*, is

$$f(x) = \frac{1}{\sigma\sqrt{2\pi}}e^{-(1/2)[(x-\mu)/\sigma]^2}$$

where
   e = the mathematical constant approximated by 2.71828
   π = the mathematical constant approximated by 3.14159
   μ = the population mean
   σ = the population standard deviation
   x = any value of the normal random variable, $-\infty < x < \infty$

You can see from the figure that the mound-shaped normal curve is symmetric about its mean μ. Furthermore, approximately 68% of the area under a normal curve lies within the interval $\mu \pm \sigma$. Approximately 95% of the area lies within the interval $\mu \pm 2\sigma$ (shaded in Figure 6.2), and almost all (99.7%) lies within the interval $\mu \pm 3\sigma$. Note that these percentages agree with the Empirical Rule of

Figure 6.2 The Normal Curve

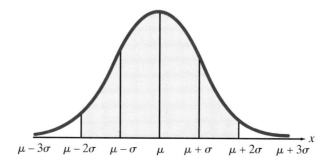

Section 3.5. (This is because the Empirical Rule is based on data that can be modeled by a normal distribution.)

Remember that areas under the normal curve have a probabilistic interpretation. Thus, if a population of measurements has approximately a normal distribution, then the probability that a randomly selected observation falls within the interval $\mu \pm 2\sigma$ is approximately .95.

Properties of the Normal Curve

1. Mound-shaped (or bell-shaped)
2. Symmetric about μ
3. $P(\mu - \sigma < x < \mu + \sigma) \approx .68$
4. $P(\mu - 2\sigma < x < \mu + 2\sigma) \approx .95$
5. $P(\mu - 3\sigma < x < \mu + 3\sigma) \approx .997$

Although always mound-shaped and symmetric, the exact shape of the normal curve will depend on the specific values of μ and σ. Several different normal curves are shown in Figure 6.3. You can see from Figure 6.3 that the mean μ measures the location of the distribution and the standard deviation σ measures its spread.

The areas under the normal curve have been computed and are given in Table B.2 of Appendix B. Since the normal curve is symmetric, we need to give areas on only one side of the mean. Consequently, the entries in Table B.2 are areas between the mean and a point x to the right of the mean.

Since the values of μ and σ vary from one normal distribution to another, the easiest way to express a distance from the mean is in terms of a z score. Recall (Section 3.6) that a z score *measures the number of standard deviations that a point x lies from its mean μ.*

Thus,

$$z = \frac{x - \mu}{\sigma} \tag{6.1}$$

is the distance between x and μ, expressed in units of σ.

Example 6.1   Computing a z Value

Suppose a population relative frequency distribution has mean $\mu = 500$ and standard deviation $\sigma = 100$. Give the z score corresponding to $x = 650$.

Figure 6.3 Three Normal Distributions with Different Means and Standard Deviations

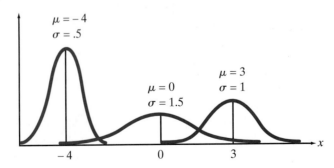

Table 6.1 Reproduction of Part of Table B.2 of Appendix B

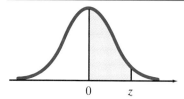

z	.00	.01	.02	.03	.04	.05	.06	.07	.08	.09
.0	.0000	.0040	.0080	.0120	.0160	.0199	.0239	.0279	.0319	.0359
.1	.0398	.0438	.0478	.0517	.0557	.0596	.0636	.0675	.0714	.0753
.2	.0793	.0832	.0871	.0910	.0948	.0987	.1026	.1064	.1103	.1141
.3	.1179	.1217	.1255	.1293	.1331	.1368	.1406	.1443	.1480	.1517
.4	.1554	.1591	.1628	.1664	.1700	.1736	.1772	.1808	.1844	.1879
.5	.1915	.1950	.1985	.2019	.2054	.2088	.2123	.2157	.2190	.2224
.6	.2257	.2291	.2324	.2357	.2389	.2422	.2454	.2486	.2517	.2549
.7	.2580	.2611	.2642	.2673	.2704	.2734	.2764	.2794	.2823	.2852
.8	.2881	.2910	.2939	.2967	.2995	.3023	.3051	.3078	.3106	.3133
.9	.3159	.3186	.3212	.3238	.3264	.3289	.3315	.3340	.3365	.3389
1.0	.3413	.3438	.3461	.3485	.3508	.3531	.3554	.3577	.3599	.3621
1.1	.3643	.3665	.3686	.3708	.3729	.3749	.3770	.3790	.3810	.3830
1.2	.3849	.3869	.3888	.3907	.3925	.3944	.3962	.3980	.3997	.4015
1.3	.4032	.4049	.4066	.4082	.4099	.4115	.4131	.4147	.4162	.4177
1.4	.4192	.4207	.4222	.4236	.4251	.4265	.4279	.4292	.4306	.4319
1.5	.4332	.4345	.4357	.4370	.4382	.4394	.4406	.4418	.4429	.4441

Source: Abridged from Table I of A. Hald, *Statistical Tables and Formulas* (New York: Wiley, 1952). Reproduced by permission of A. Hald.

Solution

The value $x = 650$ lies 150 units above $\mu = 500$. This distance, expressed in units of σ ($\sigma = 100$), is 1.5. We can get this answer directly by substituting x, μ, and σ into equation 6.1:

$$z = \frac{x - \mu}{\sigma} = \frac{650 - 500}{100} = \frac{150}{100} = 1.5$$ ❏

A partial reproduction of Table B.2 is shown in Table 6.1. The entries in the complete table give the areas to the right of the mean for distances from $z = 0.00$ to $z = 3.09$.

Example 6.2 Using the Normal Table

Find the area under a normal curve between the mean and a point $z = 1.26$ standard deviations to the right of the mean. This area represents $P(0 \le z \le 1.26)$.

Solution

To locate the proper entry, proceed down the left (z) column of the table to the row corresponding to $z = 1.2$. Then move across the top of the table to the column headed .06. The intersection of the .06 column and the 1.2 row contains the desired area, .3962 (highlighted in Table 6.1), as shown in Figure 6.4 (on page 308). ❏

Figure 6.4 The Tabulated Area in Table B.2 Corresponding to $z = 1.26$

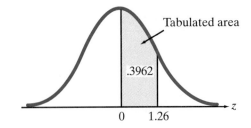

Tabulated area

.3962

0   1.26

z

The normal distribution associated with the z statistic (as shown in Figure 6.4) is called the *standard normal distribution.* The mean of a standard normal distribution is always equal to 0 (since $z = 0$ when $x = \mu$); the standard deviation is always equal to 1. Since the mean is 0, z values to the right of the mean are positive; those to the left are negative.

> **Definition 6.1**
>
> A **standard normal distribution** is a normal distribution with $\mu = 0$ and $\sigma = 1$. A **standard normal random variable**, z, is a random variable with a standard normal distribution.

Example 6.3   Area Between 0 and Negative z

Find the area beneath the standard normal curve between the mean $z = 0$ and the point $z = -1.26$. This area represents $P(-1.26 \le z \le 0)$.

Solution

The best way to solve a problem of this type is to draw a sketch of the distribution (see Figure 6.5 on page 309). Since $z = -1.26$ is negative, we know that it lies to the left of the mean, and the area that we seek is the shaded area shown.

Since the normal curve is symmetric, the area between the mean 0 and $z = -1.26$ is exactly the same as the area between the mean 0 and $z = +1.26$. We found this area in Example 6.2 to be .3962. Therefore, the area between $z = -1.26$ and $z = 0$ is .3962. ❏

Example 6.4   Area Between $-z$ and z

Find the probability that a normally distributed random variable will lie within $z = 2$ standard deviations of its mean; that is, find the probability $P(-2 \le z \le 2)$.

Solution

The probability that we seek is the shaded area shown in Figure 6.6 (page 309). Since the area between the mean and $z = 2.0$ is exactly the same as the area between the mean and $z = -2.0$, we need find only the area between the mean and $z = 2$ standard deviations to the right of the mean and multiply by 2. This area is given in Table B.2 as .4772. Therefore, the probability P that a normally distributed random variable will lie within two standard deviations of its mean is

$$P = 2(.4772) = .9544$$

❏

Figure 6.5 Standard Normal Distribution for Example 6.3

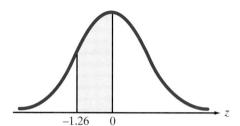

Figure 6.6 Standard Normal Distribution for Example 6.4

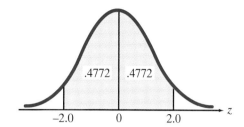

Example 6.5   Area above z

Find the probability that a normally distributed random variable x will lie more than $z = 2$ standard deviations above its mean; that is, find $P(z > 2)$.

Solution

The probability we seek is the darker shaded area shown in Figure 6.7. The total area under a standard normal curve is 1; half this area lies to the left of the mean, half to the right. Consequently, the probability P that x will lie more than two standard deviations above the mean is equal to .5 less the area A:

$$P = .5 - A$$

The area A corresponding to $z = 2.0$ is .4772. Therefore,

$$P = .5 - .4772 = .0228$$ ❏

Example 6.6   Area between Two Positive z Values

Find the area under the normal curve between $z = 1.2$ and $z = 1.6$; that is, find $P(1.2 \leq z \leq 1.6)$.

Solution

The area A that we seek lies to the right of the mean because both z values are positive. It will appear as the shaded area shown in Figure 6.8 (page 310). Let A_1 represent the area between $z = 0$ and $z = 1.2$, and A_2 represent the area between $z = 0$ and $z = 1.6$. Then the area A that we desire is $A = A_2 - A_1$. From Table B.2, we obtain:

$$A_1 = .3849 \qquad \text{and} \qquad A_2 = .4452$$

Figure 6.7 Standard Normal Distribution for Example 6.5

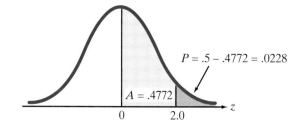

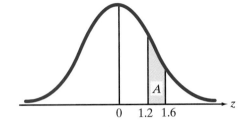

Figure 6.8 Standard Normal Distribution for Example 6.6

Then

$$A = A_2 - A_1$$
$$= .4452 - .3849 = .0603 \qquad \square$$

Example 6.7 Find a z Value for a Given Area

Find a value of z—call it z_0—such that the area to the right of z_0 is equal to .10.

Solution
The z value that we seek appears as shown in Figure 6.9. Note that we show an area to the right of z equal to .10. Since the total area to the right of the mean $z = 0$ is equal to .5, the area between the mean 0 and the unknown z value is $.5 - .1 = .4$ (as shown in the figure).

Consequently, to find z_0, we must look in the standard normal table (Table B.2) for the z value that corresponds to an area equal to .4. A reproduction of Table B.2 is shown in Table 6.2 (page 311). The area .4000 does not appear in Table 6.2. The closest values are .3997, corresponding to $z = 1.28$, and .4015, corresponding to $z = 1.29$. Since the area .3997 is closer to .4000 than is .4015, we will choose $z = 1.28$ as our answer. That is, $z_0 = 1.28$. $\square$

Examples 6.2–6.7 demonstrate how to solve the following two types of normal probability problems:

1. Examples 6.2–6.6 use Table B.2 to *find areas under the standard normal curve.* These problems may be further classified into one of three types:

   a. Finding the area between the mean $\mu = 0$, and some value of z that is located above or below $\mu = 0$ (Examples 6.2 and 6.3)

   b. Finding the area between the values z_1 and z_2, where neither z_1 nor z_2 is equal to 0 (Examples 6.4 and 6.6)

   c. Finding the area in either the upper or the lower tail of the standard normal z distribution (Example 6.5)

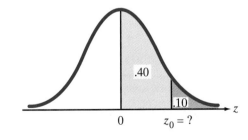

Figure 6.9 Standard Normal Distribution for Example 6.7

Table 6.2 Reproduction of Part of Table B.2 of Appendix B

z	.00	.01	.02	.03	.04	.05	.06	.07	.08	.09
.0	.0000	.0040	.0080	.0120	.0160	.0199	.0239	.0279	.0319	.0359
.1	.0398	.0438	.0478	.0517	.0557	.0596	.0636	.0675	.0714	.0753
.2	.0793	.0832	.0871	.0910	.0948	.0987	.1026	.1064	.1103	.1141
.3	.1179	.1217	.1255	.1293	.1331	.1368	.1406	.1443	.1480	.1517
.4	.1554	.1591	.1628	.1664	.1700	.1736	.1772	.1808	.1844	.1879
.5	.1915	.1950	.1985	.2019	.2054	.2088	.2123	.2157	.2190	.2224
.6	.2257	.2291	.2324	.2357	.2389	.2422	.2454	.2486	.2517	.2549
.7	.2580	.2611	.2642	.2673	.2704	.2734	.2764	.2794	.2823	.2852
.8	.2881	.2910	.2939	.2967	.2995	.3023	.3051	.3078	.3106	.3133
.9	.3159	.3186	.3212	.3238	.3264	.3289	.3315	.3340	.3365	.3389
1.0	.3413	.3438	.3461	.3485	.3508	.3531	.3554	.3577	.3599	.3621
1.1	.3643	.3665	.3686	.3708	.3729	.3749	.3770	.3790	.3810	.3830
1.2	.3849	.3869	.3888	.3907	.3925	.3944	.3962	.3980	.3997	.4015
1.3	.4032	.4049	.4066	.4082	.4099	.4115	.4131	.4147	.4162	.4177
1.4	.4192	.4207	.4222	.4236	.4251	.4265	.4279	.4292	.4306	.4319
1.5	.4332	.4345	.4357	.4370	.4382	.4394	.4406	.4418	.4429	.4441

Source: Abridged from Table I of A. Hald, *Statistical Tables and Formulas* (New York: Wiley, 1952). Reproduced by permission of A. Hald.

2. Example 6.7 uses Table B.2 to *find the z value, denoted* z_0, *corresponding to some area* in the upper tail of the standard normal z distribution. A similar procedure may be used to find a z value corresponding to a lower-tail area under the curve.

Many distributions of data that occur in the real world are approximately normal, but few are *standard* normal. However, Examples 6.8–6.10 use what you have learned about the standard normal curve to solve the same two types of problems involving *any* normal distribution:

1. Finding the probability that a normal random variable x falls between the values x_1 and x_2, or the probability that it falls in either the upper or the lower tail of the normal distribution (Examples 6.8 and 6.9)

2. Finding the value of x that places a probability P in the upper (or lower) tail of a normal distribution (Example 6.10)

Self-Test 6.1

Consider a standard normal random variable z.

a. Find $P(0 \le z \le .75)$
b. Find $P(-.28 \le z \le 1.16)$
c. Find $P(z \le -1.45)$
d. Find z_0 such that $P(z \ge z_0) = .2$

Example 6.8   Application: Finding a Normal Probability

Salt is America's second leading food additive (after sugar), both in factory-processed foods and home cooking. The average amount of salt consumed per day by an American is 15 grams (15,000 milligrams), although the actual physiological minimum daily requirement for salt is only 220 milligrams. Suppose that the

Figure 6.10 Normal Curve
Sketches for Example 6.8

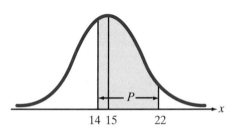

a. Amount of salt intake (in grams) per day

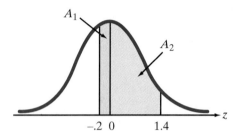

b. Standard normal

amount of salt intake per day is approximately normally distributed with a standard deviation of 5 grams. What proportion of all Americans consume between 14 and 22 grams of salt per day?

Solution

The proportion P of Americans who consume between $x = 14$ grams and $x = 22$ grams of salt is the total shaded area in Figure 6.10a. Before we can compute this area, we need to determine the z values that correspond to $x = 14$ and $x = 22$. Substituting $\mu = 15$ and $\sigma = 5$ into equation 6.1, we compute the z value for $x = 14$ as

$$z_1 = \frac{x - \mu}{\sigma} = \frac{14 - 15}{5} = \frac{-1}{5} = -.20$$

The corresponding z value for $x = 22$ is

$$z_2 = \frac{x - \mu}{\sigma} = \frac{22 - 15}{5} = \frac{7}{5} = 1.40$$

Figure 6.10b* shows these z values along with P. From this figure, we see that $P = A_1 + A_2$, where A_1 is the area corresponding to $z_1 = -.20$, and A_2 is the area corresponding to $z_2 = 1.4$. These values, given in Table B.2, are $A_1 = .0793$ and $A_2 = .4192$. Thus,

$$P = A_1 + A_2 = .0793 + .4192 = .4985$$

So 49.85% of Americans consume between 14 and 22 grams of salt per day. ❑

*Actually, the shape of the standard normal distribution shown in Figure 6.10b differs from the shape of the normal distribution in Figure 6.10a because the distributions have different variances. In fact, the normal curve associated with Figure 6.10b ($\sigma = 1$) will be more peaked and narrower than that for Figure 6.10a ($\sigma = 5$) since the variance is smaller. Nevertheless, the tail area P shaded on the two distributions is identical. For pedagogical reasons, we show the distributions with similar shapes.

The technique for finding the probability that a normal random variable falls between two values is summarized in the next box.

Finding the Probability That a Normal Random Variable Falls between Two Values, x_1 and x_2

1. Make two sketches of the normal curve, one representing the normal distribution of x and the other, the standard normal z distribution.

2. Show the approximate locations of x_1 and x_2 on the sketch of the x distribution. Be sure to locate x_1 and x_2 correctly relative to the mean μ. For example,

if x_1 is less than μ, then it should be located to the left of μ; if x_2 is greater than μ, it should be located to the right of μ.

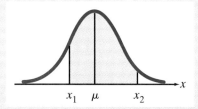

3. Find the values of z corresponding to x_1 and x_2:

$$z_1 = \frac{x_1 - \mu}{\sigma} \quad \text{and} \quad z_2 = \frac{x_2 - \mu}{\sigma}$$

4. Locate the values of z_1 and z_2 on your sketch of the z distribution.

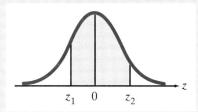

5. Use the table of areas under the standard normal curve, given in Table B.2, to find the area between z_1 and z_2. This will be the probability that x falls between x_1 and x_2.

Example 6.9   Application: Finding a Normal Probability

Refer to Example 6.8. Medical research has linked excessive consumption of salt to hypertension (high blood pressure). Physicians recommend that those Americans who want to reach a level of salt intake at which hypertension is less likely to occur should consume less than 1 gram of salt per day. What is the probability that a randomly selected American consumes less than 1 gram of salt per day?

Solution

The probability P that a randomly selected American consumes less than 1 gram of salt per day is represented by the shaded area in Figure 6.11a (page 314). The z value corresponding to $x = 1$ (shown in Figure 6.11b) is

$$z = \frac{x - \mu}{\sigma} = \frac{1 - 15}{5} = \frac{-14}{5} = -2.80$$

Since the area to the left of $z = 0$ is equal to .5, the probability that x is less than or equal to 1 is $P = .5 - A$, where A is the tabulated area corresponding to $z = -2.80$. This value, given in the standard normal table, is $A = .4974$. Then, the probability that a randomly selected American consumes no more than 1 gram of salt per day is

$$P = .5 - A$$
$$= .5 - .4974 = .0026$$

Figure 6.11 Normal Curve
Sketches for Example 6.9

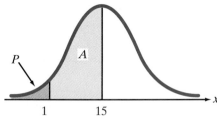

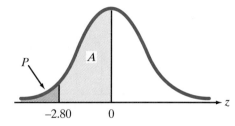

a. Amount of salt intake (in grams) per day   b. Standard normal

Self-Test

6.2

Steel used for water pipelines is often coated on the inside with cement mortar to prevent corrosion. In a study of the mortar coatings of the pipeline used in a water transmission project in California (*Transportation Engineering Journal*), researchers noted that the mortar thickness was specified to be $\frac{7}{16}$ inch. A very large sample of thickness measurements produced a mean equal to .635 inch and a standard deviation equal to .082 inch. If the thickness measurements were normally distributed, approximately what proportion were less than $\frac{7}{16}$ = .4375 inch?

Example 6.10

Application: Finding a Value of the Normal Random Variable

Psychologists traditionally treat IQ as a random variable having a normal distribution with a mean of 100 and a standard deviation of 15. Find the 10th percentile of the IQ distribution.

Solution

Let x represent the IQ of a randomly selected person. We know that x is normally distributed with $\mu = 100$ and $\sigma = 15$. We want to find the 10th percentile of the IQ distribution (call this value x_0). By definition, x_0 exceeds exactly 10 percent of the IQs in the distribution. Thus,

$$P(x < x_0) = .10$$

Unlike the two previous examples, where we were given a value of x and asked to find a corresponding probability, here we are given a probability $P = .10$ and asked to locate the corresponding value of x. We call P the **tail probability associated with x_0**. The value x_0, along with its tail probability, is shown in Figure 6.12a.

Figure 6.12 Normal Curve
Areas for Example 6.10

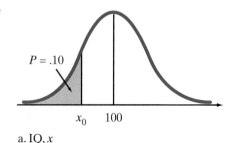

a. IQ, x

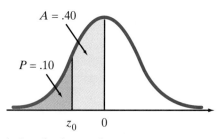

b. Standard normal, z

Now, the 10th percentile, x_0, is located to the left of (i.e., below) the mean of 100. The location of x_0 is not arbitrary, but depends on the value of P. To see this more clearly, try placing x_0 above (to the right of) the mean of 100. Now (mentally) shade in the corresponding tail probability P, i.e., the area to the left of x_0. You can see that this area cannot possibly equal .10—it is too large (.50 or greater). Thus, x_0 must lie to the left of the IQ mean of 100.

The first step, then, in solving problems of this type is to determine the location of x_0 relative to the mean μ. The second step is to find the corresponding z value for x_0—call it z_0—where

$$z_0 = \frac{x_0 - \mu}{\sigma} = \frac{x_0 - (100)}{15}$$

Once we locate z_0, we can use the equivalent relation (equation 6.2 in the next box)

$$x_0 = \mu + z_0(\sigma) = 100 + z_0(15)$$

to find x_0.

You can see from Figure 6.12b that z_0 is the z value that corresponds to the area $A = .40$ in Table B.2. In the body of the table, we see the value closest to .40 is .3997. The z value corresponding to this area is $z_0 = -1.28$. (Note that z_0 is negative because it lies to the left of 0.) Substituting into the formula above, we have

$$x_0 = 100 + z_0(15) = 100 + (-1.28)(15) = 80.8$$

Thus, 10% of the IQs fall below 80.8. In other words, 80.8 represents the 10th percentile of the IQ distribution. ❑

To help in solving problems of this type throughout the remainder of the chapter, we outline the steps leading to the solutions in the next box.

Finding a Particular Value of the Normal Random Variable Given an Associated Tail Probability

1. Make a sketch of the relative frequency distribution of the normal random variable x. Shade the tail probability P and locate the corresponding value x_0 on the sketch. Remember that x_0 will be to the right or left of the mean μ depending on the value of P.

2. Make a sketch of the corresponding relative frequency distribution of the standard normal random variable z. Locate z_0, the z value corresponding to x_0, and shade the area corresponding to P.

3. Compute the area A associated with z_0 as follows:

$$A = .5 - P$$

4. Use the area A, i.e., the area between 0 and z_0, to find z_0 in Table B.2. (If you cannot find the exact value of A in the table, use the closest value.) Note that z_0 will be negative if you place x_0 to the left of the mean in step 1.

5. Compute x_0 as follows:

$$x_0 = \mu + z_0(\sigma) \tag{6.2}$$

The preceding examples should help you to understand the use of the table of areas under the normal curve. The practical applications of this information to inference-making will become apparent in the following chapters.

PROBLEMS
6.1–6.20

Using the Tools

6.1   Find the area under the standard normal curve:

a.   Between $z = 0$ and $z = 1.2$

b.   Between $z = 0$ and $z = 1.49$

c.   Between $z = -.48$ and $z = 0$

d.   Between $z = -1.37$ and $z = 0$

e.   For values of z larger than 1.33

Show the z values and the corresponding area of interest on a sketch of the normal curve for each part of the problem.

6.2   Find the area under the standard normal curve:

a.   Between $z = 1.21$ and $z = 1.94$

b.   For values of z larger than 2.33

c.   For values of z less than -2.33

d.   Between $z = -1.50$ and $z = 1.79$

Show the z values and the corresponding area of interest on a sketch of the normal curve for each part of the problem.

6.3   Find the z value (to two decimal places) that corresponds to an area in the standard normal table (Table B.2) equal to:

a.   .1000   **b.**   .3200   **c.**   .4000   **d.**   .4500   **e.**   .4750

Show the area and corresponding value of z on a sketch of the normal curve for each part of the problem.

6.4   Find the value of z (to two decimal places) that cuts off an area in the upper tail of the standard normal curve equal to:

a.   .025   **b.**   .05   **c.**   .005   **d.**   .01   **e.**   .10

Show the area and corresponding value of z on a sketch of the normal curve for each part of the problem.

6.5   Suppose that a normal random variable x has a mean $\mu = 20.0$ and a standard deviation $\sigma = 4.0$. Find the z score corresponding to:

a. $x = 23.0$ **b.** $x = 16.0$ **c.** $x = 13.5$

d. $x = 28.0$ **e.** $x = 12.0$

For each part of the problem, locate x and μ on a sketch of the normal curve. Check to make sure that the sign and magnitude of your z score agree with your sketch.

6.6   Find the approximate value for z_0 such that the probability that z is larger than z_0 is:

a. $P = .10$ **b.** $P = .15$ **c.** $P = .20$ **d.** $P = .25$

Locate z_0 and the corresponding probability P on a sketch of the normal curve for each part of the problem.

6.7   Find the approximate value for z_0 such that the probability that z is less than z_0 is:

   a. $P = .10$ **b.** $P = .15$ **c.** $P = .30$ **d.** $P = .50$

   Locate z_0 and the corresponding probability P on a sketch of the normal curve for each part of the problem.

Applying the Concepts

6.8   A set of final examination grades in an introductory statistics course was found to be normally distributed with a mean of 73 and a standard deviation of 8.

   a. What is the probability of getting at most a grade of 91 on this exam?

   b. What percentage of students scored between 65 and 89?

   c. What percentage of students scored between 81 and 89?

   d. What is the final exam grade if only 5% of the students taking the test scored higher?

   e. If the professor "curves" (gives A's to the top 10% of the class regardless of the score), are you better off with a grade of 81 on this exam or a grade of 68 on a different exam where the mean is 62 and the standard deviation is 3? Show statistically and explain.

6.9   A statistical analysis of 1,000 long-distance telephone calls made from the headquarters of Johnson & Shurgot Corporation indicates that the length of these calls is normally distributed with $\mu = 240$ seconds and $\sigma = 40$ seconds.

   a. What percentage of these calls lasted less than 180 seconds?

   b. What is the probability that a particular call lasted between 180 and 300 seconds?

   c. How many calls lasted less than 180 seconds or more than 300 seconds?

   d. What percentage of the calls lasted between 110 and 180 seconds?

   e. What is the length of a particular call if only 1% of all calls are shorter?

6.10   The metropolitan airport commission is considering the establishment of limitations on the extent of noise pollution around a local airport. At the present time the noise level per jet takeoff in one neighborhood near the airport is approximately normally distributed with a mean of 100 decibels and a standard deviation of 6 decibels.

   a. What is the probability that a randomly selected jet will generate a noise level greater than 108 decibels in this neighborhood?

   b. What is the probability that a randomly selected jet will generate a noise level of exactly 100 decibels?

   c. Suppose a regulation is passed that requires jet noises in this neighborhood to be lower than 105 decibels 95% of the time. Assuming the standard deviation of the noise distribution remains the same, how much will the mean noise level have to be lowered to comply with the regulation?

   d. Repeat parts **a**–**c**, but assume that the standard deviation of the noise levels is 10 decibels.

*Scholz, H. "Fish Creek Community Forest: Exploratory statistical analysis of selected data." Working paper, Northern Lights College, British Columbia, Canada.

6.11 Foresters "cruising" British Columbia's boreal forest have determined that the diameter at breast height of white spruce trees in a particular community is approximately normal, with mean 17 meters and standard deviation 6 meters.*

a. Find the probability that the breast height diameter of a randomly selected white spruce in the forest community is less than 12 meters.

b. Suppose you observe a white spruce with a breast height diameter of 12 meters. Is this an unusual event? Explain.

c. Find the probability that the breast height diameter of a randomly selected white spruce in the forest community will exceed 37 meters.

d. Suppose you observe a tree in the community forest with a breast height diameter of 38 meters. Is this tree likely to be a white spruce? Explain.

e. Fifteen percent of white spruce trees in the forest will have a breast height diameter greater than what value?

6.12 The *Journal of Communication* (Summer 1997) published a study on U.S. television news coverage of major foreign earthquakes. One of the variables measured was the distance (in thousands of miles) the earthquake was from New York City—from where the ABC, NBC, and CBS nightly news broadcasts originate. The distance distribution was found to have a mean of 4.6 thousand miles and a standard deviation of 1.9 thousand miles. Assume the distribution is approximately normal. Find the probability that a randomly selected major foreign earthquake is:

a. Between 2.5 and 3.0 thousand miles from New York City

b. Between 4.0 and 5.0 thousand miles from New York City

c. Less than 6.0 thousand miles from New York City

d. Based on the mean and standard deviation reported, is it reasonable to assume that the distance distribution is approximately normal? Explain.

6.13 Competitive cyclists are thought to have a very low percentage of body fat due to intensive training methods. A group of competitive cyclists were found to have a mean percent body fat of 9% with a standard deviation of 3% (*International Journal of Sport Nutrition,* June 1995). Assume the distribution of body fat measurements for competitive cyclists is normally distributed.

a. What is the median of the percent body fat distribution?

b. What is the 3rd quartile of the percent body fat distribution?

c. What is the 95th percentile of the percent body fat distribution?

6.14 A *chewing cycle* is defined as an upward movement followed by a downward movement of the chin. Clinicians have found that the chewing cycles of "normal" children differ from the chewing cycles of children with eating difficulties. In one study (*The American Journal of Occupational Therapy,* Mar. 1984), the number of chewing cycles required for a "normal" preschool child to swallow a bite of graham cracker was found to have a mean of 15.09 and a standard deviation of 3.4. Suppose a "normal" preschool child is fed a bite of graham cracker, and assume that x, the number of chewing cycles required to swallow a bite of graham cracker, follows an approximate normal distribution.

a. Find the probability that x is less than 10.

b. Find the probability that x is between 8 and 16.

c. Find the probability that x exceeds 22.

d. One preschool child, thought to have eating difficulties, required $x = 22$ cycles to chew and swallow the bite of graham cracker. Is it likely that this child is from the group of "normal" children? Explain.

e. Find the 90th percentile of the number of chewing cycles distribution.

6.15 The Trail Making Test is frequently used by clinical psychologists to test for brain damage. Patients are required to connect consecutively numbered circles on a sheet of paper. It has been determined that the mean length of time required for a patient to perform this task is 32 seconds and the standard deviation is 4 seconds. Assume that the distribution of the lengths of time required to connect the circles is normal.

a. Find the probability that a randomly selected patient will take longer than 40 seconds to perform the task.

b. Find the probability that a randomly selected patient will take between 24 and 40 seconds to complete the task.

c. A psychologist would like to retest those persons with completion times in the highest 5% of the distribution of times required. What time would a person need to exceed on the Trail Making Test to be considered for retesting?

6.16 In a laboratory experiment, researchers at Barry University (Miami Shores, Florida) studied the rate at which sea urchins ingested turtle grass (*Florida Scientist,* Summer/Autumn 1991). The urchins, without food for 48 hours, were fed 5-cm. blades of green turtle grass. The mean ingestion time was found to be 2.83 hours and the standard deviation was .79 hour. Assume that green turtle grass ingestion time for the sea urchins has an approximate normal distribution.

a. Find the probability that a sea urchin will require 4 or more hours to ingest a 5-cm. blade of green turtle grass.

b. Find the probability that a sea urchin will require between 2 and 3 hours to ingest a 5-cm. blade of green turtle grass.

c. Ninety-nine percent of the urchins require more than how many hours to ingest the grass?

6.17 Researchers have developed sophisticated intrusion-detection algorithms to protect the security of computer-based systems. These algorithms use principles of statistics to identify unusual or unexpected data, i.e., "intruders." One popular intrusion-detection system assumes the data being monitored are normally distributed (*Journal of Information Systems,* Spring 1992). As an example, the researcher considered system data with a mean of .27, a standard deviation of 1.473, and an intrusion-detection algorithm that assumes normal data.

a. Find the probability that a data value observed by the system will fall between $-.5$ and $.5$.

b. Find the probability that a data value observed by the system exceeds 3.5.

c. Comment on whether a data value of 4 observed by the system should be considered an "intruder."

6.18   In baseball, a "no-hitter" is a regulation 9-inning game in which the pitcher yields no hits to the opposing batters. *Chance* (Summer 1994) reported on a study of no-hitters in Major League Baseball (MLB). The initial analysis focused on the total number of hits yielded per game per team for all 9-inning MLB games played between 1989 and 1993. The distribution of hits/9-innings is approximately normal with mean 8.72 and standard deviation 1.10.

a.   What percentage of 9-inning MLB games result in fewer than 6 hits?

b.   What percentage of 9-inning MLB games result in between 6 and 12 hits?

c.   Demonstrate, statistically, why a no-hitter is considered an extremely rare occurrence in MLB.

6.19   Pacemakers are used to control the heartbeat of cardiac patients. A single pacemaker is made up of several biomedical components that must be of a high quality for the pacemaker to work. One particular plastic part, called a connector module, mounts on the top of the pacemaker. Connector modules are specified to have a length between .304 inch and .322 inch to work properly. Any module with length outside these limits is *out-of-spec. Quality* (Aug. 1989) reported on one supplier of connector modules that had been shipping out-of-spec parts to the manufacturer for 12 months.

a.   The lengths of the connector modules produced by the supplier were found to follow an approximate normal distribution with mean $\mu = .3015$ inch and standard deviation $\sigma = .0016$ inch. Use this information to find the probability that the supplier produces an out-of-spec part.

b.   Once the problem was detected, the supplier's inspection crew began to employ an automated data-collection system designed to improve product quality. After two months, the process was producing connector modules with mean $\mu = .3146$ inch and standard deviation $\sigma = .0030$ inch. Find the probability that an out-of-spec part will be produced. Compare your answer to part **a**.

6.20   Behaviorists have developed a 10-item questionnaire to measure the maturity of small groups. In a study published in *Small Group Behavior* (May 1988), a class of undergraduate college students was divided into two groups, mature and immature, based on their answers to the 10-item questionnaire. A final project was then assigned and, at the end of the semester, student performances were evaluated. A summary of the grades on the project for the two groups is provided below. Assume these represent population means and standard deviations.

Group	Mean Grade	Standard Deviation
Mature	91.50	8.48
Immature	84.20	6.98

Source: Krayer, K. J. "Exploring group maturity in the classroom." *Small Group Behavior,* Vol. 19, No. 2, May 1988, p. 268.

a.   Assuming the population of project grades for the mature group is approximately normal, find the probability that a mature student will score below 80 on the final project.

b.   Repeat part **a** for the immature group.

c. Why might the assumption of normality in parts **a** and **b** be suspect? [*Hint:* Consider the fact that the highest grade that can be assigned to a project is 100.]

6.3 DESCRIPTIVE METHODS FOR ASSESSING NORMALITY

In the chapters that follow, we learn how to make inferences about the population based on information in the sample. Several of these techniques are based on the assumption that the population is approximately normally distributed. However, not all continuous random variables are normally distributed! Consequently, it will be important to determine whether the sample data come from a normal population before we can properly apply these techniques.

Several descriptive methods can be used to check for normality. In this section, we consider the four methods summarized in the box.

Determining Whether the Data Are from an Approximately Normal Distribution

1. Construct either a histogram, stem-and-leaf display, or a box-and-whisker plot for the data and note the shape of the graph. If the data are approximately normal, the shape of the histogram or stem-and-leaf display will be similar to the normal curve, Figure 6.2 (i.e., mound-shaped and symmetric about the mean) and the shape of the box-and-whisker plot will be similar to Figure 3.21a (page 183).

2. Compute the intervals $\bar{x} \pm s$, $\bar{x} \pm 2s$, and $\bar{x} \pm 3s$, and determine the percentage of measurements falling in each. If the data are approximately normal, the percentages will be approximately equal to 68%, 95%, and 100%, respectively.

3. Find the interquartile range, IQR, and standard deviation, s, for the sample, then calculate the ratio IQR/s. If the data are approximately normal, then IQR/$s \approx 1.3$.

4. Construct a *normal probability plot* for the data. If the data are approximately normal, the points will fall (approximately) on a straight line.

Why IQR/$s \approx 1.3$ for Normal Data

For normal distributions, the z values (obtained from Table B.2) corresponding to the 75th and 25th percentiles are .67 and $-.67$, respectively. Since $\sigma = 1$ for a standard normal (z) distribution,

$$\text{IQR}/\sigma = [.67 - (-.67)]/1 = 1.34$$

Example 6.11   Detecting Normality

FTC.XLS

Consider the Federal Trade Commission data on 500 cigarette brands, stored in the Excel workbook FTC.XLS. Numerical and graphical descriptive measures for the tar content data are shown on the Excel printouts, Figures 6.13a–d (page 322). Determine whether the tar contents have an approximate normal distribution.

Figure 6.13 Excel Printouts for Example 6.11

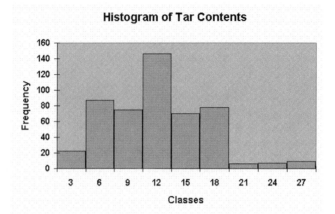

a.

	A	B
1	*Tar*	
2		
3	Mean	11.059
4	Standard Error	0.2205296
5	Median	11
6	Mode	11
7	Standard Deviation	4.931191773
8	Sample Variance	24.3166523
9	Kurtosis	0.378809998
10	Skewness	0.360265442
11	Range	26.5
12	Minimum	0.5
13	Maximum	27
14	Sum	5529.5
15	Count	500
16	Largest(1)	27
17	Smallest(1)	0.5

b.

	A	B
1	FTC Tar Content Data	
2		
3	Five-Number Summary	
4	Minimum	0.5
5	First Quartile	8
6	Median	11
7	Third Quartile	15
8	Maximum	27

c.

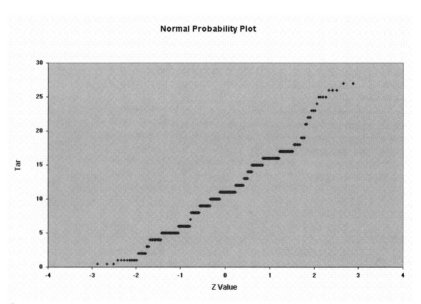

d.

Table 6.3 Describing the 500 Cigarette Tar Contents

Interval	Percentage in Interval
$\bar{x} \pm s = (6.2, 16.0)$	58.2
$\bar{x} \pm 2s = (1.3, 20.9)$	94.0
$\bar{x} \pm 3s = (-3.6, 25.8)$	99.9

Solution

As a first check, we examine the frequency histogram of the 500 tar contents shown in Figure 6.13a. Clearly, the tar contents fall in an approximately mound-shaped, symmetric distribution centered around the mean of 11.1 milligrams. Thus, from check #1 in the box, the data appear to be approximately normal.

To apply check #2, we obtain $\bar{x} = 11.1$ and $s = 4.93$ from the Excel descriptive statistics printout, Figure 6.13b. The intervals $\bar{x} \pm s$, $\bar{x} \pm 2s$, and $\bar{x} \pm 3s$ are shown in Table 6.3, as well as the percentage of tar measurements that fall in each interval. Except for the one-standard-deviation interval, these percentages agree almost exactly with those from a normal distribution.

Check #3 in the box requires that we find the interquartile range (i.e., the difference between the 75th and 25th percentiles) and the standard deviation s, and compute the ratio of these two numbers. The ratio IQR/s for a sample from a normal distribution will approximately equal 1.3. The value of s, 4.93, is highlighted in Figure 6.13b. The values of Q_1 and Q_3, 8 and 15 respectively, are highlighted in Figure 6.13c. Then IQR = $Q_3 - Q_1 = 7$ and the ratio is

$$\frac{\text{IQR}}{s} = \frac{7}{4.93} = 1.42$$

Since this value is approximately equal to 1.3, we have further confirmation that the data are approximately normal.

A fourth descriptive technique for checking normality is a *normal probability plot*. In a normal probability plot, the observations in the data set are ordered from smallest to largest and then plotted against the expected z scores of the observations calculated under the assumption that the data are from a normal distribution. When the data are, in fact, normally distributed, a linear (straight-line) trend will result. A nonlinear trend in the normal probability plot suggests that the data are nonnormal.

Definition 6.2

A **normal probability plot** for a data set is a scatterplot with the ranked data values on the vertical axis and their corresponding z values from a standard normal distribution on the horizontal axis.

Computation of the expected standard normal z scores for a normal probability plot are beyond the scope of this text. A normal probability plot for the 500 tar measurements obtained from the PHStat Add-in for Excel is shown in Figure 6.13d. Notice that the ordered measurements fall reasonably close to a straight line. Thus, check #4 also suggests that the data are likely to be approximately normally distributed. ❑

*Statistical tests of normality that provide a measure of reliability for the inference are available. However, these tests tend to be very sensitive to slight departures from normality, i.e., they tend to reject the hypothesis of normality for any distribution that is not perfectly symmetrical and mound-shaped. Consult the references (see Ramsey & Ramsey, 1990) if you want to learn more about these tests.

The checks for normality given in the box are simple, yet powerful, techniques to apply, but they are only descriptive in nature. It is possible (although unlikely) that the data are nonnormal even when the checks are reasonably satisfied. Thus, we should be careful not to claim that the 500 tar measurements in the Excel workbook FTC.XLS are, in fact, normally distributed. We can only state that it is reasonable to believe that the data are from a normal distribution.*

As we will learn in the next chapter, several inferential methods of analysis require the data to be approximately normal. If the data are clearly nonnormal, inferences derived from the method may be invalid. Therefore, it is advisable to check the normality of the data prior to conducting the analysis.

Self-Test

6.3

Which of the following suggest that the data set is approximately normal?

a.   A data set with $Q_1 = 10$, $Q_3 = 50$, and $s = 10$.

b.   A data set with the following stem-and-leaf display:

Stem	Leaves
0	1 1 2
1	1 2 5 6 7
2	0 1 1 1 3 8 8
3	4 5 6 6 6 6 7 7
4	1 1 1 2 2 3 3 3 3 4 8 9 9
5	0 0 1 2 5 5 6 8 9
6	0 1 1 1 2 3
7	1 2 5 8
8	0 7
9	2

c.   A data set with 93% of the measurements within $\bar{x} \pm s$.

d.   A data set with the following normal probability plot.

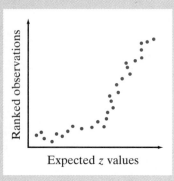

PROBLEMS
6.21–6.32

Using the Tools

6.21 If a population data set is normally distributed, what is the proportion of measurements you would expect to fall within the following intervals?

   a. $\mu \pm \sigma$ **b.** $\mu \pm 2\sigma$ **c.** $\mu \pm 3\sigma$

6.22 Consider a sample data set with the following summary statistics: $s = 95$, $Q_1 = 72, Q_3 = 195$.

   a. Calculate IQR.   **b.** Calculate IQR/s.

   c. Is the value of IQR/s approximately equal to 1.3? What does this imply?

6.23 Normal probability plots for three data sets are shown below. Which plot indicates that the data are approximately normally distributed?

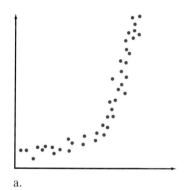
a.

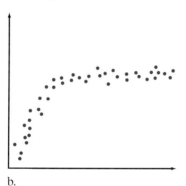
b.

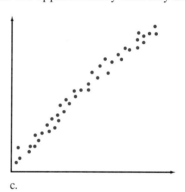
c.

6.24 The sample data for Problems 2.18 and 3.45 are reproduced below.

5.9	5.3	1.6	7.4	8.6	1.2	2.1
4.0	7.3	8.4	8.9	6.7	4.5	6.3
7.6	9.7	3.5	1.1	4.3	3.3	8.4
1.6	8.2	6.5	1.1	5.0	9.4	6.4

   a. Use the stem-and-leaf plot from Problem 2.18 to assess whether the data are from an approximately normal distribution.

   b. Compute s for the sample data.

   c. Use the values of Q_1 and Q_3 from Problem 3.45 and the value of s from part **b** to assess whether the data come from an approximately normal distribution.

   d. Generate a normal probability plot for the data and use it to assess whether the data are approximately normal.

Applying the Concepts

6.25 It is well-known that children with developmental delays (i.e., mild mental retardation) are slower, cognitively, than normally developing children. Are their social skills also lacking? A study compared the social interactions of the two groups of children in a controlled playground environment (*American Journal on Mental Retardation,* Jan. 1992). One variable of interest was the number of intervals of "no play" by each child. Children with developmental delays had a mean of 2.73 intervals of "no play" and a standard deviation of 2.58 intervals. Based on this information, is it possible for the variable of interest to be normally distributed? Explain.

6.26 How accurate are you at the game of darts? Researchers at Iowa State University attempted to develop a probability model for dart throws (*Chance,* Summer 1997). For each of 590 throws made at certain targets on a dart board, the distance from the dart to the target point was measured (to the nearest millimeter). The error distribution for the dart throws is described by the frequency table shown below.

DARTS.XLS

Distance from Target (mm)	Frequency	Distance from Target (mm)	Frequency	Distance from Target (mm)	Frequency
0	3	21	15	41	7
2	2	22	19	42	7
3	5	23	13	43	11
4	10	24	11	44	7
5	7	25	9	45	5
6	3	26	21	46	3
7	12	27	16	47	1
8	14	28	18	48	3
9	9	29	9	49	1
10	11	30	14	50	4
11	14	31	13	51	4
12	13	32	11	52	3
13	32	33	11	53	4
14	19	34	11	54	1
15	20	35	16	55	1
16	9	36	13	56	4
17	18	37	5	57	3
18	23	38	9	58	2
19	25	39	6	61	1
20	21	40	5	62	2
				66	1

Source: Stern, H. S., and Wilcox, W. "Shooting darts." *Chance,* Vol. 10, No. 3, Summer 1997, p. 17 (adapted from Figure 2).

a. Construct a histogram for the data. Is the error distribution for the dart throws approximately normal?

b. Descriptive statistics for the distances from the target for the 590 throws are given below. Use this information to decide whether the error distribution is approximately normal.

$$\bar{x} = 24.4 \text{ mm} \quad s = 12.8 \text{ mm} \quad Q_1 = 14 \text{ mm} \quad Q_3 = 34 \text{ mm}$$

c. Construct a normal probability plot for the data and use it to assess whether the error distribution is approximately normal.

6.27 Refer to the study of British Columbia's boreal forest, Problem 6.11 (page 318). The diameters at breast height (in meters) for a sample of 28 trembling aspen trees are listed at the top of page 327. Determine whether the sample data are from an approximately normal distribution.

FOREST.XLS

12.4	17.3	27.3	19.1	16.9	16.2	20.0
16.6	16.3	16.3	21.4	25.7	15.0	19.3
12.9	18.6	12.4	15.9	18.8	14.9	12.8
24.8	26.9	13.5	17.9	13.2	23.2	12.7

Source: Schotz, H. "Fish Creek Community Forest: Exploratory statistical analysis of selected data." Working paper, Northern Lights College, British Columbia, Canada.

Problem 6.27

6.28   Refer to the *Journal of Information Systems* study of intrusion-detection systems, Problem 6.17 (page 319). According to the study, many intrusion-detection systems assume that the data being monitored are normally distributed when such data are clearly nonnormal. Consequently, the intrusion-detection system may lead to inappropriate conclusions. The researcher considered the following data on input-output (I/O) units utilized by a sample of 44 users of a system.

IOUNITS.XLS

15	5	2	17	4	3	1	1	0	0	0	0
0	0	0	20	9	0	0	0	1	6	1	3
1	0	6	0	0	0	0	1	14	0	7	0
2	9	4	0	0	0	9	10				

Source: O'Leary, D. E. "Intrusion-detection systems." *Journal of Information Systems*, Spring 1992, p. 68 (Table 2).

Based on the accompanying Excel printouts here and at the top of page 328, assess whether the data are normally distributed.

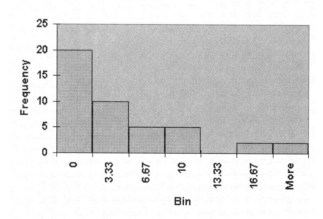

E

	A	B
1	IO Unit	
2		
3	Mean	3.431818182
4	Standard Error	0.77786348
5	Median	1
6	Mode	0
7	Standard Deviation	5.159762602
8	Sample Variance	26.62315011
9	Kurtosis	2.347878438
10	Skewness	1.730715055
11	Range	20
12	Minimum	0
13	Maximum	20
14	Sum	151
15	Count	44
16	Largest(1)	20
17	Smallest(1)	0

E

	A	B
1	I/O Units Data	
2		
3	Five-Number Summary	
4	Minimum	0
5	First Quartile	0
6	Median	1
7	Third Quartile	5.25
8	Maximum	20

Problem 6.28

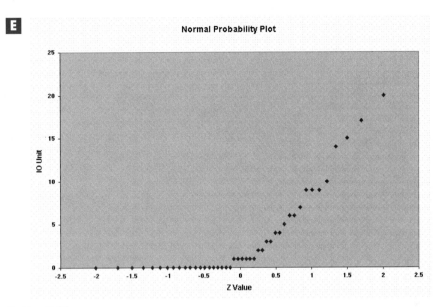

E

Normal Probability Plot

Problem 6.28

6.29　Refer to the *Small Group Behavior* study of mature and immature groups, Problem 6.20 (page 320). The grades for a sample of 25 "immature" group students and 25 "mature" group students are listed in the table, followed by an Excel stem-and-leaf display and normal probability plot (page 329) for each of the two data sets. Based on these graphs, assess whether the grade distributions are approximately normal.

SMALLGRP.XLS

Immature Group						Mature Group			
74	75	75	75	78		73	75	78	84	85
79	79	81	81	84		85	85	87	91	91
84	85	85	86	86		92	92	93	94	95
87	88	88	89	90		95	98	100	100	100
91	92	93	95	96		100	100	100	100	100

E

	A	B	C	D
1	**Stem-and-Leaf Displays**			
2	**Stem unit:**	**10**		
3				
4	**for Immature**			
5	7	4 5 5 5 8 9 9		
6	8	1 1 4 4 5 5 6 6 7 8 8 9		
7	9	0 1 2 3 5 6		
8	10			
9				
10	**for Mature**			
11	7	3 5 8		
12	8	4 5 5 5 7		
13	9	1 1 2 2 3 4 5 5 8		
14	10	0 0 0 0 0 0 0 0		
15				
1C				

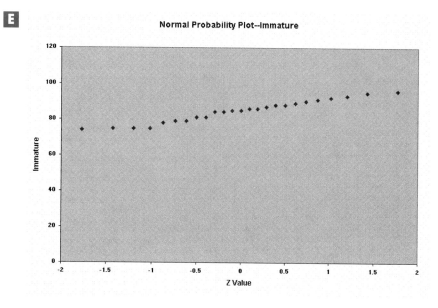

Normal Probability Plot--Immature

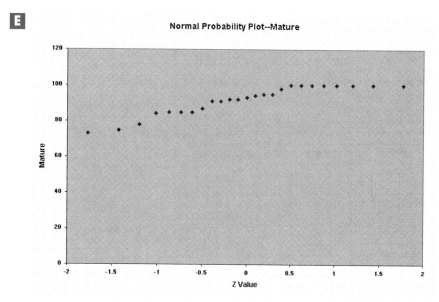

Normal Probability Plot--Mature

6.30   A chemical analysis of metamorphic rock in western Turkey was reported in *Geological Magazine* (May 1995). The trace amount (in parts per million) of the element rubidium was measured for 20 rock specimens. The data are listed below. Assess whether the sample data come from a normal population.

RUBIDIUM.XLS

164	286	355	308	277	330	323	370	241	402
301	200	202	341	327	285	277	247	213	424

Source: Bozkurt, E., et al. "Geochemistry and tetonic significance of augen gneisses from the southern Menderes Massif (West Turkey)." *Geological Magazine,* Vol. 132, No. 3, May 1995, p. 291 (Table 1)

6.31   *Meta-analysis* is a procedure to summarize and compare the findings of independent studies on similar scientific topics. *Chance* (Fall 1997) describes a meta-analysis on the relationship between social behavior and judgments based on first impressions. A literature search revealed 38 independent studies with usable results. The correlation coefficient (r) between the behavioral variables of interest was recorded for each study. A stem-and-leaf display for the 38 correlations is shown below. Comment on whether the distribution of correlations is approximately normal.

Stem	Leaf
.1	0 0 4 5 6 6
.2	1 1 1 2 3 3 4 5 6 6 7 8 9
.3	1 3 5
.4	0 0 0 1 7
.5	0 2 2 3 4 4
.6	3 8
.7	3 4
.8	7
.9	
1.0	

Source: Ambady, N., and Rosenthal, R. "Judging social behavior using 'Thin slices'." *Chance*, Vol. 10, No. 4, Fall 1997, p. 17 (Table 4).

6.32   The Harris Corporation data on voltage readings at two locations, Problem 3.38, are reproduced here. Determine whether the voltage readings at each location are approximately normal.

VOLTAGE.XLS

Old Location			New Location		
9.98	10.12	9.84	9.19	10.01	8.82
10.26	10.05	10.15	9.63	8.82	8.65
10.05	9.80	10.02	10.10	9.43	8.51
10.29	10.15	9.80	9.70	10.03	9.14
10.03	10.00	9.73	10.09	9.85	9.75
8.05	9.87	10.01	9.60	9.27	8.78
10.55	9.55	9.98	10.05	8.83	9.35
10.26	9.95	8.72	10.12	9.39	9.54
9.97	9.70	8.80	9.49	9.48	9.36
9.87	8.72	9.84	9.37	9.64	8.68

Source: Harris Corporation, Melbourne, Fla.

6.4 SAMPLING DISTRIBUTIONS

In Chapter 5 and the previous sections of this chapter, we assumed that we knew the probability distribution of a random variable, and using this knowledge we were able to find the mean, variance, and probabilities associated with the random variable. However, in most practical applications, the true mean and standard deviation are unknown quantities that will have to be estimated. Thus, our objective is to estimate a numerical characteristic of the population. These population characteristics, you will recall, are called **parameters** (see Definition 3.15, page 200). Thus, π, the probability of success in a binomial experiment, and μ and

σ, the mean and standard deviation of a normal distribution, are examples of parameters.

Recall (from Chapter 1) that inferential statistics involves using information in a sample to make inferences about unknown population parameters. Thus, we might use the sample mean $\bar{x}$ to estimate μ, or the sample standard deviation s to estimate σ. Quantities like $\bar{x}$ and s that are computed from the sample are called *sample statistics.*

Definition 6.3

A summary measure computed from the observations in a sample is called a **sample statistic**.

Although it is usually unknown to us, the value of a population parameter (e.g., μ) is constant; its value does not vary from sample to sample. However, the value of a sample statistic (e.g., $\bar{x}$) is highly dependent on the particular sample that is selected. Since statistics vary from sample to sample, any inferences based on them will necessarily be subject to some uncertainty. How, then, do we judge the reliability of a sample statistic as a tool in making an inference about the corresponding population parameter? Fortunately, the uncertainty of a statistic generally has characteristic properties that are known to us, and that are reflected in its *sampling distribution.*

Definition 6.4

The **sampling distribution** of a sample statistic (based on n observations) is the distribution of the values of the statistic theoretically generated by taking repeated random samples of size n and computing the value of the statistic for each sample.

The notion of a sampling distribution is illustrated in the next three examples.

Example 6.12   Concept of a Sampling Distribution

CHKOUT.XLS

Consider the 500 supermarket customer checkout times in the Excel workbook CHKOUT.XLS. Assume this data set is the target population and we want to estimate μ, the mean checkout time of these 500 observations. Demonstrate how to physically generate the sampling distribution $\bar{x}$, the mean of a random sample of $n = 5$ observations from the population of 500 checkout times in CHKOUT.XLS.

Solution

The sampling distribution for the statistic $\bar{x}$, based on a random sample of $n = 5$ measurements, would be generated in this manner: (1) Select a random sample of five measurements from the population of 500 checkout times; (2) Compute and record the value of $\bar{x}$ for this sample; (3) Return these five measurements to the population and repeat the procedure, i.e., draw another random sample of $n = 5$

Figure 6.14 Generating the Theoretical Sampling Distribution of the Sample Mean $\bar{x}$

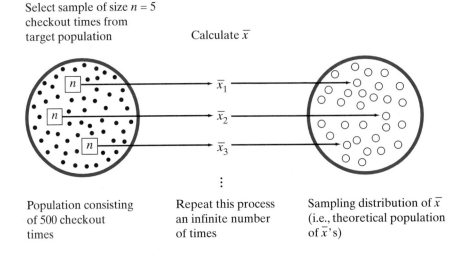

Select sample of size $n = 5$ checkout times from target population

Calculate $\bar{x}$

Population consisting of 500 checkout times

Repeat this process an infinite number of times

Sampling distribution of $\bar{x}$ (i.e., theoretical population of $\bar{x}$'s)

measurements and record the value of $\bar{x}$ for this sample; (4) Return these measurements and continue repeating the process.

If this sampling procedure could be repeated an infinite number of times, as shown in Figure 6.14, the infinite number of values of $\bar{x}$ obtained would form a new population. The data in this population could be summarized in a relative frequency distribution, called the **sampling distribution of the sample mean $\bar{x}$.** ❏

The task described in Example 6.12, which may seem impractical if not impossible, is not performed in actual practice. Instead, the sampling distribution of a statistic is obtained by applying mathematical theory or computer simulation, as illustrated in the next example.

Example 6.13   Approximating a Sampling Distribution

CHKOUT.XLS

Use computer simulation to find the approximate sampling distribution of $\bar{x}$, the mean of a random sample of $n = 5$ observations from the population of 500 checkout times in CHKOUT.XLS.

Solution
We obtained a large number—namely, 100—of computer-generated random samples of size $n = 5$ from the target population. The first ten samples are presented in Table 6.4.

For example, the first computer-generated sample contained the following measurements: 8, 15, 28, 155, 35. The corresponding value of the sample mean is

$$\bar{x} = \frac{\Sigma x}{n} = \frac{8 + 15 + 28 + 155 + 35}{5} = 48.2 \text{ seconds}$$

For each sample of five observations, the sample mean $\bar{x}$ was computed. The 100 values of $\bar{x}$ are summarized in the Excel histogram shown in Figure 6.15 (page 333). This distribution approximates the sampling distribution of $\bar{x}$ for a sample of size $n = 5$. ❏

Table 6.4 First 10 Samples of $n = 5$ Checkout Times from CHKOUT.XLS

Sample	Customer Checkout Time (seconds)				
1	8	15	28	155	35
2	16	66	45	50	35
3	50	10	50	35	15
4	31	180	53	30	353
5	150	25	15	10	20
6	53	10	113	81	97
7	93	237	6	40	23
8	12	16	35	49	100
9	27	80	50	20	20
10	66	63	50	40	50

E Figure 6.15
Sampling Distribution of $\bar{x}$: 100 Random Samples of $n = 5$ Checkout Times

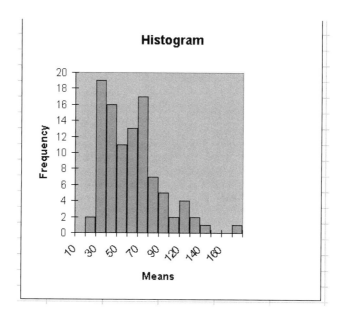

Self-Test 6.4

Perform the following experiment: Flip two fair coins and count x, the number of heads that occur. Repeat this procedure four more times, counting x each time. Record the five values of x. (These values represent a random sample of size $n = 5$.) Perform the entire experiment a total of 20 times, thus generating 20 random samples of size $n = 5$.

a. Compute $\bar{x}$ for each sample.

b. Construct a histogram or stem-and-leaf display for the 20 values of $\bar{x}$.

c. Why is the distribution of part **b** only an approximation to the sampling distribution of $\bar{x}$?

E **Figure 6.16** Histogram of the Population of 500 Customer Checkout Times

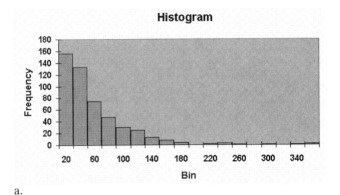

a.

	A	B
1	*Checkout times*	
2		
3	Mean	50.118
4	Standard Error	2.194224554
5	Median	35
6	Mode	30
7	Standard Deviation	49.06435261
8	Sample Variance	2407.310697
9	Kurtosis	9.731434612
10	Skewness	2.54163484
11	Range	351
12	Minimum	2
13	Maximum	353
14	Sum	25059
15	Count	500
16	Largest(1)	353
17	Smallest(1)	2

b.

A histogram for the population of 500 checkout times in CHKOUT.XLS, generated using Excel, is displayed in Figure 6.16a above, along with descriptive statistics for the data set (Figure 6.16b at left). From the printout, the population of interest in Example 6.13 has mean $\mu = 50.1$ seconds and standard deviation $\sigma = 49.1$ seconds. Let us compare the histogram for the sampling distribution of $\bar{x}$ (Figure 6.15) with the histogram for the population shown in Figure 6.16a. Note that the values of $\bar{x}$ in Figure 6.15 tend to cluster around the population mean, $\mu = 50.1$. Also, the values of the sample mean are less spread out (that is, they have less variation) than the population values shown in Figure 6.16a. These two observations are borne out by comparing the means and standard deviations of the two sets of observations, as shown in Table 6.5.

Example 6.14   Impact of *n* on Sampling Distributions

CHKOUT.XLS

Refer to Example 6.13. Simulate the sampling distribution of $\bar{x}$ for samples of size $n = 25$ from the population of 500 checkout times in CHKOUT.XLS. Compare the result with the sampling distribution of $\bar{x}$ based on samples of size $n = 5$, obtained in Example 6.13.

Solution

We obtained 100 computer-generated random samples of size $n = 25$ from the target population. An Excel histogram for the 100 corresponding values of $\bar{x}$ is shown in Figure 6.17 (page 335).

First, note that, as with the sampling distribution based on samples of size $n = 5$, the values of $\bar{x}$ tend to center about the population mean. Second, notice that the variation in the $\bar{x}$ values about their mean in Figure 6.17 is less than the

Table 6.5 Comparison of the Population Distribution and the Approximate Sampling Distribution of $\bar{x}$, Based on 100 Samples of Size $n = 5$

	Mean	Standard Deviation
Population of 500 checkout times (Figure 6.16)	$\mu = 50.1$	$\sigma = 49.1$
100 values of $\bar{x}$ based on samples of size $n = 5$ (Figure 6.15)	55.6	28.5

E **Figure 6.17**
Sampling Distribution
of $\bar{x}$: 100 Random
Samples of n = 25

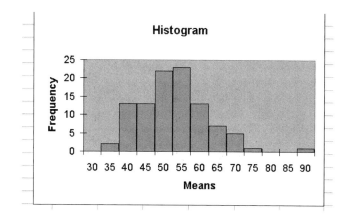

Table 6.6 Comparison of the Population Distribution and the Approximate Sampling Distribution of $\bar{x}$, Based on 100 Samples of Size n = 25

	Mean	Standard Deviation
Population of 500 checkout times (Figure 6.16)	μ = 50.1	σ = 49.1
100 values of $\bar{x}$ based on samples of size n = 5 (Figure 6.15)	55.6	28.5
100 values of $\bar{x}$ based on samples of size n = 25 (Figure 6.17)	50.3	9.4

variation in the values of $\bar{x}$ based on samples of size n = 5 (Figure 6.15). The mean and standard deviation for these 100 values of x are shown in Table 6.6 for comparison with previous results.

From Table 6.6 we observe that, as the sample size increases, there is less variation in the sampling distribution of $\bar{x}$; that is, the values of $\bar{x}$ cluster more closely about the population mean as n gets larger. This intuitively appealing result will be stated formally in the next section. ❑

PROBLEMS 6.33–6.42

Using the Tools

x	1	2	3	4	5
$p(x)$	.2	.3	.2	.2	.1

Problem 6.33

6.33 Consider the population described by the probability distribution shown at left. The random variable x is observed twice. If these observations are independent, verify that the different samples of size 2 and their probabilities are as shown at the top of page 336.

   a. Find the sampling distribution of the sample mean $\bar{x}$.

   b. Construct a probability histogram for the sampling distribution of $\bar{x}$.

   c. What is the probability that $\bar{x}$ is 4.5 or larger?

x	0	2	4	6
$p(x)$	$^1/_4$	$^1/_4$	$^1/_4$	$^1/_4$

Problem 6.34

6.34 The probability distribution shown here describes a population of measurements that can assume values of 0, 2, 4, and 6, each of which occurs with the same relative frequency:

Sample	Probability	Sample	Probability	Sample	Probability
1,1	.04	2,5	.03	4,4	.04
1,2	.06	3,1	.04	4,5	.02
1,3	.04	3,2	.06	5,1	.02
1,4	.04	3,3	.04	5,2	.03
1,5	.02	3,4	.04	5,3	.02
2,1	.06	3,5	.02	5,4	.02
2,2	.09	4,1	.04	5,5	.01
2,3	.06	4,2	.06		
2,4	.06	4,3	.04		

Problem 6.33

a. List all the different samples of $n = 2$ measurements that can be selected from this population.

b. Calculate the mean of each different sample listed in part **a**.

c. If a sample of $n = 2$ measurements is randomly selected from the population, what is the probability that a specific sample will be selected?

d. Assume that a random sample of $n = 2$ measurements is selected from the population. List the different values of $\bar{x}$ found in part **b**, and find the probability of each. Then give the sampling distribution of the sample mean $\bar{x}$ in tabular form.

6.35 The table contains 50 random samples of $n = 5$ measurements selected from a population with $\mu = 4.5$ and $\sigma^2 = 8.25$.

SAMPLE50.XLS

Sample				
1, 8, 0, 6, 6	1, 6, 0, 0, 9	3, 6, 4, 2, 0	4, 5, 3, 4, 8	2, 3, 7, 6, 3
2, 1, 7, 2, 9	6, 8, 5, 2, 8	1, 5, 0, 5, 8	5, 6, 7, 8, 2	2, 0, 6, 3, 3
4, 5, 7, 7, 1	2, 4, 9, 4, 6	4, 6, 2, 6, 2	3, 8, 6, 0, 1	1, 9, 0, 3, 2
3, 6, 1, 8, 1	6, 7, 0, 4, 3	1, 8, 8, 2, 1	1, 4, 4, 9, 0	8, 9, 2, 7, 0
9, 8, 6, 2, 9	0, 5, 9, 9, 6	9, 0, 6, 1, 7	7, 7, 9, 8, 1	1, 5, 0, 5, 1
6, 8, 8, 3, 5	4, 4, 7, 5, 6	3, 7, 3, 4, 3	9, 2, 9, 8, 7	7, 8, 7, 7, 6
9, 5, 7, 7, 9	6, 6, 5, 5, 6	4, 5, 2, 6, 6	6, 8, 9, 6, 0	9, 3, 7, 3, 9
7, 6, 4, 4, 7	5, 0, 6, 6, 5	9, 3, 7, 1, 3	3, 4, 6, 7, 0	5, 1, 1, 4, 0
6, 5, 6, 4, 2	3, 0, 4, 9, 6	1, 9, 6, 9, 2	8, 4, 7, 6, 9	2, 5, 7, 7, 9
8, 6, 8, 6, 0	3, 0, 7, 4, 1	5, 1, 2, 3, 4	6, 9, 4, 4, 2	3, 0, 6, 9, 7

a. Calculate $\bar{x}$ for each of the 50 samples.

b. Construct a histogram for the 50 sample means. This figure represents an approximation to the sampling distribution of $\bar{x}$ based on samples of size $n = 5$.

c. Compute the mean and standard deviation for the 50 sample means. Locate these values on the histogram of part **b**. Note how the sample means cluster about $\mu = 4.5$.

6.36 Refer to Problem 6.35. Combine pairs of samples (moving down the columns of the table) to obtain 25 samples of $n = 10$ measurements.

a. Calculate $\bar{x}$ for each of the 25 samples.

b. Construct a histogram for the 25 sample means. This figure represents an approximation to the sampling distribution of $\bar{x}$ based on samples of size $n = 10$. Compare with the figure constructed in Problem 6.35.

c. Compute the mean and standard deviation for the 25 sample means, and locate them on the histogram. Note how the sample means cluster about $\mu = 4.5$.

d. Compare the standard deviations of the two sampling distributions in Problems 6.35 and 6.36. Which sampling distribution has less variation?

Applying the Concepts

6.37 Consider a random sample of 25 measurements selected from each of the following populations. Describe what the sampling distribution of $\bar{x}$ would consist of for each.

a. Percent body fat measurements for all recent Olympic athletes

b. Noise levels (measured in decibels) for all passenger flights taking off at LAX airport

c. Task completion times (in minutes) for all patients who submit to a psychological exam requiring consecutively numbered circles to be connected

6.38 The following data represent the number of days absent per year in a population of six employees of a small company.

DAYSABSENT.XLS

1   3   6   7   7   12

Assuming that you sample *without* replacement:

a. Select all possible samples of size 2 and set up the sampling distribution of the mean.

b. Compute the mean of all the sample means and also compute the population mean. Are they equal?

c. Repeat parts **a** and **b** for all possible samples of size 3.

d. Compare the shape of the sampling distribution of the mean obtained in parts **a** and **c**. Which sampling distribution seems to have the least variability? Why?

6.39 Audiologists at Baylor College of Medicine fitted individuals suffering from sensorineural hearing loss with various devices (e.g., hearing aids, cochlear implants, etc.) designed to amplify sounds (*Ear & Hearing*, Apr. 1995). One subject was tested under six different experimental conditions. Under each condition, the signal intensity necessary for a 50% improvement in hearing was recorded. The six intensity scores (measured in db HL) were: 80, 85, 75, 65, 60, and 55. To test the effectiveness of the hearing device used, the audiologists were interested in the sampling distribution of $\bar{x}$, the mean intensity score for a random sample of three scores selected from the six scores recorded for the subject.

DBHL.XLS

a. List the different random samples of size $n = 3$ that can be selected from the six intensity scores.

b. For each sample, compute $\bar{x}$.

c. Use a stem-and-leaf display to portray the sampling distribution of $\bar{x}$.

d. Find the mean and the standard deviation of the sampling distribution of $\bar{x}$.

6.40 Use Excel or Table B.1 to obtain 30 random samples of size $n = 5$ from the "population" of 500 nicotine contents of cigarettes in the Excel

FTC.XLS

workbook, FTC.XLS. (Alternatively, each class member may generate several random samples, and the results can be pooled.)

a. Calculate $\bar{x}$ for each of the 30 samples. Construct a histogram for the 30 sample means.

b. Compute the average of the 30 sample means.

c. Compute the standard deviation of the 30 sample means.

d. Locate the average of the 30 sample means, computed in part **b**, on the histogram. This value could be used as an estimate for μ, the mean of the entire population of 500 nicotine measurements.

6.41 Repeat parts **a**, **b**, **c**, and **d** of Problem 6.40, using random samples of size $n = 10$. Compare the relative frequency distribution with that of Problem 6.40a. Do the values of $\bar{x}$ generated from samples of size $n = 10$ cluster more closely about μ?

FISH.XLS

6.42 Generate the approximate sampling distribution of $\bar{x}$, the mean of a random sample of $n = 15$ observations from the population of DDT measurements stored in the Excel workbook, FISH.XLS. [*Hint:* Obtain 50 random samples of size $n = 15$ from the data, compute $\bar{x}$ for each sample, and then construct a histogram for the 50 sample means. Again, each class member could generate several random samples, and the results could be pooled.]

6.5 THE SAMPLING DISTRIBUTION OF THE MEAN AND THE CENTRAL LIMIT THEOREM

Estimating the mean checkout time for all customers of a certain supermarket, or the average increase in heart rate for all human subjects after 15 minutes of vigorous exercise, or the mean yield per acre of farmland, are all examples of practical problems in which the goal is to make an inference about the mean μ of some target population. In the previous section, we have indicated that the sample mean $\bar{x}$ is often used as a tool for making an inference about the corresponding population parameter μ, and we have shown how to *approximate* its sampling distribution. The following theorem, of fundamental importance in statistics, provides information about the *actual* sampling distribution of $\bar{x}$.

The sampling distribution of $\bar{x}$, in addition to being approximately normal, has other known characteristics, which are summarized in the next box.

Properties of the Sampling Distribution of $\bar{x}$

If $\bar{x}$ is the mean of a random sample of size n from a population with mean μ and standard deviation σ, then:

1. The sampling distribution of $\bar{x}$ has a mean equal to the mean of the population from which the sample was selected. That is, if we let $\mu_{\bar{x}}$ denote the mean of the sampling distribution of $\bar{x}$, then

$$\mu_{\bar{x}} = \mu$$

2. The sampling distribution of $\bar{x}$ has a standard deviation equal to the standard deviation of the population from which the sample was selected, divided by the square root of the sample size. That is, if we let $\sigma_{\bar{x}}$ denote the standard deviation of the sampling distribution of $\bar{x}$ (also called the **standard error** of $\bar{x}$), then

$$\sigma_{\bar{x}} = \frac{\sigma}{\sqrt{n}} \qquad \qquad (6.3)$$

The Central Limit Theorem

If the sample size is *sufficiently large*, then the mean $\bar{x}$ of a random sample from a population has a sampling distribution that is approximately normal, *regardless of the shape of the relative frequency distribution of the target population*. As the sample size n increases, the sampling distribution of $\bar{x}$ moves closer and closer to the normal distribution.

Example 6.15 Properties of the Sampling Distribution of $\bar{x}$

Refer to Examples 6.13 and 6.14, where we obtained repeated random samples of sizes $n = 5$ and $n = 25$ from the population of customer checkout times stored in CHKOUT.XLS. Recall that for this target population, we know the values of the parameters μ and σ:

Population mean: $\mu = 50.1$ seconds

Population standard deviation: $\sigma = 49.1$ seconds

Show that the empirical evidence obtained in Examples 6.13 and 6.14 supports the Central Limit Theorem and the two properties of the sampling distribution of $\bar{x}$.

Solution

In Figures 6.15 and 6.17 (pages 333 and 335), we noted that the values of $\bar{x}$ cluster about the population mean, $\mu \approx 50$. This is guaranteed by property 1, which implies that, in the long run, the average of *all* values of $\bar{x}$ that would be generated in infinite repeated sampling would be equal to μ.

We also observed, from Table 6.6 (page 335), that the standard deviation of the sampling distribution of $\bar{x}$, called the *standard error of $\bar{x}$,* decreases as the sample size increases from $n = 5$ to $n = 25$. Property 2 quantifies the decrease and relates it to the sample size. For our approximate (simulated) sampling distribution based on samples of size $n = 5$, we obtained a standard deviation of 28.5, whereas property 2 tells us that, for the actual sampling distribution of $\bar{x}$, the standard deviation is equal to

$$\sigma_{\bar{x}} = \frac{\sigma}{\sqrt{n}} = \frac{49.1}{\sqrt{5}} = 22.0$$

Similarly, for samples of size $n = 25$, the sampling distribution of $\bar{x}$ actually has a standard deviation of

$$\sigma_{\bar{x}} = \frac{\sigma}{\sqrt{n}} = \frac{49.1}{\sqrt{25}} = 9.8$$

The value we obtained by simulation was 9.4.

Finally, for sufficiently large samples, the Central Limit Theorem guarantees an approximately normal distribution for $\bar{x}$, regardless of the shape of the original population. In our examples, the population from which the samples were selected is seen in Figure 6.16 (page 334) to be highly skewed to the right. Note from Figures 6.15 and 6.17 that, although the sampling distribution of $\bar{x}$ tends to be mound-shaped in each case, the normal approximation improves when the sample size is increased from $n = 5$ (Figure 6.15) to $n = 25$ (Figure 6.17). ❑

Self-Test

6.5

Let $\bar{x}$ represent the sample mean for a random sample of size $n = 100$ selected from a population with mean $\mu = 5$ and standard deviation $\sigma = 1$.

    a. Find and interpret the value of $\mu_{\bar{x}}$.

    b. Find and interpret the value of $\sigma_{\bar{x}}$.

How large must the sample size n be so that the normal distribution provides a good approximation for the sampling distribution of $\bar{x}$? The answer depends on the shape of the distribution of the sampled population, as shown by Figure 6.18 (page 341). Generally speaking, the greater the skewness of the sampled population distribution, the larger the sample size must be before the normal distribution is an adequate approximation for the sampling distribution of $\bar{x}$. For most sampled populations, sample sizes of $n \geq 30$ will suffice for the normal approximation to be reasonable. We will use the normal approximation for the sampling distribution of $\bar{x}$ when the sample size is at least 30.

Example 6.16   Application of Concepts

A study was conducted at Luton Airport (United Kingdom) to assess the suitability of concrete blocks as a surface for aircraft pavements (*Proceedings of the Institute of Civil Engineers,* Apr. 1986). The original pavement at one end of the runway was overlaid with 80-mm-thick concrete blocks. A series of tests was carried out to determine the load classification number (LCN)—a measure of breaking strength—of the new surface. Let $\bar{x}$ represent the mean LCN of a random sample of 40 concrete block sections on the repaved end of the runway.

    a. Prior to resurfacing, the mean LCN of the original pavement was known to be $\mu = 60$ and the standard deviation was $\sigma = 10$. If the mean strength of the new concrete block surface is no different from that of the original surface, give the characteristics of the sampling distribution of $\bar{x}$.

    b. If the mean strength of the new concrete block surface is no different from that of the original surface, find the probability that $\bar{x}$, the sample mean LCN of the 40 concrete block sections, exceeds 64.

    c. The tests on the new concrete block surface actually resulted in $\bar{x} = 73$. Based on this result, what can you infer about the true mean LCN of the new surface?

Figure 6.18 Sampling Distributions of $\bar{x}$ for Different Populations and Different Sample Sizes

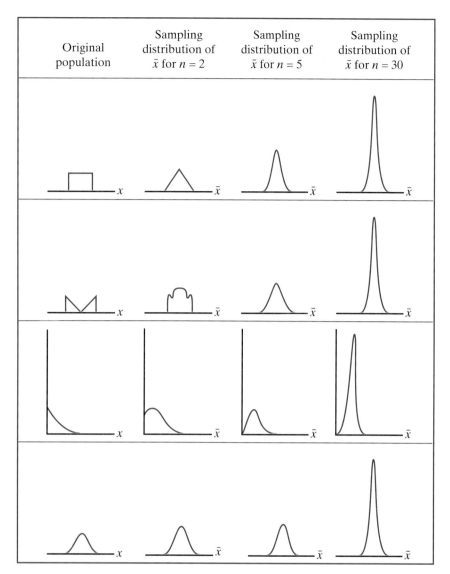

Solution

a. Although we have no information about the shape of the distribution of the breaking strengths (LCNs) for sections of the new surface, we can apply the Central Limit Theorem to conclude that the sampling distribution of $\bar{x}$, the mean LCN of the sample, is approximately normally distributed. In addition, if $\mu = 60$ and $\sigma = 10$, the mean, $\mu_{\bar{x}}$, and the standard deviation, $\sigma_{\bar{x}}$, of the sampling distribution are given by

$$\mu_{\bar{x}} = \mu = 60$$

and

$$\sigma_{\bar{x}} = \frac{\sigma}{\sqrt{n}} = \frac{10}{\sqrt{40}} = 1.58$$

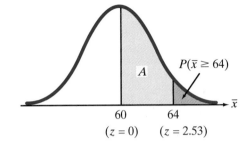

Figure 6.19 Sampling Distribution of $\bar{x}$ in Example 6.16

$P(\bar{x} \geq 64)$

A

60   64
$(z = 0)$ $(z = 2.53)$

b. If the two surfaces are of equal strength, then $P(\bar{x} \geq 64)$, the probability of observing a mean LCN of 64 or more in the sample of 40 concrete block sections, is equal to the darker shaded area shown in Figure 6.19. Since the sampling distribution is approximately normal, with mean and standard deviation as obtained in part **a**, we can compute the desired area by obtaining the z score for $\bar{x} = 64$:

$$z = \frac{\bar{x} - \mu_{\bar{x}}}{\sigma_{\bar{x}}} = \frac{64 - 60}{1.58} = 2.53$$

Thus, $P(\bar{x} \geq 64) = P(z \geq 2.53)$. This probability (area) is found using Table B.2:

$$P(\bar{x} \geq 64) = P(z \geq 2.53)$$

$$= .5 - A \qquad\qquad \text{(see Figure 6.19)}$$

$$= .5 - .4943 = .0057$$

c. If there is no difference between the true mean strengths of the new and original surfaces (i.e., $\mu = 60$ for both surfaces), the probability that we would obtain a sample mean LCN for concrete block of 64 or greater is only .0057. Observing $\bar{x} = 73$ provides strong evidence that the true mean breaking strength of the new surface exceeds $\mu = 60$. Our reasoning stems from the rare event philosophy of Chapter 4, which states that such a large sample mean ($\bar{x} = .73$) is very unlikely to occur if $\mu = 60$. ❑

Self-Test 6.6

For a random sample of size $n = 49$ selected from a population with $\mu = 17$ and $\sigma = 14$, find $P(\bar{x} < 15.5)$.

In practical terms, the Central Limit Theorem and the two properties of the sampling distribution of $\bar{x}$ assure us that the sample mean $\bar{x}$ is a reasonable statistic to use in making inferences about the population mean μ, and they allow us to compute a measure of the reliability of inferences made about μ. (This topic will be treated more thoroughly in Chapter 7.)

As we noted earlier, we will not be required to obtain sampling distributions by simulation or by mathematical arguments. Rather, for all the statistics to be used in this text, the sampling distribution and its properties (which are a matter of record) will be presented as the need arises.

Using the Tools

6.43 Suppose a random sample of n measurements is selected from a population with mean $\mu = 60$ and variance $\sigma^2 = 100$. For each of the following values of n, give the mean and standard deviation of the sampling distribution of the sample mean, $\bar{x}$.

a. $n = 10$	**b.** $n = 25$	**c.** $n = 50$	**d.** $n = 75$
e. $n = 100$	**f.** $n = 500$	**g.** $n = 1{,}000$	

6.44 Suppose a random sample of $n = 100$ measurements is selected from a population with mean μ and standard deviation σ. For each of the following values of μ and σ, give the values of $\mu_{\bar{x}}$ and $\sigma_{\bar{x}}$:

a. $\mu = 10, \sigma = 20$	**b.** $\mu = 20, \sigma = 10$
c. $\mu = 50, \sigma = 300$	**d.** $\mu = 100, \sigma = 200$

6.45 A random sample of $n = 50$ observations is selected from a population with $\mu = 21$ and $\sigma = 6$. Calculate each of the following probabilities:

a. $P(\bar{x} < 23.1)$ **b.** $P(\bar{x} > 21.7)$ **c.** $P(22.8 < \bar{x} < 23.6)$

6.46 A random sample of $n = 225$ observations is selected from a population with $\mu = 70$ and $\sigma = 30$. Calculate each of the following probabilities:

a. $P(\bar{x} > 72.5)$ **b.** $P(\bar{x} < 73.6)$
c. $P(69.1 < \bar{x} < 74.0)$ **d.** $P(\bar{x} < 65.5)$

Applying the Concepts

6.47 Medical researchers theorize that the low death rate from coronary heart disease among the Greenland Eskimos is due to their high fish consumption. The average amount of fish consumed by the Eskimos is estimated to be 400 grams per day (*New England Journal of Medicine*, May 9, 1985). Assume the standard deviation is 50 grams per day.

a. Describe the sampling distribution of $\bar{x}$, the mean amount of fish consumed per day for a sample of 25 Greenland Eskimos.

b. Find the probability that $\bar{x}$ is less than 390 grams.

c. Find the probability that $\bar{x}$ falls between 405 and 425 grams.

6.48 The National Institute for Occupational Safety and Health (NIOSH) evaluated the level of exposure of workers to the chemical dioxin, 2,3,7,8-TCDD. The distribution of TCDD levels in parts per trillion (ppt) of production workers at a Newark, New Jersey, chemical plant had a mean of 293 ppt and a standard deviation of 847 ppt (*Chemosphere*, Vol. 20, 1990). A graph of the distribution is shown on page 344. In a random sample of $n = 50$ workers selected at the New Jersey plant, let $\bar{x}$ represent the sample mean TCDD level.

a. Find the mean and standard deviation of the sampling distribution of $\bar{x}$.

b. Draw a sketch of the sampling distribution of $\bar{x}$. Locate the mean on the graph.

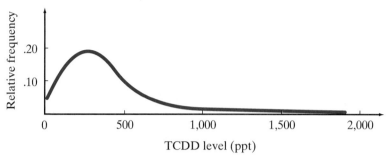

Problem 6.48

c. Find the probability that $\bar{x}$ exceeds 550 ppt.

6.49 A telephone survey of adult literacy was administered to a large number of San Diego (CA) residents (*Journal of Literacy Research,* Dec. 1996). One question asked was: "How many times during a typical week do you read a book for pleasure?" The respondents read a book for pleasure, on average, 3.10 times per week, with a standard deviation of 2.78 times per week. Assume $\mu = 3.10$ and $\sigma = 2.78$ for all American adults.

a. Explain why the distribution of responses (i.e., number of times per week you read a book for pleasure) for the population is unlikely to be normal.

b. In a random sample of 500 American adults, let $\bar{x}$ represent the mean number of times per week the sampled adults read a book for pleasure. Explain why the sampling distribution of $\bar{x}$ is normal.

c. Refer to part **b**. Find $P(\bar{x} < 3)$.

6.50 According to researchers, "One of the primary reasons relationships sour is that people stop listening to one another" (*USA Today,* Aug. 14, 1985). In a study of listening habits of 150 couples, the researchers discovered that couples in conflict listen to each other for a maximum of 14 seconds. Assume that the relative frequency distribution for the length of time couples in conflict listen to each other has a mean of $\mu = 8$ seconds and a standard deviation of $\sigma = 5$ seconds.

a. In a sample of $n = 37$ couples in conflict, let $\bar{x}$ represent the mean length of time the couples listen to each other. Describe the sampling distribution of $\bar{x}$.

b. Find the probability that $\bar{x}$ exceeds 10 seconds.

c. Within what limits would you expect approximately 95% of the sample means to fall?

6.51 Studies by neuroscientists at the Massachusetts Institute of Technology (MIT) reveal that melatonin, which is secreted by the pineal gland in the brain, functions naturally as a sleep-inducing hormone (*Tampa Tribune,* Mar. 1, 1994). Male volunteers were given various doses of melatonin or placebos and then placed in a dark room at midday and told to close their eyes and fall asleep on demand. Of interest to the MIT researchers is the time x (in minutes) required for each volunteer to fall asleep. With the placebo (i.e., no hormone), the researchers found that the mean time to fall asleep was 15 minutes. Assume that with the placebo treatment $\mu = 15$ and $\sigma = 5$.

a. Consider a random sample of $n = 20$ men who are given the sleep-inducing hormone, melatonin. Let $\bar{x}$ represent the mean time to fall asleep for this sample. If the hormone is *not* effective in inducing sleep, describe the sampling distribution of $\bar{x}$.

b. Refer to part **a**. Find $P(\bar{x} \leq 6)$.

c. In the actual study, the mean time to fall asleep for the 20 volunteers was $\bar{x} = 5$. Use this result to make an inference about the true value of μ for those taking the melatonin.

6.52 Many species of terrestrial frogs that hibernate at or near the ground surface can survive prolonged exposure to low winter temperatures. In freezing conditions, the frog's body temperature, called its *supercooling temperature,* remains relatively higher because of an accumulation of glycerol in its body fluids. Studies have shown that the supercooling temperature of terrestrial frogs frozen at $-6°C$ has a relative frequency distribution with a mean of $-2.18°C$ and a standard deviation of $.32°C$ (*Science*, May 1983). Consider the mean supercooling temperature, $\bar{x}$, of a random sample of $n = 42$ terrestrial frogs frozen at $-6°C$.

a. Find the probability that $\bar{x}$ exceeds $-2.05°C$.

b. Find the probability that $\bar{x}$ falls between $-2.20°C$ and $-2.10°C$.

6.53 Researchers at Marquette University presented MBA students (believed to be representative of entry-level managers) with decision-making situations that were clearly unethical in nature (*Journal of Business Ethics,* Vol. 6, 1987). The subjects' decisions were then rated on a scale of 1 ("definitely unethical") to 5 ("definitely ethical"). When no references to ethical concern by the "company" were explicitly stated, the ratings had a mean of 3.00 and a standard deviation of 1.03. Assume that these values represent the population mean and standard deviation, respectively, under the condition "no reference to ethical concern."

a. Suppose we present a random sample of 30 entry-level managers with a similar situation and record the ratings of each. Find the probability that $\bar{x}$, the sample mean rating, is greater than 3.40.

b. Refer to part **a**. Prior to making their decisions, the 30 entry-level managers were all read a statement from the president of the "company" concerning the company's code of business ethics. The code advocates socially responsible behavior by all employees. The researchers theorize that the population mean rating of the managers under this condition will be larger than for the "no reference to ethical concern" condition. (A higher mean indicates a more ethical response.) If the sample mean $\bar{x}$ is 3.55, what can you infer about the population mean under the "stated concern" condition?

6.54 One measure of elevator performance is cycle time. Elevator cycle time is the time between successive elevator starts, which includes the time when the car is moving and the time when it is standing at a floor. *Simulation* (Oct. 1993) published a study on the use of a microcomputer-based simulator for elevator cycle time. The simulator produced an average cycle time μ of 26 seconds when traffic intensity was set at 50 persons every five minutes. Consider a sample of 200 simulated elevator runs and let $\bar{x}$ represent the mean cycle time of this sample.

a. What do you know about the distribution of x, the time between successive elevator starts? (Give the value of the mean and standard deviation of x and the shape of the distribution, if possible.)

b. What do you know about the distribution of $\bar{x}$? (Give the value of the mean and standard deviation of $\bar{x}$ and the shape of the distribution; if possible.)

c. Assume σ, the standard deviation of cycle time x, is 20 seconds. Use this information to calculate $P(\bar{x} \geq 26.8)$.

d. Repeat part **c** but assume $\sigma = 10$.

6.55   Refer to Problem 6.54. Cycle time is related to the distance (measured by number of floors) the elevator covers on a particular run, called *running distance*. The simulated distribution of running distance, x, during a down-peak period in elevator traffic intensity is shown in the accompanying figure. The distribution has mean $\mu = 5.5$ floors and standard deviation $\sigma = 7$ floors. Consider a random sample of 80 simulated elevator runs during a down-peak in traffic intensity. Of interest is the sample mean running distance, $\bar{x}$.

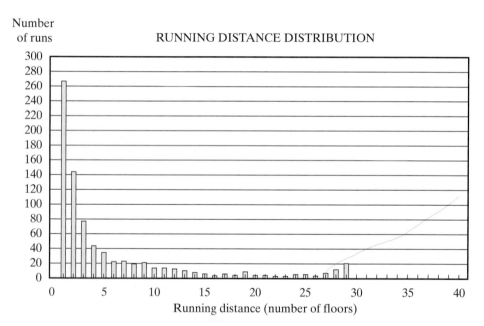

Number of runs

RUNNING DISTANCE DISTRIBUTION

Running distance (number of floors)

Source: Siikonen, M. L. "Elevator traffic simulation." *Simulation,* Vol. 61, No. 4, Oct. 1993, p. 266 (Figure 8).

a. Find $\mu_{\bar{x}}$ and $\sigma_{\bar{x}}$.

b. Is the shape of the distribution of $\bar{x}$ similar to the figure? If not, sketch the distribution.

c. During a down-peak in traffic intensity, is it likely to observe a sample mean running distance of $\bar{x} = 5.3$ floors? Explain.

FIREFRAUD.XLS

REAL-WORLD REVISITED
Detecting Fire Insurance Fraud

A wholesale furniture retailer stores in-stock items at a large warehouse located in Florida. In early 1992, a fire destroyed the warehouse and all the furniture in it. After determining the fire was an accident, the retailer sought to recover costs by submitting a claim to its insurance company.

As is typical in a fire insurance policy of this type, the furniture retailer must provide the insurance company with an estimate of "lost" profit for the destroyed items. Retailers calculate profit margin in percentage form using the Gross Profit Factor (GPF). By definition, the GPF for a single sold item is the ratio of the profit to the item's selling price measured as a percentage, i.e.,

$$\text{Item GPF} = (\text{Profit/Sales price}) \times 100$$

Of interest to both the retailer and the insurance company is the average GPF for all of the items in the warehouse. Since these furniture pieces were all destroyed, their eventual selling prices and profit values are obviously unknown. Consequently, the average GPF for all the warehouse items is unknown.

One way to estimate the mean GPF of the destroyed items is to use the mean GPF of similar, recently sold items. The retailer sold 3,005 furniture items in 1991 (the year prior to the fire) and kept paper invoices on all sales. Rather than calculate the mean GPF for all 3,005 items (the data were not computerized), the retailer sampled a total of 253 of the invoices and computed the mean GPF for these items. The 253 items were obtained by first selecting a sample of 134 items and then augmenting this sample with a second sample of 119 items. The mean GPFs for the two subsamples were calculated to be 50.6% and 51.0%, respectively, yielding an overall average GPF of 50.8%. This average GPF can be applied to the costs of the furniture items destroyed in the fire to obtain an estimate of the "lost" profit.

According to experienced claims adjusters at the insurance company, the GPF for sale items of the type destroyed in the fire rarely exceeds 48%. Consequently, the estimate of 50.8% appeared to be unusually high. (A 1% increase in GPF for items of this type equates to, approximately, an additional $16,000 in profit.) When the insurance company questioned the retailer on this issue, the retailer responded, "Our estimate was based on selecting two independent, random samples from the population of 3,005 invoices in 1991. Since the samples were selected randomly and the total sample size is large, the mean GPF estimate of 50.8% is valid."

A dispute arose between the furniture retailer and the insurance company, and a lawsuit was filed. In one portion of the suit, the insurance company accused the retailer of fraudulently representing their sampling methodology. Rather than selecting the samples randomly, the retailer was accused of selecting an unusual number of "high profit" items from the population in order to increase the average GPF of the overall sample.

To support their claim of fraud, the insurance company hired a CPA firm to independently assess the retailer's 1991 Gross Profit Factor. Through the discovery process, the CPA firm legally obtained the paper invoices for the entire population of 3,005 items sold and input the information into a computer. The selling price, profit, profit margin, and month sold for these 3,005 furniture items are available in the Excel workbook FIREFRAUD.XLS.

Your objective is to use these data to determine the likelihood of fraud. Is it likely that a random sample of 253 items selected from the population of 3,005 items would yield a mean GPF of at least 50.8%? Or is it likely that two independent, random samples of size 134 and 119 will yield mean GPFs of at least 50.6% and 51.0%, respectively? (These were the questions posed to a statistician retained by the CPA firm.) Use the ideas of probability and sampling distributions to guide your analysis.

Key Terms

Central Limit Theorem   338
Continuous probability
  distribution   304
Normal (bell) curve   305
Normal distribution   305
Normal probability plot   323

Parameter   330
Probability model   304
Sample statistic   331
Sampling distribution   331
Sampling distribution of the
  sample mean   332

Standard error of the mean   339
Standard normal distribution   308
Standard normal random
  variable   308
Tail probability   314

Key Formulas

Normal distribution:
$$z = \frac{x - \mu}{\sigma} \quad \textbf{(6.1)}, 306 \qquad x_0 = \mu + z_0\sigma \quad \textbf{(6.2)}, 315$$

Sampling distribution of $\bar{x}$:
$$\mu_{\bar{x}} = \mu \qquad \sigma_{\bar{x}} = \frac{\sigma}{\sqrt{n}} \quad \textbf{(6.3)}, 339$$

Key Symbols

Symbol	Definition
z	Standard normal random variable
z_0	Observed value of standard normal random variable
x	Normal random variable
x_0	Observed value of normal random variable
$\mu_{\bar{x}}$	Mean of the sampling distribution of $\bar{x}$
$\sigma_{\bar{x}}$	Standard deviation of the sampling distribution of $\bar{x}$ (also called standard error of $\bar{x}$)

Supplementary Problems 6.56–6.74

6.56 In order to graduate, high school students in Florida must demonstrate their competence in mathematics by scoring at least 70 (out of a possible total of 100) on a mathematics achievement test. From published accounts, it is known that the mean score on this test is 78 and the standard deviation is 7.5. Assume that the test scores are approximately normally distributed.

a. What percentage of all high school students in Florida score at least 70 on the mathematics achievement test?

b. What percentage of all high school students in Florida score below 60 on the mathematics achievement test? Locate μ and the desired probability on a sketch of the normal curve.

6.57 One measure of the water quality of a lake is its dissolved oxygen (DO) content. A marine biologist investigating a large northern lake found that the DO content at a certain time and place in the lake is approximately normally distributed with a mean of 7.8 units and a standard deviation of .4 unit.

a. What proportion of times will water samples from the lake at the designated time and place have a DO content less than 7.25 units?

b. What proportion of times will water samples from the lake at the designated time and place have a DO content greater than 6.55 units?

c. What proportion of times will water samples from the lake at the designated time and place have a DO content between 8.0 and 8.25 units?

6.58 The Taylor Manifest Anxiety Scale (TMAS) is a widely used test in psychological research. The TMAS scores of mental patients suffering from manic depression are approximately normally distributed with a mean of 47.6 and a standard deviation of 10.3.

a. Fifty percent of the mental patients have TMAS scores below what value?

b. Ninety-five percent of the mental patients have TMAS scores above what value?

c. In a sample of 50 mental patients, what is the probability that their mean TMAS score is less than 45?

6.59 Water availability is of prime importance in the life cycle of most reptiles. To determine the rate of evaporative water loss of a certain species of lizard at a particular desert site, 34 such lizards were randomly collected, weighed, and placed under the appropriate experimental conditions. After 24 hours, each lizard was removed, weighed, and its total water loss was calculated by subtracting its body weight after treatment from its initial body weight. Previous studies have shown that the relative frequency distribution of water loss for this species of lizard has a mean of 3.1 grams and a standard deviation of .8 gram.

a. Compute the probability that the 34 lizards will have a mean water loss of less than 2.7 grams.

b. Suppose the sample mean water loss for the lizards in the experiment is computed to be 2.58 grams. Based on the probability computed in part **a**, do you believe that the mean and standard deviation of the relative frequency distribution of water loss for this species of lizard may have changed since the previous studies? Explain.

6.60 In a particular population of sedentary young adults (i.e., those adults aged 20–40 who spend less than 5 hours per week in active exercise), the distribution of systolic blood pressure is approximately normal with $\mu = 125$ and $\sigma = 15$. Suppose a sedentary young adult is randomly selected from this population.

a. Find the probability that this person will have a systolic blood pressure of 90 or less.

b. Find the probability that this person will have a systolic blood pressure between 140 and 155.

c. Would you expect this person to have a systolic blood pressure below 80? Explain.

6.61 With the implantation of artificial hearts in humans there is a chance of internal infection, a problem that has occurred with implants in animals. Experiments show that calves implanted with artificial hearts can live an average of 80 days. Suppose the distribution of the number of days that a calf implanted with an artificial heart can live is approximately normally distributed with a standard deviation of 25 days.

a. What is the probability that a randomly selected calf implanted with an artificial heart will live longer than 120 days?

b. Twenty-five percent of all calves implanted with an artificial heart live longer than how many days? Show the pertinent quantities on a sketch of the normal curve for each part of the exercise.

6.62 The length of time required for a rodent to escape a noxious experimental situation was found to have a normal distribution with a mean of 6.3 minutes and a standard deviation of 1.5 minutes.

a. What is the probability that a randomly selected rodent will take longer than seven minutes to escape the noxious experimental situation?

b. Some rodents may become so disoriented in the experimental situation that they will not escape in a reasonable amount of time. Thus, the experimenter will be forced to terminate the experiment before the rodent escapes. When should each trial be terminated if the experimenter wants to allow sufficient time for 90% of the rodents to escape?

6.63 The average life of a certain steel-belted radial tire is advertised as 60,000 miles. Assume that the life of the tires is normally distributed with a standard deviation of 2,500 miles. (The life of a tire is defined as the number of miles the tire is driven before blowing out.)

a. Find the probability that a randomly selected steel-belted radial tire will have a life of 61,800 miles or less.

b. Find the probability that a randomly selected steel-belted radial tire will have a life between 62,000 miles and 66,000 miles.

c. In order to avoid a tire blowout, the company manufacturing the tires will warn purchasers to replace each tire after it has been used for a given number of miles. What should the replacement time (in miles) be so that only 1% of the tires will blow out?

d. In a sample of 100 steel-belted radial tires, what is the probability that the mean life is less than 61,800 miles? Compare your answer to part **a**.

6.64 Let $\bar{x}_{50}$ represent the mean of a random sample of size 50 obtained from a population with mean $\mu = 17$ and standard deviation $\sigma = 10$. Similarly, let $\bar{x}_{100}$ represent the mean of a random sample of size 100 selected from the same population.

a. Describe the sampling distribution of $\bar{x}_{50}$.

b. Describe the sampling distribution of $\bar{x}_{100}$.

c. Which of the probabilities, $P(15 < \bar{x}_{50} < 19)$ or $P(15 < \bar{x}_{100} < 19)$, would you expect to be the larger?

d. Calculate the two probabilities in part **c**. Was your answer to part **c** correct?

CRASH.XLS

6.65 Refer to the data on 98 crash-tested cars stored in the Excel workbook, CRASH.XLS. (The data set is described in Appendix C.) For each quantitative variable in the data set, determine whether the data are approximately normal.

6.66 This past year, an elementary school began using a new method to teach arithmetic to first graders. A standardized test, administered at the end of the year, was used to measure the effectiveness of the new method. The distribution of test scores in past years (before implementation of the new teaching method) had a mean of 75 and a standard deviation of 10. Consider the standardized test scores for a random sample of 36 first graders taught by the new method.

a. If the distribution of test scores for first graders taught by the new method is no different from that of the old method, describe the sampling distribution of $\bar{x}$, the mean test score for a random sample of 36 first graders.

b. If the sample mean test score was computed to be $\bar{x} = 79$, what would you conclude about the effectiveness of the new method of teaching arithmetic? Explain.

6.67 Birdwatchers visiting a National Wildlife Refuge located near Atlantic City, New Jersey, report a high degree of satisfaction with their visits (*Leisure Sciences,* Vol. 9, 1987). On a standard satisfaction scale that ranged from 1 to 6 (where 1 = poor rating and 6 = perfect rating), the birdwatchers have a mean score of 5.05 and a standard deviation of .98. Assume the population of satisfaction scores of birdwatchers visiting the National Wildlife Refuge is approximately normal.

a. Find the probability that a randomly selected birdwatcher visiting the refuge will have a satisfaction score of at least 5.

b. One-fourth of the birdwatchers have satisfaction scores below what value?

6.68 *Human Development* (1985) reported on a meta-analysis of the relation between race and subjective well-being based on results obtained from 54 sources. For each source, a measure of the relationship between race and well-being, called "effect size," was calculated. (Effect sizes ranged from -1 to 1, with positive values implying that whites had a higher subjective well-being than minorities and negative values implying the reverse.) A stem-and-leaf display of the 54 effect sizes is reproduced at left. The mean and standard deviation of the data were .10 and .09, respectively.

Stem	Leaf
$-.2$	4
$-.1$	6
$-.1$	2 2 1 1 1
$-.0$	9 9 8
$-.0$	1 0 0 0
.0	2
.0	5 6 6 6 6 7 7 8 8 8 9
.1	0 0 0 1 1 2 2 2 3 4 4 4
.1	5 5 6 9 9 9
.2	0 0 0 2 4
.2	5 6 7
.3	0 1

Source: Stock, W. A., Okun, M. A., Haring, M. J., and Witter, R. A. "Race and subjective well-being in adulthood: A black-white research synthesis." *Human Development,* Vol. 28, 1985, p. 195. By permission of S. Karger AG, Basel.

Problem 6.68

a. Examine the stem-and-leaf display. Is the distribution of effect sizes approximately normal?

b. Assuming normality, calculate the probability that a randomly selected study on the relationship between race and subjective well-being will have a negative effect size. (Assume $\mu = .10$ and $\sigma = .09$.)

c. Use the stem-and-leaf display to calculate the actual percentage of the 54 effect sizes that were negative. Compare the result to your answer to part **b**.

6.69 Suppose you are in charge of student ticket sales for a major college football team. From past experience, you know that the number of tickets purchased by a student standing in line at the ticket window has a distribution with a mean of 2.4 and a standard deviation of 2.0. For today's game, there are 100 eager students standing in line to purchase tickets. If only 250 tickets remain, what is the probability that all 100 students will be able to purchase the tickets they desire?

6.70 A research cardiologist wants to estimate the median rate at which a person's heartbeat increases after 15 minutes of vigorous exercise. Suppose the researcher has proposed two different statistics (call them A and B) for estimating the population median. To judge which of the statistics is more suitable, you simulated the approximate sampling distributions for each of the statistics, based on random samples of size $n = 10$ human subjects, with the results shown in the accompanying figures. Comment on the two sampling distributions. Which of the statistics, A or B, would you recommend for use?

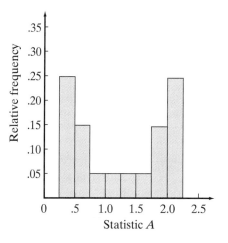
Statistic A

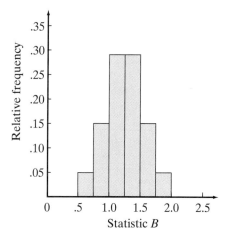
Statistic B

6.71 One of the monitoring methods used by the Environmental Protection Agency (EPA) to determine whether sewage treatment plants are conforming to standards is to take 36 1-liter specimens from the plant's discharge during the period of investigation. Chemical methods are applied to determine the percentage of sewage in each specimen. If the sample data provide evidence to indicate that the true mean percentage of sewage exceeds a limit set by the EPA, the treatment plant must undergo mandatory repair and retooling. One particular plant, at which the mean sewage discharge limit has been set at 15%, is suspected of being in violation of the EPA standard.

a. Unknown to the EPA, the relative frequency distribution of sewage percentages in 1-liter specimens at the plant in question has a mean μ of 15.7% and a standard deviation σ of 2.0%. Thus, the plant is in violation of the EPA standard. What is the probability that the EPA will obtain a sample of 36 1-liter specimens with a mean sewage percentage less than 15% even though the plant is violating the sewage discharge limit?

b. Suppose the EPA computes $\bar{x} = 14.95\%$. Does this result lead you to believe that the sample of 36 1-liter specimens obtained by the EPA was not random, but biased in favor of the sewage treatment plant? Explain.

6.72   *Cost estimation* is the term used to describe the process by which engineers estimate the cost of work contracts (e.g., road construction, building construction) that are to be awarded to the lowest bidder. The engineers' estimate is the baseline against which the low (winning) bid is compared. A study investigated the factors that affect the accuracy of engineers' estimates (*Cost Engineering*, Oct. 1988), where accuracy is measured as the percentage difference between the low bid and the engineers' estimate. One of the most important factors is number of bidders—the more bidders on the contract, the more likely the engineers are to overestimate the cost. For building contracts with five bidders, the mean percentage error was -7.02 and the standard deviation was 24.66. Consider a sample of 50 building contracts, each with five bidders.

  a.   Describe the sampling distribution of $\bar{x}$, the mean percentage difference between the low bid and the engineers' estimate, for the 50 contracts.

  b.   Find $P(\bar{x} < 0)$. (This is the probability of an overestimate.)

  c.   Suppose you observe $\bar{x} = -17.83$ for a sample of 50 building contracts. Based on the information above, are all these contracts likely to have five bidders? Explain.

6.73   The determination of the percent canopy closure of a forest is essential for wildlife habitat assessment, watershed runoff estimation, erosion control, and other forest management activities. One way in which geoscientists estimate the percent forest canopy closure is through the use of a satellite sensor called the Landsat Thematic Mapper. A study of the percent canopy closure in the San Juan National Forest (Colorado) was conducted by examining Thematic Mapper Simulator (TMS) data collected by aircraft at various forest sites (*IEEE Transactions on Geoscience and Remote Sensing*, Jan. 1986). The mean and standard deviation of the readings obtained from TMS Channel 5 were found to be 121.74 and 27.52, respectively.

  a.   Let $\bar{x}$ be the mean TMS reading for a sample of 32 forest sites. Assuming the figures given are population values, describe the sampling distribution of $\bar{x}$.

  b.   Use the sampling distribution of part **a** to find the probability that $\bar{x}$ falls between 118 and 130.

6.74   An empirical study of auditory nerve response rates was reported in the *Journal of the Acoustical Society of America* (Feb. 1986). Cats (selected because of their keen sense of hearing) were exposed to bursts of noise in a laboratory setting and the number of spikes per 200 milliseconds of noise burst received by the auditory nerve fibers of each, called the response rate, was recorded. A key question addressed by the researcher is whether the tone can be detected reliably when background noise is present. Let x represent the auditory nerve response rate under two conditions: when the stimulus is background noise only (N) and when the stimulus is a tone plus background noise (T). Empirical research has found that the probability distribution of x under either condition N or T can be approximated by a normal distribution. Based on the results of the study, the two normal distributions have means $\mu_N = 10.1$ spikes per burst and $\mu_T = 13.6$ spikes per burst, respectively, and equal variances $\sigma_N^2 = \sigma_T^2 = 2$. Given these conditions, an observer sets a threshold C and decides that a tone is present if $x \geq C$ and decides that no tone is present if $x < C$.

a. For a threshold of $C = 11$ spikes per burst, find the probability of detecting the tone given that the tone is present. That is, find $P(x \geq 11)$ under condition T. (This is known as the *detection probability*.)

b. For a threshold of $C = 11$ spikes per burst, find the probability of detecting the tone given that only background noise is present. That is, find $P(x \geq 11)$ under condition N. (This is known as the *probability of false alarm*.)

c. Usually, it is desirable to maximize detection probability while minimizing false alarm probability. Can you find a value of C that will both increase the detection probability (part **a**) and decrease the probability of false alarm (part **b**)? [*Hint:* Sketch the two probability distributions for conditions N and T, side by side, allowing some overlap between them. For any value C, shade the two probabilities in parts **a** and **b**. As you move C right or left, what happens to the probabilities?]

6.E.1 Using Microsoft Excel to Obtain Normal Probabilities

Solution Summary:

Use the PHStat add-in to answer probability questions pertaining to the normal distribution.

Example: Salt consumption data used in Chapter 6

Solution:

Use the Probability Distributions | Normal choice of the PHStat add-in to answer probability questions pertaining to the normal distribution. As an example, consider these probability questions pertaining to the salt consumption data, which have a mean of 15 grams and a standard deviation of 5 grams, that is used in several examples in this chapter:

    a. What is the probability that the salt intake will be between 14 and 22 grams? (Example 6.8 on pages 311–312)

    b. What is the probability that the salt intake will be less than 1 gram? (Example 6.9 on page 313)

    c. Ten percent of the people will have a salt intake of less than how many grams?

To answer these questions, do the following:

    1 If the PHStat add-in has not been previously loaded, load the add-in using the instructions of Section EP.3.2.

    2 Select File | New to open a new workbook (or open the existing workbook into which the normal probabilities worksheet is to be inserted).

    3 Select PHStat | Probability Distributions | Normal.

    4 In the Normal Probability Distribution dialog box (see Figure 6.E.1):

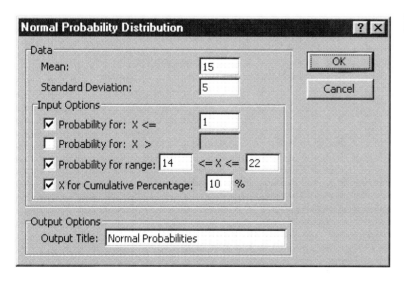

E Figure 6.E.1 PHStat Normal Probability Distribution Dialog Box

    a Enter 15 in the Mean: edit box.

    b Enter 5 in the Standard Deviation: edit box.

    c Select the Probability for: X <= check box and enter 1 in its edit box (for question b).

d Select the Probability for range: check box and enter 14 and 22, respectively, in the two edit boxes for this selection (for question a).

e Select the X for Cumulative Percentage: check box and enter 10 in its edit box (for question c).

f Enter Normal Probabilities in the Output Title: edit box.

g Click the OK button.

The add-in inserts a worksheet that reports the probabilities, *x*, and *z* values for these questions as shown in Figure 6.E.2. We can change the values for the mean and standard deviation in this worksheet to see the effects of these values on all probabilities.

 Figure 6.E.2
Normal Probabilities
Worksheet for Salt
Consumption Data of
Chapter 6

	A	B	C	D
1	NormalProbabilities			
2				
3	Mean	15		
4	Standard Deviation	5		
5				
6	Probability for X<=	1		
7	Z Value	-2.8		
8	P(X<=1)	0.002555191		
9				
10	Probability for range	14	<= X <=	22
11	Z Value for 14	-0.2		
12	Z Value for 22	1.4		
13	P(X<=14)	0.420740313		
14	P(X<=22)	0.919243289		
15	P(14<=X<=22)	0.498502976		
16				
17	Find X and Z			
18	Cumulative Percentage:	10%		
19	Z Value	-1.281550794		
20	X Value	8.592246028		

6.E.2 Using Microsoft Excel to Construct Normal Probability Plots

Solution Summary:

Use the PHStat add-in to construct a normal probability plot from a set of data.

Example: FTC cigarette data problem of Example 6.11

Solution:

Use the Probability Distributions I Normal Probability Plot choice of the PHStat add-in to construct a normal probability plot from a set of data. As an example, to construct a normal probability plot for the tar measurements of 500 cigarette brands of Example 6.11 on page 322, do the following:

1 If the PHStat add-in has not been previously loaded, load the add-in using the instructions of Section EP.3.2.

2 Open the FTC cigarette data workbook (FTC.XLS) and click the Data sheet tab. Verify that the tar measurements appear in column G.

3 Select PHStat I Probability Distributions I Normal Probability Plot.

4 In the Normal Probability Plot dialog box (see Figure 6.E.3):

a Enter G1:G501 in the Variable Cell Range: edit box.

b	Select the First cell contains label check box.
c	Enter Normal Probability Plot in the Output Title: edit box.
d	Click the OK button.

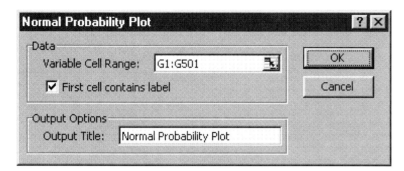

E **Figure 6.E.3** PHStat Normal Probability Plot Dialog Box

The add-in inserts a worksheet containing a table of ranks, cumulative proportions, z values, tar values, and a normal probability plot for the tar measurements similar to the one shown in Figure 6.13d on page 322.

6.E.3 Using Microsoft Excel to Simulate Sampling Distributions

Solution Summary:

Use the PHStat add-in to generate a worksheet containing a simulated sampling distribution of the mean and a chart sheet containing a histogram based on the simulated distribution.

 Simulated sampling distribution

Solution

In Figure 6.18 on page 341, we illustrated the effect of population shape and sample size on the sampling distribution of the mean. We can use the Probability Distributions | Sampling Distributions Simulation choice of the PHStat add-in to generate simulated sampling distributions from a uniformly distributed, standardized normally distributed, or discrete population. As an example, to generate a simulated sampling distribution from a uniform or standardized normal population, using 100 samples of sample size 25, do the following:

1	Select File	New to open a new workbook (or open the existing workbook into which the simulated distribution worksheet is to be inserted).
2	Select PHStat	Probability Distributions	Sampling Distributions Simulation.
3	In the Sampling Distributions Simulation dialog box (see Figure 6.E.4):

a	Enter 100 in the Number of Samples: edit box.
b	Enter 25 in the Sample Size: edit box.
c	Depending upon the type of simulation desired, select either the Uniform *or* Standardized Normal option button.
d	Enter Simulated Sampling Distribution in the Output Title: edit box.
e	Select the Histogram check box.
f	Click the OK button.

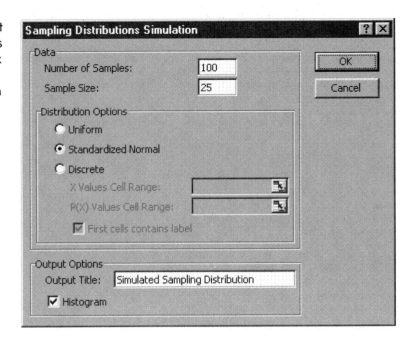

Figure 6.E.4 PHStat Sampling Distributions Simulation Dialog Box with the Standardized Normal Option Button Selected

The add-in inserts two worksheets: one containing the 100 samples of sample size 25, the other containing the histogram. A worksheet containing the histogram for a simulated sampling distribution based on a standardized normal population is shown in Figure 6.E.5. Note that this histogram is similar to, but not exactly the same as, the one shown for an actual sampling distribution for the supermarket checkout data in Figure 6.17 on page 335. To see the effects of sample size on the shape of the sampling distribution of the mean, repeat the procedure several times, each time varying the value of the sample size.

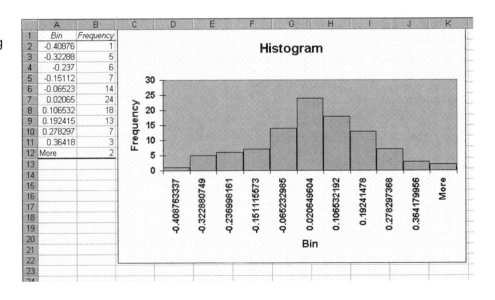

Figure 6.E.5 Worksheet Containing a Histogram for a Simulated Sampling Distribution Based on a Standardized Normal Population

The add-in can also generate a simulated sampling distribution from a discrete population. For example, to generate a simulated sampling distribution from the discrete population given in Table 6.E.1, using 100 samples of sample size 30, do the following:

1 Open the Table 6.E.1 workbook (TABLE 6.E.1.XLS) and click the Data sheet tab. Verify that the data of Table 6.E.1 have been entered into columns A and B.

2 Select PHStat | Probability Distributions | Sampling Distributions Simulation.

3 In the Sampling Distributions Simulation dialog box (see Figure 6.E.6):

a Enter 100 in the Number of Samples: edit box.

b Enter 30 in the Sample Size: edit box.

c Select the Discrete option button.

d Enter A2:B7 in the X and P(X) Values Cell Range: edit box. (Note: The cell range must not contain labels.)

e Enter Simulated Sampling Distribution in the Output Title: edit box.

f Select the Histogram check box.

g Click the OK button.

Table 6.E.1 Table of x Values and $p(x)$ Values to Simulate a Right-Skewed Discrete Distribution

x	$p(x)$
0	0.6
1	0.3
2	0.05
3	0.03
4	0.015
5	0.005

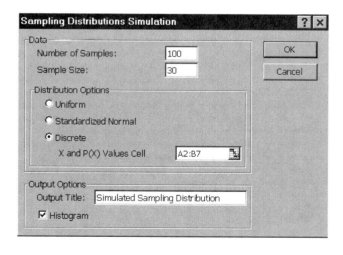

E Figure 6.E.6 PHStat Sampling Distributions Simulation Dialog Box with the Discrete Option Button Selected

Again, the add-in inserts two worksheets: one containing the 100 samples, the other containing a histogram for the simulated distribution, similar to the one shown in Figure 6.E.7. To see the effects of sample size on the shape of the sampling distribution of the mean, repeat this procedure several times, each time varying the value of the sample size. You can also change the values of $p(x)$ in the Data sheet to see their effects on the sampling distribution as well.

E **Figure 6.E.7**
Worksheet Containing
Histogram of a
Simulated Sampling
Distribution for a
Discrete Population
Based on Table 6.E.1,
Using 100 Samples of
$n = 30$

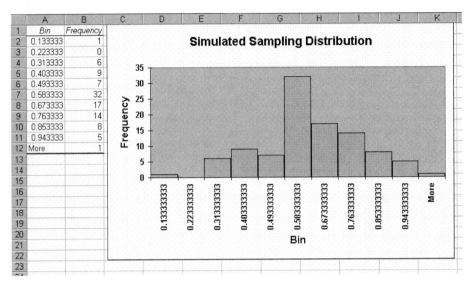

	A	B
1	Bin	Frequency
2	0.133333	1
3	0.223333	0
4	0.313333	6
5	0.403333	9
6	0.493333	7
7	0.583333	32
8	0.673333	17
9	0.763333	14
10	0.853333	8
11	0.943333	5
12	More	1

Simulated Sampling Distribution

Chapter 7

Estimation of Population Parameters Using Confidence Intervals: One Sample

CONTENTS

7.1 Point Estimators
7.2 Estimation of a Population Mean: Normal (z) Statistic
7.3 Estimation of a Population Mean: Student's t Statistic

7.4 Estimation of a Population Proportion
7.5 Choosing the Sample Size
7.6 Estimation of a Population Variance (Optional)

EXCEL TUTORIAL

7.E.1 Using Microsoft Excel to Obtain the Confidence Interval Estimate for the Mean (σ known)
7.E.2 Using Microsoft Excel to Obtain the Confidence Interval Estimate for the Mean (σ unknown)
7.E.3 Using Microsoft Excel to Obtain the Confidence Interval Estimate for the Proportion

7.E.4 Using Microsoft Excel to Determine the Sample Size for Estimating the Mean
7.E.5 Using Microsoft Excel to Determine the Sample Size for Estimating the Proportion
7.E.6 Using Microsoft Excel to Obtain the Confidence Interval Estimate for the Population Variance

REAL-WORLD APPLICATION
Scallops, Sampling, and the Law

To protect baby scallops from being harvested, a law requires that the average meat per scallop weigh at least $\frac{1}{36}$ of a pound. One fishing vessel arrived at port with 11,000 bags of scallops, from which the harbormaster randomly selected 18 bags for weighing. How should he use these data to estimate the average meat per scallop of all 11,000 bags and how accurate will it be? How does the sample size affect the accuracy of the estimate? These and other questions will be answered in this chapter. You will learn how sample size and other factors affect the behavior of sample statistics and, in Real-World Revisited, you will have an opportunity to analyze the scallop data.

7.1 POINT ESTIMATORS

In preceding chapters we learned that populations are characterized by descriptive measures (parameters), and that inferences about parameter values are based on statistics computed from the information in a sample selected from the population of interest. In this chapter, we will demonstrate how to estimate population parameters and assess the reliability of our estimates, based on knowledge of the sampling distributions of the statistics being used.

Example 7.1

Selecting a Point Estimate

Suppose you are interested in estimating the mean grade point average (GPA) of all students who attend your college or university.

    a. Identify the target population and the parameter of interest.

    b. How could one estimate the parameter of interest in this situation?

Solution

    a. The target population consists of the GPAs of all students at your college or university. Since we want to estimate the mean of this population, the parameter of interest is μ, the mean GPA of all students at your institution.

    b. An estimate of a population mean μ is the sample mean $\bar{x}$, which is computed from a random sample of n observations from the target population. Assume, for example, that we obtain a random sample of size $n = 100$ students, and then compute the value of the sample mean to be $\bar{x} = 2.63$. This value of $\bar{x}$ provides a *point estimate* of the population mean GPA. ❑

Definition 7.1

A **point estimate** of a parameter is a statistic, a single value computed from the observations in a sample, that is used to estimate the value of the target parameter.

How reliable is a point estimate of a parameter? To be truly practical and meaningful, an inference concerning a parameter (in this case, estimation of the value of μ) must consist not only of a point estimate, but also must be accompanied by a measure of the reliability of the estimate; that is, we need to be able to state how close our estimate is likely to be to the true value of the population parameter. This can be done by forming an **interval estimate** of the parameter using the characteristics of the sampling distribution of the statistic that was used to obtain the point estimate; the procedure will be illustrated in the next section.

7.2 ESTIMATION OF A POPULATION MEAN: NORMAL (z) STATISTIC

Recall from Section 6.5 that, for sufficiently large sample sizes, the sampling distribution of the sample mean $\bar{x}$ is approximately normal, as indicated in Figure 7.1.

Figure 7.1 Sampling Distribution of $\bar{x}$

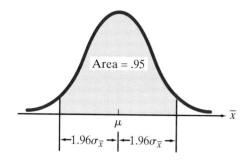

Area = .95

μ

$\leftarrow 1.96\sigma_{\bar{x}} \rightarrow | \leftarrow 1.96\sigma_{\bar{x}} \rightarrow$

$\bar{x}$

Example 7.2   Reliability of Interval Estimate

FTC.XLS

Refer to the Federal Trade Commission's cigarette data stored in the Excel workbook FTC.XLS. Suppose that our target population is the set of nicotine measurements for the 500 cigarette brands, and we are interested in estimating μ, the mean nicotine content of the cigarettes. We plan to take a sample of $n = 30$ measurements from the population of nicotine contents in FTC.XLS and construct the interval

$$\bar{x} \pm 1.96\sigma_{\bar{x}} = \bar{x} \pm 1.96\left(\frac{\sigma}{\sqrt{n}}\right)$$

where σ is the population standard deviation of the 500 nicotine measurements and $\sigma_{\bar{x}} = \sigma/\sqrt{n}$ is the **standard error** of $\bar{x}$. In other words, we will construct an interval 1.96 standard deviations around the sample mean, $\bar{x}$. How likely is it that this interval will contain the true value of the population mean μ?

Solution

We arrive at a solution by the following three-step process:

Step 1   First note that the area in the sampling distribution of $\bar{x}$ between $\mu - 1.96\sigma_{\bar{x}}$ and $\mu + 1.96\sigma_{\bar{x}}$ is approximately .95. (This area, highlighted in Figure 7.1, is obtained from the standard normal table, Table B.2 in Appendix B.) This implies that before the sample of measurements is drawn, the probability that $\bar{x}$ will fall within the interval $\mu \pm 1.96\sigma_{\bar{x}}$ is .95.

Step 2   If, in fact, the sample yields a value of $\bar{x}$ that falls within the interval $\mu \pm 1.96\sigma_{\bar{x}}$, then it is also true that the interval $\bar{x} \pm 1.96\sigma_{\bar{x}}$ will contain μ. This concept is illustrated in Figure 7.2. For a particular value of $\bar{x}$ (shown

Figure 7.2 Sampling Distribution of $\bar{x}$ in Example 7.2

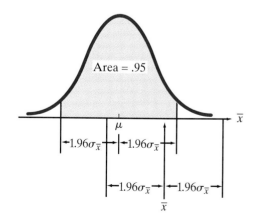

Area = .95

μ

$\leftarrow 1.96\sigma_{\bar{x}} \rightarrow | \leftarrow 1.96\sigma_{\bar{x}} \rightarrow$

$\leftarrow 1.96\sigma_{\bar{x}} \rightarrow | \leftarrow 1.96\sigma_{\bar{x}} \rightarrow$

$\bar{x}$

$\bar{x}$

with a vertical arrow) that falls within the interval $\mu \pm 1.96\sigma_{\bar{x}}$, a distance of $1.96\sigma_{\bar{x}}$ is marked off both to the left and to the right of $\bar{x}$. You can see that the value of μ must fall within $\bar{x} \pm 1.96\sigma_{\bar{x}}$.

Step 3 Steps 1 and 2 combined imply that, before the sample is drawn, the probability that the interval $\bar{x} \pm 1.96\sigma_{\bar{x}}$ will enclose μ is approximately .95. ❑

The interval $\bar{x} \pm 1.96\sigma_{\bar{x}}$ in Example 7.2 is called a large-sample 95% confidence interval for the population mean μ. The term *large-sample* refers to the sample being of a sufficiently large size that we can apply the Central Limit Theorem to determine the form of the sampling distribution of $\bar{x}$. Empirical research has found that a sample size n exceeding a value between 20 and 30 will usually yield a sampling distribution of $\bar{x}$ that is approximately normal. This result led many practitioners to adopt the conventional rule that a sample size of $n \geq 30$ is required to use large-sample confidence interval procedures. Keep in mind, though, that 30 is not a magical number, and is, in fact, quite arbitrary.

> ### Definition 7.2
>
> A **confidence interval** for a parameter is an interval within which we expect the true value of the population parameter to be contained. The endpoints of the interval are computed based on sample information.

Example 7.3 95% Confidence Interval

FTC.XLS

Refer to Example 7.2. Suppose a random sample of 30 observations from the population of nicotine measurements yielded a sample mean of $\bar{x} = .79$ mg. Construct a 95% confidence interval for μ, the population mean nicotine content of the 500 cigarettes, based on this sample information.

Solution

A 95% confidence interval for μ, based on a sample of size $n = 30$, is given by

$$\bar{x} \pm 1.96\sigma_{\bar{x}} = \bar{x} \pm 1.96\left(\frac{\sigma}{\sqrt{n}}\right) = .79 \pm 1.96\left(\frac{\sigma}{\sqrt{30}}\right)$$

To complete the calculation we need the value of σ, the standard deviation of the population of nicotine measurements. Using Excel, we obtain $\sigma = .3452$ mg. Then, our interval is

$$.79 \pm 1.96\left(\frac{.3452}{\sqrt{30}}\right) = (.666, .914)$$

We estimate that the population mean nicotine content falls within the interval from .666 mg to .914 mg. ❑

How much confidence do we have that μ, the true population nicotine content, lies within the interval (.666, .914)? Although we cannot be certain whether the sample interval contains μ (unless we calculate the true value of μ for all 500 observations), we can be reasonably sure that it does. This confidence is based on the interpretation of the confidence interval procedure: If we were to select repeated random samples of size $n = 30$ nicotine measurements and form a 1.96 standard deviation interval around $\bar{x}$ for each sample, then approximately 95% of the intervals constructed in this manner would contain μ. We say that we are "95% confident" that the particular interval (.666, .914) contains μ; this is our measure of the reliability of the point estimate $\bar{x}$.

Example 7.4   Theoretical Interpretation

FTC.XLS

To illustrate the classical interpretation of a confidence interval, we generated 100 random samples, each of size $n = 30$, from the population of nicotine measurements in FTC.XLS. For each sample, the sample mean was calculated and used to construct a 95% confidence interval for μ as in Example 7.3. Interpret the results, which are shown in Table 7.1 (page 336).

Solution

Using Excel, we found the population mean value to be $\mu = .8425$ mg. In the 100 repetitions of the confidence interval procedure described here, note that only five of the intervals (those highlighted in Table 7.1) do not contain the value of μ, whereas the remaining 95 of the 100 intervals (or 95%) do contain the true value of μ. The proportion .95 is called the *confidence coefficient* for the interval.

Keep in mind that, in actual practice, you would not know the true value of μ and you would not perform this repeated sampling; rather you would select a single random sample and construct the associated 95% confidence interval. The one confidence interval you form may or may not contain μ, but you can be fairly sure it does because of your *confidence in the statistical procedure,* the basis for which was illustrated in this example. ❑

Definition 7.3

The **confidence coefficient** is the proportion of times that a confidence interval encloses the true value of the population parameter if the confidence interval procedure is used repeatedly a very large number of times.

What "95% Confident" Means

A confidence coefficient of .95 implies that, of *all* possible 95% confidence intervals that can be formed from random samples selected from the target population, we are guaranteed 95% of them will contain μ and 5% will not.

Note that for a 95% confidence interval, the confidence coefficient of .95 is equal to the total area under the sampling distribution (1.00) less .05 of the area, which is divided equally between the two tails of the distribution (see Figure 7.1). Thus, each tail has an area of .025 and the tabulated value of z (from Table B.2 in Appendix B) that cuts off an area of .025 in the right tail of the standard normal distribution is 1.96 (see Figure 7.3 on page 367). The value $z = 1.96$ is also the distance, in terms of standard deviations, that $\bar{x}$ is from each endpoint of the 95% confidence interval.

Now, suppose we want to assign a confidence coefficient other than .95 to a confidence interval. To do this, we change the area under the sampling distribution between the endpoints of the interval, which in turn changes the tail area associated with z. Thus, this z value provides the key to constructing a confidence interval with any desired confidence coefficient. In our subsequent discussion, we will use the notation defined in Definition 7.4 (page 367).

Table 7.1 Confidence Intervals for μ for 100 Random Samples of 30 Nicotine Measurements Extracted from the Excel Workbook FTC.XLS

| Sample | $\bar{x}$ | 95% CONFIDENCE INTERVAL | | Sample | $\bar{x}$ | 95% CONFIDENCE INTERVAL | |
		Lower Limit	Upper Limit			Lower Limit	Upper Limit
1	0.8655	0.7419	0.9891	51	0.9967	0.8731	1.1203
2	0.8517	0.7281	0.9753	52	0.8033	0.6797	0.9269
3	0.8600	0.7364	0.9836	53	0.9333	0.8097	1.0569
4	0.8150	0.6914	0.9386	54	0.7967	0.6731	0.9203
5	0.8600	0.7364	0.9836	55	0.7133	0.5897	0.8369
6	0.8400	0.7164	0.9636	56	0.8667	0.7431	0.9903
7	0.8250	0.7014	0.9486	57	0.8233	0.6997	0.9469
8	0.8967	0.7731	1.0203	58	0.8567	0.7331	0.9803
9	0.8103	0.6867	0.9339	59	0.7400	0.6164	0.8636
10	0.8800	0.7564	1.0036	60	0.8333	0.7097	0.9569
11	0.9233	0.7997	1.0469	61	0.8567	0.7331	0.9803
12	0.9117	0.7881	1.0353	62	0.8600	0.7364	0.9836
13	0.8333	0.7097	0.9569	63	0.8117	0.6881	0.9353
14	0.8700	0.7464	0.9936	64	0.6867	0.5631	0.8103
15	0.9133	0.7897	1.0369	65	0.8167	0.6931	0.9403
16	0.7867	0.6631	0.9103	66	0.8400	0.7164	0.9636
17	0.8800	0.7564	1.0036	67	0.9033	0.7797	1.0269
18	0.7100	0.5864	0.8336	68	0.8133	0.6897	0.9369
19	0.9067	0.7831	1.0303	69	0.8833	0.7597	1.0069
20	0.9167	0.7931	1.0403	70	0.7667	0.6431	0.8903
21	0.8717	0.7481	0.9953	71	0.8333	0.7097	0.9569
22	0.9067	0.7831	1.0303	72	0.8700	0.7464	0.9936
23	0.8350	0.7114	0.9586	73	0.8667	0.7431	0.9903
24	0.8233	0.6997	0.9469	74	0.8633	0.7397	0.9869
25	0.8467	0.7231	0.9703	75	0.7900	0.6664	0.9136
26	0.9433	0.8197	1.0669	76	0.8900	0.7664	1.0136
27	0.7233	0.5997	0.8469	77	0.7967	0.6731	0.9203
28	0.8200	0.6964	0.9436	78	0.8433	0.7197	0.9669
29	0.9100	0.7864	1.0336	79	0.8333	0.7097	0.9569
30	0.9067	0.7831	1.0303	80	0.8417	0.7181	0.9653
31	0.8433	0.7197	0.9669	81	0.7883	0.6647	0.9119
32	0.8300	0.7064	0.9536	82	0.8267	0.7031	0.9503
33	0.7933	0.6697	0.9169	83	0.8150	0.6914	0.9386
34	0.7300	0.6064	0.8536	84	0.9500	0.8264	1.0736
35	0.8100	0.6864	0.9336	85	0.8233	0.6997	0.9469
36	0.7567	0.6331	0.8803	86	0.8967	0.7731	1.0203
37	0.8100	0.6864	0.9336	87	0.7567	0.6331	0.8803
38	0.7759	0.6523	0.8995	88	0.7333	0.6097	0.8569
39	0.8833	0.7597	1.0069	89	0.8967	0.7731	1.0203
40	0.7533	0.6297	0.8769	90	0.8100	0.6864	0.9336
41	0.9467	0.8231	1.0703	91	0.9600	0.8364	1.0836
42	0.8267	0.7031	0.9503	92	0.8800	0.7564	1.0036
43	0.8867	0.7631	1.0103	93	0.8650	0.7414	0.9886
44	0.8700	0.7464	0.9936	94	1.0333	0.9097	1.1569
45	0.8467	0.7231	0.9703	95	0.7200	0.5964	0.8436
46	0.8167	0.6931	0.9403	96	0.8733	0.7497	0.9969
47	0.7833	0.6597	0.9069	97	0.8433	0.7197	0.9669
48	0.7700	0.6464	0.8936	98	0.7750	0.6514	0.8986
49	0.8233	0.6997	0.9469	99	0.7533	0.6297	0.8769
50	0.8733	0.7497	0.9969	100	0.8655	0.7419	0.9891

Figure 7.3 Tabulated z Value Corresponding to a Tail Area of .025

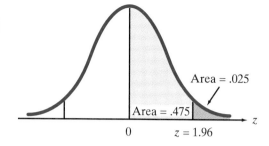

Area = .025

Area = .475

z

0

$z = 1.96$

Definition 7.4

We define $z_{\alpha/2}$ to be the z value such that an area of $\alpha/2$ lies to its right (see Figure 7.4).

Now, if an area of $\alpha/2$ lies beyond $z_{\alpha/2}$ in the right tail of the standard normal (z) distribution, then an area of $\alpha/2$ lies to the left of $-z_{\alpha/2}$ in the left tail (Figure 7.4) because of the symmetry of the distribution. The remaining area, $(1 - \alpha)$, is equal to the confidence coefficient—that is, the probability that $\bar{x}$ falls within $z_{\alpha/2}$ standard deviations of μ is $(1 - \alpha)$. Thus, a large-sample confidence interval for μ, with confidence coefficient equal to $(1 - \alpha)$, is given by

$$\bar{x} \pm z_{\alpha/2}\sigma_{\bar{x}}$$

Example 7.5   Finding z for a 90% Confidence Interval

In statistical problems using confidence interval techniques, a very common confidence coefficient is .90. Determine the value of $z_{\alpha/2}$ that would be used in constructing a 90% confidence interval for a population mean based on a large sample.

Solution
For a confidence coefficient of .90, we have

$$1 - \alpha = .90 \qquad \alpha = .10 \qquad \alpha/2 = .05$$

and we need to obtain the value of $z_{\alpha/2} = z_{.05}$ that locates an area of .05 in the upper tail of the standard normal distribution. Since the total area to the right of 0 is .50, $z_{.05}$ is the value such that the area between 0 and $z_{.05}$ is $(.50 - .05) = .45$.

Figure 7.4 Locating $z_{\alpha/2}$ on the Standard Normal Curve

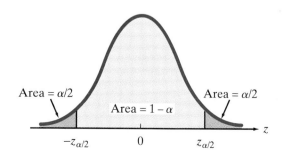

Area = $\alpha/2$

Area = $1 - \alpha$

Area = $\alpha/2$

$-z_{\alpha/2}$

0

$z_{\alpha/2}$

z

Table 7.2 Commonly Used Confidence Coefficients and Their Corresponding z Values

Confidence Coefficient $1 - \alpha$	$\alpha/2$	$z_{\alpha/2}$
.90	.05	1.645
.95	.025	1.96
.98	.01	2.33
.99	.005	2.575

Figure 7.5 Location of $z_{\alpha/2}$ for Example 7.5

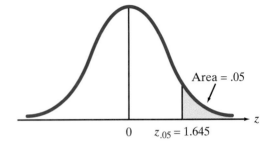

From Table B.2, we find $z_{.05} = 1.645$ (see Figure 7.5). We conclude that a large-sample 90% confidence interval for a population mean is given by $\bar{x} \pm 1.645\sigma_{\bar{x}}$ ☐

In Table 7.2 we present the values of $z_{\alpha/2}$ for the most commonly used confidence coefficients.

Self-Test 7.1

Consider a 99% confidence interval for μ.

a. What is the confidence coefficient? Interpret its value.

b. What is the value of $z_{\alpha/2}$ used to form the interval estimate?

A summary of the large-sample confidence interval procedure for estimating a population mean appears in the following box.

Large-Sample $(1 - \alpha)100\%$ Confidence Interval for a Population Mean μ

$$\bar{x} \pm z_{\alpha/2}\sigma_{\bar{x}} = \bar{x} \pm z_{\alpha/2}\left(\frac{\sigma}{\sqrt{n}}\right) \tag{7.1}$$

where $z_{\alpha/2}$ is the z value that locates an area of $\alpha/2$ to its right, σ is the standard deviation of the population from which the sample was selected, n is the sample size, and $\bar{x}$ is the value of the sample mean.

Assumption: A large random sample (usually, $n \geq 30$) is selected.

[*Note:* When the value of σ is unknown (as will usually be the case), the sample standard deviation s may be used to approximate σ in the formula for the confidence interval. The approximation is generally quite satisfactory for large samples.]

Example 7.6 Application of a 99% Confidence Interval

Twins, in their early years, tend to have lower IQs and pick up language more slowly than nontwins. Psychologists have speculated that the slower intellectual growth of twins may be caused by benign parental neglect. Suppose we want to investigate this phenomenon. A random sample of $n = 50$ sets of $2\frac{1}{2}$-year-old twin boys is selected, and the total parental attention time given to each pair during one week is recorded. The data (in hours) are listed in Table 7.3. Estimate μ, the

ATTENT.XLS

Table 7.3 Attention Time for Random Sample of $n = 50$ Sets of Twins

20.7	14.0	16.7	20.7	22.5	48.2	12.1	7.7	2.9	22.2
23.5	20.3	6.4	34.0	1.3	44.5	39.6	23.8	35.6	20.0
10.9	43.1	7.1	14.3	46.0	21.9	23.4	17.5	29.4	9.6
44.1	36.4	13.8	0.8	24.3	1.1	9.3	19.3	3.4	14.6
15.7	32.5	46.6	19.1	10.6	36.9	6.7	27.9	5.4	14.0

mean attention time given to all $2\frac{1}{2}$-year-old twin boys by their parents, using a 99% confidence interval. Interpret the interval in terms of the problem.

Solution
Applying equation 7.1 and Table 7.2, the 99% confidence interval for μ is

$$\bar{x} \pm (2.575)\frac{\sigma}{\sqrt{n}}$$

*We discuss an alternative form of the confidence interval when σ is unknown in Section 7.3.

E Figure 7.6 Excel Descriptive Statistics for $n = 50$ Sample Attention Times

	A	B
1	Time	
2		
3	Mean	20.848
4	Standard Error	1.897001372
5	Median	19.65
6	Mode	20.7
7	Standard Deviation	13.41382534
8	Sample Variance	179.9307102
9	Kurtosis	-0.672893401
10	Skewness	0.499630373
11	Range	47.4
12	Minimum	0.8
13	Maximum	48.2
14	Sum	1042.4
15	Count	50
16	Largest(1)	48.2
17	Smallest(1)	0.8

To compute the interval, we require the sample mean $\bar{x}$ and the population standard deviation σ. In most practical applications, however, the value of σ will be unknown. For large samples, the fact that σ is unknown poses only a minor problem since the sample standard deviation s provides a good approximation to σ. Consequently, we may substitute s for σ in the confidence interval formula given in the box.*

An Excel printout showing descriptive statistics for the sample of $n = 50$ attention times is displayed in Figure 7.6. The values of $\bar{x}$ and s, highlighted on the printout, are $\bar{x} = 20.85$ and $s = 13.41$. Substituting these values into the formula above, we obtain

$$20.85 \pm 2.575\left(\frac{13.41}{\sqrt{50}}\right)$$

or $(15.97, 25.73)$. We can be 99% confident that the interval $(15.97, 25.73)$ encloses the true mean weekly attention time given to $2\frac{1}{2}$-year-old twin boys by their parents. Since all the values in the interval fall below 28 hours, we conclude that there is a general tendency for $2\frac{1}{2}$-year-old twin boys to receive less than 4 hours of parental attention time per day, on the average. Further investigation would be required to relate this phenomenon to the intellectual growth of the twins. ❑

Self-Test 7.2

Based on the following summary statistics, construct a 95% confidence interval for μ: $n = 700$, $\bar{x} = 485$, $s^2 = 44,100$.

Example 7.7   Width of Confidence Interval versus $(1 - \alpha)$

Refer to Example 7.6.

a. Find a 95% confidence interval for the mean weekly attention time given to all $2\frac{1}{2}$-year-old twin boys by their parents.

b. For a fixed sample size, how is the width of the confidence interval related to the confidence coefficient?

E **Figure 7.7** 95%
Confidence Interval for
Mean Attention Time
Obtained from PHStat
Add-in for Excel

	A	B
1	Confidence Interval Estimate of the Mean	
2		
3	Population Standard Deviation	13.41
4	Sample Mean	20.85
5	Sample Size	50
6	Confidence Level	95%
7	Standard Error of the Mean	1.896460387
8	Z Value	-1.95996108
9	Interval Half Width	3.716988553
10	Interval Lower Limit	17.13301145
11	Interval Upper Limit	24.56698855

Solution

a. We used the PHStat Add-in for Excel to generate the large-sample 95% confidence interval for μ. The interval, highlighted in the Excel printout, Figure 7.7, is (17.13, 24.57).

b. The 99% confidence interval for μ was determined in Example 7.6 to be (15.97, 25.73). The 95% confidence interval in part **a** is narrower than the 99% confidence interval. This relationship holds in general, as noted in the following box. ❏

Relationship between Width of Confidence Interval and Confidence Coefficient

For a given sample size, the width of the confidence interval for a parameter increases as the confidence coefficient increases. Intuitively, the interval must become wider for us to have greater confidence that it contains the true parameter value.

Example 7.8 ### Width of a Confidence Interval versus Sample Size *n*

a. Assume that the given values of the statistics $\bar{x}$ and s were based on a sample of size $n = 100$ instead of a sample of size $n = 50$. Find a 99% confidence interval for μ, the population mean weekly attention time given to $2\frac{1}{2}$-year-old twin boys by their parents.

b. For a fixed confidence coefficient, how is the width of the confidence interval related to the sample size?

Solution

a. A 99% confidence interval for μ, generated using the PHStat Add-in for Excel, is shown and highlighted in Figure 7.8. The interval is (17.40, 24.30).

E **Figure 7.8** 99%
Confidence Interval for
Mean Attention Time
Using PHStat Add-in
for Excel

	A	B
1	Confidence Interval Estimate of the Mean	
2		
3	Population Standard Deviation	13.41
4	Sample Mean	20.85
5	Sample Size	100
6	Confidence Level	99%
7	Standard Error of the Mean	1.341
8	Z Value	-2.57583451
9	Interval Half Width	3.454194084
10	Interval Lower Limit	17.39580592
11	Interval Upper Limit	24.30419408

b. The 99% confidence interval based on a sample of size $n = 100$, constructed in part **a**, is narrower than the 99% confidence interval based on a sample of size $n = 50$, constructed in Example 7.6. This will also hold true in general, as noted in the box. ❑

Relationship between Width of Confidence Interval and Sample Size

For a fixed confidence coefficient, the width of the confidence interval decreases as the sample size increases. In other words, larger samples generally provide more information about the target population than do smaller samples.

In this section, we have introduced the concepts of point and interval estimation of the population mean μ, based on large samples. The general theory appropriate for the estimation of μ also carries over to the estimation of other population parameters. Hence, in subsequent sections we will present only the point estimate, its sampling distribution, the general form of a confidence interval for the parameter of interest, and any assumptions required for the validity of the procedure.

PROBLEMS
7.1–7.15

Using the Tools

7.1 In a large-sample confidence interval for a population mean, what does the confidence coefficient represent?

7.2 Use Table B.2 in Appendix B to determine the value of $z_{\alpha/2}$ needed to construct a large-sample confidence interval for μ, for each of the following confidence coefficients:

    a. .85   **b.** .95   **c.** .975

7.3 Suppose a random sample of size $n = 100$ produces a mean of $\bar{x} = 81$ and a standard deviation of $s = 12$.

    a. Construct a 90% confidence interval for μ.

    b. Construct a 95% confidence interval for μ.

    c. Construct a 99% confidence interval for μ.

7.4 A random sample of size n is selected from a population with unknown mean μ. Calculate a 95% confidence interval for μ for each of the following situations:

    a. $n = 35, \bar{x} = 26, \sigma = 15$ **b.** $n = 70, \bar{x} = 26, \sigma = 15$

    c. $n = 70, \bar{x} = 26, \sigma = 11$

7.5 A random sample of size 400 is taken from an unknown population with mean μ and standard deviation σ. The following values are computed:

$$\Sigma x = 2,280 \qquad \Sigma(x - \bar{x})^2 = 25,536$$

    a. Find a 90% confidence interval for μ.

    b. Find a 99% confidence interval for μ.

7.6 The mean and standard deviation of a random sample of n measurements are equal to 22 and 16, respectively.

  a. Construct a 95% confidence interval for μ if $n = 100$.

  b. Construct a 95% confidence interval for μ if $n = 500$.

7.7 Give a precise interpretation of the statement, "We are 95% confident that the interval estimate contains μ."

Applying the Concepts

DEFAULT.XLS

	A	B
1	*Default*	
2		
3	Mean	14.68181818
4	Standard Error	1.740662614
5	Median	9.6
6	Mode	11.8
7	Standard Deviation	14.14120993
8	Sample Variance	199.9738182
9	Kurtosis	5.426689724
10	Skewness	2.204234108
11	Range	74.7
12	Minimum	1.5
13	Maximum	76.2
14	Sum	969
15	Count	66
16	Largest(1)	76.2
17	Smallest(1)	1.5

Problem 7.8

7.8 Refer to the data on student-loan default rates for 66 Florida colleges, Problem 2.61 (page 115). An Excel printout showing descriptive statistics for the sample data is displayed at left.

  a. Use the descriptive statistics in the Excel printout to construct a 95% confidence interval for the true mean student-loan default rate.

  b. Interpret the interval.

  c. How could you reduce the width of the confidence interval? Are there any drawbacks to reducing the interval width? Explain.

  d. Would you feel comfortable using the 95% confidence interval to make an inference about the true mean student-loan default of all U.S. colleges? Explain.

7.9 Consider a study designed to investigate the average desired family size among low-income Hispanic women. A sample of 432 Hispanic women aged 18–50 were interviewed at an obstetrics/gynecology clinic of a large Los Angeles public hospital (*Family Planning Perspectives,* Nov./Dec. 1997).

  a. Each woman was asked, "How many sons do you desire?" The responses yielded the following summary statistics: $\bar{x} = 2.8, s = 1.4$. State the parameter of interest and construct a 90% confidence interval for its value. Interpret the results.

  b. Responses to the question, "How many daughters do you desire?" yielded the following summary statistics: $\bar{x} = .1, s = .5$. State the parameter of interest and construct a 90% confidence interval for its value. Interpret the results.

  c. Compare the results, parts **a** and **b**, and make a statement about the gender mix of the desired family size of low-income Hispanic women.

7.10 The *Journal of the American Medical Association* (Apr. 21, 1993) reported on the results of a National Health Interview Survey designed to determine the prevalence of smoking among U.S. adults. Current smokers (over 11,000 adults in the survey) were asked: "On the average, how many cigarettes do you now smoke a day?" The results yielded a mean of 20.0 cigarettes per day with an associated 95% confidence interval of (19.7, 20.3).

  a. Interpret the 95% confidence interval.

  b. State any assumptions about the target population of current cigarette smokers that must be satisfied for inferences derived from the interval to be valid.

  c. A tobacco industry researcher claims that the mean number of cigarettes smoked per day by regular cigarette smokers is less than 15. Comment on this claim.

SANIT.XLS

7.11 Refer to the Centers for Disease Control study of sanitation levels of cruise ships, Problem 2.24 (page 82). The inspection scores for a sample of 91 international cruise ships are reproduced in the table.

Ship	Score	Ship	Score	Ship	Score
Americana	89	Hanseatic Renaissance	82	Seabourn Pride	99
Amerikanis	97	Holiday	91	Seabourn Spirit	92
Azure Seas	83	Horizon	94	Seabreeze I	96
Britanis	93	Island Princess	87	Seaward	89
Caribbean Prince	84	Jubilee	89	Sky Princess	97
Caribe I	90	Mardi Gras	92	Society Explorer	66
Carla C	90	Meridian	95	Song of America	95
Carnivale	92	Nantucket Clipper	89	Song of Flower	99
Celebration	95	New Horeham II	95	Song of Norway	92
Club Med I	94	Nieuw Amsterdam	97	Southward	89
Costa Classica	91	Noordam	92	Sovereign of the Seas	93
Costa Marina	91	Nordic Empress	93	Star Princess	94
Costa Riviera	91	Nordic Prince	92	Starship Atlantic	87
Crown Monarch	94	Norway	84	Starship Majestic	94
Crown Odyssey	88	Pacific Princess	88	Starship Oceanic	97
Crown Princess	88	Pacific Star	70	Starward	96
Crystal Harmony	99	Queen Elizabeth 2	98	Stella Solaris	94
Cunard Countess	96	Regent Sea	87	Sun Viking	90
Cunard Princess	89	Regent Star	74	Sunward	95
Daphne	86	Regent Sun	95	Triton	86
Dawn Princess	86	Rotterdam	92	Topicale	93
Discovery I	93	Royal Princess	93	Universe	92
Dolphin IV	96	Royal Viking Sun	86	Victoria	96
Ecstasy	94	Sagafjord	89	Viking Princess	90
Emerald Seas	95	Scandinavian Dawn	87	Viking Serenade	96
Enchanted Isle	86	Scandinavian Song	90	Vistafjord	94
Enchanted Seas	96	Scandinavian Sun	89	Westerdam	91
Fair Princess	87	Sea Bird	86	Wind Spirit	96
Fantasy	97	Sea Goddess I	97	Yorktown Clipper	92
Festival	94	Sea Lion	91		
Golden Odyssey	89	Sea Princess	88		

Source: Center of Environmental Health and Injury Control, Miami, Florida. (reported in *Tampa Tribune,* May 17, 1992).

a. Construct a 90% confidence interval for μ, the true mean sanitation inspection score for all international ships.

b. Interpret the interval, part **a**.

c. What assumptions, if any, about the population of inspection scores are required for the confidence interval to be valid?

7.12 When a university professor attempts to publish a research article in a professional journal, the manuscript goes through a rigorous review process. Usually, anywhere from three to five reviewers read and critique the article, then pass judgment on whether or not the article should be published. A study was undertaken to seek information on how reviewers for research journals pursue their activities (*Academy of Management Journal,* Mar. 1989). A sample of 73 reviewers for the Academy of Management's *Journal* (*AMJ*) and *Review* (*AMR*) were asked how many hours they spent per paper for a typical complete review process. The sample mean and standard deviation were computed to be $\bar{x} = 5.4$ hours and $s = 3.6$ hours.

a. Find a point estimate for μ, the true mean number of hours spent by a reviewer in conducting a complete review of a paper submitted to *AMJ* or *AMR*.

b. Compute a 99% confidence interval for μ.

c. Interpret the interval, part **b**.

7.13 Tropical swarm-founding wasps, like ants and bees, rely on workers to raise their offspring. One possible explanation for this behavior is inbreeding, which increases relatedness among the wasps and makes it easier for the workers to pick out and aid their closest relatives. To test this theory, 197 swarm-founding wasps were captured in Venezuela, frozen at $-70°C$, and then subjected to a series of genetic tests (*Science*, Nov. 1988). The data were used to generate an inbreeding coefficient x for each wasp specimen, with the following results: $\bar{x} = .044$ and $s - .884$.

a. Construct a 90% confidence interval for the mean inbreeding coefficient of this species of wasp.

b. A coefficient of 0 implies that the wasp has no tendency to inbreed. Use the confidence interval, part **a**, to make an inference about the tendency for this species of wasp to inbreed.

7.14 The velocity of light emitted from a galaxy provides astronomers with valuable information on how the galaxy was formed. The *Astronomical Journal* (July 1995) reported that in a sample of 103 galaxies located in close proximity (called a galaxy cluster), the mean velocity was $\bar{x} = 27{,}117$ kilometers per second and the standard deviation was $s = 1{,}280$ kilometers per second.

a. Estimate the true mean velocity of light emitted from all galaxies in the galaxy cluster using a 95% confidence interval. Interpret the result.

b. How could you reduce the width of the confidence interval from part **a**?

7.15 Unusual rocks at "The Seven Islands," located along the lower St. Lawrence River in Canada, have attracted geologists to the area for over a century. A major geological survey of "The Seven Islands" was completed for the purpose of obtaining an accurate estimate of the rock densities in the area (*Canadian Journal of Earth Sciences*, Vol. 27, 1990). Based on samples of several varieties of rock, the following information on rock density (grams per cubic centimeter) was obtained.

Type of Rock	Sample Size	Mean Density	Standard Deviation
Late gabbro	36	3.04	.13
Massive gabbro	148	2.83	.11
Cumberlandite	135	3.05	.31

Source: Loncarevic, B. D., Feninger, T., and Lefebvre, D. "The Sept-Iles layered mafic intrusion: Geo-physical expression." *Canadian Journal of Earth Sciences,* Vol. 27, Aug. 1990, p. 505.

a. For each rock type, estimate the mean density with a 90% confidence interval.

b. Interpret the intervals, part **a**.

7.3 ESTIMATION OF A POPULATION MEAN: STUDENT'S *t* STATISTIC

In the previous section, we discussed the estimation of a population mean based on large random samples and known population standard deviation σ. However, time or cost limitations may often restrict the number of sample observations that may be obtained, so that the estimation procedures of Section 7.2 will not be applicable.

With small samples, the following two problems arise:

Problem 1   Since the Central Limit Theorem applies only to large samples, we are not able to assume that the sampling distribution of $\bar{x}$ is approximately normal. For small samples, the sampling distribution of $\bar{x}$ depends on the particular shape of the distribution of the population being sampled.

Problem 2   The sample standard deviation *s* may not be a satisfactory approximation to the population standard deviation σ if the sample size is small. Thus, replacing σ with *s* in the large-sample formula given in Section 7.2 is not appropriate.

Fortunately, we may proceed with estimation techniques based on small samples if we can make the following assumption:

Assumption Required for Estimating μ Based on Small Samples

The population from which the random sample is selected has an approximately normal distribution.

Confidence Interval for a Normal Population

When sampling is from a normal population with σ known, the appropriate confidence interval is $\bar{x} \pm z_{\alpha/2}(\sigma/\sqrt{n})$ regardless of the size of the sample. This results from the fact that the sampling distribution of $\bar{x}$ is normal whenever the population is normally distributed. (See Section 6.5.)

If this assumption is valid, then we may again use $\bar{x}$ as a point estimate for μ, and the general form of a small-sample confidence interval for μ is shown in the next box.

Small-Sample Confidence Interval for μ

$$\bar{x} \pm t_{\alpha/2}\left(\frac{s}{\sqrt{n}}\right) \tag{7.2}$$

where the distribution of *t* is based on $(n-1)$ degrees of freedom.

Now compare the large-sample confidence interval for μ (equation 7.1) to the small-sample confidence interval for μ (equation 7.2). Note that the sample standard deviation *s* replaces the population standard deviation σ in the formula. Also, the sampling distribution upon which the confidence interval is based is known as a **Student's *t* distribution**. Consequently, we must replace the value of $z_{\alpha/2}$ used in a large-sample confidence interval by a value obtained from the *t* distribution.

Student's *t* Distribution

The derivation of the *t* distribution was first published in 1908 by W. S. Gosset, who wrote under the pen name of Student. Thereafter, the distribution became known as Student's *t*.

The *t* distribution is very much like the *z* distribution. In particular, both are symmetric and mound-shaped and have a mean of 0. However, the *t* distribution is flatter, i.e., more variable (see Figure 7.9).

Also, the distribution of *t* depends on a quantity called its **degrees of freedom** (df), which is equal to $(n-1)$ when estimating a population mean based on a small

Figure 7.9 Comparison of z Distribution to the t Distribution

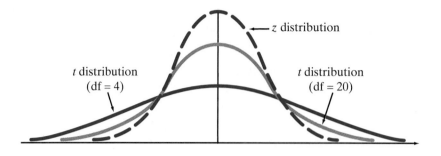

t distribution (df = 4)

t distribution (df = 20)

z distribution

Table 7.4 Reproduction of a Portion of Table B.3 of Appendix B

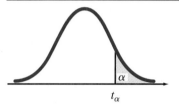

α

t_α

Degrees of Freedom	$t_{.100}$	$t_{.050}$	$t_{.025}$	$t_{.010}$	$t_{.005}$	$t_{.001}$	$t_{.0005}$
1	3.078	6.314	12.706	31.821	63.657	318.31	636.62
2	1.886	2.920	4.303	6.965	9.925	22.326	31.598
3	1.638	2.353	3.182	4.541	5.841	10.213	12.924
4	1.533	2.132	2.776	3.747	4.604	7.173	8.610
5	1.476	2.015	2.571	3.365	4.032	5.893	6.869
6	1.440	1.943	2.447	3.143	3.707	5.208	5.959
7	1.415	1.895	2.365	2.998	3.499	4.785	5.408
8	1.397	1.860	2.306	2.896	3.355	4.501	5.041
9	1.383	1.833	2.262	2.821	3.250	4.297	4.781
10	1.372	1.812	2.228	2.764	3.169	4.144	4.587
11	1.363	1.796	2.201	2.718	3.106	4.025	4.437
12	1.356	1.782	2.179	2.681	3.055	3.930	4.318
13	1.350	1.771	2.160	2.650	3.012	3.852	4.221
14	1.345	1.761	2.145	2.624	2.977	3.787	4.140
15	1.341	1.753	2.131	2.602	2.947	3.733	4.073

Source: This table is reproduced with the kind permission of the Trustees of Biometrika and Oxford University Press from E. S. Pearson and H. O. Hartley (eds.), *The Biometrika Tables for Statisticians*, Vol. 1, 3d ed., Biometrika, 1966.

sample of size n. Intuitively, we can think of the number of degrees of freedom as the amount of information available for estimating, in addition to μ, the unknown quantity σ^2. Table B.3 of Appendix B, a portion of which is reproduced in Table 7.4, gives the value of t_α that locates an area of α in the upper tail of the t distribution for various values of α and for degrees of freedom ranging from 1 to 120.

Characteristics of Student's t Distribution

1. Symmetric distribution
2. Mean of 0

3. Flatter, i.e., more area in the tails and less area in the center, than the standard normal (z) distribution

4. Depends on a quantity called *degrees of freedom* (df)

5. For large samples (i.e., large df values), the *t* and *z* distributions are nearly equivalent.

The Concept of Degrees of Freedom

Recall (from Chapter 3) that s^2 requires the computation of $\sum(x - \bar{x})^2$. Thus, to compute s^2 we must first compute $\bar{x}$. Therefore, only $n - 1$ of the sample values are *free to vary*. That is, there are $n - 1$ *degrees of freedom*.

To see this, suppose $n = 3$ and $\bar{x} = 10$. Then $\sum x = 30$. Once we know the first two measurements in the sample—say, 15 and 8—the third (and last) measurement is not free to vary. Since the sum is 30, the third measurement must be 7.

Example 7.9   Finding t Values

Use Table B.3 to determine the *t* value that would be used in constructing a 95% confidence interval for μ based on a sample of size $n = 14$.

Solution

For a confidence coefficient of .95, we have

$$1 - \alpha = .95 \qquad \alpha = .05 \qquad \alpha/2 = .025$$

We thus require the value of $t_{.025}$ for a *t* distribution based on $(n - 1) = (14 - 1) = 13$ degrees of freedom. Now in Table B.3, at the intersection of the column labeled $t_{.025}$ and the row corresponding to df = 13, we find the entry 2.160 (see Table 7.4, page 376). Hence, a 95% confidence interval for μ, based on a sample of $n = 14$ observations, would be given by

$$\bar{x} \pm 2.160\left(\frac{s}{\sqrt{14}}\right)$$ ❏

Self-Test 7.3

Find the value of $t_{\alpha/2}$ used in a 90% confidence interval for μ based on a sample of size $n = 7$.

Example 7.10   "Large" Sample

Recall that when estimating μ in Section 7.2, we used the arbitrary cutoff point of $n = 30$ for distinguishing between large and small samples. Explain why $n \geq 30$ is selected as the arbitrary cutoff.

Solution

In Table B.3, the values in the last row (corresponding to $df = \infty$) are the values from the standard normal z distribution. This phenomenon occurs because, as the sample size increases, the *t* distribution becomes more and more

SUICIDE.XLS

Table 7.5 Days in Jail Before Suicide for 14 Inmates Who Committed Suicide

10	15	3	4	7	19	15
5	22	221	85	126	29	31

like the z distribution (recall Figure 7.9). By the time n reaches 30, i.e., df = 29, there is very little difference between tabulated values of t and z. Of course, the values $n = 40$, $n = 60$, and even $n = 120$ could also have been selected as the cutoff for defining a "large" sample. However, $n = 30$ seems to be the smallest value of n for which the t values reasonably approximate the corresponding z values. ❑

Example 7.11   Small-Sample Confidence Interval for μ

Suicide is the leading cause of death of Americans incarcerated in correction facilities. On average, how much time do suicidal inmates spend in jail before they actually commit suicide? The *American Journal of Psychiatry* (June 1995) published data collected on suicides committed at a Detroit, Michigan, jail. Table 7.5 lists the number of days spent in jail before suicide for each of a sample of 14 inmates convicted of murder. Estimate the mean number of days spent in jail by inmates convicted of murder who commit suicide. Use a 90% confidence interval.

Solution

The first step in constructing the confidence interval is to compute the sample mean $\bar{x}$ and sample standard deviation s of the data in Table 7.5. These values, $\bar{x} = 42.28$ and $s = 62.14$, are highlighted in the Excel printout, Figure 7.10.

For a confidence coefficient of $1 - \alpha = .90$, we have $\alpha = .10$ and $\alpha/2 = .05$. Since the sample size is small ($n = 14$), our estimation technique requires the assumption that the distribution of days in jail before suicide for all inmates who eventually commit suicide is approximately normal.

Substituting the values for $\bar{x}, s$, and n into the formula for a small-sample confidence interval for μ (equation 7.2), we obtain

E **Figure 7.10** Excel Descriptive Statistics for Example 7.11

	A	B
1	*Days*	
2		
3	Mean	42.28571429
4	Standard Error	16.60875864
5	Median	17
6	Mode	15
7	Standard Deviation	62.14428443
8	Sample Variance	3861.912088
9	Kurtosis	5.047523202
10	Skewness	2.266427967
11	Range	218
12	Minimum	3
13	Maximum	221
14	Sum	592
15	Count	14
16	Largest(1)	221
17	Smallest(1)	3

$$\bar{x} \pm t_{\alpha/2}\left(\frac{s}{\sqrt{n}}\right) = \bar{x} \pm t_{.05}\left(\frac{s}{\sqrt{n}}\right)$$

$$= 42.28 \pm t_{.05}\left(\frac{62.14}{\sqrt{14}}\right)$$

where $t_{.05}$ is the value corresponding to an upper-tail area of .05 in the Student's t distribution based on $(n - 1) = 13$ degrees of freedom. From Table B.3, the required t value is $t_{.05} = 1.771$. Substituting this value yields

$$42.28 \pm (1.771)\left(\frac{62.14}{\sqrt{14}}\right) = 42.28 \pm 29.41$$

or 12.87 to 71.69 days. Thus, if the distribution of days spent in jail is approximately normal, then we can be 90% confident that the interval (12.87, 71.69) encloses μ, the true mean number of days spent in jail before suicide for inmates who commit suicide. ❑

Based on the following summary statistics, construct a 95% confidence interval for μ: $n = 10, \overline{x} = 485, s^2 = 44,100$.

Example 7.12   Excel-Generated Confidence Interval

Refer to Example 7.6 (page 368) and the problem of estimating μ, the mean attention time given to $2\frac{1}{2}$-year-old twin boys by their parents. Recall that a 99% confidence interval was computed based on data collected for a random sample of $n = 50$ sets of twins. An Excel printout showing a 99% confidence interval for μ is displayed in Figure 7.11. Compare the results to the interval calculated in Example 7.6.

Solution

The 99% confidence interval, highlighted on the Excel printout, is (15.76, 25.93). The interval calculated in Example 7.6 is (15.97, 25.73). The differences in the end points of the interval, although relatively minor, are because σ is unknown for the target population. As do most software packages, the PHStat add-in for Excel computes the confidence interval using the t statistic, i.e.,

$$\overline{x} \pm t_{.005}\left(\frac{s}{\sqrt{n}}\right)$$

where $t_{.005} \approx 2.68$ (based on $n - 1 = 49$ df). The confidence interval in Example 7.6, you will recall, was calculated using the z statistic, i.e.,

$$\overline{x} \pm z_{.005}\left(\frac{s}{\sqrt{n}}\right)$$

where $z_{.005} = 2.575$. Theoretically, the Excel confidence interval is the correct one, since $\overline{x}$ has a t distribution when σ is unknown. The confidence interval in Example 7.6 is approximate. But you can see the approximation is good when the sample size n is large. ❏

Before concluding this section, we will comment on the assumption that the sampled population is normally distributed. In the real world, we rarely know whether a sampled population has an exactly normal distribution. However, empirical studies indicate that moderate departures from this assumption do not seriously affect the confidence coefficients for small-sample confidence intervals.

E **Figure 7.11**
Confidence Interval
Obtained from PHStat
Add-in for Excel for
Example 7.12

	A	B	C	D
1	Confidence Interval Estimation for the Mean Weekly Attention Time			
2				
3	Sample Standard Deviation	13.4138		
4	Sample Mean	20.848		
5	Sample Size	50		
6	Confidence Level	99%		
7	Standard Error of the Mean	1.896997788		
8	Degrees of Freedom	49		
9	t Value	2.679953468		
10	Interval Half Width	5.083865802		
11	Interval Lower Limit	15.76		
12	Interval Upper Limit	25.93		

For example, if the population of days spent in jail for inmates who commit suicide of Example 7.11 has a distribution that is mound-shaped but nonnormal, it is likely that the actual confidence coefficient for the 90% confidence interval will be close to .90—at least close enough to be of practical use. As a consequence, the small-sample confidence interval given in equation 7.2 is frequently used by experimenters when estimating the population mean of a nonnormal distribution as long as the distribution is mound-shaped and only moderately skewed.

For populations that depart greatly from normality, methods that are distribution-free (called *nonparametrics*) are recommended. Nonparametric statistics are the topic of Chapter 12.

PROBLEMS
7.16–7.28

Using the Tools

7.16   Use Table B.3 in Appendix B to determine the values of $t_{\alpha/2}$ needed to construct a confidence interval for a population mean for each of the following combinations of confidence coefficient and sample size:

  a.   Confidence coefficient .99, $n = 18$

  b.   Confidence coefficient .95, $n = 10$

  c.   Confidence coefficient .90, $n = 15$

7.17   What assumptions are required for the interval estimation procedure of this section to be valid when the sample size is small, i.e., when $n < 30$?

7.18   A random sample of $n = 10$ measurements from a normally distributed population yielded $\bar{x} = 9.4$ and $s = 1.8$.

  a.   Calculate a 90% confidence interval for μ.

  b.   Calculate a 95% confidence interval for μ.

  c.   Calculate a 99% confidence interval for μ.

7.19   The following data represent a random sample of five measurements from a normally distributed population: 7, 4, 2, 5, 7.

  a.   Find a 90% confidence interval for μ.

  b.   Find a 99% confidence interval for μ.

7.20   The mean and standard deviation of n measurements randomly sampled from a normally distributed population are 33 and 4, respectively. Construct a 95% confidence interval for μ when:

  a. $n = 5$ **b.** $n = 15$ **c.** $n = 25$

7.21   How are the t distribution and the z distribution similar? How are they different?

Applying the Concepts

7.22   The *Journal of Communication* (Summer 1997) published a study on U.S. media coverage of major earthquakes that occurred in foreign countries. The number of seconds devoted to the earthquake on the nightly news broadcasts of ABC, CBS, and NBC was determined for each in a sample of 22 foreign earthquakes that caused 10 or more deaths. The mean and standard deviation were 1,780 seconds and 2,797 seconds, respectively.

a. Construct a 99% confidence interval for the average amount of time devoted to foreign earthquakes on the nightly news.

b. Interpret the interval, part **a**.

7.23 The *Journal of Applied Behavior Analysis* (Summer 1997) published a study of a treatment designed to reduce destructive behavior. A 16-year-old boy diagnosed with severe mental retardation was observed playing with toys for 10 sessions, each ten minutes in length. The number of destructive responses (e.g., head banging, pulling hair, pinching, etc.) per minute was recorded for each session in order to establish a baseline behavior pattern. The data for the 10 sessions are listed here:

3   2   4.5   5.5   4.5   3   2.5   1   3   3

DESTRUCT.XLS

Construct a 95% confidence interval for the true mean number of destructive responses per minute for the 16-year-old boy. Interpret the result.

7.24 Refer to the *Risk Management* study on fires in compartmented fire-resistive buildings, Problem 3.84 (page 202). The data shown in the table give the number of victims who died attempting to evacuate for a sample of 14 recent fires.

FIRE.XLS

Fire	Number of Victims
Las Vegas Hilton (Las Vegas)	5
Inn on the Park (Toronto)	5
Westchase Hilton (Houston)	8
Holiday Inn (Cambridge, Ohio)	10
Conrad Hilton (Chicago)	4
Providence College (Providence)	8
Baptist Towers (Atlanta)	7
Howard Johnson (New Orleans)	5
Cornell University (Ithaca, New York)	9
Westport Central Apartments (Kansas City, Missouri)	4
Orrington Hotel (Evanston, Illinois)	0
Hartford Hospital (Hartford, Connecticut)	16
Milford Plaza (New York)	0
MGM Grand (Las Vegas)	36

Source: Macdonald, J. N. "Is evacuation a fatal flaw in fire fighting philosophy?" *Risk Management,* Vol. 33, No. 2, Feb. 1986, p. 37.

a. State the assumption, in terms of the problem, that is required for a small-sample confidence interval technique to be valid.

b. Construct a 98% confidence interval for the true mean number of victims per fire who die attempting to evacuate compartmented fire-resistive buildings.

c. Interpret the interval constructed in part **b**.

7.25 The number of calories per 12-ounce serving for a sample of 22 beer brands, first presented in Problem 2.22 (page 81), is reproduced in the table on the next page.

BEERCAL.XLS

Beer	Calories	Beer	Calories
Old Milwaukee	145	Miller High Life	143
Stroh's	142	Pabst Blue Ribbon	144
Red Dog	147	Milwaukee's Best	133
Budweiser	148	Miller Genuine Draft	143
Icehouse	149	Rolling Rock	143
Molson Ice	155	Michelob Light	134
Michelob	159	Bud Light	110
Bud Ice	148	Natural Light	110
Busch	143	Coors Light	105
Coors Original	137	Miller Light	96
Gennessee Cream Ale	153	Amstel Light	95

Source: Consumer Reports, June 1996, Vol. 61, No. 6, p. 16.

a. Find a 90% confidence interval for μ, the true mean number of calories per 12-ounce serving for all beer brands.

b. Interpret the interval, part **a**.

c. What assumption about the population data is required for the confidence interval to be valid?

d. Use the stem-and-leaf display you generated in Problem 2.22 to help you decide whether the assumption of part **c** is reasonably satisfied.

7.26 According to a study reported in *Administrative Science Quarterly* (June 1988), the salary gap between a chief executive officer (CEO) of a firm and a vice president (VP) is often very large. Also, the gap appears to increase as the firm employs more VPs. Based on data collected for a sample of 105 U.S. firms drawn from *Business Week*'s Executive Compensation Scoreboard, the mean and standard deviation of the number of VPs employed by a firm are $\bar{x} = 19.4$ and $s = 10.1$. Use this information to estimate the true mean number of VPs at U.S. firms with a 90% confidence interval. Interpret the result.

7.27 The *Consumer's Research* (Nov. 1993) data on the collision-damage ratings of station wagons and minivans, Problem 3.37, is reproduced below. Recall that a higher rating implies a higher likelihood that a vehicle is in a collision.

HLDI.XLS

Vehicle Model	Collision-Damage Rating
Chevrolet Astro 4-wheel drive	50
Plymouth Voyager	59
Chevrolet Caprice	77
Oldsmobile Silhouette	72
Dodge Caravan	60
GMC Safari	60
Mazda MPV 4-wheel drive	121
Toyota Previa	77
Chevrolet Lumina APV	71
Ford Aerostar	74
Chevrolet Astro	59
Mazda MPV	114
Pontiac Trans Sport	72

a. In Problem 3.37, you computed $\bar{x} = 74.31$ and $s = 20.94$ for the sample data. Use this information to construct a 98% confidence interval for μ, the true average collision-damage rating of all station wagons and minivans.

b. Is there evidence to indicate that μ is less than 100? Explain.

7.28 Refer to the *Chemosphere* (Vol. 20, 1990) study on the TCDD levels of 20 Massachusetts Vietnam veterans who were possibly exposed to the defoliant Agent Orange, Problem 2.28 (page 84). The amounts of TCDD (measured in parts per trillion) in blood plasma drawn from each veteran are shown in the table.

TCDD.XLS

Veteran	TCDD Levels in Plasma	Veteran	TCDD Levels in Plasma
1	2.5	11	6.9
2	3.1	12	3.3
3	2.1	13	4.6
4	3.5	14	1.6
5	3.1	15	7.2
6	1.8	16	1.8
7	6.0	17	20.0
8	3.0	18	2.0
9	36.0	19	2.5
10	4.7	20	4.1

Source: Schecter, A., et al. "Partitioning of 2,3,7,8-chlorinated dibenzo-*p*-dioxins and dibenzofurans between adipose tissue and plasma lipid of 20 Massachusetts Vietnam veterans." *Chemosphere,* Vol. 20, Nos. 7–9, 1990, pp. 954–955 (Table 1).

a. Construct a 90% confidence interval for the true mean TCDD level in plasma of all Vietnam veterans exposed to Agent Orange.

b. Interpret the interval, part **a**.

c. What assumption is required for the interval estimation procedure to be valid?

d. Use one of the methods of Section 6.3 to determine whether the assumption, part **c**, is approximately satisfied.

7.4 ESTIMATION OF A POPULATION PROPORTION

We now consider the method for estimating the binomial proportion of successes— that is, the **proportion** of elements in a population that have a certain characteristic. For example, a sociologist may be interested in the proportion of urban New York City residents who are minorities; a pollster may be interested in the proportion of Americans who favor the president's new economic policy; or a television executive may be interested in the percentage of viewers tuned to a particular TV program on a given night. How would you estimate a binomial proportion π based on information contained in a sample from the population? We illustrate in the next example.

Example 7.13 **Selecting a Point Estimate**

The United States Commission on Crime is interested in estimating the proportion of crimes related to firearms in an area with one of the highest crime rates in the country. The commission selects a random sample of 300 files of recently committed crimes in the area and determines that a firearm was reportedly used in 180 of them. Estimate the true proportion π of all crimes committed in the area in which some type of firearm was reportedly used.

Figure 7.12 Sampling Distribution of p

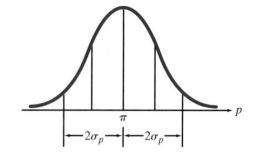

Solution

A logical candidate for a point estimate of the population proportion π is the proportion of observations in the sample that have the characteristic of interest (called a "success"); we will call this sample proportion p. In this example, the sample proportion of crimes related to firearms is given by

$$p = \frac{\text{Number of crimes in sample in which a firearm was reportedly used}}{\text{Total number of crimes in sample}}$$

$$= \frac{180}{300} = .60$$

That is, 60% of the crimes in the sample were related to firearms; the value $p = .60$ serves as our point estimate of the population proportion π. ❑

To assess the reliability of the point estimate p, we need to know its sampling distribution. This information may be derived by an application of the Central Limit Theorem (details are omitted here). Properties of the sampling distribution of p (illustrated in Figure 7.12) are given in the next box.

Sampling Distribution of the Sample Proportion p

For sufficiently large samples, the sampling distribution of $p = x/n$ is approximately normal, with

$$\text{Mean } \mu_p = \pi$$

and

$$\text{Standard deviation } \sigma_p = \sqrt{\frac{\pi(1 - \pi)}{n}} \qquad \textbf{(7.3)}$$

where
  π = true population proportion of successes
  x = number of successes in the sample
  n = sample size

A large-sample confidence interval for π may be constructed by using a procedure analogous to that used for estimating a population mean. We begin with the point estimator p, then add and subtract a certain number of standard deviations of p to obtain the desired level of confidence. The details are given in the next box.

Large-Sample $(1 - \alpha)100\%$ Confidence Interval for a Population Proportion π

$$p \pm z_{\alpha/2}\sigma_p \approx p \pm z_{\alpha/2}\sqrt{\frac{p(1-p)}{n}} \qquad (7.4)$$

where $p = x/n$ is the sample proportion of observations with the characteristic of interest.

[*Note:* The interval is approximate since we have substituted the sample value p for the corresponding population value π required for σ_p.]

Assumption: A sufficiently large random sample of size n is selected from the population.

Note that we must substitute p into the formula for

$$\sigma_p = \sqrt{\pi(1-\pi)/n}$$

here. This approximation will be valid as long as the sample size n is sufficiently large. Usually, "sufficiently large" will be satisfied if $n \geq 50$.

Example 7.14

95% Confidence Interval for π

Refer to Example 7.13. Construct and interpret a 95% confidence interval for π, the population proportion of crimes committed in the area in which some type of firearm is reportedly used.

Solution

For a confidence coefficient of .95, we have $1 - \alpha = .95$; $\alpha = .05$; $\alpha/2 = .025$; and the required z value is $z_{.025} = 1.96$. In Example 7.13, we obtained $p = 180/300 = .60$. Substituting these values into the formula for an approximate confidence interval for π (cquation 7.4) yields

$$p \pm z_{\alpha/2}\sqrt{\frac{p(1-p)}{n}} = .60 \pm 1.96\sqrt{\frac{(.60)(.40)}{300}}$$

$$= .60 \pm .06$$

or (.54, .66). We are 95% confident that the true proportion of crimes committed in the area in which firearms were used falls between .54 and .66. ❑

Example 7.15

90% Confidence Interval for π

Potential advertisers value television's well-known Nielsen ratings as a barometer of a TV show's popularity among viewers. The Nielsen rating of a certain TV program is an estimate of the proportion of viewers, expressed as a percentage, who tune their sets to the program on a given night. Suppose that in a random sample of 165 families who regularly watch television, a Nielsen survey indicated that 63 of the families were tuned to NBC's *ER* on the night of its season finale.

Estimate π, the true proportion of all TV-viewing families who watched the season finale of *ER*, using a 90% confidence interval. Interpret the interval.

Solution

In this problem the variable of interest is the response (yes or no) to the question, "Did you watch the season finale of *ER*?" The sample proportion of families that watched the season finale of *ER* is

$$p = \frac{x}{n} = \frac{\text{Number of families in sample that watched the show}}{\text{Number of families in sample}}$$

$$= \frac{63}{165} = .38$$

Using equation 7.4, the approximate 90% confidence interval is

$$p \pm z_{.05}\sqrt{\frac{p(1 - p)}{n}} = .38 \pm 1.645\sqrt{\frac{(.38)(.62)}{165}}$$

$$= .38 \pm .06$$

or (.32, .44).

We are 90% confident that the interval from .32 to .44 encloses the true proportion of TV-viewing families that watched the season finale of *ER*. If we repeatedly selected random samples of $n = 165$ families and constructed a 90% confidence interval based on each sample, then we would expect 90% of the confidence intervals constructed to contain π. ❏

Self-Test

7.5

A random sample of size 1,500 is selected from a binomial population and the number of successes is 600.

a. Find p, the sample proportion of successes.
b. Construct a 99% confidence interval for π.

Caution: Unless n is extremely large, the large-sample procedure presented in this section performs poorly when π is near 0 or 1. For example, suppose you want to estimate the proportion of people who die from a bee sting. This proportion is likely to be near 0 (say, $\pi \approx .001$). Confidence intervals for π based on a sample of size $n = 50$ will probably be misleading.

To overcome this potential problem, an *extremely* large sample size is required. Since the value of n required to satisfy "extremely large" is difficult to determine, statisticians (see Agresti & Coull, 1998) have proposed an alternative method, based on the Wilson (1927) point estimator of π. The procedure is outlined in the box. Researchers have shown that this confidence interval works well for any π even when the sample size n is very small.

Adjusted $(1 - \alpha)100\%$ Confidence Interval for a Population Proportion π

$$p^* \pm z_{\alpha/2}\sqrt{\frac{p^*(1 - p^*)}{n + 4}} \qquad \text{(7.5)}$$

where $p^* = \dfrac{x + 2}{n + 4}$ is the adjusted sample proportion of observations with the characteristic of interest, and x is the number of successes in the sample.

Example 7.16   Using the Adjusted Estimator of π

According to *True Odds: How Risk Affects Your Everyday Life* (Walsh, 1997), the probability of being the victim of a violent crime is less than .01. Suppose that in a random sample of 200 Americans, 3 were victims of a violent crime. Estimate the true proportion of Americans who were victims of a violent crime using a 95% confidence interval.

Solution

Let π represent the true proportion of Americans who were victims of a violent crime. Since π is near 0, an "extremely large" sample is required to estimate its value using the usual large-sample method. Since we are unsure whether the sample size of 200 is large enough, we will apply the adjustment outlined in the box.

The number of "successes" (i.e., number of violent crime victims) in the sample is $x = 3$. Therefore, the adjusted sample proportion is

$$p^* = \frac{x + 2}{n + 4} = \frac{3 + 2}{200 + 4} = \frac{5}{204} = .025$$

Note that this adjusted sample proportion is obtained by adding a total of 4 observations—2 "successes" and 2 "failures"—to the sample data. Substituting $p^* = .025$ into equation 7.5 for a 95% confidence interval, we obtain

$$p^* \pm 1.96\sqrt{\frac{p^*(1 - p^*)}{n + 4}} = .025 \pm 1.96\sqrt{\frac{(.025)(.975)}{204}}$$

$$= .025 \pm .021$$

or (.004, .046). Consequently, we are 95% confident that the true proportion of Americans who are victims of a violent crime falls between .004 and .046. ❑

Self-Test 7.6

A random sample of size $n = 1{,}000$ is selected from a binomial population with proportion of success $\pi \approx .005$. Use the adjustment to construct a 95% confidence interval for π if the number of successes in the sample is $x = 7$.

Using the Tools

7.29   Random samples of n measurements are selected from a population with unknown proportion of successes π. Compute an estimate of σ_p for each of the following combinations of sample size n and sample proportion of successes p.

  a. $n = 250, p = .4$ **b.** $n = 500, p = .85$ **c.** $n = 100, p = .25$

7.30   Random samples of n measurements are selected from a population with unknown proportion of successes π. Calculate a 95% confidence interval for π for each of the following combinations of sample size n and sample proportion of successes p.

  a. $n = 500, p = .38$ **b.** $n = 100, p = .45$ **c.** $n = 1,000, p = .43$

7.31   The proportion of successes in a random sample of size n is $p = .20$.

  a.   Find a 95% confidence interval for π if $n = 100$.
  b.   Find a 95% confidence interval for π if $n = 500$.

7.32   The proportion of successes in a random sample of size n is $p = .98$.

  a.   Find a 90% confidence interval for π if $n = 10,000$.
  b.   Find a 90% confidence interval for π if $n = 1,000$.
  c.   Find a 90% confidence interval for π if $n = 30$.

Applying the Concepts

7.33   A University of Minnesota survey of brand name (e.g., Levi's, Lee, and Calvin Klein) and private label jeans found a high percentage of jeans with incorrect waist and/or inseam measurements on the label. The study found that only 18 of 240 pairs of men's five-pocket, prewashed jeans sold in Minneapolis stores came within a half inch of all their label measurements (*Tampa Tribune*, May 20, 1991). Let π represent the true proportion of men's five-pocket, prewashed jeans sold in Minneapolis that have inseam and waist measurements that fall within .5 inch of the labeled measurements.

  a.   Find a point estimate of π.
  b.   Find an interval estimate of π. Use a confidence coefficient of .90.
  c.   Interpret the interval, part **b.**

7.34   An American Housing Survey (AHS) conducted by the U.S. Department of Commerce revealed that 705 of 1,500 sampled homeowners are "do-it-yourselfers"—they did most of the work themselves on at least one of their home improvements or repairs (Bureau of the Census, *Statistical Brief*, May 1992). Using a 95% confidence interval, estimate the true proportion of American homeowners who do most of their home improvement/repair work themselves. Interpret the result.

7.35   Refer to the *Journal of the American Medical Association* (Apr. 21, 1993) report on the prevalence of cigarette smoking among U.S. adults,

Problem 7.10 (page 372). Of the 43,732 survey respondents, 11,239 indicated that they were current smokers and 10,539 indicated they were former smokers.

a. Construct and interpret a 90% confidence interval for the percentage of U.S. adults who currently smoke cigarettes.

b. Construct and interpret a 90% confidence interval for the percentage of U.S. adults who are former cigarette smokers.

7.36 Refer to the *Annals of Internal Medicine* (Dec. 1, 1997) study of whether women understand basic probability, Problem 4.5 (page 215). Recall that a random sample of 287 female military veterans were asked: "How many times will a coin come up heads in 1,000 flips?" Only 155 of the females correctly answered "about 500 heads."

a. Construct a 95% confidence interval for the true proportion of female veterans who correctly answer "about 500 heads" to the coin flip question. Interpret the result.

b. Why is it important to restrict any inferences derived from the interval to female military veterans and not to women in general?

7.37 ABC News and *The Washington Post* conducted a national random-digit-dial telephone survey of 1,500 adults. One question asked: "Have you ever seen anything that you believe was a spacecraft from another planet?" Ten percent (150) of the respondents answered in the affirmative. These 150 people were then asked, "Have you personally ever been in contact with aliens from another planet?" and nine people responded "Yes." (*Chance,* Summer 1997).

a. Estimate the true proportion of adults who believe they have seen spacecraft from another planet with a 90% confidence interval.

b. Estimate the true proportion of adults who believe they have been in contact with aliens from another planet with a 90% confidence interval.

c. It is highly likely that the true value of π estimated in parts **a** and **b** is equal to 0. Given this is true, how do you explain the survey results?

7.38 The concept of tyranny in 19th-century American legal cases was investigated in the *Journal of Interdisciplinary History* (Autumn 1997). In a sample of 3,038 U.S. federal court cases that occurred in the decade 1850–1859, 27 of them used key words associated with tyranny (e.g., tyranny, despotism, dictatorship, autocracy, and absolutism).

a. Use a 99% confidence interval to estimate the true percentage of federal court cases in the decade 1850–1859 that used key words associated with tyranny.

b. Do you believe that 2% or more of all the U.S. federal court cases between 1850 and 1859 used key words associated with tyranny? Explain.

7.39 A University of South Florida research team performed a series of experiments to investigate the effects of stress on memory (*USF Magazine,* Winter 1998). In one experiment, 30 rats were trained to escape from a small pool of water by climbing onto a fixed platform hidden just below the water's surface. After learning the platform's location, all 30 rats were placed in a room with a cat. Although the cat did not touch the rats, the rats exhibited behavioral and physiological signs of stress. The rats were then immediately returned to the water tank and the researchers counted the number that forgot the location of the platform.

a. Suppose that 25 of the 30 rats forgot how to escape from the water tank. Estimate the true proportion of rats that will forget how to escape when placed in a similar experimental setting.

b. Use the result, part **a**, to form a 95% confidence interval for the true proportion of rats that will forget how to escape. Give a practical interpretation of the interval.

7.40 In zero-gravity spacecraft, astronauts rely heavily on visual information to avoid disorientation. An empirical study was conducted to assess the potential of using color brightness as a body orientation clue (*Human Factors*, Dec. 1988). Ninety college students, reclining on their backs in the dark, were disoriented when positioned on a rotating platform under a slowly rotating disk that filled their entire field of vision. Half the disk was painted with a brighter level of color than the other half. The students were asked to say "stop" when they believed they were right-side up, and the brightness level of the disk was recorded. Of the 90 students, 58 selected the brighter color level.

a. Use this information to estimate the true proportion of subjects who use the bright color level as a cue to being right-side up. Construct a 95% confidence interval for the true proportion.

b. Can you infer from the result, part **a**, that a majority of subjects would select bright color levels over dark color levels as a cue to being right-side up? Explain.

7.41 *Agoraphobia* is the psychiatric term for an abnormal fear of being in an open space. In an experiment reported in *Behavioral Research Theory* (Vol. 26, 1988), 51 agoraphobic patients were administered the Behavioral Avoidance Test (BAT). The BAT consisted of three tasks, one of which required the patient to walk unaccompanied from the hospital into a crowded urban center and to cross a wide and congested avenue. Even though most of the 51 patients experienced at least one panic attack, only seven were unable to complete the task. Construct a 90% confidence interval for the proportion of all agoraphobic patients who cannot complete the BAT task. Interpret the interval.

7.5 CHOOSING THE SAMPLE SIZE

In the preceding sections we have overlooked a problem that usually must be faced in the initial stages of an experiment. Before constructing a confidence interval for a parameter of interest, we will have to decide on the number of observations n to be included in a sample. Should we sample $n = 10$ observations, $n = 20$, or $n = 100$? To answer this question we need to decide how wide a confidence interval we are willing to tolerate and the measure of confidence—that is, the confidence coefficient—that we wish to have in the results. The following example will illustrate the method for determining the appropriate sample size for estimating a population mean.

Figure 7.13 Sampling Distribution of the Sample Mean, $\bar{x}$

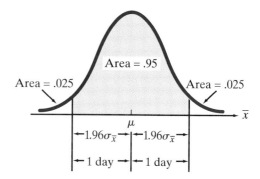

Example 7.17   Sample Size for Estimating μ

A mail-order house wants to estimate the mean length of time between shipment of an order and receipt by the customer. The management plans to randomly sample n orders and determine, by telephone, the number of days between shipment and receipt for each order. If the management wants to estimate the mean shipping time correct to within one day with confidence coefficient equal to .95, how many orders should be sampled?

Solution

We will use $\bar{x}$, the sample mean of the n measurements, to estimate μ, the mean shipping time. Its sampling distribution will be approximately normal and the probability that $\bar{x}$ will lie within

$$1.96\sigma_{\bar{x}} = 1.96\left(\frac{\sigma}{\sqrt{n}}\right)$$

of the mean shipping time μ is approximately .95 (see Figure 7.13). Therefore, we want to choose the sample size n so that $1.96\sigma/\sqrt{n}$ equals one day:

$$1.96\left(\frac{\sigma}{\sqrt{n}}\right) = 1 \tag{7.6}$$

To solve the equation $1.96\sigma/\sqrt{n} = 1$, we need to know the value of σ, a measure of variation of the population of all shipping times. Since σ is unknown (as will usually be the case in practical applications), we must approximate its value using the standard deviation of some previous sample data or deduce an approximate value from other knowledge about the population. Suppose, for example, that we know almost all shipments will be delivered within 21 days. Then the population of shipping times might appear as shown in Figure 7.14.

Figure 7.14 Hypothetical Distribution of Population of Shipping Times for Example 7.17

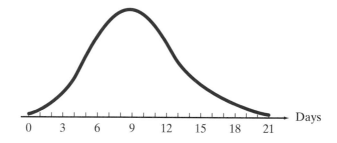

Figure 7.14 provides the information we need to find an approximation for σ. Since the Empirical Rule tells us that almost all the observations in a data set will fall within the interval $\mu \pm 3\sigma$, it follows that the range of a population is approximately 6σ. If the range of the population of shipping times is 21 days, then

$$6\sigma = 21 \text{ days}$$

and σ is approximately equal to 21/6 or 3.5 days.

The final step in determining the sample size is to substitute this approximate value of σ into equation 7.6 and solve for n. Thus, we have

$$1.96\left(\frac{3.5}{\sqrt{n}}\right) = 1$$

or

$$\sqrt{n} = \frac{1.96(3.5)}{1} = 6.86$$

Squaring both sides of this equation yields $n = 47.06$. We will follow the usual convention of rounding the calculated sample size upward. Therefore, the mail-order house needs to sample approximately $n = 48$ shipping times to estimate the mean shipping time correct to within one day with 95% confidence. ❑

In Example 7.17, we wanted our sample estimate to lie within one day of the true mean shipping time μ with 95% confidence. The value 1 (one day) represents the sampling error we desire in the confidence interval. It is the difference between $\bar{x}$, the center of the interval, and the upper (or lower) endpoint of the confidence interval.

Definition 7.5

The **sampling error**, E, in a confidence interval is the difference between the point estimator and either the upper or lower endpoint in the interval.

We could calculate the sample size for some specific sampling error E for a confidence coefficient other than .95 by changing the z value in the equation. In general, if we are willing to accept a sampling error E in estimating μ with confidence coefficient $(1 - \alpha)$, we would solve for n in the equation

$$z_{\alpha/2}\left(\frac{\sigma}{\sqrt{n}}\right) = E$$

where the value of $z_{\alpha/2}$ is obtained from Table B.2 in Appendix B. The solution is given by

$$n = \left(\frac{z_{\alpha/2}\sigma}{E}\right)^2 = \frac{(z_{\alpha/2})^2\sigma^2}{E^2}$$

For example, for a confidence coefficient of .90, we would require a sample size of

$$n = \frac{(1.645)^2\sigma^2}{E^2}$$

Choosing the Sample Size for Estimating a Population Mean μ to within E Units with Confidence Coefficient $(1 - \alpha)$

$$n = \left(\frac{z_{\alpha/2}\sigma}{E}\right)^2 = \frac{(z_{\alpha/2})^2\sigma^2}{E^2} \qquad (7.7)$$

[*Note:* The population standard deviation σ will usually have to be approximated.]

Self-Test 7.7

How large a sample is required to estimate μ to within 10 units with 90% confidence? Assume $\sigma^2 = 2,500$.

The procedures for determining the sample size needed to estimate a population proportion is analogous to the procedure for determining the sample size for estimating a population mean. We present the appropriate formula and illustrate its use with an example.

Choosing the Sample Size for Estimating a Population Proportion π to within E Units with Confidence Coefficient $(1 - \alpha)$

$$n = \frac{(z_{\alpha/2})^2\pi(1 - \pi)}{E^2} \qquad (7.8)$$

where π is the value of the population proportion that you are attempting to estimate.

[*Note:* This technique requires a previous estimate of π. If none is available, use $\pi = .5$ for a conservative choice of n.]

Example 7.18 **Sample Size for Estimating π**

In Example 7.13, the United States Commission on Crime sampled recently committed crimes to estimate the proportion in which firearms were used. Suppose the Commission wants to obtain an estimate of π that is correct to within .02 with 90% confidence. How many cases would have to be included in the Commission's sample?

Solution

Here, we desire our estimate to be within .02 of the unknown population proportion π. Thus, our sampling error is E = .02. For a confidence coefficient of $(1 - \alpha) = .90$, we have $\alpha = .10$. Therefore, to calculate n using equation 7.8, we must find the value of $z_{\alpha/2} = z_{.05}$ and find an approximation for π.

From Table B.2, the z value corresponding to an area of $\alpha/2 = .05$ in the upper tail of the standard normal distribution is $z_{.05} = 1.645$. As an approximation to π, we will use the sample estimate, $p = .60$, obtained for the sample of 300 cases in Example 7.13 (page 383).

Caution

The formulas given in this section are appropriate when the sample size n is small relative to the population size N. For situations in which n may be large relative to N, adjustments to these formulas must be made. Sample size determination for this special case (called *survey sampling*) is beyond the scope of this text. Consult the references (e.g., see Kish, 1965) if you want to learn more about this particular application.

Substituting the value of .6 for π, $z_{.05} = 1.645$, and E = .02 into equation 7.8, we have

$$n = \frac{(z_{\alpha/2})^2 \pi(1 - \pi)}{E^2} = \frac{(1.645)^2(.6)(.4)}{(.02)^2} = 1,623.6$$

Therefore, in order to estimate π to within .02 with 90% confidence, the Commission will have to sample approximately $n = 1,624$ cases. ❑

Use $\pi = .5$ When Prior Estimates Are Unavailable

In Example 7.18, we used a prior estimate of π in computing the required sample size. If such prior information were not available, we could approximate π in the sample size equation using $\pi = .5$. The nearer the substituted value of π is to .5, the larger will be the sample size obtained from the formula. Hence, if you take $\pi = .5$ as the approximation to π, you will always obtain a sample size that is at least as large as required.

PROBLEMS
7.42–7.51

Using the Tools

7.42 Determine the sample size needed to estimate μ for each of the following situations:

    a. $E = 3, \sigma = 40, (1 - \alpha) = .95$ **b.** $E = 5, \sigma = 40, (1 - \alpha) = .95$

    c. $E = 5, \sigma = 40, (1 - \alpha) = .99$

7.43 Find the sample size needed to estimate π for each of the following situations:

    a. $E = .04, \pi \approx .9, (1 - \alpha) = .90$ **b.** $E = .04, \pi \approx .5, (1 - \alpha) = .90$

    c. $E = .01, \pi \approx .5, (1 - \alpha) = .90$

7.44 Find the sample size n needed to estimate:

    a. μ to within 5 units with 99% confidence when $\sigma \approx 20$

    b. π to within .1 with 90% confidence

7.45 Find the sample size n needed to estimate:

    a. π to within .04 with 95% confidence when $\pi \approx .8$

    b. μ to within 70 units with 90% confidence when $\sigma \approx 500$

Applying the Concepts

7.46 A child psychologist wants to estimate the mean age at which a child learns to walk. How many children must be sampled if the psychologist desires an estimate that is correct to within 1 month of the true mean with 99% confidence? Assume the psychologist knows only that the age at which a child begins to walk ranges from 8 to 26 months.

7.47 *Good Housekeeping* (July 1996) reported on a study of women who shave their legs. Each woman was required to shave one leg using a cream or gel and the other leg using soap or a beauty bar. Of interest was the proportion of women who prefer shaving their legs using a cream or gel. How many women must participate in the study to estimate the proportion to within .03 with 95% confidence? Use .5 as a conservative estimate of the true proportion.

7.48 Refer to the *American Journal of Psychiatry* study of jail suicides, Example 7.11 (page 378). How many jail suicide cases must be sampled to estimate the true mean number of days the suicidal inmate spent in jail to within 10 days with 90% confidence? Use the sample standard deviation calculated in Example 7.11 as an estimate of σ.

7.49 The National Institute of Mental Health (NIMH) estimates that 3% of all American children age 3–18 commit severe acts of physical aggression against their parents. Suppose you want to estimate within .005 with 90% confidence the true proportion of American children age 3–18 who commit acts of physical aggression against their parents.

a. How many families should be included in the sample if no prior information is available about the value of the proportion?

b. What sample size is required if the NIMH proportion is used as an estimate?

7.50 The *Journal of Literacy Research* (Dec. 1996) published a study on the number of times per week a typical adult read a book for pleasure (see Problem 6.61, page 350). Suppose you want to estimate the true mean number of times per week an adult reads a book for pleasure. How many adults must be sampled to estimate with 90% confidence the mean to within .5 of its true value? Assume that the true standard deviation for the variable of interest is approximately equal to 2.8 (the value given in Problem 6.61).

7.51 Refer to the *Behavioral Research Theory* study of people who suffer from agoraphobia, Problem 7.41 (page 390). How many patients need to be studied in order to estimate the true percentage who are able to complete the BAT tasks to within .05 with 90% confidence?

7.6 ESTIMATION OF A POPULATION VARIANCE (OPTIONAL)

In the previous sections, we considered interval estimate for population means or proportions. In this optional section, we discuss a confidence interval for a population variance, σ^2.

Example 7.19 ### Estimating σ^2

Refer to the U.S. Army Corps of Engineers study of contaminated fish in the Tennessee River, Alabama. (Recall that the data are stored in FISH.XLS.) It is important for the Corps of Engineers to know how stable the weights of the contaminated fish are. That is, how large is the variation in the fish weights?

Figure 7.15 Several Chi-Square Probability Distributions

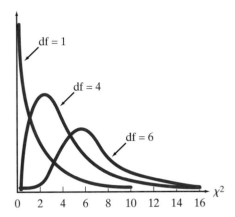

a. Identify the parameter of interest to the U.S. Army Corps of Engineers.

b. Explain how to build a confidence interval for the parameter, part **a**.

Solution

a. The Corps of Engineers is interested in the *variation* of the fish weights. Consequently, the target population parameter is σ^2, the variance of the weights of all contaminated fish inhabiting the Tennessee River.

b. Intuitively, it seems reasonable to use the sample variance s^2 to estimate σ^2 and to construct our confidence interval around this value. However, unlike sample means and sample proportions, the sampling distribution of the sample variance s^2 does not follow a normal (z) distribution or a t distribution.

Rather, when certain assumptions are satisfied (we discuss these later), the sampling distribution of s^2 possesses approximately a **chi-square (χ^2) distribution.**[*] The chi-square probability distribution, like the t distribution, is characterized by a quantity called the *degrees of freedom* associated with the distribution. Several chi-square probability distributions with different degrees of freedom are shown in Figure 7.15. Unlike z and t distributions, the chi-square distribution is not symmetric about 0. ❏

[*]Throughout this section (and this text), we will use the words *chi-square* and the Greek symbol χ^2 interchangeably.

Example 7.20   Finding Values of χ^2

Tabulated values of the χ^2 distribution are given in Table B.4 of Appendix B; a partial reproduction of this table is shown in Table 7.6 (page 397). Entries in the table give an upper-tail value of χ^2, call it χ^2_α, such that $P(\chi^2 > \chi^2_\alpha) = \alpha$. Find the tabulated value of χ^2 corresponding to 9 degrees of freedom that cuts off an upper-tail area of .05.

Solution

The value of χ^2 that we seek appears (highlighted) in the partial reproduction of Table B.4 given in Table 7.6. The columns of the table identify the value of α associated with the tabulated value of χ^2_α and the rows correspond to the degrees of freedom. For this example, we have df = 9 and α = .05. Thus, the tabulated value of χ^2 corresponding to 9 degrees of freedom is

$$\chi^2_{.05} = 16.9190$$

❏

Table 7.6 Reproduction of Part of Table B.4

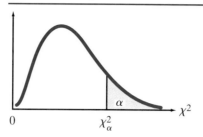

Degrees of Freedom	$\chi^2_{.100}$	$\chi^2_{.050}$	$\chi^2_{.025}$	$\chi^2_{.010}$	$\chi^2_{.005}$
1	2.70554	3.84146	5.02389	6.63490	7.87944
2	4.60517	5.99147	7.37776	9.21034	10.5966
3	6.25139	7.81473	9.34840	11.3449	12.8381
4	7.77944	9.48773	11.1433	13.2767	14.8602
5	9.23635	11.0705	12.8325	15.0863	16.7496
6	10.6446	12.5916	14.4494	16.8119	18.5476
7	12.0170	14.0671	16.0128	18.4753	20.2777
8	13.3616	15.5073	17.5346	20.0902	21.9550
9	14.6837	16.9190	19.0228	21.6660	23.5893
10	15.9871	18.3070	20.4831	23.2093	25.1882
11	17.2750	19.6751	21.9200	24.7250	26.7569
12	18.5494	21.0261	23.3367	26.2170	28.2995
13	19.8119	22.3621	24.7356	27.6883	29.8194
14	21.0642	23.6848	26.1190	29.1413	31.3193
15	22.3072	24.9958	27.4884	30.5779	32.8013
16	23.5418	26.2962	28.8454	31.9999	34.2672
17	24.7690	27.5871	30.1910	33.4087	35.7185
18	25.9894	28.8693	31.5264	34.8053	37.1564
19	27.2036	30.1435	32.8523	36.1908	38.5822

Source: From C. M. Thompson, "Tables of the Percentage Points of the χ^2-Distribution," *Biometrika,* 1941, 32, 188–189. Reproduced by permission of the *Biometrika* Trustees and Oxford University Press.

Self-Test 7.8

Find the value of $\chi^2_{.025}$ when the sample size is $n = 15$.

We use the tabulated values of χ^2 to construct a confidence interval for σ^2, as the next example illustrates.

Example 7.21 **95% Confidence Interval for σ^2**

Refer to Example 7.19. The 144 fish specimens in the U.S. Army Corps of Engineers study produced the following summary statistics: $\bar{x} = 1{,}049.7$ grams, $s = 376.6$ grams. Use this information to construct a 95% confidence interval for the true variation in weights of contaminated fish in the Tennessee River.

Figure 7.16 The Location of $\chi^2_{(1-\alpha/2)}$ and $\chi^2_{\alpha/2}$ for a Chi-Square Distribution

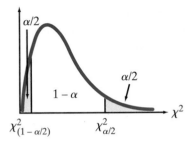

Solution

A $(1-\alpha)100\%$ confidence interval for σ^2 depends on the quantities s^2, $(n-1)$, and critical values of χ^2 as shown in the box.

A $(1-\alpha)100\%$ Confidence Interval for a Population Variance σ^2

$$\frac{(n-1)s^2}{\chi^2_{\alpha/2}} \le \sigma^2 \le \frac{(n-1)s^2}{\chi^2_{(1-\alpha/2)}} \tag{7.9}$$

where $\chi^2_{\alpha/2}$ and $\chi^2_{(1-\alpha/2)}$ are values of χ^2 that locate an area of $\alpha/2$ to the right and $\alpha/2$ to the left, respectively, of a chi-square distribution based on $(n-1)$ degrees of freedom.

Assumption: The population from which the random sample is selected has an approximate normal distribution.

Note that $(n-1)$ represents the degrees of freedom associated with the χ^2 distribution. To construct the interval, we first locate the critical values $\chi^2_{(1-\alpha/2)}$ and $\chi^2_{\alpha/2}$. These are the values of χ^2 that cut off an area of $\alpha/2$ in the lower and upper tails, respectively, of the chi-square distribution (see Figure 7.16).

For a 95% confidence interval, $(1-\alpha) = .95$ and $\alpha/2 = .025$. Therefore, we need the tabulated values $\chi^2_{.025}$ and $\chi^2_{.975}$ for $(n-1) = 143$ degrees of freedom. Looking in the df = 150 row of Table B.4 (the row with the df value closest to 143), we find $\chi^2_{.025} = 185.800$ and $\chi^2_{.975} = 117.985$. Substituting into equation 7.9, we obtain

$$\frac{(144-1)(376.6)^2}{185.800} \le \sigma^2 \le \frac{(144-1)(376.6)^2}{117.985}$$

$$109{,}156.8 \le \sigma^2 \le 171{,}898.4$$

We are 95% confident that the true variance in weights of contaminated fish in the Tennessee River falls between 109,156.8 and 171,898.4. The Army Corps of Engineers could use this interval to determine whether the weights of the fish are stable enough to allow further testing for DDT contamination. ❑

Self-Test 7.9

Find a 95% confidence interval for σ^2 based on the following summary statistics: $n = 15, \bar{x} = 72, s = 11$.

Example 7.22   Estimating σ

Refer to Example 7.21. Find a 95% confidence interval for σ, the true standard deviation of the fish weights.

Solution

A confidence interval for σ is obtained by taking the square roots of the lower and upper endpoints of a confidence interval for σ^2. Thus, the 95% confidence interval is

$$\sqrt{109{,}156.8} \le \sigma \le \sqrt{171{,}898.4}$$

$$330.4 \le \sigma \le 414.6$$

Thus, we are 95% confident that the true standard deviation of the fish weights is between 330.4 grams and 414.6 grams. ❏

Caution

The procedure for calculating a confidence interval for σ^2 in the above examples requires an assumption regardless of whether the sample size n is large or small (see previous box). We must assume that the population from which the sample is selected has an approximate normal distribution. Unlike small sample confidence intervals for μ based on the t statistic, *slight to moderate departures from normality will render the χ^2 confidence interval invalid.* It is reasonable to expect this assumption to be satisfied in Examples 7.21 and 7.22 since the histogram of the 144 fish weights in the sample, shown in Figure 3.11 (page 162), is approximately normal.

PROBLEMS 7.52–7.61

Using the Tools

7.52 For each of the following combinations of α and degrees of freedom (df), use Table B.4 in Appendix B to find the values of $\chi^2_{\alpha/2}$ and $\chi^2_{(1-\alpha/2)}$ that would be used to form a confidence interval for σ^2.

a. $\alpha = .05$, df $= 7$ b. $\alpha = .10$, df $= 16$

c. $\alpha = .01$, df $= 10$ d. $\alpha = .05$, df $= 20$

7.53 Given the following values of $\bar{x}$, s, and n, form a 90% confidence interval for σ^2.

a. $\bar{x} = 21$, $s = 2.5$, $n = 50$ b. $\bar{x} = 1.3$, $s = .02$, $n = 15$

c. $\bar{x} = 167$, $s = 31.6$, $n = 22$ d. $\bar{x} = 9.4$, $s = 1.5$, $n = 5$

What assumption about the population must be satisfied for the confidence interval to be valid?

7.54 Refer to Problem 7.53. For each part **a–d**, form a 90% confidence interval for σ.

7.55 A random sample of $n = 6$ observations from a normal distribution resulted in the following measurements: 8, 2, 3, 7, 11, 6. Form a 95% confidence interval for σ^2.

Applying the Concepts

7.56 In an *Exceptional Children* (Sept. 1984) study, a sample of 16 developmentally delayed children had a mean IQ of 80.2 and a standard deviation of 20.7. Find an estimate of the variance in IQ of developmentally delayed children using a 99% confidence interval.

7.57 A machine used to fill beer cans must operate so that the amount of beer actually dispensed varies very little. If too much beer is released, the cans will overflow, causing waste. If too little beer is released, the cans will not contain enough beer, causing complaints from customers. A random sample of the fills for 20 cans yielded a standard deviation of .07 ounce. Estimate the true variance of the fills using a 95% confidence interval.

7.58 Refer to the *Journal of Applied Behavior Analysis* study of a 16-year-old boy's destructive behavior, Problem 7.23 (page 381). The data on the boy's number of destructive responses for 10 play sessions are reproduced below.

3   2   4.5   5.5   4.5   3   2.5   1   3   3

a. Estimate the variation in the boy's number of destructive responses per session using a 99% confidence interval.

b. What assumptions are required for the interval estimate to be valid?

7.59 Refer to the *Science* study of swarm-founding wasps in Problem 7.13 (page 374). Recall that an inbreeding coefficient x was genetically determined for each of 197 wasps, resulting in a sample mean of $\bar{x} = .044$ and a sample standard deviation of $s = .884$. Use this information to construct a 90% confidence interval for σ, the true standard deviation of the inbreeding coefficients of the wasps. Interpret the interval.

7.60 *IEEE Transactions* (June 1990) presented a computer algorithm for solving a difficult mathematical programming problem. Fifty-two random problems were solved using the algorithm; the times to solution (CPU time in seconds) are listed in the accompanying table.

	A	B
1	*MathCPU*	
2		
3	Mean	0.810807692
4	Standard Error	0.208745094
5	Median	0.258
6	Mode	0.136
7	Standard Deviation	1.505282281
8	Sample Variance	2.265874747
9	Kurtosis	15.72115962
10	Skewness	3.652228455
11	Range	8.752
12	Minimum	0.036
13	Maximum	8.788
14	Sum	42.162
15	Count	52
16	Largest(1)	8.788
17	Smallest(1)	0.036

Problem 7.60

MATHCPU.XLS

.045	3.985	.506	.145	1.267	.049	.333	.379	.091	.036	.336	.219	.209
1.070	.130	.579	.045	.118	1.894	.136	1.639	.064	.258	.412	.209	.070
8.788	3.046	.179	.136	3.888	.242	.227	.182	.136	.600	.394	.258	.327
.445	1.055	.670	.088	4.170	.567	.079	.554	.912	.194	.182	.361	.258

Source: Snyder, W. S., and Chrissis J. W. "A hybrid algorithm for solving zero–one mathematical programming problems." *IEEE Transactions,* Vol. 22, No. 2, June 1990, p. 166 (Table 1).

An Excel printout giving descriptive statistics for the sample of 52 solution times is provided at left. Use this information to compute a 95% confidence interval for the variance of the solution times. Interpret the result.

7.61 *Jitter* is a term used to describe the variation in conduction time of a water power system. Low throughput jitter is critical to successful waterline technology. An investigation of throughput jitter in the opening switch of a prototype system (*Journal of Applied Physics,* Sept. 1993) yielded the following descriptive statistics on conduction time for $n =$ 18 trials.

$$\bar{x} = 334.8 \text{ nanoseconds} \qquad s = 6.3 \text{ nanoseconds}$$

(Conduction time is defined as the length of time required for the downstream current to equal 10% of the upstream current.)

a. Construct a 95% confidence interval for the true standard deviation of conduction times of the prototype system.

b. A system is considered to have low throughput jitter if the true conduction time standard deviation is less than 7 nanoseconds. Does the prototype system satisfy this requirement? Explain.

REAL-WORLD REVISITED
Scallops, Sampling, and the Law

Arnold Bennett, a Sloan School of Management professor at the Massachusetts Institute of Technology (MIT), described a recent legal case in which he served as a statistical "expert" in *Interfaces* (Mar.–Apr. 1995). The case involved a ship that fishes for scallops off the coast of New England. In order to protect baby scallops from being harvested, the U.S. Fisheries and Wildlife Service requires that "the average meat per scallop weigh at least $\frac{1}{36}$ of a pound." The ship was accused of violating this weight standard. Bennett lays out the scenario:

> The vessel arrived at a Massachusetts port with 11,000 bags of scallops, from which the harbormaster randomly selected 18 bags for weighing. From each such bag, his agents took a large scoopful of scallops; then, to estimate the bag's average meat per scallop, they divided the total weight of meat in the scoopful by the number of scallops it contained. Based on the 18 [numbers] thus generated, the harbormaster estimated that each of the ship's scallops possessed an average $\frac{1}{39}$ of a pound of meat (that is, they were about seven percent lighter than the minimum requirement). Viewing this outcome as conclusive evidence that the weight standard had been violated, federal authorities at once confiscated *95 percent* of the catch (which they then sold at auction). The fishing voyage was thus transformed into a financial catastrophe for its participants.

Bennett provided the actual scallop weight measurements for each of the 18 sampled bags in the article. The data are listed in Table 7.7 and are stored in the Excel workbook SCALLOPS.XLS.

SCALLOPS.XLS

Table 7.7 Scallop Weight Measurements for 18 Bags Sampled

.93	.88	.85	.91	.91	.84	.90	.98	.88
.89	.98	.87	.91	.92	.99	1.14	1.06	.93

Source: Bennett, A. "Misapplications review: Jail terms." *Interfaces,* Vol. 25, No. 2, March–April 1995, p. 20.

For ease of exposition, Bennett expressed each number as a multiple of $\frac{1}{36}$ of a pound, the minimum permissible average weight per scallop. Consequently, numbers below one indicate individual bags that do not meet the standard.

The ship's owner filed a lawsuit against the federal government, declaring that his vessel had fully complied with the weight standard. A Boston law firm was hired to represent the owner in legal proceedings and Bennett was retained by the firm to provide statistical litigation support and, if necessary, expert witness testimony.

Study Questions

a. Recall that the harbormaster sampled only 18 of the ship's 11,000 bags of scallops. One of the questions the lawyers asked Bennett was: "Can a reliable estimate of the mean weight of all the scallops be obtained from a sample of size 18?" Give your opinion on this issue.

b. As stated in the article, the government's decision rule is to confiscate a scallop catch if the sample mean weight of the scallops is less than $\frac{1}{36}$ of a pound. Do you see any flaws in this rule?

c. Develop your own procedure for determining whether a ship is in violation of the minimum weight restriction. Apply your rule to the data in Table 7.7. Draw a conclusion about the ship in question.

Key Terms

Starred (*) terms are from the optional section of this chapter.

*Chi-square (χ^2) distribution   396
Confidence coefficient   365
Confidence interval   364
Degrees of freedom   375

Interval estimate   362
Point estimate   362
Proportion   383
Sampling error   392

Standard error   363
Student's t distribution   375

Key Formulas

Starred (*) formulas are from the optional section of this chapter.

Large-sample confidence interval for means or proportions:

Point estimator $\pm$ ($z_{\alpha/2}$)(Standard error)   **(7.1)**, 368,   **(7.4)**, 385,   **(7.5)**, 387

Small-sample confidence interval for means or proportions:

Point estimator $\pm$ ($t_{\alpha/2}$)(Standard error)   **(7.2)**, 375

[*Note:* The respective point estimator and standard error for each parameter discussed in this chapter are provided in Table 7.8.]

Table 7.8 Summary of Estimation Procedures

Parameter	Point Estimate	Standard Error	Estimated Standard Error
μ	$\bar{x}$	$\sigma/\sqrt{n}$	$s/\sqrt{n}$
π	p	$\sqrt{\dfrac{\pi(1-\pi)}{n}}$	$\sqrt{\dfrac{p(1-p)}{n}}$

Confidence interval for variances:* $\dfrac{(n-1)s^2}{\chi^2_{\alpha/2}} \leq \sigma^2 \leq \dfrac{(n-1)s^2}{\chi^2_{(1-\alpha/2)}}$ **(7.9)**, 398

Key Symbols

Symbol	Definition
μ	Population Mean
π	Population proportion
σ^2	Population variance
$z_{\alpha/2}$	z value used in a $(1-\alpha)100\%$ large-sample confidence interval
$t_{\alpha/2}$	t value used in a $(1-\alpha)100\%$ small-sample confidence interval
$\chi^2_{\alpha/2}$ and $\chi^2_{(1-\alpha/2)}$	χ^2 values used in a $(1-\alpha)100\%$ confidence interval for a variance
$\bar{x}$	Sample mean (estimates μ)
p	Sample proportion (estimates π)
p^*	Wilson's adjusted sample proportion (estimates π)
s^2	Sample variance (estimates σ^2)
E	Distance to within what we wish to estimate a population parameter

Supplementary Problems 7.62–7.77

Starred (*) problems refer to the optional section of this chapter.

7.62 Operation Kidsafe is a nationwide safety program designed for children between the ages of 3 and 7. A study commissioned by the sponsor of Operation Kidsafe found that 340 of 500 children between the ages of 3 and 7 do not know their home phone number.

  a. Estimate π, the true percentage of children between the ages of 3 and 7 who do not know their home phone number. Use a 95% confidence interval.

  b. Give a precise interpretation of the phrase, "We are 95% confident that the interval encloses the true value of π."

  c. How would the width of the confidence interval in part **a** change if the confidence coefficient were increased from .95 to .99?

  d. How many children must be surveyed to estimate the true proportion who know their phone number to within .02 of its true value, with 95% confidence? Use the sample proportion calculated in part **a** as an estimate of π.

7.63 An experiment was conducted to estimate the mean time needed to transfer heat through sand (*Journal of Heat Transfer*, Aug. 1990). A large-sample 95% confidence interval for the mean time was found to be 20.0 ± 6.4 seconds.

  a. Give a practical interpretation of the 95% confidence interval.

  b. Give a theoretical interpretation of the 95% confidence interval.

7.64 Substance abuse problems are widespread at New Jersey businesses, according to the *Governor's Council for a Drug Free Workplace Report* (Spring/Summer 1995). A questionnaire on the issue was mailed to all New Jersey businesses that were members of the Governor's Council. Of the 72 companies that responded to the survey, 50 admitted that they had employees whose performance was affected by drugs or alcohol.

  a. Use a 95% confidence interval to estimate the proportion of all New Jersey companies with substance abuse problems.

  b. What assumptions are necessary to assure the validity of the confidence interval?

    c. Interpret the interval in the context of the problem.

    d. In interpreting the confidence interval, what does it mean to say you are "95% confident"?

    e. Would you use the interval of part **a** to estimate the proportion of all U.S. companies with substance abuse problems? Why or why not?

7.65 Twenty-four water samples were collected from Darts Lake, New York, and analyzed for concentrations of both lead and aluminum particulates (*Environmental Science & Technology,* Dec. 1985).

    a. The lead concentration measurements (in moles per liter) had a mean of 9.9 and a standard deviation of 8.4. Calculate a 99% confidence interval for the true mean lead concentration in water samples collected from Darts Lake. Interpret the interval.

    b. The aluminum concentration measurements (in moles per liter) had a mean of 6.7 and a standard deviation of 10.8. Form a 99% confidence interval for the true mean aluminum concentration in water samples collected from Darts Lake. Interpret the interval.

    *c. Refer to part **b**. Form a 90% confidence interval for the standard deviation of the aluminum concentration in water samples. Interpret the interval.

7.66 Obstructive sleep apnea is a sleep disorder that causes a person to stop breathing momentarily and then awaken briefly. These sleep interruptions, which may occur hundreds of times in a night, can drastically reduce the quality of rest and cause fatigue during waking hours. Researchers at Stanford University studied 159 commercial truck drivers and found that 124 of them suffered from obstructive sleep apnea (*Chest,* May 1995).

    a. Use the study results to estimate, with 90% confidence, the fraction of truck drivers who suffer from the sleep disorder.

    b. Sleep researchers believe that about 25% of the general population suffer from obstructive sleep apnea. Comment on whether or not this value represents the true percentage of truck drivers who suffer from the sleep disorder.

7.67 Research indicates that bicycle helmets save lives. A study reported in *Public Health Reports* (May–June 1992) was intended to identify ways of encouraging helmet use in children. One of the variables measured was the children's perception of the risk involved in bicycling. A four-point scale was used, with scores ranging from 1 (no risk) to 4 (very high risk). A sample of 797 children in grades 4–6 yielded the following results on the perception of risk variable: $\bar{x} = 3.39, s = .80$.

    a. Calculate a 90% confidence interval for the average perception of risk for all students in grades 4–6. What assumptions did you make to assure the validity of the confidence interval?

    b. If the population mean perception of risk exceeds 2.50, the researchers will conclude that students in these grades exhibit an awareness of the risk involved with bicycling. Interpret the confidence interval constructed in part **a** in this context.

ENDOW.XLS

7.68 Private and public colleges and universities rely on money contributed by individuals, corporations, and foundations for both salaries and operating expenses. Much of this money is put into a fund called an *endowment,* and the college spends only the interest earned by the fund. A random sample of eight college endowments drawn from the list of endowments in the *Chronicle of Higher Education Almanac* (Sept. 2, 1996) yielded the following endowments (in millions of dollars): 148.6, 66.1, 340.8, 500.2, 212.8, 55.4, 72.6, 83.4. Estimate the mean endowment for this population of colleges and universities using a 95% confidence interval. List any assumptions you make.

7.69 "Are 1990s undergraduates more willing to cheat in order to get good grades than those students in the 1970s?" This was the question posed to a national sample of 5,000 college professors by the Carnegie Foundation for the Advancement of Teaching (*Tampa Tribune,* Mar. 7, 1990). Forty-three percent of the professors responded "yes." Based on this survey, estimate the proportion of all college professors who feel undergraduate students in the 1990s are more willing to cheat to get good grades than undergraduate students in the 1970s. Use a confidence coefficient of .90.

7.70 The Beck Depression Inventory (BDI) is a widely used psychological test designed to measure depressive symptoms in humans. One study was conducted to assess the effect of a "life event" on the BDI scores of college undergraduates (*Journal of Human Stress,* Mar. 1983). In this experiment, the "life event" was "doing very poorly on an important exam." Thirty-three students enrolled in an introductory psychology class were identified as students who felt they performed poorly on a midterm exam. Following the exam, the students were administered the BDI. The post-exam BDI scores are summarized as follows (higher scores indicate higher levels of depression): $\bar{x} = 10.18, s = 5.26$. Estimate the mean post-exam BDI score of introductory psychology students using a 99% confidence interval.

SILICA.XLS

7.71 Researchers have experimented with generating electricity from the hot, highly saline water of the Salton Sea in southern California. Operating experience has shown that these brines leave silica scale deposits on metallic plant piping, causing excessive plant outages. In one experiment, an antiscalant was added to each of five aliquots of brine, and the solutions were filtered. A silica determination (parts per million of silicon dioxide) was made on each filtered sample after a holding time of 24 hours, with the following results (*Journal of Testing and Evaluation,* Mar. 1981): 229, 255, 280, 203, 229.

a. Estimate the mean amount of silicon dioxide present in the antiscalant solutions with a 99% confidence interval. Interpret the result.

*__b.__ Estimate the variance of the amount of silicon dioxide present in the antiscalant solutions with a 99% confidence interval. Interpret the result.

7.72 As businesses around the world rush to cash in on the popularity of the World Wide Web, questions have arisen as to what Web services users would be willing to pay for. In 1995, Georgia Institute of Technology's Graphics Visualization and Usability Center surveyed 13,000 Web users and asked them about their willingness to pay fees for access to Web sites. Of these, 2,938 were definitely not willing to pay such fees (*Inc. Technology,* No. 3, 1995).

a. Assume the 13,000 users were randomly selected. Construct a 95% confidence interval for the proportion definitely unwilling to pay fees.

b. What is the width of the interval you constructed in part **a**? For most applications, this width is unnecessarily narrow. What does that suggest about the survey's sample size?

c. How large a sample size is necessary to estimate the proportion of interest to within 2% with 95% confidence?

7.73 Accidental spillage and misguided disposal of petroleum wastes have resulted in extensive contamination of soils across the country. A common hazardous compound found in the contaminated soil is benzo(a)pyrene [B(a)p]. An experiment was conducted to determine the effectiveness of a method designed to remove B(a)p from soil (*Journal of Hazardous Materials,* June 1995). Three soil specimens contaminated with a known amount of B(a)p were treated with a toxin that inhibits microbial growth. After 95 days of incubation, the percentage of B(a)p removed from each soil specimen was measured. The experiment produced the following summary statistics: $\bar{x} = 49.3$ and $s = 1.5$.

a. Use a 99% confidence interval to estimate the mean percentage of B(a)p removed from a soil specimen in which the toxin was used.

b. Interpret the interval in terms of this application.

c. What assumption is necessary to ensure the validity of this confidence interval?

7.74 Refer to the *Human Factors* study on the use of color brightness as a body orientation clue, Problem 7.40 (page 390). How many subjects are required for a similar experiment to estimate the true proportion who use a bright color level as a cue to being right-side up to within .05 with 95% confidence? Use the sample proportion calculated in Problem 7.40 as an estimate of π.

LIQCO2.XLS

*7.75 Geologists use laser spectroscopy to measure the compositions of fluids present in crystallized rocks. An experiment was conducted to estimate the precision of this laser technique (*Applied Spectroscopy,* Feb. 1986). A chip of natural Brazilian quartz with an artificially produced fluid inclusion was subjected to laser spectroscopy. The amount of liquid carbon dioxide present in the inclusion was recorded on four different days. The data (in mole percentage) are: 86.6, 84.6, 85.5, 85.9.

*a. Obtain an estimate of the precision of the laser technique by constructing a 99% confidence interval for the variation in the carbon dioxide concentration measurements.

b. Estimate the mean amount of liquid carbon dioxide present in the inclusion with a 95% confidence interval.

IODINE.XLS

Run	Concentration
1	5.507
2	5.506
3	5.500
4	5.497
5	5.506
6	5.527
7	5.504
8	5.490
9	5.500
10	5.497

Problem 7.76

*7.76   An experiment was conducted to investigate the precision of measurements of a saturated solution of iodine. The data shown in the table at left represent $n = 10$ iodine concentration measurements on the same solution. The population variance σ^2 measures the variability—i.e., the precision—of a measurement. Use the data to find a 95% confidence interval for σ^2.

7.77   Most conventional opinion polls (e.g., Gallup, Roper, and CNN surveys) utilize a sample size of $n = 1,500$. The results of these polls are usually accompanied by the statement: "The estimated percentage is within $\pm 3\%$ of the true percentage." Consider estimating a population proportion π with a 95% confidence interval and sample size $n = 1,500$.

a.   Calculate σ_p, the standard error of the estimate, assuming (conservatively) that $\pi = .5$.

b.   Does your answer to part **a** agree with the $\pm 3\%$ margin of error stated in most polls?

Using Microsoft Excel

7.E.1 Using Microsoft Excel to Obtain the Confidence Interval Estimate for the Mean (σ known)

Solution Summary:

Use the PHStat add-in to calculate the confidence interval estimate for the mean when σ is known.

 Example: Attention time problem of Example 7.6

Solution:

Use the Confidence Intervals | Estimate for the Mean, sigma known, choice of the PHStat add-in to calculate the confidence interval estimate for the mean when σ is known. As an example, to calculate the confidence interval estimate for the mean attention time for the attention time problem of Example 7.6 on pages 368–369, do the following:

1. If the PHStat add-in has not been previously loaded, load the add-in using the instructions of Section EP.3.2.

2. Select File | New to open a new workbook (or open the existing workbook into which the confidence interval estimate worksheet is to be inserted).

3. Select PHStat | Confidence Intervals | Estimate for the Mean, sigma known.

4. In the Estimate for the Mean, sigma known, dialog box (see Figure 7.E.1):

    a. Enter 13.41 in the Population Standard Deviation: edit box.

    b. Enter 99 in the Confidence Level: edit box.

    c. Select the Sample Statistics Known option button and enter 50 in the Sample Size: edit box and 20.85 in the Sample Mean: edit box.

    d. Enter Confidence Interval Estimate for the Mean in the Output Title: edit box.

    e. Click the OK button.

E **Figure 7.E.1**
PHStat Estimate for the Mean, Sigma Known, Dialog Box

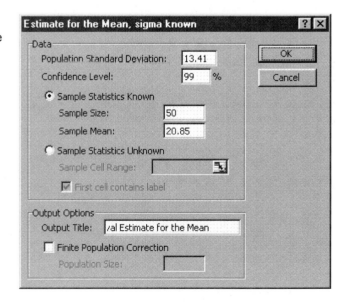

The add-in inserts a worksheet containing calculations for the confidence interval estimate for the mean similar to the one shown in Figure 7.E.2. We can change the values for the population standard deviation, sample mean, sample size, and confidence level in this worksheet to see their effects on the confidence interval estimate. (Note: For other problems in which the sample mean is not known and needs to be calculated, select the Sample Statistics Unknown option button in Step 4c and enter the cell range of the sample data in the Sample Cell Range: edit box, dimmed in Figure 7.E.1).

	A	B
1	Confidence Interval Estimate of the Mean	
2		
3	Population Standard Deviation	13.41
4	Sample Mean	20.85
5	Sample Size	50
6	Confidence Level	99%
7	Standard Error of the Mean	1.896460387
8	Z Value	-2.57583451
9	Interval Half Width	4.884968121
10	Interval Lower Limit	15.96503188
11	Interval Upper Limit	25.73496812

E Figure 7.E.2
Confidence Interval
Estimate for the Mean
for the Attention
Time Problem of
Example 7.6

7.E.2 Using Microsoft Excel to Obtain the Confidence Interval Estimate for the Mean (σ unknown)

Solution Summary:

Use the PHStat add-in to calculate the confidence interval estimate for the mean when σ is unknown.

 Example: Jail time data of Example 7.11

Solution:

Use the Confidence Intervals | Estimate for the Mean, sigma unknown, choice of the PHStat add-in to calculate a confidence interval estimate for the mean when σ is unknown. As an example, to calculate the confidence interval estimate for the mean days spent in jail for the data of Example 7.11 on page 378, do the following:

1.   If the PHStat add-in has not been previously loaded, load the add-in using the instructions of Section EP.3.2.

2.   Select File | New to open a new workbook (or open the existing workbook into which the confidence interval estimate worksheet is to be inserted).

3.   Select PHStat | Confidence Intervals | Estimate for the Mean, sigma unknown.

4.   In the Estimate for the Mean, sigma unknown, dialog box (see Figure 7.E.3):

     a.   Enter 90 in the Confidence Level: edit box.

     b.   Select the Sample Statistics Known option button and enter 14 in the Sample Size: edit box, 42.28 in the Sample Mean: edit box, and 62.14 in the Sample Standard Deviation: edit box.

     c.   Enter Confidence Interval Estimate of the Mean in the Output Title: edit box.

     d.   Click the OK button.

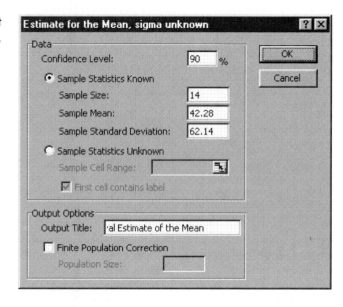

E **Figure 7.E.3** PHStat Estimate for the Mean, Sigma Unknown, Dialog Box

The add-in inserts a worksheet containing calculations for a confidence interval estimate for the mean similar to the one shown in Figure 7.E.4. We can change the values for the sample standard deviation, sample mean, sample size, and confidence level in this worksheet to see their effects on the confidence interval estimate. (Note: For other problems in which the sample mean and sample standard deviation are not known, select the Sample Statistics Unknown option button in Step 4b and enter the cell range of the sample data in the Sample Cell Range: edit box, dimmed in Figure 7.E.3).

E **Figure 7.E.4** Confidence Interval Estimate for the Mean Days Spent in Jail for the Data of Example 7.11

	A	B
1	**Confidence Interval Estimation for the Mean**	
2		
3	**Sample Standard Deviation**	62.14
4	**Sample Mean**	42.28
5	**Sample Size**	14
6	**Confidence Level**	90%
7	Standard Error of the Mean	16.60761357
8	Degrees of Freedom	13
9	*t* Value	1.770931704
10	Interval Half Width	29.41094939
11	**Interval Lower Limit**	12.86905061
12	**Interval Upper Limit**	71.69094939

7.E.3 Using Microsoft Excel to Obtain the Confidence Interval Estimate for the Proportion

Solution Summary:

Use the PHStat add-in to calculate the confidence interval estimate for the proportion.

 Crime analysis problem of Example 7.13

Solution:

Use the Confidence Intervals | Estimate for the Proportion choice of the PHStat add-in to calculate a confidence interval estimate for the proportion. As an example, to calculate the confidence interval estimate for the proportion of crimes in which a firearm was used (Example 7.13 on page 383), do the following:

1   If the PHStat add-in has not been previously loaded, load the add-in using the instructions of Section EP.3.2.

2   Select File | New to open a new workbook (or open the existing workbook into which the confidence interval estimate worksheet is to be inserted).

3   Select PHStat | Confidence Intervals | Estimate for the Proportion.

4   In the Estimate for the Proportion dialog box (see Figure 7.E.5):

     a   Enter 300 in the Sample Size: edit box.

     b   Enter 180 in the Number of Successes: edit box.

     c   Enter 95 in the Confidence Level: edit box.

     d   Enter Confidence Interval Estimate for the Proportion in the Output Title: edit box.

     e   Click the OK button.

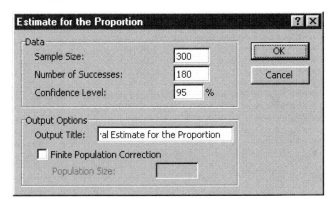

E **Figure 7.E.5**
PHStat Estimate for the Proportion Dialog Box

The add-in inserts a worksheet containing calculations for a confidence interval estimate for the proportion similar to the one shown in Figure 7.E.6. We can change the values for the sample size, number of successes, and confidence level in this worksheet to see their effects on the confidence interval estimate.

	A	B
1	Confidence Interval Estimation for the Proportion	
2		
3	Sample Size	300
4	Number of Successes	180
5	Confidence Level	95%
6	Sample Proportion	0.6
7	Z Value	-1.959961082
8	Standard Error of the Proportion	0.028284271
9	Interval Half Width	0.055436071
10	Interval Lower Limit	0.544563929
11	Interval Upper Limit	0.655436071

E **Figure 7.E.6**
Confidence Interval Estimate for the Proportion of Crimes in Which a Firearm Was Used for the Data of Example 7.13

7.E.4   Using Microsoft Excel to Determine the Sample Size for Estimating the Mean

Solution Summary:

Use the PHStat add-in to determine the sample size needed for estimating the mean.

Example: Order processing problem of Example 7.17

Solution:

Use the Sample Size | Determination for the Mean choice of the PHStat add-in to determine the sample size needed for estimating the mean. As an example, to determine the sample size needed for estimating the mean shipping time for the order processing problem of Example 7.17 on page 391, do the following:

1 If the PHStat add-in has not been previously loaded, load the add-in using the instructions of Section EP.3.2.

2 Select File | New to open a new workbook (or open the existing workbook into which the sample size determination worksheet is to be inserted).

3 Select PHStat | Sample Size | Determination for the Mean.

4 In the Determination for the Mean dialog box (see Figure 7.E.7):

    a Enter 3.5 in the Population Standard Deviation: edit box.

    b Enter 1 in the Sampling Error: edit box.

    c Enter 95 in the Confidence Level: edit box.

    d Enter Sample Size for Estimating the Mean in the Output Title: edit box.

    e Click the OK button.

E **Figure 7.E.7** PHStat
Determination for the
Mean Dialog Box

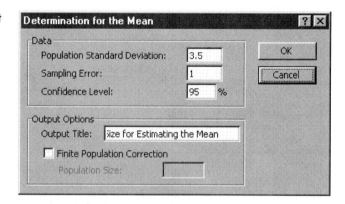

The add-in inserts a worksheet containing calculations for determining the sample size needed for estimating the mean similar to the one shown in Figure 7.E.8. We can change the values for the population standard deviation, sampling error, and confidence level in this worksheet to see their effects on the sample size needed.

E **Figure 7.E.8**
Sample Size Determination for Estimating
the Mean Shipping
Time for the Order
Processing Problem of
Example 7.17

	A	B
1	**Sample Size for Estimating the Mean**	
2		
3	**Population Standard Deviation**	3.5
4	**Sampling Error**	1
5	**Confidence Level**	95%
6	Z Value	-1.95996108
7	Calculated Sample Size	47.05773119
8	**Sample Size Needed**	48

7.E.5 Using Microsoft Excel to Determine the Sample Size for Estimating the Proportion

Solution Summary:

Use the PHStat add-in to determine the sample size needed for estimating the proportion.

 Example: Crime analysis problem of Example 7.18

Solution:

Use the Sample Size | Determination for the Proportion choice of the PHStat add-in to determine the sample size needed for estimating the proportion. As an example, to determine the sample size needed for estimating the proportion of crimes in which a firearm was used (Example 7.18 on page 393), do the following:

1. If the PHStat add-in has not been previously loaded, load the add-in using the instructions of Section EP.3.2.

2. Select File | New to open a new workbook (or open the existing workbook into which the sample size determination worksheet is to be inserted).

3. Select PHStat | Sample Size | Determination for the Proportion.

4. In the Sample Size Determination for the Proportion dialog box (see Figure 7.E.9):

   a. Enter 0.6 in the Proportion Estimation: edit box.

   b. Enter .02 in the Sampling Error: edit box.

   c. Enter 90 in the Confidence Level: edit box.

   d. Enter Sample Size for Proportion in the Output Title: edit box.

   e. Click the OK button.

Sample Size Determination for the Proportion ? X

Data
Proportion Estimation: `0.6`

Sampling Error: `.02`

Confidence Level: `90` %

Output Options

Output Title: `Sample Size for Proportion`

☐ Finite Population Correction

Population size:

OK

Cancel

E Figure 7.E.9
PHStat Sample Size Determination for the Proportion Dialog Box

The add-in inserts a worksheet containing calculations for determining the sample size needed for estimating the proportion similar to the one shown in Figure 7.E.10. We can change the values for the estimate of the true proportion, sampling error, and confidence level in this worksheet inserted by the add-in to see their effects on the sample size needed.

E **Figure 7.E.10**
Sample Size
Determination for
Estimating the
Proportion of Crimes
in Which a Firearm
Was Used for the Data
of Example 7.18

	A	B
1	**Sample Size for Estimating the Proportion**	
2		
3	**Estimate of True Proportion**	**0.6**
4	**Sampling Error**	**0.02**
5	**Confidence Level**	**90%**
6	Z Value	-1.644853
7	Calculated Sample Size	1623.324836
8	**Sample Size Needed**	**1624**

7.E.4 Using Microsoft Excel to Obtain the Confidence Interval Estimate for the Population Variance

Solution Summary:

Implement a worksheet that uses the CHIINV worksheet function to obtain the confidence interval estimate for the population variance.

 Example: Contaminated fish problem of Example 7.21

Solution:

To calculate the confidence interval estimate of the population variance for the contaminated fish problem data of Example 7.21 on pages 397–398, we can implement a worksheet using the design of Table 7.E.1. This design includes the CHIINV worksheet function, the format of which is

$$\text{CHIINV}(\textit{level of significance, degrees of freedom})$$

to return the critical value of χ^2 needed to compute the confidence interval. To implement the Table 7.E.1 design, do the following:

1. Select File | New to open a new workbook (or open the existing workbook into which the confidence interval estimate worksheet is to be inserted).

2. Rename the new worksheet Confidence.

3. Enter the title and labels for column A as shown in Table 7.E.1.

4. Enter the sample standard deviation, sample size, and confidence level in the cell range B3:B5. Enter 376.6 in cell B3, 144 in cell B4, and 0.95 in cell B5.

5. Select cell B5 and click the Percent style button on the formatting toolbar (see Section EP.2.9) to format the decimal value 0.95 as 95%.

6. Enter the formulas for the cell range B6:B13 as shown in Table 7.E.1.

Table 7.E.1 Confidence Interval Estimate Sheet Design for the Population Variance for Contaminated Fish Problem of Example 7.21

	A	B
1	Confidence Interval Estimation for the Population Variance	
2		
3	Sample Standard Deviation	xxx
4	Sample Size	xxx
5	Confidence Level	xxx
6	Degrees of Freedom	=B4−1
7	Sum of Squares	=B6*B3^2
8	Lower Chi-Square Value	=CHIINV((1−(1−B5)/2),B6)
9	Upper Chi-Square Value	=CHIINV((1−B5)/2,B6)
10	Interval Lower Limit for Variance	=B7/B9
11	Interval Upper Limit for Variance	=B7/B8
12	Interval Lower Limit for Standard Deviation	=SQRT(B10)
13	Interval Upper Limit for Standard Deviation	=SQRT(B11)

The completed worksheet will be similar to the one shown in Figure 7.E.11. Had the sample standard deviation been unknown, we could have entered a formula using the STDEV function in cell B3 to compute it from the cell range containing the data.

	A	B
1	Confidence Interval Estimation for the Population Variance	
2		
3	Sample Standard Deviation	376.6
4	Sample Size	144
5	Confidence Level	95%
6	Degrees of Freedom	143
7	Sum of Squares	20281341.08
8	Lower Chi-Square Value	111.7868951
9	Upper Chi-Square Value	177.9978476
10	Interval Lower Limit for Variance	113941.4962
11	Interval Upper Limit for Variance	181428.6108
12	Interval Lower Limit for Standard Deviation	337.5522126
13	Interval Upper Limit for Standard Deviation	425.9443752

E Figure 7.E.11
Confidence Interval Estimate for the Population Variance for the Contaminated Fish Problem of Example 7.21

Chapter 8

Testing Hypotheses about Population Parameters: One Sample

CONTENTS

8.1 The Relationship between Hypothesis Tests and Confidence Intervals

8.2 Hypothesis-Testing Methodology: Formulating Hypotheses

8.3 Hypothesis-Testing Methodology: Test Statistics and Rejection Regions

8.4 Guidelines for Determining the Target Parameter

8.5 Testing a Population Mean

8.6 Reporting Test Results: p-Values

8.7 Testing a Population Proportion

8.8 Testing a Population Variance (Optional)

8.9 Potential Hypothesis-Testing Pitfalls and Ethical Issues

EXCEL TUTORIAL

8.E.1 Using Microsoft Excel to Perform the Z test of the Hypothesis for the Mean (σ known)

8.E.2 Using Microsoft Excel to Perform the t Test of the Hypothesis for the Mean (σ unknown)

8.E.3 Using Microsoft Excel to Perform the Z test of the Hypothesis for the Proportion

8.E.4 Using Microsoft Excel to Perform the Chi-Square Test of the Hypothesis for the Variance

REAL-WORLD APPLICATION
Cellular Telephones—Road Hazards?

Many people talk on cellular telephones while driving their car. Is this practice hazardous? That is, are people who use cell phones while driving more likely to have an automobile accident? To answer this question, researchers at the University of Toronto collected data on cell phone usage for a sample of drivers who had been in a collision. In this chapter we will learn how sample data can be used to make decisions about population parameters. This methodology will be applied to the cell phone accident data in Real-World Revisited at the end of the chapter.

8.1 THE RELATIONSHIP BETWEEN HYPOTHESIS TESTS AND CONFIDENCE INTERVALS

There are two general methods available for making inferences about population parameters: We can estimate their values using confidence intervals (the subject of Chapter 7) or we can make decisions about them. Making decisions about specific values of the population parameters—*testing hypotheses* about these values—is the topic of this chapter.

It is important to know that confidence intervals and hypothesis tests are related and that either can be used to make decisions about parameters.

Example 8.1

Inference about μ from a Confidence Interval

Suppose an investigator for the Environmental Protection Agency (EPA) wants to determine whether the mean level μ of a certain type of pollutant released into the atmosphere by a chemical company meets the EPA guidelines. If 3 parts per million is the upper limit allowed by the EPA, the investigator would want to use sample data (daily pollution measurements) to decide whether the company is violating the law, i.e., to decide whether $\mu > 3$. Suppose a 99% confidence interval for μ is (3.2, 4.1). What should the EPA conclude?

Solution

The EPA is 99% confident that the mean level μ of the pollutant is between 3.2 ppm and 4.1 ppm. Since all the numbers in the interval exceed 3, the EPA is confident that μ exceeds the established limit. ❏

Example 8.2

Inference about π from a Confidence Interval

Consider a software manufacturer who purchases 3.5" computer disks in lots of 10,000. Suppose that the supplier of the disks guarantees that no more than 1% of the disks in any given lot are defective. Since the manufacturer cannot test each of the 10,000 disks in a lot, he or she must decide whether to accept or reject a lot based on an examination of a sample of disks selected from the lot. If the number x of defective disks in a sample of, say, $n = 100$ is large, the lot will be rejected and sent back to the supplier. Thus, the manufacturer wants to decide whether the proportion π of defectives in the lot exceeds .01, based on information contained in a sample. Suppose a 95% confidence interval for π is (.005, .008). What decision should the software manufacturer make?

Solution

The manufacturer is 95% confident that the proportion of defectives falls between .005 and .008. Consequently, the manufacturer will accept the lot and be confident that the proportion of defectives is less than 1%. ❏

The preceding examples illustrate how a confidence interval can be used to make a decision about a parameter. Note that both applications are one-directional: In Example 8.1 the EPA wants to determine whether $\mu > 3$; and in

Example 8.2 the manufacturer wants to know if $\pi > .01$. (In contrast, if the manufacturer is interested in determining whether $\pi > .01$ or $\pi < .01$, the inference would be two-directional.)

Recall from Chapter 7 that to find the value of z (or t) used in a $(1 - \alpha)100\%$ confidence interval, the value of α is divided in half and $\alpha/2$ is placed in both the upper and lower tails of the z (or t) distribution. Consequently, confidence intervals are designed to be two-directional. Use of a two-directional technique in a situation where a one-directional method is desired will lead the researcher (e.g., the EPA or the software manufacturer) to understate the level of confidence associated with the method. As we will explain in this chapter, hypothesis tests are appropriate for either one- or two-directional decisions about a population parameter.

The general concepts involved in hypothesis testing are outlined in the next two sections.

8.2 HYPOTHESIS-TESTING METHODOLOGY: FORMULATING HYPOTHESES

Every statistical test of hypothesis consists of five key elements: (1) null hypothesis, (2) alternative hypothesis, (3) test statistic, (4) rejection region, and (5) conclusion. In this section we focus on formulating the null and alternative hypotheses and the consequences of our decision.

Null and Alternative Hypotheses

When a researcher in any field sets out to test a new theory, he or she first formulates a hypothesis, or claim, that he or she believes to be true. In statistical terms, the hypothesis that the researcher tries to establish is called the *alternative hypothesis,* or *research hypothesis.* To be paired with the alternative hypothesis is the *null hypothesis,* which is the "opposite" of the alternative hypothesis. In this way, the null and alternative hypotheses, both stated in terms of the appropriate population parameters, describe two possible states of nature that are mutually exclusive (i.e., they cannot simultaneously be true). When the researcher begins to collect information about the phenomenon of interest, he or she generally tries to present evidence that lends support to the alternative hypothesis. As you will subsequently learn, we take an indirect approach to obtaining support for the alternative hypothesis: Instead of trying to show that the alternative hypothesis is true, we attempt to produce evidence to show that the null hypothesis (which may often be interpreted as "no change from the status quo") is false.

> **Definition 8.1**
>
> A **statistical hypothesis** is a statement about the numerical value of a population parameter.

> **Definition 8.2**
>
> The **alternative** (or research) **hypothesis**, denoted by H_a, is usually the hypothesis for which the researcher wants to gather supporting evidence.

> **Definition 8.3**
>
> The **null hypothesis**, denoted H_0, is usually the hypothesis that the researcher wants to gather evidence against (i.e., the hypothesis to be "tested").

Example 8.3   Choosing H_0 and H_a

A metal lathe is checked periodically by quality control inspectors to determine whether it is producing machine bearings with a mean diameter of 1 centimeter (cm). If the mean diameter of the bearings is larger or smaller than 1 cm, then the process is out of control and needs to be adjusted. Formulate the null and alternative hypotheses that could be used to test whether the bearing production process is out of control.

Solution

The hypotheses must be stated in terms of a population parameter. Thus, we define

μ = True mean diameter (in centimeters) of all bearings produced by the lathe

If either $\mu > 1$ or $\mu < 1$, then the metal lathe's production process is out of control. Since we wish to be able to detect either possibility, the null and alternative hypotheses are

$$H_0: \mu = 1 \text{ (the process is in control)}$$

$$H_a: \mu \neq 1 \text{ (the process is out of control)} \qquad \square$$

Example 8.4   Choosing H_0 and H_a

Printed cigarette advertisements are required by law to carry the following statement: "Warning: The surgeon general has determined that cigarette smoking is dangerous to your health." However, this warning is often located in inconspicuous corners of the advertisements and printed in small type. Consequently, a spokesperson for the Federal Trade Commission (FTC) believes that over 80% of those who read cigarette advertisements fail to see the warning. Specify the null and alternative hypotheses that would be used in testing the spokesperson's theory.

Solution

The FTC spokesperson wants to make an inference about π, the true proportion of all readers of cigarette advertisements who fail to see the surgeon general's warning. In particular, the FTC spokesperson wishes to collect evidence to support the claim that π is greater than .80; thus, the null and alternative hypotheses are

$$H_0: \pi = .80$$

$$H_a: \pi > .80 \qquad \square$$

Why an Equals Sign in H_O?

An accepted convention in hypothesis testing is to always write H_0 with an equality sign (=). In Example 8.4, we could have formulated the null hypothesis as $H_0: \pi \le .80$ to cover all situations for which $H_a: \pi > .80$ *does not* occur. However, any evidence that would cause you to reject the null hypothesis $H_0: \pi = .80$ in favor of $H_a: \pi > .80$ would also cause you to reject $H_0: \pi \le .80$. In other words, $H_0: \pi = .80$ represents the worst possible case, from the researcher's point of view, if the alternative hypothesis is *not* correct. Thus, for mathematical ease, we combine all possible situations for describing the opposite of H_a into one statement involving an equality.

An alternative hypothesis may hypothesize a change from H_0 in a particular direction, or it may merely hypothesize a change without specifying a direction. In Example 8.4, the researcher is interested in detecting a departure from H_0 in a particular direction; interest focuses on whether the proportion of cigarette advertisement readers who fail to see the surgeon general's warning is *greater than* .80. This test is called a *one-tailed* (or *one-sided*) *test*. In contrast, Example 8.3 illustrates a *two-tailed* (or *two-sided*) *test* in which we are interested in whether the mean diameter of the machine bearings differs in either direction from 1 cm., i.e., whether the process is out of control.

Definition 8.4

A **one-tailed test** of hypothesis is one in which the alternative hypothesis is directional, and includes either the "<" symbol or the ">" symbol.

Definition 8.5

A **two-tailed test** of hypothesis is one in which the alternative hypothesis does not specify a departure from H_0 in a particular direction; such an alternative will be written with the "≠" symbol.

Self-Test

8.1

A recent Food and Drug Administration (FDA) report states that 65% of all domestically produced food products are pesticide-free. Suppose an FDA spokesperson believes that less than 65% of all foreign produced food products are pesticide-free. Specify the null and alternative hypotheses that would be used to test the spokesperson's belief.

Type I and Type II Errors

The goal of any hypothesis test is to make a decision; in particular, we will decide whether to reject the null hypothesis H_0 in favor of the alternative hypothesis H_a. Although we would like to be able to always make a correct decision, we must remember that the decision will be based on sample information. Thus we can make one of two types of errors, as shown in Table 8.1.

If we reject H_0 when H_0 is true, we make a *Type I error*. The probability of a Type I error is represented by the Greek letter α. If we accept H_0 when H_0 is false, we make a *Type II error*. The probability of a Type II error is represented by the

Table 8.1 Conclusions and Consequences for Testing a Hypothesis

		TRUE STATE OF NATURE	
		H_0 true (H_a false)	H_0 false (H_a true)
Decision	Accept H_0	Correct decision	Type II error
	Reject H_0	Type I error	Correct decision

Greek letter β. There is an intuitively appealing relationship between the probabilities for the two types of error: *As α increases, β decreases; similarly, as β increases, α decreases. The only way to reduce α and β simultaneously is to increase the amount of information available in the sample, i.e., to increase the sample size.*

Definition 8.6

A **Type I error** occurs if we reject a null hypothesis when it is true. The probability of committing a Type I error is denoted by α.

$$\alpha = P(\text{Type I error}) = P(\text{Reject } H_0 \text{ when } H_0 \text{ is true})$$

Definition 8.7

A **Type II error** occurs if we accept a null hypothesis when it is false. The probability of making a Type II error is denoted by β.

$$\beta = P(\text{Type II error}) = P(\text{Accept } H_0 \text{ when } H_0 \text{ is false})$$

Example 8.5

Type I and Type II Errors

Refer to Example 8.3. The hypotheses to be tested are:

$$H_0: \mu = 1 \text{ (process is in control)}$$

$$H_a: \mu \neq 1 \text{ (process is out of control)}$$

Specify the Type I and Type II errors for the problem.

Solution

From Definition 8.6, a Type I error is that of incorrectly rejecting H_0. In our example, this would occur if we conclude that the process is out of control when, in fact, the process is in control, i.e., if we conclude that the mean bearing diameter is different from 1 cm, when the mean is equal to 1 cm. The consequence of making such an error would be that unnecessary time and effort would be expended to repair a metal lathe that is operating properly.

From Definition 8.7, a Type II error results from incorrectly accepting H_0. This occurs if we conclude that the mean bearing diameter is equal to 1 cm when, in fact, the mean differs from 1 cm. The practical significance of making a Type II error is that the metal lathe would not be repaired when, in fact, the process is out of control. ❑

Subsequently, we will see that the probability of making a Type I error is controlled by the researcher (Section 8.3); thus, it is often used as a measure of the reliability of the conclusion and is called the *significance level* of the test.

Definition 8.8

The probability, α, of making a Type I error is called the **level of significance** (or **significance level**) for a hypothesis test.

In practice, we will carefully avoid stating a decision in terms of "accept the null hypothesis H_0." Instead, if the sample does not provide enough evidence to support the alternative hypothesis H_a, we prefer to state a decision as "fail to reject H_0," or "insufficient evidence to reject H_0." This is because, if we were to "accept H_0," the reliability of the conclusion would be measured by β, the probability of a Type II error. Unfortunately, the value of β is not constant, but depends on the specific alternative value of the parameter and is difficult to compute in most testing situations.

In summary, we recommend the following procedure for formulating hypotheses and stating conclusions.

Formulating Hypotheses and Stating Conclusions

1. State the hypothesis you want to support as the alternative hypothesis H_a.

2. The null hypothesis H_0 will be the opposite of H_a and will contain an equality sign.

3. If the sample evidence supports the alternative hypothesis, you will reject the null hypothesis and will know that the probability of having made an incorrect decision (when H_0 is true) is α, a quantity that you select (prior to collecting the sample) to be as small as you wish.

4. If the sample does not provide sufficient evidence to support the alternative hypothesis, then you conclude that the null hypothesis cannot be rejected on the basis of your sample. In this situation, you may wish to obtain a larger sample to collect more information about the phenomenon under study.

Example 8.6   Hypothesis-Testing Logic

The logic used in hypothesis testing has often been likened to that used in the courtroom in which a defendant is on trial for committing a crime. Assume that the judge has issued the standard instruction to the jury: The defendant should be acquitted unless evidence of guilt is beyond a "reasonable doubt."

a. Formulate appropriate null and alternative hypotheses for judging the guilt or innocence of the defendant.

b. Interpret the Type I and Type II errors in this context.

c. If you were the defendant, would you want α to be small or large? Explain.

Solution

a. Under the American judicial system, a defendant is "innocent until proven guilty." That is, the burden of proof is *not* on the defendant to prove his or her innocence; rather, the court must collect sufficient evidence to sup-

Table 8.2 Conclusions and Consequences in Example 8.6

		TRUE STATE OF NATURE	
		Defendant Is Innocent	**Defendant Is Guilty**
Decision of Court	Defendant Is Innocent	Correct decision	Type II error
	Defendant Is Guilty	Type I error	Correct decision

port the claim that the defendant is guilty "beyond a reasonable doubt." Thus, the null and alternative hypotheses are

$$H_0: \text{Defendant is innocent}$$

$$H_a: \text{Defendant is guilty}$$

b. The four possible outcomes are shown in Table 8.2. A Type I error would occur if the court concludes that the defendant is guilty when, in fact, he or she is innocent; a Type II error would occur if the court concludes that the defendant is innocent when he or she is guilty.

c. Most would probably agree that the Type I error in this situation is far the more serious, especially the defendant. The defendant wants β, the probability of committing a Type I error, to be very small so that he or she has little or no chance of erroneously being found guilty. ❑

In the next section we provide details on how to use the sample information to decide whether or not to reject the null hypothesis.

Self-Test 8.2

Last month, a large supermarket chain received many consumer complaints about the quantity of chips in 16-ounce bags of a particular brand of potato chips. The chain decided to test the following hypotheses concerning μ, the mean weight (in ounces) of a bag of potato chips in the next shipment of chips received from their largest supplier:

$$H_0: \mu = 16$$

$$H_a: \mu < 16$$

If there is evidence that $\mu < 16$, then the shipment would be refused and a complaint registered with the supplier.

a. What is a Type I error, in terms of the problem?

b. What is a Type II error, in terms of the problem?

c. Which type of error would the chain's customers view as more serious?

d. Which type of error would the chain's supplier view as more serious?

Using the Tools

8.1 Explain the difference between an alternative hypothesis and a null hypothesis.

8.2 Explain why each of the following statements is incorrect:

a. The probability that the null hypothesis is correct is equal to α.

b. If the null hypothesis is rejected, then the test proves that the alternative hypothesis is correct.

c. In all statistical tests of hypothesis, $\alpha + \beta = 1$.

8.3 Why do we avoid stating a decision in terms of "accept the null hypothesis H_0"?

Applying the Concepts

8.4 According to *Chance* (Fall 1994), the average number of faxes transmitted in the United States each minute is 88,000. Set up the null and alternative hypotheses for testing this claim.

8.5 According to *Harpers* (Apr.–Aug. 1994), 52% of all Americans would rather spend a week in jail than be president of the United States. Set up the null and alternative hypotheses for testing this assertion.

8.6 A herpetologist wants to determine whether the egg hatching rate for a certain species of frog exceeds .5 when the eggs are exposed to ultraviolet radiation.

a. Formulate the appropriate null and alternative hypotheses.

b. Describe a Type I error. **c.** Describe a Type II error.

8.7 A manufacturer of fishing line wants to show that the mean breaking strength of a competitor's 22-pound line is really less than 22 pounds.

a. Formulate the appropriate null and alternative hypotheses.

b. Describe a Type I error. **c.** Describe a Type II error.

8.8 An individual who has experienced a long run of bad luck playing the game of craps at a casino wants to test whether the casino dice are "loaded," i.e., whether the proportion of "sevens" occurring in many tosses of the two dice is different from $\frac{1}{6}$ (if the dice are fair, the probability of tossing a "seven" is $\frac{1}{6}$).

a. Formulate the appropriate null and alternative hypotheses.

b. Describe a Type I error. **c.** Describe a Type II error.

8.9 The Environmental Protection Agency wishes to test whether the mean amount of radium-226 in soil in a Florida county exceeds the maximum allowable amount, 4 pCi/L.

a. Formulate the appropriate null and alternative hypotheses.

b. Describe a Type I error. **c.** Describe a Type II error.

8.10   Testing the thousands of compounds of a new drug to find the few that might be effective is known in the pharmaceutical industry as *drug screening*. In its preliminary stage, drug screening can be viewed in terms of a statistical decision problem. Two actions are possible: (1) to "reject" the drug, meaning to conclude that the tested drug has little or no effect, in which case it will be set aside and a new drug selected for screening; and (2) to "accept" the drug provisionally, in which case it will be subjected to further, more refined experimentation. Since it is the goal of the researcher to find a drug that effects a cure, the null and alternative hypotheses in a statistical test would take the following form:

H_0: Drug is ineffective in treating a particular disease

H_a: Drug is effective in treating a particular disease

a.   To abandon a drug when in fact it is a useful one is called a *false negative*. A false negative corresponds to which type of error, Type I or Type II?

b.   To proceed with more expensive testing of a drug that is in fact useless is called a *false positive*. A false positive corresponds to which type of error, Type I or Type II?

c.   Which of the two errors is more serious? Explain.

8.3 HYPOTHESIS-TESTING METHODOLOGY: TEST STATISTICS AND REJECTION REGIONS

Once we have formulated the null and alternative hypotheses and selected the significance level α, we are ready to carry out the test. For example, suppose we want to test the null hypothesis

$$H_0: \mu = \mu_0$$

where μ_0 is some fixed value of the population mean. The next step is to obtain a random sample from the population of interest. The information provided by this sample, in the form of a sample statistic, will help us decide whether to reject the null hypothesis. The sample statistic upon which we base our decision is called the *test statistic*. For this example, we are hypothesizing about the value of the population mean μ. Since our best guess about the value of μ is the sample mean $\bar{x}$ (see Section 7.2), it seems reasonable to use $\bar{x}$ as a test statistic.

In general, when the hypothesis test involves a specific population parameter, the test statistic to be used is the conventional point estimate of that parameter in standardized form. For example, standardizing $\bar{x}$ (i.e., converting $\bar{x}$ to its z score) leads to the test statistic:

$$z = \frac{\bar{x} - \mu_0}{\sigma_{\bar{x}}} = \frac{\bar{x} - \mu_0}{\sigma/\sqrt{n}} \tag{8.1}$$

Definition 8.9

The **test statistic** is a sample statistic, computed from the information provided by the sample, upon which the decision concerning the null and alternative hypotheses is based.

Now we need to specify the range of possible computed values of the test statistic for which the null hypothesis will be rejected. That is, what specific values of the test statistic will lead us to reject the null hypothesis in favor of the alternative hypothesis? These specific values are known collectively as the *rejection region* for the test. We learn (in the next example) that the rejection region depends on the value of α selected by the researcher.

Definition 8.10

The **rejection region** is the set of possible computed values of the test statistic for which the null hypothesis will be rejected.

To complete the test, we make our decision by observing whether the computed value of the test statistic lies within the rejection region. If the computed value falls within the rejection region, we will reject the null hypothesis; otherwise, we do not reject the null hypothesis.

A summary of the **hypothesis-testing procedure** we have developed is given in the box.

OUTLINE FOR TESTING A HYPOTHESIS

Step 1   Specify the null and alternative hypotheses, H_0 and H_a, and the significance level, α.

Step 2   Obtain a random sample from the population(s) of interest.

Step 3   Determine an appropriate test statistic and compute its value using the sample data.

Step 4   Specify the rejection region. (This will depend on the value of α selected.)

Step 5   Make the appropriate conclusion by observing whether the computed value of the test statistic lies within the rejection region. If so, reject the null hypothesis; otherwise, do not reject the null hypothesis.

Example 8.7   Rejection Region for an Upper-Tailed Test

Specify the form of the rejection region for a test of

$$H_0: \mu = 72$$

$$H_a: \mu > 72$$

at a significance level of $\alpha = .05$.

Solution

The hypothesized value of μ in H_0 is 72. Substituting this value in equation 8.1, we obtain the test statistic

$$z = \frac{\bar{x} - \mu_0}{\sigma_{\bar{x}}} = \frac{\bar{x} - 72}{\sigma/\sqrt{n}}$$

Figure 8.1 Location of Rejection Region for Example 8.7

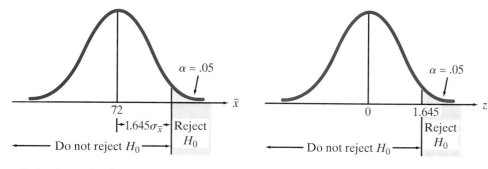

a. Rejection region in terms of $\bar{x}$

b. Rejection region in terms of z

This z score gives us a measure of how many standard deviations the observed $\bar{x}$ is from what we would expect to observe if H_0 were true. If z is "large," i.e., if $\bar{x}$ is "sufficiently greater" than 72, we will reject H_0.

Now examine Figure 8.1a and observe that the chance of obtaining a value of $\bar{x}$ more than 1.645 standard deviations above 72 is only .05, *when the true value of μ is* 72. (We are assuming that the sample size is large enough to ensure that the sampling distribution of $\bar{x}$ is approximately normal.)

Thus, if we observe a sample mean located more than 1.645 standard deviations above 72, then either H_0 is true and a relatively rare (with probability .05 or less) event has occurred, or H_a is true and the population mean exceeds 72. We would favor the latter explanation for obtaining such a large value of $\bar{x}$, and would then reject H_0.

In summary, our rejection region consists of all values of z that are greater than 1.645 (i.e., all values of $\bar{x}$ that are more than 1.645 standard deviations above 72). The *critical value* 1.645 is shown in Figure 8.1b. In this situation, the probability of a Type I error—that is, deciding in favor of H_a if in fact H_0 is true—is equal to our selected significance level, $\alpha = .05$. ❑

Definition 8.11

In specifying the rejection region for a particular test of hypothesis, the value at the boundary of the rejection region is called the **critical value**.

Example 8.8   Rejection Region for a Lower-Tailed Test

Specify the form of the rejection region for a test of

$$H_0: \mu = 72$$
$$H_a: \mu < 72$$

at significance level $\alpha = .01$.

Solution

Here, we want to be able to detect the directional alternative that μ is *less than* 72; in this case, it is "sufficiently small" values of the test statistic $\bar{x}$ that would cast doubt on the null hypothesis. As in Example 8.7, we use equation 8.1 to standardize

Figure 8.2 Location of Rejection Region for Example 8.8

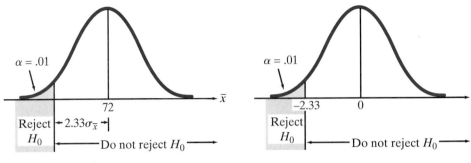

a. Rejection region in terms of $\bar{x}$ b. Rejection region in terms of $\bar{z}$

the value of the test statistic and obtain a measure of the distance between $\bar{x}$ and the null hypothesized value of 72:

$$z = \frac{\bar{x} - \mu_{\bar{x}}}{\sigma_{\bar{x}}} = \frac{\bar{x} - 72}{\sigma/\sqrt{n}}$$

This z value tells us how many standard deviations the observed $\bar{x}$ is from what would be expected *if H_0 were true.* (Again, we have assumed that the sample size n is large so that the sampling distribution of $\bar{x}$ will be approximately normal. The appropriate modifications for small samples will be discussed in Section 8.5.)

Figure 8.2a shows us that *when the true value of μ is 72,* the chance of observing a value of $\bar{x}$ more than 2.33 standard deviations below 72 is only .01. Thus, at a significance level (probability of Type I error) of $\alpha = .01$, we would reject the null hypothesis for all values of z that are less than -2.33 (see Figure 8.2b), i.e., for all values of $\bar{x}$ that lie more than 2.33 standard deviations below 72. ❑

Example 8.9   Rejection Region for Two-Tailed Test

Specify the form of the rejection region for a test of

$$H_0: \mu = 72$$

$$H_a: \mu \neq 72$$

where we are willing to tolerate a .05 chance of making a Type I error.

Solution

For this two-sided (nondirectional) alternative, $\alpha = .05$. We will reject the null hypothesis for "sufficiently small" or "sufficiently large" values of the standardized test statistic

$$z = \frac{\bar{x} - \mu_{\bar{x}}}{\sigma_{\bar{x}}} = \frac{\bar{x} - 72}{\sigma/\sqrt{n}}$$

Now, from Figure 8.3a (page 429), we note that the chance of observing a sample mean $\bar{x}$ more than 1.96 standard deviations below 72 *or* more than 1.96 standard deviations above 72, *when H_0 is true,* is only $\alpha = .05$. Thus, the rejection region consists of two sets of values: We will reject H_0 if z is either less than -1.96 or greater than 1.96 (see Figure 8.3b). For this rejection rule, the probability of a Type I error is $\alpha = .05$. ❑

Figure 8.3 Location of Rejection Region for Example 8.9

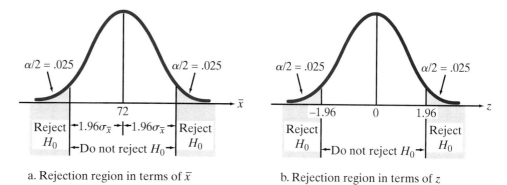

a. Rejection region in terms of $\bar{x}$

b. Rejection region in terms of z

The three previous examples all exhibit certain common characteristics regarding the rejection region, as indicated in the box.

Guidelines for Specifying the Rejection Region of a Test

1. The value of α, the probability of a Type I error, is specified in advance by the researcher. It can be made as small or as large as desired; typical values are $\alpha = .01, .02, .05$, and $.10$. For a fixed sample size, the size of the rejection region decreases as the value of α decreases (see Figure 8.4). That is, for smaller values of α, more extreme departures of the test statistic from the null hypothesized parameter value are required to permit rejection of H_0.

2. The test statistic (i.e., the point estimate of the target parameter) is standardized to provide a measure of how great its departure is from the null hypothesized value of the parameter. The standardization is based on the sampling distribution of the point estimate, assuming H_0 is true. For means and proportions, the general formula is:

$$\text{Standardized test statistic} = \frac{\text{Point estimate} - \text{Hypothesized value in } H_0}{\text{Standard error of point estimate}}$$

3. The location of the rejection region depends on whether the test is one-tailed or two-tailed, and on the prespecified significance level, α.

   a. For a one-tailed test in which the symbol ">" occurs in H_a (an **upper-tailed test**), the rejection region consists of values in the upper tail of the sampling distribution of the standardized test statistic. The critical value is selected so that the area to its right is equal to α. (See Figure 8.5a, page 430.)

   continued

Figure 8.4 Size of the Upper-Tailed Rejection Region for Different Values of α

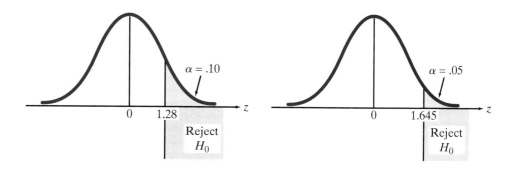

Figure 8.5 The Rejection Region for Various Alternative Hypotheses

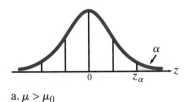

a. $\mu > \mu_0$

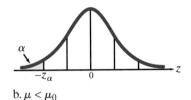

b. $\mu < \mu_0$

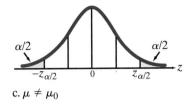

c. $\mu \neq \mu_0$

> **b.** For a one-tailed test in which the symbol "<" occurs in H_a (a **lower-tailed test**), the rejection region consists of values in the lower tail of the sampling distribution of the standardized test statistic. The critical value is selected so that the area to its left is equal to α. (See Figure 8.5b.)
>
> **c.** For a two-tailed test, in which the symbol "$\neq$" occurs in H_a, the rejection region consists of two sets of values. The critical values are selected so that the area in each tail of the sampling distribution of the standardized test statistic is equal to $\alpha/2$. (See Figure 8.5c.)

Once we have chosen a test statistic from the sample information and computed its value (step 3, page 426), we determine if its standardized value lies within the rejection region so we can decide whether to reject the null hypothesis (step 5).

Throughout this section, we have assumed that the sample size n is sufficiently large (i.e., $n \geq 30$). A summary of this large-sample test of hypothesis is provided in the box. Note that the test statistic will depend on the parameter of interest and its point estimate. In the remainder of this chapter, we cover the test statistics for a population mean μ (Section 8.5), a population proportion π (Section 8.7), and a population variance σ^2 (optional Section 8.8). First, we provide some guidelines on selecting the parameter of interest in the next section.

Large-Sample Test of Hypothesis About a Population Parameter

H_0: Parameter = Hypothesized value

Upper-Tailed Test	Two-Tailed Test	Lower-Tailed Test
H_a: Parameter > Hypothesized value	H_a: Parameter $\neq$ Hypothesized value	H_a: Parameter < Hypothesized value

$$Test \ statistic: \ = \frac{\text{Point estimate} - \text{Hypothesized value}}{\text{Standard error of point estimate}} \qquad (8.2)$$

Rejection region:	Rejection region:	Rejection region:
$z > z_\alpha$	$z < -z_{\alpha/2}$ or $z > z_{\alpha/2}$	$z < -z_\alpha$

Self-Test
8.3

At $\alpha = .10$, find the rejection region for a large-sample test of

$$H_0: \mu = 150$$
$$H_a: \mu < 150$$

PROBLEMS
8.11–8.20

Using the Tools

8.11 Suppose we want to test $H_0: \mu = 65$. Specify the form of the rejection region for each of the following (assume that the sample size will be sufficient to guarantee the approximate normality of the sampling distribution of $\bar{x}$):

  a. $H_a: \mu \neq 65, \alpha = .02$ **b.** $H_a: \mu > 65, \alpha = .05$

  c. $H_a: \mu < 65, \alpha = .01$ **d.** $H_a: \mu < 65, \alpha = .10$

8.12 Refer to Problem 8.11. Calculate the test statistic for each of the following sample results (assume s is a good approximation of σ):

  a. $n = 100, \bar{x} = 70, s = 15$ **b.** $n = 50, \bar{x} = 70, s = 15$

  c. $n = 50, \bar{x} = 60, s = 15$ **d.** $n = 50, \bar{x} = 60, s = 30$

8.13 Refer to Problems 8.11 and 8.12. Give the appropriate conclusions for each of the tests, parts **a, b, c,** and **d.**

8.14 For each of the following rejection regions, determine the value of α, the probability of a Type I error:

  a. $z < -1.96$ **b.** $z > 1.645$ **c.** $z < -2.58$ or $z > 2.58$

Applying the Concepts

8.15 Suppose you are interested in testing whether μ, the true mean score on an aptitude test used for determining admission to graduate study in your field, differs from 500. To conduct the test, you randomly select 75 applicants to graduate study and record the aptitude score, x, for each.

  a. Set up the null and alternative hypotheses for the test.

  b. Set up the rejection region for the test if $\alpha = .10$.

8.16 The Computer-Assisted Hypnosis Scale (CAHS) is designed to measure a person's susceptibility to hypnosis. CAHS scores range from 0 (no susceptibility to hypnosis) to 12 (extremely high susceptibility to hypnosis). *Psychological Assessment* (Mar. 1995) reported that University of Tennessee undergraduates had a mean CAHS score of $\mu = 4.6$. Suppose you want to test whether undergraduates at your college or university are more susceptible to hypnosis than University of Tennessee undergraduates.

  a. Set up the null and alternative hypotheses for the test.

b. Select a value of α and interpret it.

c. Find the rejection region for the test if you plan on sampling $n = 100$ undergraduates at your college.

8.17 The purchase of a coin-operated laundry is being considered by a potential entrepreneur. The present owner claims that over the past five years the average daily revenue was $675. The entrepreneur wants to test whether the true average daily revenue of the laundry is less than $675.

a. Set up the null and alternative hypotheses for the test.

b. Find the rejection region for the test if $\alpha = .05$. Assume a large sample of days will be selected from the past five years and the revenue recorded for each day.

8.18 According to the *Journal of Psychology and Aging* (May 1992), older workers (i.e., workers 45 years or older) have a mean job satisfaction rating of 4.3 on a 5-point scale. (Higher scores indicate higher levels of job satisfaction.) In a random sample of 100 older workers, suppose the sample mean job satisfaction rating is $\bar{x} = 4.1$.

a. Set up the null and alternative hypotheses for testing whether the true mean job satisfaction rating of older workers differs from 4.3.

b. At $\alpha = .05$, locate the rejection region for the test.

c. Compute the test statistic if $\sigma = .5$. What is your conclusion?

d. Repeat part **c**, but assume $\sigma = 1.5$.

8.19 *Psychological Assessment* (Mar. 1995) published the results of a study of World War II aviators who were shot down and became prisoners of war (POWs). A post-traumatic stress disorder (PTSD) score was determined for each in a sample of 33 aviators. The results, where higher scores indicate higher PTSD levels, are $\bar{x} = 9.00, s = 9.32$.

a. Psychologists have established that the mean PTSD score for Vietnam veterans is 16. Set up the null and alternative hypotheses for determining whether the true mean PTSD score μ of all World War II aviator POWs is less than 16.

b. Calculate the test statistic. (Assume $s \approx \sigma$)

c. Specify the form of the rejection region if the level of significance is $\alpha = .01$. Locate the rejection region, α, and the critical value on a sketch of the standard normal curve.

d. Use the results of parts **b** and **c** to make the proper conclusion in terms of the problem.

8.20 A certain species of beetle produces offspring with either blue eyes or black eyes. Suppose a biologist wants to determine which, if either, of the two eye colors is dominant for this species of beetle. Let π represent the true proportion of offspring that possess blue eyes. If the beetles produce blue-eyed and black-eyed offspring at an equal rate, then $\pi = .5$. Thus, the biologist desires to test

$$H_0: \pi = .5$$
$$H_a: \pi \neq .5$$

Give the form of the rejection region if the biologist is willing to tolerate a Type I error probability of $\alpha = .05$. Locate the rejection region, α, and the critical value(s) on a sketch of the normal curve. Assume that the sample size n is large. [*Hint:* The test statistic is the large-sample z statistic.]

8.4 GUIDELINES FOR DETERMINING THE TARGET PARAMETER

The key to correctly diagnosing a hypothesis test is to determine first the parameter of interest—a task that can sometimes present difficulties for the introductory statistics student, especially when the parameter is stated in words rather than symbols.

Those who are successful in diagnosing a hypothesis test generally follow a three-step process. First, identify the experimental unit (i.e., the objects upon which the measurements are taken). Second, identify the type of variable, quantitative or qualitative, measured on each experimental unit. Third, determine the target parameter based on the phenomenon of interest and the variable measured. For quantitative data, the target parameter will be either a population mean or variance; for qualitative data the parameter will be a population proportion.

Often, there are one or more key words in the statement of the problem that indicate the appropriate population parameter. In this section, we will present several examples illustrating how to determine the parameter of interest. First, we state in the box the key words to look for when conducting a hypothesis test about a single population parameter.

Diagnosing a Hypothesis Test: Determining the Target Parameter

Parameter	Key Words or Phrases
μ	Mean; average
π	Proportion; percentage; fraction; rate
σ^2	Variance; variation; spread; precision

Example 8.10   Choosing the Target Parameter

The "Pepsi Challenge" was a marketing strategy used by Pepsi-Cola. A consumer is presented with two cups of cola and asked to select the one that tastes better. Unknown to the consumer, one cup is filled with Pepsi, the other with Coke. Marketers of Pepsi claim that the true fraction of consumers who select their product will exceed .50.

   a. What is the parameter of interest to the Pepsi marketers?
   b. Set up H_0 and H_a.

Solution

a. In this problem, the experimental units are the consumers and the variable measured is *qualitative*—the consumer chooses either Pepsi or Coke. The key word in the statement of the problem is "fraction." Thus, the parameter of interest is π, where

π = True fraction of consumers who favor Pepsi over Coke in the taste test

b. To support their claim that the true fraction who favor Pepsi exceeds .5, the marketers will test

$$H_0: \pi = .5$$
$$H_a: \pi > .5 \qquad \square$$

Example 8.11   Choosing the Target Parameter

The administrator at a large hospital wants to know if the average length of stay of the hospital's patients is less than five days. To check this, lengths of stay were recorded for 100 randomly selected hospital patients.

a. What is the parameter of interest to the administrator?
b. Set up H_0 and H_a.

Solution

a. The experimental units for this problem are the hospital patients. For each patient, the variable measured is length of stay (in days)—a *quantitative* variable. Since the administrator wants to make a decision about the average length of stay, the target parameter is

μ = average length of stay of hospital patients

b. To determine whether the average length of stay is less than five days, the administrator will test

$$H_0: \mu = 5$$
$$H_a: \mu < 5 \qquad \square$$

Example 8.12   Choosing the Target Parameter

According to specifications, the variation in diameters of 2-inch bolts produced on an assembly line should be .001. A quality control engineer wants to determine whether the variation in bolt diameters differs from these specifications. What is the parameter of interest?

Solution

For the quality control engineer, the experimental units are the bolts produced by the assembly line, and the variable measured is *quantitative*—the diameter of the bolt. The key word in the statement of the problem is "variation." Therefore, the parameter of interest is σ^2. $\square$

Self-Test

8.4

Based on market surveys to determine the average number of tissues used by people when they have a cold, Kimberly-Clark Corporation, the makers of Kleenex, puts 60 tissues in a box. Suppose marketing experts at the company want to test whether the mean number of tissues used by people with colds exceeds 60.

 a.   What is the parameter of interest?

 b.   Set up the null and alternative hypotheses.

In the remaining sections of this chapter, we present a summary of the hypothesis-testing procedures for each of the parameters listed in the box on page 433. In the problems that follow each section, the target parameter can be identified by simply noting the title of the section. However, to properly diagnose a hypothesis test, it is essential to search for the key words in the statement of the problem. For practice, we strongly recommend that you read through all the Supplementary Problems at the end of the chapter and determine the parameter of interest before attempting to analyze the data.

8.5 TESTING A POPULATION MEAN

Large Samples

When testing a hypothesis about a population mean μ, the procedure that we use will depend on whether the sample size n is large (say, $n \geq 30$) or small. The accompanying box contains the elements of a large-sample hypothesis test for μ based on the z statistic. Note that for this case, the only assumption required for the validity of the procedure is that the random sample is, in fact, large so that the sampling distribution of $\bar{x}$ is normal. Technically, the true population standard deviation σ must be known in order to use the z statistic, and this is rarely, if ever, the case. However, we established in Chapter 7 that when n is large, the sample standard deviation s provides a good approximation to σ and the z statistic can be approximated as shown in the box.

Large-Sample (z) Test of Hypothesis About a Population Mean μ

Upper-Tailed Test	Two-Tailed Test	Lower-Tailed Test
$H_0: \mu = \mu_0$	$H_0: \mu = \mu_0$	$H_0: \mu = \mu_0$
$H_a: \mu > \mu_0$	$H_a: \mu \neq \mu_0$	$H_a: \mu < \mu_0$

$$\text{Test statistic: } z = \frac{\bar{x} - \mu_0}{\sigma_{\bar{x}}} \approx \frac{\bar{x} - \mu_0}{s/\sqrt{n}} \qquad (8.3)$$

Rejection region:	*Rejection region:*	*Rejection region:*
$z > z_\alpha$	$|z| > z_{\alpha/2}$	$z < -z_\alpha$

where z_α is the z value such that $P(z > z_\alpha) = \alpha$; $-z_\alpha$ is the z value such that $P(z < -z_\alpha) = \alpha$; and $z_{\alpha/2}$ is the z value such that $P(z > z_{\alpha/2}) = \alpha/2$.

continued

> *Assumption:* The random sample must be sufficiently large (say, $n \geq 30$) so that the sampling distribution of $\bar{x}$ is approximately normal and so that s provides a good approximation to σ.

Example 8.13   Large-Sample Test of μ

Humerus bones from the same species of animal tend to have approximately the same length-to-width ratios. When fossils of humerus bones are discovered, archaeologists can often determine the species of animal by examining the length-to-width ratios of the bones. It is known that species A has a mean ratio of 8.5. Suppose 41 fossils of humerus bones were unearthed at an archaeological site in East Africa, which species A is believed to have inhabited. (Assume that the unearthed bones were all from the same unknown species.) The length-to-width ratios of the bones were measured and are listed in Table 8.3.

We wish to test the hypothesis that μ, the population mean ratio of all bones of this particular species, is equal to 8.5 against the alternative that it is different from 8.5, i.e., we wish to test whether the unearthed bones are from species A.

a. Suppose we want a very small chance of rejecting H_0 if μ is equal to 8.5. That is, it is important that we avoid making a Type I error. Select an appropriate value of the significance level α.

b. Test whether μ, the population mean length-to-width ratio, is different from 8.5 using the significance level selected in part **a**.

c. What are the practical implications of the result, part **b**?

Solution

a. The hypothesis-testing procedure that we have developed gives us the advantage of being able to choose any significance level that we desire. Since the significance level α is also the probability of a Type I error, we will choose α to be very small. In general, researchers who consider a Type I error to have very serious practical consequences should perform the test at a very low α value—say, $\alpha = .01$. Other researchers may be willing to tolerate an α value as high as .10 if a Type I error is not deemed a serious error to make in practice. For this example, we will test at $\alpha = .01$.

b. We formulate the following hypotheses:

$$H_0: \mu = 8.5$$

$$H_a: \mu \neq 8.5$$

BONES.XLS

Table 8.3 Length-to-Width Ratios of a Sample of Humerus Bones

10.73	9.59	8.37	9.35	9.39	8.38
8.89	8.48	6.85	8.86	9.17	11.67
9.07	8.71	8.52	9.93	9.89	8.30
9.20	9.57	8.87	8.91	8.17	9.17
10.33	9.29	6.23	11.77	8.93	12.00
9.98	9.94	9.41	10.48	8.80	9.38
9.84	8.07	6.66	10.39	10.02	

E **Figure 8.6** Excel and PHStat Add-In for Excel Output, Example 8.13

	A	B
1	*Ratio*	
2		
3	Mean	9.257560976
4	Standard Error	0.187965286
5	Median	9.2
6	Mode	9.17
7	Standard Deviation	1.20356508
8	Sample Variance	1.448568902
9	Kurtosis	1.066260992
10	Skewness	-0.084954979
11	Range	5.77
12	Minimum	6.23
13	Maximum	12
14	Sum	379.56
15	Count	41
16	Largest(1)	12
17	Smallest(1)	6.23

a.

	A	B
1	Z Test of the Hypothesis for the Mean	
2		
3	Null Hypothesis $\mu=$	8.5
4	Level of Significance	0.01
5	Population Standard Deviation	1.2
6	Sample Size	41
7	Sample Mean	9.26
8	Standard Error of the Mean	0.187408514
9	Z Test Statistic	4.055312017
10		
11	Two-Tailed Test	
12	Lower Critical Value	-2.575834515
13	Upper Critical Value	2.575834515
14	p-Value	5.00929E-05
15	Reject the null hypothesis	

b.

The sample size is large ($n = 41$); thus, we may proceed with the large-sample (z) test about μ. The data (Table 8.3) were entered into an Excel spreadsheet; the output obtained from Excel and the PHStat Add-in are displayed in Figure 8.6. From Figure 8.6a, $\bar{x} = 9.26$ and $s = 1.20$. Substituting these values into equation 8.3, we obtain

$$z = \frac{\bar{x} - 8.5}{s/\sqrt{n}} = \frac{9.26 - 8.5}{.120/\sqrt{41}} = 4.055$$

This value is highlighted in Figure 8.6b. At $\alpha = .01$, the rejection region for this two-tailed test is

$$|z| > z_{.005} = 2.58$$

This rejection region is shown in Figure 8.7. Since $z = 4.055$ falls in the rejection region, we reject H_0 and conclude that the mean length-to-width ratio of all humerus bones of this particular species is significantly different from 8.5. If the null hypothesis is in fact true (i.e., if $\mu = 8.5$), then the probability that we have incorrectly rejected it is equal to $\alpha = .01$.

c. The *practical* implications of the result remain to be studied further. Perhaps the animal discovered at the archaeological site is of some species other than A. Alternatively, the unearthed humerus bones may have larger

Figure 8.7 Rejection Region for Example 8.13

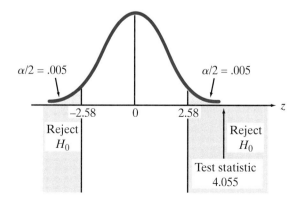

$\alpha/2 = .005$ $\alpha/2 = .005$

-2.58 0 2.58 z

Reject H_0 Reject H_0

Test statistic 4.055

than normal length-to-width ratios because of unusual feeding habits of species A. **It is not always the case that a statistically significant result implies a practically significant result.** The researcher must retain his or her objectivity and judge the practical significance using, among other criteria, knowledge of the subject matter and the phenomenon under investigation. ❑

Self-Test 8.5

Prior to the institution of a new safety program, the average number of on-the-job accidents per day at a factory was 4.5. To determine whether the safety program was effective, a factory foreman will conduct a test of

$$H_0: \mu = 4.5 \quad \text{(no change in average number of on-the-job accidents per day)}$$

$$H_a: \mu < 4.5 \quad \text{(average number of on-the-job accidents per day has decreased)}$$

where μ represents the mean number of on-the-job accidents per day at the factory after institution of the new safety program. Summary statistics for the daily number of accidents for a random sample of 120 days after institution of the new program are: $\bar{x} = 3.7$, $s = 2.6$. Conduct the test using $\alpha = .01$.

Small Samples

Time and cost considerations sometimes limit the sample size. In this case, the assumption required for a large-sample test of hypothesis about μ will be violated and s will not provide a reliable estimate of σ. We need, then, a procedure that is appropriate for use with small samples.

A hypothesis test about a population mean μ for small samples ($n < 30$) is based on a t statistic. The elements of the test are listed in the next box.

Small-Sample (t) Test of Hypothesis About a Population Mean μ

Upper-Tailed Test	Two-Tailed Test	Lower-Tailed Test
$H_0: \mu = \mu_0$	$H_0: \mu = \mu_0$	$H_0: \mu = \mu_0$
$H_a: \mu > \mu_0$	$H_a: \mu \neq \mu_0$	$H_a: \mu < \mu_0$

$$\text{Test statistic: } t = \frac{\bar{x} - \mu_0}{s/\sqrt{n}} \tag{8.4}$$

Rejection region:	Rejection region:	Rejection region:
$t > t_\alpha$	$|t| > t_{\alpha/2}$	$t < -t_\alpha$

where the distribution of t is based on $(n - 1)$ degrees of freedom, t_α is the t value such that $P(t > t_\alpha) = \alpha$; $-t_\alpha$ is the t value such that $P(t < -t_\alpha) = \alpha$; and $t_{\alpha/2}$ is the t value such that $P(t > t_{\alpha/2}) = \alpha/2$.

Assumption: The relative frequency distribution of the population from which the random sample was selected is approximately normal.

As we noticed in the development of confidence intervals (Chapter 7), when making inferences based on small samples more restrictive assumptions are required than when making inferences from large samples. In particular, this hypothesis test requires the assumption that the population from which the sample is selected is approximately normal.

Notice that the test statistic given in the box is a t statistic and is calculated exactly as our approximation to the large-sample test statistic z given earlier in this section. Therefore, just like z, the computed value of t indicates the direction and approximate distance (in units of standard deviations) that the sample mean $\bar{x}$ is from the hypothesized population mean μ_0.

Example 8.14 Small-Sample Test of μ

The building specifications in a certain city require that the sewer pipe used in residential areas have a mean breaking strength of more than 2,500 pounds per lineal foot. A manufacturer who would like to supply the city with sewer pipe has submitted a bid and provided the following additional information: An independent contractor randomly selected seven sections of the manufacturer's pipe and tested each for breaking strength. The results (pounds per lineal foot) are shown here:

PIPES.XLS

| 2,610 | 2,750 | 2,420 | 2,510 | 2,540 | 2,490 | 2,680 |

a. Compute $\bar{x}$ and s for the sample.

b. Is there sufficient evidence to conclude that the manufacturer's sewer pipe meets the required specifications? Use a significance level of $\alpha = .10$.

Solution

a. The data were entered into an Excel spreadsheet and descriptive statistics were generated. The Excel printout is shown in Figure 8.8a. The values of the sample mean breaking strength and standard deviation of breaking strengths (highlighted) are $\bar{x} = 2{,}571.4$ and $s = 115.1$.

E Figure 8.8 Excel and PHStat Add-In for Excel Output, Example 8.14

	A	B
1		*Strength*
2		
3	Mean	2571.428571
4	Standard Error	43.50306894
5	Median	2540
6	Mode	#N/A
7	Standard Deviation	115.0983017
8	Sample Variance	13247.61905
9	Kurtosis	-0.765263187
10	Skewness	0.421794298
11	Range	330
12	Minimum	2420
13	Maximum	2750
14	Sum	18000
15	Count	7
16	Largest(1)	2750
17	Smallest(1)	2420

a.

	A	B	C	D	E
1	*t* Test for the Hypothesis of the Mean				
2					
3	Null Hypothesis $\mu =$	2500			
4	Level of Significance	0.1			
5	Sample Size	7			
6	Sample Mean	2571.4			
7	Sample Standard Deviation	115.1			
8	Standard Error of the Mean	43.50371			
9	Degrees of Freedom	6			
10	T Test Statistic	1.641239			
11					
12	Upper-Tail Test			Calculations Area	
13	Upper Critical Value	1.439755		For one-tailed tests:	
14	*p*-Value	0.075928		TDIST value	0.075928
15	Reject the null hypothesis			1-TDIST value	0.924072

b.

Figure 8.9 Rejection Region for Example 8.14

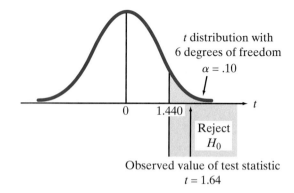

t distribution with 6 degrees of freedom

$\alpha = .10$

Observed value of test statistic $t = 1.64$

b. The relevant hypothesis test has the following elements:

$H_0: \mu = 2{,}500$ (the manufacturer's pipe does not meet the city's specifications)

$H_a: \mu > 2{,}500$ (the pipe meets the specifications)

where μ represents the true mean breaking strength (in pounds per lineal foot) for all sewer pipe produced by the manufacturer.

This small-sample ($n = 7$) test requires the assumption that the relative frequency distribution of the population values of breaking strength for the manufacturer's pipe is approximately normal. Then the test will be based on a t distribution with $(n - 1) = 6$ degrees of freedom. We will thus reject H_0 if

$$t > t_{.10} = 1.440 \text{ (see Figure 8.9)}$$

Substituting the values $\bar{x} = 2{,}571.4$ and $s = 115.1$ into equation 8.4 yields the test statistic

$$t = \frac{\bar{x} - \mu_0}{s/\sqrt{n}} = \frac{2{,}571.4 - 2{,}500}{115.1/\sqrt{7}} = 1.64$$

This value is also highlighted on the PHStat Add-in for Excel printout of the test results, Figure 8.8b. Since this value of t is larger than the critical value of 1.440, we reject H_0. There is sufficient evidence (at significance level $\alpha = .10$) that the manufacturer's pipe meets the city's building specifications. ❑

Self-Test 8.6

Refer to Self-Test 8.5 (page 438). Conduct the test $H_0: \mu = 4.5$ against $H_a: \mu < 4.5$, but assume the sample size is $n = 12$.

Remember that the small-sample test of Example 8.14 requires the assumption that the sampled population has a distribution that is approximately normal. If you know that the population is highly skewed (based on, for example, a stem-and-leaf plot of the sample data), then any inferences derived from the t test are suspect. In this case, we do not perform the t test but use a nonparametric statistical method discussed in Chapter 12.

Caution

When the sampled population is decidedly nonnormal (e.g., highly skewed), any inferences derived from the small-sample t test for μ are suspect. In this case, one alternative is to use the nonparametric sign test discussed in Section 12.2.

PROBLEMS
8.21–8.33

Using the Tools

8.21   Compute the value of the test statistic z for each of the following situations:

   a. $H_0: \mu = 9.8, H_a: \mu > 9.8, \bar{x} = 10.0, s = 4.3, n = 50$

   b. $H_0: \mu = 80, H_a: \mu < 80, \bar{x} = 75, s^2 = 19, n = 86$

   c. $H_0: \mu = 8.3, H_a: \mu \neq 8.3, \bar{x} = 8.2, s^2 = .79, n = 175$

8.22   A random sample of n observations is selected from a population with unknown mean μ and variance σ^2. For each of the following situations, specify the test statistic and rejection region.

   a. $H_0: \mu = 50, H_a: \mu > 50, n = 36, \bar{x} = 60, s^2 = 64, \alpha = .05$

   b. $H_0: \mu = 140, H_a: \mu \neq 140, n = 40, \bar{x} = 143.2, s = 9.4, \alpha = .01$

   c. $H_0: \mu = 10, H_a: \mu < 10, n = 50, \bar{x} = 9.5, s = .35, \alpha = .10$

8.23   To test the null hypothesis $H_0: \mu = 10$, a random sample of n observations is selected from a normal population. Specify the rejection region for each of the following combinations of $H_a, n,$ and α:

   a. $H_a: \mu \neq 10, n = 15, \alpha = .05$ **b.** $H_a: \mu \neq 10, n = 15, \alpha = .01$

   c. $H_a: \mu < 10, n = 15, \alpha = .05$ **d.** $H_a: \mu > 10, n = 5, \alpha = .10$

   e. $H_a: \mu > 10, n = 25, \alpha = .10$

8.24   A random sample of n observations is selected from a normal population. For each of the following situations, specify the rejection region, test statistic, and conclusion.

   a. $H_0: \mu = 3,000, H_a: \mu \neq 3,000, \bar{x} = 2,958, s = 39, n = 8, \alpha = .05$

   b. $H_0: \mu = 6, H_a: \mu > 6, \bar{x} = 6.3, s = .3, n = 7, \alpha = .01$

   c. $H_0: \mu = 22, H_a: \mu < 22, \bar{x} = 13.0, s = 6, n = 17, \alpha = .05$

8.25   A random sample of five measurements from a normally distributed population yielded the following data: 12, 4, 3, 5, 5.

   a.   Test the null hypothesis that $\mu = 4$ against the alternative hypothesis that $\mu \neq 4$. Use $\alpha = .01$.

   b.   Test the null hypothesis that $\mu = 4$ against the alternative hypothesis that $\mu > 4$. Use $\alpha = .01$.

Applying the Concepts

8.26   The 1996–1997 room and board costs for a sample of 20 American universities and colleges are listed in the accompanying table. Test the hypothesis that the true mean room and board costs of all American colleges and universities differ from $6,000. Use $\alpha = .05$.

ROOMBOARD.XLS

College/University	Room and Board Costs
Arizona State University	$4,300
Babson College	7,600
Boston University	7,000
California State University–Fresno	5,400
Case Western Reserve University	5,000
Emory University	6,500
George Washington University	6,900
Harvard University	7,000
Lafayette College	6,300
LaSalle University	6,700
Lehigh University	6,000
Montclair State University	5,300
Niagara University	5,400
New York University	7,800
Northwestern University	6,100
Notre Dame	4,800
Purdue University	4,500
Rochester Institute of Technology	6,100
Stanford University	7,300
Sienna College	5,400

Sources: Individual 1996–1997 institutional catalogues.

8.27 "Deep hole" drilling is a family of drilling processes used when the ratio of hole depth to hole diameter exceeds 10. Successful deep hole drilling depends on the satisfactory discharge of the drill chip. An experiment was conducted to investigate the performance of deep hole drilling when chip congestion exists (*Journal of Engineering for Industry,* May 1993). The length, in millimeters (mm), of 50 drill chips resulted in the following summary statistics: $\bar{x} = 81.2$ mm, $s = 50.2$ mm. Conduct a test to determine whether the true mean drill chip length μ differs from 75 mm. Use a significance level of $\alpha = .01$.

8.28 One of the most feared predators in the ocean is the great white shark. Although it is known that the great white shark grows to a mean length of 21 feet, a marine biologist believes that great white sharks off the Bermuda coast grow much longer because of unusual feeding habits. To test this claim, researchers plan to capture a number of full-grown great white sharks off the Bermuda coast, measure them, then set them free. However, because capturing sharks is difficult, costly, and very dangerous, only three are sampled. Their lengths are 24, 20, and 22 feet.

a. Do the data provide sufficient evidence to support the marine biologist's claim? Test at significance level $\alpha = .05$.

b. What assumptions are required for the hypothesis test of part **a** to be valid? Do you think these assumptions are likely to be satisfied in this particular sampling situation?

8.29 The symbol pH represents the hydrogen ion concentration in a liter of a solution. The higher the pH value, the less acidic the solution. The table lists the pH values for a sample of foods as determined by the Food and Drug Administration (FDA).

PH.XLS

Food	pH	Food	pH	Food	pH
Lima beans	6.5	Orange juice	3.9	Corn syrup	5.0
Cauliflower	5.6	Buttermilk	4.5	Cider	3.1
Figs	4.6	Crabs	7.0	Caviar	5.4
Lemons	2.3	Fresh fish	6.7	Mayonnaise	4.35
Potatoes	6.1	Blueberries	3.7	Celery	5.85
Peppers	5.15	Cheddar cheese	5.9	Artichokes	5.6
Apples	3.4	Egg whites	8.0	Parsnip	5.3
Sauerkraut	3.5	Bread	5.55	Dill pickles	3.35
Lettuce	5.9	Eclairs	4.45	Tomatoes	4.55
Sweet corn	7.3	Raisins	3.9	Bananas	4.85
Ham	6.0	Honey	3.9	Ground beef	5.55
Nectarines	3.9	Cocoa	6.3	Veal	6.0

Suppose an FDA spokesperson claims that the mean pH level of all food items is $\mu = 4.0$. Test this claim at $\alpha = .10$. Interpret the results.

8.30 *Supremism* is defined as an attitude of superiority based on race. Psychologists at Wake Forest University investigated the degree of supremism exhibited by future teachers (*Journal of Black Psychology*, Nov. 1997). Ten European-American college students who were enrolled in a 4th-year teacher education course participated in the study. Each student was given a picture of a 7-year-old African-American child and instructed to estimate the child's IQ. The IQ estimates are summarized as follows: $\bar{x} = 94.1, s = 10.3$. Is there sufficient evidence to indicate that the true mean IQ estimated by all European-American future teachers is less than 100? Test using $\alpha = .05$.

8.31 A group of researchers at the University of Texas–Houston conducted a comprehensive study of pregnant cocaine-dependent women (*Journal of Drug Issues*, Summer 1997). All the women in the study used cocaine on a regular basis (at least three times a week) for more than a year. One of the many variables measured was birthweight (in grams) of the baby delivered. For a sample of 16 cocaine-dependent women, the mean birthweight was 2,971 grams and the standard deviation was 410 grams. Test the hypothesis that the true mean birthweight of babies delivered by cocaine-dependent women is less than 3,500 grams. Use $\alpha = .05$.

8.32 A consumers' advocate group would like to evaluate the average energy efficiency rating (EER) of window-mounted, large-capacity (i.e., in excess of 7,000 Btu) air-conditioning units. A random sample of 36 such air-conditioning units is selected and tested for a fixed period of time with their EER recorded as follows:

EERAC.XLS

8.9	9.1	9.2	9.1	8.4	9.5	9.0	9.6	9.3
9.3	8.9	9.7	8.7	9.4	8.5	8.9	8.4	9.5
9.3	9.3	8.8	9.4	8.9	9.3	9.0	9.2	9.1
9.8	9.6	9.3	9.2	9.1	9.6	9.8	9.5	10.0

a. Using the .05 level of significance, is there evidence that the average EER is different from 9.0?

b. What assumptions are being made in order to perform this test?

c. What will your answer in part **a** be if the last data value is 8.0 instead of 10.0?

8.33 How does lack of sleep have an impact on one's creative ability? A British study found that loss of sleep sabotages creative faculties and the ability to deal with unfamiliar situations (*Sleep*, Jan. 1989). In the study, 12 healthy college students, deprived of one night's sleep, received an array of tests intended to measure thinking time, fluency, flexibility, and originality of thought. The overall test scores of the sleep-deprived students were compared to the average score one would expect from students who received their accustomed sleep. Suppose the overall scores of the 12 sleep-deprived students had a mean of $\bar{x} = 63$ and a standard deviation of 17. (Lower scores are associated with a decreased ability to think creatively.)

a. Test the hypothesis that the true mean score of sleep-deprived subjects is less than 80, the mean score of subjects who received sleep prior to taking the test. Use $\alpha = .05$.

b. What assumption is required for the hypothesis test of part **a** to be valid?

8.6 REPORTING TEST RESULTS: *p*-VALUES

The statistical hypothesis-testing technique that we have developed requires us to choose the significance level α (i.e., the maximum probability of a Type I error that we are willing to tolerate) before obtaining the data and computing the test statistic. By choosing α a priori, we in effect fix the rejection region for the test. Thus, no matter how large or how small the observed value of the test statistic, our decision regarding H_0 is clear-cut: Reject H_0 (i.e., conclude that the test results are statistically significant) if the observed value of the test statistic falls into the rejection region, and do not reject H_0 (i.e., conclude that the test results are insignificant) otherwise. This **"fixed" significance level** α then serves as a measure of the reliability of our inference. However, there is one drawback to a test conducted in this manner—namely, a measure of the *degree* of significance of the test results is not readily available. That is, if the value of the test statistic falls into the rejection region, we have no measure of the extent to which the data disagree with the null hypothesis.

Example 8.15 **The Degree of Disagreement between the Sample Data and H_0**

A large-sample test of $H_0: \mu = 72$ against $H_a: \mu > 72$ is to be conducted at a fixed significance level of $\alpha = .05$. Consider the following possible values of the computed test statistic:

$$z = 1.82 \quad \text{and} \quad z = 5.66$$

a. Which of the above values of the test statistic provides stronger evidence for the rejection of H_0?

b. How can we measure the extent of disagreement between the sample data and H_0 for each of the computed values?

Solution

a. The appropriate rejection region for this upper-tailed test, at $\alpha = .05$, is given by

$$z > z_{.05} = 1.645$$

Clearly, for either of the test statistic values, $z = 1.82$ or $z = 5.66$, we will reject H_0; hence, the result in each case is statistically significant. Recall, however, that the appropriate test statistic for a large-sample test concerning μ is simply the z score for the observed sample mean $\bar{x}$, calculated by using the hypothesized value of μ in H_0 (in this case, $\mu = 72$). The larger the z score, the greater the distance (in units of standard deviations) that $\bar{x}$ is from the hypothesized value of $\mu = 72$. Thus, a z score of 5.66 would present stronger evidence that the true mean is larger than 72 than would a z score of 1.82. This reasoning stems from our knowledge of the sampling distribution of $\bar{x}$; if in fact $\mu = 72$, we would certainly not expect to observe an $\bar{x}$ with a z score as large as 5.66.

b. One way of measuring the amount of disagreement between the observed data and the value of μ in the null hypothesis is to calculate the probability of observing a value of the test statistic equal to or greater than the actual computed value of z, if H_0 were true. That is, if z_c is the computed value of the test statistic, calculate

$$P(z \geq z_c)$$

assuming that the null hypothesis is true. This "disagreement" probability, or **p-value**, is calculated here for each of the computed test statistics $z = 1.82$ and $z = 5.66$, using Table B.2 in Appendix B:

$$P(z \geq 1.82) = .5 - .4656 = .0344$$
$$P(z \geq 5.66) \approx .5 - .5 = 0$$

From the discussion in part **a**, you can see that the smaller the *p*-value, the greater the extent of disagreement between the data and the null hypothesis—that is, the more significant the result. ❏

In general, *p*-values for tests based on large samples are computed as shown in the box below. (*p*-Values for small-sample tests are obtained by replacing the z statistic with the t statistic.)

Measuring the Disagreement between the Data and H_O: p-Values

Upper-tailed test: p-value $= P(z \geq z_c)$

Lower-tailed test: p-value $= P(z \leq z_c)$

Two-tailed test: p-value $= 2P(z \geq |z_c|)$

where z_c is the computed value of the test statistic and $|z_c|$ denotes the absolute value of z_c (which will always be positive).

Figure 8.10 Using p-Values to Make Conclusions

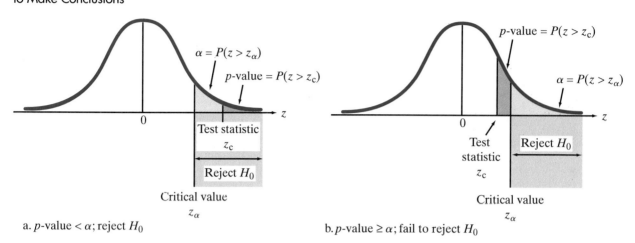

a. p-value $< \alpha$; reject H_0

b. p-value $\geq \alpha$; fail to reject H_0

Notice that the p-value for a two-tailed test is twice the probability for the one-tailed test. This is because the disagreement between the data and H_0 can be in two directions.

When publishing the results of a statistical test of hypothesis in journals, case studies, reports, etc., most researchers use p-values. Instead of selecting α a priori and then conducting a test using a rejection region, the researcher will compute and report the value of the appropriate test statistic and its associated p-value. The use of p-values has been facilitated by the routine inclusion of p-values as part of the output of statistical software and spreadsheet packages, such as Excel. Given the p-value, it is left to the reader of the report to judge the significance of the result. The reader must determine whether to reject the null hypothesis in favor of the alternative, based on the reported p-value. This p-value is often referred to as the **observed significance level** of the test. The null hypothesis will be rejected if the observed significance level is *less* than the fixed significance level α chosen by the reader (see Figure 8.10).

There are two inherent advantages of reporting test results using p-values: (1) Readers are permitted to select the maximum value of α that they would be willing to tolerate in carrying out a standard test of hypothesis in the manner outlined in this chapter; and (2) it is an easy way to present the results of test calculations performed by a statistical software or spreadsheet package.

Reporting Test Results as p-Values: How to Decide Whether to Reject H_O

1. Choose the maximum value of α that you are willing to tolerate.

2. If the observed significance level (p-value) of the test is less than α, then reject the null hypothesis.

Example 8.16 p-Value for a One-Tailed Test

Refer to Example 8.15 and the test H_0: $\mu = 72$ versus H_a: $\mu > 72$. Suppose the test statistic is $z = .42$. Compute the observed significance level of the test and interpret its value.

Solution

In this large-sample test concerning a population mean μ, the computed value of the test statistic is $z_c = .42$. Since the test is upper-tailed, the associated p-value is given by

$$P(z \geq z_c) = P(z \geq .42)$$

Figure 8.11 *p*-Value for Example 8.16

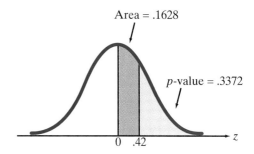

Area = .1628

p-value = .3372

From Table B.2, we obtain

$$P(z \geq .42) = .5 - .1628 = .3372 \quad \text{(see Figure 8.11)}$$

Thus, the observed significance level of the test is $p = .3372$. To reject the null hypothesis H_0: $\mu = 72$, we would have to be willing to risk a Type I error probability α of at least .3372. Most researchers would not be willing to take this risk and would deem the result insignificant (i.e., conclude that there is insufficient evidence to reject H_0). ❑

Example 8.17

p-Value for a Two-Tailed Test

Refer to Example 8.13 (page 436) and the large-sample test of H_0: $\mu = 8.5$ versus H_a: $\mu \neq 8.5$. The Excel printout for this test was given in Figure 8.6 (page 437).

a. Compute the observed significance level of the test. Compare to the value shown on the Excel printout.

b. Make the appropriate conclusion if you are willing to tolerate a Type I error probability of $\alpha = .05$.

Solution

a. The computed test statistic for this large-sample test about μ was given as $z_c = 4.055$. Since the test is two-tailed, the associated *p*-value is

$$2P(z \geq |z_c|) = 2P(z \geq |4.055|) = 2P(z \geq 4.055) \quad \text{(see Figure 8.12)}$$

Since $P(z \geq 4.055)$ is very near 0, the observed significance level of the test is approximately 0. Note that our computed value agrees with the Excel value, $p = .00005$ (5.00929E-05, in scientific notation), highlighted on Figure 8.6b.

Figure 8.12 *p*-Value for Example 8.17

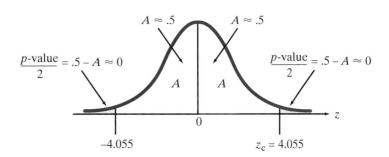

$A \approx .5$ $A \approx .5$

$\dfrac{p\text{-value}}{2} = .5 - A \approx 0$ $\dfrac{p\text{-value}}{2} = .5 - A \approx 0$

A A

-4.055 $z_c = 4.055$

b. Since the p-value is less than the maximum tolerable Type I error probability of $\alpha = .01$, we will reject H_0 and conclude that the mean length-to-width ratio of humerus bones is significantly different from 8.5. In fact, we could choose an even smaller Type I error probability (e.g., $\alpha = .001$) and still have sufficient evidence to reject H_0. Thus, the result is highly significant. ❑

**Self-Test
8.7**

Consider testing the null hypothesis H_0: $\mu = 2$. For each situation below, make the appropriate conclusion about H_0.

a. $\alpha = .01, p\text{-value} = .15$

b. $\alpha = .05, p\text{-value} = .02$

c. $\alpha = .10, p\text{-value} = .05$

Whether we conduct a test using p-values or the rejection region approach, our choice of a maximum tolerable Type I error probability becomes critical to the decision concerning H_0 and should not be hastily made. In either case, care should be taken to weigh the seriousness of committing a Type I error in the context of the problem.

Note: In this section, we demonstrated the procedure for computing p-values for a large-sample test of hypothesis based on the standard normal (z) statistic. For small samples, p-values are computed using the t distribution. However, the t table (Table B.3 in Appendix B) is not suited for this task. Consequently, it is critical to use statistical software or spreadsheet programs to find p-values for small-sample tests of hypotheses.

**PROBLEMS
8.34–8.44**

Using the Tools

8.34   For a large-sample test of

$$H_0\colon \mu = 0$$

$$H_a\colon \mu > 0$$

compute the p-value associated with each of the following computed test statistic values:

a. $z_c = 1.96$ **b.** $z_c = 1.645$

c. $z_c = 2.67$ **d.** $z_c = 1.25$

8.35   For a large-sample test of

$$H_0\colon \pi = .75$$

$$H_a\colon \pi \neq .75$$

compute the *p*-value associated with each of the following computed test statistic values:

a. $z_c = -1.01$ **b.** $z_c = -2.37$

c. $z_c = 4.66$ **d.** $z_c = -1.45$

8.36   Give the approximate observed significance level of the test H_0: $\mu = 16$ for each of the following combinations of test statistic value and H_a:

a. $z_c = 3.05, H_a: \mu \neq 16$ **b.** $z_c = -1.58, H_a: \mu < 16$

c. $z_c = 2.20, H_a: \mu > 16$ **d.** $z_c = -2.97, H_a: \mu \neq 16$

8.37   The *p*-value and the value of α for a test of H_0: $\mu = 150$ are provided for each part. Make the appropriate conclusion regarding H_0.

a. *p*-value = .217, $\alpha = .10$ **b.** *p*-value = .033, $\alpha = .05$

c. *p*-value = .001, $\alpha = .05$ **d.** *p*-value = .866, $\alpha = .01$

e. *p*-value = .025, $\alpha = .01$

Applying the Concepts

8.38   Refer to the *Psychological Assessment* study of World War II aviator POWs, Problem 8.19 (page 432). You tested whether the true mean post-traumatic stress disorder score of World War II aviator POWs is less than 16. Recall that $\bar{x} = 9.00$ and $s = 9.32$ for a sample of $n = 33$ POWs. Compute the *p*-value of the test and interpret the result on page 450.

8.39   The 1-year rate of return to shareholders in 1995 was calculated for each in a sample of 63 electric utility stocks. The data, extracted from *The Wall Street Journal* (Feb. 29, 1996), are shown in the table.

a.   Specify the null and alternative hypotheses tested for determining whether the true mean 1-year rate of return for electric utility stocks exceeded 30%.

b.   Locate the *p*-value for the test, part **a**, on the accompanying Excel printout.

	A	B
1	Z Test of the Hypothesis for the Mean	
2		
3	Null Hypothesis μ=	30
4	Level of Significance	0.05
5	Population Standard Deviation	13.47
6	Sample Size	63
7	Sample Mean	31.93
8	Standard Error of the Mean	1.697060484
9	Z Test Statistic	1.137260586
10		
11	Upper-Tail Test	
12	Upper Critical Value	1.644853
13	*p*-Value	0.127714733
14	Do not reject the null hypothesis	

c.   Interpret the result, part **b**, in the words of the problem.

8.40   Refer to Problem 8.29 (page 442) and the test to determine whether the mean pH level of food items differs from 4.0. Locate the observed

Electric Utilities	Rate of Return (%)	Electric Utilities	Rate of Return (%)
Pinnacle West Capital	52.4	Florida Progress Corp	25.5
DPL Inc	27.7	Consolidated Edison of NY	32.3
Nipsco Industries Inc	34.8	Wisconsin Energy Corp	24.6
Cipsco Inc	53.8	Potomac Electric Power	54.4
Southern Co	30.3	Dominion Resources Inc	22.6
DQE Inc	63.3	Delmarva Power & Light	35.4
Scana Corp	44.4	Northeast Utilities	20.8
Western Resources Inc	24.2	Energy Corp	43.8
Portland General Corp	59.0	Central & South West Corp	31.8
Peco Energy Co	30.5	Detroit Edison Co	41.2
Allegheny Power System	40.6	Hawaiian Electric Inds	27.8
New England Electric System	31.8	Public Service Entrp	24.5
Public Service Co of Colo	28.4	Puget Sound Power & Light	25.7
FPL Group Inc	38.1	Texas Utilities Co	38.8
CiNergy Corp	39.2	Southwestern Public Svc Co	32.9
Illinova Corp	42.8	Idaho Power Co	37.4
LG&E Energy Corp	21.0	PP&L Resources Inc	42.0
Baltimore Gas & Electric	36.8	Oklahoma Gas & Electric	39.7
American Electric Power	31.9	Pacific Gas & Electric	24.8
Kansas City Power & Light	19.8	Montana Power Co	5.5
Boston Edison Co	32.0	San Diego Gas & Electric	32.4
General Public Utilities	38.0	New York State Elec & Gas	44.8
Carolina Power & Light	37.5	Unicom Corp	44.5
Duke Power Co	30.4	MidAmerican Energy Co	31.8
Ohio Edison Co	35.9	PacifiCorp	23.4
KU Energy Corp	17.9	SCEcorp	26.1
Northern States Power	18.4	CMS Energy Corp	35.4
Ipalco Enterprises Inc	35.3	Long Island Lighting	19.0
Union Electric Co	25.9	Niagara Mohawk Power	−27.1
UtiliCorp United Inc	17.8	Centerior Energy Corp	8.6
Teco Energy Inc	32.7	AES Corp	22.4
Houston Industries Inc	46.1		

Problem 8.39

ELECUTIL.XLS

significance level (*p*-value) of the test on the PHStat Add-in for Excel printout shown on page 451. Interpret the result.

8.41 Refer to Problem 8.32 (page 443) and the test to determine whether the average energy efficiency rating is different from 9.0. Find the approximate observed significance level (*p*-value) of the test. What is your decision regarding H_0 if you are willing to risk a maximum Type I error probability of only $\alpha = .01$?

8.42 Marine scientists at the University of South Florida investigated the feeding habits of midwater fish inhabiting the eastern Gulf of Mexico (*Prog. Oceanog.,* Vol. 38, 1996). Most of the fish captured fed heavily on copepods, a type of zooplankton. The table gives the percent of shallow-living copepods in the diets of each of 35 species of myctophid fish. Use the *p*-value approach to test the hypothesis (at $\alpha = .05$) that the true mean percentage of shallow-living copepods in the diets of all myctophid fish inhabiting the Gulf of Mexico differs from 90%.

E		A	B
1		*t* Test for the Hypothesis of the Mean	
2			
3		Null Hypothesis $\mu=$	4
4		Level of Significance	0.1
5		Sample Size	36
6		Sample Mean	5.08
7		Sample Standard Deviation	1.29
8		Standard Error of the Mean	0.215
9		Degrees of Freedom	35
10		*t* Test Statistic	5.023255814
11			
12		Two-Tailed Test	
13		Lower Critical Value	-1.689572855
14		Upper Critical Value	1.689572855
15		*p*-Value	1.49544E-05
16		Reject the null hypothesis	

Problem 8.40

COPEPODS.XLS

79	100	100	98	95	96	56	51	90	95	94	93
92	71	100	100	99	100	100	100	99	80	54	56
59	92	88	81	88	59	100	85	82	74	66	

Source: Hopkins, T. L., Sutton, T. T,. and Lancraft, T. M. "The tropic structure and predation impact of low latitude midwater fish assemblage." *Prog. Oceanog.,* Vol. 38, No. 3, 1996, p. 223 (Table 6).

Problem 8.42

8.43   Radium-226 is a naturally occurring radioactive gas. Elevated levels of radium-226 in metro-Dade County (Florida) were recently investigated (*Florida Scientist,* Summer/Autumn 1991). The data in the table are radium-226 levels (measured in pCi/L) for 26 soil specimens collected in southern Dade County. The Environmental Protection Agency (EPA) has set maximum exposure levels of radium-226 at 4.0 pCi/L. Use the information in the PHStat Add-in for Excel printout (page 452) to determine whether the mean radium-226 level of soil specimens collected in southern Dade County is less than the EPA limit of 4.0 pCi/L. Use $\alpha = .10$.

RADIUM.XLS

1.46	0.58	4.31	1.02	0.17	2.92	0.91	0.43	0.91
1.30	8.24	3.51	6.87	1.43	1.44	4.49	4.21	1.84
5.92	1.86	1.41	1.70	2.02	1.65	1.40	0.75	

Source: Moore, H. E., and Gussow, D. G. "Radium and radon in Dade County ground water and soil samples." *Florida Scientist,* Vol. 54, No. 3/4, Summer/Autumn 1991, p. 155 (portion of Table 3).

8.44   During the summer of 1993, Iowa received an exceptional amount of precipitation. The abundance of standing water led to an unusually high mosquito population. Entomologists at Iowa State University collected data on the mosquito population at six Iowa sites during 1993 (*Journal of the American Mosquito Control Association,* June 1995). Using light traps, mosquitos were collected daily and their species identified and counted. A light trap index (LTI), measured as the number of mosquitos per light

E		A	B	C	D	E
1		*t* Test for the Hypothesis of the Mean				
2						
3		Null Hypothesis μ=	4			
4		Level of Significance	0.1			
5		Sample Size	26			
6		Sample Mean	2.41			
7		Sample Standard Deviation	2.08			
8		Standard Error of the Mean	0.4079216			
9		Degrees of Freedom	25			
10		*t* Test Statistic	-3.897808			
11					Calculations Area	
12		Lower-Tail Test			For one-tailed tests:	
13		Lower Critical Value	-1.316346		TDIST value	0.000322
14		*p*-Value	0.0003219		1-TDIST value	0.999678
15		Reject the null hypothesis				

Problem 8.43

trap per night, was measured at each location. At one site (Des Moines), the sample mean LTI in 1993 was $\bar{x} = 108.15$. The entomologists want to know whether the true 1993 mean LTI at Des Moines exceeds the mean LTI at Des Moines for the previous 10 years, $\mu = 47.82$.

a. Set up the null and alternative hypotheses for the appropriate test.

b. The *p*-value for the test was reported at $p = .04$. Make the appropriate conclusion at $\alpha = .05$.

c. Repeat part **b**, but use $\alpha = .01$.

8.7 TESTING A POPULATION PROPORTION

The procedure described in the box is used to test a hypothesis about a population proportion π, based on a *large* sample from the target population. (Recall that π represents the probability of success in a binomial experiment.)

Large-Sample (z) Test of Hypothesis About a Population Proportion π

Upper-Tailed Test	Two-Tailed Test	Lower-Tailed Test
$H_0: \pi = \pi_0$	$H_0: \pi = \pi_0$	$H_0: \pi = \pi_0$
$H_a: \pi > \pi_0$	$H_a: \pi \neq \pi_0$	$H_a: \pi < \pi_0$

$$\text{Test statistic: } z = \frac{p - \pi_0}{\sqrt{\pi_0(1 - \pi_0)/n}} \qquad (8.5)$$

where $p = \dfrac{x}{n}$ = sample proportion of "successes"

Rejection region:	Rejection region:	Rejection region:
$z > z_\alpha$	$|z| > z_{\alpha/2}$	$z < -z_\alpha$

Assumption: The sample size *n* is large.

Example 8.18   Large-Sample Test of π

An *American Demographics* study conducted in 1980 found that 40% of new car buyers were women. Suppose that in a random sample of $n = 120$ new car buyers in 1998, 57 were women. Does this evidence indicate that the true proportion of new car buyers in 1998 who were women is significantly larger than .40, the 1980 proportion? Use the rejection region approach, testing at significance level $\alpha = .10$.

Solution

We wish to perform a large-sample test about a population proportion π.

$$H_0: \pi = .40 \text{ (no change from 1980 to 1998)}$$

$$H_a: \pi > .40 \text{ (proportion of new car buyers who were women}$$
$$\text{was greater in 1998)}$$

where π represents the true proportion of all new car buyers in 1998 who were women.

At significance level $\alpha = .10$, the rejection region for this one-tailed test consists of all values of z for which

$$z > z_{.10} = 1.28 \quad \text{(see Figure 8.13)}$$

The test statistic requires the calculation of the sample proportion p of new car buyers who were women:

$$p = \frac{\text{Number of sampled new car buyers who were women}}{\text{Number of new car buyers sampled}}$$

$$= \frac{57}{120} = .475$$

Substituting into equation 8.5, we obtain the following value of the test statistic:

$$z = \frac{p - \pi_0}{\sqrt{\pi_0(1 - \pi_0)/n}} = \frac{.475 - .40}{\sqrt{(.40)(.60)/120}} = 1.677$$

This value of z lies within the rejection region; we thus conclude that the proportion of new car buyers in 1998 who were women increased significantly from .40. The probability of our having made a Type I error (rejecting H_0 when, in fact, it is true) is $\alpha = .10$. ❑

Figure 8.13 Rejection Region for Example 8.18

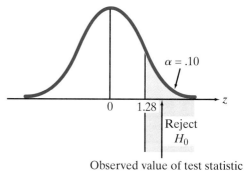

E **Figure 8.14** PHStat
Add-in for Excel
Output, Example 8.19

	A	B
1	**Z Test of Hypothesis for the Proportion**	
2		
3	**Null Hypothesis** *p*=	**0.4**
4	**Level of Significance**	**0.1**
5	**Number of Successes**	**57**
6	**Sample Size**	**120**
7	Sample Proportion	0.475
8	Standard Error	0.04472136
9	Z Test Statistic	1.677050983
10		
11	**Upper-Tail Test**	
12	**Upper Critical Value**	1.281550794
13	*p*-Value	0.046766232
14	**Reject the null hypothesis**	

Example 8.19 *p*-Value for a Test of π

Refer to Example 8.18. Find the observed significance level (*p*-value) of the test and interpret its value.

Solution

For this large-sample, upper-tailed test, the computed value of the test statistic is $z_c \approx 1.68$. Thus, the *p*-value is approximated by finding

$$P(z > z_c) = P(z > 1.68)$$

From Table B.2, this probability is $.5 - .4535 = .0465$. (The exact *p*-value for this test, .0468, is shown on the PHStat Add-in for Excel printout in Figure 8.14.

Remember, *p*-value = .0465 is the minimum α value that leads to a rejection of the null hypothesis. Consequently, we will reject H_0 for any α value exceeding .0465 (as in Example 8.18) but will fail to reject H_0 for α values below .0465. Therefore, a researcher who desires a Type I error rate of only $\alpha = .01$ or $\alpha = .02$ will have insufficient evidence to say that the proportion of new car buyers who were women in 1998 exceeds .40. However, $\alpha = .10$ (as in Example 8.18) or $\alpha = .05$ will lead the researcher to reject H_0 and claim that the proportion of new car buyers who are women has increased. ❏

Self-Test

8.8

In a nationwide telephone poll, 1,532 adults were asked: "Do you believe that people who become victims of muggings or holdups have a right to take matters into their own hands?" Forty-seven percent answered "yes." Test the hypothesis that fewer than 50% of American adults believe that crime victims have a right to take matters into their own hands. Use a significance level of $\alpha = .10$.

Although small-sample procedures are available for testing hypotheses about a population proportion, the details are omitted from our discussion. It is our experience that they are of limited utility, since most surveys of binomial populations (for example, opinion polls) performed in the real world use samples that are large enough to employ the techniques of this section.

**PROBLEMS
8.45–8.56**

Using the Tools

8.45   A random sample of n observations is selected from a binomial population to test the null hypothesis that $\pi = .40$. Specify the rejection region for each of the following combinations of H_a and α:

 a. $H_a: \pi \neq .40, \alpha = .05$ **b.** $H_a: \pi < .40, \alpha = .05$

 c. $H_a: \pi > .40, \alpha = .10$ **d.** $H_a: \pi < .40, \alpha = .01$

 e. $H_a: \pi \neq .40, \alpha = .01$

8.46   A random sample of n observations is selected from a binomial population. For each of the following situations, specify the rejection region, test statistic value, and conclusion:

 a. $H_0: \pi = .10, H_a: \pi > .10, p = .13, n = 200, \alpha = .10$

 b. $H_0: \pi = .05, H_a: \pi < .05, p = .04, n = 1,124, \alpha = .05$

 c. $H_0: \pi = .90, H_a: \pi \neq .90, p = .73, n = 125, \alpha = .01$

8.47   A random sample of 100 observations from a binomial population resulted in 45 successes.

 a.   Test the null hypothesis that the true proportion of successes in the population is .5 against the alternative hypothesis that $\pi < .5$. Use $\alpha = .05$.

 b.   Test $H_0: \pi = .35$ against $H_a: \pi \neq .35$ at $\alpha = .05$.

 c.   Test $H_0: \pi = .4$ against $H_a: \pi > .4$ at $\alpha = .05$.

8.48   Consider a test of $H_0: \pi = .75$ and let x = number of successes in a sample of size n. Compute the test statistic for each of the following:

 a. $x = 100, n = 500$ **b.** $x = 125, n = 200$

 c. $x = 1,122, n = 1,500$ **d.** $x = 31, n = 50$

Applying the Concepts

8.49   A Louis Harris poll of 1,502 working women asked: "If you had a choice, which would you prefer, more money or more time off?" The results: 706 preferred more money, 691 preferred more time off, and 105 were not sure (*Chicago Tribune*, July 27, 1995). Test the hypothesis that the true proportion of working women who prefer more time off is less than .5. Use $\alpha = .10$.

8.50   According to a survey sponsored by the American Society for Microbiology, 40% of the people using the restrooms in New York City's Penn Station do not wash their hands after going to the bathroom (*Tampa Tribune*, Sept. 17, 1996). Suppose you observe 427 people who fail to wash up in a random sample of 1,000 people who use the Penn Station restrooms. Use this information to test the claim made by the American Society for Microbiology. Test using $\alpha = .05$.

8.51   *Jeopardy!* is a popular television game show in which three contestants answer general knowledge questions on a wide variety of topics and earn money for each correct answer. The contestant with the highest total amount wins, thereby earning the right to defend his/her championship the next day. *Chance* (Spring 1994) collected data for a sample of 218 *Jeopardy!* games. In these 218 games, the defending champion won 126 times. Does this imply that the defending *Jeopardy!* champion has no better or no worse than a 50-50 chance of winning the next day? Test using $\alpha = .01$.

8.52   The *New England Journal of Medicine* (Oct. 6, 1994) published a study on the link between guns in the home and homicides. Studying 420 homicides that occurred in the homes of victims, the researchers found that 322 were killed by a spouse, family member, or acquaintance. Is there sufficient evidence (at $\alpha = .05$) to claim that a spouse, family member, or acquaintance is the killer in more than 70% of all homicides that occur in the victim's home?

8.53   According to a survey reported in the *Journal of the American Medical Association* (Apr. 22, 1998), only 608 of 1,066 randomly selected Americans recognized the warning signs of a stroke.

a.   Test the hypothesis that the true percentage of Americans who recognize the warning signs of a stroke exceeds 50%. Use $\alpha = .05$.

b.   Test the hypothesis that the true percentage of Americans who recognize the warning signs of a stroke exceeds 55%. Use the *p*-value method and $\alpha = .05$.

8.54   How attuned are mothers to their newborn babies? Researchers tested a sample of women shortly after they gave birth in a large urban hospital. Each mother, blindfolded to prevent sight and smell, felt the hands of her own infant plus two others of the same sex. Of the 68 mothers with at least an hour of exposure to their newborns, 47 correctly identified their own baby (*Development Psychology,* Jan. 1992). The researchers were "surprised," "amazed," and "excited" about the results. Do the sample results indicate that mothers have newborn recognition ability greater than chance? Test using $\alpha = .05$.

8.55   Archaeologists have found a paucity of antique grinding implements (e.g., grindstones, seed grinders) in Australia. Past excavations have indicated that 2.5% of all artifacts found in Australia are grinding implements. A recent archaeological dig at Wanmara in Central Australia found a total of 1,436 artifacts. Of these, 102 were identified as grinding implements (*Archaeol. Oceania,* July 1997). Conduct a test to determine whether the true percentage of artifacts found at Wanmara that are grinding implements differs from 2.5%. Use $\alpha = .01$.

8.56   According to a study in *Nature* (Dec. 1995), 1-year-old babies are more likely to resemble their dads in physical appearance. The researchers asked a sample of 122 people to compare photos of 1-year-old babies with photos of three possible parents (the dad, the mom, and a stranger) and to select the "parent" that most closely resembles the baby. For one particular 1-year-old baby, suppose 70 of the 122 people chose the dad's photo as most similar in appearance.

a. Conduct a test to determine whether the true proportion of people who believe the dad most resembles the baby exceeds $\frac{1}{3}$. Use $\alpha = .01$.

b. Conduct a test to determine whether the true proportion of people who believe the dad most resembles the baby exceeds .5. Use $\alpha = .01$.

8.8   TESTING A POPULATION VARIANCE (OPTIONAL)

Hypothesis tests about a population variance σ^2 are conducted using the chi-square (χ^2) distribution introduced in optional Section 7.9. The test is outlined in the box. Note that the assumption of a normal population is required regardless of whether the sample size n is large or small.

Test of Hypothesis About a Population Variance σ^2

Upper-Tailed Test	Two-Tailed Test	Lower-Tailed Test
$H_0: \sigma^2 = \sigma_0^2$	$H_0: \sigma^2 = \sigma_0^2$	$H_0: \sigma^2 = \sigma_0^2$
$H_a: \sigma^2 > \sigma_0^2$	$H_a: \sigma^2 \neq \sigma_0^2$	$H_a: \sigma^2 < \sigma_0^2$

$$\text{Test statistic: } \chi^2 = \frac{(n-1)s^2}{\sigma_0^2} \qquad (8.6)$$

Rejection region:	Rejection region:	Rejection region:
$\chi^2 > \chi_\alpha^2$	$\chi^2 < \chi_{(1-\alpha/2)}^2$ or $\chi^2 > \chi_{\alpha/2}^2$	$\chi^2 < \chi_{(1-\alpha)}^2$

where χ_α^2 and $\chi_{(1-\alpha)}^2$ are values of χ^2 that locate an area of α to the right and α to the left, respectively, of a chi-square distribution based on $(n-1)$ degrees of freedom.

Assumption: The population from which the random sample is selected has an approximate normal distribution.

Example 8.20   **Test for σ^2**

Regulatory agencies specify that the standard deviation of the amount of fill for 8-ounce cans of chicken noodle soup produced in a cannery should be less than .1 ounce. To check that specifications are being met, the quality control supervisor at the cannery sampled ten 8-ounce cans and measured the amount of fill in each. The data (in ounces) are shown here.

7.96	7.90	7.98	8.01	7.97	7.96	8.03	8.02	8.04	8.02

CANS.XLS

Is there sufficient evidence to indicate that the standard deviation σ of the fill measurements is less than .1 ounce?

a. Conduct the test using the rejection region approach.

b. Conduct the test using the p-value approach.

Solution

a. Since the null and alternative hypotheses are stated in terms of σ^2 (rather than σ), we will want to test the null hypothesis that $\sigma^2 = .01$ against the alternative that $\sigma^2 < .01$. Therefore, the elements of the test are

$$H_0: \sigma^2 = .01 \ (\sigma = .1)$$

$$H_a: \sigma^2 < .01 \ (\sigma < .1)$$

Assumption: The population of amounts of fill of the cans is approximately normal.

Test statistic:

$$\chi^2 = \frac{(n-1)s^2}{\sigma_0^2}$$

Rejection region:

The smaller the value of s^2 we observe, the stronger the evidence in favor of H_a. Thus, we reject H_0 for "small values" of the test statistic. With $\alpha = .05$ and $(n-1) = 9$ df, the critical χ^2 value is found in Table B.4 in Appendix B and illustrated in Figure 8.15. We will reject H_0 if $\chi^2 < 3.32511$. (Remember that the area given in Table B.4 is the area to the *right* of the numerical value in the table. Thus, to determine the lower-tail value that has $\alpha = .05$ to its *left*, we use the $\chi^2_{.95}$ column in Table B.4.)

To compute the test statistic, we need to find the sample standard deviation s. Numerical descriptive statistics for the sample data are provided in the Excel printout, Figure 8.16a (page 459).

The value of s, highlighted in Figure 8.16b, is $s = .043$. Substituting $s = .043$, $n = 10$, and $\sigma_0^2 = .01$ into equation 8.6, we obtain the test statistic

$$\chi^2 = \frac{(10-1)(.043)^2}{.01} = 1.66$$

Conclusion: Since the test statistic $\chi^2 = 1.66$ is less than 3.32511, the supervisor can reject H_0 and conclude that the standard deviation σ of the population of all amounts of fill is less than .1 at $\sigma = .05$. Thus, the quality control supervisor is confident in the decision that the cannery is operating within the desired limits of variability.

Figure 8.15 Rejection Region for Example 8.20

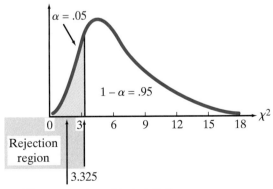

$\alpha = .05$

$1 - \alpha = .95$

Rejection region

3.325

Observed value of test statistic $\chi^2 = 1.66$

E Figure 8.16 Excel and PHStat Add-In for Excel Output, Example 8.20

	A	B
1	*Fill Amount*	
2		
3	Mean	7.989
4	Standard Error	0.013617799
5	Median	7.995
6	Mode	7.96
7	Standard Deviation	0.043063261
8	Sample Variance	0.001854444
9	Kurtosis	0.479370993
10	Skewness	-0.85380263
11	Range	0.14
12	Minimum	7.9
13	Maximum	8.04
14	Sum	79.89
15	Count	10
16	Largest(1)	8.04
17	Smallest(1)	7.9

a.

	A	B
1	Chi-Square Test of Hypothesis for the Variance	
2		
3	Null Hypothesis $\sigma^2=$	0.01
4	Level of Significance	0.05
5	Sample Size	10
6	Sample Standard Deviation	0.043
7	Degrees of Freedom	9
8	Chi-Square Test Statistic	1.6641
9		
10	Lower-Tail Test	
11	Lower Critical Value	3.32511514
12	*p*-Value	0.004263521
13	Reject the null hypothesis	

b.

b.   The *p*-value for the test is shown on the PHStat Add-in for Excel printout, Figure 8.16b. Its value (highlighted) is $p = .0042$. Since the *p*-value is less than $\alpha = .05$, we reject H_0 and conclude (as in part **a**) that $\sigma < .1$. ❑

Self-Test
8.9

A random sample of $n = 10$ observations yielded $\bar{x} = 231.7$ and $s^2 = 15.5$. Test the null hypothesis $H_0: \sigma^2 = 20$ against the alternative hypothesis $H_a: \sigma^2 < 20$. Use $\alpha = .05$. What assumptions are necessary for the test to be valid?

PROBLEMS
8.57–8.68

Using the Tools

8.57   Calculate the test statistic for $H_0: \sigma^2 = \sigma_0^2$ in a normal population for each of the following:
a. $n = 25, \bar{x} = 17, s^2 = 8, \sigma_0^2 = 10$
b. $n = 12, \bar{x} = 6.2, s^2 = 1.7, \sigma_0^2 = 1$
c. $n = 50, \bar{x} = 106, s^2 = 31, \sigma_0^2 = 50$
d. $n = 100, \bar{x} = .35, s^2 = .04, \sigma_0^2 = .01$

8.58   A random sample of n observations, selected from a normal population, is used to test the null hypothesis $H_0: \sigma^2 = 9$. Specify the appropriate rejection region for each of the following:
a. $H_a: \sigma^2 > 9, n = 20, \alpha = .01$ **b.** $H_a: \sigma^2 \neq 9, n = 20, \alpha = .01$
c. $H_a: \sigma^2 < 9, n = 12, \alpha = .05$ **d.** $H_a: \sigma^2 < 9, n = 12, \alpha = .10$

8.59   The following measurements represent a random sample of $n = 5$ observations from a normal population: 11, 7, 2, 9, 13. Is this sufficient evidence to conclude that $\sigma^2 \neq 2$? Test using $\alpha = .10$.

8.60   A random sample of $n = 35$ observations yielded $\bar{x} = .107$ and $s = .022$. Test the null hypothesis $H_0: \sigma = .04$ against the alternative hypothesis $H_a: \sigma \neq .04$. Use $\alpha = .10$. What assumptions are necessary for the test to be valid?

Applying the Concepts

8.61   A candy manufacturer must monitor the temperature at which candies are baked. Too much variation will cause inconsistency in the taste of the candy. When the process is operating correctly, the standard deviation of the baking temperature is 1.3°F. A random sample of 50 batches of candy is selected and the sample standard deviation of the baking temperature is $s = 2.1$°F. Is there evidence that the true standard deviation of the baking temperature has increased above 1.3°F? Test using $\alpha = .05$.

8.62   Polychlorinated biphenyls (PCBs) are extremely hazardous contaminants when released into the environment. The Environmental Protection Agency (EPA) is experimenting with a new device for measuring PCB concentration in fish. To check the precision of the new instrument, seven PCB readings were taken on the same fish sample. The data are recorded here (in parts per million):

6.2   5.8   5.7   6.3   5.9   5.8   6.0

PCB.XLS

Suppose the EPA requires an instrument that yields PCB readings with a variance of less than .1. Does the new instrument meet the EPA's specifications? Test using $\alpha = .05$.

8.63   Refer to the *Journal of Engineering for Industry* (May 1993) study of deep hole drilling under drill chip congestion, Problem 8.27 (page 442). Test (at $\alpha = .10$) to determine whether the true standard deviation of drill chip lengths differs from 75 mm. Recall that for $n = 50$ drill chips, $s = 50.2$.

8.64   Refer to the *Journal of Drug Issues* (Summer 1997) study of pregnant cocaine-dependent women, Problem 8.31 (page 443). Test (at $\alpha = .01$) to determine whether the variance in the birthweights of babies delivered by cocaine-dependent women is less than 200,000 grams^2. Recall that for a sample of $n = 16$ cocaine-dependent women, the babies delivered had a mean birthweight of $\bar{x} = 2{,}971$ grams and a standard deviation of $s = 410$ grams.

8.65   Refer to the study of sleep-deprived college students in Problem 8.33 (page 444). Some sleep experts believe that although most people lose their ability to think creatively when deprived of sleep, others are not nearly as affected. Consequently, they theorize that the variation in test scores for sleep-deprived subjects is larger than 225, the estimated variance in scores of those who receive their usual amount of sleep. Test the hypothesis that σ^2, the variance in test scores of sleep-deprived subjects, is larger than 225. Recall that $s = 17$ in the study of $n = 12$ sleep-deprived subjects. Use $\alpha = .01$.

8.66   A company that produces a fast-drying rubber cement in 32-ounce aluminum cans is interested in testing whether the variance of the amount of rubber cement dispensed into the cans is more than .3. If so, the dispensing machine needs adjustment. Since inspection of the canning process requires that the dispensing machines be shut down, and shut-downs for any lengthy period of time cost the company thousands of dollars in lost revenue, the company is able to obtain a random sample of only 10 cans for testing. After measuring the weights of their contents, the following summary statistics are obtained:

$$\bar{x} = 31.55 \text{ ounces} \qquad s = .48 \text{ ounce}$$

a. Does the sample evidence indicate that the dispensing machines are in need of adjustment? Test at significance level $\alpha = .05$.

b. What assumption is necessary for the hypothesis test of part **a** to be valid?

8.67 Following the Persian Gulf War of 1991, the Pentagon changed its delivery system to a "just-in-time" system. Deliveries from factories to foxholes are now expedited using bar codes, laser cards, radio tags, and databases to track supplies (*Business Week,* Dec. 11, 1995). During the Persian Gulf War, the Pentagon found the standard deviation of the order-to-delivery times for all shipments to be $\sigma = 10.5$ days. To gauge whether the standard deviation has decreased under the new system, the Pentagon randomly sampled nine shipments made during the 1995 Bosnian War. The order-to-delivery times (in days) for these nine shipments are listed below:

BOSNIA.XLS

| 15.1 | 6.4 | 5.0 | 11.4 | 6.5 | 6.5 | 3.0 | 7.0 | 5.5 |

Conduct a test to determine whether the standard deviation of the order-to-delivery times during the Bosnian War is less than 10.5 days. Use $\alpha = .05$.

8.68 To improve the signal-to-noise ratio (SNR) in the electrical activity of the brain, neurologists repeatedly stimulate subjects and average the responses —a procedure that assumes that single responses are homogeneous. A study was conducted to test the homogeneous signal theory (*IEEE Engineering in Medicine and Biology Magazine,* Mar. 1990). The null hypothesis is that the variance of the SNR readings of subjects equals the "expected" level under the homogeneous signal theory. For this study, the "expected" level was assumed to be .54. If the SNR variance exceeds this level, the researchers will conclude that the signals are nonhomogeneous.

a. Set up the null and alternative hypotheses for the researchers.

b. SNRs recorded for a sample of 41 normal children ranged from .03 to 3.0. Use this information to obtain an estimate of the sample standard deviation. [*Hint:* Assume that the distribution of SNRs is normal, and that most SNRs in the population will fall within $\mu \pm 2\sigma$, i.e., from $\mu - 2\sigma$ to $\mu + 2\sigma$. Note that the range of the interval equals 4σ.]

c. Use the estimate of s in part **b** to conduct the test of part **a**. Test using $\alpha = .10$.

8.9 POTENTIAL HYPOTHESIS-TESTING PITFALLS AND ETHICAL ISSUES

When planning to carry out a test of hypothesis based on some designed experiment or research study under investigation, several questions need to be raised to ensure that proper methodology is used:

1. What is the goal of the experiment or research? Can it be translated into null and alternative hypotheses?
2. Is the hypothesis test going to be two-tailed or one-tailed?
3. Can a random sample be drawn from the population of interest?
4. What kind of measurements will be obtained from the sample? Is the variable measured quantitative or qualitative?
5. At what significance level, or risk of committing a Type I error, should the hypothesis test be conducted?
6. Is the intended sample size large enough to use the large-sample (z) test, or should the small-sample (t) test be applied?
7. What kind of conclusions and interpretations can be drawn from the results of the hypothesis test?

Questions like these need to be raised and answered in the planning stage of a survey or designed experiment, so a person with substantial statistical training should be consulted and involved early in the process. All too often such an individual is consulted far too late in the process, after the data have been collected. Typically in such a situation, all that can be done at such a late stage is to choose the statistical test procedure that would be best for the obtained data. We are forced to assume that certain biases that have been built into the study (because of poor planning) are negligible. But this is a large assumption. Good research involves good planning. To avoid biases, adequate controls must be built in from the beginning.

In this section, we want to distinguish between what is poor research methodology and what is unethical behavior. Ethical considerations arise when a researcher is manipulative of the hypothesis-testing process. Some of the ethical issues that arise when dealing with hypothesis-testing methodology are listed below, followed by a brief description of each problem.

- Data collection method—randomization
- Informed consent from human subjects being "treated"
- Type of test—two-tailed or one-tailed
- Choice of level of significance α
- Data snooping
- Discarding of data outliers
- Reporting of findings

Data Collection Method—Randomization

To eliminate the possibility of potential biases in the results, we must use proper data collection methods. To be able to draw meaningful conclusions, the data we obtain must be the outcomes of a random sample from the target population or the outcomes from some experiment in which a **randomization** process was employed. For example, potential subjects should not be permitted to self-select for a study. In a similar manner, a researcher should not be permitted to purposely select the subjects for the study. Aside from the potential ethical issues that may be raised, such a lack of randomization can result in selection biases and destroy the value of the study.

Informed Consent from Human Subjects Being "Treated"

Ethical considerations require that any individual who is to be subjected to some "treatment" in an experiment be apprised of the research endeavor and any potential behavioral or physical side effects and provide informed consent with respect to participation. A researcher is not permitted to dupe or manipulate the subjects in a study.

Type of Test—Two-Tailed or One-Tailed

If we have prior information that leads us to test the null hypothesis against a specifically directed alternative, then a one-tailed test will be more powerful than a two-tailed test. On the other hand, we should realize that if we are interested only in *differences* from the null hypothesis, not in the *direction* of the difference, the two-tailed test is the appropriate procedure to utilize. This is an important point. For example, if previous research and statistical testing have already established the difference in a particular direction, or if an established scientific theory states that it is only possible for results to occur in one direction, then a one-tailed or directional test may be employed. However, these conditions are not often satisfied in practice, and it is recommended that one-tailed tests be used cautiously.

Choice of Level of Significance α

In a well-designed experiment or study, the level of significance α is selected in advance of data collection. One cannot be permitted to alter the level of significance, after the fact, to achieve a specific result. This would be **data snooping**. One answer to this issue of level of significance is to always report the *p*-value, not just the results of the test.

Data Snooping

Data snooping is never permissible. It would be unethical to perform a hypothesis test on a set of data, look at the results, and then select whether it should be two-tailed or one-tailed and/or choose the level of significance. These steps must be done first, as part of the planned experiment or study, before the data are collected, for the conclusions drawn to have meaning. In those situations in which a statistician is consulted by a researcher late in the process, with data already available, it is imperative that the null and alternative hypotheses be established and the level of significance chosen prior to carrying out the hypothesis test.

Discarding Data Outliers

After editing, coding, and transcribing the data, one should always look for outliers—i.e., any observations whose measurements seem to be extreme or unusual. Stem-and-leaf displays and box-and-whisker plots aid in this *exploratory data analysis* stage. In addition, the exploratory data analysis enables us to examine the data graphically with respect to the assumptions underlying a particular hypothesis test procedure.

The process of outlier detection raises a major ethical question. Should an observation be removed from a study? The answer is a qualified "yes." If it can be determined that a measurement is incomplete or grossly in error owing to some equipment problem or unusual behavioral occurrence unrelated to the study, a decision to discard the observation may be made. Sometimes there is no choice—an individual may decide to quit a particular study he has been participating in before a final measurement can be made. In a well-designed experiment or study, the researcher would plan, in advance, decision rules regarding the possible discarding of data.

Reporting of Findings

When conducting research, it is vitally important to document both good and bad results so that individuals who follow up on such research do not have to "reinvent the wheel." It would be inappropriate to report the results of hypothesis tests that show statistical significance but not those for which there was insufficient evidence in the findings.

Ethical Considerations: A Summary

Again, when discussing ethical issues concerning the hypothesis-testing methodology, the key is *intent*. We must distinguish between poor data analysis and unethical practice. Unethical behavior occurs when a researcher willfully causes a selection bias in data collection, manipulates the treatment of human subjects without informed consent, uses data snooping to select the type of test (two-tailed or one-tailed) and/or level of significance to his or her advantage, hides the facts by discarding observations that do not support a stated hypothesis, or fails to report pertinent findings.

REAL-WORLD REVISITED
Cellular Telephones—Road Hazards?

CELLPHON.XLS

Do you own a cellular telephone? If not, you probably know many others who do. Over the past several years, the market for cell phones has increased at a phenomenal rate. It is estimated that over 10% of the population now owns a cellular phone. According to a 1997 Value Line Investment Survey, the cell phone industry in North America has daily revenues of over $60 million.

With the explosion of the cell phone market, more and more people are using their cell phones while driving their cars. Safety, convenience, and increased productivity are some of the many reasons why car cell phones are so popular. But are they dangerous too? Many people claim that car phone usage leads to a greater risk of an accident or collision. In fact, some countries have enacted regulations that restrict cellular telephone usage while driving.

Is there a link between cell phone usage while driving and the risk of a motor vehicle collision? University of Toronto researchers D. A. Redelmeier and R. J. Tibshirani attempted to answer this question by conducting a controlled experiment (*Chance*, Spring 1997). The researchers identified drivers in Toronto (Ontario, Canada) who had been involved in a collision with significant damage, but no personal injury. (Data on personal injury collisions were not accessible.) For a 1-year period, 699 cases were identified in which the drivers involved in the collision had cell phones and could provide detailed cellular telephone billing records.

For each sampled case, the researchers recorded the values of two qualitative variables. The first variable was whether or not the driver made a cell phone call during the 10-minute interval immediately before the collision. The researchers called this 10-minute interval the "hazard" interval. The second variable was whether or not the driver made a cell phone call in the car during a similar interval on the day prior to the collision. The researchers called the prior day interval the "control" interval.

The data on these two qualitative variables for all 699 sampled cases are stored in the Excel workbook CELLPHON.XLS. Access the workbook and answer the following questions with a test of hypothesis.* Use a significance level α of your choice.

*The researchers analyzed the data using a sophisticated statistical method appropriate for a case-crossover design, called McNemar's test. See Maclure (1991) for a discussion of this test.

Study Questions

a. Is there evidence to indicate that the likelihood of a driver making a cell phone call during the hazard interval exceeds 5%?

b. Is there evidence to indicate that the likelihood of a driver making a cell phone call during the control interval exceeds 5%?

Key Terms

Alternative hypothesis 418	Null hypothesis 419	Test statistic 425
Critical value 427	Observed significance level 446	Two-tailed test 420
Data snooping 463	One-tailed test 420	Type I error 421
Fixed significance level 444	p-value 445	Type II error 421
Hypothesis-testing procedure 426	Randomization 462	Upper-tailed test 429
Level of significance 422	Rejection region 426	
Lower-tailed test 430	Statistical hypothesis 418	

Key Formulas

Large-sample test statistic: $z = \dfrac{\text{Estimator} - \text{Hypothesized } (H_0) \text{ value}}{\text{Standardized error}}$ **(8.3)**, 435   **(8.5)**, 452

Small-sample test statistic: $t = \dfrac{\text{Estimator} - \text{Hypothesized } (H_0) \text{ value}}{\text{Standardized error}}$ **(8.4)**, 438

Note: The respective estimators and standard errors for the population parameters μ and π are provided in Table 7.8 of Chapter 7 (page 402). [The test statistic for the parameter σ^2 from the optional section of this chapter is given below.]

Test statistic for σ^2: $\chi^2 = \dfrac{(n-1)s^2}{\sigma_0^2}$ **(8.6)**, 457

Key Symbols

Symbol	Description
H_0	Null hypothesis
H_a	Alternative hypothesis
α	Probability of a Type I error (reject H_0 when H_0 is true)
β	Probability of a Type II error (accept H_0 when H_0 is false)
μ_0	Hypothesized value of μ in H_0
π_0	Hypothesized value of π in H_0
σ_0^2	Hypothesized value of σ^2 in H_0

Supplementary Problems 8.69–8.89

Starred (*) problems refer to the optional section of this chapter.

8.69 Explain the difference between the null hypothesis and the alternative hypothesis in a statistical test.

8.70 In a test of hypothesis, is the size of the rejection region increased or decreased when the significance level α is reduced?

8.71 What are the two possible conclusions in a statistical test of hypothesis?

8.72 If the calculated value of the test statistic falls in the rejection region, we reject H_0 in favor of H_a. Does this prove that H_a is correct? Explain.

8.73 When do you risk making a Type I error? A Type II error?

8.74 Define each of the following:

  a. Type I error   **b.** Type II error   **c.** α **d.** β

  e. Critical value   **f.** *p*-value

  g. One-tailed test   **h.** Two-tailed test

8.75 Specify the form of the rejection region for a two-tailed test of hypothesis conducted at each of the following significance levels:

  a. $\alpha = .01$ **b.** $\alpha = .02$ **c.** $\alpha = .04$

  Locate the rejection region, α, and the critical values on a sketch of the standard normal curve for each part of the exercise. (Assume that the sampling distribution of the test statistic is approximately normal.)

8.76 For each of the following rejection regions, determine the value of α, the probability of a Type I error:

  a. $z > 2.58$ **b.** $z < -1.29$ **c.** $z < -1.645$ or $z > 1.645$

8.77 Suppose the observed significance level (*p*-value) of a test is .07.

  a. For what values of α would you reject H_0?

  b. For what values of α would you fail to reject H_0?

8.78 An automobile dealer provides a new car warranty that covers the engine, transmission, and drive train of all new cars for up to two years or 24,000 miles, whichever comes first. However, one sales manager believes the 2-year part of the warranty is unnecessary since μ, the true mean number of miles driven by new car owners in two years, is greater than 24,000 miles. Suppose the sales manager wishes to test

$$H_0: \mu = 24,000$$

$$H_a: \mu > 24,000$$

at a significance level of $\alpha = .01$.

a. Give the form of the rejection region for this test. Locate the rejection region, α, and the critical value on a sketch of the standard normal curve. (Assume the sample size will be sufficient to guarantee normality of the test statistic.)

b. A random sample of 32 new car owners produced the following statistics on number of miles driven after two years: $\bar{x} = 24,517$ and $s = 1,866$. Calculate the appropriate test statistic.

c. Make the appropriate conclusion in terms of the problem.

d. Describe a Type I error in terms of the problem.

e. Describe a Type II error in terms of the problem.

f. Calculate the p-value of the test and interpret its value.

8.79 A telephone survey of adult literacy was administered to a large sample of San Diego (California) residents (*Journal of Literacy Research*, Dec. 1996). Each person interviewed was presented with a list of magazine names (e.g., *New Yorker, Psychology Today*) and asked, "Do you recognize the magazine to be real?" For one of the fictitious magazines given (*American Journal Review*), 170 of the 486 respondents believed the magazine to be real. Suppose you want to know whether more than $\frac{1}{3}$ believe *American Journal Review* to be a real magazine.

a. Compute the test statistic appropriate for conducting the test.

b. Set up the rejection region for the test if you are willing to tolerate a Type I error probability of $\alpha = .10$. Locate the pertinent quantities on a sketch of the standard normal curve.

c. Give a full conclusion in terms of the problem.

d. What are the consequences of a Type I error?

e. Find the p-value of the test and interpret its value.

8.80 The branch manager of an outlet of a large, nationwide chain of pet supply stores wants to study characteristics of customers of her store. In particular, she decides to focus on two variables: the amount of money spent by customers and whether the customers own only one dog, only one cat, or more than one dog and/or one cat. The results from a sample of 70 customers are as follows:

- Amount of money spent: $\bar{x} = \$21.34$, $s = \$9.22$
- 37 customers own only a dog
- 26 customers own only a cat
- 7 customers own at least one dog and at least one cat

a. At the .05 level of significance, is there evidence that the average amount spent is different from $20?

b. At the .10 level of significance, is there evidence that the proportion of customers who own only a cat is different from .15?

c. What will your answer in part **a** be if the sample standard deviation is $6.22?

d. What will your answer in part **b** be if you want to test whether the proportion who own only a cat is different from .25 and you use the .05 level of significance?

8.81 Researchers at the University of South Florida College of Medicine conducted a study of the drug usage of U.S. physicians (*Journal of the American Medical Association,* May 6, 1992). The anonymous survey of 5,426 randomly selected physicians revealed that 7.9% (or 429) experienced substance abuse or drug dependency in their lifetime. Test the hypothesis that more than 5% of U.S. physicians have abused or depended on drugs in their lifetime. Use $\alpha = .10$.

8.82 Researchers at the University of Rochester studied the friction that occurs in the paper-feeding process of a photocopier (*Journal of Engineering for Industry,* May 1993). The experiment involved monitoring the displacement of individual sheets of paper in a stack fed through the copier. If no sheet except the top one moved more than 25% of the total stroke distance, the feed was considered successful. In a stack of 100 sheets of paper, the feeding process was successful 94 times. The success rate of the feeder is designed to be .90. Test to determine whether the true success rate of the feeder exceeds .90. Use $\alpha = .10$.

8.83 To get his or her name on the ballot of a local election, a political candidate often must obtain a petition bearing the signatures of a minimum number of registered voters. To verify that the names on the petition were signed by actual voters, election officials randomly sample the names on the list and check each sampled name for authenticity. Suppose at least 17,000 valid signatures are required to place a candidate on a ballot and the candidate submits a petition with 18,200 signatures. Election officials randomly select 100 of the 18,200 names and find only two invalid signatures. Is this sufficient evidence to believe that more than 17,000 (i.e., 93.4%) of the 18,200 signatures on the petition are valid? Test using $\alpha = .05$.

8.84 The application of adrenaline is the prevailing treatment to reduce eye pressure in glaucoma patients. Theoretically, a new synthetic drug will cause the same mean and standard deviation in blood pressure drop ($\mu = 5$ units, $\sigma = 1.2$ units) without the side effects caused by adrenaline. The new drug is given to $n = 50$ glaucoma patients, and the reduction in pressure for each patient is measured. The results are summarized as follows:

$$\bar{x} = 4.68 \qquad s = .82$$

a. Is there sufficient evidence (at $\alpha = .05$) to conclude that the mean in blood pressure reduction resulting from the new drug is different from that produced by adrenaline?

*b. Is there sufficient evidence (at $\alpha = .05$) to conclude that the standard deviation in blood pressure reduction resulting from the new drug is different from that produced by adrenaline?

8.85   Hospital patients over the age of 65 apparently face a high risk of serious treatment errors, according to a study in the *Journal of the American Geriatric Society* (Dec. 1990). The records of 122 randomly selected elderly patients were checked for errors in their prescribed medications. Of these, 73 patients were found to have at least one erroneously prescribed medication (i.e., they received an unneeded drug that might cause harmful side effects or they failed to receive a necessary drug). Prior to the study, the researcher did not expect such a high error rate. Conduct a test to determine whether the true percentage of elderly (over age 65) patients who have at least one erroneously prescribed drug exceeds .20. Use the *p*-value method and $\alpha = .05$.

8.86   How do the makers of Kleenex know how many tissues to put in a box? *The Wall Street Journal* (Sept. 21, 1984) reported that the marketing experts at Kimberly-Clark Corporation have "little doubt that the company should put 60 tissues in each pack." The researchers determined that 60 is "the average number of times people blow their nose during a cold" by asking hundreds of customers to keep count of their Kleenex use in diaries. Suppose a random sample of 250 Kleenex users yielded the following summary statistics on the number of times they blew their nose when they had a cold:

$$\bar{x} = 57 \qquad s = 26$$

Is this sufficient evidence to dispute the researcher's claim? Test at $\alpha = .05$.

8.87   Adlerian psychologists have long held that birth order affects personality. In theory, firstborn and only children tend to be self-confident, verbal, perfectionistic, anxious to please, success-oriented, and conservative. Interestingly, of the first 102 appointments to the nation's Supreme Court, 56 (or approximately 55%) have been firstborn or only children (*Presidential Studies Quarterly*, Winter 1985). This is in contrast to the population at large, in which only 37% are firstborn or only children. Assuming the 102 Supreme Court justice appointments represent a random sample of all past and future justices, test the hypothesis that the percentage of all Supreme Court justices who were firstborn or only children exceeds 37%, the national percentage. Use $\alpha = .05$. Is this a reasonable assumption?

8.88   *Marine Technology* (Jan. 1995) reported on the spillage amount and cause of puncture for 50 recent major oil spills from tankers and carriers in the United States. The data are reproduced in the table on page 470.

  a.   Test the hypothesis that the true mean spillage of all major oil spills in the U.S. exceeds 50 thousand metric tons. Use $\alpha = .10$.

  b.   Test the hypothesis that the true proportion of major oil spills in the U.S. that are caused by hull failure differs from .25. Use $\alpha = .10$.

  *c.   Test the hypothesis that the true spillage variance of all major oil spills in the U.S. differs from 3,000.

  d.   Does it appear that the spillage amounts are normally distributed? Explain.

Problem 8.88

OILSPILL.XLS

Tanker	Spillage (metric tons, thousands)	Collision	Grounding	Fire/ Explosion	Hull Failure	Unknown
				CAUSE OF SPILLAGE		
Atlantic Empress	257	X				
Castillo De Bellver	239			X		
Amoco Cadiz	221				X	
Odyssey	132			X		
Torrey Canyon	124		X			
Sea Star	123	X				
Hawaiian Patriot	101				X	
Independento	95	X				
Urquiola	91		X			
Irenes Serenade	82			X		
Khark 5	76			X		
Nova	68	X				
Wafra	62		X			
Epic Colocotronis	58		X			
Sinclair Petrolore	57			X		
Yuyo Maru No 10	42	X				
Assimi	50			X		
Andros Patria	48			X		
World Glory	46				X	
British Ambassador	46				X	
Metula	45		X			
Pericles G.C.	44			X		
Mandoil II	41	X				
Jacob Maersk	41		X			
Burmah Agate	41	X				
J. Antonio Lavalleja	38		X			
Napier	37		X			
Exxon Valdez	36		X			
Corinthos	36	X				
Trader	36				X	
St. Peter	33			X		
Gino	32	X				
Golden Drake	32			X		
Ionnis Angelicoussis	32			X		
Chryssi	32				X	
Irenes Challenge	31				X	
Argo Merchant	28		X			
Heimvard	31	X				
Pegasus	25					X
Pacocean	31				X	
Texaco Oklahoma	29				X	
Scorpio	31		X			
Ellen Conway	31		X			
Caribbean Sea	30				X	
Cretan Star	27					X
Grand Zenith	26				X	
Athenian Venture	26			X		
Venoil	26	X				
Aragon	24				X	
Ocean Eagle	21		X			

Source: Daidola, J. C. "Tanker structure behavior during collision and grounding." *Marine Technology,* Vol. 32, No. 1, Jan. 1995, p. 22 (Table 1). Reprinted with permission of The Society of Naval Architects and Marine Engineers (SNAME), 601 Pavonia Ave., Jersey City, NJ 07306, USA, (201) 798-4800. Material appearing in The Society of Naval Architect and Marine Engineers (SNAME) publications cannot be reprinted without obtaining written permission.

e.   How does your answer to part **d** impact the validity of the test, part **a**?

*f.   How does your answer to part **d** impact the validity of the test, part **c**?

8.89   How confident are you in your ability to do well? Researchers at Bowling Green University attempted to determine whether confidence is gender-related (*Psychological Reports,* Aug. 1997). In one part of the study, 87 female psychology students were given a 26-item test covering general social psychological knowledge and were asked to indicate (privately) the number of questions they were confident they had answered correctly. The number of questions the female students indicated that they were confident they had answered correctly had a mean of $\bar{x} = 9.7$ and a standard deviation of $s = 5.3$. Is there sufficient evidence to say that, on average, female students are confident on less than half (13) of the questions on the 26-item test? Use $\alpha = .10$.

Using Microsoft Excel

8.E.1 Using Microsoft Excel to Perform the Z Test of the Hypothesis for the Mean (σ known)

Solution Summary:

Use the PHStat add-in to perform a Z test of the hypothesis for the mean when σ is known.

Example: Humerus bone analysis of Example 8.13

Solution:

Use the One-Sample Tests | Z Test for the Mean, sigma known, choice of the PHStat add-in to perform a Z test of the hypothesis for the mean when σ is known. As an example, to test the hypothesis that the population mean ratio of all bones is equal to 8.5 for the humerus bone data of Example 8.13 on page 436, do the following:

1. If the PHStat add-in has not been previously loaded, load the add-in using the instructions of Section EP.3.2.

2. Select File | New to open a new workbook (or open the existing workbook into which the hypothesis testing worksheet is to be inserted).

3. Select PHStat | One-Sample Tests | Z Test for the Mean, sigma known.

4. In the Z Test for the Mean, sigma known, dialog box (see Figure 8.E.1):

E Figure 8.E.1
PHStat Z Test for the Mean, Sigma Known, Dialog Box

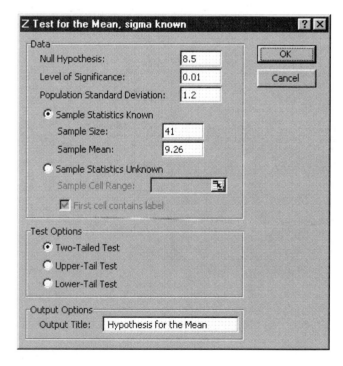

a. Enter 8.5 in the Null Hypothesis: edit box.
b. Enter 0.01 in the Level of Significance: edit box.

c Enter 1.2 in the Population Standard Deviation: edit box.

d Select the Sample Statistics Known option button and enter 41 in the Sample Size: edit box and 9.26 in the Sample Mean: edit box.

e Select the Two-Tailed Test option button.

f Enter Z Test of the Hypothesis for the Mean in the Output Title: edit box.

g Click the OK button.

The add-in inserts a worksheet containing calculations for the Z test of the hypothesis for the mean similar to the one shown in Figure 8.6b on page 437. We can change the values for the level of significance, population standard deviation, sample size, and sample mean in this worksheet to see their effects on the test of hypothesis for the mean. (Note: For problems in which the sample mean is not known, select the Sample Statistics Unknown option button in Step 4d and enter the cell range of the sample data in the Sample Cell Range: edit box, shown dimmed in Figure 8.E.1).

8.E.2 Using Microsoft Excel to Perform the *t* Test of the Hypothesis for the Mean (σ unknown)

Solution Summary:

Use the PHStat add-in to perform a *t* test of the hypothesis for the mean when σ is unknown.

Example: Sewer pipe analysis of Example 8.14

Solution:

Use the One-Sample Tests | t Test for the Mean, sigma unknown, choice of the PHStat add-in to perform a *t* test of the hypothesis for the mean when σ is unknown. As an example, to test the hypothesis whether or not there is evidence that the manufacturer's pipe has a mean breaking strength of greater than 2,500 pounds per lineal foot (Example 8.14 on page 439), do the following:

1 If the PHStat add-in has not been previously loaded, load the add-in using the instructions of Section EP.3.2.

2 Open the Example 8.14 workbook (EXAMPLE 8-14.XLS) and click the Data sheet tab. Verify that the breaking strengths appear in column A.

3 Select PHStat | One-Sample Tests | t Test for the Mean, sigma unknown.

4 In the t Test for the Mean, sigma unknown, dialog box (see Figure 8.E.2):

a Enter 2500 in the Null Hypothesis: edit box.

b Enter 0.10 in the Level of Significance: edit box.

c Select the Sample Statistics Known option button and enter 7 in the Sample Size: edit box, 2571.4 in the Sample Mean: edit box, and 115.1 in the Sample Standard Deviation: edit box.

d Select the Upper-Tail Test option button.

e Enter t Test for the Hypothesis of the Mean in the Output Title: edit box.

f Click the OK button.

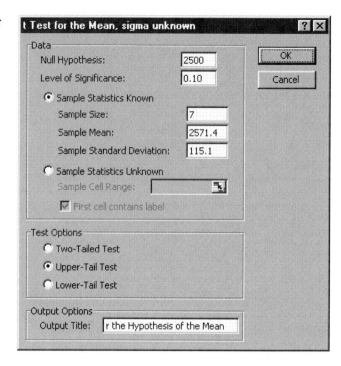

Figure 8.E.2 PHStat t Test for the Mean, Sigma Unknown, Dialog Box

The add-in inserts a worksheet containing calculations for the *t* test of the hypothesis for the mean similar to the one shown in Figure 8.8b on page 439. We can change the value for the level of significance in this worksheet to see its effect on the *t* test of hypothesis for the mean. (Note: For other problems in which the sample size, sample mean, and sample standard deviation are not known, select the Sample Statistics Unknown option button in Step 4c and enter the cell range of the sample data in the Sample Cell Range: edit box, shown dimmed in Figure 8.E.2).

8.E.3 Using Microsoft Excel to Perform the Z Test of the Hypothesis for the Proportion

Solution Summary:

Use the PHStat add-in to perform a Z test of the hypothesis for the proportion.

 Car buyers study of Example 8.18

Solution:

Use the One-Sample Tests | Z Test for the Proportion choice of the PHStat add-in to perform a Z test of the hypothesis for the proportion. As an example, to test the hypothesis whether or not there is evidence of a change in the proportion of new car buyers that were women in the years 1980 versus 1998 (Example 8.18 on page 453), do the following:

1 If the PHStat add-in has not been previously loaded, load the add-in using the instructions of Section EP.3.2.

2 Select File | New to open a new workbook (or open the existing workbook into which the hypothesis testing worksheet is to be inserted).

3 Select PHStat | One-Sample Tests | Z Test for the Proportion.

4 In the Z Test for the Proportion dialog box (see Figure 8.E.3):

a Enter 0.4 in the Null Hypothesis: edit box.

b	Enter 0.1 in the Level of Significance: edit box.
c	Enter 57 in the Number of Successes: edit box.
d	Enter 120 in the Sample Size: edit box.
e	Select the Upper-Tail Test option button.
f	Enter Z Test for the Hypothesis of the Proportion in the Output Title: edit box.
g	Click the OK button.

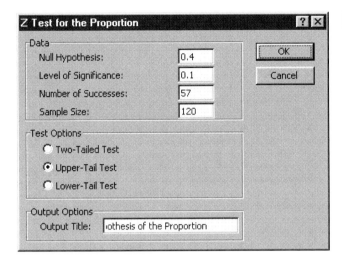

E **Figure 8.E.3**
PHStat Z Test for the Proportion Dialog Box

The add-in inserts a worksheet containing calculations for the Z test of the hypothesis for the proportion similar to the one shown in Figure 8.14 on page 454. We can change the values for the level of significance, number of successes, and sample size in this worksheet to see their effects on the test of hypothesis for the proportion.

8.E.4 Using Microsoft Excel to Perform the Chi-Square Test of the Hypothesis for the Variance

Solution Summary:

Implement a worksheet that uses the CHIINV and CHIDIST worksheet functions to perform a χ^2 test of the hypothesis for the variance.

 Quality control problem of Example 8.20

Solution:

Implement a worksheet based on the design shown in Table 8.E.1 to perform a χ^2 test of the hypothesis for the variance. This design uses the CHIINV function, first discussed in Section 7.E.4, and the CHIDIST function, the format of which is

$$\text{CHIDIST}(\textit{chi-square value, degrees of freedom})$$

to return the critical value of χ^2 (CHIINV) and the probability of exceeding a given χ^2 value for a given number of degrees of freedom (CHIDIST).

As an example, to test the hypothesis that the standard deviation of the fill measurement of the soup cans for the quality control problem of Example 8.20 on page 457 is less than 0.1 ounce, implement rows 1 through 13 of the Table 8.E.1 design. Only these rows of the design need to be implemented as only the lower-tailed test is needed for this example. To implement these rows, do the following:

	1	Select File I New to open a new workbook (or open the existing workbook into which the hypothesis testing worksheet is to be inserted).
	2	Select an unused worksheet (or select Insert I Worksheet if there are none) and rename the sheet Hypothesis.
	3	Enter the title, labels, and formulas for column A as shown in Table 8.E.1, noting the following:
	a	The Symbol font can be used to enter the σ symbol in cell A3.
	b	The formula for cell A13, typeset as two lines, should be entered as one continuous line.

Table 8.E.1 Worksheet Design for χ^2 Test of the Hypothesis for the Variance

	A	**B**
1	Chi-Square Test of Hypothesis for the Variance	
2		
3	Null Hypothesis $\sigma\text{^}2=$	xxx
4	Level of Significance	xxx
5	Sample Size	xxx
6	Sample Standard Deviation	xxx
7	Degrees of Freedom	=B5-1
8	Chi-Square Test Statistic	=B7*B6^2/B3
9		
10	Lower-Tail Test	
11	Lower Critical Value	=CHIINV(1-B4,B7)
12	p-Value	=1-CHIDIST(B8,B7)
13	=IF(B12<B4,"Reject the null hypothesis", "Do not reject the null hypothesis")	
14		
15	Upper-Tail Test	
16	Upper Critical Value	=CHIINV(B4,B7)
17	p-Value	=CHIDIST(B8,B7)
18	=IF(B17<B4,"Reject the null hypothesis", "Do not reject the null hypothesis")	
19		
20	Two-Tailed Test	
21	Lower Critical Value	=CHIINV((1-B4/2),B7)
22	Upper Critical Value	=CHIINV(B4/2,B7)
23	p-Value	=IF(B8-B21,1-CHIDIST(B8,B7),CHIDIST(B8,B7))
24	=IF(B23<B4/2,"Reject the null hypothesis", "Do not reject the null hypothesis")	

4. Enter the null hypothesis value for σ^2, the level of significance, sample size, and sample standard deviation in the cell range B3:B6. Enter 0.01 in cell B3, 0.05 in cell B4, 10 in cell B5, and 0.043 in cell B6.

5. Enter the formulas for cells B7, B8, B11, and B12.

6. Select the cell range A10:B10 and click the Merge and Center button on the formatting toolbar (see Section EP.2.9).

The completed worksheet will be similar to the one shown in Figure 8.16b on page 459.

Chapter 9

Inferences about Population Parameters: Two Samples

CONTENTS

9.1 Determining the Target Parameter

9.2 Comparing Two Population Means: Independent Samples

9.3 Comparing Two Population Means: Matched Pairs

9.4 Comparing Two Population Proportions: Independent Samples

9.5 Comparing Population Proportions: Contingency Tables

9.6 Comparing Two Population Variances (Optional)

EXCEL TUTORIAL

9.E.1 Using Microsoft Excel to Compare Two Population Means: Independent Samples

9.E.2 Using Microsoft Excel to Compare Two Population Means: Matched Pairs

9.E.3 Using Microsoft Excel to Compare Two Population Proportions

9.E.4 Using Microsoft Excel to Analyze Contingency Tables

9.E.5 Using Microsoft Excel to Perform the *F* Test for Differences in Two Variances

REAL-WORLD APPLICATION
An IQ Comparison of Identical Twins Reared Apart

Will identical twins, separated in early life and reared apart in different homes, have identical average IQs? To investigate this phenomenon, a sociologist collected IQ data for identical twins reared apart. How should we use these data to compare the twins' average IQs? In this chapter, we learn how to compare means, proportions, and variances of two populations using confidence intervals and hypothesis tests. You will have an opportunity to analyze the IQ data for identical twins reared apart in Real-World Revisited at the end of this chapter.

9.1 DETERMINING THE TARGET PARAMETER

In the previous two chapters, we covered confidence intervals and hypothesis tests applied to a single sample. In this chapter, we present confidence intervals and hypothesis tests for comparing two samples. The population parameters covered include the **difference between two population means $\mu_1 - \mu_2$, the difference between two population proportions $\pi_1 - \pi_2$,** and (optionally) **the ratio of two population variances σ_1^2/σ_2^2.** The key words or phrases that help you identify the target parameter are reported in the next box. Remember, the type of data (quantitative or qualitative) collected will aid in your decision. With quantitative data, you're interested in comparing means or variances. With qualitative data, a comparison of proportions is appropriate.

Determining the Target Parameter: Two Samples

Parameter	Key Words or Phrases
$(\mu_1 - \mu_2)$	Difference in means or averages; mean difference; unpaired comparisons of means or averages
μ_d	Mean difference; average difference; paired comparison of means or averages
$(\pi_1 - \pi_2)$	Difference in proportions, percentages, fractions, or rates; comparison of proportions, percentages, fractions, or rates
σ_1^2/σ_2^2	Ratio of variances; difference in variation; comparison of variances

Example 9.1   Choosing the Target Parameter

A sociologist wants to determine whether Democratic congressmen are older, on average, than Republican congressmen. To do this, the sociologist plans to compare the mean age of a sample of Democratic congressmen to the mean age of a sample of Republican congressmen. What is the parameter of interest to the sociologist?

Solution

In this problem, the experimental units are U.S. congressmen and the variable measured is age—a *quantitative* variable. The key words in the statement of this problem are "compare" and "mean;" thus it is clear that there are two means to be compared:

$$\mu_1 = \text{Mean age of Democratic congressmen}$$

and

$$\mu_2 = \text{Mean age of Republican congressmen}$$

Consequently, the parameter of interest is $(\mu_1 - \mu_2)$. ❑

Example 9.2   Choosing the Target Parameter

A biologist wants to compare the proportions of fish infected with tapeworms in two locations—the Mediterranean Sea and the Atlantic Ocean. Fish at both locations are captured and dissected, and the number found to be infected with the tapeworm parasite determined. What is the parameter of interest?

Solution

For this study, the experimental units are the fish captured and dissected, and the variable measured is *qualitative*—either a fish is infected with a tapeworm or it is not. The two key words in the statement of the problem are "compare" and "proportions." Therefore, the parameter of interest is $(\pi_1 - \pi_2)$, where

π_1 = Proportion of Mediterranean fish infected with tapeworms

π_2 = Proportion of Atlantic Ocean fish infected with tapeworms   ❑

Self-Test

9.1

Determine the parameter of interest for each of the following:

a. The manager of a pet supply store wants to determine whether there is a difference in the average amount of money spent by owners of cats and owners of dogs.

b. The pet store manager also wants to know if the variation in amount spent by cat owners differs from the variation in amount spent by dog owners.

9.2 COMPARING TWO POPULATION MEANS: INDEPENDENT SAMPLES

Suppose you want to use the information in two samples to compare two population means. For example, you may want to compare the mean starting salaries of college graduates with engineering and journalism degrees; or the mean gasoline consumptions that may be expected this year for drivers in two areas of the country; or the mean reaction times of men and women to a visual stimulus. The methods we present in this section are straightforward extensions of those used for estimating or testing a single population mean.

Example 9.3   Selecting a Point Estimate

Researchers at Rochester Institute of Technology investigated the use of isolation timeout as a behavioral management technique (*Exceptional Children,* Feb. 1995). Subjects for the study were 155 emotionally disturbed students enrolled in a special education facility. The students were randomly assigned to one of two types of classroom—Option I classrooms (one teacher, one paraprofessional, and a maximum of 12 students) and Option II classrooms (one teacher, one paraprofessional, and a maximum of six students). Over the academic year the number of behavioral incidents resulting in an isolation timeout was recorded for each student. Summary statistics for the two groups of students are shown in Table 9.1 (page 481). Calculate a point estimate for the difference between the mean number of timeout incidents for the two groups of students.

Solution

Let the subscript "1" refer to the Option I classrooms and the subscript "2" to the Option II classrooms; then also define the following notation:

Table 9.1 Summary of Information for Example 9.3

	Option I	Option II
Number of students	100	55
Mean number of timeout incidents	78.67	102.87
Standard deviation	59.08	69.33

Source: Costenbader, V., and Reading-Brown, M. "Isolation timeout used with students with emotional disturbance." *Exceptional Children,* Vol. 61, No. 4, Feb 1995, p. 359 (Table 3). Copyright 1995 by The Council for Exceptional Children.

μ_1 = Population mean number of timeout incidents for Option I students

μ_2 = Population mean number of timeout incidents for Option II students

Similarly, let $\bar{x}_1$ and $\bar{x}_2$ denote the respective sample means; s_1 and s_2, the respective sample deviations; and n_1 and n_2, the respective sample sizes.

Now, to estimate $(\mu_1 - \mu_2)$, it seems logical to use the difference between the sample means

$$(\bar{x}_1 - \bar{x}_2) = 78.67 - 102.87 = -24.20$$

as our point estimate of the difference between the population means. ❏

Large Samples

Confidence intervals and hypothesis tests for $(\mu_1 - \mu_2)$ are based on the sampling distribution of the point estimate $(\bar{x}_1 - \bar{x}_2)$. The properties of the sampling distribution of $(\bar{x}_1 - \bar{x}_2)$ are shown in the next box (see also Figure 9.1).

Sampling Distribution of $(\bar{x}_1 - \bar{x}_2)$

For sufficiently large sample sizes (say, $n_1 \geq 30$ and $n_2 \geq 30$), the sampling distribution of $(\bar{x}_1 - \bar{x}_2)$, based on independent random samples from two populations, is approximately normal with mean equal to the difference between the two population means and standard deviation equal to the square root of the sum of σ_1^2/n_1 and σ_2^2/n_2, i.e.,

$$\text{Mean: } \mu_{(\bar{x}_1 - \bar{x}_2)} = (\mu_1 - \mu_2) \tag{9.1}$$

$$\text{Standard deviation: } \sigma_{(\bar{x}_1 - \bar{x}_2)} = \sqrt{\frac{\sigma_1^2}{n_1} + \frac{\sigma_2^2}{n_2}} \tag{9.2}$$

where σ_1^2 and σ_2^2 are the variances of the two populations from which the samples were selected.

Figure 9.1 Sampling Distribution of $(\bar{x}_1 - \bar{x}_2)$

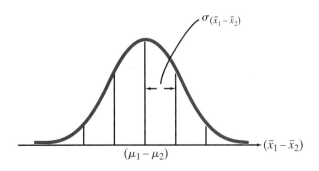

As was the case with large-sample inferences about a single population mean, the requirements of "large" sample sizes enable us to apply the Central Limit Theorem to obtain the sampling distribution of $(\bar{x}_1 - \bar{x}_2)$; they also justify the use of s_1^2 and s_2^2 as approximations to the respective population variances σ_1^2 and σ_2^2.

The procedures for testing and estimating $(\mu_1 - \mu_2)$ with large samples appear in the accompanying boxes.

Large-Sample Test of Hypothesis about $\mu_1 - \mu_2 = 0$

Upper-Tailed Test	Two-Tailed Test	Lower-Tailed Test
$H_0: (\mu_1 - \mu_2) = 0$	$H_0: (\mu_1 - \mu_2) = 0$	$H_0: (\mu_1 - \mu_2) = 0$
$H_a: (\mu_1 - \mu_2) > 0$	$H_a: (\mu_1 - \mu_2) \neq 0$	$H_a: (\mu_1 - \mu_2) < 0$

$$\text{Test statistic: } z = \frac{(\bar{x}_1 - \bar{x}_2) - 0}{\sigma_{(\bar{x}_1 - \bar{x}_2)}} = \frac{(\bar{x}_1 - \bar{x}_2) - 0}{\sqrt{\dfrac{\sigma_1^2}{n_1} + \dfrac{\sigma_2^2}{n_2}}} \approx \frac{(\bar{x}_1 - \bar{x}_2) - 0}{\sqrt{\dfrac{s_1^2}{n_1} + \dfrac{s_2^2}{n_2}}} \quad (9.3)$$

Rejection region:	Rejection region:	Rejection region:
$z > z_\alpha$	$\|z\| > z_{\alpha/2}$	$z < -z_\alpha$

Large-Sample $(1 - \alpha)100\%$ Confidence Interval for $(\mu_1 - \mu_2)$

$$(\bar{x}_1 - \bar{x}_2) \pm z_{\alpha/2}\sigma_{(\bar{x}_1 - \bar{x}_2)} = (\bar{x}_1 - \bar{x}_2) \pm z_{\alpha/2}\sqrt{\frac{\sigma_1^2}{n_1} + \frac{\sigma_2^2}{n_2}}$$

$$\approx (\bar{x}_1 - \bar{x}_2) \pm z_{\alpha/2}\sqrt{\frac{s_1^2}{n_1} + \frac{s_2^2}{n_2}} \quad (9.4)$$

[*Note:* We have used the sample variances s_1^2 and s_2^2 as approximations to the corresponding population parameters.]

Assumptions for Large-Sample Inferences about $(\mu_1 - \mu_2)$

1. The two random samples are selected in an independent manner from the target populations. That is, the choice of elements in one sample does not affect, and is not affected by, the choice of elements in the other sample.

2. The sample sizes n_1 and n_2 are sufficiently large. (We recommend $n_1 \geq 30$ and $n_2 \geq 30$.)

Example 9.4 Large-Sample Test of $(\mu_1 - \mu_2)$

Refer to Example 9.3. Is there evidence that the mean number of timeout incidents for special education students assigned to Option I classrooms is less than the corresponding mean for special education students assigned to Option II classrooms? Test using $\alpha = .05$.

Figure 9.2 Rejection Region for Example 9.4

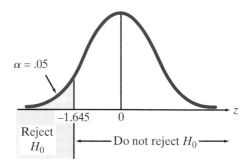

Solution

For this problem, we want to test the hypotheses

$H_0: (\mu_1 - \mu_2) = 0$ (no difference between mean number of timeout incidents for Option I and Option II students)

$H_a: (\mu_1 - \mu_2) < 0$ (mean number of timeout incidents for Option I students is less than the mean for Option II students)

This lower-tailed, large-sample (since both n_1 and n_2 exceed 30) test is based on a z statistic. Thus, we will reject H_0 if $z < -z_{.05} = -1.645$ (see Figure 9.2). We compute the test statistic by substituting the summary information in Table 9.1 into equation 9.3:

$$z = \frac{(\bar{x}_1 - \bar{x}_2) - 0}{\sqrt{\dfrac{s_1^2}{n_1} + \dfrac{s_2^2}{n_2}}} \approx \frac{(78.67 - 102.87) - 0}{\sqrt{\dfrac{(59.08)^2}{100} + \dfrac{(69.33)^2}{55}}} = -2.19$$

Since this computed value of $z = -2.19$ lies in the rejection region, there is sufficient evidence (at $\alpha = .05$) to conclude that the mean number of timeout incidents for Option I students is significantly less than the mean number of timeout incidents for Option II students.

The same conclusion can be reached by computing the observed significance level (*p*-value) of the test, described in Section 8.6. For this lower-tailed test,

$$p\text{-value} = P(z < -2.19)$$

From the standard normal table in Table B.2 in Appendix B, we find

$$
\begin{aligned}
p\text{-value} &= P(z < -2.19) \\
&= .5 - P(-2.19 < z < 0) = .5 - .4857 = .0143
\end{aligned}
$$

Since $\alpha = .05$ exceeds this *p*-value, we again reject H_0 in favor of H_a. ❏

Example 9.5   Large-Sample Confidence Interval for $(\mu_1 - \mu_2)$

Refer to Examples 9.3 and 9.4.

a. Construct a 95% confidence interval for $(\mu_1 - \mu_2)$, the difference between the mean number of timeout incidents for the two groups of special education students.

b. Interpret the interval. Compare your inference here to the results of Example 9.4.

Solution

a. The general form of a 95% confidence interval for $(\mu_1 - \mu_2)$, based on large samples from the target populations, is given by equation 9.4. Substituting $z_{.025} = 1.96$ and the summary information in Table 9.1 into the formula, we obtain:

$$(78.67 - 102.87) \pm 1.96\sqrt{\frac{\sigma_1^2}{100} + \frac{\sigma_2^2}{55}}$$

$$\approx (78.67 - 102.87) \pm 1.96\sqrt{\frac{(59.08)^2}{100} + \frac{(69.33)^2}{55}}$$

$$= -24.20 \pm 21.67$$

or $(-45.87, -2.52)$.

b. Since all the values in our interval are negative, we are 95% confident that the mean number of timeout incidents for Option II students is between 2.52 and 45.87 higher than the mean number of timeout incidents for Option I students. Option II students have, on average, more timeout incidents than Option I students.

In Example 9.4, the hypothesis test also detected a higher mean for Option II students at $\alpha = .05$. However, our confidence interval provides additional information on the magnitude of the difference. ❏

Self-Test

9.2

Independent random samples of $n_1 = 45$ blue-collar workers and $n_2 = 38$ white-collar workers yielded the following summary statistics on number of sick days taken in a recent year:

$$\bar{x}_1 = 11.5 \quad s_1 = 10.2 \quad \bar{x}_2 = 9.0 \quad s_2 = 5.6$$

a. Construct a 95% confidence interval for $(\mu_1 - \mu_2)$.
b. Test $H_0: (\mu_1 - \mu_2) = 0$ against $H_a: (\mu_1 - \mu_2) \neq 0$ at $\alpha = .05$.
c. Compare the results, parts **a** and **b**.

Small Samples

When estimating or testing the difference between two population means based on small samples from each population, we must make specific assumptions about the distributions of the two populations, as indicated in the box.

Assumptions Required for Small-Sample Inferences about $(\mu_1 - \mu_2)$

1. Both of the populations from which the samples are selected have distributions that are approximately normal.
2. The variances σ_1^2 and σ_2^2 of the two populations are equal.
3. The random samples are selected in an independent manner from the two populations.

Figure 9.3 Assumptions Required for Small-Sample Estimation of $(\mu_1 - \mu_2)$: Normal Distributions with Equal Variances

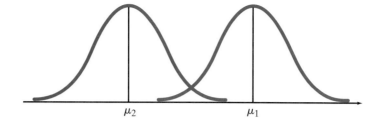

Figure 9.3 illustrates the form of the population distributions implied by assumptions 1 and 2. Observe that both populations have distributions that are approximately normal. Although the means of the two populations may differ, we require the variances σ_1^2 and σ_2^2, which measure the spread of the two distributions, to be equal. When these assumptions are satisfied, we use the Student's t distribution (specified in the next box) to construct a confidence interval for $(\mu_1 - \mu_2)$ based on small samples (say, $n_1 < 30$ or $n_2 < 30$) from the respective populations. Since we assume that the two populations have equal variances (i.e., $\sigma_1^2 = \sigma_2^2 = \sigma^2$), we construct a **pooled estimate of σ^2** (denoted s_p^2) based on the information contained in *both* samples.

Small-Sample Test of Hypothesis about $(\mu_1 - \mu_2)$

Upper-Tailed Test	Two-Tailed Test	Lower-Tailed Test
$H_0: (\mu_1 - \mu_2) = 0$	$H_0: (\mu_1 - \mu_2) = 0$	$H_0: (\mu_1 - \mu_2) = 0$
$H_a: (\mu_1 - \mu_2) > 0$	$H_a: (\mu_1 - \mu_2) \neq 0$	$H_a: (\mu_1 - \mu_2) < 0$

$$\textit{Test statistic: } t = \frac{(\bar{x}_1 - \bar{x}_2) - 0}{\sqrt{s_p^2\left(\dfrac{1}{n_1} + \dfrac{1}{n_2}\right)}} \tag{9.5}$$

where s_p^2 is given in equation 9.7 in the next box.

Rejection region:	*Rejection region:*	*Rejection region:*
$t > t_\alpha$	$|t| > t_{\alpha/2}$	$t < -t_\alpha$

where the distribution of t is based on $(n_1 + n_2 - 2)$ degrees of freedom.

Small-Sample $(1 - \alpha)100\%$ Confidence Interval for $(\mu_1 - \mu_2)$

$$(\bar{x}_1 - \bar{x}_2) \pm t_{\alpha/2}\sqrt{s_p^2\left(\frac{1}{n_1} + \frac{1}{n_2}\right)} \tag{9.6}$$

where

$$s_p^2 = \frac{(n_1 - 1)s_1^2 + (n_2 - 1)s_2^2}{n_1 + n_2 - 2} \tag{9.7}$$

and the value of $t_{\alpha/2}$ is based on $(n_1 + n_2 - 2)$ degrees of freedom.

Testing for a Nonzero Difference

Equations 9.3 and 9.5 give the test statistics for testing the null hypothesis of no difference in the means, i.e., $H_0: \mu_1 - \mu_2 = 0$. To test for a nonzero difference, i.e., $H_0: \mu_1 - \mu_2 = D$ where $D \neq 0$, use the test statistic

$$\text{Large sample: } z = \frac{(\bar{x}_1 - \bar{x}_2) - D}{\sqrt{\dfrac{\sigma_1^2}{n_1} + \dfrac{\sigma_2^2}{n_2}}} \qquad (9.8)$$

$$\text{Small sample: } t = \frac{(\bar{x}_1 - \bar{x}_2) - D}{\sqrt{s_p^2 \left(\dfrac{1}{n_1} + \dfrac{1}{n_2} \right)}} \qquad (9.9)$$

Example 9.6   Small-Sample Test for $(\mu_1 - \mu_2)$

To study the effectiveness of a new type of dental anesthetic, a dentist conducted an experiment with 10 randomly selected patients. Five patients were randomly assigned to receive the standard anesthetic (Novocaine), whereas the remaining five patients received the proposed new anesthetic. While being treated, each patient was asked to give a measure of his or her discomfort, on a scale from 0 to 100. (Higher scores indicate greater discomfort.) The discomfort scores for the 10 patients are shown in Table 9.2.

If the new anesthetic is more effective, then the mean discomfort level for Novocaine will exceed the mean discomfort level for the new treatment. Use a test of hypothesis to make an inference about the effectiveness of the new anesthetic. Test at $\alpha = .01$.

Solution

Let μ_1 and μ_2 represent the true mean discomfort levels of patients receiving Novocaine and the new anesthetic, respectively. Since the samples selected for the study are small ($n_1 = n_2 = 5$), the following assumptions are required:

1. The populations of discomfort levels of dental patients receiving Novocaine and the new anesthetic both have approximately normal distributions.
2. The variance σ^2 in the discomfort levels is the same for both groups of patients.
3. The samples were independently and randomly selected from the two target populations.

The dentist wants to test

$H_0: \mu_1 - \mu_2 = 0$ (no difference in mean discomfort levels of the two treatments)

$H_a: \mu_1 - \mu_2 > 0$ (the mean discomfort level of patients using Novocaine $[\mu_1]$ is greater than the mean discomfort level of patients on the new anesthetic $[\mu_2]$)

Table 9.2 Discomfort Scores for Example 9.6

DENTAL.XLS

Novocaine	62	71	44	50	42
New Anesthetic	38	45	27	42	30

Figure 9.4 Rejection Region for Example 9.6

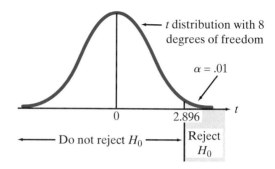

If these assumptions are valid, the test statistic will have a t distribution with $(n_1 + n_2 - 2) = (5 + 5 - 2) = 8$ degrees of freedom. With a significance level of $\alpha = .01$, we require $t_{.01}$ for this one-tailed test. From Table B.3 in Appendix B, $t_{.01} = 2.896$. Thus, the rejection region for this upper-tailed test is:

$$t > 2.896 \quad \text{(see Figure 9.4)}$$

To compute the test statistic, we need to find $\bar{x}_1, \bar{x}_2, s_1$, and s_2. These summary statistics are given in the Excel printout, Figure 9.5a, as $\bar{x}_1 = 53.8$, $\bar{x}_2 = 36.4$, $s_1 = 12.38$, and $s_2 = 7.70$. Since we have assumed that the two populations have equal variances (i.e., $\sigma_1^2 = \sigma_2^2 = \sigma^2$), the next step is to compute an estimate of this common variance. Our pooled estimate is given by equation 9.7:

E Figure 9.5
Excel Output for
Example 9.6

	A	B	C
1	Novocaine		New
2			
3	Mean	53.8	36.4
4	Standard Error	5.535341	3.443835
5	Median	50	38
6	Mode	#N/A	#N/A
7	Standard Deviation	12.3774	7.700649
8	Sample Variance	153.2	59.3
9	Kurtosis	-1.56299	-2.33196
10	Skewness	0.659365	-0.25271
11	Range	29	18
12	Minimum	42	27
13	Maximum	71	45
14	Sum	269	182
15	Count	5	5

a.

	A	B	C
1	t-Test: Two-Sample Assuming Equal Variances		
2			
3		Novocaine	New
4	Mean	53.8	36.4
5	Variance	153.2	59.3
6	Observations	5	5
7	Pooled Variance	106.25	
8	Hypothesized Mean Difference	0	
9	df	8	
10	t Stat	2.669038	
11	P(T<=t) one-tail	0.014202	
12	t Critical one-tail	2.896468	
13	P(T<=t) two-tail	0.028405	
14	t Critical two-tail	3.355381	

b.

$$s_p^2 = \frac{(n_1-1)s_1^2 + (n_2-1)s_2^2}{n_1+n_2-2} = \frac{(5-1)(12.38)^2 + (5-1)(7.70)^2}{5+5-2}$$

$$= \frac{849.98}{8} = 106.25$$

We now substitute the appropriate quantities into equation 9.9 to obtain the test statistic:

$$t = \frac{(\bar{x}_1 - \bar{x}_2) - 0}{\sqrt{s_p^2\left(\frac{1}{n_1} + \frac{1}{n_2}\right)}} = \frac{(53.8 - 36.4) - 0}{\sqrt{106.25\left(\frac{1}{5} + \frac{1}{5}\right)}} = 2.67$$

Since the test statistic $t = 2.67$ does not fall in the rejection region, we fail to reject H_0 in favor of H_a. That is, there is insufficient evidence (at $\alpha = .01$) to infer that μ_2, the mean discomfort level for patients receiving the new anesthetic, is less than μ_1, the corresponding mean for patients treated with Novocaine. Therefore, there is no evidence that the new anesthetic is more effective than the standard one.

The test could also be performed using the p-value approach. The p-value of this upper-tailed test, indicated on the Excel printout, Figure 9.5b, is .0142. Since this p-value exceeds $\alpha = .01$, we cannot reject H_0. ❑

Self-Test
9.3

The summary data for Self-Test 9.2 are repeated below:

$$\bar{x}_1 = 11.5 \quad s_1 = 10.2 \quad \bar{x}_2 = 9.0 \quad s_2 = 5.6$$

Assume independent random samples of $n_1 = 15$ blue-collar workers and $n_2 = 13$ white-collar workers produced these statistics.

a.   Compute the test statistic for testing $H_0: (\mu_1 - \mu_2) = 0$
b.   Find the rejection region for $H_a: (\mu_1 - \mu_2) > 0$ at $\alpha = .05$.

Note: The assumptions required for estimating and testing $(\mu_1 - \mu_2)$ with small samples do not have to be satisfied exactly for the procedures to be useful in practice. Slight departures from these assumptions do not seriously affect the level of confidence or α value. For situations where the variances σ_1^2 and σ_2^2 of the sampled populations are unequal, a modified t test is available (see Satterthwaite, 1946). When the sampled populations depart greatly from normality, the t distribution is invalid and any inferences derived from the procedure are suspect. For the nonnormal case it is advisable to use one of the nonparametric statistical methods discussed in Chapter 12.

Caution

When the sampled populations are decidedly nonnormal (e.g., highly skewed), any inferences derived from the small-sample t test or confidence interval for $(\mu_1 - \mu_2)$ are suspect. In this case, one alternative is to use the nonparametric Wilcoxon rank sum test of Section 12.3.

PROBLEMS
9.1–9.19

Using the Tools

9.1 Independent random samples are selected from two populations with means μ_1 and μ_2, respectively. Determine approximate values of

$$\mu_{(\bar{x}_1 - \bar{x}_2)} \qquad \text{and} \qquad \sigma_{(\bar{x}_1 - \bar{x}_2)}$$

for each of the following situations. (Assume $\sigma_1^2 = \sigma_2^2$ in each case.)

  a. $\bar{x}_1 = 150$, $s_1^2 = 36$; $\bar{x}_2 = 140$, $s_2^2 = 24$; $n_1 = n_2 = 35$

  b. $\bar{x}_1 = 125$, $s_1^2 = 225$; $n_1 = 90$; $\bar{x}_2 = 112$, $s_2^2 = 90$; $n_2 = 60$

Sample from Population 1	Sample from Population 2
$\bar{x}_1 = 15$	$\bar{x}_2 = 23$
$s_1^2 = 16$	$s_2^2 = 20$
$n_1 = 30$	$n_2 = 40$

Problem 9.2

9.2 Consider two independent random samples with 30 observations selected from population 1 and 40 from population 2. The resulting sample means and variances are shown in the accompanying table. (Assume $\sigma_1^2 = \sigma_2^2$.) Construct a confidence interval for $(\mu_1 - \mu_2)$ using a confidence coefficient of:

  a. .10   **b.** .05   **c.** .01

9.3 Refer to Problem 9.2. Find the test statistic for testing H_0: $\mu_1 - \mu_2 = 0$.

9.4 Two independent random samples, both large $(n > 30)$, are selected from populations with unknown means μ_1 and μ_2. Specify the rejection region for each of the following combinations of H_0, H_a, and α.

  a. H_0: $(\mu_1 - \mu_2) = 0$, H_a: $(\mu_1 - \mu_2) \neq 0$, $\alpha = .01$

  b. H_0: $(\mu_1 - \mu_2) = 0$, H_a: $(\mu_1 - \mu_2) \neq 0$, $\alpha = .10$

  c. H_0: $(\mu_1 - \mu_2) = 0$, H_a: $(\mu_1 - \mu_2) > 0$, $\alpha = .05$

  d. H_0: $(\mu_1 - \mu_2) = 0$, H_a: $(\mu_1 - \mu_2) < 0$, $\alpha = .05$

9.5 Two independent random samples are selected from normal populations with unknown means μ_1 and μ_2 and variances $\sigma_1^2 = \sigma_2^2$. Specify the rejection region for each of the following combinations of H_a, n_1, n_2, and α when testing the null hypothesis H_0: $(\mu_1 - \mu_2) = 0$.

  a. H_a: $(\mu_1 - \mu_2) \neq 0$, $n_1 = 10$, $n_2 = 10$, $\alpha = .05$

  b. H_a: $(\mu_1 - \mu_2) > 0$, $n_1 = 8$, $n_2 = 4$, $\alpha = .10$

  c. H_a: $(\mu_1 - \mu_2) < 0$, $n_1 = 6$, $n_2 = 5$, $\alpha = .01$

9.6 The following tables show summary statistics for random samples selected from two normal populations that are assumed to have the same variance. In each case, find s_p^2, the pooled estimate of the common variance.

a.

Sample from Population 1	Sample from Population 2
$\bar{x}_1 = 552$	$\bar{x}_2 = 369$
$s_1^2 = 4,400$	$s_2^2 = 7,481$
$n_1 = 6$	$n_2 = 7$

b.

Sample from Population 1	Sample from Population 2
$\bar{x}_1 = 10.8$	$\bar{x}_2 = 8.4$
$s_1^2 = .313$	$s_2^2 = .499$
$n_1 = 8$	$n_2 = 8$

9.7 Independent random samples from two normal populations with equal variances produced the sample means and sample variances listed in the

Sample from Population 1	Sample from Population 2
$n_1 = 14$	$n_2 = 7$
$\bar{x}_1 = 53.2$	$\bar{x}_2 = 43.4$
$s_1^2 = 96.8$	$s_2^2 = 102.0$

Problem 9.7

accompanying table. Construct a confidence interval for $(\mu_1 - \mu_2)$ using a confidence coefficient of:

a. .10 **b.** .05 **c.** .01

9.8 Refer to Problem 9.7. Find the test statistic for testing $H_0: \mu_1 - \mu_2 = 0$.

9.9 To use the t statistic in making small-sample inferences about the difference between the means of two populations, what assumptions must be made about the two populations? About the two samples?

Applying the Concepts

9.10 A Johns Hopkins University medical researcher compared the health preferences of 14 female healthy volunteers to those of 13 male healthy volunteers (*Women & Therapy*, Vol. 16, 1995). The volunteers rated several health conditions on a 5-point scale (1 = very desirable, ..., 5 = very undesirable). In response to the condition "body pains," the women had an average rating of 4.64 while the men had an average rating of 4.07.

a. Based on only this information, can a reliable inference be made about the true difference in the mean "body pain" ratings of men and women? Explain.

b. A test of the null hypothesis $H_0: \mu_{women} = \mu_{men}$ resulted in an observed significance level of .006. Interpret this result.

9.11 Students enrolled in music classes at the University of Texas (Austin) participated in a study to compare the teacher evaluations of music education majors and nonmusic education majors (*Journal of Research in Music Education*, Winter 1991). Independent random samples of 100 music majors and 100 nonmajors rated the overall performance of their teacher using a 6-point scale, where 1 = lowest rating and 6 = highest rating.

	Music Majors	Nonmusic Majors
Sample size	100	100
Mean "overall" rating	4.26	4.59
Standard deviation	.81	.78

Source: Duke, R. A., and Blackman, M. D. "The relationship between observers' recorded teacher behavior and evaluation of music instruction." *Journal of Research in Music Education*, Vol. 39, No. 4, Winter 1991 (Table 2).

a. Conduct a test of hypothesis to compare the mean teacher ratings of the two groups of music students. Use $\alpha = .05$.

b. Repeat part **a**, but use a 95% confidence interval. Interpret the result.

9.12 In developing countries such as India, working women are often the target of violence. Thus, in theory, these women are under greater stress. *Collegium Antropologicum* (June 1997) reported on a study to compare the anxiety levels of working and nonworking mothers in India. A random sample of 94 working mothers had a mean anxiety level of 33.44 (on a 50-point scale) with a standard deviation of 12.15. An independent random sample of 94 nonworking mothers had a mean anxiety level of 35.14 with a standard deviation of 11.09.

a. Is there sufficient evidence to support the theory that working mothers in India are under greater stress than nonworking mothers? Use $\alpha = .10$.

b. Estimate the difference between the true mean anxiety levels of working and nonworking mothers in India with a 95% confidence interval.

9.13 As a result of recent advances in educational telecommunications, many colleges and universities are utilizing instruction by interactive television for "distance" education. For example, each semester Ball State University televises six graduate business courses to students at remote off-campus sites (*Journal of Education for Business,* Jan./Feb. 1991). To compare the performance of the off-campus MBA students at Ball State (who take the televised classes) to the on-campus MBA students (who have a "live" professor), a comprehensive test was administered to a sample of both groups of students. The test scores (50 points maximum) are summarized in the table. Based on these results, the researchers report that "there was no significant difference between the two groups of students."

	Mean	Standard Deviation
On-campus students	41.93	2.86
Off-campus TV students	44.56	1.42

Source: Arndt, T. L., and LaFollette, W. R. "Interactive television and the nontraditional student," *Journal of Education for Business,* Jan./Feb. 1991, p. 184.

a. Note that the sample sizes were not given in the journal article. Assuming 50 students are sampled from each group, perform the desired analysis. Do you agree with the researchers' findings?

b. Repeat part **a**, but assume 15 students are sampled from each group.

9.14 Intensive supervision (ISP) officers are parole officers who receive extensive training on the principles of effective intervention in order to help the criminal offender on parole or probation. The *Prison Journal* (Sept. 1997) published a study comparing the attitudes of ISP officers to regular parole officers who had not received any intervention training. Independent random samples of 11 ISP parole officers and 61 regular parole officers were recruited for the study. The officers' responses to their subjective role (7–42 point scale) and strategy (4–24 point scale) are summarized in the table.

		Sample Size	SUBJECTIVE ROLE		STRATEGY	
			Mean	Standard Deviation	Mean	Standard Deviation
Group	Regular	61	26.15	5.51	12.69	2.94
	ISP	11	19.50	4.35	9.00	2.00

Source: Fulton, B., et al. "Moderating probation and parole officer attitudes to achieve desired outcomes." *The Prison Journal,* Vol. 77, No. 3, Sept. 1997, p. 307 (Table 3).

a. At $\alpha = .05$, conduct a test to detect a difference between the mean subjective role responses for the two groups of parole officers.

b. Construct and interpret a 95% confidence interval for the difference between the mean strategy responses for the two groups of parole officers.

c. What assumptions are necessary for the validity of the intervals in parts **a** and **b**?

9.15 *Psychological Reports* (Aug. 1997) published a study on whether confidence in your ability to do well is gender-related. Undergraduate students in a social psychology class were given a 26-item pretest early in the semester. Each student was asked to indicate the number of questions they were confident that they had answered correctly. A summary of the results, by gender, is given in the table.

		Sample Size	NUMBER OF "CONFIDENT" ANSWERS		ACTUAL NUMBER OF CORRECT ANSWERS	
			Mean	Standard Deviation	Mean	Standard Deviation
Gender	Men	29	14.1	6.0	12.1	2.5
	Women	87	9.7	5.3	12.1	3.2

Source: Yarab, P. E., Sensibaugh, C. C., and Allgeier, E. R. "Over-confidence and under-performance: Men's perceived accuracy and actual performance in a course." *Psychological Reports,* Vol. 81, No. 1, Aug. 1997, p. 77 (Table 1).

a. Compare the mean number of "confident" answers for men to the corresponding mean for women with a test of hypothesis. Use $\alpha = .01$.

b. Compare the mean number of actual correct answers for men to the corresponding mean for women with a test of hypothesis. Use $\alpha = .01$.

9.16 The *Florida Scientist* (Summer/Autumn 1991) reported on a study of the feeding habits of sea urchins. A sample of 20 urchins were captured from Biscayne Bay (Miami), placed in marine aquaria, then food was withheld for 48 hours. Each sea urchin was then fed a 5-cm blade of turtle grass. Ten of the urchins received only green blades, whereas the other half received only decayed blades. (Assume that the two samples of sea urchins—10 urchins per sample—were randomly and independently selected.) The ingestion time, measured from the time the blade first made contact with the urchin's teeth to the time the urchin had finished eating the blade, was recorded. A summary of the results is provided in the table.

	Green Blades	Decayed Blades
Number of sea urchins	10	10
Mean ingestion time (hours)	3.35	2.36
Standard deviation (hours)	.79	.47

Source: Montague, J. R., et al. "Laboratory measurement of ingestion rate for the sea urchin *Lytechinus variegatus,*" *Florida Scientist,* Vol. 54, Nos. 3/4, Summer/Autumn 1991 (Table 1).

a. Construct a 90% confidence interval for the difference between the mean ingestion times of sea urchins feeding on green and decayed turtle grass.

b. According to the researchers, "the difference in rates at which the urchins ingested the blades suggest that green, unblemished turtle grass may not be a particularly palatable food compared with decayed turtle grass. If so, urchins in the field may find it more profitable to selectively graze on decayed portions of the leaves." Does the result, part **a**, support this conclusion?

9.17 Magnetotherapy involves the use of electromagnetic fields in the healing process of muscles. Rehabilitation researchers in Croatia experimented with magnetotherapy in the treatment of the quadriceps in the leg muscle (*Collegium Antropologicum,* June 1997). Two groups of 30 patients each were studied. Both groups received the standard exercise treatment; however, magnetotherapy was added to the treatment of the second group. Treatment ended when therapeutic results were achieved. The length of time (in days) each patient spent in therapy was recorded. The data are summarized in the table at left. Is there sufficient evidence to conclude that the average time spent in therapy for patients in the exercise group is greater than the average time for patients treated with magnetotherapy? Test at $\alpha = .05$.

Exercise Group	Magnetotherapy Group
$n_1 = 30$	$n_2 = 30$
$\bar{x}_1 = 39.83$	$\bar{x}_2 = 33.50$
$s_1 = 7.01$	$s_2 = 6.87$

Problem 9.17

9.18 Refer to the *Dalton Transactions* (Dec. 1997) study of bond lengths of new metal compounds, Problem 2.67 (page 118). The data, reproduced in the accompanying table, are bond lengths for samples of 36 copper–selenium compounds and 7 copper–phosphorus compounds.

BOND.XLS

Copper–Selenium					
256.1	259.0	256.0	264.7	262.5	239.0
238.9	237.4	265.9	240.2	239.7	238.3
273.8	246.4	246.0	247.4	290.6	260.9
255.6	253.1	283.7	261.4	271.9	267.9
238.7	237.6	240.0	241.1	240.3	240.8
246.5	248.3	246.0	250.3	250.1	254.6

Copper–Phosphorus					
223.8	222.7	223.3	226.1	226.3	226.0
225.0					

Source: Deveson, A., et al. "Syntheses and structures of four new copper (I)–selenium clusters: Size dependence of the cluster on the reaction conditions." *Dalton Transactions,* No. 23, Dec. 1997, p. 4492 (Table 1).

a. Find a point estimate for the true difference between the mean bond lengths of copper–selenium and copper–phosphorus metal compounds.

b. Form a 99% confidence interval around the point estimate, part **a**. Interpret the result.

c. Compare the mean bond lengths of the two metal compounds with a two-tailed test of hypothesis. Use $\alpha = .01$. Does your inference agree with that of part **b**?

9.19 Refer to the Harris Corporation/University of Florida study to determine whether a manufacturing process performed at a remote location can be established locally, Problem 3.38 (page 170). Test devices (pilots) were set up at both the old and new locations and voltage readings on 30 production runs at each location were obtained. The data are reproduced in the table on page 494. Summary results are displayed in the accompanying Excel printout. [*Note:* Smaller voltage readings are better than larger voltage readings.]

VOLTAGE.XLS

Old Location			New Location		
9.98	10.12	9.84	9.19	10.01	8.82
10.26	10.05	10.15	9.63	8.82	8.65
10.05	9.80	10.02	10.10	9.43	8.51
10.29	10.15	9.80	9.70	10.03	9.14
10.03	10.00	9.73	10.09	9.85	9.75
8.05	9.87	10.01	9.60	9.27	8.78
10.55	9.55	9.98	10.05	8.83	9.35
10.26	9.95	8.72	10.12	9.39	9.54
9.97	9.70	8.80	9.49	9.48	9.36
9.87	8.72	9.84	9.37	9.64	8.68

Source: Harris Corporation, Melbourne, Fla.

	A	B	C
1	t-Test: Two-Sample Assuming Equal Variances		
2			
3		Old	New
4	Mean	9.803667	9.422333
5	Variance	0.29259	0.229322
6	Observations	30	30
7	Pooled Variance	0.260956	
8	Hypothesized Mean Difference	0	
9	df	58	
10	t Stat	2.891126	
11	P(T<=t) one-tail	0.002697	
12	t Critical one-tail	1.671553	
13	P(T<=t) two-tail	0.005394	
14	t Critical two-tail	2.001716	

a. Compare the mean voltage readings at the two locations using a test of hypothesis ($\alpha = .05$).

b. Based on the test results, does it appear that the manufacturing process can be established locally?

9.3 COMPARING TWO POPULATION MEANS: MATCHED PAIRS

The procedures for comparing two population means presented in Section 9.2 were based on the assumption that the samples were randomly and independently selected from the target populations. Sometimes we can obtain more information about the difference between population means ($\mu_1 - \mu_2$) by selecting paired observations or repeated measurements on the same experimental unit.

Example 9.7 **Drawback to Using Independent Samples**

An elementary education teacher wants to compare two methods for teaching reading skills to first graders. One way to design the experiment is to randomly select 20 students from all available first graders, and then randomly assign 10 students to method 1 and 10 students to method 2. The reading achievement test scores obtained after completion of the experiment would represent independent random samples of scores attained by students taught reading skills by the two different methods. Consequently, the procedures described in Section 9.2 could be used to make inferences about $(\mu_1 - \mu_2)$, the difference between the mean achievement test scores of the two methods.

    a. Comment on the potential drawbacks of using independent random samples to make inferences about $(\mu_1 - \mu_2)$.

    b. Propose a better method of sampling, one that will yield more information on the parameter of interest.

Solution

    a. Assume that method 1 is truly more effective than method 2 in teaching reading skills to first graders. A potential drawback to the independent sampling plan is that the differences in the reading skills of first graders due to IQ, learning ability, socioeconomic status, and other factors are not taken into account. For example, by chance the sampling plan may assign the 10 "worst" students to method 1 and the 10 "best" students to method 2. This unbalanced assignment may mask the fact that method 1 is more effective than method 2, i.e., the resulting test of hypothesis or confidence interval on $(\mu_1 - \mu_2)$ may fail to show that μ_1 exceeds μ_2.

    b. A better method of sampling is one that attempts to remove the variation in achievement test scores due to extraneous factors such as IQ, learning ability, and socioeconomic status. One way to do this is to match the first graders in pairs, where the students in each pair have similar IQ, socioeconomic status, etc. From each pair, one member would be randomly selected to be taught by method 1; the other member would be assigned to the class taught by method 2 (see Figure 9.6). The differences between the **matched pairs** of achievement test scores should provide a clearer picture of the true difference in achievement for the two reading methods because the matching would tend to cancel the effects of the extraneous factors that formed the basis of the matching. ❑

    The sampling plan shown in Figure 9.6 is commonly known as a *matched-pairs experiment*. In the box on page 496 we give the procedures for estimating and testing the difference between two population means based on matched-pairs data. You can see that once the differences in the paired observations are obtained, the analysis proceeds as a one-sample problem. That is, a confidence interval or test for a single mean (the mean of the difference μ_d) is computed.

Figure 9.6 Matched Pairs Experiment, Example 9.7 (Pair Members Identified as Student A and Student B)

First-Grader Pair	Assignment Method 1	Method 2
1	A	B
2	B	A
⋮	⋮	⋮
10	A	B

Matched Pairs Test of Hypothesis about $\mu_d = (\mu_1 - \mu_2)$

Let $d_1, d_2, ..., d_n$ represent the differences between the pairwise observations in a *random sample* of n matched pairs, $\bar{d}$ = mean of the n sample differences, and s_d = standard deviation of the n sample differences.

Upper-Tailed Test	Two-Tailed Test	Lower-Tailed Test
$H_0: \mu_d = 0$	$H_0: \mu_d = 0$	$H_0: \mu_d = 0$
$H_a: \mu_d > 0$	$H_a: \mu_d \neq 0$	$H_a: \mu_d < 0$

Large Sample:

$$\text{Test statistic: } z = \frac{\bar{d} - 0}{\sigma_d/\sqrt{n}} \approx \frac{\bar{d} - 0}{s_d/\sqrt{n}} \qquad (9.10)$$

Rejection region:	Rejection region:	Rejection region:
$z > z_\alpha$	$|z| > z_{\alpha/2}$	$z < -z_\alpha$

Small Sample:

$$\text{Test statistic: } t = \frac{\bar{d} - 0}{s_d/\sqrt{n}} \qquad (9.11)$$

Rejection region:	Rejection region:	Rejection region:
$t > t_\alpha$	$|t| > t_{\alpha/2}$	$t < -t_\alpha$

$(1 - \alpha)100\%$ Confidence Interval for $\mu_d = (\mu_1 - \mu_2)$ Matched Pairs

Large Sample:

$$\bar{d} \pm z_{\alpha/2}\left(\frac{\sigma_d}{\sqrt{n}}\right) \quad (9.12)$$

where σ_d is the population standard deviation of differences.

Small Sample:

$$\bar{d} \pm t_{\alpha/2}\left(\frac{s_d}{\sqrt{n}}\right) \quad (9.13)$$

where $t_{\alpha/2}$ is based on $(n - 1)$ degrees of freedom.

Matched Pairs Assumptions with a Small Sample ($n < 30$)

1. The relative frequency distribution of the population of differences is approximately normal.
2. The paired differences are randomly selected from the population of differences.

Example 9.8 Hypothesis Test for μ_d

In the comparison of two methods for teaching reading discussed in Example 9.7, suppose that the $n = 10$ pairs of achievement test scores were as shown in Table 9.3 (page 497). At $\alpha = .05$, test for a difference between the mean achievement scores of students taught by the two methods.

READING.XLS

Table 9.3 Reading Achievement Test Scores for Example 9.8

	STUDENT PAIR									
	1	**2**	**3**	**4**	**5**	**6**	**7**	**8**	**9**	**10**
Method 1 Score	78	63	72	89	91	49	68	76	85	55
Method 2 Score	71	44	61	84	74	51	55	60	77	39
Pair Difference	7	19	11	5	17	−2	13	16	8	16

E **Figure 9.7**
Excel Printout for
Example 9.8

	A	B	C
1	t-Test: Paired Two Sample for Means		
2			
3		Method 1	Method 2
4	Mean	72.6	61.6
5	Variance	198.0444	217.8222
6	Observations	10	10
7	Pearson Correlation	0.89842	
8	Hypothesized Mean Difference	0	
9	df	9	
10	t Stat	5.325352	
11	P(T<=t) one-tail	0.000239	
12	t Critical one-tail	1.833114	
13	P(T<=t) two-tail	0.000478	
14	t Critical two-tail	2.262159	

Solution

We want to test H_0: $\mu_d = 0$ against H_a: $\mu_d \neq 0$, where $\mu_d = (\mu_1 - \mu_2)$. The differences between the $n = 10$ matched pairs of reading achievement test scores are computed as

$$d = (\text{Method 1 score}) - (\text{Method 2 score})$$

and are shown in the third row of Table 9.3. Since $n = 10$ (i.e., a small sample), we must assume that these differences are from an approximately normal population in order to conduct the small-sample test. The mean and standard deviation of these sample differences are $\bar{d} = 11.0$ and $s_d = 6.53$. Substituting these values into the formula for the small-sample test statistic (equation 9.11), we obtain

$$t = \frac{\bar{d} - 0}{s_d/\sqrt{n}} = \frac{11.0 - 0}{6.53/\sqrt{10}} = 5.33$$

This test statistic is shown on the Excel printout, Figure 9.7, as well as the two-tailed p-value of the test. Since p-value $= .000478$ is smaller than $\alpha = .05$, there is sufficient evidence to reject H_0 in favor of H_a. Our conclusion is that method 1 produces a mean achievement test score that is statistically different than the mean score for method 2. ❑

Self-Test
9.4

A random sample of $n = 200$ paired observations yielded the following summary information: $\bar{d} = 115$, $s_d = 250$.

a. Is there evidence that μ_1 differs from μ_2? Test at $\alpha = .05$.
b. Find a 95% confidence interval for $\mu_d = \mu_1 - \mu_2$.

Often, the pairs of observations in a matched-pairs experiment arise naturally by recording two measurements on the same experimental unit at two different points in time. For example, a professor at Brandeis University investigated the effectiveness of a prescription drug in improving the SAT scores of nervous test takers (*Newsweek,* Nov. 16, 1987). The experimental units (the objects upon which the measurements are taken) were 22 high school juniors who took the SAT twice, once without and once with the drug. Thus, two measurements were taken on each junior: (1) SAT score with no drug and (2) SAT score with the drug. These two observations, taken for all students in the study, formed the "matched pairs" of the experiment.

In an analysis of matched-pairs observations, it is important to stress that the pairing of the experimental units must be performed *before* the data are collected. Recall that the objective is to compare two methods of "treating" the experimental units. By using matched pairs of units that have similar characteristics, we are able to cancel out the effects of the variables used to match the pairs.

As in the previous section, we close with a caution.

Caution

It is inappropriate to apply the small-sample matched-pairs t procedure when the population of differences is decidedly nonnormal (e.g., highly skewed). In this case, an alternative procedure is the nonparametric Wilcoxon signed ranks test of Section 12.4.

PROBLEMS
9.20–9.33

Using the Tools

9.20   The data for a random sample of four paired observations are shown in the accompanying table.

Pair	Observation from Population A	Observation from Population B
1	2	0
2	5	7
3	10	6
4	8	5

a.   Calculate the difference within each pair, subtracting observation B from observation A. Use the differences to calculate $\bar{d}$ and s_d.

b.   If μ_1 and μ_2 are the means of populations A and B, respectively, express μ_d in terms of μ_1 and μ_2.

c.   Compute the test statistic for testing H_0: $\mu_d = 0$ against H_a: $\mu_d > 0$.

d.   Find the rejection region ($\alpha = .05$) for the test, part c.

e.   What is the conclusion of the test, parts c and d?

f.   Construct a 95% confidence interval for μ_d.

9.21   A matched-pairs experiment is used to test the hypothesis H_0: $(\mu_1 - \mu_2) = 0$. For each of the following situations, specify the rejection region, test statistic value, and conclusion.

a. $H_a: (\mu_1 - \mu_2) \ne 0, \bar{d} = 400, s_d = 435, n = 100, \alpha = .01$

b. $H_a: (\mu_1 - \mu_2) > 0, \bar{d} = .48, s_d = .08, n = 5, \alpha = .05$

c. $H_a: (\mu_1 - \mu_2) < 0, \bar{d} = -1.3, s_d^2 = .95, n = 5, \alpha = .10$

9.22   A random sample of 10 paired observations yielded the following summary information: $\bar{d} = 2.3, s_d = 2.67$.

a.   Find a 90% confidence interval for μ_d.

b.   Find a 95% confidence interval for μ_d.

c.   Find a 99% confidence interval for μ_d.

9.23   A random sample of $n = 50$ paired observations yielded the following summary statistics: $\bar{d} = 19.3, s_d = 5.2$.

a.   Compute the test statistic for testing $H_0: \mu_d = 0$ against $H_a: \mu_d \ne 0$.

b.   Find the p-value for the test, part **a**.

c.   What is the conclusion of the test, parts **a** and **b**?

d.   Construct a 95% confidence interval for μ_d.

9.24   The data for a random sample of seven paired observations are shown in the accompanying table. Let $\mu_d = (\mu_A - \mu_B)$.

Pair	Observation from Population A	Observation from Population B
1	48	54
2	50	56
3	47	50
4	50	55
5	63	64
6	65	65
7	55	61

a.   Test $H_0: \mu_d = 0$ against $H_a: \mu_d < 0$ at $\alpha = .05$.

b.   Test $H_0: \mu_d = 0$ against $H_a: \mu_d \ne 0$ at $\alpha = .05$.

c.   Find a 90% confidence interval for μ_d.

9.25   List the assumptions required to make valid inferences about μ_d based on:

a.   a large sample of matched pairs

b.   a small sample of matched pairs

Applying the Concepts

Subject	Before	After
1	136.9	130.2
2	201.4	180.7
3	166.8	149.6
4	150.0	153.2
5	173.2	162.6
6	169.3	160.1

Problem 9.26

BIOFEED.XLS

9.26   Recent experimentation suggests that it may be possible for a person to control certain body functions if he or she is trained in a program of *biofeedback* exercises. An experiment is conducted to determine whether blood pressure levels can be consciously reduced in people trained in this program. The blood pressure measurements (in millimeters of mercury) listed in the table represent readings before and after the biofeedback training of six subjects.

a.   At $\alpha = .10$, conduct a test to determine whether the mean blood pressure is reduced after biofeedback training.

b.   Construct a 90% confidence interval for the difference in mean blood pressure measurements before and after the biofeedback training. Interpret the result.

9.27 A supermarket advertisement in the *Tampa Tribune* (FL) states: "Winn-Dixie offers you the lowest total food bill! Here's the proof!" The "proof" (shown here) is a side-by-side listing of the prices of 60 grocery items purchased at Winn-Dixie and at Publix on the same day.

SUPRMKT.XLS

Item	Winn-Dixie	Publix	Item	Winn-Dixie	Publix
Big Thirst Towel	1.21	1.49	Keb Graham Crust	.79	1.29
Camp Crm/Broccoli	.55	.67	Spiffits Glass	1.98	2.19
Royal Oak Charcoal	2.99	3.59	Prog Lentil Soup	.79	1.13
Combo Chdr/Chz Snk	1.29	1.29	Lipton Tea Bags	2.07	2.17
Sure Sak Trash Bag	1.29	1.79	Carnation Hot Coco	1.59	1.89
Dow Handi Wrap	1.59	2.39	Crystal Hot Sauce	.70	.87
White Rain Shampoo	.96	.97	C/F/N Coffee Bag	1.17	1.15
Post Golden Crisp	2.78	2.99	Soup Start Bf Veg	1.39	2.03
Surf Detergent	2.29	1.89	Camp Pork & Beans	.44	.49
Sacramento T/Juice	.79	.89	Sunsweet Pit Prune	.98	1.33
SS Prune Juice	1.36	1.61	DM Vgcls Grdn Duet	1.07	1.13
V-8 Cocktail	1.18	1.29	Argo Corn Starch	.69	.89
Rodd Kosher Dill	1.39	1.79	Sno Drop Bowl Clnr	.53	1.15
Bisquick	2.09	2.19	Cadbury Milk Choc	.79	1.29
Kraft Italian Drs	.99	1.19	Andes Crm/De Ment	1.09	1.30
BC Hamburger Helper	1.46	1.75	Combat Ant & Roach	2.33	2.39
Comstock Chrry Pie	1.29	1.69	Joan/Arc Kid Bean	.45	.56
Dawn Liquid King	2.59	2.29	La Vic Salsa Pican	1.22	1.75
DelMonte Ketchup	1.05	1.25	Moist N Beef/Chz	2.39	3.19
Silver Floss Kraut	.77	.81	Ortega Taco Shells	1.08	1.33
Trop Twist Beverag	1.74	2.15	Fresh Step Cat Lit	3.58	3.79
Purina Kitten Chow	1.09	1.05	Field Trial Dg/Fd	3.49	3.79
Niag Spray Starch	.89	.99	Tylenol Tablets	5.98	5.29
Soft Soap Country	.97	1.19	Rolaids Tablets	1.88	2.20
Northwood Syrup	1.13	1.37	Plax Rinse	2.88	3.14
Bumble Bee Tuna	.58	.65	Correctol Laxative	3.44	3.98
Mueller Elbow/Mac	2.09	2.69	Tch Scnt Potpourri	1.50	1.89
Kell Nut Honey Crn	2.95	3.25	Chld Enema 2.250	.98	1.15
Cutter Spray	3.09	3.95	Gillette Atra Plus	5.00	5.24
Lawry Season Salt	2.28	2.97	Colgate Shave	.94	1.10

a. Explain why the data should be analyzed as matched pairs.

b. An Excel printout showing the results of a hypothesis test of $(\mu_{Winn} - \mu_{Publix})$, the difference between the mean prices of grocery items purchased at the two supermarkets, is displayed on page 501. Interpret the results.

9.28 According to *Webster's New World Dictionary,* a tongue-twister is "a phrase that is hard to speak rapidly." Do tongue-twisters have an effect on the length of time it takes to read silently? To answer this question, 42 undergraduate psychology students participated in a reading experiment (*Memory & Cognition,* Sept. 1997). Two lists, each comprised of 600 words, were constructed. One list contained a series of tongue-twisters and the other list (called the *control*) did not contain any tongue-twisters. Each student read both lists and the length of time (in minutes) required to complete the lists was recorded. The researchers used a test of hypothesis to compare the mean reading response times for the tongue-twister and control lists.

a. Set up the null hypothesis for the test.

	A	B	C
1	t-Test: Paired Two Sample for Means		
2			
3		Winn-Dixie	Publix
4	Mean	1.6655	1.9195
5	Variance	1.20170653	1.217473
6	Observations	60	60
7	Pearson Correlation	0.96895919	
8	Hypothesized Mean Difference	0	
9	df	59	
10	t Stat	-7.1773633	
11	P(T<=t) one-tail	6.7738E-10	
12	t Critical one-tail	1.67109192	
13	P(T<=t) two-tail	1.3548E-09	
14	t Critical two-tail	2.00099748	

Problem 9.27

b. Use the information in the accompanying table to find the test statistic and p-value of the test.

c. Give the appropriate conclusion. Use $\alpha = .05$.

		RESPONSE TIME (MINUTES)	
		Mean	**Standard Deviation**
List Type	Tongue-twister	6.59	1.94
	Control	6.34	1.92
	Difference	.25	.78

Source: Robinson, D. H., and Katayama, A. D. "At-lexical, articulatory interference in silent reading: The 'upstream' tongue-twister effect." *Memory & Cognition,* Vol. 25, No. 5, Sept. 1997, p. 663.

9.29 Refer to the *Chemosphere* study of Vietnam veterans' exposure to Agent Orange, Problem 7.28 (page 383). In addition to the amount of TCDD (measured in parts per million) in blood plasma, the TCDD in fat tissue drawn from 20 exposed Vietnam veterans was recorded. The data are shown in the table. Compare the mean TCDD level in plasma to the mean TCDD level in fat tissue for Vietnam veterans exposed to Agent Orange using a test of hypothesis at $\alpha = .05$.

TCDDFAT.XLS

Veteran	TCDD Levels in Plasma	TCDD Levels in Fat Tissue	Veteran	TCDD Levels in Plasma	TCDD Levels in Fat Tissue
1	2.5	4.9	11	6.9	7.0
2	3.1	5.9	12	3.3	2.9
3	2.1	4.4	13	4.6	4.6
4	3.5	6.9	14	1.6	1.4
5	3.1	7.0	15	7.2	7.7
6	1.8	4.2	16	1.8	1.1
7	6.0	10.0	17	20.0	11.0
8	3.0	5.5	18	2.0	2.5
9	36.0	41.0	19	2.5	2.3
10	4.7	4.4	20	4.1	2.5

Source: Schecter, A., et al. "Partitioning of 2,3,7,8-chlorinated dibenzo-*p*-dioxins and dibenzofurans between adipose tissue and plasma lipid of 20 Massachusetts Vietnam veterans." *Chemoshpere,* Vol 20, Nos. 7–9, 1990, pp. 954–955 (Tables I & II).

9.30 Refer to the *Journal of STAR Research* (July 1994) study of traffic counts on Florida highways, Problem 3.40 (page 171). The table below gives the traffic counts at the 30th highest hour and the 100th highest hour of a recent year for 20 randomly selected Department of Transportation (DOT) permanent count stations. Recall that the Florida DOT bases roadway design on the 100th highest hourly traffic volume in a year rather than the traditional 30th highest hourly volume. Is there a significant difference between the mean traffic count for the 30th highest hour and the mean traffic count for the 100th highest hour on Florida roadways? Use $\alpha = .10$.

DOTHOUR.XLS

Station	30th Highest Hour	100th Highest Hour
0117	1,890	1,736
0087	2,217	2,069
0166	1,444	1,345
0013	2,105	2,049
0161	4,905	4,815
0096	2,022	1,958
0145	594	548
0149	252	229
0038	2,162	2,048
0118	1,938	1,748
0047	879	811
0066	1,913	1,772
0094	3,494	3,403
0105	1,424	1,309
0113	4,571	4,425
0151	3,494	3,359
0159	2,222	2,137
0160	1,076	989
0164	2,167	2,039
0165	3,350	3,123

Source: Ewing, R. "Roadway levels of service in an era of growth management." *Journal of STAR Research,* Vol. 3, July 1994, p. 103 (Table 2).

9.31 The drug clonidine has been shown to be beneficial in the treatment of Tourette's Syndrome. Since patients with Tourette's disorder often stutter, clonidine may be useful in treating stuttering. The *American Journal of Psychiatry* (July 1995) reported on a study of 25 stuttering children who were treated with clonidine. Prior to treatment, a speech sample was recorded for each child and the frequency of speech repetitions was measured. The experiment was repeated eight weeks later after the children were treated with a daily dose of clonidine. The difference in stuttering frequency before and after treatment was recorded for each child. These differences are summarized here:

$$\bar{d} = -2.33 \qquad s_d = 16.98$$

a. The journal reported a 95% confidence interval for the true mean difference in stuttering frequencies as $(-9.3, 4.7)$. Do you agree with the result?

b. The journal reported that "clonidine was not found to have a positive effect on stuttering." Do you agree? Explain.

SWIMMAZE.XLS

Litter	Male	Female
1	8	5
2	8	4
3	6	7
4	6	3
5	6	5
6	6	3
7	3	8
8	5	10
9	4	4
10	4	4
11	6	5
12	6	3
13	12	5
14	3	8
15	3	4
16	8	12
17	3	6
18	6	4
19	9	5

Source: Thomas E. Bradstreet, Merck Research Labs, BL 3-2, West Point, Penn. 19486.

Problem 9.32

c. Conduct a test of hypothesis (at $\alpha = .05$) to confirm the result, part **b**.

9.32 Merck Research Labs conducted an experiment to evaluate the effect of a new drug using the Single-T swim maze. Nineteen impregnated dam rats were captured and allocated a dosage of 12.5 milligrams of the drug. One male and one female pup were randomly selected from each resulting litter to perform in the swim maze. Each rat pup is placed in the water at one end of the maze and allowed to swim until it successfully escapes at the opposite end. If the rat pup fails to escape after a certain period of time, it is placed at the beginning of the maze and given another attempt to escape. The experiment is repeated until three successful escapes are accomplished by each rat pup. The number of swims required by each pup to perform three successful escapes is reported in the accompanying table. Is there sufficient evidence of a difference between the mean number of swims required by male and female rat pups? Test using $\alpha = .10$.

9.33 Refer to the *Journal of Geography* (May/June 1997) study of the effectiveness of computer-based interactive maps in the classroom, Problem 2.27 (page 84). Each of 45 seventh grade geography students was given two assignments. For the first assignment, the students used traditional library resources (books, atlases, and encyclopedias) to research a particular African region. For the second assignment, the students used computer resources (electronic atlases and encyclopedias, interactive maps) to research a different African region. At the end of each assignment, a test is administered to the students. The test scores are listed in the accompanying table.

GEOSCORE.XLS

Student	Test #1	Test #2	Student	Test #1	Test #2	Student	Test #1	Test #2
1	9	9	16	56	59	31	38	47
2	16	9	17	25	31	32	94	91
3	9	3	18	59	75	33	94	100
4	56	56	19	81	66	34	75	75
5	38	47	20	59	47	35	50	66
6	31	63	21	81	63	36	81	91
7	50	47	22	41	56	37	75	78
8	53	59	23	91	75	38	88	63
9	69	72	24	53	69	39	78	78
10	63	56	25	56	63	40	44	63
11	41	63	26	38	56	41	72	69
12	56	47	27	25	72	42	59	63
13	50	47	28	63	72	43	69	81
14	59	53	29	47	69	44	84	88
15	53	63	30	50	63	45	69	91

Source: Linn, S. E. "The effectiveness of interactive maps in the classroom: A selected example in studying Africa." *Journal of Geography,* Vol. 96, No. 3, May/June 1997, p. 167 (Table 1).

a. Is there evidence of a difference between the two mean test scores at $\alpha = .10$?

b. What assumptions (if any) are necessary for the inference in part **a** to be valid? Explain.

9.4 COMPARING TWO POPULATION PROPORTIONS: INDEPENDENT SAMPLES

In this section we present methods for making inferences about the difference between two population proportions when the samples are collected independently. For example, one may be interested in comparing the proportions of married and unmarried persons who are overweight, or the proportions of homes in two states that are heated by natural gas, or the proportions of teenagers today and 20 years ago who smoke, etc.

Example 9.9 **Selecting a Point Estimate**

In a study of ethical management decision making, 48 of 50 female managers responded that concealing one's on-the-job errors was very unethical. In contrast, only 30 of 50 male managers responded in a similar manner (*Journal of Business Ethics,* Aug. 1987). Construct a point estimate for the difference between the proportions of female and male managers who believe that concealing one's errors is very unethical.

Solution

For this example, define

π_1 = Population proportion of female managers who believe that concealing on-the-job errors is unethical

π_2 = Population proportion of male managers who believe that concealing on-the-job errors is unethical

x_1 = Number of females in the sample who believe that concealing on-the-job errors is unethical

x_2 = Number of males in the sample who believe that concealing on-the-job errors is unethical

As a point estimate of $(\pi_1 - \pi_2)$, we will use the difference between the corresponding sample proportions $(p_1 - p_2)$, where

$$p_1 = \frac{x_1}{n_1} = \frac{48}{50} = .96$$

and

$$p_2 = \frac{x_2}{n_2} = \frac{30}{50} = .60$$

Thus, the point estimate of $(\pi_1 - \pi_2)$ is

$$(p_1 - p_2) = .96 - .60 = .36 \qquad \square$$

To judge the reliability of the point estimate $(p_1 - p_2)$, we need to know the characteristics of its performance in repeated independent sampling from two binomial populations. This information is provided by the sampling distribution of $(p_1 - p_2)$, illustrated in Figure 9.8 and described in the accompanying box.

Figure 9.8 Sampling Distribution of $p_1 - p_2$

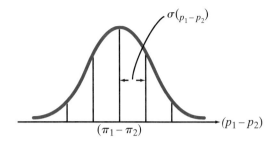

Sampling Distribution of $p_1 - p_2$

For sufficiently large sample sizes n_1 and n_2, the sampling distribution of $(p_1 - p_2)$, based on independent random samples from two binomial populations, is approximately normal with

$$\text{Mean: } \mu_{(p_1 - p_2)} = (\pi_1 - \pi_2) \tag{9.14}$$

and

$$\text{Standard deviation: } \sigma_{(p_1 - p_2)} = \sqrt{\frac{\pi_1(1 - \pi_1)}{n_1} + \frac{\pi_2(1 - \pi_2)}{n_2}} \tag{9.15}$$

Large-sample confidence intervals and tests of hypothesis for $(\pi_1 - \pi_2)$ are derived from the sampling distribution of $(p_1 - p_2)$. Details are presented in the next three boxes.

Large-Sample Test of Hypothesis about $(\pi_1 - \pi_2) = 0$

Upper-Tailed Test	Two-Tailed Test	Lower-Tailed Test
$H_0: (\pi_1 - \pi_2) = 0$	$H_0: (\pi_1 - \pi_2) = 0$	$H_0: (\pi_1 - \pi_2) = 0$
$H_a: (\pi_1 - \pi_2) > 0$	$H_a: (\pi_1 - \pi_2) \neq 0$	$H_a: (\pi_1 - \pi_2) < 0$

$$\text{Test statistic: } z = \frac{(p_1 - p_2) - 0}{\sigma_{(p_1 - p_2)}} \approx \frac{(p_1 - p_2) - 0}{\sqrt{p(1 - p)\left(\frac{1}{n_1} + \frac{1}{n_2}\right)}} \tag{9.16}$$

where

$$p_1 = \frac{x_1}{n_1} = \frac{\text{Number of successes in sample 1}}{n_1}$$

$$p_2 = \frac{x_2}{n_2} = \frac{\text{Number of successes in sample 2}}{n_2}$$

$$p = \frac{x_1 + x_2}{n_1 + n_2} = \frac{\text{Total number of successes}}{\text{Total sample size}}$$

Rejection region:	*Rejection region:*	*Rejection region:*
$z > z_\alpha$	$|z| > z_{\alpha/2}$	$z < -z_\alpha$

> ## Large-Sample $(1 - \alpha)100\%$ Confidence Interval for $(\pi_1 - \pi_2)$
>
> $$(p_1 - p_2) \pm z_{\alpha/2}\sigma_{(p_1 - p_2)} \approx (p_1 - p_2) \pm z_{\alpha/2}\sqrt{\frac{p_1(1 - p_1)}{n_1} + \frac{p_2(1 - p_2)}{n_2}} \quad \textbf{(9.17)}$$
>
> where p_1 and p_2 are the sample proportions of observations with the characteristic of interest, $q_1 = 1 - p_1$, and $q_2 = 1 - p_2$.

> ## Assumptions for Making Inferences about $(\pi_1 - \pi_2)$
> 1. Independent random samples are collected.
> 2. The samples are sufficiently large (i.e., $n_1 \geq 30$ and $n_2 \geq 30$).

Example 9.10

Hypothesis Test for $(\pi_1 - \pi_2)$

Refer to Example 9.9. Conduct a test to determine whether the proportion of female managers who believe that concealing one's errors is very unethical exceeds the corresponding proportion for male managers. Use $\alpha = .05$.

Solution

Since we want to know whether π_1, the proportion for female managers, exceeds π_2, the proportion for male managers, we will test

$$H_0: \pi_1 - \pi_2 = 0$$
$$H_a: \pi_1 - \pi_2 > 0$$

For this upper-tailed test, the null hypothesis will be rejected at $\alpha = .05$ if

$$z > z_{.05} = 1.645 \quad \text{(see Figure 9.9)}$$

From Example 9.9, we have $n_1 = 50$, $n_2 = 50$, $p_1 = .96$ and $p_2 = .60$. Since we are testing whether the two proportions are equal, the best estimate of this unknown π is obtained by "pooling" the information from both samples into a single estimator p:

$$p = \frac{x_1 + x_2}{n_1 + n_2} = \frac{48 + 30}{50 + 50} = \frac{78}{100} = .78$$

Substituting the values of p_1, p_2, and p into equation 9.16, we obtain the test statistic:

Figure 9.9 Rejection Region for Example 9.10

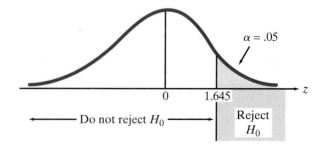

$$z = \frac{(p_1 - p_2) - 0}{\sqrt{p(1 - p)\left(\dfrac{1}{n_1} + \dfrac{1}{n_2}\right)}} = \frac{(.96 - .60) - 0}{\sqrt{(.78)(.22)\left(\dfrac{1}{50} + \dfrac{1}{50}\right)}} = 4.35$$

Since the test statistic $z = 4.35$ is greater than 1.645, it falls in the rejection region. Consequently, there is sufficient evidence (at $\alpha = .05$) to conclude that the proportion of female managers who believe that concealing on-the-job errors is very unethical exceeds the corresponding proportion of male managers.

Using the p-value approach, we will reach the same conclusion. The p-value of this upper-tailed test is, by definition,

$$p\text{-value} = P(z > 4.35) \approx 0$$

Since this p-value is less than $\alpha = .05$, we reject H_0. ❑

**Self-Test
9.5**

Independent random samples of size $n_1 = n_2 = 500$ from populations 1 and 2 produced sample successes $x_1 = 220$ and $x_2 = 260$.

a. Test $H_0: (\pi_1 - \pi_2) = 0$ against $H_a: (\pi_1 - \pi_2) \neq 0$ using $\alpha = .05$.
b. Find a 95% confidence interval for $(\pi_1 - \pi_2)$.

The procedures presented in this section require large, independent random samples from the populations. When the samples are not "sufficiently large," we must resort to another statistical technique, as stated in the box.

Caution
_____
The z test for comparing π_1 and π_2 is inappropriate when the sample sizes are not "sufficiently large." In this case, π_1 and π_2 can be compared using a statistical technique to be discussed in Section 9.5.

**PROBLEMS
9.34–9.47**

Using the Tools

9.34 Suppose you want to estimate $\pi_1 - \pi_2$, the difference between the proportions of two populations. Based on independent random samples, you can compute the sample proportions of successes, p_1 and p_2, respectively. Estimate $\mu_{(p_1 - p_2)}$ and $\sigma_{(p_1 - p_2)}$ for each of the following results:

a. $n_1 = 150, p_1 = .3; n_2 = 130, p_2 = .4$
b. $n_1 = 100, p_1 = .10; n_2 = 100, p_2 = .05$
c. $n_1 = 200, p_1 = .76; n_2 = 200, p_2 = .96$

Sample from Population 1	Sample from Population 2
$n_1 = 400$	$n_2 = 350$
$x = 200$	$x = 210$

Problem 9.35

9.35 Independent random samples are taken from two populations. The accompanying table shows the sample sizes and the number of observations with the characteristic of interest.

    a. Calculate the test statistic for testing $H_0: (\pi_1 - \pi_2) = 0$ against H_a: $(\pi_1 - \pi_2) \neq 0$.

    b. Find the rejection region for the test, part **a**, at $\alpha = .05$.

    c. Make the proper conclusion for the test, parts **a** and **b**.

    d. Find a 90% confidence interval for $(\pi_1 - \pi_2)$.

    e. Find a 95% confidence interval for $(\pi_1 - \pi_2)$.

    f. Find a 99% confidence interval for $(\pi_1 - \pi_2)$.

9.36 Independent random samples of size n_1 and n_2 are selected from two binomial populations to test the null hypothesis $H_0: (\pi_1 - \pi_2) = 0$. For each of the following situations, specify the rejection region, test statistic value, and conclusion.

    a. $H_a: (\pi_1 - \pi_2) \neq 0, x_1 = 600, x_2 = 600, n_1 = 1,500, n_2 = 2,000, \alpha = .01$

    b. $H_a: (\pi_1 - \pi_2) > 0, x_1 = 50, x_2 = 20, n_1 = n_2 = 1,000, \alpha = .05$

    c. $H_a: (\pi_1 - \pi_2) < 0, x_1 = 72, x_2 = 78, n_1 = n_2 = 120, \alpha = .01$

9.37 Consider independent random samples from populations 1 and 2. Construct a confidence interval for $(\pi_1 - \pi_2)$ if:

    a. $\alpha = .10, p_1 = .20, p_2 = .15, n_1 = 70, n_2 = 95$

    b. $\alpha = .05, p_1 = .73, p_2 = .68, n_1 = 1,025, n_2 = 640$

    c. $\alpha = .01, p_1 = .51, p_2 = .23, n_1 = 500, n_2 = 500$

9.38 Independent random samples of 250 observations each are selected from two populations. The samples from populations 1 and 2 produced, respectively, 100 and 75 observations possessing the characteristic of interest.

    a. Test $H_0: (\pi_1 - \pi_2) = 0$ against $H_a: (\pi_1 - \pi_2) > 0$ at $\alpha = .05$.

    b. Test $H_0: (\pi_1 - \pi_2) = 0$ against $H_a: (\pi_1 - \pi_2) \neq 0$ at $\alpha = .01$.

    c. Construct a 90% confidence interval for $(\pi_1 - \pi_2)$.

    d. Construct a 99% confidence interval for $(\pi_1 - \pi_2)$.

Applying the Concepts

9.39 *Sports Illustrated* (Dec. 8, 1997) asked a sample of middle school and high school students whether they agreed or disagreed with the statement: "African-American players have become so dominant in sports such as football and basketball that many whites feel they can't compete at the same level as blacks." Forty-five percent of the white males in the sample and 27% of the black males in the sample disagreed with the statement. Assume that 500 white males and 500 black males participated in the study.

    a. Using a 95% confidence interval, compare the percentages of white and black males who disagreed with the statement.

    b. Make the comparison, part **a**, using a test of hypothesis. Use $\alpha = .05$.

9.40 Economists at the Federal Reserve Bank of Cleveland examined the relationship between earnings and education (*Economic Review,* Qtr. 4, 1996). One of the variables investigated was high school dropout rate. A comparison was made between the high school dropout rate in 1963 and 1993. In 1963, 42% of the 25,000 adults interviewed were high school dropouts. In 1993, 11% of the 60,000 adults interviewed were high school dropouts.

a. Is there evidence that the high school dropout rate has significantly decreased from 1963 to 1993? Test using $\alpha = .01$.

b. Repeat part **a**, but assume that 40% of the 60,000 adults interviewed in 1993 were high school dropouts.

c. Refer to part **b**. Comment on the difference between "statistical" significance and "practical" significance.

9.41 According to an American Heart Association (AHA) researcher, "people who are hostile and mistrustful are more likely to die young or develop life-threatening heart disease than those with more 'trusting hearts'" (*Tampa Tribune*, Jan. 17, 1989). A sample of 118 male doctors, lawyers, and workers in a large industrial firm in Chicago were divided into two groups based on a standard psychological test designed to measure hostility. Of the 35 men who scored high in hostility, 7 died at a relatively early age (i.e., between the ages of 25 and 50). In contrast, only 4 of the 83 men whose hostility rating was low died at an early age.

a. Test the AHA researcher's claim using $\alpha = .05$.

b. Estimate the true difference between the proportion of men with high hostility scores who die at an early age and the corresponding proportion of men with low hostility scores with a 95% confidence interval. Does your result agree with that of part **a**?

9.42 The herbal medicine EGb, made from the extract of the Ginkgo biloba tree, may improve the condition of Alzheimer's patients (*Journal of the American Medical Association*, Oct. 21, 1997). Patients with mild to moderately severe dementia were randomly divided into two groups. One group received a daily dosage of EGb while the other group received a placebo. (The experiment was "double blind"—neither the patients nor the researchers knew who received which treatment.) Assume there were 30 patients in each group and that 21 of the EGb patients demonstrated improved cognitive ability compared to eight of the placebo patients. Use a test of hypothesis to compare the proportion of patients with dementia who showed improved cognitive ability for the two treatments. Use $\alpha = .10$.

9.43 The *Journal of Fish Biology* (Aug. 1990) reported on a study to compare the prevalence of parasites (tapeworms) found in species of Mediterranean and Atlantic fish. In the Mediterranean Sea, 588 brill were captured and dissected; 211 were found to be infected by the parasite. In the Atlantic Ocean, 123 brill were captured and dissected; 26 were found to be infected. Compare the proportions of infected brill at the two capture sites using a 90% confidence interval. Interpret the interval.

9.44 Since the early 1990s marketers have wrestled with the appropriateness of using ads that appeal to children to sell adult products. One controversial advertisement campaign is Camel cigarettes' use of the cartoon character "Joe Camel" as its brand symbol. The Federal Trade Commission has considered banning ads featuring Joe Camel because they supposedly encourage young people to smoke. Lucy L. Henke, a marketing professor at the University of New Hampshire, assessed young children's abilities to recognize cigarette brand advertising symbols. She found that 15 out of 28 children under the age of 6 and 46 out of 55 children age 6 and over recognized Joe Camel, the brand symbol of Camel cigarettes (*Journal of Advertising,* Winter 1995).

a. Use a 95% confidence interval to estimate the proportion of all children that recognize Joe Camel. Interpret the interval.

b. Do the data indicate that recognition of Joe Camel increases with age? Test using $\alpha = .05$.

9.45 A new mood stabilizer drug, called Depakote, has been developed for treating mania. In advertisements for Depakote published in medical journals (e.g., the *American Journal of Psychiatry,* Jan. 1998), the effectiveness of the new drug is described. In a double-blind clinical trial, 120 manic patients were randomly assigned to one of two treatment groups for 21 days. (Assume 60 patients were in each group.) The first group received a daily dosage of Depakote and the second group received a placebo (i.e., no drug) every day. Each patient was administered the Manic Rating Scale (MRS) prior to and following the clinical trial. (MRS is designed to measure a patient's mania level.) The results are shown in the table. Conduct a test of hypothesis to compare the proportions of manic patients with significant improvement in MRS score for the Depakote and placebo treatment groups. Use $\alpha = .05$.

Drug	Proportion of Patients with Significant Improvement in MRS Score
Depakote	.58
Placebo	.29

Source: Abbot Laboratories, Inc., North Chicago, IL, Nov. 1997.

9.46 The city of Niagara Falls, New York, and the surrounding county are known to contain a large number of toxic-waste disposal sites (dump sites). Following the negative publicity of the "Love Canal" (located in Niagara Falls), the New York State Department of Health funded a death-certificate study to determine whether lung cancer in Niagara County might be associated with exposure to pollution from the dump sites (*Environmental Research,* Feb. 1989). The study involved a comparison of two samples. The first sample (called cases) comprised the $n_1 = 327$ residents of Niagara County who died of cancer of the trachea, bronchus, or lung. The second sample (called controls) consisted of $n_2 = 667$ residents of Niagara County who died from causes other than respiratory cancers. Of the 327 cases, 50 resided near dump sites containing lung carcinogens; of the 667 controls, 102 resided near dump sites containing lung carcinogens.

a. Compare the percentages of cases and controls with residences near dump sites with a 90% confidence interval.

b. If the confidence coefficient were decreased to .80, would you expect the width of the interval to increase, decrease, or stay the same? Would you recommend decreasing the confidence coefficient to .80? Explain.

9.47   Sports fans often argue that the outcomes of professional basketball games are not decided until the fourth (and last) quarter. In contrast, the conventional "wisdom" in Major League Baseball (MLB) suggests that most nine-inning games are "over" by the seventh inning. Can the outcome of a baseball game really be predicted based on the team leading after seven innings? And is it really true that the team leading after three quarters of a National Basketball Association (NBA) game is no more likely to win than the trailing team? To answer these and other questions, University of Missouri researchers collected and analyzed data for 200 NBA games and 100 MLB games played during the 1990 regular season (*Chance*, Vol. 5, 1992). The accompanying table reports the number and percentage of games in which the team leading after three quarters or seven innings lost. [*Note:* Tied games after three quarters or seven innings were removed from the data set.] Use a test of hypothesis to compare the loss percentage of NBA teams leading after three quarters to the loss percentage of MLB teams leading after seven innings. Test using $\alpha = .05$.

Sports	Games	Leader Wins	Leader Loses	% Reversals
NBA	189	150	39	20.6
MLB	92	86	6	6.5

Source: Cooper, H., DeNeve, K. M., and Mosteller, F. "Predicting professional sports game outcomes from intermediate game scores." *Chance: New Directions for Statistics and Computing,* Vol. 5, Nos. 3–4, 1992, pp. 18–22 (Table 1).

9.5 COMPARING POPULATION PROPORTIONS: CONTINGENCY TABLES

Data are often categorized according to two qualitative variables for the purpose of determining whether the two variables are related. For example, we may want to determine whether religious affiliation and marital status are related, or whether socioeconomic status is related to race. Recall from Section 2.4 that such data can be summarized in the form of a cross-classification table. In this section, we present a method for making inferences about the proportions in a cross-classification table.

A cross-classification table is often called a **contingency table** because the objective of the study is to investigate whether the proportions associated with the categories of one of the qualitative variables *depend* (or are *contingent*) on the proportions associated with the categories of the other variable. We illustrate with an example.

Example   9.11   Independence of Two Qualitative Variables

Refer to the *American Sociological Review* study of whether men and women interrupt a speaker equally often, Example 2.6 (page 85). Undergraduate students were organized into discussion groups and their interactions recorded on video. Recall that the researchers recorded two variables each time the group speaker was interrupted: (1) the gender of the speaker and (2) the gender of the interrupter. A cross-classification (or contingency) table for a sample of $n = 40$

Table 9.4 Contingency Table for Example 9.11

		INTERRUPTER		
		Male	**Female**	**TOTALS**
SPEAKER	Male	15	5	20
	Female	10	10	20
	TOTALS	25	15	40

interruptions is shown in Table 9.4. In terms of proportions, what does it mean to say that speaker gender and interrupter gender are independent?

Solution

From our discussion of independent events in Chapter 4, we know that the probability of one event occurring (e.g., a male interrupter) does not depend on the other event (e.g., a female speaker) if the events are independent. Consequently, if the two variables, speaker gender and interrupter gender, are independent, then the proportion of times a male interrupts will be the same for both male and female speakers.

If we let π_1 represent the true proportion of male speakers who are interrupted by a male and π_2 represent the true proportion of female speakers who are interrupted by a male, then independence implies $\pi_1 = \pi_2$. ❑

A test of the null hypothesis that two qualitative variables are independent uses a test statistic that compares the actual number of observations in each cell of the table to the number expected if H_0 is true. The following examples illustrate the procedure for conducting the test.

Example 9.12

Computing Expected Numbers in a Contingency Table

Refer to Example 9.11. Based on the assumption that the two qualitative variables, speaker gender and interrupter gender, are independent, how many observations would you expect to fall in each of the cells of Table 9.4?

Solution

Note that 20 of the 40 interruptions sampled fell in row 1 of Table 9.4 and 20 fell in row 2. If speaker gender is independent of interrupter gender, we would expect $20/40 = \frac{1}{2}$ of the 25 male interruptions to fall in the first row of Table 9.4 and $\frac{1}{2}$ to fall in the second row. Therefore, the expected number of observations falling in row 1, column 1, is

$$E_{11} = \left(\frac{20}{40}\right)(25) = \left(\frac{\text{Row 1 total}}{n}\right)(\text{Column 1 total}) = 12.5$$

and the expected number of observations falling in row 2, column 1, is

$$E_{21} = \left(\frac{20}{40}\right)(25) = \left(\frac{\text{Row 2 total}}{n}\right)(\text{Column 1 total}) = 12.5$$

Table 9.5 Expected Number if Speaker Gender and Interrupter Gender Are Independent

| | | INTERRUPTER | | |
		Male	Female	TOTALS
SPEAKER	Male	12.5	7.5	20
	Female	12.5	7.5	20
	TOTALS	25	15	40

We now move to column 2 and note that the column total is 15, i.e., there were 15 total female interruptions. Again, we expect $20/40 = \frac{1}{2}$ of these 15 observations to fall in the first row and $\frac{1}{2}$ to fall in the second row. Thus, the expected number for row 1, column 2, is

$$E_{12} = \left(\frac{20}{40}\right)(15) = \left(\frac{\text{Row 1 total}}{n}\right)(\text{Column 2 total}) = 7.5$$

and the expected number for row 2, column 2, is

$$E_{22} = \left(\frac{20}{40}\right)(15) = \left(\frac{\text{Row 2 total}}{n}\right)(\text{Column 2 total}) = 7.5$$

These expected numbers are shown in Table 9.5.

The formula for calculating any expected value in a contingency table can be deduced from the values previously calculated. Each **expected cell count** is equal to the product of its respective row and column totals divided by the total sample size n. The general formula for calculating the expected cell count for a cell in any row and column of a contingency table is given in the box. ❏

Formula for Computing Expected Cell Counts in a Contingency Table

$$E_{ij} = \frac{(R_i)(C_j)}{n} \tag{9.18}$$

where
E_{ij} = expected cell count for the cell in row i and column j
R_i = row total corresponding to row i
C_j = column total corresponding to column j
n = sample size

Self-Test 9.6

Calculate the expected cell counts for the following contingency table:

| | | Columns (Variable 2) | |
		1	2
Rows	1	20	10
(Variable 1)	2	80	50

Example 9.13   Computing the Test Statistic

Refer to Examples 9.11 and 9.12. To test the null hypothesis that the two qualitative variables, speaker gender and interrupter gender, are independent, we compare the observed number and expected number of each cell in the contingency table in the following manner:

$$\frac{(\text{Observed cell count} - \text{Expected cell count})^2}{\text{Expected cell count}} \qquad \textbf{(9.19)}$$

The sum of these quantities over all cells, called χ^2, represents the test statistic for the test of independence. (The χ^2 statistic was first presented in Section 7.6.) Calculate χ^2 using the information in Tables 9.4 and 9.5.

Solution

We substitute the observed cell count from Table 9.4 and the expected cell count from Table 9.5 into equation 9.19 for each cell to obtain the following quantities:

Row 1, Column 1: $\qquad \dfrac{(15 - 12.5)^2}{12.5} = .5$

Row 2, Column 1: $\qquad \dfrac{(5 - 7.5)^2}{7.5} = .833$

Row 1, Column 2: $\qquad \dfrac{(10 - 12.5)^2}{12.5} = .5$

Row 2, Column 2: $\qquad \dfrac{(10 - 7.5)^2}{7.5} = .833$

Summing these values, we obtain the test statistic:

$$\chi^2 = .5 + .833 + .5 + .833 = 2.66 \qquad \qquad \square$$

Example 9.14   Finding the Rejection Region

Specify the rejection region for the test described in Example 9.13. Use $\alpha = .05$. Then test to determine whether speaker gender and interrupter gender are independent.

Solution

If the two qualitative variables are, in fact, independent, then the observed cell counts should be nearly equal to the expected cell counts. Consequently, the greater the difference between the observed and expected cell counts, the greater is the evidence to indicate a lack of independence. Since the value of chi-square increases as the differences between the observed and expected cell counts increase, we will reject

H_0: Speaker gender and interrupter gender are independent

for values of χ^2 *larger* than the critical value $\chi^2_{.05}$.

The appropriate degrees of freedom for this χ^2 test will always be $(r - 1)(c - 1)$, where r is the number of rows and c is the number of columns in the contingency

Figure 9.10 Rejection Region for the Contingency Table, Example 9.14

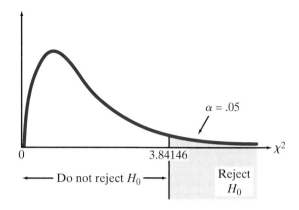

table. For these data, we have $r = 2$ rows and $c = 2$ columns; hence, the appropriate number of degrees of freedom for χ^2 is

$$\text{df} = (r - 1)(c - 1) = (1)(1) = 1$$

The tabulated value of $\chi^2_{.05}$ corresponding to 1 degree of freedom, obtained from Table B.4 in Appendix B, is 3.84146. Therefore, the rejection region (highlighted in Figure 9.10) is

Rejection region: $\chi^2 > 3.84146$

Since the computed value $\chi^2 = 2.66$ falls below the critical value 3.84146, we fail to reject the null hypothesis of independence at $\alpha = .05$. There is not sufficient evidence to claim that the proportion of times the speaker is interrupted by a male depends on the gender of the speaker. ❑

The elements of a χ^2 test for a contingency table are summarized in the box.

A Test of Hypothesis for Independence of Two Qualitative Variables in a Contingency Table

H_0: The two qualitative variables comprising the contingency table are independent

H_a: The two qualitative variables comprising the contingency table are dependent

$$\text{Test statistic: } \chi^2 = \sum_{i=1}^{r} \sum_{j=1}^{c} \frac{(O_{ij} - E_{ij})^2}{E_{ij}} \qquad (9.20)$$

where
  r = number of rows in the table
  c = number of columns in the table
  O_{ij} = observed number of responses in the cell in row i and column j
  E_{ij} = expected number of responses in cell ij = $\dfrac{(R_i)(C_j)}{n}$

$$\text{Rejection region: } \chi^2 > \chi^2_\alpha$$

where χ^2_α is the tabulated value of the chi-square distribution based on $(r - 1)(c - 1)$ degrees of freedom such that $P(\chi^2 > \chi^2_\alpha) = \alpha$.

Assumption: The sample size n is "large." This will be satisfied if the expected count for each of the $r \times c$ cells is at least 1. (See the next box.)

"Large" n in Contingency Tables

For the χ^2 test to give accurate results, the expected counts of the $r \times c$ cells must be large. There has been much debate among statisticians as to the definition of "large." Some researchers adopt a conservative approach and require each expected count to be at least 5. Others have found that the test is accurate if no more than 20% of the cells contain expected frequencies less than 5 and no cells have expected frequencies less than 1. In the case of a $2 \times c$ table, researchers have found the test gives accurate results as long as all expected cell counts equal at least 0.5. *A reasonable compromise between these points of view is to be sure that all expected counts are at least 1.*

Self-Test 9.7

The contingency table for Self-Test 9.6 is reproduced below:

		Columns (Variable 2)	
		1	**2**
Rows	1	20	10
(Variable 1)	2	80	50

a. Calculate the χ^2 statistic for testing whether the two qualitative variables are independent.

b. Give the rejection region when $\alpha = .01$.

c. Conduct the test.

The contingency table of Examples 9.11–9.14 is a 2×2 table—that is, there are two rows (representing the two outcomes for the first qualitative variable) and two columns (representing the two outcomes for the second qualitative variable). For this special case, we can assume the rows represent two different populations. In the examples above, the two populations would be (1) male speakers and (2) female speakers. Now, consider the proportion of male interrupters for each population and define

π_1 = Proportion of male interrupters for the population of male speakers

π_2 = Proportion of female interrupters for the population of female speakers

As we learned in Example 9.14, if the two variables are independent, $\pi_1 = \pi_2$. Consequently, for the 2×2 table, a test of independence is equivalent to testing

$$H_0: (\pi_1 - \pi_2) = 0$$

This null hypothesis is equivalent to that tested when comparing two population proportions with a large-sample z test (Section 9.4). Therefore, both the z test and the χ^2 test for a 2×2 contingency table can be used to test for the difference between two population proportions. The z test, however, requires "large" samples (i.e., $n_1 \geq 30$ and $n_2 \geq 30$). In contrast, the χ^2 test requires only that n be large

Table 9.6 Contingency Table for Example 9.15

| | | FOREIGN LANGUAGE HIRING PREFERENCE | | | |
		Yes	Neutral	No	TOTALS
FIRM TYPE	U.S.	50	57	19	126
	Foreign	60	22	7	89
	TOTALS	110	79	26	215

Source: Cornick, M. F., et al. "The value of foreign language skills for accounting and business majors." *Journal of Education for Business,* Jan./Feb. 1991, p. 162 (Table 2).

enough so that the expected number in each cell is at least 1. Thus, the χ^2 test may be used as an alternative to the z test when the samples are small, as long as the expected cell counts are all at least 1.

For contingency tables with more than two rows or two columns, the χ^2 test is usually performed using computer software because of the numerous calculations involved, as the next example illustrates.

Example 9.15

2 × 3 Contingency Table Analysis

In Example 2.8 (page 89), we considered a study published in the *Journal of Education for Business.* Each in a sample of 215 personnel directors was classified according to two qualitative variables: foreign language hiring preference (yes, no, or neutral) and type of business (foreign or domestic). The response data are summarized in the contingency table, Table 9.6. Is there sufficient evidence to claim that foreign language hiring preference depends on type of business? Test using $\alpha = .01$.

Solution

We want to test

H_0: Foreign language hiring preference and type of business are independent

H_a: Foreign language hiring preference and type of business are dependent

The results of the test are shown in the Excel printout, Figure 9.11 (page 518). Note that the expected counts are given in the cells of the contingency table shown on the printout, and that each expected count is at least 1. Thus, it is appropriate to apply the χ^2 test.

The test statistic and corresponding p-value are

$$\chi^2 = 16.061 \qquad p\text{-value} = .00033$$

Since the p-value is less than $\alpha = .01$, there is sufficient evidence to reject H_0 and claim that foreign language hiring preference does depend on type of business. In fact, in our solution to Example 2.8, we showed that the estimated proportion of U.S. firms that favor foreign language hiring was 50/126 = .397, while the estimated proportion of foreign firms that give foreign language hiring preference was 60/89 = .674—almost twice as high. ❏

E Figure 9.11 PHStat Add-In for Excel Output, Example 9.15

	A	B	C	D	E	F	G
3	Observed Frequencies:			Foreign Language Hiring Preference			
4			Firm Type	Yes	Neutral	No	Totals
5			U.S.	50	57	19	126
6			Foreign	60	22	7	89
7			Totals	110	79	26	215
8							
9	Expected Frequencies:			Foreign Language Hiring Preference			
10			Firm Type	Yes	Neutral	No	Total
11			U.S.	64.46511628	46.29767442	15.2372093	126
12			Foreign	45.53488372	32.70232558	10.7627907	89
13			Total	110	79	26	215
14							
15							
16	Level of Significance	0.01					
17	Number of Rows	2					
18	Number of Columns	3					
19	Degrees of Freedom	2					
20	Critical Value	9.21035					
21	Chi-Square Test Statistic	16.0609					
22	p-Value	0.00033					
23	Reject the null hypothesis						

Caution

Inferences derived from χ^2 tests that do not satisfy the sample size requirement outlined earlier (i.e., expected cell counts of at least 1) are suspect. To make sure the sample size requirement is satisfied, it may be necessary to collapse two or more low-frequency categories into one category in the contingency table prior to performing the test. Alternative procedures are available (see Marascuilo and McSweeney, 1977) if the combining or pooling of categories is undesirable or nonsensical.

PROBLEMS 9.48–9.60

Using the Tools

9.48 Refer to the 2 × 2 contingency table shown here.

		COLUMNS		
		1	2	TOTALS
ROWS	1	133	219	352
	2	201	247	448
	TOTALS	334	466	800

a. Calculate the expected cell counts for the contingency table.

b. Calculate the chi-square statistic for the table.

c. At $\alpha = .05$, are the rows and columns independent?

9.49 Refer to the accompanying 2 × 3 (two rows and three columns) contingency table.

a. Calculate the expected cell counts for the contingency table.

b. Calculate the chi-square statistic for the table.

c. At $\alpha = .05$, are the rows and columns independent?

	COLUMNS			
	1	**2**	**3**	**TOTALS**
ROWS 1	14	37	23	74
2	21	32	38	91
TOTALS	35	69	61	165

Problem 9.49

	Variable 2	
	0	**1**
Variable 1 0	6	24
1	16	36

Problem 9.51

9.50 Give the degrees of freedom for a test of independence of the two qualitative variables comprising a contingency table with:

    a. $r = 2$ rows and $c = 2$ columns   **b.** $r = 4$ rows and $c = 2$ columns

    c. $r = 3$ rows and $c = 3$ columns   **d.** $r = 3$ rows and $c = 4$ columns

9.51 Test the null hypothesis of independence of the two qualitative variables for the 2×2 contingency table shown here. Use $\alpha = .05$.

9.52 Refer to the 3×4 contingency table given here.

		COLOR				
		Red	**White**	**Blue**	**Yellow**	**TOTALS**
LENGTH	Short	18	12	21	37	88
	Medium	7	10	15	31	63
	Long	9	6	14	30	59
	TOTALS	34	28	50	98	210

    a. Calculate the expected cell counts for the contingency table.

    b. Calculate the chi-square statistic for the table.

    c. Test the hypothesis of independence of the two qualitative variables for the 3×4 table. Use $\alpha = .01$.

Applying the Concepts

9.53 A University of Florida sociologist claimed that, "If you kill a white person (in Florida), the chances of getting the death penalty is greater than if you kill a black person" (*Gainesville Sun*, Oct. 20, 1986). Concentrating only on crimes against strangers, the sociologist classified the data of 326 murder cases in Florida according to race of the victim and death sentence, as shown in the accompanying table.

		DEATH SENTENCE		
		Yes	**No**	**TOTALS**
RACE OF VICTIM	White	30	184	214
	Black	6	106	112
	TOTALS	36	290	326

    a. Conduct a test to determine whether the victim's race and death sentence are independent. Use $\alpha = .05$.

    b. Do the results, part **a**, support the statement made by the sociologist?

    c. Construct a 95% confidence interval for $(\pi_1 - \pi_2)$, where π_1 is the percentage of murderers of whites who received the death penalty and π_2 is the percentage of murderers of blacks who received the death penalty. Interpret the interval. Do the results agree with the test, part **a**?

9.54 Refer to the *American Journal on Mental Retardation* (Jan. 1992) study of the social interactions of two groups of children, Problem 2.38 (page 94). Each in a sample of 30 children was classified according to whether or not they are diagnosed with developmental delays and whether or not they exhibit disruptive behavior. The contingency table describing the data is reproduced below. Conduct a test to determine whether the proportion of children who exhibit disruptive behavior depends on the diagnosis. Use $\alpha = .10$.

		BEHAVIOR		
		Disruptive	**Nondisruptive**	**TOTALS**
DIAGNOSIS	With developmental delays	12	3	15
	Without developmental delays	5	10	15
	TOTALS	17	13	30

9.55 Refer to the *Industrial Marketing Management* (Feb. 1993) study of humor in trade magazine advertisements, Problem 2.35 (page 92). Each in a sample of 665 ads was classified according to nationality (British, German, American) of the trade magazine in which they appeared and whether or not they were considered humorous by a panel of judges. The numbers of ads falling into each of the categories are provided in the accompanying table. Conduct an analysis to determine whether nationality of trade magazine and humor in advertisements are independent. Use $\alpha = .01$.

		HUMOROUS	
		Yes	**No**
	British	52	151
NATIONALITY	German	44	148
	American	56	214

Source: McCullough, L. S., and Taylor, R. K. "Humor in American, British, and German ads." *Industrial Marketing Management,* Vol. 22, No. 1, Feb. 1993, p. 22 (Table 3).

9.56 Refer to the *Psychological Science* (Mar. 1995) "water-level task" experiment, Problem 2.39 (page 94). Recall that groups of subjects completed a task in which they were asked to judge the location of the water line in a glass tilted at a 45° angle. The group and the angle of the judged line were recorded for each subject. The contingency table for the study is shown below.

		GROUP					
		Students	**Waitresses**	**Housewives**	**Bartenders**	**Bus Drivers**	**TOTALS**
JUDGE LINE	More than 5° above or below surface	11	15	14	12	5	57
	Within 5° of surface	29	5	6	8	15	63
	TOTALS	40	20	20	20	20	120

a. Test the hypothesis that the judged location of the water line depends on group. Use $\alpha = .05$.

b. Repeat part **a**, but only use the data for waitresses and bartenders.

9.57 Refer to the *Journal of the American Geriatrics Society* (Jan. 1998) study of age-related macular degeneration (AMD), Problem 2.40 (page 95). Recall that the researchers investigated the link between AMD incidence and alcohol consumption. Each individual in a national sample of 3,072 adults was asked about the most frequent alcohol type (beer, wine, liquor, or none) consumed in the past year. In addition, each adult was clinically tested for AMD. The results are displayed in the accompanying table.

Most Frequent Alcohol Type Consumed	Total Number	Number Diagnosed with AMD*
None	965	87
Beer	430	30
Wine	1,284	51
Liquor	393	20
TOTALS	3,072	188

* Numbers based on percentages given in Figure 1.

Source: Obisean, T. O., et al. "Moderate wine consumption is associated with decreased odds of developing age-related macular degeneration in NHANES-1." *Journal of the American Geriatrics Society,* Vol. 46, No. 1, Jan. 1998, p. 3 (Figure 1).

a. In Problem 2.40**e** you constructed a graph to compare the incidence of AMD among the four alcohol-type response groups. Specify the null and alternative hypotheses for testing whether the incidence of AMD depends on alcohol type consumed.

b. An Excel printout for the test, part **a**, is displayed below. Interpret the results of the test.

	A	B	C	D	E	F	G	H
1	Chi-Square Test of Independence for Type of Alcohol and AMD Incidence							
2								
3	Observed Frequencies:			Most Frequent Type of Alcohol Consumed				
4			Diagnosed with AMD	None	Beer	Wine	Liquor	Totals
5			Yes	87	30	51	20	18
6			No	878	400	1233	373	288
7			Totals	965	430	1284	393	307
8								
9	Expected Frequencies:			Most Frequent Type of Alcohol Consumed				
10			Diagnosed with AMD	None	Beer	Wine	Liquor	Total
11			Yes	59.05598958	26.31510417	78.578125	24.0507813	18
12			No	905.9440104	403.6848958	1205.421875	368.949219	288
13			Total	965	430	1284	393	307
14								
15								
16	Level of Significance	0.05						
17	Number of Rows	2						
18	Number of Columns	4						
19	Degrees of Freedom	3						
20	Critical Value	7.8147247						
21	Chi-Square Test Statistic	25.64881						
22	p-Value	1.1178E-05						
23	Reject the null hypothesis							

9.58 According to *Nature* (Sept. 1993), biologists define a "hotspot" as a species-rich, 10-kilometer-square geographical area. Do rare flying insects tend to congregate in hotspots of their own species? To answer this question, 105 rare British flying insect species were captured and the following two variables measured for each: (1) type of species captured and (2) type of hotspot where they were found. The data are summarized in the accompanying table. Is there evidence to indicate that hotspot type depends on rare species type? Use $\alpha = .10$.

		HOTSPOT TYPE			
		Butterfly	**Dragonfly**	**Bird**	**TOTALS**
RARE SPECIES TYPE	Butterfly	31	17	20	68
	Dragonfly	11	13	13	37
	TOTALS	42	30	33	105

MOVIES.XLS

9.59   Each week on their syndicated television show *At The Movies,* reviewers Gene Siskel *(Chicago Tribune)* and Roger Ebert *(Chicago Sun-Times)* rate the latest film releases. University of Florida statisticians examined data on 160 movie reviews by Siskel and Ebert during the period 1995–1996 *(Chance,* Spring 1997). Each critic's review was categorized as pro ("thumbs up"), con ("thumbs down"), or mixed. Consequently, each movie has a Siskel rating (pro, con, or mixed) and an Ebert rating (pro, con, or mixed). The accompanying Excel printout gives the results of a test of hypothesis to determine whether the movie reviews of the two critics are independent.

a.   Locate the *p*-value of the test.

b.   Interpret the results at $\alpha = .01$.

	A	B	C	D	E	F	G
1	Chi-Square Test of Independence for Siskel Ebert Movie Ratings						
2							
3	Observed Frequencies:				Ebert		
4			Siskel	Con	Mix	Pro	Total
5			Con	24	8	13	45
6			Mix	8	13	11	32
7			Pro	10	9	64	83
8			Total	42	30	88	160
9							
10	Expected Frequencies:				Ebert		
11			Siskel	Con	Mix	Pro	Total
12			Con	11.8125	8.4375	24.75	45
13			Mix	8.4	6	17.6	32
14			Pro	21.7875	15.5625	45.65	83
15			Total	42	30	88	160
16							
17							
18	Level of Significance	0.05					
19	Number of Rows	3					
20	Number of Columns	3					
21	Degrees of Freedom	4					
22	Critical Value	9.48773					
23	Chi-Square Test Statistic	#NUM!					
24	*p* -Value	3.4E-09					
25	Reject the null hypothesis						

Note: Due to a fault in the Microsoft Excel worksheet function used to generate the chi-square test statistic, the message #NUM!—and not the value of the statistic—appears in cell B23. This fault only appears when the *p*-value approaches zero and does not affect the decision displayed in row 25.

9.60   *Dear enemy recognition* is the term used by naturalists and ecologists for the aggressive behavior of birds, mammals, and ants when their territorial boundaries are violated by one of their own species. The red-backed salamander, for example, will attempt to bite an opponent's snout—an injury that could reduce a salamander's ability to locate prey, mates, and territorial competitors. In one study, 144 salamanders were collected from a forest and inspected for scar tissue in the snout (*The American Naturalist*, June 1981). The results are shown in the table.

	Male	Female	TOTALS
Scar tissue in snout	5	12	17
No scar tissue in snout	76	51	127
TOTALS	81	63	144

Source: Jaeger, R. G. "Dear enemy recognition and the costs of aggression between salamanders." *The American Naturalist*, June 1981, Vol. 117, pp. 962–973. © 1981 The University of Chicago.

a. Use a chi-square test to determine if there is a difference between the proportions of males and females with scar tissue in the snout. Use $\alpha = .01$.

b. Conduct the test of part **a**, but use the large-sample z test.

c. Compare the results, parts **a** and **b**. Do they agree?

9.6 COMPARING TWO POPULATION VARIANCES (OPTIONAL)

In this optional section, we present a test of hypothesis for comparing two population variances, σ_1^2 and σ_2^2. Variance tests have broad applications. For example, a production manager may be interested in comparing the variation in the length of eye-screws produced on each of two assembly lines. A line with a large variation produces too many individual eye-screws that do not meet specifications (either too long or too short), even though the mean length may be satisfactory. Similarly, an investor might want to compare the variation in the monthly rates of return for two different stocks that have the same mean rate of return. In this case, the stock with the smaller variance may be preferred because it is less risky—that is, it is less likely to have many very low and very high monthly return rates.

Variance tests can also be applied before conducting a small-sample t test for $(\mu_1 - \mu_2)$, discussed in Section 9.2. Recall that the t test requires the assumption that the variances of the two sampled populations are equal. If the two population variances are greatly different, any inferences derived from the t test are suspect. Consequently, it is important that we detect a significant difference between the two variances, if it exists, before applying the small-sample t test.

Example 9.16   **Setting Up the Hypothesis Test**

Suppose you want to conduct a test of hypothesis to compare two population variances. Let σ_1^2 represent the variance of population 1 and σ_2^2 represent the variance of population 2. Assume independent random samples of size $n_1 = 13$ and $n_2 = 18$ are selected and that the two populations are both normal.

a. Set up the null and alternative hypotheses for testing whether σ_1^2 exceeds σ_2^2.

b. Give the form of the test statistic.

c. Specify the rejection region for $\alpha = .05$.

Solution

a. To detect whether σ_1^2 exceeds σ_2^2, we test

$$H_0: \sigma_1^2 = \sigma_2^2$$

$$H_a: \sigma_1^2 > \sigma_2^2$$

However, the common statistical procedure for comparing the two variances makes an inference about the ratio σ_1^2/σ_2^2. This is because the sampling distribution of the estimator of σ_1^2/σ_2^2 is well known when the *samples are randomly and independently selected from two normal populations.* Consequently, we typically write the null and alternative hypotheses as follows:

$$H_0: \sigma_1^2/\sigma_2^2 = 1$$

$$H_0: \sigma_1^2/\sigma_2^2 > 1$$

b. Since s_1^2 and s_2^2 are the best estimates of σ_1^2 and σ_2^2, respectively, the best estimate of σ_1^2/σ_2^2 is s_1^2/s_2^2. Consequently, the test statistic, called an F statistic, is

$$F = \frac{s_1^2}{s_2^2} \tag{9.21}$$

c. Under the assumption that both samples come from normal populations, the F statistic, $F = s_1^2/s_2^2$, possesses an F *distribution* with $\nu_1 = (n_1 - 1)$ numerator degrees of freedom and $\nu_2 = (n_2 - 1)$ denominator degrees of freedom. An F distribution can be symmetric about its mean, skewed to the left, or skewed to the right; its exact shape depends on the values of ν_1 and ν_2. In this example, $\nu_1 = (n_1 - 1) = 12$ and $\nu_2 = (n_2 - 1) = 17$. An F distribution with $\nu_1 = 12$ numerator df and $\nu_2 = 17$ denominator df is shown in Figure 9.12. You can see that this particular F distribution is skewed to the right.

Figure 9.12 Rejection Region for Example 9.16

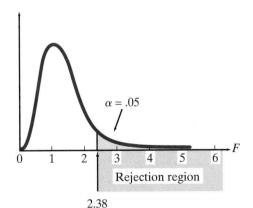

$\alpha = .05$

Rejection region

2.38

Table 9.7 Reproduction of Part of Table B.5 of Appendix B ($\alpha = .05$)

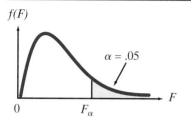

ν_1	NUMERATOR DEGREES OF FREEDOM									
ν_2	**10**	**12**	**15**	**20**	**24**	**30**	**40**	**60**	**120**	**∞**
1	241.9	243.9	245.9	248.0	249.1	250.1	251.1	252.2	253.3	254.3
2	19.40	19.41	19.43	19.45	19.45	19.46	19.47	19.48	19.49	19.50
3	8.79	8.74	8.70	8.66	8.64	8.62	8.59	8.57	8.55	8.53
4	5.96	5.91	5.86	5.80	5.77	5.75	5.72	5.69	5.66	5.63
5	4.74	4.68	4.62	4.56	4.53	4.50	4.46	4.43	4.40	4.36
6	4.06	4.00	3.94	3.87	3.84	3.81	3.77	3.74	3.70	3.67
7	3.64	3.57	3.51	3.44	3.41	3.38	3.34	3.30	3.27	3.23
8	3.35	3.28	3.22	3.15	3.12	3.08	3.04	3.01	2.97	2.93
9	3.14	3.07	3.01	2.94	2.90	2.86	2.83	2.79	2.75	2.71
10	2.98	2.91	2.85	2.77	2.74	2.70	2.66	2.62	2.58	2.54
11	2.85	2.79	2.72	2.65	2.61	2.57	2.53	2.49	2.45	2.40
12	2.75	2.69	2.62	2.54	2.51	2.47	2.43	2.38	2.34	2.30
13	2.67	2.60	2.53	2.46	2.42	2.38	2.34	2.30	2.25	2.21
14	2.60	2.53	2.46	2.39	2.35	2.31	2.27	2.22	2.18	2.13
15	2.54	2.48	2.40	2.33	2.29	2.25	2.20	2.16	2.11	2.07
16	2.49	2.42	2.35	2.28	2.24	2.19	2.15	2.11	2.06	2.01
17	2.45	2.38	2.31	2.23	2.19	2.15	2.10	2.06	2.01	1.96

Source: From M. Merrington and C. M. Thompson, "Tables of Percentage Points of the Inverted Beta (F)-Distribution." *Biometrika,* 1943, 33, 73–88. Reproduced by permission of the *Biometrika* Trustees and Oxford University Press.

Upper-tail critical values of F are found in Table B.5 of Appendix B. Table 9.7, a partial reproduction of Table B.5, gives F values that correspond to $\alpha = .05$ upper-tail areas for different pairs of degrees of freedom. The columns of the table correspond to various numerator degrees of freedom (ν_1), while the rows correspond to various denominator degrees of freedom (ν_2).

Thus, if the numerator degrees of freedom are $\nu_1 = 12$ and the denominator degrees of freedom are $\nu_2 = 17$, we find the F value,

$$F_{.05} = 2.38$$

As shown in Figure 9.12, $\alpha = .05$ is the tail area to the right of 2.38 in the F distribution with 12 numerator df and 17 denominator df. Thus, the probability that the F statistic will exceed 2.38 is $\alpha = .05$.

Given this information on the F distribution, we are now able to find the rejection region for this test. Logically, large values of $F = s_1^2/s_2^2$ will lead us to reject H_0 in favor of H_a. (If H_a is, in fact, true, then $\sigma_1^2 > \sigma_2^2$. Consequently, s_1^2 will very likely exceed s_2^2 and F will be large.) How large? The critical F value depends on our choice of α. The general form of the rejection region is $F > F_\alpha$. For $\alpha = .05$, we have $F_{.05} = 2.38$ (based on $\nu_1 = 12$ and $\nu_2 = 17$ df). Thus, the rejection region is

$$\text{Reject } H_0 \text{ if } F > 2.38 \qquad \square$$

Self-Test

9.8

Consider independent random samples of size $n_1 = 10$ and $n_2 = 25$ selected from normal populations. Give the rejection region for testing $H_0: \sigma_1^2/\sigma_2^2 = 1$ against $H_a: \sigma_1^2/\sigma_2^2 \neq 1$ at $\alpha = .10$.

The elements of both one-tailed and two-tailed hypothesis tests for the ratio of two population variances σ_1^2/σ_2^2 are summarized in the box.

Test of Hypothesis for the Ratio of Two Population Variances σ_1^2/σ_2^2

Upper-Tailed Test	Two-Tailed Test	Lower-Tailed Test
$H_0: \sigma_1^2/\sigma_2^2 = 1$ $(\sigma_1^2 = \sigma_2^2)$	$H_0: \sigma_1^2/\sigma_2^2 = 1$ $(\sigma_1^2 = \sigma_2^2)$	$H_0: \sigma_1^2/\sigma_2^2 = 1$ $(\sigma_1^2 = \sigma_2^2)$
$H_a: \sigma_1^2/\sigma_2^2 > 1$ $(\sigma_1^2 > \sigma_2^2)$	$H_a: \sigma_1^2/\sigma_2^2 \neq 1$ $(\sigma_1^2 \neq \sigma_2^2)$	$H_a: \sigma_1^2/\sigma_2^2 < 1$ $(\sigma_1^2 < \sigma_2^2)$
Test statistic:	*Test statistic:*	*Test statistic:*
$F = s_1^2/s_2^2$	$F = \dfrac{\text{Larger sample variance}}{\text{Smaller sample variance}}$	$F = s_2^2/s_1^2$
Rejection region:	*Rejection region:*	*Rejection region:*
$F > F_\alpha$	$F > F_{\alpha/2}$	$F > F_\alpha$
where	where	where
$\nu_1 = n_1 - 1$	$\nu_1 = n - 1$ for sample variance in numerator	$\nu_1 = n_2 - 1$
$\nu_2 = n_2 - 1$	$\nu_2 = n - 1$ for sample variance in denominator	$\nu_2 = n_1 - 1$

Assumptions:
1. Both the populations from which the samples are selected have relative frequency distributions that are approximately normal.
2. The random samples are selected in an independent manner from the two populations.

Example 9.17   **Conducting the Test for Comparing Variances**

Ethylene oxide (ETO) is used quite frequently in sterilizing hospital supplies. A study was conducted to investigate the effect of ETO on hospital personnel involved with the sterilization process. Thirty subjects were randomly selected and randomly assigned to one of two tasks: 19 subjects were assigned the task of opening the sterilization package that contains ETO (task 1); the remaining 11 subjects were assigned the task of opening and unloading the sterilizer gun filled with ETO (task 2). After the tasks were performed, researchers measured the

Table 9.8 Data on ETO Levels, Example 9.17

	Task 1	Task 2
Sample size	19	11
Mean	5.90	5.60
Standard deviation	1.93	4.10

amount of ETO (in milligrams) present in the bloodstream of each subject. A summary of the results appears in Table 9.8. Do the data provide sufficient evidence to indicate a difference in the variability of the ETO levels in subjects assigned to the two tasks? Test using $\alpha = .10$.

Solution

Let

$$\sigma_1^2 = \text{Population variance of ETO levels in subjects assigned task 1}$$

$$\sigma_2^2 = \text{Population variance of ETO levels in subjects assigned task 2}$$

For this test to yield valid results, we must assume that both samples of ETO levels come from normal populations and that the samples are independent.

Now, following the guidelines in the box, the elements of this two-tailed hypothesis test follow:

$$H_0: \sigma_1^2/\sigma_2^2 = 1 \qquad (\sigma_1^2 = \sigma_2^2)$$
$$H_a: \sigma_1^2/\sigma_2^2 \neq 1 \qquad (\sigma_1^2 \neq \sigma_2^2)$$

$$\textit{Test statistic: } F = \frac{\text{Larger } s^2}{\text{Smaller } s^2} = \frac{s_2^2}{s_1^2} = \frac{(4.10)^2}{(1.93)^2} = 4.51$$

Rejection region: For this two-tailed test, $\alpha = .10$ and $\alpha/2 = .05$. Thus, the rejection region is

$$F > F_{.05} = 2.41 \quad \text{(from Table B.5 in Appendix B)}$$

where the distribution of F is based on $\nu_1 = (n_2 - 1) = 10$ and $\nu_2 = (n_1 - 1) = 18$ degrees of freedom.

Conclusion: Since the test statistic $F = 4.51$ falls in the rejection region, we reject H_0. Therefore, the data provide sufficient evidence to indicate that the population variances differ. It appears that hospital personnel involved with opening the sterilization package (task 1) have less variable ETO levels than those involved with opening and unloading the sterilizer gun (task 2). ❑

Example 9.17 illustrates the technique for calculating the test statistic and rejection region for a two-tailed F test. The reason we place the larger sample variance in the numerator of the test statistic is that only upper-tail values of F are shown in the F tables of Appendix B—no lower-tail values are given. By placing the larger sample variance in the numerator, we make certain that only the upper tail of the rejection region is used. The fact that the upper-tail area is $\alpha/2$ reminds us that the test is two-tailed.

The problem of not being able to locate an F value in the lower tail of the F distribution is avoided in a one-tailed test because we can control how we specify the ratio of the population variances in H_0 and H_a. That is, we can always make a one-tailed test an upper-tailed test. For example, if we want to test whether σ_1^2 is greater than σ_2^2, then we write the alternative hypothesis as

$$H_a: \frac{\sigma_1^2}{\sigma_2^2} > 1 \qquad (\sigma_1^2 > \sigma_2^2)$$

and the appropriate test statistic is $F = s_1^2/s_2^2$. Conversely, if we want to test whether σ_1^2 is less than σ_2^2 (i.e., whether σ_2^2 is greater than σ_1^2), we write

$$H_a: \frac{\sigma_2^2}{\sigma_1^2} > 1 \qquad (\sigma_2^2 > \sigma_1^2)$$

and the corresponding test statistic is $F = s_2^2/s_1^2$, as shown in the box.

**Self-Test
9.9**

Consider independent random samples of size $n_1 = 6$ and $n_2 = 10$ selected from normal populations. Find the rejection region for testing $H_0: \sigma_1^2 = \sigma_2^2$ against $H_a: \sigma_1^2 < \sigma_2^2$ at $\alpha = .05$.

Caution

The F test is much less robust (i.e., much more sensitive) to departures from normality than the t test for comparing population means (Section 9.2). If you have doubts about the normality of the two populations, avoid using the F test for comparing the two variances. Rather, use a nonparametric test (see Hollander and Wolfe, 1973).

**PROBLEMS
9.61–9.71**

Using the Tools

9.61   Find F_α for an F distribution with 15 numerator df and 12 denominator df for the following values of α:

    a. $\alpha = .025$ **b.** $\alpha = .05$ **c.** $\alpha = .10$

9.62   Find $F_{.05}$ for an F distribution with:

    a.   Numerator df = 7, denominator df = 25

    b.   Numerator df = 10, denominator df = 8

    c.   Numerator df = 30, denominator df = 60

    d.   Numerator df = 15, denominator df = 4

9.63   Calculate the value of the test statistic for testing $H_0: \sigma_1^2/\sigma_2^2 = 1$ in each of the following cases:

    a. $H_a: \sigma_1^2/\sigma_2^2 > 1$; $s_1^2 = 1.75$; $s_2^2 = 1.23$

    b. $H_a: \sigma_1^2/\sigma_2^2 < 1$; $s_1^2 = 1.52$; $s_2^2 = 5.90$

    c. $H_a: \sigma_1^2/\sigma_2^2 \neq 1$; $s_1^2 = 2,264$; $s_2^2 = 4,009$

9.64   Under what conditions does the sampling distribution of s_1^2/s_2^2 have an F distribution?

Applying the Concepts

9.65 A study published in the *Journal of the American Concrete Institute* (Jan.–Feb. 1986) investigated the cracking strength of reinforced concrete T-beams. Several different types of T-beams were used in the experiment, each type having a different flange width. Cracking torsion moments for eight beams with 70-cm slab widths and eight beams with 100-cm slab widths are recorded here:

70-cm slab width: 6.00, 7.20, 10.20, 13.20, 11.40, 13.60, 9.20, 11.20
100-cm slab width: 6.80, 9.20, 8.80, 13.20, 11.20, 14.90, 10.20, 11.80

TBEAMS.XLS

a. Is there evidence of a difference in the variation in the cracking torsion moments of the two types of T-beams? Use $\alpha = .10$.

b. What assumptions are required for the test to be valid?

9.66 An *Environmental Science & Technology* (Oct. 1993) study used the data in the table to compare the mean oxon/thion ratios at a California orchard under two weather conditions: foggy and clear/cloudy. Test the assumption of equal variances required for the comparison of means to be valid. Use $\alpha = .02$.

OTRATIO.XLS

Date	Condition	Thion	Oxon	Oxon/Thion Ratio
Jan. 15	Fog	38.2	10.3	.270
17	Fog	28.6	6.9	.241
18	Fog	30.2	6.2	.205
19	Fog	23.7	12.4	.523
20	Clear	74.1	45.8	.618
21	Fog	88.2	9.9	.112
21	Clear	46.4	27.4	.591
22	Fog	135.9	44.8	.330
23	Fog	102.9	27.8	.270
23	Cloudy	28.9	6.5	.225
25	Fog	46.9	11.2	.239
25	Clear	44.3	16.6	.375

Source: Sclber, J. N., et al. "Air and fog deposition residues of four organophosphate insecticides used on dormant orchards in the San Joaquin Valley, California." *Environmental Science & Technology*, Vol. 27, No. 10, Oct. 1993, p. 2240 (Table V).

9.67 Refer to the *Prison Journal* (Sept. 1997) study of two groups of parole officers, Problem 9.14 (page 491). Summary statistics on subjective role and strategy for regular officers and for intensive supervision (ISP) officers are reproduced in the accompanying table.

			SUBJECTIVE ROLE		STRATEGY	
		Sample Size	Mean	Standard Deviation	Mean	Standard Deviation
Group	Regular	61	26.15	5.51	12.69	2.94
	ISP	11	19.50	4.35	9.00	2.00

Source: Fulton, B., et al. "Moderating probation and parole officer attitudes to achieve desired outcomes." *The Prison Journal*, Vol. 77, No. 3, Sept. 1997, p. 307 (Table 3).

a. In Problem 9.14**a**, you made an inference about the difference between the mean subjective role response for the two groups of parole officers, assuming the variances for the two groups were equal. Conduct a test to determine if this assumption is violated. Use $\alpha = .05$.

b. In Problem 9.14**b**, you made an inference about the difference between the mean strategy responses for the two groups of parole officers, assuming the variances for the two groups were equal. Conduct a test to determine if this assumption is violated. Use $\alpha = .05$.

9.68 Hermaphrodites are animals that possess the reproductive organs of both sexes. *Genetical Research* (June 1995) published a study of the mating systems of hermaphroditic snail species. The mating habits of the snails were classified into two groups: (1) self-fertilizing (selfing) snails that mate with snails of the same sex and (2) cross-fertilizing (outcrossing) snails that mate with snails of the opposite sex. One variable of interest in the study was the effective population size of the snail species. The means and standard deviations of the effective population size for independent random samples of 17 outcrossing snail species and five selfing snail species are given in the accompanying table. Geneticists are often more interested in comparing the variation in population size of the two types of mating systems. Conduct this analysis for the researcher using $\alpha = .10$. Interpret the result.

| | | Sample Size | EFFECTIVE POPULATION SIZE | |
			Mean	Standard Deviation
Snail Mating System	Outcrossing	17	4,894	1,932
	Selfing	5	4,133	1,890

Source: Jarne, P. "Mating system, bottlenecks, and generic polymorphism in hermaphroditic animals." *Genetical Research*, Vol. 65, No. 3, June 1995, p. 197 (Table 4). Copyright 1995 Genetical Research.

9.69 A study in the *Journal of Occupational and Organizational Psychology* (Dec. 1992) investigated the relationship of employment status and mental health. A sample of working and unemployed people was selected, and each person was given a mental health examination using the General Health Questionnaire (GHQ), a widely recognized measure of mental health. Although the article focused on comparing the mean GHQ levels, a comparison of the variability of GHQ scores for employed and unemployed men and women is of interest as well.

a. In general terms, what does the amount of variability in GHQ scores tell us about the group?

b. What are the appropriate null and alternative hypotheses to compare the variability of the mental health scores of the employed and unemployed groups? Define any symbols you use.

c. The standard deviation for a sample of 142 employed men was 3.26, while the standard deviation for 49 unemployed men was 5.10. Conduct the test you set up in part **b** using $\alpha = .05$. Interpret the results.

d. What assumptions are necessary to assure the validity of the test?

9.70 Refer to the *Florida Scientist* study of the feeding habits of sea urchins, Problem 9.16 (page 492). Recall that ten urchins were fed green turtle grass blades and ten were fed decayed blades. A summary of the ingestion times (in hours) for the two groups of sea urchins is reproduced in the table.

	Green Blades	Decayed Blades
Number of sea urchins	10	10
Mean ingestion time (hours)	3.35	2.36
Standard deviation (hours)	.79	.47

Source: Montague, J. R., et al. "Laboratory measurement of ingestion rate for the sea urchin *Lytechinus variegatus.*" *Florida Scientist,* Vol. 54, Nos. 3/4, Summer/Autumn 1991 (Table 1).

a. Use a hypothesis test to compare the variances of the ingestion times for the two groups of sea urchins. Test at $\alpha = .01$.

b. How does the result, part **a**, impact your inference about the difference between the mean ingestion times made in Problem 9.16**b**? Explain.

9.71 An experiment was conducted to compare the variability in the sentence perception of normal-hearing individuals with no prior experience in speechreading to those with experience in speechreading (*Journal of the Acoustical Society of America,* Feb. 1986). The sample consisted of 24 inexperienced and 12 experienced subjects. All subjects were asked to verbally reproduce sentences while speechreading. A summary of the results (percentage of correct syllables) for the two groups is given in the table. Conduct a test to determine whether the variance in the percentage of correctly reproduced syllables differs between the two groups of speechreaders. Test using $\alpha = .10$.

Inexperienced Speechreaders	Experienced Speechreaders
$n_1 = 24$	$n_2 = 12$
$\bar{x}_1 = 87.1$	$\bar{x}_2 = 86.1$
$s_1 = 8.7$	$s_2 = 12.4$

Source: Breeuwer, M., and Plomp, R. "Speechreading supplemented with auditorily presented speech parameters." *Journal of the Acoustical Society of America,* Vol. 79, No. 2, Feb. 1986, p. 487.

REAL-WORLD REVISITED
An IQ Comparison of Identical Twins Reared Apart

How much of our personality, our likes and dislikes, our individuality, is predetermined by our genes? And which of our traits are shaped and changed by our environment? Identical twins, because they share an identical genotype, make ideal subjects for investigating the degree to which various environmental conditions may instigate change. The classical method of studying this phenomenon and the subject of an interesting book by Susan Farber (1981) is the study of identical twins separated early in life and reared apart.*

Identical twins are formed when a single egg, fertilized by a single sperm, splits into two parts and each develops into a separate embryo. In contrast, fraternal twins develop when two eggs from one ovary or one egg from each ovary is released and each is fertilized by a different sperm. Although they have been the subjects in many studies of twins, fraternal twins are no more genetically similar than two siblings. Therefore, a study of fraternal twins would leave unanswered the question of whether a given individual trait is due to hereditary or environmental differences. Likewise, identical twins reared in the same family share almost identical environments and make it difficult to separate the two factors. Thus, claims Farber, "in theory at least, the clearest demarcation of heredity and environment is found when identical twins have been separated early in life and reared apart in different homes, by different parents, and often in widely varying socioeconomic and geographic circumstances."

Over the years, several studies of identical twins have been conducted. Farber's book contains a chronicle and reanalysis of 95 pairs of identical twins reared apart. Much of her discussion focuses on a comparison of IQ scores.

The question of concern is, "Are there significant differences between the IQ scores of identical twins, where one member of the pair is reared by the natural parents and the other member of the pair is not?" The data for this analysis (extracted from Table E6 of Farber's book) appear in Table 9.9 and are available in the Excel workbook TWINS.XLS. One member (A) of each of the $n = 32$ pairs of twins was reared by a natural parent, whereas the other member (B) was reared by a relative or some other person.

TWINS.XLS

Table 9.9 IQ Scores of Identical Twins Reared Apart

Pair ID	Twin A	Twin B	Pair ID	Twin A	Twin B
112	113	109	228	100	88
114	94	100	232	100	104
126	99	86	236	93	84
132	77	80	306	99	95
136	81	95	308	109	98
148	91	106	312	95	100
170	111	117	314	75	86
172	104	107	324	104	103
174	85	85	328	73	78
180	66	84	330	88	99
184	111	125	338	92	111
186	51	66	342	108	110
202	109	108	344	88	83
216	122	121	350	90	82
218	97	98	352	79	76
220	82	94	416	97	98

Study Questions

a. How should the data be analyzed, as matched pairs or independent samples? Explain.

b. Conduct a test of hypothesis to determine whether there is evidence of a difference between mean IQ scores of identical twins reared apart. Test using $\alpha = .10$.

c. Construct a 90% confidence interval for the mean difference in IQ scores, $\mu_A - \mu_B$.

d. Based on the interval constructed in part **c**, is there evidence of a difference in mean IQ scores between twins reared by a natural parent and twins reared by a relative or some other person? Does your inference agree with that of part **b**?

e. What assumptions are necessary for the validity of the procedures used in parts **b** and **c**? Check if these assumptions are satisfied.

f. Repeat parts **b–e**, but analyze the data using the alternative method. In other words, if you analyzed the data as matched pairs, now use the independent samples method. If you analyzed the data as independent samples, now use the matched pairs method. Compare the results of the two methods.

Key Terms

Starred (*) terms are from the optional section of this chapter.

Contingency table 511
Difference between means 479
Difference between proportions 479

Expected cell counts 513
Independence of two qualitative variables 511
Independent samples 480

Matched pairs 495
Pooled estimate of variance 485
Ratio of two variances* 479

Key Formulas

Starred (*) formulas are from the optional section of this chapter.

Confidence Interval for Comparing Means or Proportions

Large Sample:

Estimator $\pm\ (z_{\alpha/2})$(Standard error)

(9.4), 482
(9.12), 496
(9.17), 506

Small Sample:

Estimator $\pm\ (t_{\alpha/2})$(Standard error)

(9.6), 485
(9.13), 496

Test Statistic for Comparing Means or Proportions

$$z = \frac{\text{Estimator} - \text{Hypothesized value}}{\text{Standard error}}$$

(9.3), 482
(9.10), 496
(9.16), 505

$$t = \frac{\text{Estimator} - \text{Hypothesized value}}{\text{Standard error}}$$

(9.5), 485
(9.11), 496

[*Note:* The respective point estimator and standard error for each parameter are provided in Table 9.10.]

Table 9.10 Estimators and Standard Errors for Population Parameters

Parameter	Point Estimate	Standard Error	Estimated Standard Error
$(\mu_1 - \mu_2)$	$(\bar{x}_1 - \bar{x}_2)$	$\sqrt{\dfrac{\sigma_1^2}{n_1} + \dfrac{\sigma_2^2}{n_2}}$	$\sqrt{\dfrac{s_1^2}{n_1} + \dfrac{s_2^2}{n_2}}$
μ_d	$\bar{d}$	$\sigma_d/\sqrt{n}$	$s_d/\sqrt{n}$
$(\pi_1 - \pi_2)$	$(p_1 - p_2)$	$\sqrt{\dfrac{\pi_1(1 - \pi_1)}{n_1} + \dfrac{\pi_2(1 - \pi_2)}{n_2}}$	$\sqrt{\dfrac{p_1(1 - p_1)}{n_1} + \dfrac{p_2(1 - p_2)}{n_2}}$
$^*\sigma_1^2/\sigma_2^2$	s_1^2/s_2^2	(not necessary)	(not necessary)

Test Statistic for χ^2 Test of Independence

$$\chi^2 = \sum\sum \frac{(O_{ij} - E_{ij})^2}{E_{ij}} \quad \textbf{(9.20)}, 515$$

where $E_{ij} = \dfrac{R_i C_j}{n}$ **(9.18)**, 513

*Test Statistic for Comparing Variances

$$F = s_1^2/s_2^2 \quad \textbf{(9.21)}, 524$$

Key Symbols

Symbol	Description
$\mu_1 - \mu_2$	Difference between two population means (independent samples)
$\pi_1 - \pi_2$	Difference between two population proportions (independent samples)
μ_d	Population mean difference (matched pairs)
σ_d	Population standard deviation of differences (matched pairs)
$\bar{x}_1 - \bar{x}_2$	Difference between two sample means (estimates $\mu_1 - \mu_2$)
$p_1 - p_2$	Difference between two sample proportions (estimates $\pi_1 - \pi_2$)
$\bar{d}$	Sample mean difference, matched pairs (estimates μ_d)
s_p^2	Pooled sample variance (estimates $\sigma_1^2 = \sigma_2^2$)
s_d	Sample standard deviation of differences, matched pairs (estimates σ_d)
σ_1^2/σ_2^2*	Ratio of two population variances
s_1^2/s_2^2*	Ratio of two sample variances (estimates σ_1^2/σ_2^2)
F_α*	Critical F value used to compare variances
R_i	Total for row i
C_j	Total for column j
E_{ij}	Expected count for cell in row i and column j

Supplementary Problems 9.72–9.89

Starred (*) problems refer to the optional section of this chapter.

9.72 A study compared a traditional approach to teaching basic nursing skills with an innovative approach (*Journal of Nursing Education,* Jan. 1992). Forty-two students enrolled in an upper-division nursing course participated in the study. Half (21) were randomly assigned to labs that utilized the innovative approach. After completing the course, all students were given short-answer questions about scientific principles underlying each of 10 nursing skills. The objective of the research is to compare the mean scores of the two groups of students.

    a. What is the appropriate test to use to compare the two groups?

    b. Are any assumptions required for the test?

    c. One question dealt with the use of clean/sterile gloves. The mean scores for this question were 3.28 (traditional) and 3.40 (innovative). Is there sufficient information to perform the test?

    d. Refer to part **c.** The *p*-value for the test was reported as $p = .79$. Interpret this result.

    e. Another question concerned the choice of a stethoscope. The mean scores of the two groups were 2.55 (traditional) and 3.60 (innovative) with an associated *p*-value of .02. Interpret these results.

9.73 Did you know that the use of aspirin to alleviate the symptoms of viral infections in children may lead to serious complications (Reyes' syndrome)? A random sample of 500 children with viral infections received no aspirin to alleviate symptoms, and 12 developed Reyes' syndrome. In a random sample of 450 children with viral infections who were given aspirin, 23 developed Reyes' syndrome.

    a. Is there sufficient evidence to indicate that the proportion of children with viral infections who develop Reyes' syndrome is greater for those who take aspirin than for those who do not? Test at $\alpha = .05$.

b. Construct a 95% confidence interval for the difference in the proportions of children who develop Reyes' syndrome between those who receive no aspirin and those who receive aspirin during a viral infection. Interpret the interval.

9.74 Propranolol, a class of heart drugs called *beta blockers,* have been used in an attempt to reduce anxiety in students taking the Scholastic Aptitude Test (SAT). In one study, 22 high school juniors who had not performed as well as expected on the SAT were administered a beta blocker one hour prior to retaking the test in their senior year. The junior and senior year SAT scores for these 22 students are shown in the table. Nationally, students who retake the test without special preparations will increase their score by an average of 38 points (reported in *Newsweek,* Nov. 16, 1987). Test the hypothesis that the true mean increase in SAT scores for those students who take a beta blocker prior to the exam exceeds the national average increase of 38. Use $\alpha = .05$.

SATSCORES.XLS

	SAT SCORES			SAT SCORES	
Student	Junior Year	Senior Year	Student	Junior Year	Senior Year
1	810	984	12	933	945
2	965	1015	13	811	897
3	707	1006	14	774	875
4	652	995	15	780	923
5	983	997	16	913	884
6	822	963	17	655	931
7	874	860	18	906	1136
8	900	915	19	737	854
9	693	847	20	788	927
10	1115	1202	21	878	872
11	749	910	22	912	1054

Note: SAT scores are simulated based on information provided in *Newsweek.*

9.75 Do carbohydrate-electrolyte drinks (e.g., Gatorade, Powerade, All Sport) improve performance in sports? To answer this question, researchers at Springfield College (Mass.) recruited eight trained male competitive cyclists to participate in a study (*International Journal of Sport Nutrition,* June 1995). During one cycling trial, each cyclist was given a carbohydrate-electrolyte drink. During a second trial, the cyclists were given a placebo drink (Crystal Light). At the end of each trial, several physiological variables were measured. These variables were analyzed using matched pairs. A summary of the results is shown in the table.

Variable	PLACEBO Mean	Std. Dev.	CARBOHYDRATE-ELECTROLYTE Mean	Std. Dev.	MATCHED PAIRS p-Value
1. Perceived exertion rating	5.60	1.01	4.99	1.11	.003
2. Heart rate (beats/minute)	161.9	6.8	160.1	10.0	>.05
3. Peak power (WAT)	747.6	97.3	794.0	83.5	.02
4. Fatigue index (%)	21.3	12.4	20.9	10.4	>.05

Source: Ball, T. C., et al. "Periodic carbohydrate replacement during 50 min. of high intensity cycling improves subsequent sprint performance." *International Journal of Sport Medicine,* Vol. 5, No. 2, June 1995, p. 155 (Table 3).

a. Interpret the *p*-values shown in the table. Make a practical conclusion for each variable.

b. Sufficient information is provided in the table to analyze the data as independent random samples. Perform these tests and compare the results to those of part **a**.

c. Explain why the matched pairs tests are more appropriate than the independent samples tests.

9.76 *Social Science Quarterly* (Sept. 1993) reported on a study of gender differences among workers in the computer software industry. Questionnaires were administered to a sample of 298 females and 264 males who were employed full-time in software-related jobs. In the female sample, 89 had professional occupations (e.g., programmers, analysts, computer scientists) and 209 had nonprofessional jobs (computer and peripheral equipment operators). In contrast, the male sample included 150 professionals and 114 nonprofessionals. Conduct a test of hypothesis to compare the proportions of male and female software workers who hold professional positions. Use $\alpha = .10$.

9.77 Epidemiologists have theorized that the risk of coronary heart disease can be reduced by an increased consumption of fish. One study monitored the diet and health of a random sample of middle-age men who live in The Netherlands. The men were divided into groups according to the number of grams of fish consumed per day. Twenty years later, the level of dietary cholesterol (one of the risk factors for coronary disease) present in each was recorded. The results for two groups of subjects, the "no fish consumption" group (0 grams per day) and the "high fish consumption" group (greater than 45 grams per day), are summarized in the table. (Dietary cholesterol is measured in milligrams per 1,000 calories.)

	No Fish Consumption (0 Grams/day)	High Fish Consumption (> 45 Grams/day)
Sample size	159	79
Mean	146	158
Standard deviation	66	75

Source: Kromhout, D., Bosschieter, E. B., and Coulander, C. L. "The inverse relationship between fish consumption and 20-year mortality from coronary heart disease." *New England Journal of Medicine,* May 9, 1985, Vol. 312, No. 19, pp. 1205–1209.

a. Calculate a 99% confidence interval for the difference between the mean levels of dietary cholesterol present in the two groups.

b. Based on the interval constructed in part **a**, what can you infer about the true difference? Explain.

*c. Is there evidence of a difference between the variances of the cholesterol levels for the two groups of men? Test using $\alpha = .05$.

d. How does the result, part **c**, impact the validity of the inference made in part **b**? Explain.

9.78 Pesticides applied to an extensively grown crop can result in inadvertent area-wide air contamination. *Environmental Science & Technology* (Oct. 1993) reported on air deposition residues of the insecticide diazinon used on dormant orchards in the San Joaquin Valley, California. Air samples were collected and analyzed at an orchard site for each of 11 days during the most intensive period of spraying. The levels of diazinon residue (in ng/m^3) during the day and at night are recorded in the table. The researchers want to know whether the mean diazinon residue levels differ from day to night.

DIAZINON.XLS

Date	DIAZINON RESIDUE Day	Night	Date	DIAZINON RESIDUE Day	Night
Jan. 11	5.4	24.3	Jan. 17	6.1	104.3
12	2.7	16.5	18	7.7	96.9
13	34.2	47.2	19	18.4	105.3
14	19.9	12.4	20	27.1	78.7
15	2.4	24.0	21	16.9	44.6
16	7.0	21.6			

Source: Selber, J. N., et al. "Air and fog deposition residues for organophospate insecticides used on dormant orchards in the San Joaquin Valley, California." *Environmental Science & Technology,* Vol. 27, No. 10, Oct. 1993, p. 2240 (Table IV).

a. Conduct the appropriate test at $\alpha = .10$.

b. What assumption is necessary for the validity of the test? Check this assumption with a graphical method.

9.79 Geneticists at Duke University Medical Center have identified the E2F1 transcription factor as an important component of cell proliferation control (*Nature,* Sept. 23, 1993). The researchers induced DNA synthesis in two batches of serum-starved cells. Each cell in one batch was micro-injected with the E2F1 gene, while the cells in the second batch (the controls) were not exposed to E2F1. After 30 hours, the number of cells in each batch that exhibited altered growth was determined. The results of the experiment are summarized in the table.

	Control	E2F1-Treated Cells
Total Number of Cells	158	92
Number of Growth-Altered Cells	15	41

Source: Johnson, D. G., et al. "Expression of transcription factor E2F1 induces quiescent to enter S phase." *Nature,* Vol. 365, No. 6444, Sept. 23, 1993, p. 351 (Table 1).

a. Compare the percentages of cells exhibiting altered growth in the two batches with a test of hypothesis. Use $\alpha = .10$.

b. Use the results, part **a**, to make an inference about the ability of the E2F1 transcription factor to alter cell growth.

9.80 To reduce nonresponse in mail surveys, several different techniques for formatting questionnaires have been proposed. An experiment was conducted to study the effect of questionnaire layout and page size on response rates in a mail survey. Approximately 850 students enrolled at the University of Leyden (The Netherlands) were questioned about their attitudes toward suicide. Four different questionnaire formats were used: (1) typewriting on small (15 × 21 cm) page; (2) typewriting on large (18.5 × 25.5 cm) page; (3) typeset on small page; and (4) typeset on large page. The numbers of students mailed each type of questionnaire and the numbers responding are given in the table on page 538.

		Number of Responses	Number of Nonresponses	TOTAL NUMBER MAILED
Questionnaire Format	Typewritten, small page	86	57	143
	Typewritten, large page	191	97	288
	Typeset, small page	72	69	141
	Typeset, large page	192	92	284
	TOTALS	541	315	856

Source: Reprinted with permission of author and publisher from: Jansen, J. H. "Effect of questionnaire layout and size and issue-involvement on response rates in mail surveys." *Perceptual and Motor Skills,* Vol. 61, 1985, pp. 139–142.

a. Scan the data. Do the response rates appear to differ among the four questionnaire formats?

b. Why is a statistical test useful in answering part **a**?

c. Calculate the number of students you would expect to fall in each of the eight cells of the contingency table if, in fact, the response rates are identical for the four questionnaire types.

d. Find the difference between the observed and the (estimated) expected numbers for each of the eight cells.

e. Calculate and interpret the value of the chi-square statistic for the contingency table.

9.81 An experiment was conducted in England to examine the diet metabolizable energy (ME) content of commercial cat foods. The researchers monitored the diets of 57 adult domestic short-haired cats; 28 cats were fed a diet of commercial canned cat food, whereas 29 cats were fed a diet of dry cat food over a 3-week period. At the end of the trial, the ME content was determined for each cat, with the results shown in the table.

	Canned Food	Dry Food
Sample size	28	29
Mean ME content	.96	3.70
Standard deviation	.26	.48

Source: Kendall, P. T., Burger, I. N., and Smith, P. M. "Methods of estimation of the metabolizable energy content of cat foods." *Feline Practice,* Vol. 15, No. 2, Feb. 1986, pp. 38–44.

a. Conduct a test to determine whether the average ME content of cats fed canned food differs from the average ME content of cats fed dry food. Use $\alpha = .10$.

b. List the assumptions necessary for the test, part **a**, to be valid.

*c. If possible, test the assumptions of part **b**. How does this impact the result, part **a**?

9.82 Researchers at Mount Sinai Medical Center in New York believe that ALS, a neurological disorder (also known as "Lou Gehrig's" disease) that slowly paralyzes and kills its victims, may be linked to household pets, especially small dogs. A five-member medical team found that 72% of the afflicted patients studied had small household dogs at least 20 years before contracting ALS. In contrast, only 33% of a healthy control group had pet dogs early in life. (Assume that 100 afflicted patients and 100 healthy controls were studied.) Is there sufficient evidence to indicate that the percentage of all patients with ALS who had pet dogs early in life is larger than the corresponding percentage for nonafflicted people? Test using $\alpha = .01$.

Fast Muscle	Slow Muscle
$n_1 = 12$	$n_2 = 12$
$\bar{x}_1 = .57$	$\bar{x}_2 = .37$
$s_1 = .104$	$s_2 = .035$

Source: Ushio, H., and Watabe, S. "Ultra-structural and biochemical analysis of the sarcoplasmic reticulum from crayfish fast and slow striated muscles." *The Journal of Experimental Zoology,* Vol. 267, Sept. 1993, p. 16 (Table 1).

9.83 Marine biochemists at the University of Tokyo studied the properties of crustacean skeletal muscles (*The Journal of Experimental Zoology,* Sept. 1993). It is well known that certain muscles contract faster than others. The main purpose of the experiment was to compare the biochemical properties of fast and slow muscles of crayfish. Using crayfish obtained from a local supplier, twelve fast muscle fiber bundles were extracted and each fiber bundle tested for uptake of a certain protein. Twelve slow muscle fiber bundles were extracted from a second sample of crayfish, and protein uptake measured. The results of the experiment are summarized at left. (All measurements are in moles per milligram.)

a. Analyze the data using a 95% confidence interval. Make an inference about the difference between the protein uptake means of fast and slow muscles.

*b. Conduct a test to determine whether the variation in protein uptake differs for fast and slow muscles. Use $\alpha = .05$.

c. How does the result, part **b**, impact the validity of the inference made in part **a**? Explain.

9.84 A federal traffic safety researcher was hired to ascertain the effect of wearing safety devices (shoulder harnesses, seat belts) on reaction times to peripheral stimuli. To investigate this question, he randomly selected 15 subjects from the students enrolled in a driver education program. Each subject performed a simulated driving task that allowed reaction times to be recorded under two conditions, wearing a safety device (restrained condition) and no safety device (unrestrained condition). Thus, each subject received two reaction-time scores, one for the restrained condition and one for the unrestrained condition. The data (in hundredths of a second) are shown in the accompanying table, followed by an Excel printout of the analysis.

REACTIME.XLS

		DRIVER													
	1	2	3	4	5	6	7	8	9	10	11	12	13	14	15
Restrained	36.7	37.5	39.3	44.0	38.4	43.1	36.2	40.6	34.9	31.7	37.5	42.8	32.6	36.8	38.0
Unrestrained	36.1	35.8	38.4	41.7	38.3	42.6	33.6	40.9	32.5	30.7	37.4	40.2	33.1	33.6	37.5

	A	B	C
1	t-Test: Paired Two Sample for Means		
2			
3		*Restrained*	*Unrestrained*
4	Mean	38.00666667	36.82666667
5	Variance	12.78495238	13.07352381
6	Observations	15	15
7	Pearson Correlation	0.945244008	
8	Hypothesized Mean Difference	0	
9	df	14	
10	t Stat	3.838637004	
11	P(T<=t) one-tail	0.000903799	
12	t Critical one-tail	1.76130925	
13	P(T<=t) two-tail	0.001807598	
14	t Critical two-tail	2.144788596	
15			

a. Is there evidence of a difference between mean reaction-time scores for the restrained and unrestrained drivers? Test using $\alpha = .05$.

b. What assumptions are necessary for the validity of the test of part **a**?

c. Based on the test of part **a**, what would you infer about the mean reaction times for the driving conditions?

9.85 The label "Machiavellian" was derived from the sixteenth century Florentine writer Niccolo Machiavelli, who wrote on ways of manipulating others to accomplish one's objective. Critics often accuse marketers of being manipulative and unethical, or "Machiavellian," in nature. An article in the *Journal of Marketing* (Summer 1984) explored the question of whether "marketers are more Machiavellian than others." The Machiavellian scores (measured by the Mach IV scale) for a sample of marketing professionals were recorded and compared to the Machiavellian scores for other groups of people, including a sample of college students in an earlier study. The results are summarized in the accompanying table. (Higher scores are associated with Machiavellian attitudes.) Is there sufficient evidence of a mean difference in Machiavellian scores between marketing professionals and college students? Test using $\alpha = .01$.

	Marketing Professionals	College Students
Sample size	1,076	1,782
Mean score	85.7	90.7
Standard deviation	13.2	14.3

9.86 *Chance* (Fall 1988) reported on the results of two studies designed to investigate the effectiveness of aspirin in the prevention of heart attacks. Both studies used physicians as subjects. The U.S. Physicians' Health Study involved a clinical trial in which about half of the physicians were assigned at random to receive one Bufferin brand aspirin tablet every other day. The other half received a placebo, a harmless and ineffective substitute. (The study was double-blind, i.e., designed so that neither the participants nor the medical scientists who were conducting the research knew which tablet, the Bufferin or the placebo, was being administered.) A British study also used doctors, two-thirds of whom were randomly chosen to take daily aspirin. The remaining physicians were not given a placebo, but instead were instructed "to avoid aspirin and products containing aspirin unless some specific indication for aspirin was thought to have developed." The results of both studies are summarized in the table.

U.S. Physicians' Health Study	Aspirin Group	Placebo Group
Sample size	11,037	11,034
Number of fatal heart attacks	5	18

British Physicians' Study	Aspirin Group	Placebo Group
Sample size	3,429	1,710
Number of fatal heart attacks	89	47

Source: Greenhouse, J. B., and Greenhouse, S. M. "An aspirin a day ...?"
Chance: New Directions for Statistics and Computing, Vol. 1, No. 4, Fall 1988, pp. 24–31, New York, Springer-Verlag.

a. Consider the results of the U.S. Physicians' Health Study. Compare the fatal heart attack rates of the aspirin group and the placebo group using a test of hypothesis at $\alpha = .05$.

b. Repeat part **a**, but use the results of the British study.

c. Refer to parts **a** and **b**. Does your inference about the beneficial effect of aspirin in the prevention of heart attacks depend on which study you consider?

d. Why might the two studies yield contrasting results? [*Hint:* Consider one or more of the following issues: sample size; the fact that the U.S. study used physicians who had extraordinarily low cardiovascular mortality rates; double-blind study versus unblinded study; placebo versus "no aspirin."]

9.87 It is estimated that one of every six adult Americans is afraid to fly. To determine whether fear of flying is a significant problem for the airline industry, a series of surveys was conducted. One question was posed to determine whether anxiety with respect to flying is dependent on flight experience on a commercial aircraft. Respondents were first classified as either flyers (those who have flown at least once), nonflyers likely to fly (those who have never flown but consider themselves likely to fly in the future), or nonflyers not likely to fly (those who have never flown and are not likely to fly in the future). The numbers of each group falling into each of three levels of anxiety with respect to flying are shown in the table.

		FLIGHT EXPERIENCE		
		Flyers	**Nonflyers Likely to Fly**	**Nonflyers Not Likely to Fly**
Anxiety Level	No anxiety	1,043	128	113
	Anxious	189	46	6
	Afraid	140	47	141

Source: Dean, R. D., and Whitaker, K. M. "Fear of flying: Impact on the U.S. air travel industry." *Journal of Travel Research,* Summer 1982, p. 10.

a. Does the anxiety level with respect to flying depend on flight experience? Use $\alpha = .05$.

b. Find a 95% confidence interval for the difference between the proportions of flyers and nonflyers likely to fly who have no anxiety toward flying. Interpret the interval.

9.88 The director of training for a company manufacturing electronic equipment is interested in determining whether different training methods have an effect on the productivity of assembly-line employees. She randomly assigns 42 recently hired employees into two groups of 21, of which the first group receives a computer-assisted, individual-based training program and the other receives a team-based training program. Upon completion of the training, the employees are evaluated on the time (in seconds) it took to assemble a part. The results are shown on page 542.

a. Using a .05 level of significance, is there evidence of a difference in the average time to assemble a part between the two programs?

*b. Using the .05 level of significance, is there evidence of a difference in the variances between the two programs?

TRAINING.XLS

Computer-Assisted, Individual-Based Program		Team-Based Program	
19.4	16.7	22.4	13.8
20.7	19.3	18.7	18.0
21.8	16.8	19.3	20.8
14.1	17.7	15.6	17.1
16.1	19.8	18.0	28.2
16.8	19.3	21.7	20.8
14.7	16.0	30.7	24.7
16.5	17.7	23.7	17.4
16.2	17.4	23.2	20.1
16.4	16.8	12.3	15.2
18.5		16.0	

Problem 9.88

9.89 Suppose you are investigating allegations of gender discrimination in the hiring practice of a particular firm. An equal-rights group claims that females are less likely to be hired than males with the same background, experience, and other qualifications. Data on hiring status and gender were collected on 28 former applicants and are shown in the table.

DISCRIM.XLS

Hiring Status	Gender	Hiring Status	Gender
Not hired	Female	Hired	Male
Not hired	Male	Not hired	Female
Hired	Male	Not hired	Male
Hired	Male	Hired	Male
Not hired	Female	Not hired	Female
Hired	Female	Not hired	Male
Not hired	Male	Not hired	Female
Not hired	Female	Not hired	Male
Not hired	Female	Hired	Male
Hired	Female	Not hired	Female
Not hired	Male	Not hired	Female
Not hired	Female	Not hired	Male
Not hired	Female	Hired	Male
Not hired	Female	Hired	Female

a. Summarize the data in a contingency table.

b. Test the hypothesis that hiring status and gender are independent. Use $\alpha = .01$.

9.E.1 Using Microsoft Excel to Compare Two Population Means: Independent Samples

Solution Summary:

If sample data are available, use the Data Analysis t-Test: Two-Sample Assuming Equal Variances tool to perform the pooled-variance *t* test for differences in two means using sample data. If only summary data are available, use the PHStat add-in to perform this test.

Using Sample Data: Unstacked

 Example: Dental anesthetic analysis of Example 9.6 on page 486

Solution:

To perform the pooled-variance *t* test for differences in two means using sample data, use the Data Analysis t-Test: Two-Sample Assuming Equal Variances tool. As an example, consider the dental anesthetic data. To determine whether or not there is evidence that the average discomfort level is higher for Novocaine than for the new anesthetic, do the following:

1.   Open the DENTAL.XLS workbook and click the Data sheet tab. Verify that the Novocaine and the New Anesthetic data of Table 9.2 on page 487 have been entered into columns A and B, respectively.

2.   Select Tools | Data Analysis. Select t-Test: Two-Sample Assuming Equal Variances from the Analysis Tools list box in the Data Analysis dialog box. Click the OK button.

3.   In the t-Test: Two-Sample Assuming Equal Variances dialog box (see Figure 9.E.1):

    a   Enter A1:A6 in the Variable 1 Range: edit box.

    b   Enter B1:B6 in the Variable 2 Range: edit box.

    c   Enter 0 (zero) in the Hypothesized Mean Difference: edit box.

    d   Select the Labels check box.

    e   Enter 0.01 in the Alpha: edit box.

    f   Select the New Worksheet Ply option button and enter Data Analysis t Test as the name of the new sheet.

    g   Click the OK button.

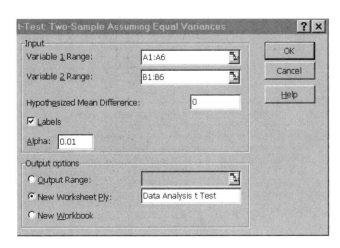

E Figure 9.E.1 Data Analysis t-Test: Two-Sample Assuming Equal Variances Tool Dialog Box

The Data Analysis tool add-in inserts a worksheet containing calculations for the *t* test similar to the one shown in Figure 9.E.2. This worksheet is *not* dynamically changeable, so any changes made to the underlying dental anesthetic data would require using the Data Analysis tool a second time in order to produce updated test results.

E **Figure 9.E.2** *t* Test for the Differences in Two Means for the Dental Anesthetic Data

	A	B	C
1	t-Test: Two-Sample Assuming Equal Variances		
2			
3		*Novocaine*	*New*
4	Mean	53.8	36.4
5	Variance	153.2	59.3
6	Observations	5	5
7	Pooled Variance	106.25	
8	Hypothesized Mean Difference	0	
9	df	8	
10	t Stat	2.669038	
11	P(T<=t) one-tail	0.014202	
12	t Critical one-tail	2.896468	
13	P(T<=t) two-tail	0.028405	
14	t Critical two-tail	3.355381	

Using Sample Data: Stacked

The Data Analysis t-Test tool requires that the data for each group be in separate columns, an arrangement known as unstacked data. Some data sets are "stacked," in which the data for both groups appear in a single column, identified by a grouping variable in another column. To use such stacked data with the Data Analysis t-Test tool, select the Data Preparation | Unstack Data choice of the PHStat add-in to unstack the data as part of Step 1.

Using Summary Data

To perform the pooled-variance *t* test for differences in two means using summary data use the Two-Sample Tests | t Test for Differences in Two Means choice of the PHStat add-in. As an example, consider the dental anesthetic data. To determine whether or not there is evidence that the average discomfort level is higher for Novocaine than for the new anesthetic, do the following:

1 If the PHStat add-in has not been previously loaded, load the add-in using the instructions of Section EP.3.2.

2 Open the DENTAL.XLS workbook and click the Data sheet tab. Verify that the Novocaine and the New Anesthetic data of Table 9.2 on page 487 have been entered into columns A and B, respectively.

3 Select PHStat | Two-Sample Tests | t Test for Differences in Two Means.

4 In the t Test for Differences in Two Means dialog box (see Figure 9.E.3):

a Enter 0 (zero) in the Hypothesized Difference: edit box.

b Enter 0.01 in the Level of Significance: edit box.

c Enter the summary data for Table 9.2 on page 487 for both samples. Enter 5, 53.8, and 12.4 as the sample size, sample mean, and sample standard deviation for the population 1 sample. Enter 5, 36.4, and 7.7 as the sample size, sample mean, and sample standard deviation for the population 2 sample.

d Select the Upper-Tail Test option button.

e Enter t Test for Differences in Means for Dental Anesthetics in the Output Title: edit box.

f Click the OK button.

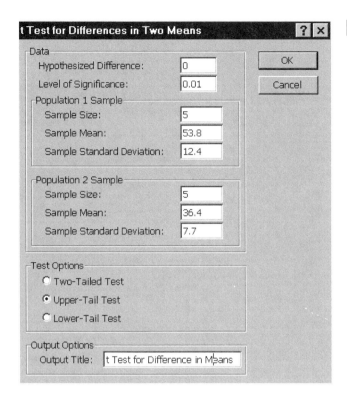

E **Figure 9.E.3** PHStat t Test for Differences in Two Means Dialog Box

9.E.2 Using Microsoft Excel to Compare Two Population Means: Matched Pairs

Solution Summary:

Use the Data Analysis t-Test: Paired Two Sample for Means tool to perform the *t* test for matched pairs.

 Reading scores data of Examples 9.7 and 9.8 on pages 494–497

Solution:

To perform the *t* test for matched pairs, use Data Analysis t-Test: Paired Two Sample for Means tool. As an example, consider the reading scores data of Examples 9.7 and 9.8. To determine whether or not there is evidence of a significant difference in the average scores for the two methods, do the following:

1   Open the READING.XLS workbook and click the Data sheet tab. Verify that the reading scores data for method 1 and method 2 of Table 9.3 on page 497 have been entered into columns A and B, respectively.

2   Select Tools | Data Analysis. Select t-Test: Paired Two Sample for Means from the Analysis Tools list box in the Data Analysis dialog box. Click the OK button.

3   In the t-Test: Paired Two Sample for Means dialog box (see Figure 9.E.4):

         a   Enter A1:A11 in the Variable 1 Range: edit box.

         b   Enter B1:B11 in the Variable 2 Range: edit box.

         c   Enter 0 (zero) in the Hypothesized Mean Difference: edit box.

         d   Select the Labels check box.

         e   Enter 0.05 in the Alpha: edit box.

f   Select the New Worksheet Ply option button and enter Data Analysis t Test as the name of the new sheet.

g   Click the OK button.

E **Figure 9.E.4** Data Analysis t-Test: Paired Two Sample for Means Tool Dialog Box

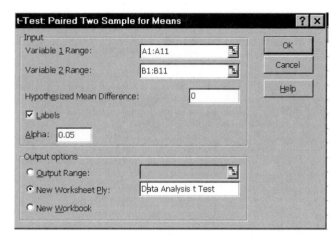

The Data Analysis tool add-in inserts a worksheet containing calculations for the *t* test similar to the one shown in Figure 9.7 on page 497. This worksheet is *not* dynamically changeable, so any changes made to the underlying reading score data would require using the Data Analysis tool a second time in order to produce updated test results.

Using Stacked Data

The Data Analysis t-Test tool requires that the data for each group be in separate columns, an arrangement known as unstacked data. Some data sets are "stacked," in which the data for both groups appear in a single column, identified by a grouping variable in another column. To use such stacked data with the Data Analysis t-Test tool, select the Data Preparation | Unstack Data choice of the PHStat add-in to unstack the data as part of Step 1.

9.E.3   Using Microsoft Excel to Compare Two Population Proportions

Solution Summary:

Use the PHStat add-in to perform the Z test for the difference in two proportions.

Example: Ethical management decision-making analysis of Example 9.9 on pages 504–507

Solution:

To perform the Z test for the difference in two proportions, use the Two-Sample Tests | Z Test for the Difference in Two Proportions choice of the PHStat add-in. As an example, consider the ethical management decision-making example. To determine whether or not there is evidence of a significant difference in the proportion of males and females who believe it is very unethical to conceal one's on-the-job errors, do the following:

1   If the PHStat add-in has not been previously loaded, load the add-in using the instructions of Section EP.3.2.

2   Select File | New to open a new workbook (or open the existing workbook into which the hypothesis testing worksheet is to be inserted).

3   Select PHStat | Two-Sample Tests | Z Test for Differences in Two Proportions.

4   In the Z Test for the Difference in Two Proportions dialog box (see Figure 9.E.5):

a. Enter 0 in the Hypothesized Difference: edit box.
b. Enter 0.05 in the Level of Significance: edit box.
c. Enter the number of successes and sample size for both samples. Enter 48 in the number of successes and 50 as the sample size for the population 1 sample. Enter 30 as the number of successes and 50 as the sample size for the population 2 sample.
d. Select the Upper-Tail Test option button.
e. Enter Z Test for Difference in Two Proportions in the Output Title: edit box.
f. Click the OK button.

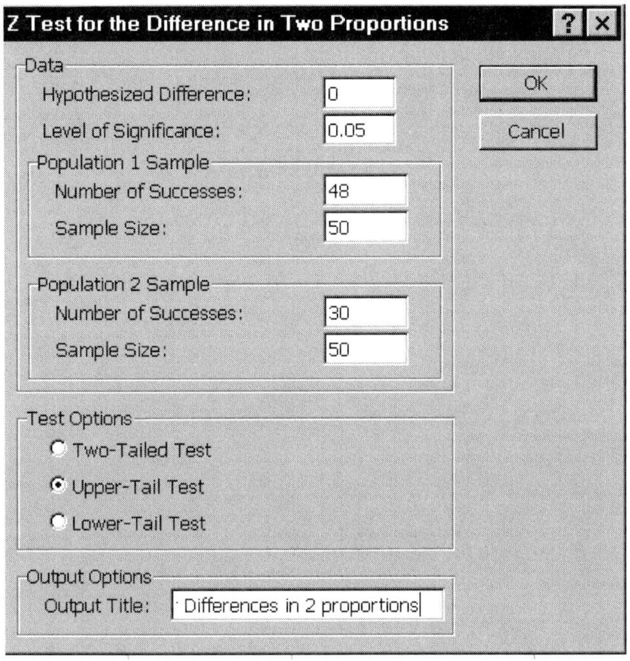

E **Figure 9.E.5** PHStat Z Test for the Difference in Two Proportions Dialog Box

The add-in inserts a worksheet containing calculations for the Z test for the difference in two proportions similar to the one shown in Figure 9.E.6.

9-3E.xls

	A	B
1	**Z Test for Differences in Two Proportions for Ethical Decision Making**	
2		
3	**Hypothesized Difference**	**0**
4	**Level of Significance**	**0.05**
5	**Group 1**	
6	**Number of Successes**	**48**
7	**Sample Size**	**50**
8	**Group 2**	
9	**Number of Successes**	**30**
10	**Sample Size**	**50**
11	Group 1 Proportion	0.96
12	Group 2 Proportion	0.6
13	Difference in Two Proportions	0.36
14	Average Proportion	0.78
15	Z Test Statistic	4.345240946
16	**Upper-Tailed Test**	
17	**Upper Critical Value**	**1.644853**
18	***p*-Value**	**6.96143E-06**

E **Figure 9.E.6** Z Test for Difference in Two Proportions for the Ethical Management Decision-Making Example Obtained from the PHStat Add-in for Microsoft Excel

We can change the values for the level of significance or the number of successes or sample size of each group in the worksheet inserted by the add-in to see their effects on the Z test. For example, if we changed the group 2 number of successes in cell B9 from 30 to 45, we would observe that the two-tailed *p*-value changes from .00000696 to .1198, thereby causing us not to reject the null hypothesis.

9.E.4 Using Microsoft Excel to Analyze Contingency Tables

Solution Summary:

    A.   Use the PHStat add-in to perform the χ^2 test for differences in two proportions.

    B.   Use the PHStat add-in to perform the χ^2 test for differences in *c* proportions or as a test of independence.

Example: Speaker interruption data of Examples 9.11–9.14 on pages 511–515

Part A Solution: Using the PHStat add-in to perform the χ^2 test for differences in two proportions.

To perform the χ^2 test for differences in two proportions, use the Two-Sample Tests | Chi-Square Test for Differences in Two Proportions choice of the PHStat add-in. As an example, consider the speaker interruption data of Examples 9.11–9.14. To determine whether or not there is evidence of a relationship between speaker gender and interrupter gender, do the following:

[1]   If the PHStat add-in has not been previously loaded, load the add-in using the instructions of Section EP.3.2.

[2]   Select File | New to open a new workbook (or open the existing workbook into which the hypothesis testing worksheet is to be inserted).

[3]   Select PHStat | Two-Sample Tests | Chi-Square Test for Differences in Two Proportions.

[4]   In the Chi-Square Test for Differences in Two Proportions dialog box (see Figure 9.E.7):

        [a] Enter 0.05 in the Level of Significance: edit box.

        [b] Enter Chi-Square Test for Differences in Two Proportions for Speaker Interruption in the Output Title: edit box.

        [c] Click the OK button.

E **Figure 9.E.7** PHStat Chi-Square Test for Differences in Two Proportions Dialog Box

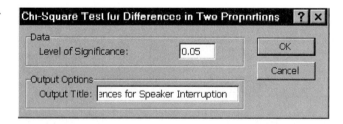

The add-in inserts a worksheet that contains tables for observed and expected frequencies and an area for test statistics. Because there are no observed frequencies, many cells display the message #DIV/0! at this point in the procedure. This is not an error.

[5]   In the newly inserted worksheet:

        [a] Enter replacement labels for the row and column variables. Enter Speaker in cell C4 and Interrupter in cell D3.

        [b] Enter replacement labels for the row and column categories. Enter Male in cell C5 and Female in cell C6. Enter Male in cell D4 and Female in cell E4.

c Enter the values from Table 9.4 (see page 512) for the Male and Female counts for the Speaker and Interrupter in the cell range D5:E6. Enter 15 in cell D5, 10 in cell D6, 5 in cell E5, and 10 in cell E6.

The completed worksheet will be similar to the one shown in Figure 9.E.8.

Observed Frequencies:			Interrupter		
		Speaker	Male	Female	Total
		Male	15	5	20
		Female	10	10	20
		Total	25	15	40
Expected Frequencies:			Interrupter		
		Speaker	Male	Female	Total
		Male	12.5	7.5	20
		Female	12.5	7.5	20
		Total	25	15	40
Level of Significance	0.05				
Number of Rows	2				
Number of Columns	2				
Degrees of Freedom	1				
Critical Value	3.841455				
Chi-Square Test Statistic	2.666671				
p-Value	0.10247				
Do not reject the null hypothesis					

E Figure 9.E.8 PHStat Add-in for Microsoft Excel Output for the Speaker–Interrupter Example

Due to a fault in the Microsoft Excel worksheet function used to generate the chi-square test statistic, the message #NUM!—and not the value of the statistic—may appear. This fault only appears when the p-value approaches zero and does not affect the decision displayed, which is based on the p-value.

We can change the values for the level of significance or any of the observed frequency counts in the worksheet inserted by the add-in to see their effects on the χ^2 test. For example, if we changed the count for male speaker–male interrupter (cell D5) from 15 to 20, and female speaker–male interrupter from 10 to 5, we would observe that the two-tailed p-value changes from .10247 to .003163, thereby causing us to reject the null hypothesis.

Example: Foreign language hiring preference data of Example 9.15 on pages 517–518

Part B Solution: Using the PHStat add-in to perform the χ^2 test for differences in c proportions or as a test of independence

To perform the χ^2 test for differences in c proportions or a χ^2 test of independence, use the c-Sample Tests I Chi-Square Test choice of the PHStat add-in. As an example, consider the foreign language hiring preference data. To determine whether or not the foreign language hiring preference is different for different types of businesses, do the following:

1 If the PHStat add-in has not been previously loaded, load the add-in using the instructions of Section EP.3.2.

2 Select File I New to open a new workbook (or open the existing workbook into which the hypothesis testing worksheet is to be inserted).

3 Select PHStat I c-Sample Tests I Chi-Square Test.

4 In the Chi-Square Test dialog box (see Figure 9.E.9):

   a Enter 0.01 in the Level of Significance: edit box.

   b Enter 2 in the Number of Rows: edit box.

   c Enter 3 in the Number of Columns: edit box.

   d Enter Chi-Square Test for Differences in c Proportions for Foreign Language Hiring Preference in the Output Title: edit box.

   e Click the OK button.

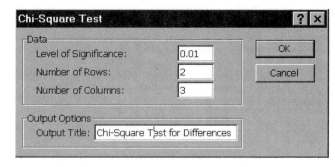

E **Figure 9.E.9** PHStat Chi-Square Test Dialog Box

The add-in inserts a worksheet that contains tables for observed and expected frequencies and an area for test statistics. Because there are no observed frequencies, many cells display the message #DIV/0! at this point in the procedure. This is not an error. Continue with the next step:

5 In the newly inserted worksheet:

a Enter replacement labels for the row and column variables. Enter Firm Type in cell C4 and Foreign Language Hiring Preference in cell D3.

b Enter replacement labels for the row and column categories. Enter U. S. in cell C5 and Foreign in cell C6. Enter Yes in cell D4, Neutral in cell E4, and No in cell F4.

c Enter the values from Table 9.6 (see page 517) for the firm types and foreign language hiring preferences in the cell range D5:F6. Enter 50 in cell D5, 60 in cell D6, 57 in cell E5, 22 in cell E6, 19 in cell F5, and 7 in cell F6.

The completed worksheet will be similar to the one shown in Figure 9.11 on page 518. Due to a fault in the Microsoft Excel worksheet function used to generate the chi-square test statistic, the message #NUM!—and not the value of the statistic—may appear. This fault only appears when the p-value approaches zero and does not affect the decision displayed, which is based on the p-value.

As discussed earlier, we can change the values for the level of significance or any of the observed frequency counts in the worksheet inserted by the add-in to see their effects on the χ^2 test.

9.E.5 Using Microsoft Excel to Perform the *F* Test for Differences in Two Variances

Solution Summary:

If sample data are available, use the Data Analysis F-Test Two-Sample for Variances tool to perform the *F* test for differences in the variances of two populations. If only summary data is available, use the PHStat add-in to perform this test.

Using Summary Data

Example: Hospital sterilization data of Example 9.17 on pages 526–527

Solution:

To perform the *F* test for differences in the variances of two populations using summary data, use the Two-Sample Tests | F Test for Differences in Two Variances choice of the PHStat add-in. As an example, consider the hospital sterilization data. To determine whether or not there is evidence of a significant difference in the variability in the ethylene oxide level under two different tasks, do the following:

1 If the PHStat add-in has not been previously loaded, load the add-in using the instructions of Section EP.3.2.

550

2   Select File | New to open a new workbook (or open the existing workbook into which the F test worksheet is to be inserted).

3   Select PHStat | Two-Sample Tests | F Test for Differences in Two Variances.

4   In the F Test for Differences in Two Variances dialog box (see Figure 9.E.10):

    a   Enter 0.05 in the Level of Significance: edit box.

    b   Enter the summary data of Table 9.8 on page 527 for both samples. Enter 11 and 4.10 as the sample size and sample standard deviation for population 1 (which has the larger variance). Enter 19 and 1.93 as the sample size and sample standard deviation for population 2.

    c   Select the Upper-Tail Test option button.

    d   Enter F Test for Differences in Variances for Tasks in the Output Title: edit box.

    e   Click the OK button.

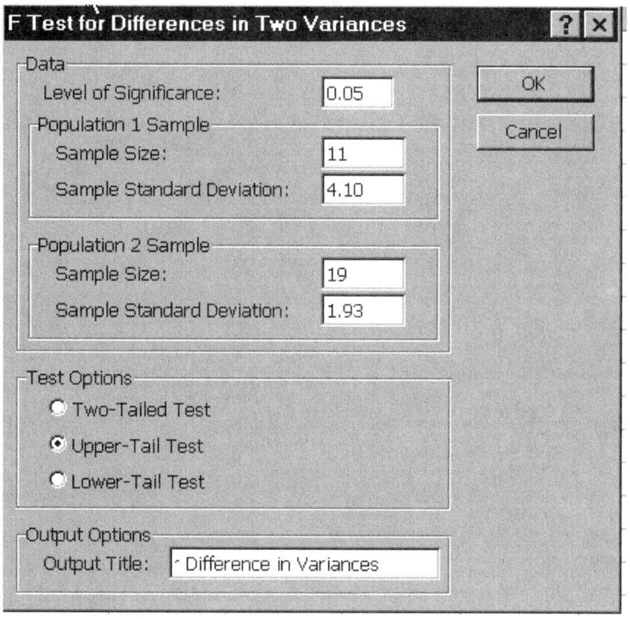

E **Figure 9.E.10**
PHStat F Test for Differences in Two Variances Dialog Box

The add-in inserts a worksheet containing calculations for the F test similar to the one shown in Figure 9.E.11.

F-Test for Differences in Variances for Tasks	
Level of Significance	0.05
Population 1 Sample	
Sample Size	11
Sample Standard Deviation	4.1
Population 2 Sample	
Sample Size	19
Sample Standard Deviation	1.93
F-Test Statistic	4.512873
Population 1 Sample Degrees of Freedom	10
Population 2 Sample Degrees of Freedom	18
Upper-Tail Test	
Upper Critical Value	2.411703
p-Value	0.00274
Reject the null hypothesis	

E **Figure 9.E.11**
PHStat Add-In for Microsoft Excel Output of the F Test for the Differences in Two Variances for the Hospital Sterilization Summary Data of Example 9.17

We can change the values for the sample size and sample standard deviation from each population as well as level of significance in the worksheet inserted by the add-in to see their effects on the test of hypothesis for the differences between two variances.

Using Sample Data: Unstacked

Example: Cracking strength data of Problem 9.65 on page 529

Solution:

To perform the *F* test for differences in the variances of two populations using sample data, use the Data Analysis F-Test Two-Sample for Variances tool. As an example, consider the cracking strength data of Problem 9.65. To determine whether or not there is evidence of a significant difference in the variability in the cracking strength for two different T-beams, do the following:

1. Open the TBEAMS.XLS workbook and click the Data sheet tab. Verify that the 70-cm slab and the 100-cm slab data have been entered into columns A and B, respectively.

2. Select Tools | Data Analysis. Select F-Test Two-Sample for Variances from the Analysis Tools list box in the Data Analysis dialog box. Click the OK button.

3. In the F-Test Two-Sample for Variances dialog box (see Figure 9.E.12):

   a. Enter A1:A9 in the Variable 1 Range: edit box.
   b. Enter B1:B9 in the Variable 2 Range: edit box.
   c. Select the Labels check box.
   d. Enter 0.10 in the Alpha: edit box.
   e. Select the New Worksheet Ply option button and enter Data Analysis F Test as the name of the new sheet.
   f. Click the OK button.

E **Figure 9.E.12** Data Analysis F-Test Two-Sample for Variances Tool Dialog Box

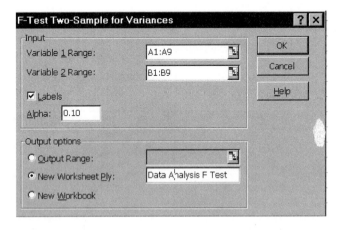

The Data Analysis tool add-in inserts a worksheet containing calculations for the F test similar to the one shown in Figure 9.E.13. This worksheet is *not* dynamically changeable, so any changes made to the underlying concrete strength data would require using the Data Analysis tool a second time in order to produce updated test results.

	A	B	C
1	F-Test Two-Sample for Variances		
2			
3		70-cm	100-cm
4	Mean	10.25	10.7625
5	Variance	7.231429	6.662679
6	Observations	8	8
7	df	7	7
8	F	1.085364	
9	P(F<=f) one-tail	0.458363	
10	F Critical one-tail	2.78493	

E Figure 9.E.13 Data Analysis F-Test Two-Sample for Variances Output for the Concrete Strength Data of Problem 9.65

Using Sample Data: Stacked

The Data Analysis F-Test tool requires that the data for each group be in separate columns, an arrangement known as unstacked data. Some data sets are "stacked," in which the data for both groups appear in a single column, identified by a grouping variable in another column. To use such stacked data with the Data Analysis F-Test tool, select the Data Preparation | Unstack Data choice of the PHStat add-in to unstack the data as part of Step 1.

Chapter 10

Regression Analysis

CONTENTS

10.1 Introduction to Regression Models

10.2 The Straight-Line Model: Simple Linear Regression

10.3 Estimating and Interpreting the Model Parameters

10.4 Model Assumptions

10.5 Measuring Variability around the Least Squares Line

10.6 Inferences about the Slope

10.7 Inferences about the Correlation Coefficient (Optional)

10.8 The Coefficient of Determination

10.9 Using the Model for Estimation and Prediction

10.10 Computations in Simple Linear Regression (Optional)

10.11 Residual Analysis: Checking the Assumptions (Optional)

10.12 Multiple Regression Models (Optional)

10.13 Pitfalls in Regression and Ethical Issues

EXCEL TUTORIAL

10.E.1 Using Microsoft Excel to Generate Scatter Diagrams and a Regression Line

10.E.2 Using Microsoft Excel for Simple Linear Regression

10.E.3 Using Microsoft Excel to Test for the Existence of Correlation

10.E.4 Using Microsoft Excel for Multiple Regression (Optional)

REAL-WORLD APPLICATION
The S.O.B. Effect among College Administrators

At major colleges and universities, administrators (e.g., deans, chairpersons, provosts, presidents) are among the highest paid state employees. Is there a link between the raises administrators receive and their performance on the job? This and other questions concerning the relationship between two variables will be discussed in this chapter. You will have the opportunity to analyze the raise–performance relationship of college administrators in Real-World Revisited at the end of this chapter.

10.1 INTRODUCTION TO REGRESSION MODELS

Suppose a university administrator wants to predict the annual merit raise (in dollars) awarded to professors employed by the university. To do this, the administrator will build a mathematical equation, i.e., a **model**, for the annual merit raise for any particular professor. The process of finding a mathematical model to predict the value of a variable is part of a statistical method known as **regression analysis**.

In regression analysis, the variable y to be predicted is called the *dependent* (or *response*) *variable*. In this example, y = annual merit raise for a professor.

Definition 10.1

The variable to be predicted (or modeled), y, is called the **dependent** (or **response**) **variable**.

The administrator knows that the actual value of y will vary from professor to professor depending on the professor's rank, teaching rating, tenure status, and numerous other factors. These factors used to predict y are called *independent* (or *explanatory*) *variables*.

Definition 10.2

The variables used to predict (or model) y are called **independent** (or **explanatory**) **variables** and are denoted by the symbols x_1, x_2, x_3, etc.

A **general regression model** takes the form shown in the next box.

General Regression Model for y

$$y = \beta_0 + \beta_1 x_1 + \beta_2 x_2 + \cdots + \beta_k x_k + \varepsilon \tag{10.1}$$

where $x_1, x_2, \ldots, x_k$ are the independent variables, $\beta_0, \beta_1, \ldots, \beta_k$ are the unknown model parameters, and ε is unexplainable (or random) error.

Suppose the administrator wants to use teaching rating x as the only predictor of merit raise y. When a single independent variable is used to predict y, the model simplifies to

$$y = \beta_0 + \beta_1 x + \varepsilon$$

The values β_0 and β_1 are unknown population parameters that need to be estimated from the sample data and ε represents unexplainable (or random) error. The process involves obtaining a sample of professors and recording teaching rating x as well as merit raise y. Subjecting this sample data to a regression analysis will yield estimates of the model parameters and enable the administrator to predict the merit raise y for a particular professor. The prediction equation takes the form

$$\hat{y} = b_0 + b_1 x$$

where $\hat{y}$ is the predicted value of y and b_0 and b_1 are estimates of the model parameters β_0 and β_1, respectively.

Example 10.1   Identifying the Regression Variables

The National Collegiate Athletic Association (NCAA) attempts to predict the academic success of college athletes using a regression model. The NCAA wants to predict an athlete's GPA as a function of several variables including high school GPA, Scholastic Assessment Test (SAT) score, and number of hours tutored during an academic year.

a. Identify the dependent variable y of interest to the NCAA.

b. Identify the independent variables.

c. Give the equation of the regression model for y if the NCAA uses only high school GPA as a predictor.

d. Give the equation of the regression model for y if the NCAA uses all the independent variables as predictors.

Solution

a. The dependent variable is the variable to be predicted. Since the NCAA wants to predict the GPA of an athlete, the dependent variable is

$$y = \text{GPA of a college athlete}$$

b. The independent variables the NCAA wants to use to predict y are:

x_1 = High school GPA of a college athlete

x_2 = SAT score of a college athlete

x_3 = Number of hours the athlete is tutored during the year

c. Using only x_1 as a predictor variable, the regression model is

$$y = \beta_0 + \beta_1 x + \varepsilon$$

where ε represents random error.

d. Applying equation 10.1 shown in the box (page 555), the full model takes the form

$$y = \beta_0 + \beta_1 x_1 + \beta_2 x_2 + \beta_3 x_3 + \varepsilon \qquad \square$$

Self-Test 10.1

A real estate appraiser wants to use square footage of living area, assessed value, and number of rooms to predict the sale price of a home.

a. Identify the dependent and independent variables.

b. Write the equation of a regression model that can be used to make the prediction.

An overview of the steps involved in a regression analysis is given in the box.

Steps in a Regression Analysis

1. Propose a model for y as a function of the independent variables $x_1, x_2, ..., x_k$.
2. Collect the sample data.
3. Use the sample data to estimate unknown parameters in the model.
4. Make specific assumptions about the probability distribution of the random error term ε and estimate any unknown parameters of this distribution.
5. Statistically check the usefulness of the model.
6. Check whether the assumptions on the random error term (step 4) are satisfied. If not, make model modifications.
7. When satisfied that the (modified) model is useful, use it for prediction, for estimation, and for making inferences about the model parameters.

We present the simplest of all regression models—called the straight-line model—in Sections 10.2–10.11. More complex models are discussed in optional Section 10.12.

10.2 THE STRAIGHT-LINE MODEL: SIMPLE LINEAR REGRESSION

Recall that the Federal Trade Commission collects data on the levels of tar, nicotine, and carbon monoxide in smoke emitted from domestic cigarettes. (These data are stored in the Excel workbook FTC.XLS.) Suppose you are interested in predicting a cigarette brand's carbon monoxide (CO) measurement from its nicotine content. Then, the dependent variable is

y = Carbon monoxide (CO) content (measured in milligrams)

and the only independent variable is

x = Nicotine content (also measured in milligrams)

In the following examples, we develop a model for y called the **straight-line model**.

Example 10.2 **Interpreting a Scatterplot**

Table 10.1 is a list of the nicotine and CO measurements for a hypothetical sample of 18 different cigarette brands. Examine the relationship between CO content y and nicotine content x. What type of regression model is suggested by this relationship?

CONICOTINE.XLS

Table 10.1 Carbon Monoxide–Nicotine Data for Example 10.2

CO y	Nicotine x	CO y	Nicotine x	CO y	Nicotine x
6	0.4	9	0.8	12	1.2
8	0.4	15	0.8	18	1.2
6	0.5	11	0.9	13	1.3
9	0.5	15	0.9	17	1.3
9	0.7	13	1.1	14	1.4
11	0.7	16	1.1	22	1.4

E **Figure 10.1**
Excel Scatterplot of
CO Content versus
Nicotine Content,
Example 10.2

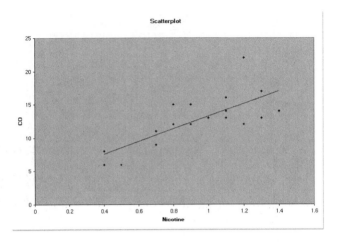

Solution

Recall (Section 3.8) that we can gain insight into the relationship between two quantitative variables by constructing a scatterplot for the sample data. Figure 10.1 shows a scatterplot for the sample data in Table 10.1 generated with Excel.

You can see that the level of CO y increases as nicotine content x increases. Although the points do not fall in a perfect pattern (e.g., a straight line or a curve), such a result is expected with regression models. The **random error component ε** in a regression model allows for random fluctuation of the points around some relationship between x and y. The points in Figure 10.1 appear to fluctuate around a straight line (highlighted on the graph). Therefore, it is reasonable to propose a straight-line relationship between y and x. ❏

Example 10.3 **Equation of a Straight-Line Model**

Refer to Example 10.2. Write the straight-line equation relating CO content y to nicotine content x.

Solution

A straight-line equation involves two parameters, the y-intercept and the slope of the line. In regression, the Greek symbols β_0 and β_1 represent the y-intercept and slope, respectively, since they are population parameters that will be known only if we have access to the entire population of (x, y) measurements. Therefore, the equation of the model is:

$$y = \beta_0 + \beta_1 x + \varepsilon$$

Note that this is the simple, one-variable model discussed in Section 10.1. A graph of this model is shown in Figure 10.2. Note that β_0 (the **y-intercept**) is the point at which the line cuts through the y axis, and β_1 (the **slope**) is the amount of change in y for every 1-unit increase in x. ❏

In this chapter, we will primarily consider the simplest of regression models—the straight-line model. The analysis of the straight-line model is commonly

Figure 10.2 The Straight-Line Model

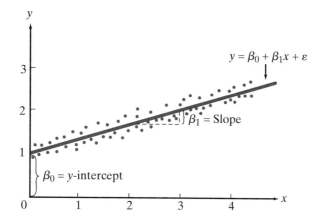

known as **simple linear regression** analysis. The elements of the simple linear regression model are summarized in the box.

The Simple Linear Regression (Straight-Line) Model

$$y = \beta_0 + \beta_1 x + \varepsilon \qquad (10.2)$$

where

y = quantitative dependent variable
x = quantitative independent variable
ε = random error component
β_0 (beta zero) = y-intercept of the line (point at which the line intercepts or cuts through the y axis—see Figure 10.2)
β_1 (beta one) = slope of the line (amount of increase, or decrease, in the deterministic component of y for every 1-unit increase in x—see Figure 10.2)

Example 10.4 **Identifying the Model Parameters**

Suppose it is known that y is related to x by the straight-line model

$$y = 17 + 4x + \varepsilon$$

Give the y-intercept and slope of the line.

Solution

From equation 10.2 in the box, the slope β_1 is the number multiplied by x in the model. Thus, $\beta_1 = 4$. The y-intercept β_0 is the number that is not multiplied by x. Therefore, $\beta_0 = 17$. ❏

Self-Test

10.2

Give the y-intercept and slope of the straight-line model

$$y = -100 + 67x + \varepsilon$$

PROBLEMS 10.1–10.6

Using the Tools

10.1   Suppose that y is exactly related to x by the straight-line equation $y = 1.5 + 2x$.

    a. Find the y-intercept for the line.   **b.** Find the slope of the line.

    c. If you increase x by one unit, how much will y increase or decrease?

    d. If you decrease x by one unit, how much will y increase or decrease?

    e. What is the value of y when $x = 0$?

10.2   Suppose that y is exactly related to x by the straight-line equation $y = 1.5 - 2x$.

    a. Find the y-intercept for the line.   **b.** Find the slope of the line.

    c. If you increase x by one unit, how much will y increase or decrease?

    d. What is the value of y when $x = 0$?

    e. What does this line have in common with the line in Problem 10.1? How do the two lines differ?

10.3   Graph the lines corresponding to each of the following equations:

    a. $y = 1 + 3x$ **b.** $y = 1 - 3x$ **c.** $y = -1 + .5x$

    d. $y = -1 - 3x$ **e.** $y = 2 - .5x$ **f.** $y = -1.5 + x$

    g. $y = 3x$ **h.** $y = -2x$

10.4   Give the values of β_0 and β_1 corresponding to each of the lines in Problem 10.3.

Applying the Concepts

10.5   In a straight-line model, what do the values β_0 and β_1 represent?

10.6   If a straight-line model accurately represents the true relationship between y and an independent variable x, does it imply that every value of y will always fall exactly on the line? Explain.

10.3 ESTIMATING AND INTERPRETING THE MODEL PARAMETERS

The following example illustrates the technique we will use to *fit the straight-line model to the data*, i.e., to estimate the slope and y-intercept of the line using information provided by the sample data.

Example 10.5   **How to Estimate β_0 and β_1**

Suppose a psychologist wants to model the relationship between the creativity score y and the flexibility score x of a mentally retarded child. Based on practical experience, the psychologist hypothesizes the straight-line model

$$y = \beta_0 + \beta_1 x + \varepsilon$$

CREATIVITY.XLS

Table 10.2 Creativity–Flexibility Scores for Example 10.5

Child	Flexibility Score x	Creativity Score y
1	2	2
2	6	11
3	4	7
4	5	10
5	3	5
6	5	8
7	3	4
8	3	3
9	7	11
10	9	13

If the psychologist were able to obtain the flexibility and creativity scores of *all* mentally retarded children, i.e., the entire population of (x, y) measurements, then the values of the population parameters β_0 and β_1 could be determined exactly. Collecting this mass of data would, of course, be impossible. The problem, then, is to estimate the unknown population parameters based on the information contained in a sample of (x, y) measurements. Suppose the psychologist tests 10 randomly selected mentally retarded children. The creativity and flexibility scores (both measured on a scale of 1 to 20) are given in Table 10.2. How can we best use the sample information to estimate the unknown y-intercept β_0 and the slope β_1?

Solution

Estimates of the unknown parameters β_0 and β_1 are obtained by finding the "best-fitting" straight line through the sample data points of Table 10.2. (These points are graphed in the Excel scatterplot, Figure 10.3.) We denote the estimates as b_0 and b_1, respectively. Then the "best-fitting" line can be written

$$\hat{y} = b_0 + b_1 x$$

where $\hat{y}$ is the predicted value of the creativity score y. ❑

E **Figure 10.3**
Excel Scatterplot of
Creativity–Flexibility
Score Data, Table 10.2

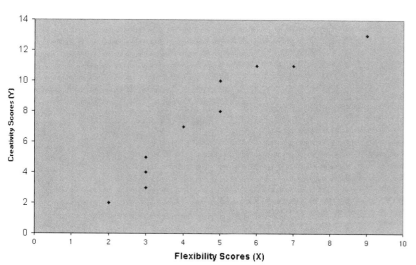

Although "best fit" can be defined in a variety of ways, the simplest approach involves making the difference between the actual y and predicted y, i.e., $y - \hat{y}$, as small as possible. We call the difference, $y - \hat{y}$, the error of prediction. Because the error of prediction may be positive for some observations and negative for others, it is better, mathematically, to minimize the sum of the squared errors, denoted SSE. A mathematical technique that finds the estimate of β_0 and β_1 and minimizes SSE is known as the **method of least squares**.

Least Squares Criterion for Finding the "Best-Fitting" Line

Choose the line that minimizes the sum of squared errors,

$$\text{SSE} = \sum (y - \hat{y})^2 \tag{10.3}$$

This is called the **least squares line**, or the **least squares prediction equation**.

The sum of the errors for the least squares line will always equal 0, i.e., $\sum(y - \hat{y}) = 0$. Since there are many other fitted lines that also have this property, we do not use this as the only criterion for choosing the "best-fitting" line.

With the least squares method, we find b_0 and b_1 using the formulas given below.

Formulas for the Least Squares Estimates

$$\textit{Slope: } b_1 = \frac{\text{SSxy}}{\text{SSxx}} \tag{10.4}$$

$$\textit{y-Intercept: } b_0 = \bar{y} - b_1\bar{x} \tag{10.5}$$

where

$$\text{SSxy} = \sum xy - \frac{(\sum x)(\sum y)}{n} \tag{10.6}$$

$$\text{SSxx} = \sum x^2 - \frac{(\sum x)^2}{n} \tag{10.7}$$

n = sample size

Example 10.6 Finding b_0 and b_1

Refer to Example 10.5. Use the least squares method to estimate β_0 and β_1 in the straight-line model.

Solution

Almost all spreadsheet and statistical software perform the least squares computations in simple linear regression. Consequently, we use Microsoft Excel to find the estimates of β_0 and β_1. (For those who are interested, these computations are illustrated in optional Section 10.10.)

Figure 10.4 (page 563) is an Excel printout of the simple linear regression analysis of the data in Table 10.2.

The estimates of β_0 and β_1 are indicated in Figure 10.4 under the column labeled **Coefficients**. The estimate of β_0, the y-intercept, is given in the row labeled **Intercept** and the estimate of β_1, the slope, is given in the row labeled **Flexibility Score**. (*Note:* The estimate of the slope will always be in the row named for the independent variable in the model.) These estimates are:

$$b_0 = -.437 \qquad b_1 = 1.667$$

E Figure 10.4
Excel Simple Linear
Regression Analysis of
Data in Table 10.2

R^2

s

SSR

SSE

SST

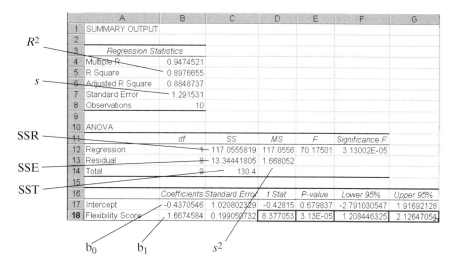

b_0 b_1 s^2

E Figure 10.5 Excel
Plot of Least Squares
Line, Example 10.6

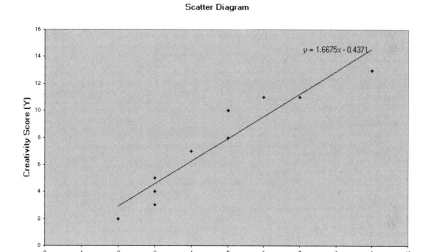

Consequently, the "best-fitting" line—also known as the least squares line—is

$$\hat{y} = -.437 + 1.667x$$

This line is plotted in Figure 10.5 using Excel in order for you to see how well the least squares line fits the data. ❏

Example 10.7   **Interpreting b_O and b_1**

Refer to Example 10.6. Interpret the values of the estimated y-intercept and estimated slope.

Solution
First, we'll interpret the estimated slope b_1. Recall from Section 10.2 that the slope represents the change in y for every 1-unit increase in x. Therefore, $b_1 = 1.667$ is the estimated change in creativity score y for every 1-point increase in

In simple linear regression, the estimated y-intercept will often not have a practical interpretation. It will, however, be practical if the value $x = 0$ is meaningful and within the range of the x-values in the sample.

flexibility score x. Since b_1 is positive, we estimate creativity score to *increase* 1.667 points for every 1-point increase in flexibility score.

The y-intercept is the point at which the least squares line crosses the y axis. This point has an x value of 0. Consequently, the estimated y-intercept b_0 can be interpreted as the predicted creativity score y when flexibility score x is 0. However, in this application b_0 is meaningless since it is impractical for a mentally retarded child to have a flexibility score of 0. In fact, none of the children in the sample had flexibility scores this low. ❑

Example 10.8   Interpreting b_0 and b_1

CRASH.XLS

Refer to the National Car Assessment Program (NCAP) crash test data stored in the Excel workbook CRASH.XLS. Two of the many variables measured are severity of driver's head injury (DRIVHEAD) and overall driver injury rating (DRIVSTAR). Recall that overall driver injury rating ranges from 1 to 5; the higher the rating, the less the chance of an injury. Suppose we want to model the overall driver injury rating y as a straight-line function of the driver's head injury score x. The straight-line model is fit to the NCAP data collected for 98 crash-tested cars. The Excel printout of the simple linear regression analysis is displayed in Figure 10.6. Interpret the estimates of β_0 and β_1 shown on the printout.

Solution
The estimates of β_0 and β_1, indicated on Figure 10.6, are

$$b_0 = 5.67177 \qquad b_1 = -.00289$$

The estimated y-intercept, $b_0 = 5.67177$, will have a practical interpretation only if the value $x = 0$ is meaningful. In this problem, a driver's head injury score of $x = 0$ is very unlikely to occur when a car is crashed into a wall at 35 mph. Even if $x = 0$ was theoretically possible, none of the 98 cars crash-tested in the sample had a driver's head injury score of 0. In fact, the lowest x-value in the sample data is 238! Therefore, the value $b_0 = 5.67177$ has no practical interpretation.

The estimated slope, $b_1 = -.00289$, represents the change in overall driver's injury rating y for every 1-point increase in driver's head injury score x. The

E Figure 10.6
Partial Simple Linear Regression Output Obtained from Excel, Example 10.8

SUMMARY OUTPUT					
Regression Statistics					
Multiple R	0.741739				
R Square	0.5501767				
Adjusted R Square	0.5454911				
Standard Error	0.486442				
Observations	98				
ANOVA					
	df	*SS*	*MS*	*F*	*Significance F*
Regression	1	27.78392506	27.78393	117.4172	2.40925E-18
Residual	96	22.71607494	0.236626		
Total	97	50.5			
	Coefficients	*Standard Error*	*t Stat*	*P-value*	*Lower 95%*	*Upper 95%*
Intercept	5.6717676	0.168209252	33.71852	5.14E-55	5.337874597	6.005660559
Drive head	-0.002887	0.000266457	-10.8359	2.41E-18	-0.003416219	-0.002358393

b_0
b_1

negative value implies that the overall injury rating *decreases* by .00289 point as the head injury score increases by 1 point. We also detected this negative relationship between the two variables in Example 3.22 (page 193). ❏

Example 10.9   Predicting y

Refer to Example 10.8. Use the least squares line to predict the overall driver's injury rating y when severity of head injury is $x = 1,000$.

Solution

The least squares prediction equation, obtained in Example 10.8, is $\hat{y} = 5.67177 - .00289x$. To obtain the predicted value of y, we substitute $x = 1,000$ into the least squares prediction equation:

$$\hat{y} = 5.67177 - .00289(1,000) = 2.78$$

Consequently, we predict a driver involved in a head-on crash at 35 mph to have an overall injury rating of 2.78 (a fairly low rating) when the severity of his or her head injury rating is 1,000. We show how to provide a measure of reliability for this inference in Section 10.9. ❏

Self-Test
10.3

Determining the nutritive quality of a new food product is often accomplished by feeding the food product to animals whose metabolic processes are very similar to our own (e.g., rats). This technique was used to evaluate the protein efficiency of two forms of a new product—one solid and the other liquid. Ten rats were randomly assigned a solid diet and 10 a liquid diet. At the end of the feeding period, the total protein intake x (in grams) and the weight gain y (in grams) were recorded for each of the rats. A straight line was fit to the 10 (x, y) data points for each diet by the method of least squares. The least squares prediction equations follow:

Liquid: $\hat{y} = 110 - .75x$

Solid: $\hat{y} = 84 + 3.66x$

a. Interpret the estimated y-intercept and slope for each diet.

b. For each diet, find the predicted weight gain of a rat with a protein intake of 10 grams.

PROBLEMS
10.7–10.14

Applying the Concepts

10.7   The data on retail price and energy cost for nine large side-by-side refrigerators, Problems 2.45 (page 101) and 3.76 (page 195), are reproduced at the top of page 566.

REFRIG.XLS

Brand	Price ($)	Energy Cost per Year ($)
KitchenAid Superba KSRS25QF	1,600	73
Kenmore (Sears) 5757	1,200	73
Whirlpool ED25DQXD	1,550	78
Amana SRD25S3	1,350	85
Kenmore (Sears) 5647	1,700	93
GE Profile TPX24PRY	1,700	93
Frigidaire Gallery FRS26ZGE	1,500	95
Maytag RSW2400EA	1,400	96
GE TFX25ZRY	1,200	94

Source: "The Kings of Cool." Copyright 1997 by Consumers Union of United States, Inc., Yonkers, NY 10703. Adapted from *Consumer Reports,* Jan. 1998, p. 52.

a. Give the equation of a straight-line model relating retail price y to energy cost x.

b. The model, part **a**, was fit to the data using Excel. The Excel printout is shown below. Locate the estimated slope and y-intercept of the line on the printout.

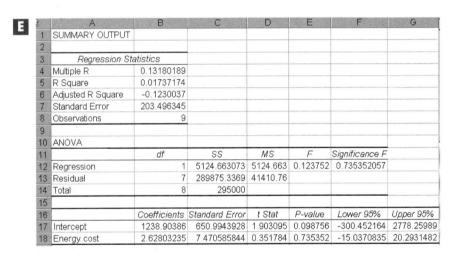

c. Write the equation of the least squares line.

d. Why is the line, part **c**, the "best" fitting line through the data?

e. Interpret the values of b_0 and b_1.

f. Use the least squares line to predict the retail price of a refrigerator with an energy cost of $80 per year.

10.8 Refer to *The American Statistician* (Feb. 1991) study of a process for drilling rock, Problems 2.46 (page 101) and 3.77 (page 195). Recall that experimenters wanted to determine whether the time y it takes to dry drill a distance of five feet in rock increases with depth x at which drilling begins. The data are reproduced on page 567.

a. Write a straight-line model relating drill time y to depth x.

b. Use Excel to fit the simple linear regression model, part **a**, to the data.

c. Refer to part **b**. Interpret the estimated slope of the line.

d. Refer to part **b**. Interpret the estimated y-intercept of the line.

DRILL.XLS

Depth at which Drilling Begins x (feet)	Time to Drill 5 Feet y (minutes)
0	4.90
25	7.41
50	6.19
75	5.57
100	5.17
125	6.89
150	7.05
175	7.11
200	6.19
225	8.28
250	4.84
275	8.29
300	8.91
325	8.54
350	11.79
375	12.12
395	11.02

Source: Penner, R., and Watts, D. G. "Mining information." *The American Statistician,*
Vol. 45, No. 1, Feb. 1991, p. 6 (Table 1).

Problem 10.8

10.9   Carbon dioxide–baited traps are typically used by entomologists to monitor mosquito populations. An article in the *Journal of the American Mosquito Control Association* (Mar. 1995) investigated whether temperature influences the number of mosquitos caught in a trap. Six mosquito samples were collected on each of nine consecutive days. For each day, two variables were measured: average temperature (degrees Centigrade) and mosquito catch ratio (the number of mosquitos caught in each sample divided by the largest sample caught).

MOSQUITO.XLS

Date	Average Temperature	Catch Ratio
July 24	16.8	.66
25	15.0	.30
26	16.5	.46
27	17.7	.44
28	20.6	.67
29	22.6	.99
30	23.3	.75
31	18.2	.24
Aug. 1	18.6	.51

Source: Petric, D., et al. "Dependence of CO_2-baited suction
trap captures on temperature variations." *Journal of the
Mosquito Control Association,* Vol. 11, No. 1, Mar. 1995, p. 8.

a.   Construct a scatterplot relating catch ratio to average temperature. Is there a linear trend in the data?

b.   Use Excel to fit a straight-line model relating catch ratio y to average temperature x. Interpret the β estimates.

10.10   Refer to the *American Journal of Psychiatry* (July 1995) study relating verbal memory retention y and right hippocampal volume x of 21 Vietnam veterans, Problems 2.48 (page 102) and 3.80 (page 196). The data, reproduced in the table, were subjected to a simple linear regression analysis using Excel. The Excel printout is also shown here.

BRAIN.XLS

Veteran	Verbal Memory Retention y	Right Hippocampal Volume x
1	26	960
2	22	1,090
3	30	1,180
4	65	1,000
5	70	1,010
6	60	1,070
7	60	1,080
8	58	1,040
9	65	1,040
10	83	1,030
11	69	1,045
12	60	1,210
13	62	1,220
14	66	1,215
15	76	1,220
16	65	1,300
17	66	1,350
18	84	1,350
19	90	1,370
20	102	1,400
21	102	1,460

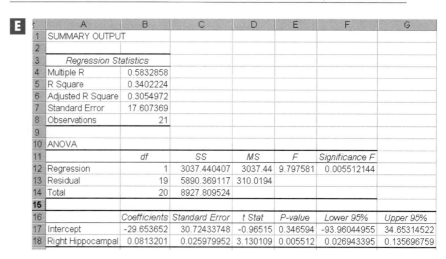

	A	B	C	D	E	F	G
1	SUMMARY OUTPUT						
2							
3	Regression Statistics						
4	Multiple R	0.5832858					
5	R Square	0.3402224					
6	Adjusted R Square	0.3054972					
7	Standard Error	17.607369					
8	Observations	21					
9							
10	ANOVA						
11		df	SS	MS	F	Significance F	
12	Regression	1	3037.440407	3037.44	9.797581	0.005512144	
13	Residual	19	5890.369117	310.0194			
14	Total	20	8927.809524				
15							
16		Coefficients	Standard Error	t Stat	P-value	Lower 95%	Upper 95%
17	Intercept	-29.653652	30.72433748	-0.96515	0.346594	-93.96044955	34.65314522
18	Right Hippocampal	0.0813201	0.025979952	3.130109	0.005512	0.026943395	0.135696759

a.   Locate the estimated slope and y-intercept on the printout.

b.   Write the equation of the least squares line.

c.   Interpret the value of the estimated slope.

d.   Interpret the value of the estimated y-intercept.

e.   Predict the verbal memory retention of a Vietnam veteran with a right hippocampal volume of 1,000.

10.11   Is it possible to predict the number of wins of a professional basketball team during the regular National Basketball Association (NBA) season? Charles Fisk, a California data analyst, collected performance data on NBA teams from 1979 to 1996. A regression model was developed to predict *y*, the number of wins for a team in a particular year, standardized using a *z* score calculation. Independent variables (all standardized using *z* scores) that were found to be the best predictors included:

x_1 = The difference between a team's 2-point field goal percentage and its opponents' percentage during the year

x_2 = The difference between a team's rebounds and its opponents' rebounds during the year

x_3 = The difference between opponents' turnovers and a team's turnovers during the year

The values of these variables for each of the 29 NBA teams during the 1995–96 season are listed in the table below.

NBA.XLS

Team	Wins y	2-Point Field Goals x_1	Rebounds x_2	Turnovers x_3
Chicago	2.21	0.77	2.14	2.02
Seattle	1.64	1.70	0.29	0.67
Orlando	1.35	1.28	−0.34	0.69
San Antonio	1.28	1.02	−0.23	0.54
Utah	1.00	1.49	1.11	1.61
LA Lakers	0.85	1.10	−0.65	1.50
Indiana	0.78	0.73	0.86	−0.67
Houston	0.50	0.50	−0.99	−0.18
New York	0.43	0.81	−0.56	0.18
Cleveland	0.43	0.18	−0.47	1.84
Detroit	0.36	0.29	0.36	−0.54
Atlanta	0.36	−0.89	−0.06	1.56
Portland	0.21	0.58	1.93	−1.63
Miami	0.07	0.92	0.78	−0.93
Phoenix	0.00	−0.05	0.49	−0.14
Charlotte	0.00	−0.60	−0.27	−0.23
Washington	−0.14	0.54	−0.86	0.31
Sacramento	−0.14	−0.29	0.36	−0.76
Golden State	−0.36	−0.90	0.21	0.28
Denver	0.43	−0.23	0.89	−1.19
Boston	−0.57	−0.93	−0.60	0.11
New Jersey	−0.78	−1.14	1.86	−0.93
LA Clippers	−0.85	−0.10	−0.90	0.02
Dallas	−1.07	−2.26	−0.10	0.69
Minnesota	−1.07	−0.44	−0.38	−0.73
Milwaukee	−1.14	−0.31	−0.66	−0.73
Toronto	−1.42	−0.44	−0.35	−1.92
Philadelphia	−1.64	−1.66	−1.78	−1.11
Vancouver	−1.85	−1.65	−2.07	0.67

a. Plot each of the independent variables against the dependent variable *y*. Which independent variable appears to have the strongest relationship with standardized number of wins?

b. Use Excel to fit a straight-line model relating standardized wins *y* to x_1.

c. Interpret the values of b_0 and b_1 obtained in part **b**.

d. Repeats parts **b** and **c** for independent variable x_2.

e. Repeats parts **b** and **c** for independent variable x_3.

10.12 Civil engineers often use a straight-line equation to model the relationship between the shear strength y of masonry joints and precompression stress x. To test this theory, a series of stress tests was performed on solid bricks arranged in triplets and joined with mortar (*Proceedings of the Institute of Civil Engineers,* Mar. 1990). The precompression stress was varied for each triplet; the ultimate shear load just before failure (called the shear strength) was recorded. The stress results for seven triplets (measured in N/mm^2) are shown in the accompanying table.

TRIPLETS.XLS

	TRIPLET TEST						
	1	2	3	4	5	6	7
Shear strength y	1.00	2.18	2.21	2.41	2.59	2.82	3.06
Precompression stress x	0	.60	1.20	1.33	1.43	1.75	1.75

Source: Riddington, J. R., and Ghazali, M. Z. "Hypothesis for shear failure in masonry joints." *Proceedings of the Institute of Civil Engineers, Part 2,* Mar. 1990, Vol. 89, p. 96 (Figure 7).

a. Construct a scatterplot for the seven data points. Does the relationship between shear strength and precompression stress appear to be linear?

b. Use the method of least squares (and Excel) to estimate the parameters of the linear model.

c. Interpret the values of b_0 and b_1.

d. Predict the shear strength for a triplet with a precompression stress of 1 N/mm^2.

10.13 The electroencephalogram (EEG) is a device used to measure brain waves. Neurologists have found that the peak EEG frequency in normal children increases with age. In one study (reported in *Science,* 1982), 287 normal children ranging from 2 to 16 years old were instructed to hold a 65-gram weight in the palm of their outstretched hand for a brief but unspecified time. The peak EEG frequency (measured in hertz) was then recorded for each child. The children were grouped according to age and the average peak frequency for each age group was recorded. The data appear in the accompanying table.

EEG.XLS

Age x (years)	Average Peak EEG Frequency y (hertz)	Age x (years)	Average Peak EEG Frequency y (hertz)
2	5.33	10	7.28
3	5.75	11	7.06
4	5.80	12	7.60
5	5.60	13	7.45
6	6.00	14	8.23
7	5.78	15	8.50
8	5.90	16	9.38
9	6.23		

Source: Tryon, W. W. "Development equation for postural tremor." *Science,* Vol. 215, No. 2, 1982, pp. 300–301. Copyright 1982 by the AAAS.

a. Construct a scatterplot for the data. Do you observe a trend?

b. Write the equation of a straight-line model relating y to x.

c. Find the least squares prediction equation using Excel.

d. Graph the least squares line on the scatterplot.

e. Interpret the estimated slope of the line.

f. Interpret the estimated y-intercept of the line.

g. Use the least squares line to predict the peak EEG frequency of a 6-year-old child.

10.14 The Mechanics Baseline and the Force Concept Inventory are two standardized tests designed to measure the competency of high school students in physics. The *American Journal of Physics* (July 1995) reported the inventory and baseline scores for 27 students in an honors physics course taught by one of the developers of the tests. These scores are given in the accompanying table. Suppose you are interested in predicting a student's memory score y from his or her baseline score x.

PHYSICS.XLS

Student	Inventory Score y (%)	Baseline Score x (%)	Student	Inventory Score y (%)	Baseline Score x (%)
1	97	100	15	83	68
2	97	93	16	83	52
3	90	84	17	76	52
4	90	80	18	70	52
5	90	75	19	70	58
6	97	72	20	80	58
7	86	75	21	62	58
8	86	65	22	76	61
9	86	61	23	62	38
10	83	65	24	42	38
11	76	65	25	67	41
12	72	65	26	52	45
13	95	68	27	45	33
14	95	52			

Source: Wells, M., Hestenes, D., and Swackhamer, G. "A modeling method for high school physics instruction." *American Journal of Physics,* Vol. 63, No. 7, July 1995, p. 612 (adapted from Fig. 3), p. 162.

a. An Excel plot of the data is shown below. Does baseline score appear to be a good predictor of inventory score?

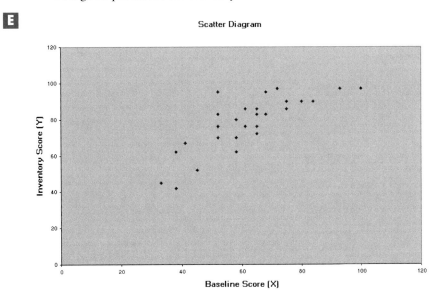

b. Write the equation of a straight-line model relating the two variables.

c. An Excel simple linear regression printout for the data is shown below. Locate and interpret the estimates of β_0 and β_1 on the printout.

E

	A	B	C	D	E	F	G
1	SUMMARY OUTPUT						
2							
3	*Regression Statistics*						
4	Multiple R	0.7952947					
5	R Square	0.6324936					
6	Adjusted R Square	0.6177933					
7	Standard Error	9.5728377					
8	Observations	27					
9							
10	ANOVA						
11		*df*	*SS*	*MS*	*F*	*Significance F*	
12	Regression	1	3942.871321	3942.871	43.02602	7.16126E-07	
13	Residual	25	2290.98053	91.63922			
14	Total	26	6233.851852				
15							
16		*Coefficients*	*Standard Error*	*t Stat*	*P-value*	*Lower 95%*	*Upper 95%*
17	Intercept	31.314697	7.362792722	4.2531	0.000258	16.15075243	46.47864211
18	Baseline Score	0.7541835	0.114977118	6.559423	7.16E-07	0.517383856	0.990983138

d. What is the predicted inventory score of a physics student with a baseline score of 50%?

10.4 MODEL ASSUMPTIONS

In the following sections, we describe the statistical methods (e.g., tests of hypotheses and confidence intervals) appropriate for making inferences from a simple linear regression analysis. As with most statistical procedures, the validity of the inferences depends on certain assumptions being satisfied. These assumptions, made about the random error term in the straight-line probabilistic model, are summarized in the box.

> ### Assumptions about the Random Error ε Required for a Simple Linear Regression Analysis
>
> **1.** The mean of the probability distribution of ε is 0.[*]
>
> **2.** The variance of the probability distribution of ε is constant for all values of the independent variable x and is equal to σ^2.
>
> **3.** The probability distribution of ε is normal.
>
> **4.** The errors associated with any two observations are independent. That is, the error associated with one value of y has no effect on the errors associated with other y values.

*The assumption that mean error is 0 is *not* guaranteed by the method of least squares. The method of least squares yields a *sample* mean error of 0, i.e., $[\Sigma(y - \hat{y})]/n = 0$, for all models regardless of whether the model is the correct one. The zero mean assumption in simple linear regression is equivalent to assuming that the form of the correct model is, in fact, a straight line.

Figure 10.7 shows a pictorial representation of the assumptions given in the box. For each value of x shown in the figure, the relative frequency distribution of the errors is normal with mean 0 and with a constant variance (all the distributions shown have the same amount of spread or variability) equal to σ^2.

Figure 10.7 The Probability Distribution of the Random Error Component ε

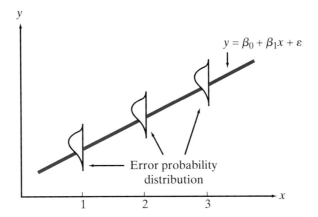

Statistical techniques are available for detecting when one or more of the assumptions are grossly violated. These methods are based on an analysis of the least squares errors of prediction, or *residuals*. (We present an analysis of residuals in optional Section 10.11.) In practice, however, you will never know whether the data exactly satisfy the four assumptions. Fortunately, the estimators and test statistics used in a simple linear regression have sampling distributions that remain relatively stable for minor departures from the assumptions.

> In regression analysis, the symbol $E(y)$ is used in place of μ to represent the true mean (or *expected value*) of y.

There is an additional assumption that is implied in a simple linear regression analysis, but often forgotten—namely, the assumption that the relationship between the mean value of y, denoted $E(y)$, and the independent variable x is correctly modeled by a *straight line*. In a real application, the relationship between $E(y)$ and x may possess some curvature. Therefore, when we conduct a simple linear regression analysis, we are assuming that this curvature is minimal over the set of values for which x is measured. The implications of this assumption can be seen in Figure 10.8. If x is measured over the interval between two points, say x_L and x_U, a simple linear regression analysis may produce a very good prediction equation for predicting y for values of x between x_L and x_U, but very poor estimates and predictions for values of x outside this range. The rule, then, is *never to attempt to predict values of y for values of x outside the range of values used in the regression analysis*.

Figure 10.8 Hypothetical Comparison of the True Relationship between $E(y)$ and x with the Simple Linear Regression Model

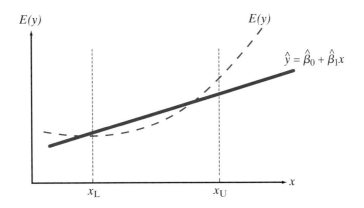

10.5 MEASURING VARIABILITY AROUND THE LEAST SQUARES LINE

In Example 10.5, is the creativity score y really related to the flexibility score x, or is the linear relation that we seem to see a result of chance? That is, could it be the case that x and y are completely unrelated, and that the apparent linear configuration of the data points in the scatterplot of Figure 10.3 is due to random variation? The statistical method that will answer this question requires that we know how much y will vary for a given value of x. That is, we must know the value of the quantity, called σ^2, that measures the variability of the y values about the least squares line. This value is the variance σ^2 of the random error identified in Section 10.4.

Figure 10.9a shows a hypothetical situation where the value of σ^2 is small. Note that there is little variation in the y values about the least squares line. In contrast, Figure 10.9b illustrates a situation where σ^2 is large; the y values deviate greatly about the least squares line.

Since the variance σ^2 will rarely be known, we estimate its value using the sum of squared errors, SSE, and the formulas shown in the box. However, we can rely on computer software such as Excel to compute an estimate of σ^2.

Degrees of Freedom for Estimating σ^2

The divisor of s^2, $n - 2$, represents the *number of degrees of freedom available for estimating σ^2*. The value results from the fact that we "lose" two degrees of freedom for estimating the model parameters β_0 and β_1.

Estimation of σ^2, a Measure of the Variability of the y Values about the Least Squares Line

An estimate of σ^2 is given by

$$s^2 = \frac{\text{SSE}}{n - 2} \qquad \text{(10.8)}$$

where

$$\text{SSE} = \sum (y - \hat{y})^2 = \text{SSyy} - b_1 \text{SSxy} \qquad \text{(10.9)}$$

$$\text{SSyy} = \sum (y - \hat{y})^2 = \sum y^2 - \frac{(\sum y)^2}{n} \qquad \text{(10.10)}$$

$$\text{SSxy} = \sum (x - \bar{x})(y - \bar{y}) = \sum xy - \frac{(\sum x)(\sum y)}{n} \qquad \text{(10.6)}$$

Figure 10.9 Illustration of Error Variance σ^2

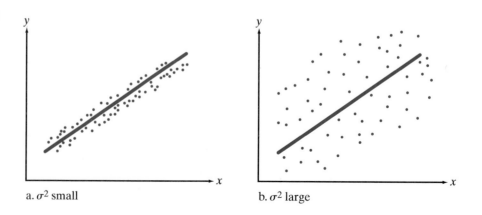

a. σ^2 small   b. σ^2 large

Example 10.10   Estimating σ^2 and σ

Refer to the simple linear regression relating creativity score to flexibility score, Example 10.5.

a. Estimate the value of the error variance σ^2.

b. Estimate and interpret the value of σ.

Solution

a. Refer to the Excel output for the simple linear regression relating creativity score y to flexibility score x, Figure 10.4 (page 563). The value of SSE (indicated on the printout) is 13.3444 and the estimate of σ^2 (also indicated on the printout) is 1.66805.

b. An estimate of the standard deviation σ of the random error is obtained by taking the square root of s^2. Therefore, our estimate is

$$s = \sqrt{s^2} = \sqrt{1.66805} = 1.29153$$

This value is also indicated on Figure 10.4. Since s measures the spread of the distribution of the y values about the least squares line, we should not be surprised to find that most of the observations lie within $2s = 2(1.29) = 2.58$ of the least squares line. From Figure 10.10 we see that, for this example, all 10 data points have y values that lie within $2s$ of $\hat{y}$, the least squares predicted value. Since the distribution of errors is assumed to be normal, we expect about 95% of the y values to lie within $2s$ of the least squares line. ❑

Practical Interpretation of s, the Estimated Standard Deviation of σ

Given the assumptions outlined in the box in Section 10.4, we expect most (about 95%) of the observed y values to lie within $2s$ of their respective least squares predicted values y.

Figure 10.10
Observations within $2s$ of the Least Squares Line

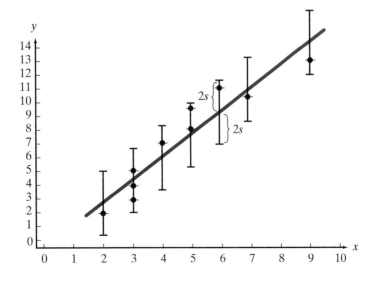

Self-Test

10.4

Refer to Self-Test 10.3 (page 565). Recall that simple linear regression was used to fit the model $y = \beta_0 + \beta_1 x + \varepsilon$ to $n = 10$ data points, where y is the weight gain for a rat on a diet of a new food product and x is the total protein intake of the rat. For rats on a solid diet, SSE = 300.

    a. Compute an estimate of σ.

    b. Interpret the estimate in part **a.**

PROBLEMS

10.15–10.23

Using the Tools

10.15 Suppose you fit a least squares line to 10 data points and find SSE = .22.

    a. Find s^2, the estimate of σ^2.   **b.** Find s, the estimate of σ.

10.16 Repeat Problem 10.15, but assume $n = 100$.

10.17 Repeat Problem 10.15, but assume SSE = 22.

10.18 Suppose you fit a straight line to 25 data points. Find s^2 and s for each of the following values of SSE.

    a. 100   **b.** 6.20   **c.** 8,865

Applying the Concepts

10.19 Locate and interpret the values of SSE, s^2, and s on the Excel printout, Problem 10.7 (page 566).

10.20 Locate and interpret the values of SSE, s^2, and s on the Excel printout, Problem 10.10 (page 568).

10.21 Refer to the three straight-line models for number of NBA wins y, Problem 10.11 (page 569). For each model, find the estimate of σ on the Excel printout and interpret it. Which of the three independent variables, x_1, x_2, or x_3, yields the linear prediction equation with the smallest error variance?

10.22 Refer to Problem 10.13 (page 570).

    a. Find SSE and s^2.   **b.** Find s, the estimate of σ.

    c. Interpret the value of s.

10.23 Locate and interpret the values of SSE, s^2, and s on the Excel printout, Problem 10.14 (page 571).

10.6 INFERENCES ABOUT THE SLOPE

After fitting the model to the data and computing an estimate of σ^2, we can statistically check the usefulness of the model. That is, we can use a statistical inferential procedure (a test of hypothesis or confidence interval) to determine whether the least squares straight-line (linear) model is a reliable tool for predicting y for a given value of x.

Example 10.11

How to Test the Model

Consider the straight-line model

$$y = \beta_0 + \beta_1 x + \varepsilon$$

How do we determine statistically whether this model is useful for prediction purposes? In other words, how could we test whether x provides useful information for the prediction of y?

Solution

Suppose x is *completely unrelated* to y. What could we say about the values of β_0 and β_1 in the probabilistic model, if in fact x contributes no information for the prediction of y? For y to be independent of x, the true slope β_1 of the line must be equal to 0. Therefore, to test the null hypothesis that x contributes no information for the prediction of y against the alternative that these variables are linearly related with a slope differing from 0, we test

$$H_0: \beta_1 = 0$$

$$H_a: \beta_1 \neq 0$$

If the data support the alternative hypothesis, we will conclude that x does contribute information for the prediction of y using the straight-line model, although the true relationship between $E(y)$ and x could be more complex than a straight line. ❏

It is well known that the sampling distribution of b_1, the estimator of the slope β_1, has a t distribution with standard error

$$s_{b_1} = \frac{s}{\sqrt{SSxx}} \tag{10.11}$$

Using the hypothesis-testing technique developed in Chapter 8, we set up the test as shown in the next box.

Test of Hypothesis for the Slope of the Straight-Line Model

Upper-Tailed Test	Two-Tailed Test	Lower-Tailed Test
$H_0: \beta_1 = 0$	$H_0: \beta_1 = 0$	$H_0: \beta_1 = 0$
$H_a: \beta_1 > 0$	$H_a: \beta_1 \neq 0$	$H_a: \beta_1 < 0$

$$Test\ statistic: t = \frac{b_1}{s_{b_1}} \approx \frac{b_1}{s/\sqrt{SSxx}} \tag{10.12}$$

Rejection region:	Rejection region:	Rejection region:
$t > t_\alpha$	$|t| > t_{\alpha/2}$	$t < -t_\alpha$

where the distribution of t is based on $(n - 2)$ degrees of freedom, t_α is the t value such that $P(t > t_\alpha) = \alpha$, and $t_{\alpha/2}$ is the t value such that $P(t > t_{\alpha/2}) = \alpha/2$.

Assumptions: See Section 10.4.

Inferences based on this hypothesis test require the standard least squares assumptions about the random error term listed in the box in Section 10.4. However, the test statistic has a sampling distribution that remains relatively stable for minor departures from the assumptions. That is, our inferences remain valid for practical cases in which the assumptions are nearly, but not completely, satisfied.

Example 10.12   Testing the Model

Refer to the simple linear regression relating creativity score y to flexibility score x, Example 10.5 (page 561). At a significance level $\alpha = .05$, test whether the straight-line model is useful for predicting y.

Solution

Testing the usefulness of the model requires testing the hypotheses

$$H_0: \beta_1 = 0$$

$$H_a: \beta_1 \neq 0$$

With $n = 10$ and $\alpha = .05$, the critical value based on $(10 - 2) = 8$ df is obtained from Table B.3 in Appendix B:

$$t_{\alpha/2} = t_{.025} = 2.306$$

Thus, we will reject H_0 if $|t| > 2.306$ (see Figure 10.11).

Rather than compute the test statistic using equation 10.12, we'll use the results obtained from Excel. The Excel simple linear regression analysis is reproduced in Figure 10.12 (page 579).

The estimate $b_1 = 1.667$ and its standard error $s_{b_1} = .199$ are indicated in the printout as well as the value of the test statistic $t = 8.38$. Since this calculated t value falls in the upper tail of the rejection region (see Figure 10.11), we reject the null hypothesis and conclude that the slope β_1 is not 0.

At the $\alpha = .05$ level of significance, the sample data provide sufficient evidence to conclude that flexibility score *does* contribute useful information for the prediction of creativity score using the straight-line model.

The same conclusion can be reached by using the p-value approach. The p-value for this two-tailed test (indicated on Figure 10.12) is near 0.* Since this value is smaller than $\alpha = .05$, we reject H_0 and conclude that the slope β_1 differs from 0. ❏

*Scientific Notation: The p-value on the Excel printout is expressed in scientific notation as 3.13E-05. This means to move the decimal point 5 places to the left. Therefore, the p-value is .0000313.

Figure 10.11 Rejection Region for Example 10.12

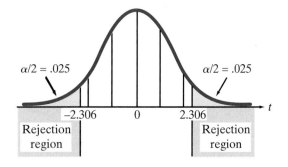

E **Figure 10.12** Excel Simple Linear Regression Output, Example 10.12

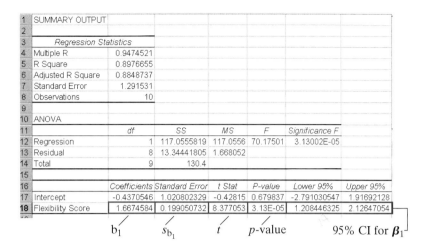

SUMMARY OUTPUT					
Regression Statistics					
Multiple R	0.9474521				
R Square	0.8976655				
Adjusted R Square	0.8848737				
Standard Error	1.291531				
Observations	10				
ANOVA					
	df	*SS*	*MS*	*F*	*Significance F*
Regression	1	117.0555819	117.0556	70.17501	3.13002E-05
Residual	8	13.34441805	1.668052		
Total	9	130.4			
	Coefficients	*Standard Error*	*t Stat*	*P-value*	*Lower 95%*	*Upper 95%*
Intercept	-0.4370546	1.020802329	-0.42815	0.679837	-2.791030547	1.91692128
Flexibility Score	1.6674584	0.199050732	8.377053	3.13E-05	1.208446325	2.12647054

b_1 s_{b_1} t p-value   95% CI for β_1

If the test in Example 10.12 had resulted in "fail to reject H_0," would we have concluded that $\beta_1 = 0$? The answer to this question is "no" (recall the discussion on Type II errors in Chapter 8). Rather, we acknowledge that additional data might indicate that β_1 differs from 0, or that a more complex relationship (other than a straight line) may exist between y and x.

Self-Test

10.5

A consumer investigator obtained the following least squares straight-line model (based on a sample of $n = 100$ families) relating the yearly food cost y (thousands of dollars) for a family of four to annual income x:

$$\hat{y} = 467 + .26x$$

In addition, the investigator computed the standard error $s_{b_1} = .22$. Conduct a test to determine whether mean yearly food cost y increases as annual income x increases, i.e., whether the slope β_1 of the line is positive. Use $\alpha = .05$.

In addition to testing whether the slope β_1 is 0, we may also be interested in estimating its value with a confidence interval. The procedure is illustrated in the following example.

Example 10.13   Confidence Interval for the Slope

Refer to Example 10.12. Construct a 95% confidence interval for the slope β_1 in the straight-line model relating creativity score y to flexibility score x.

Solution

The method of Chapter 7 can be used to construct a confidence interval for β_1. The interval, derived from the sampling distribution of b_1, is given in the box (page 580).

A $(1 - \alpha)100\%$ Confidence Interval for the Slope β_1

$$b_1 \pm (t_{\alpha/2})s_{b_1} = b_1 \pm t_{\alpha/2}\left(\frac{s}{\sqrt{SSxx}}\right) \qquad \textbf{(10.13)}$$

where the distribution of t is based on $(n - 2)$ degrees of freedom and $t_{\alpha/2}$ is the value of t such that $P(t > t_{\alpha/2}) = \alpha/2$.

For a 95% confidence interval, $\alpha = .05$. Therefore, we need to find the value of $t_{.025}$ based on $(10 - 2) = (10 - 2) = 8$ df. In Example 10.12, we found that $t_{.025} = 2.306$. From the Excel printout, Figure 10.12, we have

$$b_1 = 1.667 \qquad \text{and} \qquad s_{b_1} = .199$$

Now we apply equation 10.13 to obtain a 95% confidence interval for the slope in the model relating creativity score to flexibility score:

$$b_1 \pm (t_{.025})s_{b_1} = 1.667 \pm (2.306)(.199)$$
$$= 1.667 \pm .459$$

Our interval estimate of the slope parameter β_1 is then 1.208 to 2.126. This confidence interval also appears on the Excel printout, Figure 10.12, as indicated. ❏

Example 10.14   Interpreting the Confidence Interval for β_1

Interpret the interval estimate of β_1 derived in Example 10.13.

Solution

Since all the values in the interval $(1.208, 2.126)$ are positive, we say that we are 95% confident that the slope β_1 is positive. That is, we are 95% confident that the mean creativity score $E(y)$ increases as flexibility score x increases. In addition, we can say that for every 1-point increase in flexibility score x, the increase in mean creativity score $E(y)$ could be as small as 1.208 or as large as 2.126 points. However, the relatively large width of the interval reflects the small number of data points (and, consequently, a lack of information) in the experiment. We could expect a narrower interval if the sample size were increased. ❏

Self-Test
10.6

Use the information in Self-Test 10.5 (page 579) to find a 90% confidence interval for β_1. Interpret the results.

PROBLEMS
10.24–10.35

Using the Tools

10.24   Suppose you want to test $H_0: \beta_1 = 0$ versus $H_a: \beta_1 \neq 0$ in a simple linear regression model. Give the degrees of freedom associated with the value of the test statistic for each of the following sample sizes.

a. $n = 6$ **b.** $n = 10$ **c.** $n = 25$ **d.** $n = 50$

10.25 For each of the following, specify the rejection region for testing the null hypothesis $H_0: \beta_1 = 0$ in simple linear regression.

a. $H_a: \beta_1 \neq 0, n = 5, \alpha = .05$

b. $H_a: \beta_1 > 0, n = 5, \alpha = .10$

c. $H_a: \beta_1 < 0, n = 7, \alpha = .01$

10.26 A simple linear regression analysis of $n = 10$ data points yielded the following results:

$$b_1 = 56 \qquad s_{b_1} = 16$$

a. Test the null hypothesis that the slope β_1 of the line equals 0 against the alternative hypothesis that β_1 is not equal to 0. Use $\alpha = .10$.

b. Find a 90% confidence interval for the slope β_1.

10.27 A simple linear regression analysis of $n = 45$ data points yielded the following results:

$$b_1 = -300.7 \qquad s_{b_1} = 193.4$$

a. Test the null hypothesis that the slope β_1 of the line equals 0 against the alternative hypothesis that β_1 is negative. Use $\alpha = .10$.

b. Find a 95% confidence interval for the slope β_1.

Applying the Concepts

10.28 The data on temperature at flight time and O-ring damage index for 23 launches of the space shuttle *Challenger,* first presented in Problem 2.50 (page 104), are reproduced in the table on page 582.

a. Hypothesize a straight-line model relating O-ring damage y to flight time temperature x.

b. Fit the model, part **a**, to the data using Excel.

c. Theoretically, O-ring damage will increase as flight time temperatures decrease. Test this theory at $\alpha = .01$.

DRILL.XLS

10.29 Refer to *The American Statistician* investigation of dry drilling in rock, Problem 10.8 (page 567). Is there evidence to indicate that dry drill time y increases with depth x? Test using $\alpha = .10$.

MOSQUITO.XLS

10.30 Refer to the *Journal of the American Mosquito Control Association* study of the influence of temperature on number of mosquitos caught in carbon dioxide–baited traps, Problem 10.9 (page 567).

a. Construct a 95% confidence interval for β_1, the slope of the line relating catch ratio y to temperature x.

b. Interpret the result, part **a**.

EEG.XLS

10.31 Refer to the *Science* study of EEG frequency in normal children, Problem 10.13 (page 570). Do the data provide sufficient evidence to indicate that age x contributes information for the prediction of average peak frequency y? Test using $\alpha = .05$.

ORING.XLS

Flight Number	Temperature (°F)	O-Ring Damage Index
1	66	0
2	70	4
3	69	0
5	68	0
6	67	0
7	72	0
8	73	0
9	70	0
41-B	57	4
41-C	63	2
41-D	70	4
41-G	78	0
51-A	67	0
51-B	75	0
51-C	53	11
51-D	67	0
51-F	81	0
51-G	70	0
51-I	67	0
51-J	79	0
61-A	75	4
61-B	76	0
61-C	58	4

Note: Data for flight number 4 are omitted due to an unknown O-ring condition.

Primary Sources: Report of the Presidential Commission on the Space Shuttle Challenger *Accident,* Washington, D.C., 1986, Vol. II (pp. H1–H3) and Volume IV (p. 664). *Post-*Challenger *Evaluation of Space Shuttle Risk Assessment and Management,* Washington, D.C., 1988, pp. 135–136.

Secondary Source: Tufte, E. R. *Visual and Statistical Thinking: Displays of Evidence for Making Decisions.* Graphics Press, 1997.

Problem 10.28

PHYSICS.XLS

10.32 Refer to the *American Journal of Physics* study of two standardized physics competency tests, Problem 10.14 (page 571). Is there evidence of a positive linear relationship between the Mechanics Baseline score x and Force Concept Inventory score y? Use $\alpha = .01$.

10.33 Refer to the *Scientific American* (Dec. 1993) study of fertility rates in developing countries, Problem 3.81 (page 196). The data on fertility rate y and percentage of married women who use contraception x for each of 27 developing countries are reproduced on page 583. The simple linear regression model was fit to the data using Excel. The Excel printout follows.

   a. Give the least squares prediction equation.

   b. Conduct a test to determine whether fertility rate y and contraceptive prevalence x are linearly related. Use $\alpha = .05$.

   c. Construct a 95% confidence interval for β_1. Interpret the interval.

   d. Does the confidence interval, part **c**, support the results of the test, part **b**? Explain.

FERTIL.XLS

Country	Contraceptive Prevalence x	Fertility Rate y
Mauritius	76	2.2
Thailand	69	2.3
Colombia	66	2.9
Costa Rica	71	3.5
Sri Lanka	63	2.7
Turkey	62	3.4
Peru	60	3.5
Mexico	55	4.0
Jamaica	55	2.9
Indonesia	50	3.1
Tunisia	51	4.3
El Salvador	48	4.5
Morocco	42	4.0
Zimbabwe	46	5.4
Egypt	40	4.5
Bangladesh	40	5.5
Botswana	35	4.8
Jordan	35	5.5
Kenya	28	6.5
Guatemala	24	5.5
Cameroon	16	5.8
Ghana	14	6.0
Pakistan	13	5.0
Senegal	13	6.5
Sudan	10	4.8
Yemen	9	7.0
Nigeria	7	5.7

Source: Robey, B., et al. "The fertility decline in developing countries." *Scientific American,* Dec. 1993, p. 62. [*Note:* The data values are estimated from a scatterplot.]

Problem 10.33

	A	B	C	D	E	F	G
1	SUMMARY OUTPUT						
2							
3	*Regression Statistics*						
4	Multiple R	0.86501984					
5	R Square	0.74825933					
6	Adjusted R Square	0.7381897					
7	Standard Error	0.69571072					
8	Observations	27					
9							
10	ANOVA						
11		*df*	*SS*	*MS*	*F*	*Significance F*	
12	Regression	1	35.96633167	35.9663	74.30855	5.8648E-09	
13	Residual	25	12.10033499	0.48401			
14	Total	26	48.06666667				
15							
16		*Coefficients*	*Standard Error*	*t Stat*	*P-value*	*Lower 95%*	*Upper 95%*
17	Intercept	6.73192924	0.290342518	23.1862	2.02E-18	6.13395805	7.3299
18	Contraceptive Prevalence	-0.05461028	0.006335123	-8.62024	5.86E-09	-0.0676577	-0.04156

Problem 10.33

10.34   *Chemosphere* (Vol. 20, 1990) published a study of Vietnam veterans exposed to Agent Orange. The blood plasma and fat tissue levels of several types of dioxin (called cogeners) were measured for each of the 20 Vietnam veterans. For each cogener, a simple linear regression analysis was conducted to predict (1) fat tissue level from blood plasma level and (2) blood plasma level from fat tissue level. The results for three of these cogeners are shown in the table.

Cogener	y = Fat Tissue Level x = Blood Plasma Level	y = Blood Plasma Level x = Fat Tissue Level	t Value for Testing β_1
2,3,4,7,8,-P$_n$ CDF	$\hat{y} = .8109 + .9713x$	$\hat{y} = .9855 + .7605x$	7.13
H$_x$ CDD	$\hat{y} = 18.1565 + .7377x$	$\hat{y} = 5.2009 + .9018x$	5.98
OCDD	$\hat{y} = 118.6057 + .3679x$	$\hat{y} = 167.723 + 1.5752x$	4.98

Source: Schecter, A., et al. "Partitioning of 2,3,7,8-chlorinated dibenzo-p-dioxins and dibenzofurans between adipose tissue and plasma lipid of 20 Massachusetts Vietnam veterans." *Chemosphere,* Vol. 20, Nos. 7–9, 1990, pp. 954–955 (Table III).

a.   For the cogener 2,3,4,7,8-P$_n$ CDF, are the two regression models statistically adequate for predicting y? Test both using $\alpha = .05$.

b.   Repeat part **a** for the cogener H$_x$CDD.

c.   Repeat part **a** for the cogener OCDD.

10.35   A Florida real estate appraisal company regularly employs a paired sales technique to evaluate land sales. When two parcels of land are sold, the ratio of the acreage of the larger parcel to the smaller parcel is calculated as well as the difference in sale prices (measured as a percentage of the higher sale price and called a downward adjustment). The variables land sale ratio x and downward price adjustment y were measured for a random sample of 14 pairs of land sales in Seminole County, Florida. The data (provided by a confidential source) are shown in the accompanying table. An Excel simple linear regression analysis of the data is shown in the printout on page 585.

LANDSALE.XLS

Pair	Land Sale Ratio x	Downward Price Adjustment y (%)
1	1.7	8
2	2.4	7
3	3.3	8
4	6.8	23
5	5.7	30
6	18.2	52
7	26.2	51
8	33.6	52
9	43.2	57
10	48.4	51
11	59.2	55
12	80.0	57
13	85.3	55
14	140.9	61

a.   Give the least squares prediction equation.

b.   Conduct a test (at $\alpha = .05$) to determine whether downward price adjustment y is linearly related to land sale ratio x.

E

	A	B	C	D	E	F	G
1	SUMMARY OUTPUT						
2							
3	*Regression Statistics*						
4	Multiple R	0.7319503					
5	R Square	0.53575125					
6	Adjusted R Square	0.49706385					
7	Standard Error	14.6154869					
8	Observations	14					
9							
10	ANOVA						
11		*df*	*SS*	*MS*	*F*	*Significance F*	
12	Regression	1	2958.150516	2958.151	13.84821	0.002919831	
13	Residual	12	2563.349484	213.6125			
14	Total	13	5521.5				
15							
16		*Coefficients*	*Standard Error*	*t Stat*	*P-value*	*Lower 95%*	*Upper 95%*
17	Intercept	25.7957004	5.556198606	4.642689	0.000568	13.68978383	37.901617
18	Land Sale Ratio	0.37098611	0.099692114	3.721319	0.00292	0.153775659	0.58819657

Problem 10.35

10.7 INFERENCES ABOUT THE CORRELATION COEFFICIENT (OPTIONAL)

In the previous section, we discovered that b_1, the least squares slope, provides useful information on the linear relationship between two variables y and x. You may recall that another way to measure this association is to compute the **correlation coefficient** r. The correlation coefficient, defined in Section 3.9 (page 190), provides a quantitative measure of the strength of the linear relationship between x and y in the sample, just as does the least squares slope b_1. The values of r and b_1 will always have the same sign (either both positive or both negative) in simple linear regression. However, unlike the slope, the correlation coefficient r is not expressed in units along a scale. (Recall that r is always between -1 and $+1$, no matter what the units of x and y are.)

Just as b_1 estimates the population parameter β_1 (the true slope of the line), r estimates the parameter that represents the coefficient of correlation for the population of all (x, y) data points. The **population correlation coefficient** is denoted by the Greek letter ρ (rho). A test of the null hypothesis that there is no linear association between x and y in the population is described in the box on page 586.

The next example illustrates the test for correlation. In addition, it demonstrates how the correlation coefficient r may be a misleading measure of the strength of the association between x and y in situations where the true relationship is nonlinear.

Example 10.15 Test for Zero Correlation

Underinflated or overinflated tires can increase tire wear and decrease gas mileage. A manufacturer of a new tire tested the tire for wear at different pressures with the results shown in Table 10.3.

a. Test the hypothesis of no correlation between mileage y and tire pressure x. Use $\alpha = .05$.

b. Interpret the result.

TIRES.XLS

Table 10.3 Data for Example 10.15

Pressure x (pounds per sq. inch)	Mileage y (thousands)	Pressure x (pounds per sq. inch)	Mileage y (thousands)
30	29.5	33	37.6
30	30.2	34	37.7
31	32.1	34	36.1
31	34.5	35	33.6
32	36.3	35	34.2
32	35.0	36	26.8
33	38.2	36	27.4

Test of Hypothesis for Linear Correlation

Upper-Tailed Test	Two-Tailed Test	Lower-Tailed Test
$H_0: \rho = 0$	$H_0: \rho = 0$	$H_0: \rho = 0$
$H_a: \rho > 0$	$H_a: \rho \neq 0$	$H_a: \rho < 0$

$$\text{Test statistic: } t = \frac{r}{\sqrt{\frac{1 - r^2}{n - 2}}} \tag{10.14}$$

where

$$r = \frac{SSxy}{\sqrt{(SSxx)(SSyy)}} \tag{10.15}$$

Rejection region:	Rejection region:	Rejection region:
$t > t_\alpha$	$|t| > t_{\alpha/2}$	$t < -t_\alpha$

where the distribution of t depends on $(n - 2)$ df, and $t_\alpha, t_{\alpha/2},$ and $-t_\alpha$ are the critical values obtained from Table B.3 in Appendix B.

Assumptions: The sample of (x, y) values is randomly selected from a (bivariate) normal population.*

*A bivariate normal population implies that the probability distributions of both x and y are normal.

Solution

a.   An Excel printout of the correlation analysis is shown in Figure 10.13 (page 587). The value of r, highlighted on Figures 10.13a and 10.13b, is $r = -.114$. This relatively small value for r describes a weak linear relationship between pressure x and mileage y.

The p-value for conducting the test $H_0: \rho = 0$ against $H_a: \rho \neq 0$, highlighted on Figure 10.13b, is .698. Since this p-value exceeds $\alpha = .05$, there is insufficient evidence to reject H_0 and conclude that pressure x and mileage y are linearly related.

b.   Although the test of part **a** was nonsignificant, the manufacturer would be remiss in concluding that tire pressure has little or no impact on wear of the tire. On the contrary, the relationship between pressure and wear is fairly strong, as the Excel scatterplot in Figure 10.14 (page 588) illustrates. The relationship shown is curvilinear; both underinflated tires (low pres-

E **Figure 10.13** Excel Printout of Correlation Analysis of Data in Table 10.3

	A	B	C	D
1	Pressure	Mileage		
2	30	29.5		
3	30	30.2		
4	31	32.1	Correlation	-0.114
5	31	34.5		
6	32	36.3		
7	32	35		
8	33	38.2		
9	33	37.6		
10	34	37.7		
11	34	36.1		
12	35	33.6		
13	35	34.2		
14	36	26.8		
15	36	27.4		

a.

	A	B
1	t test for the Existence of Correlation	
2		
3	Level of Significance	0.05
4	Sample Size	14
5	Sample Correlation Coefficient	-0.114
6	Degrees of Freedom	12
7	t Test Statistic	-0.3975
8		
9	Two-Tailed Test	
10	Lower Critical value	-2.17881
11	Upper Critical Value	2.178813
12	p-Value	0.697979
13	Do not reject the null hypothesis	

b.

sure values) and overinflated tires (high pressure values) lead to low mileages. ❑

Example 10.15 points out the danger of using r to determine how well x predicts y: The correlation coefficient r describes only the *linear* relationship between x and y. For nonlinear relationships, the value of r may be misleading and we must resort to other methods for describing and testing such a relationship. Regression models for curvilinear relationships are presented in optional Section 10.12.

We conclude this section by pointing out that the null hypothesis $H_0: \rho = 0$ is equivalent to the null hypothesis $H_0: \beta_1 = 0$. Consequently, a test for zero correlation in the population is equivalent to testing for a zero slope in a straight-line model. For example, when we tested the null hypothesis $H_0: \beta_1 = 0$ in the straight-line model relating creativity score y to flexibility score x (Example 10.12), the data led to a rejection of the hypothesis for $\alpha = .05$. This implies that the null

E Figure 10.14
Scatterplot of Data in
Table 10.3

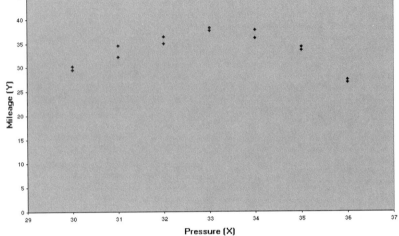

Scatter Diagram

hypothesis of a zero linear correlation between the two variables (creativity score and flexibility score) can also be rejected at $\alpha = .05$. The only real difference between the least squares slope b_1 and the coefficient of correlation r is the measurement scale. Therefore, the information they provide about the utility of the least squares model is redundant and either test can be used. If, however, you want to know how much y increases (or decreases) for every 1-unit increase in x, you need to construct a confidence interval for the slope β_1; the coefficient of correlation r does not provide this information.

Self-Test

10.7

A simple linear regression analysis on $n = 40$ (x, y) data points yielded the following results:

$$b_1 = 35.5 \qquad s_{b_1} = 10.0 \qquad r = .50$$

Show that the test statistic used to test $H_0: \rho = 0$ is equivalent to the test statistic used to test $H_0: \beta_1 = 0$.

PROBLEMS
10.36–10.45

Using the Tools

10.36 What value does r assume if all the sample points fall on the same straight line and if the line has:

a. A positive slope?   **b.** A negative slope?

10.37 For each of the following situations, specify the rejection region and state your conclusion for testing the null hypothesis H_0: There is no linear correlation between x and y:

a. H_a: The variables x and y are positively correlated; $r = .68, n = 10, \alpha = .01$

b. H_a: The variables x and y are negatively correlated; $r = -.68$, $n = 52$, $\alpha = .01$

c. H_a: The variables x and y are linearly correlated; $r = -.84, n = 10, \alpha = .10$

Applying the Concepts

10.38 Can teaching effectiveness be gauged from very brief observations of behavior? In an attempt to answer this question, Harvard psychologists N. Ambady and R. Rosenthal collected data on a random sample of 13 graduate student teachers (*Chance*, Fall 1997). Two variables were measured for each: (1) the teacher's end-of-semester average student evaluation rating (on a scale of 1 to 5); and (2) the average nonverbal behavior rating of the teacher (on a scale of 1 to 9) as judged by nine undergraduates who viewed short video clips of each teacher's behavior in the classroom.

a. The coefficient of correlation relating the two variables is $r = .76$. Interpret this result.

b. Use the result, part **a**, to calculate the test statistic for testing for a positive correlation between teacher effectiveness and nonverbal behavior.

c. Is there evidence (at $\alpha = .05$) of a positive linear relationship between teacher effectiveness and nonverbal behavior?

10.39 Refer to the *Small Group Behavior* (May 1988) study of the maturity levels of MIS students, Problem 3.79 (page 195). Recall that the maturity level y of each of 58 students was measured on a scale of 0–100, where more mature students received higher scores. The correlation coefficient relating y to the number x of meetings held with their groups outside of regular class sessions was found to be $r = .46$. Is this sufficient evidence to indicate a positive correlation between group maturity and outside-of-class meetings? Test using $\alpha = .01$.

10.40 In the business world, the term *Machiavellian* is often used to describe one who employs aggressive, manipulative, exploitative, and devious moves to achieve personal and corporate objectives. Do young marketing professionals tend to be more Machiavellian than older marketing professionals? To answer this question, a sample of 1,076 members of the American Marketing Association were administered a questionnaire that measured tendency toward Machiavellianism. (The higher the score, the greater the tendency toward Machiavellianism.) The sample correlation coefficient between age and Machiavellianism score was found to be $r = -.20$ (*Journal of Marketing,* Summer 1984). Is there evidence of a negative linear relationship between age of marketers and Machiavellianism score? Test using $\alpha = .05$.

10.41 Researchers at Boston College investigated the tactics used by sociology professors to learn students' names (*Teaching Sociology,* July 1995). Questionnaire data were obtained from a sample of 138 sociology professors. Two of the many variables measured were (1) the number of years the professor had been employed as a full-time faculty member and (2) the total number of name-learning tactics used by the professor. [*Note:* 18 tactics were listed on the questionnaire (e.g., name tags, seating chart) and the respondent circled those he or she used in the past two years.]

a. Set up the null and alternative hypotheses for testing whether years of full-time teaching are correlated with total number of name-learning tactics used.

b. The researchers reported that the test, part **a**, was nonsignificant at $\alpha = .10$, but did not report the value of the correlation coefficient, r. Show that r, for this problem, was less than .14.

10.42 Passive exposure to environmental tobacco has been associated with growth suppression and an increased frequency of respiratory tract infections in normal children. Is this association more pronounced in children with cystic fibrosis? To answer this question, 43 children (18 girls and 25 boys) attending a 2-week summer camp for cystic fibrosis patients were studied (*New England Journal of Medicine*, Sept. 20, 1990). Among several variables measured were the child's weight percentile y and the number of cigarettes smoked per day in the child's home x.

a. For the 18 girls, the coefficient of correlation between y and x was reported as $r = -.50$. Interpret this result.

b. Refer to part **a**. The p-value for testing $H_0: \rho = 0$ against $H_a: \rho \neq 0$ was reported as $p = .03$. Interpret this result.

c. For the 25 boys, the coefficient of correlation between y and x was reported as $r = -.12$. Interpret this result.

d. Refer to part **c**. The p-value for testing $H_0: \rho = 0$ against $H_a: \rho \neq 0$ was reported as $p = .57$. Interpret this result.

10.43 Refer to the *Journal of Literacy Research* (Dec. 1996) study of adult literacy, Problem 8.79 (page 467). Total literacy for each of the over 500 adults surveyed was measured by summing the scores of four separate literacy tests—Author Recognition, Magazine Recognition, Cultural Literacy, and Vocabulary Literacy Tests—administered over the telephone. Several other variables were measured in addition to total literacy, including years of education.

a. For the fathers who participated in the study, the correlation between total literacy and years of education was $r = .10$. A test for positive correlation resulted in a significance level of $p = .025$. Interpret the results.

b. For the mothers who participated in the study, the correlation between total literacy and years of education was found to be nonsignificant at $\alpha = .10$. Interpret this result.

c. Refer to part **b**. For mothers, the correlation between years of education and score on the Magazine Recognition test was $r = .18$ (with p-value $= .002$). Interpret this result.

10.44 A Global Positioning System (GPS) is a satellite system designed to locate wetland boundaries for regulatory purposes. Although the GPS can locate boundaries much faster than a traditional land survey, its accuracy varies with the number of satellites used in the system. Simple linear regression was used to examine the relationship between the accuracy of a GPS and the number of satellites used (*Wetlands*, Mar. 1995). Multiple GPS readings were collected for each of sixteen points located on a wetland boundary in South Carolina. The dependent variable is the distance y between the average GPS reading and the actual location (in meters).

The independent variable is the percentage x of readings obtained when three satellites were in use. The data and an output of the simple linear regression, generated by Excel, is shown here.

GPS.XLS

Survey Location	Distance Error y	Percent of Readings with Three Satellites x
1	7.8	94
2	4.2	83
3	4.5	23
4	2.8	58
5	1.5	78
6	3.0	100
7	1.4	100
8	14.9	100
9	0.4	28
10	18.9	100
11	5.8	75
12	0.6	2
13	30.9	100
14	30.3	100
15	31.4	100
16	7.5	53

Source: Hook, D. D., et al. "Locating delineated wetland boundaries in coastal South Carolina using Global Positioning Systems." *Wetlands,* Vol. 15, No. 1, Mar. 1995, p. 34 (Table 2).

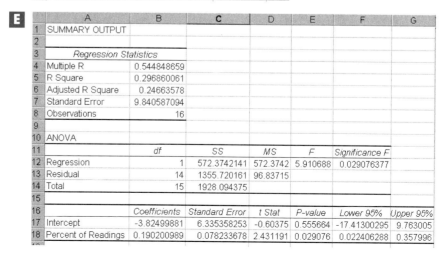

    a. Locate the coefficient of correlation r on the printout and interpret it.

    b. Locate the p-value for a test of correlation between distance error y and percent x of readings with three satellites. Is there evidence of a correlation between y and x in the population at $\alpha = .10$?

    c. Locate the p-value for a test of $H_0: \beta_1 = 0$ against $H_a: \beta_1 \neq 0$. Conduct the test at $\alpha = .10$.

    d. Do the results, parts **b** and **c**, agree?

10.45 In cotherapy two or more therapists lead a group. An article in the *American Journal of Dance Therapy* (Spring/Summer 1995) examined the use of cotherapy in dance/movement therapy. Two of several variables measured on each of a sample of 136 professional dance/movement therapists were years of formal training x and reported success rate y (measured as a percentage) of coleading dance/movement therapy groups.

    a. Propose a linear model relating y to x.

    b. The researcher hypothesized that dance/movement therapists with more years in formal dance training will report higher perceived success rates in cotherapy relationships. State the hypothesis in terms of the parameter of the model, part **a**.

    c. The correlation coefficient for the sample data was reported as $r = -.26$. Interpret this result.

    d. Does the value of r in part **c** support the hypothesis in part **b**? Test using $\alpha = .05$.

10.8   THE COEFFICIENT OF DETERMINATION

In this section, we define a numerical descriptive measure of how well the least squares line fits the sample data. This measure, called the *coefficient of determination,* is very useful for assessing how much the errors of prediction of y can be reduced by using the information provided by x. We develop this descriptive measure of model adequacy in the next three examples.

Example 10.16   **Best Predictor of y without x**

Refer to the creativity–flexibility data, Table 10.2 (page 561). Suppose you do not use x, the flexibility score, to predict y, the creativity score.

    a. If you have access to the sample of 10 mentally retarded children's creativity scores only, what quantity would you use as the best predictor for any y value?

    b. Use this predictor to compute the sum of squared errors.

Solution

    a. If we have no information on the distribution of the y values other than that provided by the sample, then $\bar{y}$, the sample average creativity score, would be the best predictor for *any* y value. From Table 10.2, we find $\bar{y} = 7.4$.

Figure 10.15 Errors of Prediction: Using $\bar{y}$ to Predict Creativity Score y

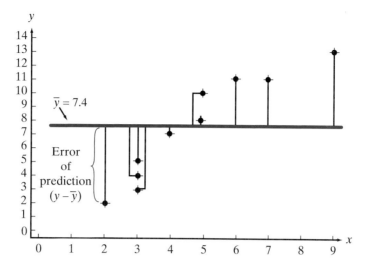

b. The errors of prediction using $\bar{y}$ as the predictor of y for the data of Table 10.2 are illustrated in Figure 10.15. With $\bar{y}$ as our predictor, the sum of squared errors (denoted SSTotal) is:

$$\text{SSTotal} = \sum(\text{Actual } y - \text{Predicted } y)^2 = \sum(y - \bar{y})^2$$
$$= (2 - 7.4)^2 + (11 - 7.4)^2 + \cdots + (13 - 7.4)^2 = 130.4$$

The magnitude of SSTotal is an indicator of how well $\bar{y}$ behaves as a predictor of y. It is also commonly described as the total variation of the sample y values around their mean. ❑

Example 10.17   Best Predictor of y Using x

Refer to Example 10.16. Suppose now that you use a straight-line model with flexibility score x to predict creativity score y. How do we measure the additional information provided by using the value of x in the least squares prediction equation, rather than $\bar{y}$, to predict y?

Solution

If we use the model with x to predict y, then $\hat{y}$ is our predicted value, where $\hat{y} = b_0 + b_1 x$. The errors of prediction using $\hat{y}$ for the data of Table 10.2 are illustrated in Figure 10.16 (page 594). Notice how much smaller these errors are relative to the errors shown in Figure 10.15.

We already know (from Example 10.10) that the sum of squared errors for the least squares line is

$$\text{SSE} = \sum(y - \hat{y})^2 = 13.3444$$

A convenient way of measuring how well the least squares equation performs as a predictor of y is to compare the sum of squared errors using $\bar{y}$ as a predictor (i.e., the total sample variation, SSTotal) to the sum of squared errors using $\hat{y}$ as a predictor (i.e., SSE). The difference between these, SSTotal $-$ SSE, represents the sum of squares *explained* by the linear regression between x and y, and is called

Figure 10.16 Errors of Prediction: Using the Least Squares Model

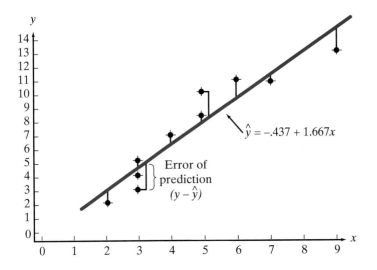

SSR. The ratio of SSR to SSTotal is called the *coefficient of determination,* and is denoted by the symbol R^2. ❑

Definition 10.3

The **coefficient of determination** is

$$R^2 = \frac{\text{SSTotal} - \text{SSE}}{\text{SSTotal}} = \frac{\text{SSR}}{\text{SSTotal}} \qquad (10.16)$$

where

$$\text{SSTotal} = \sum (y - \bar{y})^2$$
$$\text{SSE} = \sum (y - \hat{y})^2$$
$$\text{SSR} = \text{SSTotal} - \text{SSE}$$

and $0 \le R^2 \le 1$. R^2 represents the proportion of the sum of squares of deviations of the y values about their mean that can be attributed to a linear relationship between y and x. [*Note:* In simple linear regression, it may also be computed as the square of the coefficient of correlation, i.e., $R^2 = (r)^2$.]

Since r is between -1 and $+1$, the coefficient of determination R^2 will always be between 0 and 1. Thus, $R^2 = .75$ means that 75% of the sum of squares of deviations of the y values about their mean is attributable to the linear relationship between y and x. In other words, the error of prediction can be reduced by 75% when the least squares equation, rather than $\bar{y}$, is used to predict y.

Example 10.18   Computing R^2

Find the coefficient of determination for the flexibility–creativity score data and interpret its value. The data are repeated in Table 10.4 for convenience.

Table 10.4 Creativity–Flexibility Score Data

Child	Flexibility Score x	Creativity Score y
1	2	2
2	6	11
3	4	7
4	5	10
5	3	5
6	5	8
7	3	4
8	3	3
9	7	11
10	9	13

Solution

From previous examples, we have SSTotal = 130.4 and SSE = 13.3444. Applying equation 10.16, we obtain

$$\frac{\text{SSR}}{\text{SSTotal}} = \frac{(103.4 - 13.444)}{130.4} = \frac{117.0556}{130.4} = .8977$$

This value is also indicated on the Excel printout, Figure 10.12 (page 579), next to the cell labeled **R Square**. We interpret this value as follows: The use of flexibility score x to predict creativity score y with the least squares line

$$\hat{y} = -.44 + 1.667x$$

accounts for 89.8% of the total variation of the 10 sample creativity scores about their mean. That is, we can reduce the sum of squares errors by nearly 90% by using the least squares equation $\hat{y} = -.44 + 1.667x$, instead of $\bar{y}$, to predict y. ❏

Practical Interpretation of the Coefficient of Determination R^2

About $100(R^2)\%$ of the total variation in the sample y values about their mean $\bar{y}$ can be explained by (or attributed to) using x to predict y in the straight-line model.

Self-Test

10.8

The coefficient of determination for a straight-line model relating annual salary y to years of experience x for factory workers is $R^2 = .62$. Interpret this value.

Obviously, the coefficient of correlation r and the coefficient of determination R^2 are very closely related. Hence, there may be some confusion as to when each should be used. Our recommendations are as follows: If you are interested only in measuring the strength of the linear relationship between two variables x and y, use the coefficient of correlation r. However, if you want to determine how well the least squares straight-line model fits the data, use the coefficient of determination R^2.

**PROBLEMS
10.46–10.56**

Using the Tools

10.46 For a set of $n = 20$ data points, SSTotal $= 210$ and SSE $= 31$. Find R^2.

10.47 For a set of $n = 30$ data points, $\Sigma(y - \bar{y})^2 = 12$ and $\Sigma(y - \hat{y})^2 = 2.74$. Find R^2.

10.48 The coefficient of correlation between x and y is $r = .72$. Calculate the coefficient of determination R^2.

10.49 The coefficient of correlation between x and y is $r = -.35$. Calculate the coefficient of determination R^2.

Applying the Concepts

10.50 Refer to the simple linear regression relating drill time y and depth at which drilling begins x, Problem 10.8 (page 567). Find and interpret R^2.

DRILL.XLS

10.51 Refer to the simple linear regression relating peak EEG frequency y and age x, Problem 10.13 (page 570). Find and interpret the coefficient of determination R^2.

EEG.XLS

10.52 Refer to the simple linear regression relating Force Concept Inventory score y to Mechanics Baseline score x, Problem 10.14 (page 571). Find and interpret R^2.

PHYSICS.XLS

10.53 Refer to the simple linear regression relating O-ring damage y to flight time temperature x, Problem 10.28 (page 581). Find and interpret R^2.

ORING.XLS

10.54 Refer to the simple linear regression relating fertility rate y to contraceptive prevalence x, Problem 10.33 (page 582). Find and interpret R^2.

FERTIL.XLS

10.55 In Problem 10.38 (page 589), the coefficient of correlation between teacher effectiveness and nonverbal behavior rating was $r = .76$. Find the coefficient of determination R^2 for a straight-line model relating teacher effectiveness y to nonverbal behavior rating x. Interpret the result.

10.56 In Problem 10.41 (page 589), assume the coefficient of correlation between total number of name-learning techniques and years of full-time teaching was $r = .12$. Find the coefficient of determination R^2 for a straight-line model relating number of name-learning techniques y to years of teaching x. Interpret the result.

10.9 USING THE MODEL FOR ESTIMATION AND PREDICTION

After we have tested our straight-line model and are satisfied that x is a statistically useful linear predictor of y, we are ready to accomplish our original objective—using the model for prediction and estimation.

The most common uses of a regression model for making inferences can be divided into two categories, which are listed in the box.

Uses of the Regression Model for Making Inferences

1. Use the model for estimating $E(y)$, the mean value of y, for a specific value of x.
2. Use the model for predicting a particular y value for a given value of x.

In the first case, we want to estimate the mean value of y for a very large number of experiments at a given x value. For example, a psychologist may want to estimate the mean creativity score for *all* mentally retarded children with flexibility scores of 3. In the second case, we wish to predict the outcome of a single experiment (predict an individual value of y) at the given x value. For example, the psychologist may want to predict the creativity score of a particular mentally retarded child who scored 3 on the flexibility test.

In Section 10.3, we showed how to use the least squares line

$$\hat{y} = b_0 + b_1 x$$

to predict a particular value of y for a given x. The least squares line is also used to obtain $E(y)$, an estimate of the mean value of y, as demonstrated in the next example.

Example 10.19   **Point Estimator of $E(y)$**

Refer to Example 10.6 (page 562). We found the least squares model relating creativity score y to flexibility score x to be

$$\hat{y} = -.437 + 1.667x$$

Give a point estimate for the mean creativity score of all mentally retarded children that have a flexibility score of 5.

Solution

Just as we used $\hat{y}$ to predict an individual value of y (see Example 10.9, page 565), we use $\hat{y}$ to estimate $E(y)$ for a particular x value. When $x = 5$, we have

$$\hat{y} = -.437 + 1.667(5) = 7.90$$

Thus, the estimated mean creativity score for all mentally retarded children with flexibility score 5 is 7.90. ❑

Self-Test
10.9

The least squares line relating annual salary y to years of experience x for factory workers is $\hat{y} = 22,000 + 1,000x$.

a. Find a point estimate of y for a worker who has 10 years of experience.

b. Find a point estimate of $E(y)$ for all workers who have eight years of experience.

Since the least squares model is used to obtain both the estimator of $E(y)$ and the predictor of y, how do the two inferences differ? The difference lies in the accuracies with which the estimate and the prediction are made. These accuracies are best measured by the repeated *sampling errors* of the least squares line when it is used as an estimator and predictor, respectively. These sampling errors are given in the next box.

Sampling Errors for the Estimator of the Mean of *y* and the Predictor of an Individual *y*: Simple Linear Regression

1. The standard deviation of the sampling distribution of the estimator $\hat{y}$ of the mean value of y at a fixed x is

$$\sigma_{\hat{y}} = \sigma\sqrt{\frac{1}{n} + \frac{(x - \bar{x})^2}{\Sigma(x - \bar{x})^2}} \qquad (10.17)$$

where σ is the square root of σ^2, the measure of variability discussed in Section 10.4.

2. The standard deviation of the prediction error for the predictor $\hat{y}$ of an individual y value at a fixed x is

$$\sigma_{(y - \hat{y})} = \sigma\sqrt{1 + \frac{1}{n} + \frac{(x - \bar{x})^2}{\Sigma(x - \bar{x})^2}} \qquad (10.18)$$

where σ is the square root of σ^2, the measure of variability discussed in Section 10.4.

Since the true value of σ will rarely be known, we estimate σ by s. The sampling errors are then used in estimation and prediction intervals as shown below.

Estimation and Prediction Intervals in Simple Linear Regression

$(1 - \alpha)100\%$ Confidence Interval for $E(y)$ at a Fixed x:

$$\hat{y} \pm (t_{\alpha/2})s\sqrt{\frac{1}{n} + \frac{(x - \bar{x})^2}{\Sigma(x - \bar{x})^2}} \qquad (10.19)$$

$(1 - \alpha)100\%$ Prediction Interval for an Individual y at a Fixed x:

$$\hat{y} \pm (t_{\alpha/2})s\sqrt{1 + \frac{1}{n} + \frac{(x - \bar{x})^2}{\Sigma(x - \bar{x})^2}} \qquad (10.20)$$

Example 10.20   Confidence Interval for E(y)

Refer to the simple linear regression relating creativity score y to flexibility score x.

a. Find and interpret a 95% confidence interval for the mean creativity score of all mentally retarded children that have a flexibility score of 5.

b. Using a 95% prediction interval, predict the creativity score of a particular retarded child if his flexibility score is 5. Interpret the interval.

E Figure 10.17
Output from PHStat
Add-In for Excel 95%
Confidence Interval
for $E(y)$

	A	B
1	Regression Analysis for Creativity Scores	
2		
3	X Value	5
4	Confidence Level	95%
5	Sample Size	10
6	Degrees of Freedom	8
7	*t* Value	2.3060056
8	Sample Mean	4.7
9	Sum of Squared Difference	42.10
10	Standard Error of the Estimate	1.291531
11	*h* Statistic	0.1021378
12	Average Predicted Y (YHat)	7.9002375
13		
14		
15	For Average Predicted Y (YHat)	
16	Interval Half Width	0.9518278
17	Confidence Interval Lower Limit	6.9484098
18	Confidence Interval Upper Limit	8.8520653
19		
20	For Individual Response Y	
21	Interval Half Width	3.1266778
22	Prediction Interval Lower Limit	4.7735597
23	Prediction Interval Upper Limit	11.026915

95% CI for $E(y)$ (rows 17–18)

95% PI for y (rows 22–23)

Solution

a. We used the PHStat add-in for Microsoft Excel to obtain the 95% confidence interval. (See Section 10.10 for the calculations required to obtain the interval.) The 95% confidence interval for $E(y)$ when $x = 5$ is indicated on the printout, Figure 10.17. The interval is (6.95, 8.85). Hence, we are 95% confident that the mean creativity score $E(y)$ for all mentally retarded children with a flexibility score of $x = 5$ ranges from 6.95 to 8.85.

b. The 95% prediction interval is also indicated on the Excel printout, Figure 10.17. (Again, consult Section 10.10 for details on the calculations.) The interval is (4.77, 11.03). Thus, with 95% confidence, we predict that the creativity score for a retarded child with a flexibility score of 5 will fall between 4.77 and 11.03. ❑

In comparing the results in Example 10.20, it is important to note that the prediction interval for an individual creativity score y is wider than the corresponding confidence interval for the mean creativity score $E(y)$. By examining the formulas for the two intervals, you can see that this will always be true.

Additionally, over the range of the sample data, the widths of both intervals increase as the value of x gets farther from $\bar{x}$. (See Figure 10.18, page 600.) Thus, the more x deviates from $\bar{x}$, the less useful the interval will be in practice. In fact, when x is selected far enough away from $\bar{x}$ so that it falls outside the range of the sample data, it is dangerous to make any inferences about $E(y)$ or y, as the following box explains.

Caution

Using the least squares prediction equation to estimate the mean value of y or to predict a particular value of y for values of x that fall *outside the range* of the values of x contained in your sample data may lead to errors of estimation or prediction that are much larger than expected. Although the least squares model may provide a very good fit to the data over the range of x values contained in the sample, *it could give a poor representation of the true model for values of x outside this region.*

Figure 10.18 Comparison of Widths of 95% Confidence Interval and Prediction Interval

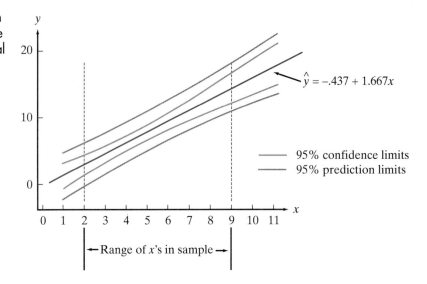

$\hat{y} = -.437 + 1.667x$

95% confidence limits
95% prediction limits

← Range of x's in sample →

Self-Test

10.10

Refer to Self-Test 10.9 (page 597) and the line relating annual salary y to years of experience x.

a. A 95% confidence interval for E(y) when x = 8 is (27,500, 32,500). Interpret this interval.

b. A 95% prediction interval for y when x = 8 is (23,000, 37,000). Interpret this interval.

**PROBLEMS
10.57–10.67**

Using the Tools

10.57 A simple linear regression analysis based on $n = 20$ data points produced the following results:

$$\hat{y} = 2.1 + 3.4x \qquad s = .475 \qquad \bar{x} = 2.5 \qquad \Sigma(x - \bar{x})^2 = 4.77$$

a. Obtain $\hat{y}$ when $x = 2.5$. What two quantities does $\hat{y}$ estimate?

b. Find a 95% confidence interval for $E(y)$ when $x = 2.5$. Interpret the interval.

c. Find a 95% confidence interval for $E(y)$ when $x = 2.0$. Interpret the interval.

d. Find a 95% confidence interval for $E(y)$ when $x = 3.0$. Interpret the interval.

e. Examine the widths of the confidence intervals obtained in parts **b, c,** and **d.** What happens to the width of the confidence interval for $E(y)$ as the value of x moves away from the value of $\bar{x}$?

f. Find a 95% prediction interval for a value of y to be observed in the future when $x = 2.5$. Interpret the interval.

g. Compare the width of the confidence interval from part **c** with the width of the prediction interval from part **f**. Explain why for a given value of x, a prediction interval for y is always wider than a confidence interval for $E(y)$.

10.58 In fitting a least squares line to $n = 22$ data points, the following results were obtained:

$$\hat{y} = 1.4 + .8x \qquad s = .22 \qquad \bar{x} = 2 \qquad \Sigma(x - \bar{x})^2 = 25$$

a. Obtain $\hat{y}$ when $x = 1$. What two quantities does $\hat{y}$ estimate?

b. Find a 95% confidence interval for $E(y)$ when $x = 1$.

c. Find a 95% prediction interval for y when $x = 1$.

10.59 A simple linear regression analysis on $n = 85$ data points yielded the following results:

$$\hat{y} = 67.6 - 14.3x \qquad s_{\hat{y}} = 3.5 \qquad s_{(y - \hat{y})} = 4.7$$

a. Estimate the mean value of y when $x = 2$, using a 90% confidence interval. Interpret the interval.

b. Suppose you plan to observe the value of y for a particular experimental unit with $x = 2$. Find a 90% prediction interval for the value of y that you will observe. Interpret the interval.

c. Which of the two intervals constructed in parts **a** and **b** is wider? Why?

10.60 A simple linear regression analysis on $n = 40$ data points yielded the following results:

$$\hat{y} = -1{,}867 + 45x \qquad s_{\hat{y}} = 104 \qquad s_{(y - \hat{y})} = 198$$

a. Estimate the mean value of y when $x = 100$, using a 95% confidence interval. Interpret the interval.

b. Suppose you plan to observe the value of y for a particular experimental unit with $x = 100$. Find a 95% prediction interval for the value of y that you will observe. Interpret the interval.

c. Which of the two intervals constructed in parts **a** and **b** is wider? Why?

E

	A	B
1	Regression Analysis for Refrigerators	
2		
3	X Value	90
4	Confidence Level	95%
5	Sample Size	9
6	Degrees of Freedom	7
7	t Value	2.3646226
8	Sample Mean	86.666667
9	Sum of Squared Difference	742.00
10	Standard Error of the Estimate	203.49634
11	h Statistic	0.1260857
12	Average Predicted Y (YHat)	1475.4268
13		
14		
15	For Average Predicted Y (YHat)	
16	Interval Half Width	170.86428
17	Confidence Interval Lower Limit	1304.5625
18	Confidence Interval Upper Limit	1646.2911
19		
20	For Individual Response Y	
21	Interval Half Width	510.62745
22	Prediction Interval Lower Limit	964.79933
23	Prediction Interval Upper Limit	1986.0542

Problem 10.61

Applying the Concepts

10.61 The PHStat add-in for Excel printout for Problem 10.7 (page 565) is given at left. Locate a 95% prediction interval for the retail price of a large side-by-side refrigerator that has an energy cost of $x = \$90$ per year. Interpret the interval.

DRILL.XLS

10.62 Refer to Problem 10.8 (page 566). Use the PHStat add-in for Excel to find a 95% prediction interval for drill time y when drilling begins at a depth of $x = 300$ feet. Interpret your result.

EEG.XLS

10.63 Refer to Problem 10.13 (page 570) and the simple linear regression relating average peak EEG frequency y to age x.

a. Use the PHStat add-in for Excel to find a 95% confidence interval for $E(y)$ when $x = 7$ and interpret it.

b. Use the PHStat add-in for Excel to find a 95% prediction interval for y when $x = 7$ and interpret it.

10.64 In forestry, the diameter of a tree at breast height (which is fairly easy to measure) is used to predict the height of the tree (a difficult measurement to obtain). Silviculturalists working in British Columbia's boreal forest attempted to predict the heights of several species of trees. The data in the accompanying table are the breast height diameters (in centimeters) and heights (in meters) for a sample of 36 white spruce trees.

FOREST1.XLS

Breast Height Diameter x (cm)	Height y (m)	Breast Height Diameter x (cm)	Height y (m)
18.9	20.0	16.6	18.8
15.5	16.8	15.5	16.9
19.4	20.2	13.7	16.3
20.0	20.0	27.5	21.4
29.8	20.2	20.3	19.2
19.8	18.0	22.9	19.8
20.3	17.8	14.1	18.5
20.0	19.2	10.1	12.1
22.0	22.3	5.8	8.0
23.6	18.9	20.7	17.4
14.8	13.3	17.8	18.4
22.7	20.6	11.4	17.3
18.5	19.0	14.4	16.6
21.5	19.2	13.4	12.9
14.8	16.1	17.8	17.5
17.7	19.9	20.7	19.4
21.0	20.4	13.3	15.5
15.9	17.6	22.9	19.2

Source: Scholz, H., Northern Lights College, British Columbia.

a. Construct a scatterplot for the data. Do you detect a trend?
b. A printout of a simple linear regression analysis of the data using Excel is shown below. Find the least squares line.
c. Plot the least squares line on your graph, part **a.**

	A	B	C	D	E	F	G
1	SUMMARY OUTPUT						
2							
3	*Regression Statistics*						
4	Multiple R	0.8105027					
5	R Square	0.6569146					
6	Adjusted R Square	0.6468239					
7	Standard Error	1.6777331					
8	Observations	36					
9							
10	ANOVA						
11		*df*	*SS*	*MS*	*F*	*Significance F*	
12	Regression	1	183.2446941	183.2447	65.1007	2.08926E-09	
13	Residual	34	95.70280593	2.814788			
14	Total	35	278.9475				
15							
16		*Coefficients*	*Standard Error*	*t Stat*	*P-value*	*Lower 95%*	*Upper 95%*
17	Intercept	9.146839	1.121313098	8.157257	1.63E-09	6.868058138	11.4256199
18	Breast Height Diameter	0.4814743	0.059673328	8.0685	2.09E-09	0.360203558	0.60274498

d. Do the data provide sufficient evidence to indicate that the breast height diameter x contributes information for the prediction of tree height y? Test using $\alpha = .05$.

e. The PHStat add-in for Excel printout is shown below. Locate a 95% confidence interval for the average height of white spruce trees with a breast height diameter of 20 cm on the printout. Interpret the interval.

	A	B
1	Regression Analysis for Tree Height	
2		
3	X Value	20
4	Confidence Level	95%
5	Sample Size	36
6	Degrees of Freedom	34
7	t Value	2.032243174
8	Sample Mean	18.19722222
9	Sum of Squared Difference	790.47
10	Standard Error of the Estimate	1.677733116
11	h Statistic	0.031889267
12	Average Predicted Y (YHat)	18.77632445
13		
14	For Average Predicted Y (YHat)	
15	Interval Half Width	0.608864732
16	Confidence Interval Lower Limit	18.16745972
17	Confidence Interval Upper Limit	19.38518918
18		
19	For Individual Response Y	
20	Interval Half Width	3.463499251
21	Prediction Interval Lower Limit	15.3128252
22	Prediction Interval Upper Limit	22.2398237

ORING.XLS

10.65 Refer to Problem 10.28 (page 581). On January 28, 1986, the day the space shuttle *Challenger* exploded and seven astronauts died, the launch temperature was 32°F.

a. Use the PHStat add-in for Excel to find a 95% prediction interval for the O-ring damage of a space shuttle when flight time temperature is 32 degrees. Interpret the interval.

b. Comment on the validity of the prediction interval, part **a**. (Consider the range of temperatures in the sample data used in the simple linear regression analysis.)

10.66 The PHStat add-in for Excel printout for Problem 10.33 (page 583) is shown at the top of page 604. Locate a 95% confidence interval for the mean fertility rate of all developing countries with a contraceptive prevalence of 30%. Interpret the interval.

10.67 The PHStat add-in for Excel printout for Problem 10.44 (page 591) is reproduced on page 604. Locate a 95% prediction interval for GPS distance error of a wetland when 100% of the readings are obtained with three satellites. Interpret the interval.

	A	B
1	**Regression Analysis for Fertility Rate**	
2		
3	**X Value**	**30**
4	**Confidence Level**	**95%**
5	Sample Size	27
6	Degrees of Freedom	25
7	*t* Value	2.05953711
8	Sample Mean	40.66666667
9	Sum of Squared Difference	12060.00
10	Standard Error of the Estimate	0.695710716
11	*h* Statistic	0.046471347
12	Average Predicted Y (YHat)	5.093620785
13		
14		
15	**For Average Predicted Y (YHat)**	
16	Interval Half Width	0.308880822
17	**Confidence Interval Lower Limit**	**4.784739963**
18	**Confidence Interval Upper Limit**	**5.402501607**
19		
20	**For Individual Response Y**	
21	Interval Half Width	1.465757027
22	**Prediction Interval Lower Limit**	**3.627863758**
23	**Prediction Interval Upper Limit**	**6.559377812**

Problem 10.66

	A	B
1	**Regression Analysis for GPS Distance Error**	
2		
3	**X Value**	**100**
4	**Confidence Level**	**95%**
5	Sample Size	16
6	Degrees of Freedom	14
7	*t* Value	2.144788596
8	Sample Mean	74.625
9	Sum of Squared Difference	15821.75
10	Standard Error of the Estimate	9.840587094
11	*h* Statistic	0.103196549
12	Average Predicted Y (YHat)	15.1951001
13		
14		
15	**For Average Predicted Y (YHat)**	
16	Interval Half Width	6.780131052
17	**Confidence Interval Lower Limit**	**8.414969047**
18	**Confidence Interval Upper Limit**	**21.97523115**
19		
20	**For Individual Response Y**	
21	Interval Half Width	22.16827746
22	**Prediction Interval Lower Limit**	**-6.973177359**
23	**Prediction Interval Upper Limit**	**37.36337756**

Problem 10.67

10.10 COMPUTATIONS IN SIMPLE LINEAR REGRESSION (OPTIONAL)

In our development of the simple linear (straight-line) model, we have primarily focused on using the output of software such as Microsoft Excel. In this optional section, we demonstrate the computations that were involved in obtaining the numerical results used in a simple linear regression analysis. Although many of the computing formulas were given earlier in the chapter, they are repeated here for convenience.

Consider, again, the simple linear regression analysis of the creativity–flexibility data, Table 10.2 (page 561). The computations involved in each step of the analysis are illustrated in the examples that follow. Recall that we want to model creativity score y as a straight-line function of flexibility score x. The initial step is to hypothesize the model

$$y = \beta_0 + \beta_1 x + \varepsilon$$

and the second step is to collect the sample data. We begin with the third step in the analysis.

Example 10.21   **Estimating β_0 and β_1 (Step 3)**

The formulas for computing estimates of the y-intercept β_0 and the slope β_1 were given in equations 10.4–10.7. These are repeated in the box. Apply these formulas to the creativity–flexibility data of Table 10.2.

Formulas for the Least Squares Estimates
___

$$Slope: \mathrm{b_1} = \frac{\mathrm{SSxy}}{\mathrm{SSxx}} \tag{10.4}$$

$$y\text{-}Intercept: \mathrm{b_0} = \bar{y} - \mathrm{b_1}\bar{x} \tag{10.5}$$

where

$$\mathrm{SSxy} = \sum xy - \frac{(\sum x)(\sum y)}{n} \tag{10.6}$$

$$\mathrm{SSxx} = \sum x^2 - \frac{(\sum x)^2}{n} \tag{10.7}$$

Solution

According to the formulas, we need to compute the sum of the x values ($\sum x$), the sum of the y values ($\sum y$), the sums of squares of the x values ($\sum x^2$), and the sums of squares of the cross-products of the x and y values ($\sum xy$). A table of these sums is shown in Table 10.5 (page 606). (We have also included $\sum y^2$ in the table. This quantity will be used in step 4.)

Substituting the values of $\sum x$, $\sum x^2$, and $\sum xy$ into the formulas for SSxy (equation 10.6) and SSxx (equation 10.7), we obtain

$$\mathrm{SSxy} = \sum xy - \frac{(\sum x)(\sum y)}{n} = 418 - \frac{(47)(74)}{10} = 70.2$$

$$\mathrm{SSxx} = \sum x^2 - \frac{(\sum x)^2}{n} = 263 - \frac{(47)^2}{10} = 42.1$$

Now we can compute $\mathrm{b_1}$ using equation 10.4:

$$\mathrm{b_1} = \frac{\mathrm{SSxy}}{\mathrm{SSxx}} = \frac{70.2}{42.1} = 1.667$$

Table 10.5 Sums of Squares for Data of Table 10.2

x	y	x^2	xy	y^2
2	2	4	4	4
6	11	36	66	121
4	7	16	28	49
5	10	25	50	100
3	5	9	15	25
5	8	25	40	64
3	4	9	12	16
3	3	9	9	9
7	11	49	77	121
9	13	81	117	169
TOTALS $\sum x = 47$	$\sum y = 74$	$\sum x^2 = 263$	$\sum xy = 418$	$\sum y^2 = 678$

To obtain b_0, we substitute the values of b_1, $\sum x$, and $\sum y$ into equation 10.5:

$$b_0 = \bar{y} - b_1\bar{x} = \left(\frac{\sum y}{n}\right) - b_1\left(\frac{\sum x}{n}\right)$$

$$= \frac{74}{10} - (1.667)\left(\frac{47}{10}\right) = -.437$$

Note that these estimates, $b_0 = -.437$ and $b_1 = 1.667$, agree (except for rounding) with the values shown on the Excel printout, Figure 10.4 (page 563). ❑

Self-Test

10.11

Consider the data listed in the table.

x	−1	0	1	2	3
y	−1	1	2	4	5

Find the least squares line relating y to x.

Example 10.22 **Computing SSE and s (Step 4)**

The formulas for computing SSE and an estimate of σ^2 are repeated in the next box. Apply these formulas to the creativity–flexibility data of Table 10.2. Also, compute an estimate of σ.

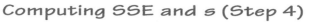

Estimation of σ^2, a Measure of the Variability of the y Values about the Least Squares Line

$$s^2 = \frac{SSE}{n-2} \qquad (10.8)$$

where

$$SSE = \sum(y - \hat{y})^2 = SSyy - b_1(SSxy) \qquad (10.9)$$

Solution

The formula for computing SSE is given in equation 10.9. Note that it requires the values of SSxy and b_1 (calculated in Example 10.21) and SSyy. To obtain SSyy, we use equation 10.10 and the values of $\sum y^2$ and $\sum y$ calculated in Table 10.5:

$$SSyy = \sum y^2 - \frac{(\sum y)^2}{n} = 678 - \frac{(74)^2}{10} = 130.4$$

Substituting SSyy, SSxy, and b_1 into equation 10.9, we obtain the value of SSE:

$$SSE = SSyy - b_1 SSxy = 130.4 - (1.667)(70.2) = 13.37$$

Except for rounding, this value agrees with the value of SSE shown in the Excel printout, Figure 10.4 (page 563).

To obtain s^2, the estimate of σ^2, we use equation 10.8:

$$s^2 = \frac{SSE}{n-2} = \frac{13.37}{8} = 1.67$$

Then s, the estimate of σ, is

$$s = \sqrt{s^2} = \sqrt{1.67} = 1.29$$

Again, except for rounding, both $s^2 = 1.67$ and $s = 1.29$ match the values shown in the Excel printout, Figure 10.4. ❏

Self-Test

10.12

Refer to the data, Self-Test 10.11. Compute SSE and s for the simple linear regression.

Example 10.23

Testing the Slope (Step 5)

Refer to the creativity–flexibility data. Apply equation 10.12 (repeated below) to obtain the test statistic for testing $H_0: \beta_1 = 0$.

Computing the Test Statistic for Testing $H_0: \beta_1 = 0$ in Simple Linear Regression

$$t = \frac{b_1}{s/\sqrt{SSxx}} \qquad \text{(10.12)}$$

Solution

To compute the test statistic t we require the values of b_1, s, and SSxx, computed previously. Substitution into equation 10.12 yields

$$t = \frac{b_1}{s/\sqrt{SSxx}} = \frac{1.667}{1.29/\sqrt{42.1}} = \frac{1.667}{.199} = 8.37$$

This value agrees (except for rounding) with that computed by Excel in Figure 10.4 (page 563). ❏

Self-Test

10.13

Refer to Self-Tests 10.11 and 10.12. Compute the test statistic for testing $H_0: \beta_1 = 0$.

Example 10.24   Confidence Interval for the Slope (Step 5)

Construct a 95% confidence interval for the slope of the line relating creativity score y to flexibility score x.

Solution

The formula for computing the confidence interval is repeated in the next box.

Computing a Confidence Interval for β_1 in Simple Linear Regression

$$b_1 \pm (t_{\alpha/2})\frac{s}{\sqrt{SS_{xx}}} \tag{10.13}$$

where $t_{\alpha/2}$ is based on $(n - 2)$ degrees of freedom.

From previous examples, we found

$$b_1 = 1.667 \qquad s = 1.29 \qquad SS_{xx} = 42.1$$

For a 95% confidence interval, $\alpha = .05$ and $\alpha/2 = .025$. We need to find the value of $t_{.025}$ in Table B.3 in Appendix B. For $n - 2 = 8$ df, $t_{.025} = 2.306$. Substituting these values into equation 10.13, we obtain

$$b_1 \pm (t_{\alpha/2})\frac{s}{\sqrt{SS_{xx}}} = 1.667 \pm (2.306)\left(\frac{1.29}{\sqrt{42.1}}\right)$$
$$= 1.667 \pm .459$$
$$= (1.208, 2.126) \qquad \Box$$

Self-Test

10.14

Refer to Self-Tests 10.11–10.13. Compute a 95% confidence interval for β_1.

Example 10.25   Computing R^2 (Step 5)

Compute the coefficient of determination R^2 for the straight-line model relating creativity score y to flexibility score x.

Solution

The formula for computing R^2 is repeated in the next box.

Computing R^2 in Simple Linear Regression

*SSyy = SSTotal

$$R^2 = \frac{SSyy - SSE}{SSyy} = \frac{SSR}{SSyy} \qquad \textbf{(10.16)}^{*}$$

where $SSyy = \Sigma(y - \bar{y})^2$

From previous examples, we found SSE = 13.37 and SSyy = 130.4. Substituting these values into equation 10.16, we obtain

$$R^2 = \frac{130.4 - 13.37}{130.4} = \frac{117.03}{130.4} = .897 \qquad \square$$

Self-Test

10.15

Refer to Self-Tests 10.11–10.14. Compute R^2.

Step 6 (check model assumptions) is covered in the next section. We continue with Step 7 in the next example.

Example 10.26

Estimation and Prediction Computations (Step 7)

Refer to the straight-line model relating creativity score y to flexibility score x. Use the data in Table 10.2 to compute

a.   a 95% confidence interval for $E(y)$ when $x = 5$

b.   a 95% prediction interval for y when $x = 5$

Solution

As we demonstrated in Example 10.19 (page 597), the point estimate for both y and $E(y)$ is $\hat{y}$, where $\hat{y}$ is obtained by substituting $x = 5$ into the prediction equation:

$$\hat{y} = b_0 + b_1 x = -.437 + 1.667(5) = 7.90$$

The formulas for computing the confidence interval for $E(y)$ and the prediction interval for y are repeated in the box below.

Computing a Confidence Interval for E(y) or a Prediction Interval for y for x = x_p

$(1 - \alpha)100\%$ Confidence Interval for $E(y)$:

† SSxx = $\Sigma(x - \bar{x})^2$

$$\hat{y} \pm (t_{\alpha/2})s\sqrt{\frac{1}{n} + \frac{(x_p - \bar{x})^2}{SSxx}} \qquad \textbf{(10.19)}^{\dagger}$$

continued

*SSxx $= \Sigma(x - \bar{x})^2$

(1 − α)100% Prediction Interval for an Individual y:

$$\hat{y} \pm (t_{\alpha/2})s\sqrt{1 + \frac{1}{n} + \frac{(x_p - \bar{x})^2}{\text{SSxx}}} \qquad (10.20)^*$$

where x_p is the value of x used to predict y or estimate $E(y)$, and $t_{\alpha/2}$ is based on $(n - 2)$ degrees of freedom.

From previous examples, we found $n = 10$, $s = 1.29$, $\bar{x} = 4.7$, SSxx $= 42.1$, and $t_{\alpha/2} = t_{.025} = 2.306$. The value of x used to predict y, denoted x_p, is $x_p = 5$. Now we substitute these values into the formulas:

a.   95% confidence interval for $E(y)$:

$$\hat{y} \pm (t_{\alpha/2})s\sqrt{\frac{1}{n} + \frac{(x_p - \bar{x})^2}{\text{SSxx}}}$$

$$= 7.90 \pm (2.306)(1.29)\sqrt{\frac{1}{10} + \frac{(5 - 4.7)^2}{42.1}}$$

$$= 7.90 \pm .951$$

$$= (6.95, 8.85)$$

b.   95% prediction interval for y:

$$\hat{y} \pm (t_{\alpha/2})s\sqrt{1 + \frac{1}{n} + \frac{(x_p - \bar{x})^2}{\text{SSxx}}}$$

$$= 7.90 \pm (2.306)(1.29)\sqrt{1 + \frac{1}{10} + \frac{(5 - 4.7)^2}{42.1}}$$

$$= 7.90 \pm 3.12$$

$$= (4.78, 11.02)$$

Except for rounding, these intervals agree with those obtained from the PHStat add-in for Excel in Example 10.20 (page 599). ❑

**Self-Test
10.16**

Refer to Self-Tests 10.11–10.15.

a.   Find a 95% prediction interval for y when $x = 0$.
b.   Find a 95% confidence interval for $E(y)$ when $x = 2$.

**PROBLEMS
10.68–10.72**

x	−2	0	1	4	3
y	−1	3	5	9	5

Problem 10.68

Using the Tools

10.68   Consider the data listed in the table. Compute each of the following.

a.   SSxy   b.   SSxx   c. $\bar{y}$ d. $\bar{x}$ e. b_1

f. b_0 g.   SSyy   h.   SSE   i. s j. s_{b_1}

k. t for testing $H_0: \beta_1 = 0$ **l.**   a 95% confidence interval for β_1

m. R^2 **n.** $\hat{y}$ when $x = 2$

o.   a 95% prediction interval for y when $x = 2$

10.69   Consider the five data points:

x	−1	0	1	2	3
y	−1	1	1	2.5	3.5

a.   Construct a scatterplot for the data.

b.   Use the computing formulas to find the least squares prediction equation.

c.   Graph the least squares line on the scatterplot and visually confirm that it provides a good fit to the data points.

d.   Compute the test statistic for testing $H_0: \beta_1 = 0$ against $H_a: \beta_1 > 0$.

e.   Give the conclusion for the test of part **d**.

f.   Compute and interpret R^2.

g.   Compute a 95% confidence interval for $E(y)$ when $x = 0$. Interpret the result.

x	1	1.5	1.9	2.5
y	3.1	2.2	1.0	.3

Problem 10.70

10.70   Consider the four data points shown in the table at left.

a.   Construct a scatterplot for the data.

b.   Use the computing formulas to find the least squares prediction equation.

c.   Graph the least squares line on the scatterplot and visually confirm that it provides a good fit to the data points.

d.   Compute the test statistic for testing $H_0: \beta_1 = 0$ against $H_a: \beta_1 < 0$.

e.   Give the conclusion for the test of part **d**.

f.   Compute and interpret R^2.

g.   Compute a 95% confidence interval for $E(y)$ when $x = 1.5$. Interpret the result.

10.71   Consider the following seven data points:

x	−5	−3	−1	0	1	3	5
y	.8	1.1	2.5	3.1	5.0	4.7	6.2

a.   Construct a scatterplot for the data.

b.   Using the computing formulas, find the least squares prediction equation.

c.   Graph the least squares line on the scatterplot and visually confirm that it provides a good fit to the data points.

d.   Compute the test statistic for testing $H_0: \beta_1 = 0$ against $H_a: \beta_1 \neq 0$.

e.   Give the conclusion for the test of part **d**.

f.   Compute and interpret R^2.

g.   Compute a 90% prediction interval for y when $x = -4$. Interpret the result.

10.72   Consider the four data points in the accompanying table.

x	−3.0	2.4	−1.1	2.0
y	2.7	.4	1.3	.5

a. Construct a scatterplot for the data.

b. Using the computing formulas, find the least squares prediction equation.

c. Graph the least squares line on the scatterplot and visually confirm that it provides a good fit to the data points.

d. Compute the test statistic for testing $H_0: \beta_1 = 0$ against $H_a: \beta_1 \neq 0$.

e. Give the conclusion for the test of part **d**.

f. Compute and interpret R^2.

g. Compute a 90% confidence interval for $E(y)$ when $x = 1.0$. Interpret the result.

10.11 RESIDUAL ANALYSIS: CHECKING THE ASSUMPTIONS (OPTIONAL)

In Section 10.4 we listed four assumptions about the random error ε in the model required in a simple linear regression analysis. For convenience, they are repeated here.

Assumptions Required in Regression

1. Mean of ε is 0 (i.e., $E(\varepsilon) = 0$)
2. Variance of ε is constant for all values of x (i.e., $\text{Var}(\varepsilon) = \sigma^2$)
3. Probability distribution of ε is normal
4. ε's are independent

It is unlikely that these assumptions are ever satisfied exactly in a practical application of regression analysis. Fortunately, experience has shown that least squares regression produces reliable statistical tests, confidence intervals, and prediction intervals as long as departures from the assumptions are not too great. In this section, we present some methods for determining whether the data indicate significant departures from the assumptions.

The methods of this section are based on an analysis of *residuals*.

Definition 10.4

A regression **residual** is defined as the difference between an observed y value and its corresponding predicted value:

$$\text{Residual} = (y - \hat{y}) \tag{10.21}$$

For each assumption, we show how a graph of the residuals can be used to detect violations and we provide some guidance on how to remedy the problem.

Mean Error of Zero

The assumption that the average of the errors is 0 will be violated if an inappropriate regression model is used. For example, if you hypothesize a straight-line relationship between y and x, when the true relationship is more complex (e.g., curvilinear), then you have misspecified the model and this assumption will be violated.

One way to detect model misspecification is to plot the value of each residual versus the corresponding value of the independent variable, x. (If the model contains more than one independent variable, a plot would be constructed for each of the independent variables.) If the plot shows a random pattern of residuals, the assumption is likely to be satisfied. However, if the plot reveals a strong pattern, the assumption is violated. The following example illustrates this method.

Example 10.27   Detecting Model Misspecification

Fit the straight-line model $y = \beta_0 + \beta_1 x + \varepsilon$ to the data shown in Table 10.6. Then calculate the residuals, plot them versus x, and analyze the plot.

Solution

We obtained the least squares equation for the data using Excel. The Excel printout for the simple linear regression is shown in Figure 10.19a. You can see that the resulting prediction equation is

$$\hat{y} = 3.167 + 1.0952x$$

Substituting each value of x into this prediction equation, we can calculate $\hat{y}$ and the corresponding residual $(y - \hat{y})$. The predicted value $\hat{y}$ and the residual $(y - \hat{y})$ are shown in Figure 10.19b for each of the data points.

Table 10.6 Data for Example 10.27

x	y	x	y
0	1	4	9
1	4	5	10
2	6	6	10
3	8	7	8

E Figure 10.19 Excel Printout for the Simple Linear Model, Example 10.27

a.

b.

E **Figure 10.20** Excel
Residual Plot for Example
10.27

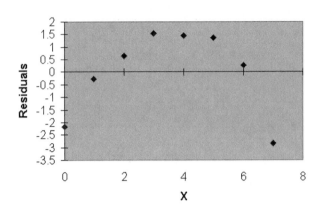

A plot of these residuals versus the independent variable x is shown in Figure 10.20. Instead of varying in a random pattern as x increases, the values of the residuals cycle from negative to positive to negative. This cyclical behavior is caused because we have fit a straight-line model to data for which a more complex relationship is appropriate—that is, we have misspecified the model.

Figure 10.20 shows why fitting the wrong model to a set of data can produce patterns in the residuals when they are plotted versus an independent variable. For this example, the nonrandom (in this case, curvilinear) behavior of the residuals can be eliminated by fitting the curvilinear model

$$y = \beta_0 + \beta_1 x + \beta_2 x^2 + \varepsilon$$

to the data. In general, certain patterns in the values of the residuals may suggest a need to modify the regression model, but the exact change that is needed may not always be obvious. We consider more complex models (such as the curvilinear model above) in optional Section 10.12. ❑

Constant Error Variance

To detect a violation of assumption #2, that the variation in the error is constant for all x, we plot the regression residuals against $\hat{y}$, the predicted value of y. For example, a plot of the residuals versus the predicted value $\hat{y}$ may display a pattern as shown in Figure 10.21. In the figure, the range in values of the residuals increases as $\hat{y}$ increases, indicating that the variance of the random error ε becomes larger as $\hat{y}$ increases in value. Since according to the model, increasing x will increase (or decrease) $\hat{y}$, this implies that the variance of ε is *not* constant for all x—a violation of assumption #2.

Figure 10.21 Residual Plot
Showing Changes in the
Variance of ε

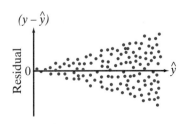

Residual plots of the type shown in Figure 10.21 are not uncommon because the variance of y often depends on the mean value of y. Dependent variables that represent counts per unit of area, volume, time, etc. (e.g., Poisson random variables—see Section 5.4) are cases in point. Other types of dependent variables that tend to violate this assumption are business and economic data (e.g., prices, salaries) and proportions or percentages.

Example 10.28   Detecting a Nonconstant Error Variance

The data in Table 10.7 are the salaries y and years of experience x for a sample of 50 auditors. The straight-line model was fit to the data using Excel. The Excel printout is shown in Figure 10.22, followed by an Excel plot of the residuals versus $\hat{y}$ in Figure 10.23 (page 616). Interpret the results.

Table 10.7 Salary Data for Example 10.28

AUDITOR.XLS

Years of Experience x	Salary y	Years of Experience x	Salary y	Years of Experience x	Salary y
7	$26,075	21	$43,628	28	$99,139
28	79,370	4	16,105	23	52,624
23	65,726	24	65,644	17	50,594
18	41,983	20	63,022	25	53,272
19	62,308	20	47,780	26	65,343
15	41,154	15	38,853	19	46,216
24	53,610	25	66,537	16	54,288
13	33,697	25	67,447	3	20,844
2	22,444	28	64,785	12	32,586
8	32,562	26	61,581	23	71,235
20	43,076	27	70,678	20	36,530
21	56,000	20	51,301	19	52,745
18	58,667	18	39,346	27	67,282
7	22,210	1	24,833	25	80,931
2	20,521	26	65,929	12	32,303
18	49,727	20	41,721	11	38,371
11	33,233	26	82,641		

Figure 10.22 Excel Output for Example 10.28

	A	B	C	D	E	F	G
1	SUMMARY OUTPUT						
2							
3	*Regression Statistics*						
4	Multiple R	0.887079601					
5	R Square	0.786910218					
6	Adjusted R Square	0.782470848					
7	Standard Error	8642.441396					
8	Observations	50					
9							
10	ANOVA						
11		*df*	*SS*	*MS*	*F*	*Significance F*	
12	Regression	1	13239655469	1.32E+10	177.2572	9.8691E-18	
13	Residual	48	3585206077	74691793			
14	Total	49	16824861546				
15							
16		*Coefficients*	*Standard Error*	*t Stat*	*P-value*	*Lower 95%*	*Upper 95%*
17	Intercept	11368.72114	3160.316978	3.597336	0.000758	5014.4817	17722.96057
18	Years	2141.380732	160.8392325	13.3138	9.87E-18	1817.99197	2464.769494

E **Figure 10.23**
Excel Residual Plot
for the Data of
Example 10.28

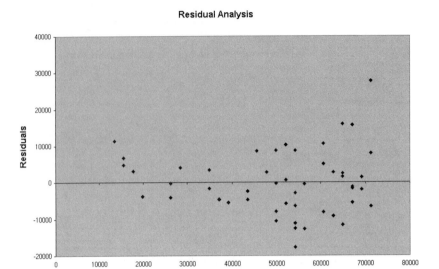

Solution

The Excel printout, Figure 10.22, suggests that the straight-line model provides an adequate fit to the data. The R^2 value indicates that the model explains 78.7% of the sample variation in salaries. The t value for testing β_1, 13.31, is highly significant (p-value ≈ 0) and indicates that the model contributes information for the prediction of y. However, an examination of the residuals plotted against $\hat{y}$ (Figure 10.23) reveals a potential problem. Note the "cone" shape of the residual variability; the size of the residuals increases as the estimated mean salary increases.

This residual plot strongly suggests that a nonconstant error variance exists. Thus, assumption #2 appears to be violated. ❑

Statisticians have found that certain transformations on the dependent variable y can stabilize the error variance. For example, when the data are prices or salaries (as in Example 10.28), the error variance will remain constant if the dependent variable in the model is expressed as the natural logarithm of y [i.e., $\log(y)$]. Similarly, with Poisson data, the error variance can be stabilized by using the square root of y (i.e., $\sqrt{y}$) as the dependent variable. The transformations $\log(y)$ and $\sqrt{y}$ are called **variance-stabilizing transformations** on the dependent variable y because they lead to regression models that satisfy the assumption of a constant error variance.

Normal Errors

Of the four regression assumptions, the assumption that the random error is normally distributed is the least restrictive when we apply regression analysis in practice. That is, moderate departures from the assumption of normality have very little effect on the validity of the statistical tests, confidence intervals, and prediction intervals. In this case we say that regression is **robust** with respect to nonnormality. However, great departures from normality cast doubt on any inferences derived from the regression analysis.

The simplest way to determine whether the data grossly violate the assumption of normality is to construct either a histogram, a stem-and-leaf display, or a normal probability plot of the residuals, as illustrated in the next example.

E **Figure 10.24**
Excel Histogram of the
Regression Residuals,
Example 10.29

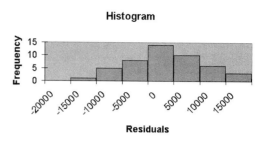

Histogram

Example **10.29**   **Checking for Normal Errors**

Refer to the straight-line model relating salary to experience, fit in Example 10.28.
A histogram of the residuals of the model, generated by Excel, is shown in Figure
10.24. Interpret the plot.

Solution

You can see from Figure 10.24 that the distribution of the residuals is mound-
shaped and reasonably symmetric about 0. Consequently, it is unlikely that the
normality assumption would be violated using these data.

When gross nonnormality of the random error term is detected, it can often be
rectified by applying a transformation on the dependent variable. For example, if
the relative frequency distribution of the residuals is highly skewed to the right (as
it is for Poisson data), the square-root transformation on y will stabilize (approximately) the variance and, at the same time, will reduce skewness in the distribution of residuals. Nonnormality may also be due to outliers, discussed next. ❏

Outliers

Residual plots can also be used to detect *outliers,* values of y that appear to be in
disagreement with the model. Since almost all values of y should lie within 3σ of
$E(y)$, the mean value of y, we would expect most of them to lie within $3s$ of $\hat{y}$. If a
residual is larger than $3s$ (in absolute value), we consider it an outlier and seek
background information that might explain the reason for its large value.

Definition 10.5

A residual that is larger than $3s$ (in absolute value) is considered to be an
outlier.

To detect outliers we can construct horizontal lines located a distance of $3s$
above and below 0 (see Figure 10.25 on page 618) on a residual plot. Any residual
falling outside the band formed by these lines would be considered an outlier. We
would then initiate an investigation to seek the cause of the departure of such
observations from expected behavior.

Although some analysts advocate eliminating outliers, regardless of whether
cause can be assigned, others encourage correcting only those outliers that can be
traced to specific causes. The best philosophy is probably a compromise between
these extremes. For example, before deciding the fate of an outlier, you may want
to determine how much influence it has on the regression analysis. When an accurate outlier (i.e., an outlier that is not due to recording or measurement error) is

Figure 10.25 3s Lines
Used to Locate Outliers

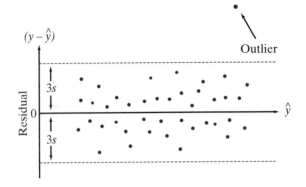

found to have a dramatic effect on the regression analysis, it may be the model
and not the outlier that is suspect. Omission of important variables could be the
reason why the model is not predicting well for the outlying observation. Several
sophisticated numerical techniques are available for identifying outlying influential observations. Consult the references given at the end of this chapter (e.g.,
Belsley, Kuh, and Welsch [1980], Berenson and Levine [1999], Mendenhall and
Sincich [1996]) for a discussion of these techniques.

Independent Errors

The assumption that the random errors are independent (uncorrelated) is most
often violated when the data employed in a regression analysis are a **time series**.
With time series data, the experimental units in the sample are time periods (e.g.,
years, months, or days) in consecutive time order.

For most business and economic time series, there is a tendency for the regression residuals to have positive and negative runs over time. For example, consider
fitting a straight-line regression model to yearly time series data. The model takes
the form

$$y = \beta_0 + \beta_1 t + \varepsilon$$

where y is the value of the time series in year t. A plot of the yearly residuals may
appear as shown in Figure 10.26. Note that if the residual for year t is positive (or
negative), there is a tendency for the residual for year $(t + 1)$ to be positive (or
negative). That is, neighboring residuals tend to have the same sign and appear to
be correlated. Thus, the assumption of independent errors is likely to be violated,
and any inferences derived from the model are suspect.

Consult the references at the end of this chapter (e.g., Neter et al. [1996]) to
learn how to fit and apply time series models.

Figure 10.26 Residual Plot
for Yearly Time Series Model

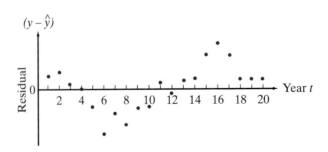

A Summary of Steps to Follow in a Residual Analysis

1. Check for a *misspecified model* by plotting the residuals $(y - \hat{y})$ against each independent variable in the model. A curvilinear trend detected in a plot implies that a quadratic term for that particular x variable will probably improve model adequacy.

2. Check for *unequal variances* by plotting the residuals against the predicted values $\hat{y}$. If you detect a cone-shape pattern, refit the model using an appropriate variance-stabilizing transformation on y.

3. Check for *nonnormal errors* by constructing a stem-and-leaf display, histogram, or normal probability plot of the residuals. If you detect extreme skewness in the data, look for one or more outliers.

4. Check for *outliers* by locating residuals that lie a distance of $3s$ or more above or below 0 on a residual plot versus $\hat{y}$. Before eliminating an outlier from the analysis, you should conduct an investigation to determine its cause. If the outlier is found to be the result of a coding or recording error, fix it or remove it. Otherwise, you may want to determine how influential the outlier is before deciding its fate.

5. Check for *correlated errors* by plotting the residuals in time order. If you detect runs of positive and negative residuals, propose a time series model to account for the residual correlation.

PROBLEMS
10.73–10.83

Using the Tools

10.73   Identify the problem(s) with each of the following residual plots:

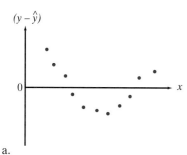

a.

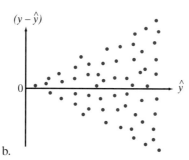

b.

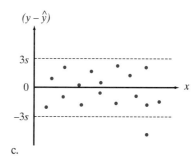

c.

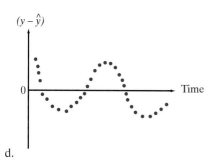

d.

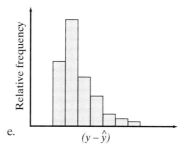

e.

10.74 A straight-line model is fit to the data shown in the table, with the following results: $\hat{y} = 2.588 + .541x, s = .356$.

x	-2	-2	-1	-1	0	0	1	1	2	2	3	3
y	1.1	1.3	2.0	2.1	2.7	2.8	3.4	3.6	4.0	3.9	3.8	3.6

   a. Calculate the residuals for the model.

   b. Plot the residuals versus x. Do you detect any trends? If so, what does the pattern suggest about the model?

   c. Plot the residuals versus $\hat{y}$. Identify any outliers on the plot.

   d. Refer to the residual plot constructed in part **c**. Do you detect any trends? If so, what does the pattern suggest about the model?

10.75 A straight-line model is fit to the data shown in the table, with the following results: $\hat{y} = -3.179 + 2.491x, s = 4.154$.

x	2	4	7	10	12	15	18	20	21	25
y	5	10	12	22	25	27	39	50	47	65

   a. Calculate the residuals for the model.

   b. Plot the residuals versus x. Do you detect any trends? If so, what does the pattern suggest about the model?

   c. Plot the residuals versus $\hat{y}$. Identify any outliers on the plot.

   d. Refer to the residual plot constructed in part **c**. Do you detect any trends? If so, what does the pattern suggest about the model?

Applying the Concepts

10.76 Refer to the *American Journal of Psychiatry* simple linear regression relating verbal memory retention y to right hippocampal volume of the brain x, Problem 10.10 (page 568). Several Excel residual plots are shown on page 621. Interpret these plots. Do the least squares assumptions appear to be satisfied?

PHYSICS.XLS

10.77 Refer to the *American Journal of Physics* simple linear regression relating inventory score y to baseline score x for 27 physics students, Problem 10.14 (page 571). Use Excel to conduct a residual analysis.

LANDSALE.XLS

10.78 Refer to Problem 10.35 (page 584) and the simple linear regression relating downward price adjustment y and land sale ratio x. The real estate appraisal company has been advised that a curvilinear relationship exists between y and x and, therefore, that the straight-line model is inappropriate for predicting price adjustment. Plot the residuals of the straight-line model against x. Does the plot support this claim?

10.79 Refer to *The New England Journal of Medicine* study of passive exposure to environmental tobacco smoke in children with cystic fibrosis, Problem 10.42 (page 590). Recall that the researchers investigated the correlation between a child's weight percentile y and the number of cigarettes smoked per day in the child's home x. The table on page 621 lists the data for the 25 boys.

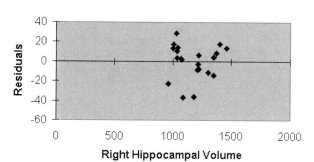

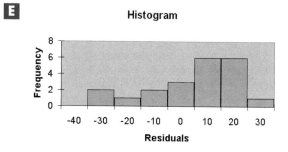

Problem 10.76

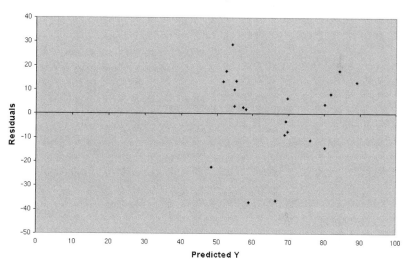

	Number of Cigarettes		Number of Cigarettes
Weight Percentile y	Smoked per Day x	Weight Percentile y	Smoked per Day x
6	0	43	0
6	15	49	0
2	40	50	0
8	23	49	22
11	20	46	30
17	7	54	0
24	3	58	0
25	0	62	0
17	25	66	0
25	20	66	23
25	15	83	0
31	23	87	44
35	10		

Source: Rubin, B. K. "Exposure of children with cystic fibrosis to environmental tobacco smoke." *The New England Journal of Medicine,* Sept. 20, 1990, Vol. 323, No. 12, p. 785 (data extracted from Figure 3).

Problem 10.79

a.   Use Excel to fit the straight-line model relating y to x and obtain the residuals.

b.   Verify that the sum of the residuals is 0.

c.   Use Excel to plot the residuals against the number of cigarettes smoked per day x.

d.   Do you detect any patterns, trends, or unusual observations on the plot, part **c**?

e.   Suggest a remedy for any problems identified in part **d**.

10.80   Breakdowns of machines that produce steel cans are very costly. The more breakdowns, the fewer cans produced, and the smaller the company's profits. To help anticipate profit loss, the owners of a can company would like to find a model that will predict the number of breakdowns on the assembly line. One model proposed by the company's statisticians is

$$y = \beta_0 + \beta_1 x + \varepsilon$$

where y is the number of breakdowns per 8-hour shift and x is the number of inexperienced personnel working on the assembly line. After the model is fit using the least squares procedure, the residuals are plotted against $\hat{y}$, as shown in the figure at left.

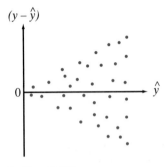

$(y - \hat{y})$

Problem 10.80

a.   Do you detect a pattern in the residual plot? What does this suggest about the least squares assumptions?

b.   Given the nature of the response variable y and the pattern detected in part **a**, what model adjustments would you recommend?

10.81   A certain type of rare gem serves as a status symbol for many of its owners. In theory, as the price of the gem increases, the demand will decrease at low prices, level off at moderate prices, and increase at high prices as a result of the status the owners believe they gain by obtaining the gem. Although a nonlinear model would seem to match the theory, the model proposed to explain the demand for the gem by its price is the straight-line model

$$y = \beta_0 + \beta_1 x + \varepsilon$$

where y is the demand (in thousands) and x is the retail price per carat (dollars). Consider the following data.

GEM.XLS

x	100	700	450	150	500	800	70	50	300	350	750	700
y	130	150	60	120	50	200	150	160	50	40	180	130

a.   Use Excel to fit the model to the 12 data points given in the table. Obtain the residuals.

b.   Plot the residuals against retail price per carat x.

c.   Can you detect any trends in the residual plot? What does this imply?

10.82   In 1919, the Cincinnati Reds defeated the heavily favored Chicago White Sox in the World Series, five games to three. A year later, eight Chicago players (known as the "Black Sox") were banned from major league baseball for conspiring to lose the 1919 World Series. The data in the table, published in *The American Statistician* (Nov. 1993), are performance statistics for the 16 players who played in every game of the 1919 World Series. Batting player game percentage (PGP) y represents the player's ability to hit in the "clutch" when the outcome of the game is at stake. (The higher the number, the better the player's performance in the clutch.) Slugging average x is total bases accumulated by a player's hits divided by number of at bats. The players marked by an asterisk in the table are "Black Sox" players.

BLACKSOX.XLS

Player	Batting PGP y	Slugging Average x
1	−1.6	.05
*2	−3.6	.17
3	−0.8	.20
*4	−1.8	.23
5	−1.4	.24
6	−0.8	.26
*7	−0.5	.30
8	1.2	.30
9	−1.0	.31
10	−0.8	.32
11	−0.2	.35
12	0.4	.36
23	1.0	.38
14	−0.1	.46
*15	−0.3	.50
*16	2.1	.56

Source: Bennett, J. "Did Shoeless Joe Jackson throw the 1919 World Series?" *The American Statistician,* Vol. 47, No. 4, Nov. 1993, p. 246 (adapted from Figure 2).

Consider the straight-line model $y = \beta_0 + \beta_1 x + \varepsilon$.

a. Use Excel to obtain the least squares prediction equation.

b. Use Excel to obtain the regression residuals.

c. Do you detect any outliers in the data? If so, are these outliers Black Sox players? [*Note:* Player #16 in the table is Shoeless Joe Jackson. His alleged participation in the "fix" is the only factor preventing his election into baseball's Hall of Fame.]

10.83 PCBs make up a family of hazardous chemicals that are often dumped, illegally, by industrial plants into the surrounding streams, rivers, or bays. The table on page 624 reports the 1984 and 1985 concentrations of PCBs (measured in parts per billion) in water samples collected from 37 U.S. bays and estuaries. An official from the Environmental Protection Agency (EPA) wants to model the 1985 PCB concentration y of a bay as a function of the 1984 PCB concentration x.

a. Use Excel to fit a straight-line model to the data.

b. Is the model adequate for predicting y? Explain.

c. Construct a residual plot for the data. Do you detect any outliers? If so, identify them.

PCBWATER.XLS

Bay	State	PCB CONCENTRATION 1984	1985
Casco Bay	ME	95.28	77.55
Merrimack River	MA	52.97	29.23
Salem Harbor	MA	533.58	403.1
Boston Harbor	MA	17104.86	736
Buzzards Bay	MA	308.46	192.15
Narragansett Bay	RI	159.96	220.6
East Long Island Sound	NY	10	8.62
West Long Island Sound	NY	234.43	174.31
Raritan Bay	NJ	443.89	529.28
Delaware Bay	DE	2.5	130.67
Lower Chesapeake Bay	VA	51	39.74
Pamlico Sound	NC	0	0
Charleston Harbor	SC	9.1	8.43
Sapelo Sound	GA	0	0
St. Johns River	FL	140	120.04
Tampa Bay	FL	0	0
Apalachicola Bay	FL	12	11.93
Mobile Bay	AL	0	0
Round Island	MS	0	0
Mississippi River Delta	LA	34	30.14
Barataria Bay	LA	0	0
San Antonio Bay	TX	0	0
Corpus Christi Bay	TX	0	0
San Diego Harbor	CA	422.1	531.67
San Diego Bay	CA	6.74	9.3
Dana Point	CA	7.06	5.74
Seal Beach	CA	46.71	46.47
San Pedro Canyon	CA	159.56	176.9
Santa Monica Bay	CA	14	13.69
Bodega Bay	CA	4.18	4.89
Coos Bay	OR	3.19	6.6
Columbia River Mouth	OR	8.77	6.73
Nisqually Beach	WA	4.23	4.28
Commencement Bay	WA	20.6	20.5
Elliott Bay	WA	329.97	414.5
Lutak Inlet	AK	5.5	5.8
Nahku Bay	AK	6.6	5.08

Source: Environmental Quality, 1987–1988.

d. Refer to part **c**. Although the residual for Boston Harbor is not, by definition, an outlier, the EPA believes that it strongly influences the regression because of its large y value. Remove the observation for Boston Harbor from the data and refit the model using Excel. Has model adequacy improved?

10.12 MULTIPLE REGRESSION MODELS (OPTIONAL)

Many practical applications of regression analysis require models that are more complex than the simple straight-line model. For example, a realistic probabilistic model for the carbon monoxide rating y of an American-made cigarette would include more variables than the nicotine content x of the cigarette (discussed in Example 10.2). Additional variables such as tar content, length, filter type, and menthol flavor might also be related to carbon monoxide ranking. Thus, we would want to incorporate these and other potentially important independent variables into the model if we needed to make accurate predictions of the carbon monoxide ranking y.

A more complex model relating y to various independent variables, say $x_1, x_2, x_3, \dots$, is called a **multiple regression model**. The general form of a multiple regression model is shown in the box.

General Multiple Regression Model

$$y = \beta_0 + \beta_1 x_1 + \beta_2 x_2 + \cdots + \beta_k x_k + \varepsilon$$

where

y is the dependent variable (the variable to be predicted)

$x_1, x_2, \dots, x_k$ are the independent variables

β_i determines the contribution of the independent variable x_i

ε is the random error component of the model

[*Note:* The symbols $x_1, x_2, \dots, x_k$ may represent functions of other x's. For example, $x_3 = x_1 x_2$ and $x_4 = x_1^2$.]

The steps to follow in a multiple regression analysis are similar to those in a simple linear regression, as outlined in the box.

Steps to Follow in a Multiple Regression Analysis

1. Hypothesize the form of the model.

2. State assumptions about the random error. (These are the same assumptions as those listed in Section 10.4.)

3. Estimate the unknown parameters $\beta_0, \beta_1, \beta_2, \dots, \beta_k$ using the method of least squares.

4. Check whether the fitted model is useful for predicting y.

5. Check that the assumptions of step 2 are satisfied using a residual analysis. Make modifications to the model, if necessary.

6. If we decide that the model is useful and the assumptions are satisfied, use it to estimate the mean value of y or to predict a particular value of y for given values of the independent variables.

In addition to those outlined in the box, there are other similarities. A t test on an individual β parameter in multiple regression is performed identically to the t test on β_1 in simple linear regression. Interpretations of s, the estimate of σ, and R^2, the coefficient of determination, are also identical. There are some key differences, however, between analyzing a multiple and simple linear regression model. For example, the interpretations of the β estimates in the multiple regression model are not always the same; they will depend on the form of the model. Also, the methodology for testing the overall adequacy of the multiple regression model differs from that in simple linear regression. We illustrate these concepts in the next several examples.

The Basic Multiple Regression Model

We will start by considering the simplest of all multiple regression models, called a **first-order model**. This model is described in the box for the two independent variables case.

A First-Order Multiple Regression Model with Two Independent Variables

$$y = \beta_0 + \beta_1 x_1 + \beta_2 x_2 + \varepsilon$$

where β_i represents the slope of the line relating y to x_i when the other x is held fixed (i.e., β_i measures the change in y for every 1-unit increase in x_i, holding the other x fixed).

[*Note:* Both independent variables in the model are quantitative.]

Example 10.30

Interpreting β Estimates in the First-Order Model

A property appraiser wants to model the relationship between the sale price y (in dollars) of a residential property in a midsize city and the following two independent variables: x_1 = appraised value of improvements (i.e., home value in dollars) on the property and x_2 = area of living space on the property (i.e., home size in square feet). Consider the first-order multiple regression model

$$y = \beta_0 + \beta_1 x_1 + \beta_2 x_2 + \varepsilon$$

To fit the model, the appraiser selected a random sample of $n = 20$ properties from the thousands of properties that were sold in a particular year. The resulting data are given in Table 10.8.

APPRAISAL.XLS

Table 10.8 Real Estate Appraisal Data for 20 Properties

Property #	Sale Price y	Improvements Value x_1	Area x_2
1	68,900	44,967	1,873
2	48,500	27,860	928
3	55,500	31,439	1,126
4	62,000	39,592	1,265
5	116,500	72,827	2,214
6	45,000	27,317	912
7	45,000	29,856	899
8	83,000	47,752	1,803
9	59,000	39,117	1,204
10	47,500	29,349	1,725
11	40,500	40,166	1,080
12	40,000	31,679	1,529
13	97,000	58,510	2,455
14	45,500	23,454	1,151
15	40,900	20,897	1,173
16	80,000	56,248	1,960
17	56,000	20,859	1,344
18	37,000	22,610	988
19	50,000	35,948	1,076
20	22,400	5,779	962

E **Figure 10.27**
Excel Output for
Sale Price Model,
Example 10.30

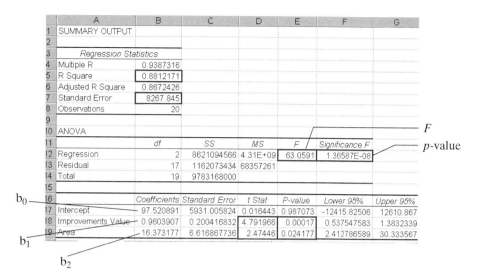

a. Use the method of least squares to estimate the unknown parameters β_0, β_1, and β_2, in the model.

b. Interpret the estimates of the β parameters.

Solution

a. The model hypothesized in this example is fit to the data of Table 10.8 using Excel. The Excel printout is reproduced in Figure 10.27. The least squares estimates of the β parameters appear in the column labeled **Coefficients**. You can see that $b_0 = 97.52$, $b_1 = .96$, and $b_2 = 16.37$. Therefore, the equation that minimizes SSE for this data set (i.e., the least squares prediction equation) is

$$\hat{y} = 97.52 + .96x_1 + 16.37x_2$$

b. Recall that in the straight-line model, β_0 represents the y-intercept of the line and β_1 represents the slope of the line. From our earlier discussion, β_1 has a practical interpretation—it represents the mean change in y for every 1-unit increase in x. When the independent variables are *quantitative*, the β parameters in the first-order model have similar interpretations. The difference is that when we interpret the β that multiplies one of the variables (e.g., x_1), we must be certain to hold the values of the remaining independent variables (e.g., x_2) fixed. Therefore, β_1 measures the change in y for every 1-unit increase in x_1 when the other variable in the model, x_2, is held fixed. A similar statement can be made about β_2: β_2 measures the change in y for every 1-unit increase in x_2 when the other variable in the model, x_1, is held fixed. Consequently, we obtain the following interpretations:

$b_1 = .96$: We estimate the sale price of a property to increase .96 dollar for every \$1 increase in appraised improvements x_1 when living area x_2 is held fixed.

$b_2 = 16.37$: We estimate the sale price of a property to increase \$16.37 per additional square foot of living area x_2 when appraised value of improvements x_1 is held fixed.

The value $b_0 = 97.52$ does not have a meaningful interpretation in this example. To see this, note that $\hat{y} = b_0$ when $x_1 = x_2 = 0$. Thus, $b_0 = 97.52$ represents the predicted sale price when the values of both independent variables are set equal to 0. Since a residential property with these characteristics—appraised improvements of $0 and 0 square feet of living area—is not practical, the value of b_0 has no meaningful interpretation. In general, b_0 will not have a practical interpretation unless it makes sense to set the values of the x's simultaneously equal to 0. ❏

Caution

The interpretation of the β parameters in multiple regression will depend on the terms specified in the model. The interpretations given in Example 10.30 are for a first-order model only. In practice, you should be sure that a first-order model is the correct model for y before making these β interpretations.

Test of Overall Model Adequacy

One of the goals in a multiple regression analysis is to conduct a test of the adequacy of the model—that is, test to determine whether the model is really useful for predicting y (step 4). In simple linear regression we performed this step by conducting a t test on the slope parameter β_1. In multiple regression, conducting t tests on each β parameter in a model is generally *not* an appropriate way to determine whether a model is contributing information for the prediction of y. This is due to the fact that a t test for, say, β_1 in a multiple regression model tests the contribution of x_1, holding all the other x's in the model fixed. To test the overall model, we need to conduct a test involving *all* the β parameters (except β_0) simultaneously. The null and alternative hypotheses for this overall test of model utility are given in the box.

Hypotheses for Testing Whether a Multiple Regression Model Is Useful for Predicting y

$$H_0: \beta_1 = \beta_2 = \cdots = \beta_k = 0$$

H_a: At least one of the β parameters in H_0 is nonzero

Example 10.31 Conducting the F Test

Refer to Example 10.30. Test (using $\alpha = .05$) whether the first-order model in the two quantitative independent variables is useful for predicting sale price y by testing the null hypothesis

$$H_0: \beta_1 = \beta_2 = 0$$

against the alternative hypothesis

H_a: At least one of the model parameters, β_1 and/or β_2, differs from 0

Solution

The test statistic used in the test for model utility is an F statistic, as described in the next box.

Procedure for Testing Whether the Overall Multiple Regression Model Is Useful for Predicting y

$$H_0: \beta_1 = \beta_2 = \cdots = \beta_k = 0$$

H_a: At least one of the parameters, $\beta_1, \beta_2, \ldots, \beta_k$, differs from 0

$$Test\ statistic:\ F = \frac{\text{Mean square for model}}{\text{Mean square for error}} = \frac{\text{SS(Model)}/k}{\text{SSE}/[n - (k + 1)]} \quad (10.22)$$

$$Rejection\ region:\ F > F_\alpha$$

where

n = number of observations

k = number of parameters in the model (excluding β_0)

and the distribution of F depends on k numerator degrees of freedom and $n - (k + 1)$ denominator degrees of freedom.

The F value for the sale price model, indicated on the Excel printout shown in Figure 10.27 (page 627), is $F = 63.06$. The p-value of the test, also indicated on the printout, is approximately 0. Since $\alpha = .05$ exceeds the p-value, there is sufficient evidence to reject H_0 and to conclude that at least one of the model coefficients, β_1 and/or β_2, is nonzero. This test result implies that the first-order model is useful for predicting the sale price of residential properties. ❏

After we have determined that the overall multiple regression model is useful for predicting y using the F test, we may elect to conduct one or more t tests on the individual β parameters. However, when the model includes a large number of independent variables, we should limit the number of t tests conducted to avoid a potential problem of making too many Type I errors. In addition to these tests, we should examine the value of s (the estimate of σ) and R^2. Ideally, we want s to be small and R^2 to be near 1. Once we are satisfied with the results of the F test (and possible t tests) and the values of R^2 and s, we will be comfortable using the model for estimation and prediction (step 6 on page 625).

Guidelines for Checking the Utility of a Multiple Regression Model

1. Conduct a test of overall model adequacy using the F test, i.e., test

$$H_0: \beta_1 = \beta_2 = \cdots = \beta_k$$

If the model is deemed adequate (i.e., if you reject H_0), then proceed to step 2. Otherwise, you should hypothesize and fit another model. The new model may include more independent variables or more complex terms.

2. Conduct t tests on those β parameters that you are particularly interested in (i.e., the "most important" β's). However, it is a safe practice to limit the number of β's tested. Conducting a series of t tests leads to a high probability of making at least one Type I error.

3. Examine the values of R^2 and s. Models with high R^2 values and low s values are preferred.

Example 10.32   Testing Individual β Parameters

Refer to the first-order model relating sale price y to appraised improvements x_1 and living area x_2, Examples 10.30 and 10.31. Conduct the tests $H_0: \beta_1 = 0$ and $H_0: \beta_2 = 0$, each at $\alpha = .05$. Interpret the results.

Solution

The test statistics and corresponding p-values of the two tests are highlighted on the Excel printout, Figure 10.27. The results are shown below.

$$H_0: \beta_1 = 0, t = 4.79, \text{p-value} = .0002$$

$$H_0: \beta_2 = 0, t = 2.47, \text{p-value} = .0242$$

Note that both p-values are less than $\alpha = .05$; hence, we reject H_0 in both cases.

The test for β_1 implies that there is sufficient evidence that appraised improvements x_1 contributes to the prediction of y when living area x_2 is held fixed. Similarly, the test for β_2 implies that living area x_2 contributes to the prediction of y when appraised value of improvements x_1 is held fixed. Both variables appear to be contributing to the overall model's ability to predict sale price. ❏

Example 10.33   Interpreting s and R^2

Refer to the multiple regression model of Examples 10.30–10.32, and the Excel printout, Figure 10.27 (page 627).

a. Locate s, the estimate of σ, on the Excel printout and interpret its value.

b. Locate R^2, the coefficient of determination, on the Excel printout and interpret its value.

Solution

a. The value of s is shown at the top of the printout, Figure 10.27, in the row labeled **Standard Error**. The value $s = 8{,}267.84$ is interpreted as follows: We expect to predict sale price y to within $2s = 2(8{,}267.84) \approx 16{,}536$ dollars of its true value using the first-order multiple regression model.

b. The value of R^2 is also shown at the top of Figure 10.27. The value $R^2 = .8812$ implies that about 88% of the sample variation in sale price y can be explained by the multiple regression model with improvement value x_1 and living area x_2 as independent variables. ❏

Adjusted Coefficient of Determination

The value of R^2 in Example 10.33 is high (near 90%), indicating the model fits the data well. We already know (from the F test in Example 10.31) that the model is statistically useful for predicting sale price y. Consequently, one might assume that a model with a high R^2 will always be useful for predicting y. Unfortunately, this is not always the case.

For example, consider a first-order model with 10 parameters (i.e., nine independent variables):

$$y = \beta_0 + \beta_1 x_1 + \beta_2 x_2 + \cdots + \beta_9 x_9 + \varepsilon$$

Now suppose (unknown to us) that none of the nine x's are useful predictors of y. If we fit the model to a data set with $n = 10$ data points (equal to the number of β parameters in the model), the value of R^2 will equal 1. That is, we will obtain a "perfect fit" using a model with independent variables that are unrelated to y! The problem stems from the fact that for a fixed sample size n, R^2 will increase artificially as independent variables (no matter how useful) are added to the model.

Caution

In a multiple regression analysis, the value of R^2 will increase as more independent variables are added to the model. Consequently, use the value of R^2 as a measure of model utility only if the sample contains substantially more data points than the number of β parameters in the model.

As an alternative to using R^2 as a measure of model adequacy, the *adjusted coefficient of determination*, denoted R_a^2, is often reported. The formula for R_a^2 is shown in the box.

The Adjusted Coefficient of Determination

The **adjusted coefficient of determination** is given by

$$R_a^2 = 1 - \left[\frac{n-1}{n-(k+1)}\right](1 - R^2) \tag{10.23}$$

Unlike R^2, R_a^2 takes into account ("adjusts" for) both the sample size n and the number of β parameters in the model. R_a^2 will always be smaller than R^2, and more important, R_a^2 cannot be "forced" to 1 by simply adding more and more independent variables to the model. Consequently, analysts prefer the more conservative R_a^2 when choosing a measure of model adequacy in multiple regression.

Example 10.34   Interpreting R_a^2

Refer to Examples 10.30–10.33. Locate the value of R_a^2 on the Excel printout, Figure 10.27 (page 627). Interpret its value.

Solution

The value of R_a^2 is shown on the Excel printout directly underneath the value of R^2. Note that $R_a^2 = .8672$, a value only slightly smaller than R^2. Our interpretation is that after adjusting for sample size and the number of parameters in the model, approximately 87% of the sample variation in sale price can be "explained" by the first-order model.   ❑

As stated in the beginning of this section, multiple regression models can be more complex than the first-order model. These models may include **quadratic** (squared) terms (e.g., x_1^2); *cross-product* terms (e.g., $x_1 x_2$), commonly known as **interaction** terms; and terms for a qualitative independent variable (called **dummy variables**). Some of these other multiple regression models and their properties

Figure 10.28 Graphs of Some Multiple Regression Models

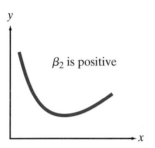
β_2 is positive

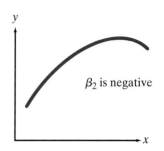
β_2 is negative

a. Quadratic model: $y = \beta_0 + \beta_1 x + \beta_2 x^2 + \varepsilon$

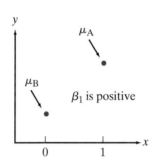
μ_A μ_B β_1 is positive

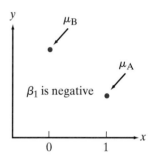
μ_B μ_A β_1 is negative

b. Dummy variable model: $y = \beta_0 + \beta_1 x + \varepsilon$, where $x = \begin{cases} 1 \text{ if A} \\ 0 \text{ if B} \end{cases}$

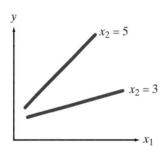
$x_2 = 5$ $x_2 = 3$

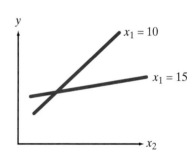
$x_1 = 10$ $x_1 = 15$

c. Interaction model: $y = \beta_0 + \beta_1 x_1 + \beta_2 x_2 + \beta_3 x_1 x_2 + \varepsilon$

are listed below. Graphical representations of these models are displayed in Figure 10.28. Details are beyond the scope of this text. Consult the references (e.g., Mendenhall and Sincich [1996]) for Chapter 10 to learn more about these models.

> **Quadratic (Second-Order) Model for a Single Quantitative Independent Variable**
>
> $y = \beta_0 + \beta_1 x + \beta_2 x_2 + \varepsilon$ (see Figure 10.28a)
>
> $\beta_0 = y$-intercept of the curve
>
> $\beta_1 = $ Shift parameter
>
> $\beta_2 = $ Rate of curvature

Dummy Variable Model for a Single Qualitative Independent Variable at Two Levels

$$y = \beta_0 + \beta_1 x + \varepsilon, \text{ where } x = \begin{cases} 1 \text{ if level A} \\ 0 \text{ if level B} \end{cases} \quad \text{(see Figure 10.28b)}$$

$$\beta_0 = \mu_B$$

$$\beta_1 = \mu_A - \mu_B$$

$\mu_A = E(y)$ when $x = 1$ (mean of y for level A)

$\mu_B = E(y)$ when $x = 0$ (mean of y for level B)

Interaction Model for Two Quantitative Independent Variables

$$y = \beta_0 + \beta_1 x_1 + \beta_2 x_2 + \beta_3 x_1 x_2 + \varepsilon \quad \text{(see Figure 10.28c)}$$

$(\beta_1 + \beta_3 x_2) = $ Slope of line relating y to x_1

$(\beta_2 + \beta_3 x_1) = $ Slope of line relating y to x_2

PROBLEMS
10.84–10.101

Using the Tools

10.84 Give the equation of a first-order model relating y to three quantitative independent variables.

10.85 Give the equation of an interaction model relating y to two quantitative independent variables.

10.86 Give the equation of a model relating y to a qualitative independent variable with two levels, A and B.

10.87 Give the equation of a curvilinear model relating y to a quantitative independent variable x.

10.88 Suppose y is related to four quantitative independent variables x_1, x_2, x_3, and x_4 by the first-order model

$$y = \beta_0 + \beta_1 x_1 + \beta_2 x_2 + \beta_3 x_3 + \beta_4 x_4 + \varepsilon$$

You fit this model to a set of $n = 15$ data points and find $R^2 = .74$, SS(Total) = 1.690, and SSE = .439.

    a. Calculate s^2, the estimate of the variance of the random error.

    b. Calculate the F statistic for testing $H_0: \beta_1 = \beta_2 = \beta_3 = \beta_4 = 0$.

    c. Do the data provide sufficient evidence to indicate that the model contributes information for predicting y? Test using $\alpha = .05$.

10.89   Suppose you fit the first-order multiple regression model

$$y = \beta_0 + \beta_1 x_1 + \beta_2 x_2 + \varepsilon$$

to $n = 25$ data points and obtain the prediction equation

$$\hat{y} = 6.4 + 3.1x_1 + .92x_2$$

The estimated standard deviations of the sampling distributions of b_1 and b_2 are 2.3 and .27, respectively.

a.   Test $H_0: \beta_1 = 0$ against $H_a: \beta_1 > 0$. Use $\alpha = .05$.

b.   Test $H_0: \beta_2 = 0$ against $H_a: \beta_2 \neq 0$. Use $\alpha = .05$.

c.   Find a 90% confidence interval for β_1. Interpret the interval.

d.   Find a 99% confidence interval for β_2. Interpret the interval.

10.90   Suppose you fit the first-order multiple regression model

$$y = \beta_0 + \beta_1 x_1 + \beta_2 x_2 + \beta_3 x_3 + \varepsilon$$

to $n = 20$ data points and obtain $R^2 = .2623$. Test the null hypothesis $H_0: \beta_1 = \beta_2 = \beta_3 = 0$ against the alternative hypothesis that at least one of the β parameters is nonzero. Use $\alpha = .05$.

Applying the Concepts

10.91   Backpacks are commonly seen in many places, especially college campuses, shopping malls, airplanes, and hiking trails. In the Aug. 1997 issue of *Consumer Reports,* information was provided concerning different features of backpacks—including their price (dollars), volume (cubic inches), and number of 5-inch by 7-inch books that the backpack can hold. The results are listed in the table.

BACKPACKS.XLS

Price	Volume	Books	Price	Volume	Books	Price	Volume	Books
48	2,200	59	35	1,950	49	35	1,519	43
45	1,670	49	32	1,385	45	95	1,102	73
50	2,200	48	40	1,700	38	40	1,316	55
42	1,700	52	35	2,000	51	40	1,760	43
29	1,875	52	28	1,500	46	25	1,844	42
50	1,500	49	40	1,950	46	50	2,150	52
48	1,874	50	40	1,810	44	35	1,810	50
38	1,586	47	45	1,910	48	50	2,180	46
33	1,910	53	27	1,875	42	35	1,635	40
40	1,500	49	25	1,450	42	15	1,245	47

Source: "Packs for town and country." Copyright 1997 by Consumers Union of United States, Inc., Yonkers, N.Y. 10703. Adapted by permission from *Consumer Reports,* August 1997, pp. 20–21. Although these data sets originally appeared in *Consumer Reports,* the selective adaptation and resulting conclusions presented are those of the authors and are not sanctioned or endorsed in any way by Consumers Union, the publisher of *Consumer Reports.*

Suppose that we want to develop a first-order multiple regression model to predict the price of a backpack y based on the volume x_1 and the number of books it can hold x_2.

a.   Write the equation of the model.

b.   Use Excel to obtain the multiple regression results and then give the least squares prediction equation.

c.   Interpret the meaning of the estimated model parameters.

d. Predict the price of a backpack that has a volume of 2,000 cubic inches and can hold 50 5" $\times$ 7" books.

e. Determine whether there is a significant relationship between price and the two explanatory variables (volume and number of books) at $\alpha = .05$.

f. Find and interpret s.

g. Find and interpret adjusted R^2.

h. At $\alpha = .05$, determine whether each explanatory variable makes a significant contribution to the regression model.

i. Do the results, parts **d–h**, support the use of this model in predicting backpack price? Explain.

10.92 UCLA political science professor A. F. Simon investigated the predictors of the amount of U.S. media coverage given to foreign earthquakes (*Journal of Communication*, Summer 1997). Three independent variables were used to predict y, the total number of seconds devoted to the earthquake on the national news television broadcasts of ABC, CBS, and NBC:

x_1 = Number of people killed by the earthquake

x_2 = Number of people affected by the earthquake

x_3 = Distance (in thousands of miles) from the earthquake to New York City

The first-order model $y = \beta_0 + \beta_1 x_1 + \beta_2 x_2 + \beta_3 x_3 + \varepsilon$ was fit to data collected for $n = 22$ recent foreign earthquakes.

a. The adjusted R^2 for the model is .4924. Interpret this value.

b. The F value for testing $H_0: \beta_1 = \beta_2 = \beta_3 = 0$ is $F = 7.79$. Use this value to make an inference about the utility of the model at $\alpha = .05$.

c. The t values (and associated p-values) for each independent variable in the model are shown below. Interpret these results.

Independent Variable	t Value	p-Value
x_1	3.73	.002
x_2	0.53	.603
x_3	-3.00	.008

10.93 Perfectionists are persons who set themselves standards and goals that cannot be reasonably met or accomplished. One theory suggests that those individuals who are depressed have a tendency toward perfectionism. To study this phenomenon, 76 members of an introductory psychology class completed four questionnaires: (1) the ASO scale, designed to measure self-acceptance; (2) the Burns scale, designed to measure perfectionism; (3) the Zung scale, designed to measure depression; and (4) the Rotter scale, designed to measure perceptions between actions and reinforcement (*The Journal of Adlerian Theory, Research, and Practice*, Mar. 1986).

a. Give the equation of a first-order model relating depression (Zung scale) to self-acceptance (ASO scale), perfectionism (Burns scale), and reinforcement (Rotter scale).

b. The model, part **a**, was fit to the $n = 76$ points and resulted in a coefficient of determination of $R^2 = .70$. Interpret this value.

c. The model resulted in $F = 56$. Is there sufficient evidence to indicate that the model is useful for predicting depression (Zung scale) score? Test using $\alpha = .05$.

d. A t test for the perfectionism (Burns scale) variable resulted in a (two-tailed) p-value of .87. Interpret this value.

10.94 A marketing analyst for a major shoe manufacturer wants to determine which variables can be used in predicting the durability of running shoes. Two independent variables are to be considered, x_1 (FOREIMP), a measurement of the forefoot shock-absorbing capability, and x_2 (MID-SOLE), a measurement of the change in impact properties over time. The dependent variable y (LTIMP) is a measure of the long-term ability to absorb shock after a repeated impact test. A random sample of 15 types of currently manufactured running shoes was selected for testing and the data used to fit a first-order multiple regression model. Using Excel, the following (partial) output is provided:

ANOVA	df	SS	MS	F	Significance F
Regression	2	12.61020	6.30510	97.69	0.0001
Residual	12	0.77453	0.06454		
Total	14	13.38473			

Variable	Coefficients	Standard Error	t Stat	p-Value
Intercept	−0.02686	.06905	−0.39	
Foreimp	0.79116	.06295	12.57	.0000
Midsole	0.60484	.07174	8.43	.0000

a. Give the equation of the model fit to the data.

b. Interpret the estimates of the model parameters shown on the printout.

c. Use the information on the printout to find R^2. Interpret the result.

d. Find and interpret s, the estimated standard deviation of the model.

e. Determine whether there is a significant relationship between long-term impact and the two explanatory variables at the $\alpha = .05$ level.

f. Find a 95% confidence interval for the slope of the relationship between long-term impact and forefoot impact. Interpret the result.

g. At $\alpha = .05$, determine whether each explanatory variable makes a significant contribution to the regression model.

10.95 Refer to the *Prison Journal* (Sept. 1997) study of parole officer attitudes, Problem 9.14 (page 491). Multiple regression analysis was used to model the subjective role y of a parole officer (measured on a scale from 7 to 42). Data collected on 72 parole officers were used to fit the model

$$y = \beta_0 + \beta_1 x_1 + \beta_2 x_2 + \beta_3 x_3 + \beta_4 x_4 + \beta_5 x_5 + \varepsilon$$

where

$$x_1 = \begin{cases} 1 \text{ if female} \\ 0 \text{ if male} \end{cases}$$

$$x_2 = \begin{cases} 1 \text{ if intensive supervision officer} \\ 0 \text{ if regular officer} \end{cases}$$

$$x_s = \begin{cases} 1 \text{ if work in midwestern agency} \\ 0 \text{ if work in northeastern agency} \end{cases}$$

x_4 = length of time (in months) worked as a probation officer

x_5 = age (in years)

The results are displayed in the following table.

Independent Variable	β Estimate	Standard Error	t Value	p-Value
x_1	−.540	1.432	−.377	.7087
x_2	−7.410	1.930	−3.839	.0003
x_3	−1.060	1.434	.739	.4630
x_4	−.0017	.010	−.170	.8639
x_5	−.136	.117	−1.162	.2501

$R^2 = .212$ $R_a^2 = .144$ $F = 3.12$ (p-value = .0146)

a. Identify the type (quantitative or qualitative) of the independent variables.

b. At $\alpha = .05$, is the model adequate for predicting subjective role y?

c. Interpret the estimate of β_4.

d. At $\alpha = .05$, is there sufficient evidence that age x_5 is a useful predictor of subjective role y?

e. Interpret R^2 and R_a^2. Why do these values differ?

10.96 A manufacturer of boiler drums wants to use regression to predict the number of man-hours needed to erect the drums in future projects. To accomplish this, data for 35 boilers were collected. In addition to man-hours y, the variables measured were boiler capacity (x_1 = pounds per hour), boiler design pressure (x_2 = pounds per square inch), boiler type ($x_3 = 1$ if industry field erected, 0 if utility field erected), and drum type ($x_4 = 1$ if steam, 0 if mud). The data are provided in the accompanying table on page 638. An Excel printout for the model $y = \beta_0 + \beta_1 x_1 + \beta_2 x_2 + \beta_3 x_3 + \beta_4 x_4 + \varepsilon$ appears below.

	A	B	C	D	E	F	G
3	Regression Statistics						
4	Multiple R	0.950241938					
5	R Square	0.90295974					
6	Adjusted R Square	0.890438416					
7	Standard Error	894.6031853					
8	Observations	36					
9							
10	ANOVA						
11		df	SS	MS	F	Significance F	
12	Regression	4	230854854.1	57713714	72.11376	2.97665E-15	
13	Residual	31	24809760.63	800314.9			
14	Total	35	255664614.8				
15							
16		Coefficients	Standard Error	t Stat	P-value	Lower 95%	Upper 95%
17	Intercept	-3783.43295	1205.489975	-3.1385	0.003711	-6242.04734	-1324.819
18	Boiler Capacity	0.008749011	0.000903468	9.683809	6.86E-11	0.006906375	0.010592
19	Design Pressure	1.926477177	0.648906909	2.968804	0.005723	0.603022072	3.249932
20	Boiler Type	3444.254644	911.7282884	3.77772	0.000675	1584.771503	5303.738
21	Drum Type	2093.353564	305.6336847	6.849224	1.12E-07	1470.009206	2716.698

a. Conduct a test of the overall adequacy of the model. Use $\alpha = .05$.

BOILER.XLS

Man-Hours y	Boiler Capacity x_1	Design Pressure x_2	Boiler Type x_3	Drum Type x_4
3,137	120,000	375	Industrial	Steam
3,590	65,000	750	Industrial	Steam
4,526	150,000	500	Industrial	Steam
10,825	1,073,877	2,170	Utility	Steam
4,023	150,000	325	Industrial	Steam
7,606	610,000	1,500	Utility	Steam
3,748	88,200	399	Industrial	Steam
2,972	88,200	399	Industrial	Steam
3,163	88,200	399	Industrial	Steam
4,065	90,000	1,140	Industrial	Steam
2,048	30,000	325	Industrial	Steam
6,500	441,000	410	Industrial	Steam
5,651	441,000	410	Industrial	Steam
6,565	441,000	410	Industrial	Steam
6,387	441,000	410	Industrial	Steam
6,454	627,000	1,525	Utility	Steam
6,928	610,000	1,500	Utility	Steam
4,268	150,000	500	Industrial	Steam
14,791	1,089,490	2,170	Utility	Steam
2,680	125,000	750	Industrial	Steam
2,974	120,000	375	Industrial	Mud
1,965	65,000	750	Industrial	Mud
2,566	150,000	500	Industrial	Mud
1,515	150,000	250	Industrial	Mud
2,000	150,000	500	Industrial	Mud
2,735	150,000	325	Industrial	Mud
3,698	610,000	1,500	Utility	Mud
2,635	90,000	1,140	Industrial	Mud
1,206	30,000	325	Industrial	Mud
3,775	441,000	410	Industrial	Mud
3,120	441,000	410	Industrial	Mud
4,206	441,000	410	Industrial	Mud
4,006	441,000	410	Industrial	Mud
3,728	627,000	1,525	Utility	Mud
3,211	610,000	1,500	Utility	Mud
1,200	30,000	325	Industrial	Mud

Source: Dr. Kelly Uscategui, University of Connecticut.

Problem 10.96

    b. Test the hypothesis that boiler capacity x_1 is positively linearly related to man-hours y. Use $\alpha = .05$.

    c. Test the hypothesis that boiler pressure x_2 is positively linearly related to man-hours y. Use $\alpha = .05$.

    d. Construct a 95% confidence interval for β_3.

    e. It can be shown that β_3 represents the difference between the mean number of man-hours required for industrial and utility field erection boilers. Use this information to interpret the confidence interval of part **d**.

    f. Construct a 95% confidence interval for β_4 and interpret the result. [*Hint:* $\beta_4 = \mu_{\text{Steam}} - \mu_{\text{Mud}}$, where μ_i represents the mean number of man-hours required for drum type *i*.]

10.97 Furman University researchers used multiple regression to determine the most important indicators of educational achievement in South Carolina (*Furman Studies*, June 1996). The following multiple regression model was fit to data collected for $n = 91$ South Carolina school districts during the 1992–1993 academic year:

$$y = \beta_0 + \beta_1 x_1 + \beta_2 x_2 + \beta_3 x_3 + \beta_4 x_4 + \beta_5 x_5 + \beta_6 x_6 + \beta_7 x_7 + \beta_8 x_8 + \beta_9 x_9 + \varepsilon$$

where

y = percentage of eleventh grade students who performed above the 50th national percentile on the Stanford-8 Achievement Test.

x_1 = average number of years of total education of teachers

x_2 = average number of years of total education of the administrative staff

x_3 = operation expenditures per pupil

x_4 = ratio of students to teaching staff

x_5 = percentage of students classified as nonwhite

x_6 = percentage of population 20 years or over with less than a twelfth-grade education

x_7 = per capita personal income

x_8 = percentage of total school membership enrolled in private schools

x_9 = percentage of population meeting all reading and math standards

a. The analysis yielded $R^2 = .82$. Interpret this result.

b. The global F value for the model is $F = 47.07$ with an associated p-value of $p \approx 0$. Interpret these results.

c. The estimate of β_1 in the model is $b_1 = 1.98$. Interpret this result.

d. The test statistic for testing H_0: $\beta_1 = 0$ is $t = 4.644$ with an associated p-value of $p \approx 0$. Interpret these results.

e. The test statistic for testing H_0: $\beta_8 = 0$ is $t = .88$ with an associated p-value of $p = .381$. Interpret these results.

10.98 Unions in the United States have officially opposed a free trade agreement with Mexico, fearing a decrease in wages. However, an article in the *Journal of Labor Research* (Spring 1993) used regression analysis to demonstrate that a Mexican free trade agreement will have little influence on union wages and should increase nonunion wages. The model for union wages y included the independent variable x_1 = years of completed education, among others.

a. The researcher hypothesized a curvilinear relationship between union wages y and education x_1. Write a model for y as a function of x_1 that incorporates this hypothesis.

b. Refer to the model, part **a**. The researcher also hypothesized that education x_1 will have a positive net impact on union wages y. Specifically, wages y will increase with education x_1, but at a decreasing rate. Explain how to test this hypothesis.

10.99 Most educators agree that effective parent involvement in school programs can be facilitated by the actions and attitudes of the school principal. Principals usually employ three strategies when interacting with parents: (1) cooperation (attempting to achieve a common goal), (2) socialization (molding parents' attitudes to those of the school), and

(3) formalization (adopting formal measures that weaken parental demands). A study was conducted to investigate the influence of these three strategies on the degree of parent involvement in public elementary schools in Taipei, China (*Proceedings of the National Science Council, Republic of China,* July 1997). Using survey data collected for 172 elementary school principals, the following variables were measured:

$$y = \text{Overall parent involvement score}$$

$$x_1 = \text{Level of principal's cooperation strategy}$$

$$x_2 = \text{Level of principal's socialization strategy}$$

$$x_3 = \text{Level of principal's formalization strategy}$$

a. A first-order model was fit to the data. The resulting F statistic was $F = 11.09$ with a corresponding p-value of $p \approx 0$. Interpret these results.

b. The coefficient of determination for the model, part **a**, was $R^2 = .165$. Interpret this result.

c. The β estimates and corresponding t values (and p-values) are listed below. Give a practical interpretation of each β estimate.

Variable	β Estimate	t Value	p-Value
x_1	.341	4.77	.000
x_2	.070	1.00	.319
x_3	.050	1.26	.209

d. Refer to part **c**. Conduct a test to determine whether principal's cooperation x_1 is linearly related to parent involvement y. Use $\alpha = .05$.

e. Refer to part **c**. Conduct a test to determine whether principal's formalization x_3 is linearly related to parent involvement y. Use $\alpha = .05$.

10.100 Because the coefficient of determination R^2 always increases when a new independent variable is added to the model, it may be tempting to include many variables in a model to force R^2 to be near 1. However, doing so reduces the degrees of freedom available for estimating σ^2, which adversely affects our ability to make reliable inferences. As an example, suppose you want to predict the selling price of a used car using 18 independent variables (such as make, model, year, and odometer reading). You fit the model

$$y = \beta_0 + \beta_1 x_1 + \beta_2 x_2 + \cdots + \beta_{17} x_{17} + \beta_{18} x_{18} + \varepsilon$$

where $y = $ selling price and $x_1, x_2, \ldots, x_{18}$ are the predictor variables. Using the relevant information on $n = 20$ used cars to fit the model, you obtain $R^2 = .95$ and $F = 1.056$. Test to determine whether this value of R^2 is large enough for you to infer that this model is useful—i.e., that at least one term in the model is important for predicting used car selling price. Use $\alpha = .05$.

10.101 According to the 1990 census, the number of homeless people in the U.S. is more than a quarter of a million. Yet, little is known about what causes homelessness. Economists at the City University of New York used multiple regression to assist in determining the factors that cause homelessness in American cities (*American Economic Review,* Mar. 1993). Data on the number y of homeless per 100,000 population in

$n = 50$ metropolitan areas were obtained from the Department of Housing and Urban Development. In addition, the 16 independent variables listed in the accompanying table were measured for each city and a multiple regression analysis performed by fitting the first-order model

$$y = \beta_0 + \beta_1 x_1 + \beta_2 x_2 + \cdots + \beta_{16} x_{16} + \varepsilon$$

Independent Variable	β Estimate	t Value
Intercept	307.54	(–)
Rental price (10th percentile)	2.87	3.93
Vacancy rate (10th percentile)	−872.9	−1.58
Rent-control law (yes or no)	−15.50	−.23
Employment growth	−859.09	−1.58
Share of employment in service industries	−347.69	−1.33
Size of low-skill labor market	−1,003.87	−.38
Households (per 100,000) below poverty level	.013	1.22
Public welfare expenditures	.11	.59
AFDC benefits	−.95	−2.58
SSI benefits	1.07	2.14
Percent reduction in AFDC (nonpoor percents)	146.62	1.49
AFDC accuracy rate	98.15	.13
Mental health in-patients (per 100,000)	−.83	−1.50
Fraction of births to teenage mothers	−1,173.00	−1.39
Blacks (per 100,000)	.004	1.78
1984 population (100,000s)	1.22	1.44

Source: Honig, M., and Filer, R. K. "Cause of intercity variation in homelessness." *American Economic Review,* Vol. 83, No. 1, Mar. 1993, p. 251 (Table 2).

a. Interpret the β estimate for the independent variable rental price.

b. Test the hypothesis that the incidence of homelessness decreases as employment growth increases. Use $\alpha = .05$.

c. What is the danger in performing t tests for all 16 independent variables to determine model adequacy?

d. For this model, $R_a^2 = .83$. Interpret this result.

10.13 PITFALLS IN REGRESSION AND ETHICAL ISSUES

Regression analysis is perhaps the most widely used and, unfortunately, the most widely misused statistical technique applied to data. A few of the most frequently encountered difficulties are discussed below.

Pitfalls

1. *Assumptions:* Many users of regression lack an awareness of the standard least squares regression assumptions. (This problem has been magnified by the ease with which computer software can be employed to fit a regression model. Nowhere on the printout are the four assumptions listed as a reminder to the user!) Of those users who are aware of the assumptions, many do not know how to evaluate them or what to do if, in fact, any of the assumptions are violated. Consequently, the user with this lack of knowledge risks drawing inferences and making decisions from an invalid model.

To avoid this pitfall, residual plots are of vital importance to a regression analysis. Always check the assumptions with a residual analysis before using the model statistics to make inferences.

2. *Parameter Interpretations:* In any regression analysis, it is important to interpret the estimates of the β parameters correctly. A typical misconception is that β_i *always* measures the effect of x_i on $E(y)$, *independent* of the other x variables in the model. This may be true for some models, but it is not true in general. Generally, the interpretation of an individual β parameter becomes increasingly more difficult as the model becomes more complex. (See the boxes, pages 632–633.)

Another misconception about parameter estimates is that a statistically significant b_i value establishes a *cause-and-effect* relationship between $E(y)$ and x_i. That is, if b_i is found to be significantly greater than 0, then some practitioners would infer that an increase in x_i *causes* an increase in the mean response $E(y)$. Unfortunately, it is dangerous to infer a causal relationship between two variables based on a regression analysis. There may be many other independent variables (some of which we may have included in our model, some of which we may have omitted) that affect the mean response. Unless we can control the values of these other variables, we are uncertain about what is actually causing the observed increase in y. In Chapter 11, we introduce the notion of *designed experiments,* where the values of the independent variables are set in advance before the value of y is observed. Only with such an experiment can a cause-and-effect relationship be established.

3. *Multicollinearity:* Often, two or more of the independent variables used in the model for $E(y)$ will contribute redundant information. That is, the independent variables will be correlated with each other. For example, suppose we want to construct a model to predict the gasoline mileage rating y of a truck as a function of the weight of its load x_1 and the horsepower x_2 of its engine. In general, you would expect heavier loads to require greater horsepower and to result in lower mileage ratings. Thus, although both x_1 and x_2 contribute information for the prediction of mileage rating, some of the information is overlapping because x_1 and x_2 are correlated. When the independent variables are correlated, we say that **multicollinearity** exists. In practice, it is not uncommon to observe correlations among the independent variables. However, a few problems arise when serious multicollinearity is present in the regression analysis.

First, high correlations among the independent variables increase the likelihood of rounding errors in the calculations of the β estimates, standard errors, and so forth. Second, the regression results may be confusing and misleading.

To illustrate, if the gasoline mileage rating model

$$y = \beta_0 + \beta_1 x_1 + \beta_2 x_2 + \varepsilon$$

were fit to a set of data, we might find that the t values for both b_1 and b_2 (the least squares estimates) are nonsignificant. However, the F test for $H_0: \beta_1 = \beta_2 = 0$ would probably be highly significant. The tests may seem to be contradictory, but really they are not. The t tests indicate that the contribution of one variable, say x_1 = load, is not significant after the effect of x_2 = horsepower has been discounted (because x_2 is also in the model). The significant F test, on the other hand, tells us that at least one of the two variables is making a contribution to the prediction of y (i.e., either β_1, β_2, or both differ from 0). In fact, both are probably contributing, but the contribution of one overlaps with that of the other.

Multicollinearity can also have an effect on the signs of the parameter estimates. More specifically, a value of b_i may have the opposite sign from what is expected. For example, we expect the signs of both of the parameter estimates for the gasoline mileage rating model to be negative, yet the regression analysis for the model might yield the estimates $b_1 = .2$ and $b_2 = -.7$. The positive value of b_1 seems to contradict our expectation that heavy loads will result in lower mileage ratings. However, it is dangerous to interpret a β coefficient when the independent variables are correlated. Because the variables contribute redundant information, the effect of load weight x_1 on mileage rating is measured only partially by b_1.

Several methods are available for detecting multicollinearity in regression. A simple technique is to examine the coefficient of correlation r between each pair of independent variables in the model. If one or more of the r values is near 1 (in absolute value), the variables in question are highly correlated and a severe multicollinearity problem may exist. Other indications of the presence of multicollinearity include those mentioned above—namely, nonsignificant t tests for the individual β parameters when the F test for overall model adequacy is significant, and parameter estimates with opposite signs from what is expected. More formal methods for detecting multicollinearity (such as variance-inflation factors) are available. Consult the references given at the end of the chapter (e.g., Mendenhall and Sincich [1996]) if you want to learn more about these methods.

One way to avoid the problems of multicollinearity in regression is to conduct a designed experiment (Chapter 11) so that the levels of the x variables are uncorrelated. Unfortunately, time and cost constraints may prevent you from collecting data in this manner. Most analysts, when confronted with highly correlated independent variables, choose to include only one of the correlated variables in the model.

4. *Predicting Outside the Experimental Region:* The fitted regression model enables us to construct a confidence interval for $E(y)$ and a prediction interval for y for values of the independent variable only within the region of experimentation, i.e., within the range of values of the independent variables used in the experiment. For example, suppose that you conduct experiments on the mean strength of plastic molded at several different temperatures in the interval 200°F to 400°F. The regression model that you fit to the data is valid for estimating $E(y)$ or for predicting values of y for values of temperature x in the range 200°F to 400°F. However, if you attempt to extrapolate beyond the experimental region, you risk the possibility that the fitted model is no longer a good approximation to the mean strength of the plastic (see Figure 10.29). For example, the plastic may

Figure 10.29 Using a Regression Model Outside the Experimental Region

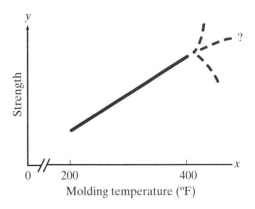

become too brittle when formed at 500°F and possess no strength at all. Estimating and predicting outside of the experimental region are sometimes necessary. If you do so, keep in mind the possibility of a large extrapolation error.

5. *Model Building:* In many regression applications, only a single model (usually a first-order model) is fit and evaluated. Even if the model is deemed adequate (based on statistics such as adjusted R^2, s, and the F statistic), the model is not guaranteed to be the best predictor of y. A more complex model—one with squared terms, interactions, or other independent variables—may be a significantly better predictor of y in practice. A user unaware of this alternative model will risk making erroneous conclusions about the relationship between y and the independent variables.

In addition to the initial model fit, the most successful regression analysts always consider one or more alternative models. Model comparisons are made and the model judged "best" is used to make inferences and predictions.

Ethical Issues

Ethical considerations arise when a user of regression manipulates the process of developing the regression model. The key here is intent. Some examples of unethical behavior on the part of the regression user are listed below.

1. Predicting the dependent variable of interest with the willful intent of possibly excluding certain independent variables from consideration in the model.

2. Deleting observations from the model to obtain a better model without giving reasons for deleting these observations.

3. Making inferences about the model without providing an evaluation of the assumptions when he or she knows that the assumptions of least squares regression are violated.

4. Willfully failing to remove independent variables from the model that exhibit a high degree of multicollinearity.

All of these situations emphasize the importance of following the steps of regression (Sections 10.1 and 10.12).

REAL-WORLD REVISITED
The S.O.B. Effect among College Administrators

At major colleges and universities, administrators (e.g., deans, chairpersons, provosts, vice presidents, and presidents) are among the highest-paid state employees. Is there a relationship between the raises administrators receive and their performance on the job? This was the question of interest to a group of faculty union members at the University of South Florida called the United Faculty of Florida (UFF).

The UFF compared the annual ratings of 15 University of South Florida administrators (as determined by faculty in a survey) to their subsequent raises. The data for the analysis, listed in Table 10.9, are saved in the Excel workbook

USFRAISE.XLS. [*Note:* Ratings are measured on a 5-point scale, where 1 = very poor and 5 = very good.]

USFRAISE.XLS

Table 10.9 Raises and Ratings of University of South Florida Administrators

Administrator	Raise[a]	Average Rating (5-point scale)[b]
1	$18,000	2.79
2	16,700	1.52
3	15,787	4.40
4	10,608	3.10
5	10,268	3.83
6	9,795	2.84
7	9,513	2.10
8	8,459	2.38
9	6,099	3.59
10	4,557	4.11
11	3,571	3.41
12	3,718	3.64
13	3,652	3.36
14	3,227	2.92
15	2,808	3.00

Sources: [a]Faculty and A&P Salary Report, University of South Florida, Resource Analysis and Planning, 1990.

[b]Administrative Compensation Survey, *Chronicle of Higher Education,* Jan. 1991.

According to the union, the "relationship is inverse, i.e., the lower the rating by the faculty, the greater the raise. Apparently, bad administrators are more valuable than good administrators."* (With tongue in cheek, the UFF refers to this phenomenon as "the S.O.B. effect.") The UFF based its conclusions on a simple linear regression analysis of the data in Table 10.9, where y = administrator's raise and x = average rating of administrator.

UFF Faculty Forum, University of South Florida Chapter, Vol. 3, No. 5, May 1991.

Study Questions

a. Initially, the UFF conducted the analysis using all 15 data points in Table 10.9. Fit a straight-line model to the data of Table 10.9. Is there evidence (at $\alpha = .05$) to support the union's claim of an inverse relationship between raise and rating?

b. A second simple linear regression was performed with only 14 of the data points in Table 10.9. The data for administrator #3 were eliminated based on the fact that he was promoted to dean in the middle of the academic year. (No other reason was given for removing this data point from the analysis.) Statistically, is this data point an outlier? Explain.

c. Remove the data point for administrator #3 and rerun the simple linear regression analysis. Now is there evidence to support the union's claim of an inverse relationship between raise and rating?

d. From our discussion in Section 10.11, we know that it is extremely dangerous to remove a data point from an analysis simply to improve the fit of the model. (In fact, the outlying data point may be an indicator of a "bad" model. By eliminating this point, we might be throwing away information that could aid in developing a better model.) Suppose, however, that the union argues that because administrator #3 was promoted to an administration position (dean) in mid-year, his salary and/or rating is not indicative of administrators in general, but of professors who have different responsibilities (e.g., teaching and research). On that basis, would you recommend that data point #3 be removed from the analysis?

e. Now suppose that the administrator's raise and rating were determined after the promotion to dean and that data point #3 is "typical" of administrators. What is your recommendation concerning the removal of this data point? Explain.

f. Based on the results of the regression, part c, the union computed estimated raises for selected faculty ratings of administrators. These are shown in Table 10.10. Verify the predicted raises in the table.

Table 10.10 Estimated Raises for Selected Ratings

Ratings		Raise
Very poor	1.00	$15,939
	1.50	13,960
Poor	2.00	11,980
	2.50	10,001
Average	3.00	8,021
	3.50	6,042
Good	4.00	4,062
	4.50	2,083
Very good	5.00	103

g. What problems do you perceive with using Table 10.10 to estimate an administrator's raise at the University of South Florida? Explain.

h. The ratings of administrators listed in Table 10.9 were determined by surveying the faculty at the University of South Florida. All faculty are mailed the survey each year, but the response rate is typically low (approximately 10%–20%). The danger with such a survey is that only disgruntled faculty, who are more apt to give a low rating to an administrator, will respond. Many of these faculty also think they are underpaid and that the administrators are overpaid. Comment on how such a survey could bias the results shown in Table 10.10.

i. Based on your answers to the previous questions, would you support the UFF's claim?

Key Terms

Starred (*) terms are from the optional sections of this chapter.

Adjusted coefficient of determination* 631
Coefficient of determination 594
Correlation coefficient 585
Dependent (response) variable 555

Dummy variable* 633
First-order model* 626
General regression model 555
Independent (explanatory) variable 555
Interaction model* 633

Least squares line 562
Least squares prediction equation 562
Method of least squares 562
Model 555
Multicollinearity 642

Multiple regression model*   625
Outlier*   617
Population correlation
  coefficient   585
Quadratic model*   632
Random error component   558

Regression analysis   555
Residual   612
Robust*   616
Simple linear regression   559
Slope   558
Straight-line model   557

Time series*   618
Variance-stabilizing
  transformation*   616
y-intercept   558

Key Formulas

Starred (*) formulas are from the optional sections of this chapter.

Simple Linear Regression

Least squares estimates of the β's:

$$b_1 = \frac{SSxy}{SSxx} \quad \textbf{(10.4)}, 562$$

$$b_0 = \bar{y} - b_1\bar{x} \quad \textbf{(10.5)}, 562$$

where

$$SSxy = \sum xy - \frac{(\sum x)(\sum y)}{n} \quad \textbf{(10.6)}, 562$$

$$SSxx = \sum x^2 - \frac{(\sum x)^2}{n} \quad \textbf{(10.7)}, 562$$

$$SSyy = \sum y^2 - \frac{(\sum y)^2}{n} \quad \textbf{(10.10)}, 574$$

Least squares line:
$$\hat{y} = b_0 + b_1 x$$

Sum of squared errors:
$$SSE = \sum(y - \hat{y})^2 = SSyy - b_1 SSxy \quad \textbf{(10.9)}, 574$$

Estimated variance of ε:
$$s^2 = \frac{SSE}{n-2} = MSE \quad \textbf{(10.8)}, 574$$

Estimated standard error of b_1:
$$s_{b_1} = s/\sqrt{SSxx} \quad \textbf{(10.11)}, 577$$

Test statistic for testing $H_0: \beta_1 = 0$:
$$t = \frac{b_1}{s_{b_1}} \quad \textbf{(10.12)}, 577$$

$(1-\alpha)100\%$ confidence interval for β_1:
$$b_1 \pm (t_{\alpha/2})s_{b_1} \quad \textbf{(10.13)}, 580$$

Coefficient of determination:
$$R^2 = \frac{SSR}{SSyy} = \frac{SSyy - SSE}{SSyy} \quad \textbf{(10.16)}, 594$$

$(1-\alpha)100\%$ confidence interval for $E(y)$ when $x = x_p$:
$$\hat{y} \pm (t_{\alpha/2})s\sqrt{\frac{1}{n} + \frac{(x_p - \bar{x})^2}{SSxx}} \quad \textbf{(10.19)}, 598$$

$(1-\alpha)100\%$ prediction interval for y when $x = x_p$:
$$\hat{y} \pm (t_{\alpha/2})s\sqrt{1 + \frac{1}{n} + \frac{(x_p - \bar{x})^2}{SSxx}} \quad \textbf{(10.20)}, 598$$

Multiple Regression

Estimated variance of ε:*
$$s^2 = \frac{SSE}{n-(k+1)} = MSE$$

Test statistic for testing $H_0: \beta_i = 0$:*
$$t = \frac{b_i}{s_{b_i}}$$

$(1-\alpha)100\%$ confidence interval for β_i:*
$$b_i \pm (t_{\alpha/2})s_{b_i}$$

Test statistic for testing $H_0: \beta_1 = \beta_2 = \cdots = \beta_k = 0$:*
$$F = \frac{MS(Model)}{MSE} \quad \textbf{(10.22)}, 629$$

Residual:*
$$y - \hat{y} \quad \textbf{(10.21)}, 612$$

Key Symbols

Symbol	Description
y	Dependent variable (variable to be predicted or modeled)
x	Independent (predicator) variable
$E(y)$	Expected (mean) value of y
β_0	y-intercept of true line
β_1	Slope of true line
b_0	Least squares estimate of y-intercept
b_1	Least squares estimate of slope
ε	Random error
$\hat{y}$	Predicted value of y
$(y - \hat{y})$	Error of prediction (or residual)
SSE	Sum of squared errors (will be smallest for least squares line)
SSxx	Sum of squares of x values
SSyy	Sum of squares of y values
SSxy	Sum of squares of cross products, $x \cdot y$
r	Coefficient of correlation
R^2	Coefficient of determination
R_a^2	Adjusted coefficient of determination
x_p	Value of x used to predict y
x_1^2	Quadratic term that allows for curvature in the relationship between y and x
$x_1 x_2$	Interaction term
β_i	Coefficient of x_i in the multiple regression model
b_i	Least squares estimate of β_i
s_{b_i}	Estimated standard error of b_i
F	Test statistic for testing global usefulness of model

Supplementary Problems 10.102–10.115

Starred (*) problems refer to one of the optional sections of this chapter.

10.102   A medical item used to administer to a hospital patient is called a factor. For example, factors can be intravenous (I.V.) tubing, I.V. fluid, needles, shave kits, bedpans, diapers, dressings, medications, and even code carts. The coronary care unit at Bayonet Point Hospital (St. Petersburg, Florida) investigated the relationship between the number of factors per patient x and the patient's length of stay (in days) y. The data for a random sample of 50 coronary care patients are given in the table (page 649), followed by an Excel printout of the simple linear regression analysis.

   a.   Construct a scatterplot of the data.

   b.   Find the least squares line for the data and sketch it on your scatterplot.

   c.   Define β_1 in the context of this problem.

   d.   Test the hypothesis that the number of factors per patient x contributes no information for the prediction of the patient's length of stay y when a linear model is used (use $\alpha = .05$). Draw the appropriate conclusions.

   e.   Find and interpret a 90% confidence interval for β_1.

   f.   Find and interpret the coefficient of correlation for the data.

   g.   Find and interpret the coefficient of determination for the model.

FACTORS.XLS

Number of Factors x	Length of Stay y (days)	Number of Factors x	Length of Stay y (days)
231	9	354	11
323	7	142	7
113	8	286	9
208	5	341	10
162	4	201	5
117	4	158	11
159	6	243	6
169	9	156	6
55	6	184	7
77	3	115	4
103	4	202	6
147	6	206	5
230	6	360	6
78	3	84	3
525	9	331	9
121	7	302	7
248	5	60	2
233	8	110	2
260	4	131	5
224	7	364	4
472	12	180	7
220	8	134	6
383	6	401	15
301	9	155	4
262	7	338	8

Source: Bayonet Point Hospital, Coronary Care Unit.

Problem 10.102

E

	A	B	C	D	E	F	G
1	SUMMARY OUTPUT						
2							
3	*Regression Statistics*						
4	Multiple R	0.6115913					
5	R Square	0.3740439					
6	Adjusted R Square	0.3610031					
7	Standard Error	2.100774					
8	Observations	50					
9							
10	ANOVA						
11		df	SS	MS	F	Significance F	
12	Regression	1	126.5839332	126.5839	28.68269	2.37754E-06	
13	Residual	48	211.8360668	4.413251			
14	Total	49	338.42				
15							
16		Coefficients	Standard Error	t Stat	P-value	Lower 95%	Upper 95%
17	Intercept	3.3060317	0.672974255	4.912568	1.09E-05	1.952927102	4.659136373
18	Number of Factors	0.0147549	0.002755021	5.355623	2.38E-06	0.009215514	0.020294188

Problem 10.102

h. Find a 95% prediction interval for the length of stay of a coronary care patient who is administered a total of $x = 200$ factors.

i. Explain why the prediction interval obtained in part **h** is so wide. How could you reduce the width of the interval?

j. Conduct a residual analysis of the data. What model modifications (if any) do you recommend?

10.103   A study was conducted to model the thermal performance of tubes used in the refrigeration and process industries (*Journal of Heat Transfer,* Aug. 1990). Twenty-four specially manufactured copper tubes were used in the experiment. Vapor was released downward into each tube, and the heat transfer ratio was measured. Theoretically, heat transfer will be related to the area at the top of the tube that is "unflooded" by condensation of the vapor. The data in the table are the unflooded area ratio x and heat transfer ratio y values recorded for the 24 copper tubes. Use Excel to perform a complete simple linear regression analysis of the data. Be sure to check the regression assumptions with a residual analysis.

HEAT.XLS

Unflooded Area Ratio x	Heat Transfer Ratio y	Unflooded Area Ratio x	Heat Transfer Ratio y
1.93	4.4	2.00	5.2
1.95	5.3	1.77	4.7
1.78	4.5	1.62	4.2
1.64	4.5	2.77	6.0
1.54	3.7	2.47	5.8
1.32	2.8	2.24	5.2
2.12	6.1	1.32	3.5
1.88	4.9	1.26	3.2
1.70	4.9	1.21	2.9
1.58	4.1	2.26	5.3
2.47	7.0	2.04	5.1
2.37	6.7	1.88	4.6

Source: Marto, P. J., et al. "An experimental study of R-113 film condensation on horizontal integral-fin tubes." *Journal of Heat Transfer,* Vol. 112, Aug. 1990, p. 763 (Table 2).

10.104   The long jump is a track and field event in which a competitor attempts to jump a maximum distance into a sand pit after obtaining a running start. At the edge of the sand pit is a takeoff board, and jumpers usually try to plant their toe at the front edge of the board to maximize jumping distance. The absolute distance between the front edge of the takeoff board and where the toe actually lands on the board prior to jumping is called "takeoff error." Is takeoff error in the long jump linearly related to best jumping distance? To answer this question, kinesiology researchers videotaped the performances of 18 novice long jumpers at a high school track meet (*Journal of Applied Biomechanics,* May 1995). The average takeoff error x and best jumping distance (out of three jumps) y for each jumper are recorded in the table on page 651. Use Excel to conduct a complete simple linear regression analysis of the data. Assume one of the researchers' goals is to predict the best jumping distance for a jumper with an average takeoff error of .05 meter.

10.105   The Consumer Attitude Survey, performed by the University of Florida Bureau of Economic and Business Research (BEBR), is conducted using random-digit telephone dialings of Florida households. The reliability of a telephone survey such as this depends on the *refusal rate,* i.e., the percentage of dialed households that refuse to take part in the study. One factor thought to be related to refusal rate is personal income. The table on page 651 gives the refusal rate y and personal income per capita x for 12 randomly selected Florida counties from a recent BEBR survey.

LONGJUMP.XLS

Jumper	Best Jumping Distance y (meters)	Average Takeoff Error x (meters)
1	5.30	.09
2	5.55	.17
3	5.47	.19
4	5.45	.24
5	5.07	.16
6	5.32	.22
7	6.15	.09
8	4.70	.12
9	5.22	.09
10	5.77	.09
11	5.12	.13
12	5.77	.16
13	6.22	.03
14	5.82	.50
15	5.15	.13
16	4.92	.04
17	5.20	.07
18	5.42	.04

Source: Reprinted by permission from W. P. Berg and N. L. Greer, "A kinematic profile of the approach run of novice long jumpers." *Journal of Applied Biomechanics,* Vol. 11, No. 2, May 1995, p. 147 (Table 1).

Problem 10.104

BEBR.XLS

County	Refusal Rate y	Per Capita Income x	County	Refusal Rate y	Per Capita Income x
1	.296	$ 7,737	7	.429	$11,466
2	.498	12,330	8	.422	10,000
3	.386	12,058	9	.441	10,052
4	.327	9,927	10	.191	8,636
5	.500	6,904	11	.526	7,445
6	.333	9,463	12	.405	9,059

Source: Bureau of Economic and Business Research, University of Florida.

Problem 10.105

a. Estimate the coefficient of correlation between refusal rate and per capita income.

b. Do the data provide sufficient evidence to indicate a correlation between refusal rate y and per capita income x? Use $\alpha = .05$.

Variable	Coefficient of Correlation r
x_1 and x_2	.33
x_1 and x_3	.08
x_2 and x_3	.16

Problem 10.106

10.106 Refer to the *Proceedings of the National Science Council, Republic of China* study of the variables related to y, parent involvement in public elementary schools, Problem 10.99. Three independent variables were included in the regression model: level of principal's cooperation x_1, level of principal's socialization x_2, and level of principal's formalization x_3. The intercorrelations among these variables are listed at left. Do you detect extreme multicollinearity in the data? What impact does this have, if any, on the analysis of the multiple regression model, Problem 10.99 (page 640)?

10.107 At temperatures approaching absolute zero (273 degrees below zero Celsius), helium exhibits traits that defy many laws of conventional physics. An experiment has been conducted with helium in solid form at various temperatures near absolute zero. The solid helium is placed in a

HELIUM.XLS

Proportion of Impurity Passing through Helium y	Temperature x (°C)
.315	−262
.202	−265
.204	−256
.620	−267
.715	−270
.935	−272
.957	−272
.906	−272
.985	−273
.987	−273

Problem 10.107

dilution refrigerator along with a solid impure substance, and the proportion (by weight) of the impurity passing through the solid helium is recorded. (This phenomenon of solids passing directly through solids is known as *quantum tunneling*.) The data are given in the table at left.

a. Use Excel to conduct a complete simple linear regression analysis of the data. Be sure to check the assumptions and make any model modifications, if necessary.

b. Would you recommend using the fitted model, part **a**, to predict the proportion of impurity passing through helium at a temperature of −250°C? Explain.

10.108 A company that has the distribution rights to home video sales of previously released movies would like to be able to predict the number of units that it can expect to sell (based on the box office gross sales of the movie). Data are available for 30 movies that indicate the box office gross (in millions of dollars) and the number of units sold (in thousands) of home videos. The results are as follows:

VIDEOS.XLS

Movie	Box Office Gross ($ millions)	Home Video Units Sold (thousands)
1	1.10	57.18
2	1.13	26.17
3	1.18	92.79
4	1.25	61.60
5	1.44	46.50
6	1.53	85.06
7	1.53	103.52
8	1.69	30.88
9	1.74	49.29
10	1.77	24.14
11	2.42	115.31
12	5.34	87.04
13	5.70	128.45
14	6.43	126.64
15	8.59	107.28
16	9.36	190.80
17	9.89	121.57
18	12.66	183.30
19	15.35	204.72
20	17.55	112.47
21	17.91	162.95
22	18.25	109.20
23	23.13	280.79
24	27.62	229.51
25	37.09	277.68
26	40.73	226.73
27	45.55	365.14
28	46.62	218.64
29	54.70	286.31
30	58.51	254.58

Using Excel, conduct a complete simple linear regression analysis of the data. One of your goals is to predict the video unit sales for a movie that had a box office gross of $20 million.

*10.109 Refer to the *Journal of Communications* (Summer 1997) study of homicide newsworthiness, Problem 1.5 (page 33). The researchers fit the following multiple regression model to data collected for $n = 100$ Milwaukee homicides:

$$y = \beta_0 + \beta_1 x_1 + \beta_2 x_2 + \beta_3 x_3 + \beta_4 x_4 + \beta_5 x_5 + \beta_6 x_6 + \beta_7 x_7 + \beta_8 x_8 + \beta_9 x_1 x_2 + \varepsilon$$

where

y = average length of published news stories on a homicide

$x_1 = \begin{cases} 1 \text{ if white victim or suspect} \\ 0 \text{ if not} \end{cases}$

$x_2 = \begin{cases} 1 \text{ if female suspect} \\ 0 \text{ if not} \end{cases}$

$x_3 = \begin{cases} 1 \text{ if female victim} \\ 0 \text{ if not} \end{cases}$

$x_4 = \begin{cases} 1 \text{ if white child or senior citizen} \\ 0 \text{ if not} \end{cases}$

x_5 = per capita income of census tract in which crime occurred

$x_6 = \begin{cases} 1 \text{ if suspect and victim had any type of prior relationship} \\ 0 \text{ if not} \end{cases}$

$x_7 = \begin{cases} 1 \text{ if victim was involved with drugs, gambling, gangs, or prostitution} \\ 0 \text{ if not} \end{cases}$

$x_8 = \begin{cases} 1 \text{ if police information ban} \\ 0 \text{ if not} \end{cases}$

The regression results are summarized in the accompanying table.

Variable	β Estimate	p-Value
Race (x_1)	.42	$p < .001$
Gender of suspect (x_2)	$-.33$	$.001 < p < .01$
Gender of victim (x_3)	.26	$.01 < p < .05$
Age of victim (x_4)	.33	$p < .001$
Per capita income (x_5)	$-.02$	$p > .10$
Relationship (x_6)	$-.11$	$p > .10$
Risky behavior (x_7)	$-.06$	$p > .10$
Information ban (x_8)	$-.15$	$p > .10$
Race $\times$ Gender ($x_1 x_2$)	$-.25$	$.01 < p < .05$

Overall model: $R^2 = .29, p < .001$

Source: Pritchard, D., and Hughes, K. D. "Patterns of deviance in crime news." *Journal of Communication*, Vol. 47, No. 3, Summer 1997, p. 59 (Table 2).

a. Identify the qualitative independent variables used in the model.

b. Give the null hypothesis for testing whether the overall model is useful for predicting the average length of published news stories on a homicide.

c. Conduct the test, part **b**. Use $\alpha = .01$.

d. Interpret the value of R^2.

e. Interpret the estimate of β_3.

f. Interpret the estimate of β_5.

g. Is there evidence of interaction between race x_1 and gender of suspect x_2? Test using $\alpha = .05$.

*10.110 Does exercise improve the human immune system? An experiment was conducted by a physiologist at the University of Florida to determine whether such a relationship exists. Thirty subjects volunteered to participate in the study. The amount of immunoglobulin known as IgG (an indicator of long-term immunity) and the maximal oxygen uptake (a measure of aerobic fitness level) were recorded for each subject. The resulting data are given in the accompanying table.

IGG.XLS

Subject	IgG y	Maximal Oxygen Uptake x	Subject	IgG y	Maximal Oxygen Uptake x	Subject	IgG y	Maximal Oxygen Uptake x
1	881	34.6	11	752	33.0	21	1,158	37.4
2	1,290	45.0	12	1,687	52.0	22	965	35.1
3	2,147	62.3	13	1,782	61.4	23	1,456	43.0
4	1,909	58.9	14	1,529	50.2	24	1,273	44.1
5	1,282	42.5	15	969	34.1	25	1,418	49.8
6	1,530	44.3	16	1,660	52.5	26	1,743	54.4
7	2,067	67.9	17	2,121	69.9	27	1,997	68.5
8	1,982	58.5	18	1,382	38.8	28	2,177	69.5
9	1,019	35.6	19	1,714	50.6	29	1,965	63.0
10	1,651	49.6	20	1,959	69.4	30	1,264	43.2

a. Construct a scatterplot for the data. What model does the data suggest?

b. Excel printouts for the quadratic model $y = \beta_0 + \beta_1 x + \beta_2 x^2 + \varepsilon$ are shown below and on page 655. Conduct a complete analysis of the model including a residual analysis.

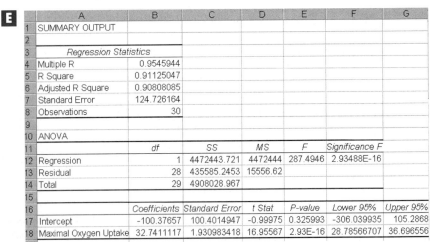

*10.111 Does extensive media coverage of a military crisis influence public opinion on how to respond to the crisis? Political scientists at UCLA researched this question and reported their results in *Communication Research* (June 1993). The military crisis of interest was the 1990 Persian Gulf War, precipitated by Iraqi leader Saddam Hussein's invasion of Kuwait. The researchers used multiple regression analysis to model the level y of support Americans had for a military (rather than

E

Maximal Oxygen Uptake Residual Plot

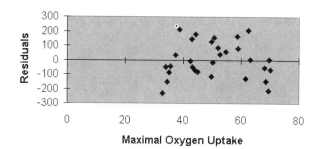

E

Residual Analysis

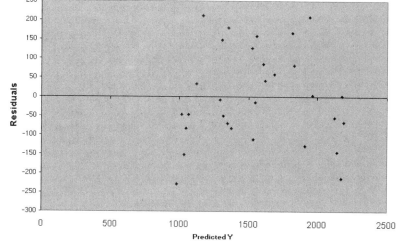

E

Histogram

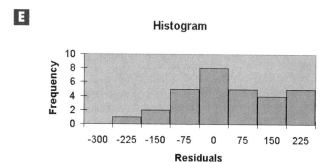

Problem 10.110

a diplomatic) response to the crisis. Values of y ranged from 0 (preference for a diplomatic response) to 4 (preference for a military response).
The following independent variables were used in the model:

x_1 = Level of TV news exposure in a selected week (number of days)

x_2 = Knowledge of 7 political figures (1 point for each correct answer)

x_3 = Gender (1 if male, 0 if female)

x_4 = Race (1 if nonwhite, 0 if white)

x_5 = Partisanship (0–6 scale, where 0 = strong Democrat and 6 = strong Republican)

x_6 = Defense spending attitude (1–7 scale, where 1 = greatly decrease spending and 7 = greatly increase spending)

x_7 = Education level (1–7 scale, where 1 = less than eight grades and 7 = college)

Data from a survey of 1,763 Americans were used to fit the model

$$y = \beta_0 + \beta_1 x_1 + \beta_2 x_2 + \beta_3 x_3 + \beta_4 x_4 + \beta_5 x_5 + \beta_6 x_6 + \beta_7 x_7 + \beta_8 x_2 x_3 + \beta_9 x_2 x_4 + \varepsilon$$

The regression results are shown in the accompanying table.

Variable	β Estimate	Standard Error	Two-Tailed p-Value
TV news exposure (x_1)	.02	.01	.03
Political knowledge (x_2)	.07	.03	.03
Gender (x_3)	.67	.11	<.001
Race (x_4)	−.76	.13	<.001
Partisanship (x_5)	.07	.01	<.001
Defense spending (x_6)	.20	.02	<.001
Education (x_7)	.07	.02	<.001
Knowledge × Gender ($x_2 x_3$)	−.09	.04	.02
Knowledge × Race ($x_2 x_4$)	.10	.06	.08

Source: Iyengar, S., and Simon, A. "News coverage of the Gulf Crisis and public opinion." *Communication Research,* Vol. 20, No. 3, June 1993, p. 380 (Table 2). Copyright 1993 by Sage Publications. Reprinted by permission of Sage Publications, Inc.

a. Interpret the β estimate for the variable x_1, TV news exposure.

b. Conduct a test to determine whether an increase in TV news exposure is associated with an increase in support for a military resolution of the crisis. Use α = .05.

c. Is there sufficient evidence to indicate that the relationship between support for a military resolution y and gender x_3 depends on political knowledge x_2? Test using α = .05.

d. Is there sufficient evidence to indicate that the relationship between support for a military resolution y and race x_4 depends on political knowledge x_2? Test using α = .05.

e. The coefficient of determination for the model was R^2 = .194. Interpret this value.

f. For this model, F = 46.88. Conduct a global test for model utility. Use α = .05.

10.112 Is there a link between the loneliness of parents and their offspring? Research psychologists examined this question in the *Journal of Marriage and the Family* (Aug. 1986). The participants in the study were 130 female college undergraduates and their parents. Each triad of daughter, mother, and father completed the UCLA Loneliness Scale, a 20-item questionnaire designed to assess loneliness and several variables theoretically related to loneliness, such as social accessibility to others, difficulty in making friends, and depression. Pearson product moment correlations relating daughter's loneliness to parent's loneliness score as well as the other variables were calculated. The results are summarized in the table.

Variable	CORRELATION r BETWEEN DAUGHTER'S LONELINESS AND PARENTAL VARIABLES	
	Mother	Father
Loneliness	.26	.19
Depression	.11	.06
Self-esteem	−.14	−.06
Assertiveness	−.05	.01
Number of friends	−.21	−.10
Quality of friendships	−.17	.01

Source: Lobdell, J., and Perlman, D. "The intergenerational transmission of loneliness. A study of college females and their parents." *Journal of Marriage and the Family,* Vol. 48, No. 8, Aug. 1986, p. 592. Copyright 1986 by the National Council on Family Relations, 3989 Central Ave., N.E., Suite #550, Minneapolis, MN 55421.

a. The researchers concluded that "mother and daughter loneliness scores were (positively) significantly correlated at $\alpha = .01$." Do you agree?

b. Determine which, if any, of the other sample correlations are large enough to indicate (at $\alpha = .01$) that linear correlation exists between daughter's loneliness score and the variable measured.

c. Explain why it would be dangerous to conclude that a causal relationship exists between mother's loneliness and daughter's loneliness.

d. Explain why it would be dangerous to conclude that the variables with nonsignificant correlations in the table are unrelated.

*10.113 Residential property appraisers make extensive use of multiple regression in their evaluation of property. Typically, the sale price y of a property is modeled as a function of several home-related conditions (e.g., gross living area, location, number of bedrooms). However, appraisers are not interested in the predicted price. Rather, they use the regression model as a tool for making value adjustments to the property. These adjustments are derived from the parameter estimates of the model. The *Real Estate Appraiser* (Apr. 1992) reported the results of a multiple regression on the price y of $n = 157$ residential properties recently sold in a northern Virginia subdivision. A table showing the results of the multiple regression analysis is reproduced on page 658. Note that there are 27 independent variables in the model.

a. Interpret the values of **F**, **s**, **R-Square**, and **Adj R-Sq** shown in the table.

b. One of the independent variables in the model is gross living area (G.L.A.), measured in square feet. A 95% confidence interval for the β coefficient associated with G.L.A. is shown in the table. Interpret this interval.

c. Note that the independent variables with β coefficients significantly different from 0 (at $\alpha = .05$) are highlighted in bold in the table. The nonsignificant variables are not highlighted. Would you advise the property appraiser to ignore any value adjustments based on nonsignificant independent variables? Explain.

OVERALL MODEL STATISTICS

$s = 6544.324$ R-Square $= .8140$
$F = 20.914$ (p $= .0001$)   Adj R-Sq $= .7751$

PARAMETER ESTIMATES

Variable	Parameter Estimate	Std Error	95% Confidence Interval (@ 129df = 1.98)	T for H_0: Parameter = 0	Prob>|T|
Intercept	96,603	12,530	(71,794 to 121,412)	7.710	.0001
Time	150	123	(−94 to 394)	1.220	.2248
Lot Size	**.60**	**.30**	**(0.01 to 1.19)**	2.022	**.0452***
Age	381	502	(−613 to 1,375)	.758	.4501
G.L.A.	**22.40**	**3.67**	**(15.13 to 29.67)**	6.099	**.0001***
Bedrooms	2,263	1,609	(−923 to 5,499)	1.407	.1619
Half Baths	**5,962**	**2,934**	**(153 to 11,771)**	2.032	**.0442***
Corner Lot	−1,481	1,692	(−4,831 to 1,869)	−.876	.3829
Cul-de-Sac	−56	2,557	(−5,119 to 5,007)	−.022	.9825
Back to Woods	**4,086**	**2,044**	**(39 to 8,133)**	1.999	**.0477***
Deck	2,408	2,167	(−1,883 to 6,699)	1.111	.2686
Fence	**2,896**	**1,271**	**(379 to 5,413)**	2.279	**.0243***
Shed	70	1,343	(−2,589 to 2,729)	.052	.9588
Patio	2,377	1,671	(−932 to 5,686)	1.423	.1572
Portico	−906	2,963	(−6,773 to 4,961)	−.306	.7603
Screen Porch	**5,021**	**2,038**	**(986 to 9,056)**	2.463	**.0151***
In-grnd Pool	**7,570**	**3,028**	**(1,575 to 13,565)**	2.500	**.0137***
Garage	**2,989**	**1,446**	**(126 to 5,852)**	2.068	**.0407***
Driveway	−1,844	3,222	(−8,224 to 4,536)	−.572	.5681
Fireplace	1,290	1,277	(−1,238 to 3,818)	1.010	.3144
Brick Facade	−2,140	2,369	(−6,381 to 2,551)	−.903	.3680
Updated Kit.	**4,171**	**1,470**	**(1,260 to 7,082)**	2.837	**.0053***
Remodel Kit.	**6,091**	**2,367**	**(1,404 to 10,778)**	2.574	**.0112***
Intercom	1,933	2,146	(−2,316 to 6,182)	.901	.3693
Cen. Vacuum	**−4,636**	**2,166**	**(−8,925 to −347)**	−2.140	**.0342***
Skylights	**7,744**	**2,622**	**(2,552 to 12,936)**	−2.954	**.0037***
Air Filter	874	2,506	(−4,088 to 5,836)	−.349	.7280
Bay Window	−3,174	2,086	(−7,304 to 956)	−1.522	.1305

*Indicates significance at the 5% significance level.

Source: Gilson, S. J. "A Case Study—Comparing the results: Multiple regression analysis vs. matched pairs in residential subdivision." *The Real Estate Appraiser,* Apr. 1992, p. 37 (Table 4).

Problem 10.113

FTC.XLS

*10.114 Refer to the FTC cigarette data stored in the Excel workbook FTC.XLS. Select a random sample of $n = 50$ cigarette brands from the data set. Suppose you want to use the data to fit the model $y = \beta_0 + \beta_1 x_1 + \beta_2 x_2 + \varepsilon$, where y = carbon monoxide content, x_1 = tar content, and x_2 = nicotine content.

a. Use Excel to obtain the correlation between x_1 and x_2. What does this imply?

b. Using Excel, fit the model $y = \beta_0 + \beta_1 x_1 + \beta_2 x_2 + \varepsilon$. Do you detect any signs of multicollinearity?

*10.115 An article published in *Geography* (July 1980) used multiple regression to predict annual rainfall levels in California. Data on the average annual precipitation y, altitude x_1, latitude x_2, and distance from the

RAIN.XLS

Pacific coast x_3 for 30 meteorological stations scattered throughout California are listed in the accompanying table. Initially, the first-order model $y = \beta_0 + \beta_1 x_1 + \beta_2 x_2 + \beta_3 x_3 + \varepsilon$ was fit to the data. Excel print-outs of the analysis follow.

Station	Average Annual Precipitation y (inches)	Altitude x_1 (feet)	Latitude x_2 (degrees)	Distance from Coast x_3 (miles)
1. Eureka (W)	39.57	43	40.8	1
2. Red Bluff (L)	23.27	341	40.2	97
3. Thermal (L)	18.20	4,152	33.8	70
4. Fort Bragg (W)	37.48	74	39.4	1
5. Soda Springs (W)	49.26	6,752	39.3	150
6. San Francisco (W)	21.82	52	37.8	5
7. Sacramento (L)	18.07	25	38.5	80
8. San Jose (L)	14.17	95	37.4	28
9. Giant Forest (W)	42.63	6,360	36.6	145
10. Salinas (L)	13.85	74	36.7	12
11. Fresno (L)	9.44	331	36.7	114
12. Pt. Piedras (W)	19.33	57	35.7	1
13. Pasa Robles (L)	15.67	740	35.7	31
14. Bakersfield (L)	6.00	489	35.4	75
15. Bishop (L)	5.73	4,108	37.3	198
16. Mineral (W)	47.82	4,850	40.4	142
17. Santa Barbara (W)	17.95	120	34.4	1
18. Susanville (L)	18.20	4,152	40.3	198
19. Tule Lake (L)	10.03	4,036	41.9	140
20. Needles (L)	4.63	913	34.8	192
21. Burbank (W)	14.74	699	34.2	47
22. Los Angeles (W)	15.02	312	34.1	16
23. Long Beach (W)	12.36	50	33.8	12
24. Los Banos (L)	8.26	125	37.8	74
25. Blythe (L)	4.05	268	33.6	155
26. San Diego (W)	9.94	19	32.7	5
27. Daggett (L)	4.25	2,105	34.09	85
28. Death Valley (L)	1.66	−178	36.5	194
29. Crescent City (W)	74.87	35	41.7	1
30. Colusa (L)	15.95	60	39.2	91

Source: Taylor, P. J. "A pedagogic application of multiple regression analysis." *Geography,* July 1980, Vol. 65, pp. 203–212.

E

	A	B	C	D	E	F	G
3	*Regression Statistics*						
4	Multiple R	0.7747871					
5	R Square	0.60029505					
6	Adjusted R Square	0.55417524					
7	Standard Error	11.097986					
8	Observations	30					
9							
10	ANOVA						
11		*df*	*SS*	*MS*	*F*	*Significance F*	
12	Regression	3	4809.355964	1603.119	13.01599	2.20513E-05	
13	Residual	26	3202.297623	123.1653			
14	Total	29	8011.653587				
15							
16		*Coefficients*	*Standard Error*	*t Stat*	*P-value*	*Lower 95%*	*Upper 95%*
17	Intercept	-102.35743	29.20548173	-3.50473	0.001676	-162.3901959	-42.324662
18	Altitude	0.00409052	0.001218311	3.357532	0.002431	0.001586242	0.00659479
19	Latitude	3.45107976	0.794863124	4.341728	0.000191	1.817214135	5.08494538
20	Distance from Coast	-0.1428578	0.036340056	-3.93114	0.000559	-0.21755588	-0.0681597

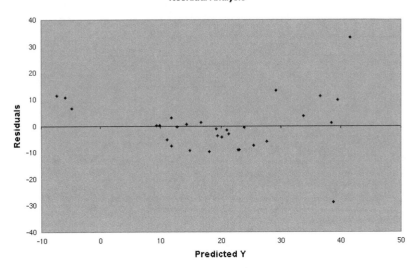

E

Residual Analysis

a.   Conduct a complete analysis of the multiple regression model. Is the model useful for predicting precipitation level y?

b.   Conduct an analysis of the regression residuals. Are model modifications necessary? Explain.

c.   Refer to your plot of the residuals against $\hat{y}$ in part **b**. The *Geography* researcher noted that stations located on the westward-facing slopes of the California mountains (identified by the symbol "W" in the table) invariably had positive residuals (i.e., the least squares model underpredicted the level of precipitation) whereas stations on the leeward side of the mountains (identified by the symbol "L" in the table) had negative residuals (i.e., the least squares model overpredicted the level of precipitation). In the researcher's words, "This suggests a very clear shadow effect of the mountains, for which California is known." Verify this observation on your residual plot.

d.   The shadow effect detected in part **c** suggests that a dummy variable for leeward/westward will improve the fit of the model. Let $x_4 = 1$ if leeward side and $x_4 = 0$ if westward side and then fit the modified model $y = \beta_0 + \beta_1 x_1 + \beta_2 x_2 + \beta_3 x_3 + \beta_4 x_4 + \varepsilon$ to the data using Excel. Does the addition of the β_4 term improve the fit of the model? Explain.

Using Microsoft Excel

10.E.1 Using Microsoft Excel to Generate Scatter Diagrams and a Regression Line

Solution Summary:

A. Use the Chart Wizard to generate a scatter diagram to plot the relationship between an X variable and a Y variable.

B. Modify the scatter diagram generated by the wizard to include a trend line, the regression equation, and the value of R^2.

Example: Creativity scores analysis of Examples 10.5 and 10.6 on pages 560–563.

Part A Solution: Using the Chart Wizard to generate a scatter diagram

Use the XY (Scatter) choice of the Chart Wizard (see Section 2.E.4 on page 141) to generate a scatter diagram to plot the relationship between an X variable and a Y variable. As an example, to generate a scatter diagram that explores the relationship between the flexibility score and the creativity score for the creativity scores analysis of Examples 10.5 and 10.6, do the following:

1 Open the CREATIVITY.XLS workbook and click the Data sheet tab. Verify that child, flexibility score, and creativity score data of Table 10.2 on page 561 have been entered into columns A, B, and C, respectively.

2 Select Insert | Chart.

3 In the Step 1 dialog box:

  a Select the Standard Types tab and then select XY (Scatter) from the Chart type: list box. Select the first (top) choice from the Chart subtypes, which is identified as "Scatter: Compares pairs of values" when selected.

  b Click the Next button.

4 In the Step 2 dialog box:

  a Select the Data range tab. Enter B1:C11 in the Data range: edit box.

  b Select the Columns option button in the Series in: group.

  c Click the Next button.

5 In the Step 3 dialog box:

  a Select the Titles tab. Enter Regression Analysis for Creativity Scores in the Chart title: edit box, Flexibility Score in the Value (X) axis: edit box, and Creativity Score ($000) in the Value ($Y$) axis: edit box.

  b Select, in turn, the Gridlines, Axes, Legend, and Data Labels tabs and verify that their settings match those given in Table 2.E.2 on page 131.

  c Click the Next button.

6 In the Step 4 dialog box:

  a Select the As new sheet: option button and enter Scatter in the edit box to the right of the option button.

  b Click the Finish button.

The Chart Wizard inserts a chart sheet containing the scatter diagram for the data of Table 10.2 similar to one shown in Figure 10.3 on page 561.

Data Order and Scatter Diagrams

When generating scatter diagrams, the Chart Wizard always assumes that the first column (or row) of data of the data range entered in the Step 2 dialog box contains values for the X variable (as it does in the preceding example). Had the X variable data been located in the second column of the data range, it would have been necessary to select the Series tab of the Step 2 dialog box and change the cell ranges in the X Values: and Y Values: edit boxes. Furthermore, due to a quirk in this wizard, the revised cell ranges must be entered into those edit boxes as formulas that include sheet names, e.g., =Data!B1:B11. Using the simple cell range form (B1:B11) would cause Microsoft Excel to display the misleading "The formula you typed contains an error" error message.

Part B Solution: Modifying the scatter diagram

After the Chart Wizard generates a scatter diagram, we can modify the chart by adding a trend line, the regression equation, and the value of R^2. As an example, consider the chart produced in Figure 10.3 for the creativity scores problem. To add a trend line to this scatter diagram, do the following:

1.   Click the sheet tab of the Scatter chart sheet generated in Part A Solution of this procedure.

2.   Select Chart | Add Trendline. (Note: the Chart choice appears on the Microsoft Excel menu bar only when a chart or chart sheet is selected.)

3.   In the Add Trendline dialog box:

     a   Select the Type tab and select the Linear choice in the Trend/Regression type group (see Figure 10.E.1a).

     b   Select the Options tab and select the Automatic option button and the Display equation on chart and Display R-squared value on chart check boxes (see Figure 10.E.1b).

     c   Click the OK button.

E **Figure 10.E.1a**
Type Tab of the Add
Trendline Dialog Box

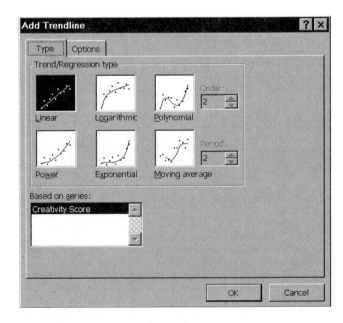

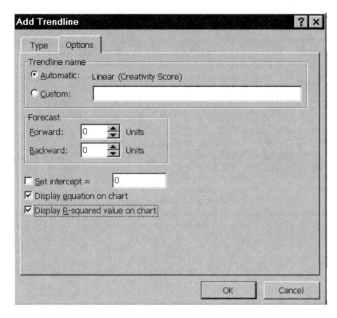

E **Figure 10.E.1b**
Options Tab of
the Add Trendline
Dialog Box

Microsoft Excel modifies the scatter diagram by adding a line of regression, the regression equation, and the value of R^2. The modified scatter diagram will be similar to the one shown in Figure 10.5 on page 563. We can change the values for the creativity score or flexibility score in the Data sheet to see their effects on the fit of the regression model as represented by the line of regression and regression equation on the scatter diagram.

10.E.2 Using Microsoft Excel for Simple Linear Regression

Solution Summary:

Use the PHStat add-in to calculate the coefficients of the simple linear regression equation, to obtain residual plots, and to estimate predicted values.

Example: Creativity scores analysis of Examples 10.5 and 10.6 on pages 560–563.

Solution:

Use the Regression | Simple Linear Regression choice of the PHStat add-in to perform a simple linear regression analysis on a set of data. This add-in modifies and extends the output generated by the Data Analysis Regression tool that could also be used to calculate the regression coefficients and obtain residual plots. As an example, to calculate the coefficients of the simple linear regression equation that represents the relationship between flexibility score and creativity score, to obtain residual plots, and to estimate predicted values for the creativity scores analysis of Examples 10.5 and 10.6, do the following:

1 Open the CREATIVITY.XLS workbook and click the Data sheet tab. Verify that child, flexibility score, and creativity score data of Table 10.2 on page 561 have been entered into columns A, B, and C, respectively.

2 Select PHStat | Regression | Simple Linear Regression.

3 In the Simple Linear Regression dialog box (see Figure 10.E.2):

    a Enter C1:C11 in the Y Variable Cell Range: edit box.

    b Enter B1:B11 in the X Variable Cell Range: edit box.

    c Select the First cells in both ranges contain label check box.

d   Enter 95 in the Confidence Lvl. for regression: edit box.

e   Select all four check boxes of the Regression Tool Output Options group.

f   Enter Regression for Creativity Scores in the Output Title: edit box.

g   Select the Scatter Diagram check box if desired.

h   Select the Confidence & Prediction Interval for X check box and enter 5 in its edit box. Enter 95 in the Confidence level for int.: edit box.

i   Click the OK button.

The add-in inserts a worksheet that contains the regression coefficients and other summary information for the simple linear regression for the Table 10.2 data, similar to the worksheet shown in Figure 10.4 on page 563. In addition, the worksheet contains a table of predicted Y values and residuals and a residual plot of the residuals on the vertical Y axis and the X values on the horizontal axis. This worksheet is *not* dynamically changeable, so any changes made to the underlying creativity data would require using the PHStat add-in a second time to produce updated results.

To obtain a plot of the residuals and the predicted Y values, follow the instructions for obtaining a scatter diagram provided in Section 10.E.1, substituting the cell range for the predicted Y values and the residuals for the cell range of the X and Y variables.

E   **Figure 10.E.2**
PHStat Simple Linear
Regression Dialog Box

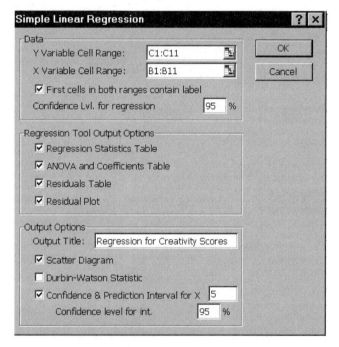

10.E.3 Using Microsoft Excel to Test for the Existence of Correlation

Solution Summary:

Implement a worksheet that uses the TINV and TDIST worksheet functions to perform a t test of the hypothesis of no correlation.

Example: Tire pressure problem of Example 10.15 on page 586

Solution:

To perform a t test of the hypothesis of no correlation, implement a worksheet based on the design shown in Table 10.E.1 on page 666. This design uses the TINV function and the TDIST function. The format of these functions is

$$TINV(level\ of\ significance,\ degrees\ of\ freedom)$$

$$TDIST(absolute\ value\ of\ the\ t\ test\ statistic,\ degrees\ of\ freedom,\ tails)$$

where
   $tails = 1$ for a one-tailed test and 2 for a two-tailed test

TINV returns the critical value of the t statistic. TDIST returns the probability of exceeding a given t value for a given number of degrees of freedom and type of test.

   As an example, to test the null hypothesis that the correlation coefficient for the tire pressure problem of Example 10.15 is zero, implement rows 1 through 13 of the Table 10.E.1 design. Only those rows need to be implemented as only the two-tailed test is needed for this example. To implement these rows, do the following:

1 Select File | Open to open a new workbook (or open the existing workbook into which the hypothesis testing worksheet is to be inserted).

2 Select an unused worksheet (or select Insert | Worksheet if there are none) and rename the sheet hypothesis.

3 Enter the title, labels, and formulas for column A as shown in Table 10.E.1. Enter the formulas in cells A13, A18, and A23, which have been typeset as two lines, as one continuous line.

4 Enter the level of significance, sample size, and sample correlation coefficient in the cell range B3:B5. If the correlation coefficient needs to be computed, use the CORREL function explained in Section 3.E.1 on page 210. For the tire pressure example, enter 0.05 in cell B3, 14 in cell B4, and –0.114 in cell B5.

5 Enter the formulas for cells B6, B7, B10:B12.

The resulting worksheet will be similar to the one shown in Figure 10.13b on page 587.

Table 10.E.1 Worksheet Column A and B Design for Testing for the Existence of Correlation

	A	B
1	T Test for the Existence of Correlation	
2		
3	Level of Significance	.xx
4	Sample Size	xxx
5	Sample Correlation Coefficient	xxx
6	Degrees of Freedom	=B4-2
7	*t* Test Statistic	=B5/SQRT((1-B5^2))
8		
9	Two-Tailed Test	
10	Lower Critical Value	=-(TINV(B3,B6))
11	Upper Critical Value	=(TINV(B3,B6)
12	*p*-Value	=TDIST(ABS(B7),B6,2)
13	=IF(B12<$B$3,"Reject the null hypothesis", Do not reject the null hypothesis")	
14		
15	Lower-Tail Test	
16	Lower Critical Value	=-(TINV(2*B3,B6))
17	*p*-Value	=IF(B7<0,E19,E20)
18	=IF(B17<$B$3,"Reject the null hypothesis", Do not reject the null hypothesis")	
19		
20	Upper-Tail Test	
21	Upper Critical Value	=(TINV(2*B3,B6))
22	*p*-Value	=IF(B7<0,E20,E19)
23	=IF(B22<$B$3,"Reject the null hypothesis", Do not reject the null hypothesis")	

Table 10.E.2 Worksheet Column D and E Design for Testing for the Existence of Correlation

	D	E
17	Calculations Area	
18	For one-tail tests	
19	TDIST value	=TDIST(ABS(B7),B6,1)
20	1-TDIST value	1 – E19

Note: Column C has been left blank.

10.E.4 Using Microsoft Excel for Multiple Regression (Optional)

Solution:

A. Use the PHStat add-in to perform a multiple regression analysis.

B. (Optional) Use Excel to create a second-order variable.

C. (Optional) Use Excel to create a dummy variable with two categories.

Example: Real estate sale price analysis of Examples 10.30–10.34 on pages 626–633

Part A Solution: Using the PHStat add-in to perform a multiple regression analysis

Use the Regression | Multiple Regression choice of the PHStat add-in to perform a multiple regression analysis. This choice modifies and extends the output generated by the Data Analysis Regression tool, which could also be used to calculate the regression coefficients and obtain residual plots. As an example, consider the real estate sale price analysis of Examples 10.30–10.34. To calculate the coefficients of the multiple regression equation that represent the relationship between the sale price and the appraised improvements and the area, and to obtain residual plots, do the following:

1 Open the APPRAISAL.XLS workbook and click the Data sheet tab. Verify that the sale price, appraised improvements, and the area data of Table 10.8 on page 626 have been entered into columns B, C, and D, respectively.

2 Select PHStat | Regression | Multiple Regression.

3 In the Multiple Regression dialog box (see Figure 10.E.3):

 a Enter B1:B21 in the Y Variable Cell Range: edit box.

 b Enter C1:D21 in the X Variables Cell Range: edit box. Note that all *X* variables must be in contiguous columns.

 c Select the First cells in both ranges contain label check box.

 d Enter 95 in the Confidence Lvl. for regression: edit box.

 e Select all four check boxes of the Regression Tool Output Options group.

 f Enter Regression for Sales Price in the Output Title: edit box.

 g Click the OK button.

The add-in inserts a worksheet that contains the regression coefficients and other summary information for the simple linear regression for the Table 10.8 data, similar to the worksheet shown in Figure 10.27 on page 627. In addition, the worksheet contains a table of predicted *Y* values and residuals and residual plots for each of the *X* variables. This worksheet is *not* dynamically changeable, so any changes made to the data would require using the PHStat add-in a second time to produce updated results.

To obtain a plot of the residuals and the predicted *Y* values, follow the instructions for obtaining a scatter diagram provided in Section 10.E.1, substituting the cell range for the predicted *Y* values and the residuals for the cell range of the *X* and *Y* variables.

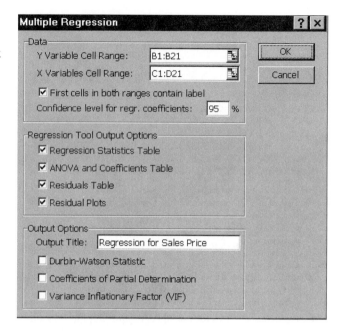

E **Figure 10.E.3**
PHStat Multiple
Regression Dialog Box

Part B Solution: Using Excel to create a second-order variable

If a second-order (quadratic) regression model is to be used, we can use simple formulas to create the square of an explanatory variable. For example, if the cell range for an explanatory variable is B2:B20, a variable that is the square of this X variable can be created by entering the formula =B2^2 in cell C2, and copying this formula through cell C20.

Part C Solution: Using Excel to create a dummy variable

If a regression model with dummy variables is to be used, and the category values for a dummy variable are provided rather than numerical codes, the Excel find and replace procedure can be used to code the category names into 0 and 1 codes. Suppose that the category codes Yes and No were contained in the cell range C1:C51. To change these categorical labels into the numerical codes of 0 for No and 1 for Yes, do the following:

1.   Select the range C1:C51 containing the categorical responses Yes and No, then select Edit | Replace.
2.   In the Replace dialog box:
   - **a**   Enter Yes in the Find what: edit box.
   - **b**   Enter 1 in the Replace with: edit box.
   - **c**   Click the Replace All button.
3.   With the cell range C1:C51 still selected, select Edit | Replace a second time.
4.   In the Replace dialog box:
   - **a**   Enter No in the Find what: edit box.
   - **b**   Enter 0 in the Replace with: edit box.
5.   Click the Replace All button.

Chapter 11

Analysis of Variance

CONTENTS

11.1 Experimental Design

11.2 ANOVA Fundamentals

11.3 Completely Randomized Designs: One-Way ANOVA

11.4 Factorial Designs: Two-Way ANOVA

11.5 Follow-Up Analysis: Multiple Comparisons of Means

11.6 Checking ANOVA Assumptions

11.7 Calculation Formulas for ANOVA (Optional)

EXCEL TUTORIAL

11.E.1 Using Microsoft Excel for a One-Way ANOVA

11.E.2 Using Microsoft Excel for a Two-Way ANOVA

11.E.3 Using Microsoft Excel for Multiple Comparisons

REAL-WORLD APPLICATION
Reluctance to Transmit Bad News—The MUM Effect

The reluctance to deliver bad news has been termed the "MUM effect." Forty undergraduates at Duke University participated in a study designed to determine what factors influence the MUM effect. The subjects were equally divided into four groups, and each group was treated under a different experimental condition. The length of time it took each to deliver the bad news was recorded. The goal was to determine if the average length of time required to deliver the bad news differed for the four groups. In this chapter, we consider the general problem of comparing more than two population means. We examine the MUM effect in greater detail in Real-World Revisited.

11.1 EXPERIMENTAL DESIGN

As we have seen in Chapters 7, 8, and 9, the solutions to many statistical problems are based on inferences about population means. In Chapters 7 and 8 we considered inferences about a single population mean, and in Chapter 9 we developed inferences for the difference between two population means. In this chapter we expand our discussion to a comparison of three or more means. The procedure for selecting sample data for a comparison of several population means is called the *design of the experiment,* and the statistical procedure for comparing the means is called an *analysis of variance.* The objective of this chapter is to introduce some aspects of experimental design and the analysis of data from such experiments using an analysis of variance.

The study of experimental design originated in England and, in its early years, was associated solely with agricultural experimentation. The need for experimental design in agriculture was very clear: It takes a full year to obtain a single observation on the yield of a new variety of wheat. Consequently, the need to save time and money led to a study of ways to obtain more information using smaller samples. Similar motivation led to its subsequent acceptance and wide use in fields such as biology, engineering, psychology, and sociology. Despite this fact, the terminology associated with experimental design clearly indicates its early association with the biological sciences.

We call the process of collecting sample data an *experiment* and the variable to be measured the *response.* (In this chapter we consider only experiments in which the response is a quantitative variable. Experiments with qualitative responses are beyond the scope of this text.) The planning of the sampling procedure is called the *design* of the experiment. The object upon which the response measurement is taken is called an *experimental* (or *sampling*) unit.

Definition 11.1

The process of collecting sample data is called an **experiment**.

Definition 11.2

The plan for collecting the sample is called the **design** of the experiment.

Definition 11.3

The variable measured in the experiment is called the **response variable**. (In this chapter, all response variables will be quantitative variables.) The response will be denoted with the symbol y.

Definition 11.4

The object upon which the response variable is measured is called an **experimental** (or **sampling**) **unit**.

Variables that may be related to a response variable are called *factors.* The value assumed by a factor in an experiment is called a *level.* The combinations of levels of the factors for which the response will be observed are called *treatments.*

Definition 11.5

The variables, quantitative or qualitative, that are related to a response variable are called **factors**.

Definition 11.6

The intensity setting of a factor (i.e., the value assumed by a factor in an experiment) is called a **factor level**.

Definition 11.7

A **treatment** is a particular combination of levels of the factors involved in an experiment.

Example 11.1 A Designed Experiment

A marketing study is conducted to investigate the effects of brand and shelf location on mean weekly coffee sales. Coffee sales are recorded for each of the two brands (brand A and brand B) and three shelf locations (bottom, middle, and top). The $2 \times 3 = 6$ combinations of brand and shelf location were varied each week for a period of 18 weeks. A layout of the design is displayed in Figure 11.1. For this experiment, identify

   a. the experimental unit   **b.** the response y

   c. the factors   **d.** the factor levels

   e. the treatments   **f.** the objective of the analysis

Solution

   a. Since our data will be collected each week for a period of 18 weeks, the experimental unit is a week.

   b. The variable of interest, i.e., the response, is $y =$ weekly coffee sales. Note that weekly coffee sales is a quantitative variable.

   c. Since we are interested in investigating the effect of brand and shelf location on mean sales, *brand* and *shelf location* are the factors. Note that both factors are qualitative variables, although, in general, they may be quantitative or qualitative.

Figure 11.1 Layout for Designed Experiment of Example 11.1

Brand	Shelf Location		
	Bottom	Middle	Top
A	Week 1 9 14	Week 2 7 16	Week 4 12 17
B	Week 5 10 13	Week 3 8 18	Week 6 11 15

d. For this experiment, brand is measured at two levels (A and B) and shelf location at three levels (bottom, middle, and top).

e. Since coffee sales are recorded for each of the six brand-shelf location combinations (brand A, bottom), (brand A, middle), (brand A, top), (brand B, bottom), (brand B, middle), and (brand B, top), the experiment involves six treatments. The term *treatments* is used to describe the factor-level combinations to be included in an experiment because many experiments involve "treating" or doing something to alter the nature of the experimental unit. Thus, we might view the six brand-shelf location combinations as treatments on the experimental units in the marketing study involving coffee sales.

f. The objective of the analysis is to compare the mean weekly coffee sales associated with the six treatments. ❏

Now that you understand some of the terminology, it is helpful to think of the design of an experiment in four steps.

Designing an Experiment

1. Select the factors to be included in the experiment and identify the parameters that are the object of the study. Usually, the target parameters are the population means associated with the factor-level combinations (i.e., treatments).

2. Choose the treatments (the factor-level combinations) to be included in the experiment.

3. Determine the number of observations (sample size) to be made for each treatment. (This will usually depend on the standard error(s) that you desire. See Section 7.5.)

4. Plan how the treatments will be assigned to the experimental units. That is, decide on which design to employ.

**Self-Test
11.1**

The United States Golf Association (USGA) regularly tests golf equipment to ensure that it conforms to USGA standards. Suppose it wishes to compare the mean distance traveled by four different brands of golf balls when struck by a driver (the club used to maximize distance). The following experiment is conducted: 10 balls of each brand are randomly selected. Each is struck by "Iron Byron" (the USGA's golf robot named for the famous golfer Byron Nelson) using a driver, and the distance traveled is recorded. Identify each of the following elements in this experiment: response, factors, levels, treatments, and experimental units.

Entire texts are devoted to properly executing these steps for various experimental designs (e.g., Cochran and Cox [1957]). The main objective of this chapter, however, is to show how to analyze the data that are collected in a designed experiment.

In Sections 11.3 and 11.4, we consider two widely used experimental designs and demonstrate how to analyze the data for each design. First, in Section 11.2, we present a short discussion of the logic behind the analysis of data collected from such experiments.

11.2 ANOVA FUNDAMENTALS

Once the data for a designed experiment have been collected, we use the sample information to make inferences about the population means associated with the various treatments. The method used to compare the treatment means is known as **analysis of variance**, or **ANOVA**. The concept behind an analysis of variance can be explained using the following simple example.

Example 11.2   *Concept of ANOVA*

Suppose we want to compare the means (μ_1 and μ_2) of two populations using independent random samples of size $n_1 = n_2 = 5$ from each of the populations. The sample observations and the sample means are listed in Table 11.1 and shown on a line (dot) plot in Figure 11.2.

a. Do you think these data provide sufficient evidence to indicate a difference between the population means μ_1 and μ_2?

b. Now look at two more samples of $n_1 = n_2 = 5$ measurements from the populations, as listed in Table 11.2 and plotted in Figure 11.3 (page 674). Do these data appear to provide evidence of a difference between μ_1 and μ_2?

Solution

a. One way to determine whether a difference exists between the population means μ_1 and μ_2 is to examine the difference *between* the sample means, $(\bar{y}_1 - \bar{y}_2)$, and to compare it to a measure of variability *within* the samples. The greater the disparity in these two measures, the greater will be the evidence to indicate a difference between μ_1 and μ_2.

Table 11.1 Data for Example 11.2a

Sample from Population 1	Sample from Population 2
6	8
−1	1
0	3
4	7
1	6
$\bar{y}_1 = 2$	$\bar{y}_2 = 5$

Figure 11.2 Line Plot of Data in Table 11.1

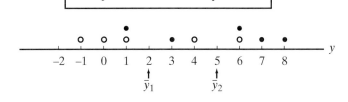

Table 11.2 Data for Example 11.2b

Sample from Population 1	Sample from Population 2
2	5
3	5
2	5
2	4
1	6
$\bar{y}_1 = 2$	$\bar{y}_2 = 5$

Figure 11.3 Line Plot of Data in Table 11.2

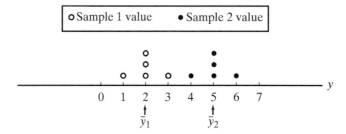

For the data of Table 11.1, you can see in Figure 11.2 that *the difference between the sample means is small relative to the variability within the sample observations.* Thus, we think you will agree that the difference between $\bar{y}_1$ and $\bar{y}_2$ is probably not large enough to indicate a difference between μ_1 and μ_2.

b. Notice that the difference between the sample means for the data of Table 11.2 is identical to the difference shown in Table 11.1. However, since there is now very little variability within the sample observations (see Figure 11.3), *the difference between the sample means is large compared to the variability within the sample observations.* Thus, the data appear to give evidence of a difference between μ_1 and μ_2. ❑

To conduct a formal test of hypothesis, we need to quantify these two measures of variation. The **within-sample variation** is measured by the pooled s^2 that we computed for the independent random samples t test of Section 9.2, namely

$$\text{Within-sample variation: } s^2 = \frac{\sum\limits_{i=1}^{n_1}(y_{i1} - \bar{y}_1)^2 + \sum\limits_{i=1}^{n_2}(y_{i2} - \bar{y}_2)^2}{n_1 + n_2 - 2} \qquad \textbf{(11.1)}$$

$$= \frac{\text{SSE}}{n_1 + n_2 - 2} = \frac{\text{SSE}}{n - 2} = \text{MSE}$$

where y_{i1} is the *i*th observation in sample 1, y_{i2} is the *i*th observation in sample 2, and $n = (n_1 + n_2)$ is the total sample size. The quantity in the numerator of s^2 is often denoted SSE, the sum of squared errors. SSE measures unexplained variability—that is, it measures variability *unexplained* by the differences between the sample means. The ratio $\text{SSE}/(n - 2)$ is also called **mean square for error**, or **MSE**. Thus, MSE is another name for s^2.

A measure of the **between-sample variation** is given by the weighted sum of squares of deviations of the individual sample means about the mean for all n observations, $\bar{y}$, divided by the number of samples minus 1, i.e.,

$$\text{Between-sample variation: } s^2 = \frac{n_1(\bar{y}_1 - \bar{y})^2 + n_2(\bar{y}_2 - \bar{y})^2}{2 - 1} \quad \textbf{(11.2)}$$

$$= \frac{\text{SST}}{1} = \text{MST}$$

The quantity in the numerator is often denoted SST, the sum of squares for treatments, because it measures the variability *explained* by the differences between the sample means of the two treatments. The ratio SST/(2 − 1) is also called **mean square for treatments**, or **MST**.

For this experimental design, SSE and SST sum to a known total, namely,

$$\text{SS(Total)} = \sum (y_i - \bar{y})^2$$

In the next section we will see that

$$F = \frac{\text{Between-sample variation}}{\text{Within-sample variation}} = \frac{\text{MST}}{\text{MSE}} \quad \textbf{(11.3)}$$

has an F distribution with $\nu_1 = 1$ and $\nu_2 = n_1 + n_2 - 2$ degrees of freedom (df) and therefore can be used to test the null hypothesis of no difference between the treatment means. Large values of F lead us to conclude that a difference between the treatment (population) means exists.

The additivity property of the sums of squares led early researchers to view this analysis as a *partitioning* of SS(Total) $= \Sigma(y_i - \bar{y})^2$ into sources corresponding to the factors included in the experiment and to SSE. The formulas for computing the sums of squares, the additivity property, and the form of the test statistic made it natural for this procedure to be called an *analysis of variance*. We demonstrate the analysis of variance procedures for the general problem of comparing k population means for two special types of experimental designs in Sections 11.3 and 11.4.

11.3 COMPLETELY RANDOMIZED DESIGNS: ONE-WAY ANOVA

The simplest experimental design employed in practice is called a *completely randomized design*. This experiment involves a comparison of the means for a number, say k, of treatments, based on independent random samples of $n_1, n_2, \ldots, n_k$ observations, drawn from populations associated with treatments 1, 2, $\ldots, k$, respectively. A layout for a completely randomized design is shown in Figure 11.4

Figure 11.4 Layout for a Completely Randomized Design

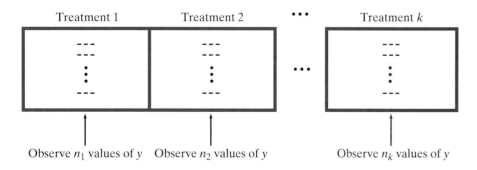

Definition 11.8

A **completely randomized design** to compare k treatment (population) means is one in which the treatments are randomly assigned to the experimental units, or in which independent random samples are drawn from each of the k target populations.

A key feature of a completely randomized design is that there is only a single factor of interest. (For this reason, the analysis of the data is often called a *one-way analysis of variance.*) The levels of this factor represent the k treatments (or populations) to be compared. If we define μ_i as the mean of the population of measurements associated with treatment i, for $i = 1, 2, ..., k$, then the null hypothesis to be tested is that the k treatment means are equal, i.e.,

$$H_0: \mu_1 = \mu_2 = \cdots = \mu_k$$

and the alternative hypothesis is that at least two of the treatment means differ.

Example 11.3   Setting Up the One-Way ANOVA

Sociologists often conduct experiments to investigate the relationship between socioeconomic status and college performance. Socioeconomic status is generally partitioned into three groups: lower class, middle class, and upper class. Consider the problem of comparing the mean grade point averages of those college freshmen associated with the lower class, those associated with the middle class, and those associated with the upper class. The grade point averages for random samples of seven college freshmen associated with each class—a total of 21 freshmen—were selected from a university's files at the end of the academic year. The data are recorded in Table 11.3.

a. Verify that the experimental design is completely randomized, and identify the treatments.

b. State the null and alternative hypotheses for a test to determine whether there is a difference among the true mean grade point averages (GPAs) for the three socioeconomic classes.

GPACLASS.XLS

Table 11.3 Grade Point Averages for Three Socioeconomic Groups, Example 11.3

	(1) Lower Class	(2) Middle Class	(3) Upper Class
	2.87	3.23	2.25
	2.16	3.45	3.13
	3.14	2.78	2.44
	2.51	3.77	3.27
	1.80	2.97	2.81
	3.01	3.53	1.36
	2.16	3.01	2.53
Sample Means:	$\bar{y}_1 = 2.52$	$\bar{y}_2 = 3.25$	$\bar{y}_3 = 2.54$

Solution

a. Since the data are collected from independent random samples of college freshmen from each population (socioeconomic class), this is a completely randomized design. The three socioeconomic classes—lower class, middle class, and upper class—represent the three treatments (or populations).

b. Our objective is to determine whether differences exist among the three population (socioeconomic class) GPA means. Consequently, we will test the null hypothesis "All three population means—μ_1, μ_2, and μ_3—are equal." That is,

$$H_0: \mu_1 = \mu_2 = \mu_3$$

where μ_1 is the mean GPA for those college freshmen associated with the lower socioeconomic class, μ_2 is the mean GPA for those associated with the middle class, and μ_3 is the mean GPA for those associated with the upper class. The alternative hypothesis of interest is

$$H_a: \text{At least two of the population means differ}$$

If an analysis of variance of the data in Table 11.3 shows that the differences among the sample means are large enough to indicate differences among the corresponding population means, we will reject the null hypothesis H_0 in favor of the alternative hypothesis H_a; otherwise, we will not reject H_0 and will conclude that there is insufficient evidence to indicate differences among the population means. ❑

An analysis of variance for a completely randomized design partitions SS(Total) into two components, SSE and SST (see Figure 11.5). Recall that the quantity SST denotes the sum of squares for treatments and measures the variation explained by the differences between the treatment means. The sum of squares for error, SSE, is a measure of the unexplained variability, obtained by calculating a pooled measure of the variability *within* the k samples. If the treatment means truly differ, then SSE should be substantially smaller than SST. We compare the two sources of variability by forming an F statistic:

$$F = \frac{\text{SST}/(k-1)}{\text{SSE}/(n-k)} = \frac{\text{MST}}{\text{MSE}} \tag{11.4}$$

where n is the total number of measurements.

Under certain conditions, the F statistic has an F *distribution* that depends on ν_1 numerator degrees of freedom and ν_2 denominator degrees of freedom. For the completely randomized design, F is based on $\nu_1 = (k-1)$ and $\nu_2 = (n-k)$ degrees of freedom. If the computed value of F exceeds the upper critical value F_α, we reject H_0 and conclude that at least two of the treatment means differ.

Figure 11.5
The Partitioning of SS(Total) for a Completely Randomized Design

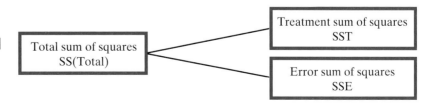

ANOVA F Test to Compare k Population Means for a Completely Randomized Design

$H_0: \mu_1 = \mu_2 = \cdots = \mu_k$ (i.e., no difference in the treatment [population] means)

H_a: At least two treatment means differ

$$\text{Test statistic: } F = \frac{\text{MST}}{\text{MSE}}$$ (11.4)

Rejection region: $F > F_\alpha$

where the distribution of F is based on $(k - 1)$ numerator df and $(n - k)$ denominator df, and F_α is the F value found in Table B.5 of Appendix B such that $P(F > F_\alpha) = \alpha$.

Assumptions:
1. All k population probability distributions are normal.
2. The k population variances are equal.
3. The samples from each population are random and independent.

The results of an analysis of variance are usually summarized and presented in an **analysis of variance (ANOVA) table**. Such a table shows the sources of variation, their respective degrees of freedom, sums of squares, mean squares, and computed F statistic. The general form of the ANOVA table for a completely randomized design is shown in the box. In practice, the entries in the ANOVA table are obtained using spreadsheet or statistical software. (For those who are interested, we provide the calculation formulas in optional Section 11.7.)

General ANOVA Table for a Completely Randomized Design

Source	df	SS	MS	F
Treatments	$k - 1$	SST	$\text{MST} = \dfrac{\text{SST}}{k - 1}$	$F = \dfrac{\text{MST}}{\text{MSE}}$
Error	$n - k$	SSE	$\text{MSE} = \dfrac{\text{SSE}}{n - k}$	
TOTAL	$n - 1$	SS(Total)		

Example 11.4   Conducting the One-Way ANOVA

Refer to Example 11.3 and the data presented in Table 11.3. Conduct the test of the hypotheses

$$H_0: \mu_1 = \mu_2 = \mu_3$$

H_a: At least two population means are different

where μ_1 is the true mean GPA for lower-class college freshmen, μ_2 is the true mean GPA for middle-class college freshmen, and μ_3 is the true mean GPA for upper-class college freshmen. Use $\alpha = .05$.

E **Figure 11.6**
Excel Output for
One-Way ANOVA,
Example 11.4

	A	B	C	D	E	F	G
1	Anova: Single Factor						
2							
3	SUMMARY						
4	*Groups*	*Count*	*Sum*	*Average*	*Variance*		
5	Lower Class	7	17.65	2.521429	0.254114		
6	Middle Class	7	22.74	3.248571	0.124348		
7	Upper Class	7	17.79	2.541429	0.406748		
8							
9							
10	ANOVA						
11	*Source of Variation*	*SS*	*df*	*MS*	*F*	*P-value*	*F crit*
12	Between Groups	2.401438095	2	1.200719	4.587511	0.024543	3.554561
13	Within Groups	4.711257143	18	0.261737			
14	Total	7.112695238	20				

Treatments — 12
Error — 13

SST   SSE   MST   MSE

Solution
An Excel printout of the analysis is shown in Figure 11.6. The sums of squares (SST and SSE) and mean squares (MST and MSE) in the resulting ANOVA table are indicated on the printout. The test statistic, $F = 4.59$, is highlighted on the printout. To carry out the test, we can use either the rejection region (at $\alpha = .05$) or the p-value of the test. Both the critical F value and p-value of the test are also highlighted in Figure 11.6.

The critical F value for the $\alpha = .05$ rejection region, given as **F Crit** in Figure 11.6, is $F_{.05} = 3.55$. Thus, we will reject the null hypothesis that the three means are equal if

$$F > F_{.05} = 3.55$$

Since the computed value of the test statistic, $F = 4.59$, exceeds the critical value, we have sufficient evidence (at $\alpha = .05$) to conclude that the true mean freshmen GPAs differ for at least two of the three socioeconomic classes.

The same conclusion is obtained using the p-value of the test. Since $p = .0245$ is less than $\alpha = .05$, we have sufficient evidence to reject H_0. ❏

Self-Test
11.2

Suppose the USGA wants to compare the mean distances associated with four different brands of golf balls when struck with a driver. A completely randomized design is employed, with Iron Byron, the USGA's robotic golfer, using a driver to hit a random sample of 10 balls of each brand in a random sequence. The ANOVA table is shown below.

Source	df	SS	MS	*F*	*p*-value
Brand	3	2,794	931.5	43.99	.0001
Error	36	762	21.2		
TOTAL	39	3,556			

    a. State the null hypothesis.

    b. What is the value of the test statistic?

    c. Make the appropriate conclusion at $\alpha = .05$.

An analysis of variance for a completely randomized design may include the construction of confidence intervals for a single mean or for the difference between two means. Because the independent sampling design involves the selection of independent random samples, we can find a confidence interval for a single mean using the method of Section 7.3 and for the difference between two population means using the method of Section 9.2.*

However, the researcher is usually interested in making all possible pairwise comparisons of the means. Consequently, several intervals need to be constructed, inflating the overall probability of a Type I error. We discuss a better, alternative approach to constructing confidence intervals for means in ANOVA in Section 11.5.

Before ending our discussion of completely randomized designs, we make the following comment. The proper application of the ANOVA procedure requires that certain assumptions be satisfied, i.e., all k populations are approximately normal with equal variances. In Section 11.6, we provide details on how to use the data to determine if these assumptions are satisfied to a reasonable degree. If the analysis, for example, reveals that one or more of the populations are nonnormal (e.g., highly skewed), then any inferences derived from the ANOVA of the data are suspect. In this case, we can apply a nonparametric technique (Chapter 12).

*The only modification we will make in these two procedures is that we will use an estimate of σ^2 based on the information contained in all k samples—namely, the pooled measure of variability within the k samples:

$$\text{MSE} = s^2 = \frac{\text{SSE}}{(n-k)}.$$

Caution

When the assumptions for analyzing data collected from a completely randomized design are violated, any inferences derived from the ANOVA are suspect. An alternative technique to use for the completely randomized design is the nonparametric Kruskal-Wallis test (see Section 12.5).

PROBLEMS 11.1–11.12

Using the Tools

11.1 Independent random samples were selected from two populations, with the results shown in the table at left.

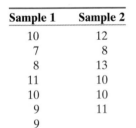

Sample 1	Sample 2
10	12
7	8
8	13
11	10
10	10
9	11
9	

Problem 11.1

a. Construct a line plot of the data similar to Figures 11.2 and 11.3. Do you think the data provide evidence of a difference between the population means?

b. Calculate the within-sample variation (MSE) for the data using equation 11.1 (page 674).

c. Calculate the between-sample variation (MST) for the data using equation 11.2 (page 675).

d. Using equation 11.3 (page 675), compute the test statistic appropriate for testing $H_0: \mu_1 = \mu_2$ against the alternative hypothesis that the two means differ.

e. Summarize the results of parts **b–d** in an ANOVA table.

f. Specify the rejection region, using a significance level of $\alpha = .05$.

g. Make the proper conclusion. How does this compare to your answer to part **a**?

11.2   Problem 11.1 involves a test of the null hypothesis H_0: $\mu_1 = \mu_2$ based on independent random sampling. Recall that a test of this hypothesis was conducted in Section 9.2 using a t statistic.

a.   Use the t test to test H_0: $\mu_1 = \mu_2$ against the alternative hypothesis H_a: $\mu_1 \neq \mu_2$. Test using $\alpha = .05$.

b.   It can be shown (proof omitted) that an F statistic with $\nu_1 = 1$ numerator degree of freedom and ν_2 denominator degrees of freedom is equal to t^2, where t is a t statistic based on ν_2 degrees of freedom. Square the value of t that you calculated in part **a** and show that it is equal to the value of F calculated in Problem 11.1.

c.   Is the analysis of variance F test for comparing two population means a one- or a two-tailed test of H_0: $\mu_1 = \mu_2$? [*Hint:* Although the t test can be used to test for either H_a: $\mu_1 > \mu_2$ or H_a: $\mu_1 < \mu_2$, the alternative hypothesis for the F test is H_a: The two means are different.]

11.3   Independent random samples were selected from three populations. The data are shown in the accompanying table, followed by an Excel printout of the analysis of variance.

Sample 1	Sample 2	Sample 3
2.1	4.4	1.1
3.3	2.6	.2
.2	3.0	2.0
	1.9	

E

1	Anova: Single Factor						
2							
3	SUMMARY						
4	*Groups*	*Count*	*Sum*	*Average*	*Variance*		
5	Sample 1	3	5.6	1.866667	2.443333		
6	Sample 2	3	10	3.333333	0.893333		
7	Sample 3	3	3.3	1.1	0.81		
8							
9							
10	ANOVA						
11	*Source of Variation*	*SS*	*df*	*MS*	*F*	*P-value*	*F crit*
12	Between Groups	7.726667	2	3.863333	2.795016	0.138739	5.143249
13	Within Groups	8.293333	6	1.382222			
14							
15	Total	16.02	8				

a.   Locate the value of MST. What type of variability is measured by this quantity?

b.   Locate the value of MSE. What type of variability is measured by this quantity?

c.   How many degrees of freedom are associated with MST?

d.   How many degrees of freedom are associated with MSE?

e.   Locate the value of the test statistic for testing H_0: $\mu_1 = \mu_2 = \mu_3$ against the alternative hypothesis that at least one population mean is different from the other two.

f.   Summarize the results of parts **a–e** in an ANOVA table.

g.   Specify the rejection region, using a significance level of $\alpha = .05$.

h. State the proper conclusion.

i. Locate and interpret the *p*-value for the test of part **e**. Does this agree with your answer to part **h**?

11.4 A partially completed ANOVA table for a completely randomized design is shown here.

Source	df	SS	MS	F
Treatment	4	24.7	—	—
Error	—	—	—	
TOTAL	34	62.4		

a. Complete the ANOVA table.

b. How many treatments are involved in the experiment?

c. Do the data provide sufficient evidence to indicate a difference among the population means? Test using $\alpha = .10$.

Applying the Concepts

11.5 *Bulimia nervosa* is an eating disorder characterized by episodes of binge eating, extreme efforts to counteract ingested calories, and a morbid fear of becoming fat. In a study published in the *American Journal of Psychiatry* (July 1995), a random sample of 75 subjects was selected from females clinically diagnosed with bulimia nervosa. The subjects were divided into three groups based on when they were born: (1) before 1950, (2) between 1950 and 1959, and (3) after 1959. For one part of the study, the variable of interest was age (in years) at onset of the eating disorder. The goal was to compare the mean ages at onset of the three groups.

a. Identify the response variable in the experiment.

b. Identify the treatments in the experiment.

c. State the null and alternative hypotheses to be tested.

d. The observed significance level of the ANOVA *F* test was less than .01. Interpret this result.

11.6 Refer to the Harris Corporation/University of Florida study to compare the mean voltage emitted by a manufacturing process established at two locations, Problems 3.38 (page 170) and 9.19 (page 493). The voltage readings of 30 production runs at the old location and the new location are reproduced in the table. A one-way analysis of variance was performed on the data using Excel. The Excel printout is displayed on page 683.

VOLTAGE.XLS

Old Location			New Location		
9.98	10.12	9.84	9.19	10.01	8.82
10.26	10.05	10.15	9.63	8.82	8.65
10.05	9.80	10.02	10.10	9.43	8.51
10.29	10.15	9.80	9.70	10.03	9.14
10.03	10.00	9.73	10.09	9.85	9.75
8.05	9.87	10.01	9.60	9.27	8.78
10.55	9.55	9.98	10.05	8.83	9.35
10.26	9.95	8.72	10.12	9.39	9.54
9.97	9.70	8.80	9.49	9.48	9.36
9.87	8.72	9.84	9.37	9.64	8.68

Source: Harris Corporation, Melbourne, Fla.

E	A	B	C	D	E	F	G
1	Anova: Single Factor						
2							
3	SUMMARY						
4	Groups	Count	Sum	Average	Variance		
5	Old Location	30	294.11	9.803667	0.29259		
6	New Location	30	282.67	9.422333	0.229322		
7							
8							
9	ANOVA						
10	Source of Variation	SS	df	MS	F	P-value	F crit
11	Between Groups	2.181227	1	2.181227	8.358608	0.005394	4.006864
12	Within Groups	15.13543	58	0.260956			
13							
14	Total	17.31666	59				

a. Locate on the printout the F statistic for testing whether the mean voltage readings at the two locations are equal.

b. Is there evidence (at $\alpha = .05$) to indicate that the mean voltage readings at the two locations are different?

c. In Problem 9.19**a** you compared the two voltage means using a two-sample t test. Compare the t statistic to the value of F obtained in part **a**. Verify that $F = t^2$ and that the two tests yield the same conclusion at $\alpha = .05$.

11.7 The data in the following table represent actual sale prices (in thousands of dollars) of residential properties recently sold in Tampa, Florida. Independent random samples of 10 properties were selected from each of seven different neighborhoods (A, B, C, D, E, F, and G). Suppose you want to compare the mean sale prices of properties in the seven Tampa neighborhoods.

TAMSALES.XLS

NEIGHBORHOOD						
A	B	C	D	E	F	G
191.5	100.5	35.3	86.0	69.9	46.0	14.5
208.5	147.5	50.0	140.0	159.5	42.0	50.0
225.0	115.0	160.0	120.0	84.5	48.0	39.0
375.0	210.0	59.9	72.9	57.7	57.0	20.0
205.0	155.0	38.5	72.0	45.4	34.5	32.2
191.5	127.5	105.9	68.7	43.9	74.5	54.0
189.0	115.0	312.0	90.0	205.0	66.9	21.5
139.0	110.0	84.5	100.0	59.9	54.5	55.0
138.0	183.0	68.5	97.0	37.0	46.8	43.0
162.0	107.5	160.0	84.0	68.0	6.9	13.3

Source: Hillsborough County (Fla.) property appraiser's office.

a. What type of experimental design is employed?

b. Identify the response variable and the treatments.

c. Give the null hypothesis to be tested.

d. Use Excel to perform the analysis of variance. Summarize the results in an ANOVA table.

e. Is there evidence of differences among the mean sale price populations for the seven neighborhoods? Use $\alpha = .01$.

11.8 A statistics professor wants to study four different strategies of playing the game of Blackjack (Twenty-One). The four strategies are

1. Dealer's strategy
2. Five-count strategy
3. Basic ten-count strategy
4. Advanced ten-count strategy

A calculator that is programmed to play Blackjack is utilized and data from five sessions of each strategy are collected. The profits (or losses) from each session are given in the accompanying table. The professor wants to know whether there is evidence of a difference among the four Blackjack strategies. Use Excel to conduct the analysis at $\alpha = .01$.

BLCKJACK.XLS

	STRATEGY		
Dealer's	**Five Count**	**Basic Ten Count**	**Advanced Ten Count**
−$56	−$26	+$16	+$60
−$78	−$12	+$20	+$40
−$20	+$18	−$14	−$16
−$46	−$8	+$6	+$12
−$60	−$16	−$25	+$4

FISH.XLS

11.9 Consider the data on DDT measurements of contaminated fish in the Tennessee River (Alabama) stored in the Excel workbook FISH.XLS. Recall that the qualitative variable, SPECIES, stored in the workbook represents the species of fish captured—channel catfish, smallmouth buffalo, or largemouth bass. According to the U.S. Army Corps of Engineers, who collected the data, it is important to know whether the mean DDT levels of the three species of fish differ and, if so, which species has the highest mean level. Using Excel, run the proper analysis. Summarize the results in an ANOVA table.

11.10 Speech recognition technology has advanced to the point that it is now possible to communicate with a computer through verbal commands. A study was conducted to evaluate the value of speech recognition in human interactions with computer systems (*Special Interest Group on Computer-Human Interaction Bulletin,* July 1993.) A sample of 45 subjects was randomly divided into three groups (15 subjects per group) and each subject was asked to perform tasks on a basic voice mail system. A different interface was employed in each group: (1) touch-tone, (2) human operator, or (3) simulated speech recognition. One of the variables measured was overall time (in seconds) to perform the assigned tasks. An analysis was conducted to compare the mean overall performance times of the three groups.

a. Identify the experimental design employed in this study.

b. Identify the treatments in this study.

c. State the appropriate null hypothesis to be tested.

d. The sample mean performance times for the three groups are given on page 685. Despite differences among the sample means, the null hypothesis of part **c** could not be rejected at $\alpha = .05$. Explain how this is possible.

11.11 Do Hispanic visitors to a national forest in the U.S. have the same environmental concerns as Anglo visitors? To answer this question, 334

Group	Mean Performance Time (seconds)
Touch-tone	1,400
Human operator	1,030
Speech recognition	1,040

Problem 11.10

people touring the Angeles National Forest and the San Bernardino National Forest were surveyed on several environmental issues (e.g., oil spills, car emissions, and wildfires). Each visitor was classified into one of four ethnic categories: (1) U.S.-born Anglos, (2) U.S.-born Hispanics, (3) Mexican-born Hispanics, and (4) Central American-born Hispanics. Environmental concern for each issue was measured on a 5-point scale where 1 = not harmful, 2 = somewhat harmful, 3 = neutral, 4 = harmful, and 5 = very harmful. A summary of the results on the issues of car emissions and wildfires, published in the *Journal of Environmental Education* (Spring 1995), is provided below.

Ethnic Group	Sample Size	MEAN RESPONSE Car Emissions	Wildfires
U.S.-born Anglo	39	3.61	3.61
U.S.-born Hispanic	79	3.32	4.00
Mexican-born Hispanic	173	3.13	3.55
Central American-born Hispanic	43	2.61	3.35

a. State the null and alternative hypotheses appropriate for determining whether the four ethnic groups differ regarding their mean responses on the issue of car emissions.

b. At $\alpha = .05$, the test of part **a** was statistically significant. Interpret this result.

c. State the null and alternative hypotheses appropriate for determining whether the four ethnic groups differ regarding their mean responses on the issue of wildfires.

d. At $\alpha = .05$, the test of part **c** was not statistically significant. Interpret this result.

11.12 The Minnesota Multiphasic Personality Inventory (MMPI) is a questionnaire used to gauge personality type. Several scales are built into the MMPI to assess response distortion; these include the Infrequency (I), Obvious (O), Subtle (S), Obvious-subtle (O-S), and Dissimulation (D) scales. *Psychological Assessment* (Mar. 1995) published a study that investigated the effectiveness of these MMPI scales in detecting deliberately distorted responses. A completely randomized design with four treatments was employed. The treatments consisted of independent random samples of females in the following four groups: nonforensic psychiatric patients ($n_1 = 65$), forensic psychiatric patients ($n_2 = 28$), college students who were requested to respond honestly ($n_3 = 140$), and college students who were instructed to provide "fake bad" responses ($n_4 = 45$). All 278 participants were given the MMPI and the I, O, S, O-S, and D scores were recorded for each. Each scale was treated as a response variable and an analysis of variance conducted. The ANOVA F values are reported in the table.

Response Variable	ANOVA F Value
Infrequency (I)	155.8
Obvious (O)	49.7
Subtle (S)	10.3
Obvious-subtle (O-S)	45.4
Dissimulation (D)	39.1

a. For each response variable, determine whether the mean scores of the four groups completing the MMPI differ significantly. Use $\alpha = .05$ for each test.

b. If the MMPI is effective in detecting distorted responses, then the mean score for the "fake bad" treatment group will be largest. Based on the information provided, can the researchers make an inference about the effectiveness of the MMPI? Explain.

11.4 FACTORIAL DESIGNS: TWO-WAY ANOVA

In Section 11.3 we presented a one-way ANOVA for analyzing data from an experiment on a single factor. Suppose we want to investigate the effect of two factors on the mean value of a response variable. That is, we want to conduct a two-way ANOVA. The design appropriate for a two-way ANOVA is a **factorial design**.

In a factorial design, experimental units are measured for various combinations of the factor levels. For example, suppose an experiment involves two factors, one at three levels and the other at two levels. If the response is measured for each of the $2 \times 3 = 6$ factor-level combinations, the design is called a *complete 2 × 3 factorial design* (since all $2 \times 3 = 6$ possible treatments are included in the experiment).

Definition 11.9

A **complete factorial design** is an experiment that includes *all* possible factor-level combinations (treatments).

If one observation on the response variable y is taken for each of the six factor-level combinations, we say that we conducted one replication of a 2×3 factorial experiment. Stating that we conducted one replication of the experiment means that we obtained one measurement on y for each of the six factor-level combinations. As we will see, most factorial designs involve two or more replications.

Definition 11.10

A single **replication** of a factorial experiment is one in which the response variable is observed once for every possible factor-level combination.
[*Note:* In practice, the number of replications r for a factorial experiment is almost always chosen to be two or larger.]

The data for a two-factor factorial experiment are presented in a two-way table, with rows corresponding to the levels of one factor and columns corresponding to the levels of the other factor. For each combination of factor levels, the data fall in one of the row-column cells of the table. To illustrate, consider the following example.

Example 11.5

2 × 3 Factorial Design

A company that stamps gaskets out of sheets of rubber, plastic, and cork wants to compare the mean number of gaskets produced per hour for two different types of stamping machines. Practically, the manufacturer wants to determine whether one machine is more productive than the other and, even more important, whether one machine is more productive in producing rubber gaskets while the other is more productive in producing plastic or cork gaskets. To answer these questions, the manufacturer decides to conduct an experiment using the three types of gasket material, cork (C), rubber (R), and plastic (P), with each of the two stamping machines, (1) and (2). Each machine is operated for three 1-hour time periods for each of the gasket materials, with the 18 1-hour time periods assigned to the six machine-material combinations in a random order. (The purpose of the randomization is to eliminate the possibility that uncontrolled environmental factors might bias the results.) The data for the experiment, the number of gaskets (in thousands) produced per hour, are shown in Table 11.4.

a. Identify the response y, factors, and treatments (factor-level combinations) in the experiment.

b. Is the design selected a complete factorial design?

c. How many replications are in the experiment?

Solution

a. Since the company will produce gaskets using three types of material with each of two types of stamping machine, the experiment involves two factors, machine and material, with machine at two levels (1, 2) and material at three levels (C, R, P). Each of the $2 \times 3 = 6$ combinations of machine and material—(1, C), (1, R), (1, P), (2, C), (2, R), and (2, P)—represents the treatments in the experiment. For this reason the experiment is referred to as a 2×3 factorial design.

GASKET.XLS

Table 11.4 Data for the 2 × 3 Factorial Experiment of Example 11.5

| | | \multicolumn{3}{c}{GASKET MATERIAL} | | |
		Cork (C)	Rubber (R)	Plastic (P)
Stamping Machine	1	4.31	3.36	4.01
		4.27	3.42	3.94
		4.40	3.48	3.89
	2	3.94	3.91	3.48
		3.81	3.80	3.53
		3.99	3.85	3.42

For each treatment, the response (i.e., quantitative variable) measured is y = number of gaskets (in thousands) produced per hour. One goal of the experiment will be to compare the mean responses of the six treatments.

b. Since all possible treatments are included in the experiment, a complete factorial design is employed.

c. Note that there are three observations on y recorded for each treatment in Table 11.4. Hence, the experiment consists of $r = 3$ replications. ❏

Self-Test

11.3

Refer to Self-Test 11.2 (page 679). Suppose the USGA tests four brands of golf balls (A, B, C, D) and two different clubs (driver, 5-iron) using a complete factorial design. Iron Byron (the robotic golfer) will hit four balls of each brand with the driver, then four balls of each brand with the 5-iron—a total of 32 hits (four hits for each of the $4 \times 2 = 8$ brand-club combinations). The goal is to compare the mean distances of the eight brand-club combinations.

a. Identify the response y, factors, and treatments in this experiment.

b. How many replications are in the experiment?

As stated in Example 11.5a, one objective of the factorial experiment is to compare the means for the six treatments. One way to accomplish this is to investigate how the means for one factor (material) differ when the levels of the other factor (machine) are varied. The next example illustrates this idea.

Example 11.6   Factor Interaction

Refer to the data of Table 11.4. Suppose that we have calculated the six cell (treatment) means. Consider two possible outcomes. Figure 11.7 shows two hypothetical plots of the six means. For both plots, the three means for stamping machine 1 are connected by dark-color line segments. The corresponding three means for machine 2 are connected by lighter-color line segments. What do these plots imply?

Figure 11.7
Hypothetical Plots of the
Means for the Six Machine-
Material Combinations

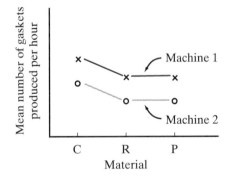

a. No interaction

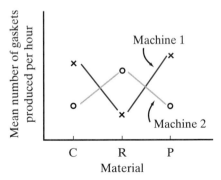

b. Interaction

Solution

Figure 11.7a suggests that machine 1 produces a larger number of gaskets per hour, regardless of the gasket material, and is therefore superior to machine 2. On the average, machine 1 stamps more cork (C) gaskets per hour than rubber or plastic, but the *difference* in the mean numbers of gaskets produced by the two machines remains approximately the same, regardless of the gasket material. Thus, the *difference* between the mean numbers of gaskets produced by the two machines is *independent* of the gasket material used in the stamping process.

In contrast, Figure 11.7b shows the productivity of machine 1 to be larger than for machine 2 when the gasket material is cork (C) or plastic (P). But the means are reversed for rubber (R) gasket material. For this material, machine 2 produces, on the average, more gaskets per hour than machine 1. Thus, Figure 11.7b illustrates a situation where the mean value of the response variable *depends* on the combination of the factor levels. When this situation occurs, we say that the factors *interact*. Thus, one of the most important objectives of a factorial experiment is to detect factor interaction if it exists. ❑

Definition 11.11

In a factorial design with two factors A and B, when the difference between the mean levels of factor A depends on the different levels of factor B, we say there is **interaction** between factors A and B. If the difference is independent of the levels of B, then there is **no interaction** between factors A and B.

The analysis of variance for a two-factor factorial design (i.e., a two-way ANOVA) involves a partitioning of SS(Total) into four parts as shown in Figure 11.8. These sums of squares allow us to investigate and test for factor interaction as well as for **factor main effects**; i.e., an overall effect of factor A (called the main effect for A) and an overall effect of factor B (called the main effect for B). (The formulas for the sums of squares and the corresponding mean squares for a factorial experiment arc given in optional Section 11.7.) As in the previous section,

Figure 11.8 Partitioning of the Total Sum of Squares for a Two-Factor Factorial Design

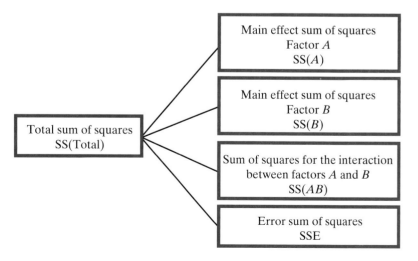

we will use Excel to perform the analysis of the data and to summarize the ANOVA results in a table. In general, the ANOVA table for a two-factor factorial design, with factor A at a levels, factor B at b levels, and with r replications, appears as shown in the box. A summary of the ANOVA F tests for a factorial design follows.

General ANOVA Table for a Two-Factor Factorial Design

Source	df	SS	MS	F
Main effect, A	$a - 1$	SS(A)	$MS(A) = \dfrac{SS(A)}{a - 1}$	$F = \dfrac{MS(A)}{MSE}$
Main effect, B	$b - 1$	SS(B)	$MS(B) = \dfrac{SS(B)}{b - 1}$	$F = \dfrac{MS(B)}{MSE}$
AB interaction	$(a - 1)(b - 1)$	SS(AB)	$MS(AB) = \dfrac{SS(AB)}{(a - 1)(b - 1)}$	$F = \dfrac{MS(AB)}{MSE}$
Error	$ab\,(r - 1)$	SSE	$MSE = \dfrac{SSE}{ab(r - 1)}$	
TOTAL	$abr - 1$	SS(Total)		

ANOVA F Tests for a Two-Factor Factorial Design

Test for Factor Interaction

H_0: No interaction between factors A and B

H_a: Factors A and B interact

$$\text{Test statistic: } F = \frac{MS(AB)}{MSE} = \frac{MS(AB)}{s^2} \tag{11.5}$$

Rejection region: $F > F_\alpha$, where F is based on $\nu_1 = (a - 1)(b - 1)$ and $\nu_2 = ab(r - 1)$ df

Test for Main Effects for Factor A

H_0: There are no differences among the means for main effect A

H_a: At least two of the main effect A means differ

$$\text{Test statistic: } F = \frac{MS(A)}{MSE} = \frac{MS(A)}{s^2} \tag{11.6}$$

Rejection region: $F > F_\alpha$, where F is based on $\nu_1 = (a - 1)$ and $\nu_2 = ab(r - 1)$ df

Test for Main Effects for Factor B

H_0: There are no differences among the means for main effect B

H_a: At least two of the main effect B means differ

$$\text{Test statistic: } F = \frac{MS(B)}{MSE} = \frac{MS(B)}{s^2} \tag{11.7}$$

Rejection region: $F > F_\alpha$, where F is based on $\nu_1 = (b - 1)$ and $\nu_2 = ab(r - 1)$ df

Assumptions:

1. The population distribution of the observations for any factor-level combination is approximately normal.

2. The variance of the probability distribution is constant and the same for all factor-level combinations.

3. The treatments (factor-level combinations) are randomly assigned to the experimental units.

4. The observations for each factor-level combination represent independent random samples.

Caution

When the assumptions for analyzing data collected from a factorial experiment are violated, any inferences derived from the ANOVA are suspect. Nonparametric methods are available for analyzing factorial experiments, but they are beyond the scope of this text. Consult the references given at the end of Chapter 12 if you want to learn about such techniques.

Example 11.7 Two-Way ANOVA

Refer to Example 11.5. Excel printouts for the 2 × 3 factorial ANOVA of the data in Table 11.4 (page 867) are shown in Figure 11.9a below and Figure 11.9b (page 692). Summarize the results in an ANOVA table.

Solution

Figure 11.9a gives the sample means for each of the 2 × 3 = 6 treatments. The sums of squares and mean squares for the sources of variation, machine main effect, material main effect, machine × material interaction, and error, are indicated on the Excel printout, Figure 11.9b. These values (highlighted) make up the ANOVA table for the experiment. ❑

E **Figure 11.9** Excel Output for Factorial ANOVA, Example 11.7

	A	B	C	D	E
1	Anova: Two-Factor With Replication				
2					
3	SUMMARY	C	R	P	Total
4	*Machine 1*				
5	Count	3	3	3	9
6	Sum	12.98	10.26	11.84	35.08
7	Average	4.326667	3.42	3.946667	3.897778
8	Variance	0.004433	0.0036	0.003633	0.158394
9					
10	*Machine 2*				
11	Count	3	3	3	9
12	Sum	11.74	11.56	10.43	33.73
13	Average	3.913333	3.853333	3.476667	3.747778
14	Variance	0.008633	0.003033	0.003033	0.045694
15					
16	*Total*				
17	Count	6	6	6	
18	Sum	24.72	21.82	22.27	
19	Average	4.12	3.636667	3.711667	
20	Variance	0.05648	0.058987	0.068937	

a.

E Figure 11.9
Continued

	A	B	C	D	E	F	G
	ANOVA						
	Source of Variation	SS	df	MS	F	P-value	F crit
Machines —	Sample	0.10125	1	0.10125	23.04046	0.000434	4.747221
Materials —	Columns	0.811944	2	0.405972	92.38306	5.15E-08	3.88529
	Interaction	0.768033	2	0.384017	87.38685	7.03E-08	3.88529
Error —	Within	0.052733	12	0.004394			
	Total	1.733961	17				

b.

Example 11.8   F Test for Interactions

Refer to Example 11.7. A plot of the six means corresponding to the six machine-material combinations is shown in Figure 11.10 (page 693).

a. Based on the plot, do you think that the factors machine and material interact?

b. Perform a test of hypothesis to determine whether the data provide sufficient evidence to indicate that the more productive stamping machine depends on the gasket material. Test using $\alpha = .05$.

Solution

a. Figure 11.10 shows that the difference between the means for machine 1 and machine 2 changes depending on the material stamped. For example, when cork is stamped, machine 1 is more productive (i.e., has the higher mean). But when rubber is stamped, machine 2 is more productive. Consequently, it appears that the two factors interact. Remember, however, that the plot is based on *sample* means. A statistical test is required to determine whether the pattern of sample means shown in Figure 11.10 implies interaction in the population.

b. The test statistic, critical F value, and p-value for testing the null hypothesis of no interaction between machines and materials, highlighted in the Excel printout, Figure 11.9b, are

$$F = 87.387 \qquad F_{.05} = 3.885 \qquad p\text{-value} \approx 0$$

Since the p-value is less than $\alpha = .05$ (or the test statistic F exceeds the critical value), we conclude that there is an interaction between machines and materials. Therefore, there is sufficient evidence to indicate that neither machine is the more productive for all three materials. If the differences in mean productivity are large enough, the manufacturer should use both machines, selecting the machine that gives the greater productivity for a specific material. ❑

   Tests for differences in the mean levels of the main effects in a factorial experiment are relevant *only when the factors do not interact*. When there is no factor interaction, the differences in the mean levels of factor A are the same for all levels of factor B. The test for main effect A tests the significance of these differences. In the presence of interaction, however, the main effect test is irrelevant since the differences in the mean levels of factor A are not the same at each level

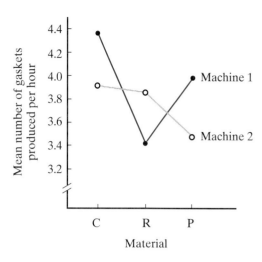

E **Figure 11.10**
Plot of the Means for
the Six Machine-
Material Combinations

of factor B. The following example demonstrates the tests for main effects and discusses their implications.

Example 11.9   Testing for Main Effects

A 2×2 factorial design was employed to determine the effects of work scheduling (factor A) and method of payment (factor B) on attitude toward the job (y). Two types of scheduling were employed, the standard 8:00 A.M.–5:00 P.M. work day and a modification whereby the worker was permitted to vary the starting time and the length of the lunch hour. The two methods of payment were a standard hourly rate and a reduced hourly rate with an added piece rate based on worker production. Four workers were randomly assigned to each of the four scheduling-payment combinations; each completed an attitude test after one month on the job. An ANOVA table for the test scores data is shown in Table 11.5. Interpret the results using $\alpha = .05$.

Solution

First notice that the test for factor interaction is not statistically significant (i.e., there is no evidence of factor interaction) since the p-value for the test, .8899, exceeds $\alpha = .05$. Therefore, we may focus on the main effects for the two factors, schedule (A) and payment (B).

Note that the p-values for the tests of main effect A (schedule) and main effect B (payment) are .0189 and .0002, respectively, and are both less than $\alpha = .05$.

Table 11.5 ANOVA Table for the 2×2 Factorial Design, Example 11.9

Source	df	SS	MS	F	p-value
Schedule (A)	1	361	361	7.37	.0189
Payment (B)	1	1,444	1,444	29.47	.0002
Schedule-Payment Interaction (AB)	1	1	1	.02	.8899
Error	12	588	49		
TOTAL	15	2,394			

Thus, there is evidence (at $\alpha = .05$) of differences in the mean levels of the respective main effects. Practically, this implies that the mean worker attitude scores differ for the two schedules, but that this difference does not depend on method of payment (due to lack of interaction). Similarly, the mean worker attitude scores for the two methods of payment differ, but the difference does not depend on the work schedule. ❏

An analysis of variance for a factorial experiment is based on the assumption that the observations in the cells of the two-way table represent independent random samples. Under this assumption, formulas for constructing confidence intervals for the difference between two cell means are exactly the same as for the completely randomized design. We present one approach to constructing such confidence intervals in Section 11.5.

> **Caution**
> In a factorial experiment, the F test for factor interaction should always be conducted first because the F tests for factor main effects are usually relevant only when factor interaction is *not* significant. If interaction is detected, *do not* perform the F tests for main effects.

**Self-Test
11.4**

Refer to the factorial experiment described in Self-Test 11.3 (page 688). The p-values for the ANOVA F tests are listed below.

Factor	p-Value
Club Main Effect	.0001
Brand Main Effect	.3420
Club $\times$ Brand Interaction	.2603

a. Which ANOVA F test should be conducted first?

b. Is there evidence (at $\alpha = .01$) of interaction between club and brand?

c. Based on the results of the test, part **b**, should tests for main effects be conducted? If so, perform these tests using $\alpha = .01$.

**PROBLEMS
11.13–11.25**

Using the Tools

11.13   Suppose you conduct a 3 $\times$ 4 factorial experiment with two observations per treatment. Construct an analysis of variance table for the experiment showing the sources of variation and degrees of freedom for each.

11.14   The analysis of variance for a 3×2 factorial design with four observations per treatment produced the ANOVA table entries shown here.

Source	df	SS	MS	F
A	—	100	—	—
B	1	—	—	—
AB interaction	2	—	2.5	—
Error	—	—	2.0	
TOTAL	—	700		

   a.   Complete the ANOVA table.

   b.   Test for interaction between factor A and factor B. Use $\alpha = .05$.

   c.   Test for differences in main effect means for factor A. Use $\alpha = .05$.

   d.   Test for differences in main effect means for factor B. Use $\alpha = .05$.

11.15   The data for a 3×4 factorial design with two observations per treatment are shown in the accompanying table. An analysis of variance was conducted on the data using Excel. The results are reported in the accompanying printout.

		FACTOR B			
		1	2	3	4
Factor A	1	5 4	7 9	6 5	5 7
	2	6 4	10 9	5 8	9 7
	3	8 10	7 6	5 8	6 5

E

ANOVA						
Source of Variation	SS	df	MS	F	P-value	F crit
A —— Sample	6.583333	2	3.291667	1.837209	0.201344	3.88529
B —— Columns	13.79167	3	4.597222	2.565891	0.103331	3.4903
Interaction	35.08333	6	5.847222	3.263566	0.038502	2.996117
Within	21.5	12	1.791667			
Total	76.95833	23				

   a.   Do the data provide sufficient evidence of interaction between factor A and factor B? Test using $\alpha = .05$.

   b.   Given the results of the test in part **a**, would you recommend that tests for main effects be conducted? Explain.

Applying the Concepts

11.16   A videocassette recorder (VCR) repair service wished to study the effect of VCR brand and service center on the repair time measured in minutes. Three VCR brands (A, B, C) and three service centers were specifically selected for analysis. Each service center repaired two VCRs of each brand. The results are shown on page 696.

VCR.XLS

		VCR BRANDS		
		A	**B**	**C**
Service Centers	1	52	48	59
		57	39	67
	2	51	61	58
		43	52	64
	3	37	44	65
		46	50	69

a. Use Excel to obtain an ANOVA table for this 3×3 factorial.

b. Conduct the test for factor interaction at $\alpha = .05$. Interpret the results.

c. Is there a main effect due to service centers? Test at $\alpha = .05$.

d. Is there a main effect due to VCR brand? Test at $\alpha = .05$.

11.17 In the *Journal of Nutrition* (July 1995), University of Georgia researchers examined the impact of a vitamin-B supplement (nicotinamide) on the kidney. The experimental "subjects" were 28 Zucker rats—a species that tends to develop kidney problems. Half of the rats were classified as obese and half as lean. Within each group, half were randomly assigned to receive a vitamin-B-supplemented diet and half were not. Thus, a 2×2 factorial experiment was conducted with seven rats assigned to each of the four combinations of size (lean or obese) and diet (supplement or not). One of the response variables measured was weight (in grams) of the rat's kidney at the end of a 20-week feeding period. The data (simulated from summary information provided in the journal article) are shown in the table.

KIDNEY.XLS

		DIET			
		Regular		**Vitamin-B Supplement**	
Rat Size	Lean	1.62	1.47	1.51	1.63
		1.80	1.37	1.65	1.35
		1.71	1.71	1.45	1.66
		1.81		1.44	
	Obese	2.35	2.84	2.93	2.63
		2.97	2.05	2.72	2.61
		2.54	2.82	2.99	2.64
		2.93		2.19	

a. Use Excel to conduct an analysis of variance on the data. Summarize the results in an ANOVA table.

b. Conduct the appropriate ANOVA F tests at $\alpha = .01$. Interpret the results.

11.18 Do women enjoy the thrill of a close basketball game as much as men? To answer this question, male and female undergraduate students were recruited to participate in an experiment (*Journal of Sport & Social Issues*, Feb. 1997). The students watched one of eight live televised games of a recent NCAA basketball tournament. (None of the games involved a home team to which the students could be considered emotionally committed.) The "suspense" of each game was classified into one of four categories according to the closeness of scores at the game's conclusion: minimal (15 point or greater differential), moderate (10–14 point dif-

	GENDER	
Suspense	**Male**	**Female**
Minimal	1.77	2.73
Moderate	5.38	4.34
Substantial	7.16	7.52
Extreme	7.59	4.92

Source: Gan, Su-lin, et al. "The thrill of a close game: Who enjoys it and who doesn't?" *Journal of Sport & Social Issues,* Vol. 21, No. 1, Feb. 1997, pp. 59–60.

Problem 11.18

ferential), substantial (5–9 point differential), and extreme (1–4 point differential). After the game, each student rated his or her enjoyment on an 11-point scale ranging from 0 (not at all) to 10 (extremely). The enjoyment rating data were analyzed as a 4×2 factorial design, with suspense (four levels) and gender (two levels) as the two factors. The $4 \times 2 = 8$ treatment means are shown in the table at left.

a. Plot the treatment means in a graph similar to Figure 11.7. Does the pattern of means suggest interaction between suspense and gender? Explain.

b. The ANOVA F test for interaction yielded the following results: numerator df = 3, denominator df = 68, $F = 4.42$, p-value = .007. What can you infer from these results?

c. Based on the test, part **b**, is the difference between the mean enjoyment levels of males and females the same, regardless of the suspense level of the game?

11.19 The *Accounting Review* (Jan. 1991) reported on a study of the effect of two factors, confirmation of accounts receivable and verification of sales transactions, on account misstatement risk by auditors. Both factors were held at the same two levels: completed or not completed. Thus, the experimental design is a 2×2 factorial design.

a. Identify the factors, factor levels, and treatments for this experiment.

b. Explain what factor interaction means for this experiment.

c. A graph of the hypothetical mean misstatement risks for each of the $2 \times 2 = 4$ treatments is displayed here. In this hypothetical case, does it appear that interaction exists?

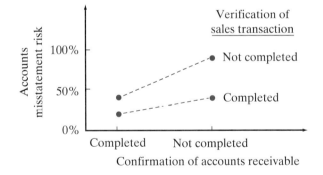

Source: Brown, C. E., and Solomon, I. "Configural information processing in auditing: The role of domain-specific knowledge." *The Accounting Review,* Vol. 66, No. 1, Jan. 1991, p. 105 (Figure 1).

11.20 The citrus mealybug is an insect that feeds on citrus fruit, especially grapefruit. An experiment was conducted to determine whether the density of mealybugs in an orchard depends on time of year (*Environmental Entomology,* June 1995). Thirty-five grapefruits were randomly sampled from each of two orchards (A and B) in each of four time periods (June, July, September, and October), and the number of mealybugs per fruit was recorded. The data were subjected to a 2×4 factorial ANOVA, with the results shown in the ANOVA table on page 698. Using $\alpha = .05$, conduct the appropriate ANOVA F tests. What conclusions can you draw from the ANOVA?

Source	df	SS	MS	F	p-value
Orchard	1	1.1535	1.1535	3.44	.06
Period	3	20.6278	6.8759	20.52	.001
Orchard × Period	3	2.9707	.9902	2.96	.03
Error	272	91.1437	.3351		

Problem 11.20

11.21 Many temperate-zone animal species exhibit physiological and morphological changes when the hours of daylight begin to decrease during autumn months. A study was conducted to investigate the "short day" traits of collared lemmings (*The Journal of Experimental Zoology*, Sept. 1993). A total of 124 lemmings were bred in a colony maintained with a photoperiod of 22 hours of light per day. At weaning (19 days of age), the lemmings were weighed and randomly assigned to live under one of two photoperiods: 16 hours or less of light per day, more than 16 hours of light per day. (Each group was assigned the same number of males and females.) After 10 weeks, the lemmings were weighed again. The response variable of interest was the gain in body weight (measured in grams) over the 10-week experimental period. The researchers analyzed the data using an ANOVA for a 2 × 2 factorial design, where the two factors are photoperiod (at two levels) and gender (at two levels).

a. Construct an ANOVA table for the experiment, listing the sources of variation and associated degrees of freedom.

b. The F test for interaction was not significant. Interpret this result.

c. The p-values for testing for photoperiod and gender main effects were both smaller than .001. Interpret these results.

11.22 A hospital administrator wished to examine postsurgical hospitalization periods following knee surgery. A random sample of 30 patients was selected, five for each combination of age group and type of surgery. The results, in number of postsurgical hospitalization days, are listed in the accompanying table. Conduct a complete analysis of the data using Excel. Interpret the results.

KNEESURG.XLS

	AGE GROUP		
Type of Knee Surgery	**Under 30**	**30 to 50**	**Over 50**
	1	4	3
	3	3	5
Arthroscopy	2	2	2
	6	3	3
	2	2	3
	3	4	4
	10	5	8
Arthrotomy	6	11	12
	7	5	10
	8	6	3

11.23 The impact of a fictional story on real-world beliefs was investigated in *Psychonomic Bulletin & Review* (Sept. 1997). The researchers theorized that readers unfamiliar with the setting of a fictional story would be vulnerable to its assertion, while readers familiar with the setting would not. In a designed experiment, each in a group of Yale University students was assigned to read a story that took place either at Yale (familiar setting) or at Princeton University (unfamiliar setting). Also, the student read either a story about a real-world truth (e.g., sunlight is bad for your skin) or a story that contained a false assertion (e.g., sunlight is good for your skin). After reading the story, the student was asked to rate the believability of a true statement about the story item (e.g., sunlight's effect on skin). Thus, there were $2 \times 2 = 4$ experimental conditions, one for each of the combinations of setting (Yale or Princeton) and story (assertion true or false). The students' standardized agreement ratings were analyzed as a 2×2 factorial design ANOVA.

a. Identify the factors, factor levels, and treatments in this experiment.

b. Identify the response variable y.

c. The ANOVA F test for interaction resulted in a p-value of less than .01. Interpret this result.

d. Based on the result, part **c**, would you recommend conducting F tests on the main effects? Explain.

e. The sample means for the four experimental conditions are shown in the table. [*Note:* Positive standardized agreement ratings reflect the tendency for the students to agree with the statement. Negative ratings reflect the tendency for the students to disagree with the statement.] Do the sample means support the researchers' theory? Explain.

| | | STORY ASSERTION | |
		True	False
SETTING	Yale (familiar)	−.16	−.26
	Princeton (unfamiliar)	.69	−.56

11.24 What is the optimal method of directing newcomers to a specific location in a complex building? Researchers at Ball State University (Indiana) investigated this "wayfinding" problem and reported their results in *Human Factors* (Mar. 1993). Subjects met in a starting room on a multi-level building and were asked to locate the "goal" room as quickly as possible. (Some of the subjects were provided directional aids, while others were not.) Upon reaching their destination, the subjects returned to the starting room and were given a second room to locate. (One of the goal rooms was located in the east end of the building, the other in the west end.) The experimentally controlled variables in the study were aid type at three levels (signs, map, no aid) and room order at two levels (east/west, west/east). Subjects were randomly assigned to each of the $3 \times 2 = 6$ experimental conditions and the travel time (in seconds) recorded. The results of the analysis of the east room data for this 3×2 factorial design are provided in the table on page 700. Interpret the results.

11.25 Those who care for a spouse with Alzheimer's disease often experience both physical and mental health problems. Four strategies caregivers use to cope with the stress are *wishfulness* ("wished you were a stronger person"), *acceptance* ("accepted the situation"), *intrapsychic* ("fantasies about how things will turn out"), and *instrumental* ("followed a plan of action"). In one experiment, 78 caregivers were classified into one of

Source	df	MS	F	p-Value
Aid type	2	511,323.06	76.67	<.001
Room order	1	13,005.08	1.95	>.10
Aid × Order	2	8,573.13	1.29	>.10
Error	46	6,668.94		

Source: Butler, D. L., et al. "Wayfinding by newcomers in a complex building." *Human Factors,* Vol. 35, No. 1, Mar. 1993, p. 163 (Table 2).

Problem 11.24

the four coping strategies and as having a high or low level of stress. All subjects then completed a standardized test designed to measure level of coping (*Journal of Applied Gerontology,* Mar. 1997). The mean coping scores were calculated for subjects in each of the $4 \times 2 = 8$ strategy × stress classes. A graph of these eight means is shown in the accompanying figure. An ANOVA for a 4×2 factorial design was conducted on the data.

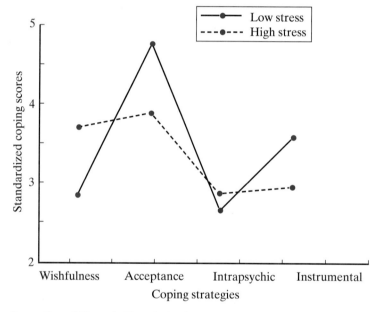

Source: Rose, S. K., et al. "The relationship of self-restraint and distress to coping among spouses caring for persons with Alzheimer's disease." *Journal of Applied Gerontology,* Vol. 16, No. 1, Mar. 1997, p. 97 (Figure 1).

a. Based on the graph, do you believe that strategy and stress interact?

b. In practical terms, what does it mean to say that strategy and stress interact?

11.5 FOLLOW-UP ANALYSIS: MULTIPLE COMPARISONS OF MEANS

Many designed experiments are conducted with the ultimate goal of determining the largest (or the smallest) treatment mean. For example, suppose a drugstore is considering five floor displays for a new product. After conducting an ANOVA on the data and discovering differences among the mean weekly sales associated with the five displays, the drugstore wants to determine which display yields the greatest mean weekly sales of the product. Similarly, a psychologist might want to determine which among six behavior types achieves the highest mean score on an anxiety exam. An entomologist might want to choose one species of ant, from among several, that produces the highest mean number of offspring, and so on.

Once differences among treatment means have been detected in an ANOVA, choosing the treatment with the largest population mean μ might appear to be a simple matter. For example, if there were three treatment means, $\bar{y}_1$, $\bar{y}_2$, and $\bar{y}_3$, we could compare them by constructing a $(1 - \alpha)100\%$ confidence interval for the difference between each pair of means. In general, if there are k treatment means, there are

$$c = k(k - 1)/2$$

pairs of means to be compared. In the case of $k = 3$ means, there are $c = 3(3 - 1)/2 = 3$ confidence intervals to construct, one for each of the following differences: $(\mu_1 - \mu_2)$, $(\mu_1 - \mu_3)$, and $(\mu_2 - \mu_3)$.

However, there is a problem associated with this procedure. If you construct a series of $(1 - \alpha)100\%$ confidence intervals, the risk of making *at least one* Type I error in the series of inferences—called the *experimentwise error rate*—will be larger than the value of α specified for a single interval. For example, if 10 95% confidence intervals are constructed, the probability of at least one Type I error is greater than .05 and may be as high as .40! In this case, there is a good chance that we will conclude that a difference between some pair of means exists, when, in fact, the means are equal.

> **Number of Pairwise Comparisons c When Comparing k Treatment Means**
>
> $$c = k(k - 1)/2$$

Definition 11.12

The **experimentwise error rate (EER)** in a designed experiment to compare k means is the probability of making at least one Type I error in a series of inferences about the population means, based on $(1 - \alpha)100\%$ confidence intervals.

$$\text{EER} = P(\text{at least one Type I error})$$

where $P(\text{Type I error}) = P(\text{Reject } H_0\colon \mu_i = \mu_j \mid \mu_i = \mu_j) = \alpha$.

There are a number of alternative procedures for comparing and ranking a group of treatment means as part of a *follow-up analysis* to the ANOVA. One method that is useful for making pairwise comparisons in many experimental designs is called the **Bonferroni multiple comparisons procedure.**[*] Bonferroni's method is designed to control the experimentwise error rate EER. The user selects a small value of EER (say, EER $\leq$ .10) so that inferences derived from Bonferroni's method can be made with an overall confidence level of $(1 - \text{EER})$. (See Miller [1981] for proof of this result.)

The formula for constructing confidence intervals for the differences between treatment means using Bonferroni's procedure is provided in the box (page 702). Note that confidence intervals for the difference between two treatment means for all possible pairs of treatments are produced based on the EER selected by the analyst.

[*]Other procedures, such as Tukey, Tukey-Kramer, Duncan, and Scheffé, may be more appropriate in certain sampling situations. Consult the Chapter 11 references for details on how to use these multiple comparisons methods.

Bonferroni's Multiple Comparisons Procedure for All Pairwise Comparisons of k Treatment Means

1. Select the experimentwise error rate EER.

2. Determine c, the number of pairwise comparisons (i.e., confidence intervals), where

$$c = k(k - 1)/2 \qquad \text{(11.8)}$$

3. Compute

$$\alpha = \frac{\text{EER}}{c} \qquad \text{(11.9)}$$

   Each interval constructed will have a confidence level of $(1 - \alpha)$.

4. For each treatment pair (i, j) calculate the critical difference:

$$B_{ij} = (t_{\alpha/2})\sqrt{\text{MSE}\left(\frac{1}{n_i} + \frac{1}{n_j}\right)} \qquad \text{(11.10)}$$

   where

$$
\begin{aligned}
t_{\alpha/2} &= \text{the critical } t \text{ value with tail area } \alpha/2 \text{ based on } \nu \text{ degrees of freedom} \\
\nu &= \text{df(Error) in the ANOVA table} \\
\text{MSE} &= \text{mean square for error in the ANOVA table} \\
n_i &= \text{sample size for treatment } i \\
n_j &= \text{sample size for treatment } j
\end{aligned}
$$

5. For each treatment pair (i, j), compute the confidence interval

$$(\bar{y}_i - \bar{y}_j) \pm B_{ij} \qquad \text{(11.11)}$$

6. a. If the interval contains 0, then conclude that the two treatment means are *not* significantly different.

   b. If the interval contains all positive numbers, then conclude that $\mu_i > \mu_j$.

   c. If the interval contains all negative numbers, then conclude that $\mu_j > \mu_i$.

[*Note:* The level of confidence associated with all inferences drawn from the analysis simultaneously is at least $(1 - \text{EER})$.]

Example 11.10   Comparing Means in a One-Way Table

Refer to the one-way ANOVA for the completely randomized design discussed in Examples 11.3 and 11.4 (page 679). Recall that we rejected the null hypothesis of no differences among the mean GPAs of freshmen in three socioeconomic classes. Use Bonferroni's method to perform pairwise comparisons of the means for the three groups of freshmen. Use an experimentwise error rate of .09.

Solution

Step 1   We selected EER = .09. This value represents the probability of making at least one Type I error in the analysis. Also, the overall level of confidence associated with all the inferences derived from the analysis is $(1 - .09) = .91$.

Step 2   Since there are three treatments (i.e., socioeconomic classes) in this experiment, $k = 3$. Then the number of possible pairwise comparisons is (from equation 11.8) $c = 3(3 - 1)/2 = 3$. In other words, we will construct $c = 3$ confidence intervals, one for each of the following treatment pairs: (lower class, middle class), (lower class, upper class), and (middle class, upper class).

Step 3   The value of α used to construct each confidence interval is (from equation 11.9)

$$\alpha = \frac{\text{EER}}{c} = \frac{.09}{3} = .03$$

Since $1 - \alpha = .97$, we will construct three 97% confidence intervals in order to obtain an experimentwise error rate of .09.

Step 4   To find the critical difference for each pair of treatment means, we require the quantities $t_{\alpha/2}$, df(Error), and MSE. From the ANOVA table on the Excel printout, Figure 11.6 (page 679), $\nu = $ df(Error) $= 18$ and MSE $= .262$. We know from Step 3 that $\alpha = .03$ and therefore $\alpha/2 = .015$. We need to find $t_{\alpha/2} = t_{.015}$ where the distribution of t is based on $\nu = 18$ df. If you examine Table B.3 of Appendix B, you will not find a column for $t_{.015}$. However, Excel's TINV function can provide this information. Using Excel, we obtain $t_{.015} = 2.356$.

Since all three treatments have seven observations, i.e., $n_1 = n_2 = n_3 = 7$, the Bonferroni critical difference B_{ij} will be the same for any pair of treatment means. In other words, $B_{12} = B_{13} = B_{23}$. Applying equation 11.10, we find the critical difference

$$B_{ij} = t_{\alpha/2} \sqrt{\text{MSE}\left(\frac{1}{n_i} + \frac{1}{n_j}\right)}$$

$$= 2.356 \sqrt{.262\left(\frac{1}{7} + \frac{1}{7}\right)} = .64$$

Step 5   From the Excel printout, Figure 11.6, we obtain the following sample means:

$$\bar{y}_1 = 2.52 \qquad \bar{y}_2 = 3.25 \qquad \bar{y}_2 = 2.54$$

Substituting these values and B_{ij} into equation 11.11 produces the following three confidence intervals:

Confidence interval for $(\mu_1 - \mu_2)$: $(\bar{y}_1 - \bar{y}_2) \pm B_{12} = (2.52 - 3.25) \pm .64$
$$= -.73 \pm .64$$
$$= (-1.39, -.09)$$

Confidence interval for $(\mu_1 - \mu_3)$: $(\bar{y}_1 - \bar{y}_3) \pm B_{13} = (2.52 - 2.54) \pm .64$
$$= -.02 \pm .64$$
$$= (-.66, .62)$$

Confidence interval for $(\mu_2 - \mu_3)$: $(\bar{y}_2 - \bar{y}_3) \pm B_{12} = (3.25 - 2.54) \pm .64$
$$= .71 \pm .64$$
$$= (.07, 1.35)$$

Overall, we are 91% confident that the intervals *collectively* contain the differences between the true mean GPAs.

Step 6   We demonstrated how to interpret confidence intervals for the difference between two means in Section 9.2. If zero is included in the interval, then

E **Figure 11.11**
Excel Output for the
Bonferroni Analysis,
Example 11.10

1	Bonferroni Procedure for GPA			
2				
3	Group 1		Group 1 to Group 2 Comparison	
4	Sample Mean	2.52	Absolute Difference	0.73
5	Sample Size	7	Standard Error of Difference	0.2736003
6	Group 2		Bonferroni Critical Difference	0.6446523
7	Sample Mean	3.25	Means are different	
8	Sample Size	7		
9	Group 3		Group 1 to Group 3 Comparison	
10	Sample Mean	2.54	Absolute Difference	0.02
11	Sample Size	7	Standard Error of Difference	0.2736003
12	MSE	0.262	Bonferroni Critical Difference	0.6446523
13	Error DF	18	Means are not different	
14	Experimentwise Error Rate	0.09		
15	Number of Groups	3	Group 2 to Group 3 Comparison	
16	Treatment pairs	3	Absolute Difference	0.71
17	α value for comparisons	0.03	Standard Error of Difference	0.2736003
18	t statistic for comparisons	2.356182	Bonferroni Critical Difference	0.6446523
19			Means are different	

there is no evidence of a difference between the means compared. Note that the confidence interval for $(\mu_1 - \mu_3)$ contains zero. Consequently, there is no significant difference between the lower class and upper class GPA means μ_1 and μ_3.

The confidence intervals for $(\mu_1 - \mu_2)$ and $(\mu_2 - \mu_3)$ do not contain zero. Since the first interval, $(\mu_1 - \mu_2)$, contains all negative numbers, we infer that the mean GPA for the middle class (μ_2) exceeds the mean for the lower class (μ_1). Similarly, the third interval, $(\mu_2 - \mu_3)$, contains all positive numbers, implying the middle class mean (μ_2) exceeds the upper class mean (μ_3).

These inferences can also be derived by comparing the absolute difference between the treatment means, $|\bar{y}_i - \bar{y}_j|$, to the Bonferroni critical difference B_{ij}. These results are shown on an Excel printout of the Bonferroni analysis, Figure 11.11.

The absolute differences and Bonferroni critical differences are highlighted on the printout for each pair of treatment means. If $|\bar{y}_i - \bar{y}_j|$ exceeds B_{ij}, then Excel indicates that the treatment means are significantly different. If $|\bar{y}_i - \bar{y}_j|$ is less than B_{ij}, then Excel states that the treatments are not significantly different. You can see from Figure 11.11 that Excel found no difference between the GPA means for the lower (1) and upper (3) classes.

A convenient summary of the results of the Bonferroni multiple comparisons procedure is a listing of the treatment means from lowest to highest, with a solid line connecting those means that are *not* significantly different. This summary is shown in Figure 11.12. Our interpretation is that the mean GPA for the middle class of freshmen students exceeds that of either of the other two mean GPAs; but the means of the lower and upper classes are not significantly different. Again, all these inferences are made with an overall confidence level of 91% since our experimentwise error rate is .09. ❏

Figure 11.12 Summary of
Bonferroni's Multiple
Comparisons, Example 11.10

Mean GPA	2.52	2.54	3.25
Treatment (Class)	Lower	Upper	Middle

Self-Test

11.5

Refer to the one-way ANOVA, Self-Test 11.2 (page 679). Bonferroni's procedure was used to rank the four golf ball brand mean distances with EER = .06. The results are summarized below.

Mean	249.3	250.8	261.1	270.0
Brand	D	A	B	C

a. How many pairwise comparisons of the four brand means are included in the analysis?

b. Find the value of α used to form the confidence intervals.

c. Find the value of the Bonferroni critical difference used to rank the means.

d. Which brand has the highest ranked mean? The smallest?

e. Interpret the meaning of EER = .06.

In addition to a completely randomized design, Bonferroni's multiple comparisons of means procedure can be applied to a wide variety of experimental designs in ANOVA (including the factorial design of Section 11.4). Keep in mind, however, that many other methods of making multiple comparisons are available and one or more of these techniques may be more appropriate to use in your particular application. Consult the Chapter 11 references for details on other techniques (e.g., Hsu [1996]).

In closing, we remind you that multiple comparisons of treatment means should be performed only as a follow-up analysis to the ANOVA, i.e., only after you have conducted the appropriate analysis of variance F test(s) and determined that sufficient evidence exists of differences among the treatment means. Be wary of conducting multiple comparisons when the ANOVA F test indicates no evidence of a difference among a small number of treatment means—this may lead to confusing and contradictory results.*

*When a large number of treatments are to be compared, a borderline, nonsignificant F value (e.g., .05 < p-value < .10) may mask differences between some of the means. In this situation, it is better to ignore the F test and proceed directly to a multiple comparisons procedure.

Caution

In practice, it is advisable to avoid conducting multiple comparisons of a small number of treatment means when the corresponding ANOVA F test is nonsignificant; otherwise, confusing and contradictory results may occur.

**PROBLEMS
11.26–11.43**

Using the Tools

11.26 Consider a completely randomized design with k treatments. For each of the following values of k, find the number of pairwise comparisons of treatments to be made in a Bonferroni analysis.

a. $k = 3$ **b.** $k = 5$ **c.** $k = 4$ **d.** $k = 8$

11.27 Consider a two-factor factorial design with factor A at a levels and factor B at b levels. For each of the following, find the number of pairwise comparisons of treatments to be made in a Bonferroni analysis.

   a. $a = 2, b = 3$ **b.** $a = 2, b = 2$

   c. $a = 4, b = 2$ **d.** $a = 3, b = 2$

11.28 Refer to Problem 11.26. For each part, find the value of α used to construct the Bonferroni confidence intervals if EER = .10.

11.29 Refer to Problem 11.27. For each part, find the value of α used to construct the Bonferroni confidence intervals if EER = .05.

11.30 Find the Bonferroni critical difference B_{ij} for comparing k treatment means in each of the following situations:

   a. $n_i = 10, n_j = 10$, EER = .06, $k = 3$, MSE = 25

   b. $n_i = 8, n_j = 12$, EER = .05, $k = 5$, MSE = 100

   c. $n_i = 25, n_j = 20$, EER = .10, $k = 5$, MSE = 4.8

11.31 Bonferroni's multiple comparisons procedure was applied to data from a completely randomized design with five treatments ($A, B, C, D,$ and E). The findings are summarized as follows:

Mean	10.2	11.3	14.7	14.9	20.6
Treatment	B	C	A	E	D

   a. Identify the treatment(s) in the group with the statistically largest mean(s).

   b. Identify the treatment(s) in the group with the statistically smallest mean(s).

11.32 Bonferroni's multiple comparisons procedure was applied to data from a 2×2 factorial design. The findings are summarized as follows:

Mean	100	145	155	160
Treatment	A_1B_2	A_2B_2	A_2B_1	A_1B_1

   a. Identify the treatment(s) in the group with the statistically largest mean(s).

   b. Identify the treatment(s) in the group with the statistically smallest mean(s).

11.33 In a design with four treatments ($A, B, C,$ and D), Bonferroni's multiple comparisons procedure found that (1) treatment A had a statistically smaller mean than the other three treatments, (2) the means for treatments C and D are not statistically different, and (3) the mean for treatment B is statistically larger than those for treatments C and D. Summarize these results by using lines to connect treatment means that are not statistically different.

Applying the Concepts

11.34 *Science Education* (Jan. 1995) investigated whether the level of chemistry education impacts performance on a test about conservation of matter. A sample of 120 students was divided into six groups of 20 subjects each according to age and level of chemistry instruction. The six experimental groups were (1) seventh graders, (2) ninth graders, (3) eleventh grade science students, (4) eleventh grade nonscience students, (5) college psychology majors, and (6) college chemistry majors. The mean proportion of correct answers on the exam for each group is listed in the accompanying table.

Group	Mean Score
College chemistry majors	.891
College psychology majors	.674
11th grade nonscience students	.660
11th grade science students	.588
9th graders	.519
7th graders	.384

a. An ANOVA F test for group differences resulted in a p-value less than .001. Interpret this result.

b. A multiple comparisons of treatment means was performed using an experimentwise error rate of .05. The vertical lines in the table summarize the results of the multiple comparisons. Which group(s) has the largest mean proportion of correct answers? The smallest?

c. Interpret the value of EER = .05.

11.35 Are all videocassette recorder (VCR) users alike or can they be segmented into subgroups with different motives and behaviors? This question was the topic of research reported in the *Journal of Advertising Research* (Apr./May 1988). A sample of over 300 members of a large videotape rental club in a southeastern city were surveyed about their VCR use. Based on their responses, each member was categorized into one of five groups as shown in the table below. One of the dependent variables measured was degree to which the user "zipped" (i.e., fast-forwarded) through commercials while replaying a taped TV program. This "ad avoidance" variable was measured on a 7-point scale, where 1 = almost always and 7 = never.

GROUP	
(1) Videophile:	Record TV programs often, and rent/buy videotapes often
(2) Time shifter:	Record TV programs often, rarely rent/buy videotapes
(3) Source Shifter:	Rarely record TV programs, rent/buy videotapes often
(4) Low user:	Rarely record TV programs or rent/buy videotapes
(5) Regular user:	Periodically record TV programs and/or rent/buy videotapes

a. The ANOVA F test for testing H_0: $\mu_1 = \mu_2 = \mu_3 = \mu_4 = \mu_5$ resulted in a p-value of less than .01. Interpret this result.

b. The mean "ad avoidance" levels of the five groups are listed here, as well as the results of a Bonferroni multiple comparisons analysis (using EER = .05). Interpret the results.

Mean "ad avoidance"	1.8	2.6	2.8	2.9	3.4
VCR user segment	Time shifter	Videophile	Regular user	Low user	Source shifter

11.36 Refer to the *American Journal of Psychiatry* study of bulimia nervosa, Problem 11.5 (page 682). Recall that females diagnosed with bulimia nervosa were divided into three age groups based on year of birth. A multiple comparisons procedure was used to rank the mean age (in years) at onset of the eating disorder for the three groups, using an experimentwise error rate of .01. The results are summarized below. What conclusions can you draw from this analysis?

32.6	18.8	17.5
Before	Between	After
1950	1950 and 1959	1959

11.37 Refer to the one-way ANOVA on the Blackjack profit data, Problem 11.8 (page 684). Apply the Bonferroni method (and Excel) to rank the four Blackjack strategies. Use EER = .10.

11.38 Refer to the analysis of variance on the DDT data stored in FISH.XLS, Problem 11.9 (page 684). Excel was used to make a Bonferroni multiple comparisons of the mean DDT levels of the three species of fish using an experimentwise error rate of .05. Interpret the results shown on the Excel printout.

Bonferroni Procedure for DDT			
Group 1		Group 1 to Group 2 Comparison	
Sample Mean	33.29938	Absolute Difference	25.137713
Sample Size	96	Standard Error of Difference	19.197631
Group 2		Bonferroni Critical Difference	46.513474
Sample Mean	8.161667	Means are not different	
Sample Size	36		
Group 3		Group 1 to Group 3 Comparison	
Sample Mean	1.38	Absolute Difference	31.91938
Sample Size	12	Standard Error of Difference	30.076908
MSE	9649.284	Bonferroni Critical Difference	72.872609
Error DF	141	Means are not different	
Experimentwise Error Rate	0.05		
Number of Groups	3	Group 2 to Group 3 Comparison	
Treatment pairs	3	Absolute Difference	6.781667
α value for comparisons	0.016667	Standard Error of Difference	32.74359
t statistic for comparisons	2.422876	Bonferroni Critical Difference	79.333648
		Means are not different	

11.39 Refer to the *Journal of Environmental Education* study of environmental concerns, Problem 11.11 (page 685). On the issue of oil spills, the ANOVA F test comparing the means of the four ethnic groups was significant at $\alpha = .05$. A Bonferroni follow-up analysis yielded the following results at an experimentwise error rate of .05. What conclusions can you draw from the analysis?

Mean response	4.74	4.53	4.04	3.79
Ethnic group	U.S.-born Anglos	U.S.-born Hispanics	Mexican-born Hispanics	Central American-born Hispanics

11.40 Refer to the *Psychological Assessment* (Mar. 1995) study of the effectiveness of the Minnesota Multiphasic Personality Inventory (MMPI) in detecting distorted responses, Problem 11.12 (page 685). You conducted an ANOVA on each of five MMPI scales and determined that the mean scores of four groups of females differed significantly for each scale. Recall that the four groups were nonforensic psychiatric patients (NFP), forensic psychiatric patients (FP), college students who responded honestly (CSH), and college students who "faked bad" (CSFB) responses. The Bonferroni method was used to rank the means of the four groups using EER = .05. The results for each response variable are shown in the table. [*Note:* Overbars connect means that are not significantly different.] Interpret the results for each response variable. Recall that the MMPI will be deemed effective in detecting distorted responses if the mean score of the "fake bad" group is largest.

Response Variable					
Infrequency (F)	Mean	7.1	11.3	14.6	33.6
	Group	CHS	NFP	FP	CSFB
Obvious (O)	Mean	240.9	270.7	287.5	341.9
	Group	CSH	FP	NFP	CSFB
Subtle (S)	Mean	231.4	244.1	244.5	259.7
	Group	CSFB	CSH	FP	NFP
Obvious-subtle (O-S)	Mean	51.3	88.2	93.4	198.8
	Group	CSH	NFP	FP	CSFB
Dissimulation (D)	Mean	12.0	12.1	20.8	21.0
	Group	FP	NFP	CSH	CSFB

Source: Bagby, R. M., Burs, T., and Nicholson, R.A. "Relative effectiveness of the standard validity scales in detecting fake-bad and fake-good responding: Replication and extension." *Psychological Assessment,* Vol. 7, No. 1, Mar. 1995, p. 86 (Table 1).

11.41 Refer to the *Journal of Sport & Social Issues* study of enjoyment of college basketball games, Problem 11.18 (page 696). Recall that the 4×2 factorial experiment resulted in a significant interaction between suspense and gender. Because of this finding, the researchers compared the mean enjoyment ratings of the four suspense levels separately for males and females using a Bonferroni multiple comparisons technique.

a. The experimentwise error rate for the analysis was .05. Interpret this value.

b. The multiple comparisons of the four suspense means for male viewers are summarized below. Interpret the results.

1.77	5.38	7.16	7.59
Minimal	Moderate	Substantial	Extreme

c. The multiple comparisons of the four suspense means for female viewers are summarized below. Interpret the results.

2.73	4.34	4.92	7.52
Minimal	Moderate	Extreme	Substantial

11.42 Refer to the *Journal of Nutrition* study of a vitamin-B supplement, Problem 11.17 (page 696). Use Bonferroni's method (and Excel) to rank the mean weights of rats in the four treatment conditions. Use an experimentwise error rate of .01.

11.43   Refer to the *Environmental Entomology* study of citrus mealybugs, Problem 11.20 (page 697). The means for number of mealybugs found per fruit for each of the four time periods were compared using an experimentwise error rate of .05. The results are summarized here. Interpret these results.

Mean number per fruit	1.91	1.87	1.58	0.09
Period	Oct.	July	Sept.	June

11.6   CHECKING ANOVA ASSUMPTIONS

For both of the experimental designs discussed in this chapter, we listed the assumptions underlying the analysis in the relevant boxes. The assumptions for a completely randomized design are as follows: (1) the k probability distributions of the response y corresponding to the k treatments are normal; and (2) the population variances of the k treatments are equal. Similarly, for factorial designs, the data for the treatments must come from normal probability distributions with equal variances.

Checks on these ANOVA assumptions can be performed using graphs of the response variable y. A brief overview of these techniques is given in the box.

Checking ANOVA Assumptions

Detecting Nonnormal Populations

1. For each treatment, construct a histogram, stem-and-leaf display, box-and-whisker plot, or normal probability plot of the response (or dependent) variable y. Look for highly skewed distributions. [*Note:* ANOVA, like regression, is robust with respect to the normality assumption. That is, slight departures from normality will have little impact on the validity of the inferences derived from the analysis. However, if the sample size for each treatment is small, then these graphs will probably be of limited use.]

2. Formal statistical tests of normality are also available. The null hypothesis is that the probability distribution of the response y is normal. These tests, however, are sensitive to slight departures from normality. Since in most scientific applications the normality assumption will not be satisfied exactly, these tests will likely result in a rejection of the null hypothesis and, consequently, are of limited use in practice. Consult the references for more information on these formal tests.

3. If the distribution of the response y departs greatly from normality, a *normalizing transformation* may be necessary. For example, for highly skewed distributions, transformations such as $\log(y)$ or $\sqrt{y}$ tend to "normalize" the data since these functions "pull" the observations in the tail of the distribution back toward the mean.

Detecting Unequal Variances

1. For each treatment, construct a line plot for the values of the response y and look for differences in the spread (variability) shown in the plots. (See Figure 11.13 [page 711] for an example.)

2. When the sample sizes are small for each treatment, only a few points are plotted on the frequency plots, making it difficult to detect differences in variation. In this situation, you may want to use one of several formal statistical tests of homogeneity of variances that are available. Consult the Chapter 11 references for information on these tests (e.g., Gujarati [1995]).

3. When unequal variances are detected, use a *variance stabilizing transformation* such as $\log(y)$ or $\sqrt{y}$.

Example 11.11   Checking Assumptions

Refer to the completely randomized ANOVA, Examples 11.3 and 11.4. Check the ANOVA assumptions for this experiment.

Solution

For this completely randomized design, the three treatments are the three socio-economic classes (lower, middle, and upper) and the response y is grade point average (GPA). Thus, we require the GPA distribution for each class to be normally distributed with equal variances. Recall that the sample size for each treatment was seven.

With such a small number of observations per treatment, a stem-and-leaf display to check for normality will not be very revealing. Not knowing whether the GPAs are normally distributed is not a major concern, since ANOVA, like regression, is robust with respect to nonnormal data. That is, the procedure yields valid inferences even when the distribution of the response y deviates from normality.

To check for equal variances, we plotted the GPAs for each socioeconomic class in Figure 11.13. The line plot shows the spread of the GPAs for each class. Except for one low GPA (1.36) in the upper class, the variability of the response is about the same for each treatment; thus, the assumption of equal variances appears to be reasonably satisfied. ❑

Figure 11.13 Line Plot for Example 11.11

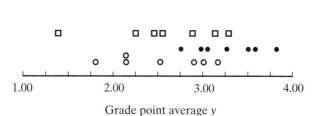

Self-Test
11.6

Refer to the one-way ANOVA, Self-Test 11.2 (page 679). Line plots of the distances for each brand of golf ball are illustrated in the accompanying figure. Does it appear that the brands have approximately the same variation in distance?

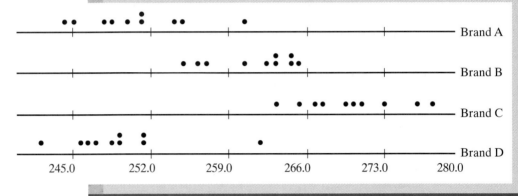

PROBLEMS
11.44–11.48

Applying the Concepts

11.44   Check the assumptions for the completely randomized design ANOVA of Problem 11.6 (page 682).

11.45   Check the assumptions for the completely randomized design ANOVA of Problem 11.7 (page 683).

11.46   Check the assumptions for the completely randomized design ANOVA of Problem 11.8 (page 684).

11.47   Check the assumptions for the completely randomized design ANOVA of Problem 11.9 (page 685).

11.48   Check the assumptions for the factorial design ANOVA of Problem 11.17 (page 696).

11.7 CALCULATION FORMULAS FOR ANOVA (OPTIONAL)

In this optional section we demonstrate the calculation formulas for constructing an ANOVA table for both experimental designs discussed in this chapter. Except for rounding, these formulas will produce results identical to those obtained from statistical or spreadsheet software such as Excel.

ANOVA Calculation Formulas for a Completely Randomized Design

1. CM = Correction for the mean

$$= \frac{(\text{Total of all observations})^2}{\text{Total number of observations}} = \frac{(\Sigma y)^2}{n} \qquad (11.12)$$

k = number of means (treatments) to be compared

n_i = number of observations for treatment i

n = total sample size = $n_1 + n_2 + \cdots + n_k$

2. $SS(\text{Total})$ = Total sum of squares

$$= (\text{Sum of squares of all observations}) - CM$$

$$= \Sigma y^2 - CM \qquad (11.13)$$

3. SST = Sum of squares for treatments

$$= \left(\begin{array}{c} \text{Sum of squares of treatment totals} \\ \text{with each square divided by the} \\ \text{number of observations in that treatment} \end{array} \right) - CM$$

$$= \frac{T_1^2}{n_1} + \frac{T_2^2}{n_2} + \cdots + \frac{T_k^2}{n_k} - CM \qquad (11.14)$$

where T_i is the total of all observations for treatment $i, i = 1, 2, ..., k$.

4. SSE = Sum of squares for error

$$= SS(\text{Total}) - SST \qquad (11.15)$$

5. MST = Mean square for treatment = $\dfrac{SST}{k-1} \qquad (11.16)$

6. MSE = Mean square for error = $\dfrac{SSE}{n-k} \qquad (11.17)$

7. F = Test statistic = $\dfrac{MST}{MSE} \qquad (11.18)$

ANOVA Calculation Formulas for a Two-Factor Factorial Design

1. CM = Correction for the mean

$$= \frac{(\text{Total of all } n \text{ observations})^2}{n} = \frac{(\Sigma y)^2}{n} \qquad (11.19)$$

where $n = abr$

a = number of levels of factor A

b = number of levels of factor B

r = number of replications of the factorial experiment

2. $SS(\text{Total})$ = (Sum of squares of all observations) $- CM$

$$= \Sigma y^2 - CM \qquad (11.20)$$

continued

3. Main effect sum of squares for factor A:

$$SS(A) = \frac{A_1^2 + A_2^2 + \cdots + A_a^2}{br} - CM \qquad (11.21)$$

where A_i is the total of all observations at level i for factor A.

4. Main effect sum of squares for factor B:

$$SS(B) = \frac{B_1^2 + B_2^2 + \cdots + B_b^2}{ar} - CM \qquad (11.22)$$

where B_i is the total of all observations at level i for factor B.

5. Interaction sum of squares:

$$SS(AB) = \sum_{i,j} \frac{(AB_{ij})^2}{r} - SS(A) - SS(B) - CM \qquad (11.23)$$

where AB_{ij} is the sum of all observations in the cell corresponding to the ith level of factor A and the jth level of factor B. [*Note:* To find SS(AB), square each cell total, sum the squares for all cell totals, divide by r, and then subtract SS(A), SS(B), and CM.]

6. $$SSE = SS(Total) - SS(A) - SS(B) - SS(AB) \qquad (11.24)$$

7. $$MS(A) = \frac{SS(A)}{a - 1} \qquad (11.25)$$

$$MS(B) = \frac{SS(B)}{b - 1} \qquad (11.26)$$

$$MS(AB) = \frac{SS(AB)}{(a - 1)(b - 1)} \qquad (11.27)$$

$$MSE = s^2 = \frac{SSE}{ab(r - 1)} \qquad (11.28)$$

8. F statistic for testing AB interaction: $F = \dfrac{MS(AB)}{MSE} \qquad (11.29)$

9. F statistic for testing A main effect: $F = \dfrac{MS(A)}{MSE} \qquad (11.30)$

10. F statistic for testing B main effect: $F = \dfrac{MS(B)}{MSE} \qquad (11.31)$

Example 11.12   Using Calculation Formulas for a One-Way ANOVA

The data for the completely randomized design of Example 11.3 are reproduced in Table 11.6 (page 715). Use the calculation formulas shown in the box to construct an ANOVA table.

Solution

Step 1   The first step is to find the sum of all $n = 21$ sample observations, or $\sum y$. The sums of the GPAs for each class (treatment) are shown at the bottom of Table 11.6. Adding these three sums yields

$$\sum y = T_1 + T_2 + T_3 = 17.65 + 22.74 + 17.79 = 58.18$$

Table 11.6 Data (GPAs) and Sums for the Completely Randomized ANOVA, Example 11.12

	Lower Class	Middle Class	Upper Class
	2.87	3.23	2.25
	2.16	3.45	3.13
	3.14	2.78	2.44
	2.51	3.77	3.27
	1.80	2.97	2.81
	3.01	3.53	1.36
	2.16	3.01	2.53
Sums:	$T_1 = 17.65$	$T_2 = 22.74$	$T_3 = 17.79$

Then we compute CM—called the correction for the mean—using equation 11.12:

$$\text{CM} = \frac{(\Sigma y)^2}{n} = \frac{(58.18)^2}{21} = 161.1863$$

Step 2   SS(Total) is obtained using equation 11.13. (Note that this requires that we sum the squares of all $n = 21$ sample GPA values.)

$$\begin{aligned} \text{SS(Total)} &= \sum y^2 - \text{CM} \\ &= (2.87)^2 + (2.16)^2 + (3.14)^2 + \cdots + (2.53)^2 - 161.1863 \\ &= 168.299 - 161.1863 = 7.1127 \end{aligned}$$

Step 3   Applying equation 11.14, the sum of squares for treatments, SST, is

$$\begin{aligned} \text{SST} &= \frac{T_1^2}{n_1} + \frac{T_2^2}{n_2} + \frac{T_3^2}{n_3} - \text{CM} \\ &= \frac{(17.65)^2}{7} + \frac{(22.74)^2}{7} + \frac{(17.79)^2}{7} \\ &= 163.5877 - 161.1863 = 2.4014 \end{aligned}$$

Step 4   The sum of squared errors, SSE, is obtained using equation 11.15:

$$\begin{aligned} \text{SSE} &= \text{SS(Total)} - \text{SST} \\ &= 7.1127 - 2.4014 = 4.7113 \end{aligned}$$

Step 5   Applying equation 11.16, we obtain the mean square for treatments:

$$\text{MST} = \frac{\text{SST}}{k-1} = \frac{2.4014}{3-1} = 1.2007$$

Step 6   Applying equation 11.17, we obtain the mean square for error:

$$\text{MSE} = \frac{\text{SSE}}{n-k} = \frac{4.7113}{21-3} = .2617$$

Step 7   Finally, from equation 11.18, the test statistic is

$$F = \frac{\text{MST}}{\text{MSE}} = \frac{1.2007}{.2617} = 4.59$$

Table 11.7 ANOVA Table for Example 11.12

Source	df	SS	MS	F
Class (treatments)	2	2.4014	1.2007	4.59
Error	18	4.7113	.2617	
TOTAL	20	7.1127		

These computations are summarized in the ANOVA table shown in Table 11.7. Except for rounding, these numbers agree with those shown in the Excel printout, Figure 11.6 (page 679). ❑

Example 11.13 **Using Calculation Formulas for a Two-Way ANOVA**

The data for the factorial design of Example 11.5 are reproduced in Table 11.8. Use the calculation formulas shown in the box to construct an ANOVA table for the data.

Solution

For this 2×3 factorial design, let machine represent factor A and material represent factor B. Then the number of levels of factors A and B, respectively, are $a = 2$ and $b = 3$. Also, there are $r = 3$ replications in this experiment.

Step 1   First, we calculate the correction factor for the mean, CM. From Table 11.8, the sum of all $n = 18$ sample observations is 68.81. Substituting into equation 11.19, we find:

$$CM = \frac{(\Sigma y)^2}{n} = \frac{(68.81)^2}{18} = 263.04534$$

Step 2   Applying equation 11.20, we obtain SS(Total):

$$SS(Total) = \sum y^2 - CM$$
$$= (4.31)^2 + (4.27)^2 + (4.40)^2 + \cdots + (3.42)^2 - 263.04534$$
$$= 264.7793 - 263.04534 = 1.73396$$

Table 11.8 Data (Thousands of Gaskets) and Sums for the Factorial ANOVA, Example 11.13

		C	R	P	Sums
		MATERIAL			
MACHINE	**1**	4.31	3.36	4.01	$A_1 = 35.08$
		4.27	3.42	3.94	
		4.40	3.48	3.89	
		Sum = 12.98	Sum = 10.26	Sum = 11.84	
	2	3.94	3.91	3.48	$A_2 = 33.73$
		3.81	3.80	3.53	
		3.99	3.85	3.42	
		Sum = 11.74	Sum = 11.56	Sum = 10.43	
Sums		$B_1 = 24.72$	$B_2 = 21.82$	$B_3 = 22.27$	TOTAL = 68.81

Step 3   The main effect sum of squares for factor A (machine) requires that we first find the totals for each level of A. These sums, $A_1 = 35.08$ and $A_2 = 33.73$, are shown in Table 11.8. Substituting these values into equation 11.21, we obtain:

$$SS(A) = \frac{A_1^2 + A_2^2}{br} - CM$$

$$= \frac{(35.08)^2 + (33.73)^2}{(3)(3)} - 263.04534$$

$$= 263.14659 - 263.04534 = .10125$$

Step 4   To find the main effect sum of squares for factor B (material), we need the totals for each level of B. These sums, $B_1 = 24.72$, $B_2 = 21.82$, and $B_3 = 22.27$, are also shown in Table 11.8. Substituting into equation 11.22, we find

$$SS(B) = \frac{B_1^2 + B_2^2 + B_3^2}{ar} - CM$$

$$= \frac{(24.72)^2 + (21.82)^2 + (22.27)^2}{(2)(3)} - 263.04534$$

$$= 263.85728 - 263.04534 = .81194$$

Step 5   To calculate the interaction sum of squares, we need the sum of responses in each combination of the levels of the two factors. These sums are shown in the six cells of Table 11.8. Applying equation 11.23 yields

$$SS(AB) = \frac{(A_1B_1)^2 + (A_1B_2)^2 + (A_1B_3)^2 + (A_2B_1)^2 + (A_2B_2)^2 + (A_2B_3)^2}{r} - SS(A) - SS(B) - CM$$

$$= \frac{(12.98)^2 + (10.26)^2 + (11.84)^2 + (11.74)^2 + (11.56)^2 + (10.43)^2}{3} - .10125 - .81194 - 263.04534$$

$$= 264.72657 - .10125 - .81194 - 263.04534$$

$$= .76803$$

Step 6   The sum of squares for error, SSE, is obtained using equation 11.24:

$$SSE = SS(Total) - SS(A) - SS(B) - SS(AB)$$

$$= 1.73396 - .10125 - .81194 - .76803 = .05274$$

Step 7   Using equations 11.25–11.28, we obtain the following mean squares:

$$MS(A) = \frac{SS(A)}{a - 1} = \frac{.10125}{2 - 1} = .10125$$

$$MS(B) = \frac{SS(B)}{b - 1} = \frac{.81194}{3 - 1} = .40597$$

$$MS(AB) = \frac{SS(AB)}{(a - 1)(b - 1)} = \frac{.76803}{(2 - 1)(3 - 1)} = .384015$$

$$MSE = \frac{SSE}{ab(r - 1)} = \frac{.05274}{(2)(3)(3 - 1)} = .004395$$

Table 11.9 ANOVA Table for Example 11.13

Source	df	SS	MS	F
Machine (A)	1	.10125	.10125	23.04
Material (B)	2	.81194	.40597	92.37
A × B	2	.76803	.384015	87.38
Error	12	.05274	.004395	
TOTAL	15	1.73396		

Steps 8–10   The ANOVA *F* statistics are calculated using equations 11.29–11.31:

$$\text{Interaction: } F = \frac{MS(AB)}{MSE} = \frac{.384015}{.004395} = 87.38$$

$$\text{Main Effect } A \text{ (machine): } F = \frac{MS(A)}{MSE} = \frac{.10125}{.004395} = 23.04$$

$$\text{Main Effect } B \text{ (machine): } F = \frac{MS(B)}{MSE} = \frac{.40597}{.004395} = 92.37$$

These computations are summarized in the ANOVA table shown in Table 11.9. Except for rounding, these numbers agree with those in the Excel printout, Figure 11.9 (page 692). ❑

PROBLEMS 11.49–11.53

Using the Tools

11.49   Use the ANOVA calculation formulas for a completely randomized design to construct the ANOVA summary table for Problem 11.3 (page 681).

11.50   Use the ANOVA calculation formulas for a completely randomized design to construct the ANOVA summary table for Problem 11.6 (page 682).

11.51   Use the ANOVA calculation formulas for a completely randomized design to construct the ANOVA summary table for Problem 11.7 (page 683).

11.52   Use the ANOVA calculation formulas for a factorial design to construct the ANOVA summary table for Problem 11.15 (page 695).

11.53   Use the ANOVA calculation formulas for a factorial design to construct the ANOVA summary table for Problem 11.17 (page 696).

*Rosen, S., and Tesser, A. "On reluctance to communicate undesirable information: The MUM effect." *Journal of Communication,* Vol. 22, 1970, pp. 124–141.

REAL-WORLD REVISITED
Reluctance to Transmit Bad News—The MUM Effect

In a 1970 experiment, psychologists S. Rosen and A. Tesser found that people were reluctant to transmit bad news to peers. Rosen and Tesser termed this phenomenon the "MUM effect."* Since that time, numerous studies have been conducted to determine why people have a tendency to keep mum when given the opportunity to transmit bad news to others. Two theories have emerged from this research. The first maintains that the MUM effect is an aversion to private discomfort. To avoid discomforts such as empathy with the victim's distress or guilt feelings for their own good fortune, would-be communicators of bad news keep mum. The second theory is that the MUM effect is a public display. People experience little or no discomfort when transmitting bad news, but keep mum to avoid an unfavorable impression or to pay homage to a social norm.

Here, we consider an article by C. F. Bond and E. L. Anderson published in the *Journal of Experimental Social Psychology*. Bond and Anderson conducted a designed experiment to determine which of the two explanations for the MUM effect is more plausible. "If the MUM effect is an aversion to private discomfort," they state, "subjects should show the effect whether or not they are visible [to the victim]. If the effect is a public display, it should be stronger if the subject is visible than if the subject cannot be seen."*

Forty undergraduates at Duke University participated in the experiment. Each subject was asked to administer an IQ test to another student and then provide the test taker with his or her percentile score. Unknown to the subject, the test taker was a confederate student who was working with the researchers. The experiment manipulated two factors, *subject visibility* and *confederate success,* each at two levels. Subject visibility was either visible (i.e., the subject was visible to the test taker) or not visible. Confederate success was either success (i.e., the test taker supplied a set of answers that placed him in the top 20% of all Duke undergraduates) or failure (i.e., the bottom 20%). Ten subjects were randomly assigned to each of the $2 \times 2 = 4$ experiment conditions; thus, a 2×2 factorial design with 10 replications was employed.

One of several behavioral variables that were measured during the experiment was *latency to feedback,* defined as the time (in seconds) between the end of the test and the delivery of feedback (i.e., the percentile score) from the subject to the test taker. The data on latency to feedback for the 40 subjects, listed in Table 11.10, are stored in the Excel workbook MUM.XLS.

*Bond, C. F., and Anderson, E. L. "The reluctance to transmit bad news: Private discomfort or public display?" *Journal of Experimental Social Psychology,* Vol. 23, 1987, pp. 176–187.

MUM.XLS

Table 11.10 Data on Latency to Feedback for the 2 × 2 Factorial Design

| | | CONFEDERATE SUCCESS | | | |
		Success		Failure	
Subject Visibility	Visible	67.6	46.3	118.7	169.4
		83.8	75.9	153.6	171.2
		75.7	51.4	127.1	156.2
		84.9	87.0	141.3	137.6
		54.8	104.3	153.9	143.2
	Not Visible	79.1	65.7	76.2	62.0
		106.2	99.6	90.3	85.0
		87.2	105.3	83.3	64.8
		104.5	93.3	46.2	37.9
		100.5	55.0	83.4	96.4

Note: Data are simulated based on summary statistics reported in Bond and Anderson (1987).

Study Questions

a. Conduct a complete analysis of variance for the data, including a multiple comparisons of treatment means (if necessary).

b. Bond and Anderson conclude that "subjects appear reluctant to transmit bad news—but only when they are visible to the news recipient." Do you agree?

Key Terms

Analysis of variance
(ANOVA) 673
ANOVA table 678
Between-sample variation 674
Bonferroni multiple comparisons
procedure 701
Complete factorial design 686
Completely randomized
design 676
Design 670

Experiment 670
Experimental (sampling) unit 670
Experimentwise error rate 701
Factor 671
Factor interaction 689
Factor level 671
Factor main effect 689
Factorial design 686
Mean square for error 674
Mean square for treatments 675

Multiple comparisons of
means 701
Replication 686
Response variable 670
Sum of squares for
treatments 675
Treatment 671
Within-sample variation 674

Key Formulas

Completely randomized design:

Testing treatments $\qquad F = \dfrac{\text{MST}}{\text{MSE}}$ **(11.4)**, 678

Factorial design with 2 factors:

Testing $A \times B$ interaction $\qquad F = \dfrac{\text{MS}(AB)}{\text{MSE}}$ **(11.5)**, 690

Testing main effect A $\qquad F = \dfrac{\text{MS}(A)}{\text{MSE}}$ **(11.6)**, 690

Testing main effect B $\qquad F = \dfrac{\text{MS}(B)}{\text{MSE}}$ **(11.7)**, 690

Number of pairwise comparisons for k treatment means $\quad c = \dfrac{k(k-1)}{2}$ **(11.8)**, 702

Bonferroni critical difference $\qquad B_{ij} = (t_{\alpha/2})\sqrt{\text{MSE}\left(\dfrac{1}{n_i} + \dfrac{1}{n_j}\right)}$ **(11.10)**, 702

Key Symbols

Symbol	Description
ANOVA	Analysis of variance
SS(Total)	Total sum of squares of deviations
SST	Sum of squares for treatments
SSB	Sum of squares for blocks
SSE	Sum of squared errors
MST	Mean square for treatments
MSB	Mean square for blocks

MSE	Mean square for errors
F	Ratio of mean squares
SS(A)	Sum of squares for main effect A
SS(AB)	Sum of squares for interaction between factors A and B
MS(A)	Mean square for main effect A
MS(AB)	Mean square for AB interaction

Supplementary Problems 11.54–11.67

11.54 Bell Communications Research (Bellcore) publishes, on average, 150 technical memos (TMs) each month. Since not all TMs are relevant to all employees, Bellcore recognizes that "information filtering" will naturally occur—each employee will retain only the information that is relevant to him or her. Bellcore researchers conducted a study of four automated information filtering methods (*Communications of the Association for Computing Machinery,* Dec. 1992). These methods are (1) keyword match-word profile, (2) latent semantic indexing (LSI)-word profile, (3) keyword match-document profile, and (4) LSI-document profile. A sample of TM abstracts was filtered by each method and the relevance of the filtered abstracts was rated on a 7-point scale by a panel of Bellcore employees. In addition, a subset of TM abstracts was randomly selected and rated by the panel. The mean relevance ratings of the five filtering methods (four automated methods plus the random selection method) were compared using an analysis of variance for a completely randomized design.

   a. Identify the treatments in the experiment.

   b. Identify the response variable.

   c. The ANOVA resulted in a test statistic of $F = 117.5$, based on 4 numerator degrees of freedom and 132 denominator degrees of freedom. Interpret this result at a significance level of $\alpha = .05$.

11.55 Turnover among truck drivers is a major problem for both carriers and shippers. Since knowledge of driver-related job attitudes is valuable for predicting and controlling future turnover, a study of the work-related attitudes of truck drivers was conducted (*Transportation Journal,* Fall 1993). The two factors considered in the study were career stage and time spent on the road. Career stage was set at three levels: early (less than 2 years), mid-career (between 2 and 10 years), and late (more than 10 years). Road time was dichotomized as short (gone for one weekend or less) and long (gone for longer than one weekend). Data were collected on job satisfaction for drivers sampled in each of the $3 \times 2 = 6$ combinations of career stage and road time. (Job satisfaction was measured on a 5-point scale, where 1 = really dislike and 5 = really like.)

   a. Identify the response variable for this experiment.

   b. Identify the factors for this experiment.

   c. Identify the treatments for this experiment.

   d. The ANOVA table for the analysis appears on page 722. Fully interpret the results.

   e. The researchers theorized that the impact of road time on job satisfaction may be different depending on the career stage of the driver. Do the results support this theory?

Source	F Value	p-Value
Career stage (CS)	26.67	$p \le .001$
Road time (RT)	.19	$p > .05$
CS × RT	1.59	$p > .05$

Source: McElroy, J. C., et al. "Career stage, time spent on the road, and truckload driver attitudes." *Transportation Journal,* Vol. 33, No. 1, Fall 1993, p. 10 (Table 2).

Problem 11.55d

f. The researchers also theorized that career stage impacts the job satisfaction of truck drivers. Do the results support this theory?

11.56 When marketing its products in a foreign country, should a company use its own salespeople or salespeople from the target market country? To research this question, a study was designed to investigate the effect of salesperson nationality on buyer attitudes (*Journal of Business Research,* Vol. 22, 1991). A sample of American MBA students was divided into two groups and shown a videotape of an advertisement for forklift trucks made in India. For group 1, an Indian sales representative made the presentation; for group 2, an American sales representative made the presentation. After viewing the tape, the subjects were asked whether the salesperson was trustworthy (measured on a 5-point scale). The mean scores were compared using an ANOVA.

a. The ANOVA resulted in an F value of 2.32, with an observed significance level of .13. Is there evidence of a difference between the mean trustworthiness scores of the two groups of MBA students? Use $\alpha = .10$.

b. The sample mean scores for the two groups are $\bar{y}_1 = 3.12$ and $\bar{y}_2 = 3.49$. Suppose you were to test $H_0: \mu_1 = \mu_2$ against $H_a: \mu_1 < \mu_2$ at $\alpha = .10$. Use the result, part **a**, to make the proper conclusion. [*Hint:* Use the fact that the p-value for a two-tailed t test is double the p-value for a one-tailed test.]

11.57 Studies conducted at the University of Melbourne (Australia) indicate that there may be a difference in the pain thresholds of blonds and brunettes. Men and women of various ages were divided into four categories according to hair color: light blond, dark blond, light brunette, and dark brunette. The purpose of the experiment was to determine whether hair color is related to the amount of pain produced by common types of mishaps and assorted types of trauma. Each person in the experiment was given a pain threshold score based on his or her performance in a pain sensitivity test (the higher the score, the lower the person's pain tolerance). Consider the results shown in the accompanying table and an Excel printout of the analysis of variance on page 723.

HAIRPAIN.XLS

Light Blond	Dark Blond	Light Brunette	Dark Brunette
62	63	42	32
60	57	50	39
71	52	41	51
55		37	30
48			35

E

	A	B	C	D	E	F	G
1	Anova: Single Factor						
2							
3	SUMMARY						
4	*Groups*	*Count*	*Sum*	*Average*	*Variance*		
5	LightBlond	5	296	59.2	72.7		
6	DarkBlond	3	172	57.33333	30.33333		
7	LightBrunette	4	170	42.5	29.66667		
8	DarkBrunette	5	187	37.4	69.3		
9							
10							
11	ANOVA						
12	*Source of Variation*	*SS*	*df*	*MS*	*F*	*P-value*	*F crit*
13	Between Groups	1566.569	3	522.1895	9.459077	0.001397	3.410534
14	Within Groups	717.6667	13	55.20513			
15							
16	Total	2284.235	16				
17							

a. Is there evidence of a difference among the mean pain thresholds for people with the four hair color types? Use $\alpha = .05$.

b. Which of the four hair-color categories shows the lowest mean pain threshold? Use an EER of .05.

11.58 In social psychology, researchers have found that when individuals have positive attitudes toward an issue, a less credible source has a higher impact on positive attitude change than a highly credible source. Does this phenomenon exist in a personal selling situation? The *Journal of Personal Selling & Sales Management* (Fall 1990) reported on a study to examine the effects of salesperson credibility on buyer persuasion. The experiment involved two factors, each at two levels: brand quality (high versus low) and salesperson credibility (high versus low). Each of 64 undergraduate students was randomly assigned to one of the $2 \times 2 = 4$ experimental treatments (8 students per treatment). After viewing a presentation on laptop computers, the students' intentions to buy were measured with a questionnaire.

a. A portion of the ANOVA table for this experiment is shown here. Interpret the results.

Source of Variation	df	SS	MS	F	*p*-Value
Brand quality (A)	1	59.30	59.30	39.74	.001
Salesperson credibility (B)	1	3.51	3.51	2.35	.130
$A \times B$	1	15.13	15.13	10.14	.002

Source: Sharma, A. "The persuasive effect of salesperson credibility: conceptual and empirical examination." *Journal of Personal Selling & Sales Management,* Fall 1990, Vol. 10, pp. 71–80 (Table 2).

b. Bonferroni's multiple comparisons procedure (with EER = .01) was used to compare the mean buyer intentions for the $2 \times 2 = 4$ experimental conditions. Since the two factors (brand quality and salesperson credibility) interact, the means for high and low salesperson credibility were compared for each of the two levels of brand quality. The results for high brand quality are shown on page 724. Interpret the results.

Mean buyer intention	4.88	5.27
Sales credibility	High	Low

Problem 11.58b

c. Refer to part **b**. The results for low brand quality are shown below. Interpret the results.

Mean buyer intention	2.35	3.92
Sales credibility	High	Low

11.59 A clinical trial was conducted to study the physiological effects on patients of five active exercises used in the earliest stages of cardiac rehabilitation (*Physical Therapy,* Aug. 1986). Six treatments were investigated in the study—the five exercises (exercises A, B, C, D, E) and the control (rest). The response variable was patient's heart rate (in beats per minute) after exercising. Bonferroni's multiple comparisons procedure was conducted on the six treatment means. The ranked treatment means are listed below, with overbars connecting the means that are not significantly different at EER = .05. Interpret the results.

Treatment	Rest	A	B	C	D	E
Mean heart rate (beats/minute)	61	62	66	68	69	69

11.60 The *American Journal of Psychology* (Winter 1991) reported on a study designed to investigate the way in which adolescents with low reading ability comprehend simple text-based problems. Fourteen-year-old students from New York City schools participated in the experiment. Based on expert evaluation, the students were divided into four groups: (1) learning disabled, low socioeconomic status; (2) nondisabled, low socioeconomic status; (3) learning disabled, high socioeconomic status; and (4) nondisabled, high socioeconomic status. Each student was asked to read and retell a "story problem"; however, the way in which the problem was presented was varied. Some students read a "no-priority" problem (i.e., a problem with no clear goals and/or objectives), while others read a "priority" problem (i.e., a problem with a clear statement of the character's priority). The experiment was designed as a 4 × 2 factorial, with group at four levels and problem type (priority or no-priority) at two levels. One of the dependent variables measured was proportion of ideas recalled correctly.

a. The test for group by problem type interaction was nonsignificant at $\alpha = .01$. Interpret this result.

b. The test for problem type main effects was nonsignificant at $\alpha = .01$. Interpret this result.

c. The test for group main effects was statistically significant at $\alpha = .01$. Interpret this result.

d. The mean proportions of ideas recalled correctly for the four groups were compared with a multiple comparisons procedure (EER = .05); the results are shown below. Interpret the results.

Mean	.252	.361	.379	.589
Group	LD-Low SES	ND-Low SES	LD-High SES	ND-High SES

11.61 The cattle raised on the Biological Reserve of Doñana (Spain) live under free-range conditions, with virtually no human interference. The cattle population is organized into four herds (LGN, MTZ, PLC, and QMD). The *Journal of Zoology* (July 1995) investigated the ranging behavior of the four herds across the four seasons. Thus, a 4×4 factorial experiment was employed, with herd and season representing the two factors. Three animals from each herd during each season were sampled and the home range of each individual was measured (in square kilometers). The data were subjected to an ANOVA, with the results shown in the table below.

Source	df	F	p-Value
Herd (H)	3	17.2	$p < .001$
Season (S)	3	3.0	$p < .05$
H × S	9	1.2	$p > .05$
Error	32		
TOTAL	47		

  a. Conduct the appropriate ANOVA *F* tests and interpret the results.

  b. The researcher ranked the four herd means independently of season. Do you agree with this strategy? Explain.

  c. Refer to part **b**. The Bonferroni rankings of the four herd means (EER = .05) are shown below. Interpret the results.

Mean home range (km²)	.75	1.0	2.7	3.8
Herd	PLC	LGN	QMD	MTZ

11.62 In business, the prevailing theory is that companies can be categorized into one of four types based on their strategic profile: reactors (marginal competitors, unstable, victims of industry forces); defenders (specialize in established products, lower costs while maintaining quality); prospectors (develop new/improved products); and analyzers (operate in two product areas—one stable, one dynamic). The *American Business Review* (Jan. 1990) reported on a study that proposes a fifth organization type, balancers, who operate in three product spheres—one stable and two dynamic. Each firm in a sample of 78 glassware firms was categorized into one of these five types; the level of performance (process research and development ratio) of each was measured.

  a. A completely randomized design ANOVA of the data resulted in a significant (at $\alpha = .05$) *F* value for treatments (organization types). Interpret this result.

  b. Bonferroni multiple comparisons of the five performance levels (using EER = .05) are summarized in the following table. Interpret the results.

Mean	.138	.235	.820	.826	.911
Type	Reactor	Prospector	Defender	Analyzer	Balancer

Source: Wright, P., et al. "Business performance and conduct of organization types: A study of select special-purpose and laboratory glassware firms." *American Business Review,* Jan. 1990, p. 95 (Table 4).

11.63   Vanadium (V) is an essential trace element. An experiment was conducted to compare the concentrations of V in biological materials using isotope dilution mass spectrometry. The accompanying table gives the quantities of V (measured in nanograms per gram) in dried samples of oyster tissue, citrus leaves, bovine liver, and human serum.

YANADIUM.XLS

Oyster Tissue	Citrus Leaves	Bovine Liver	Human Serum
2.35	2.32	.39	.10
1.30	3.07	.54	.17
.34	4.09	.30	.14

Source: Fassett, J. D., and Kingston, H. M. "Determination of nanogram quantities of vanadium in biological material by isotope dilution thermal ionization mass spectrometry with ion counting detection." *Analytical Chemistry,* Vol. 57, No. 13, Nov. 1985, p. 2474 (Table II). Copyright 1985 American Chemical Society.

a.   Use Excel to analyze the data. Summarize the results in an ANOVA table.

b.   Is there sufficient evidence (at $\alpha = .05$) to indicate that the mean V concentrations differ among the four biological materials?

c.   Use Bonferroni's method (and Excel) to rank the mean V concentrations of the four biological materials. Use an experimentwise error rate of .05.

CM	CS	TE	TI
4	6	5	8
7	9	5	4
5	5	7	8
6	7	8	10
8	6	7	3

Source: Data are simulated values based on the group means reported in *Human Factors,* Feb. 1984. Copyright 1984 by the Human Factors Society, Inc.

Problem 11.64

DECODE.XLS

11.64   Operators of sonar display consoles must learn to decode abbreviated words quickly and accurately. In one experiment, 20 Navy and civilian personnel participated in a study of abbreviations on a sonar console. Five of these subjects were highly familiar with the sonar system. The 15 subjects unfamiliar with the system were randomly divided into three groups of five. Thus, the study consisted of a total of four groups (one experienced and three inexperienced groups), with five subjects per group. The experienced group and one inexperienced group (denoted TE and TI, respectively) were assigned to learn the simple method of abbreviation. One of the remaining inexperienced groups was assigned the conventional single abbreviation method (denoted CS), whereas the other was assigned the conventional multiple abbreviation method (denoted CM). Each subject was then given a list of 75 abbreviations to learn, one at a time, through the display console of a minicomputer. The number of trials until the subject accurately decoded at least 90% of the words on the list was recorded as shown in the table at left. Use Excel to determine (at $\alpha = .05$) if differences exist among the mean numbers of trials required for the four groups. If so, which group performed the best?

11.65   The chemical element antimony is sometimes added to tin-lead solder to replace the more expensive tin and to reduce the cost of soldering. A factorial experiment was conducted to determine how antimony affects the strength of the tin-lead solder joint (*Journal of Materials Science,* May 1986). Tin-lead solder specimens were prepared using one of four possible cooling methods (water-quenched, WQ; oil-quenched, OQ; air-blown, AB; and furnace-cooled, FC) and with one of four possible amounts of antimony (0%, 3%, 5%, and 10%) added to the composition. Three solder joints were randomly assigned to each of the $4 \times 4 = 16$ treatments and the shear strength of each measured. The experiment results, shown in the accompanying table, were subjected to an ANOVA using Excel. The Excel printouts are also reproduced here and on page 728.

TINLEAD.XLS

Amount of Antimony (% weight)	Cooling Method	Shear Strength (MPa)
0	WQ	17.6, 19.5, 18.3
0	OQ	20.0, 24.3, 21.9
0	AB	18.3, 19.8, 22.9
0	FC	19.4, 19.8, 20.3
3	WQ	18.6, 19.5, 19.0
3	OQ	20.0, 20.9, 20.4
3	AB	21.7, 22.9, 22.1
3	FC	19.0, 20.9, 19.9
5	WQ	22.3, 19.5, 20.5
5	OQ	20.9, 22.9, 20.6
5	AB	22.9, 19.7, 21.6
5	FC	19.6, 16.4, 20.5
10	WQ	15.2, 17.1, 16.6
10	OQ	16.4, 19.0, 18.1
10	AB	15.8, 17.3, 17.1
10	FC	16.4, 17.6, 17.6

Source: Tomlinson, W. J., and Cooper, G. A. "Fracture mechanism of brass/Sn-Pb-Sb solder joints and the effect of production variables on the joint strength." *Journal of Materials Science,* Vol. 21, No. 5, May 1986, p. 1731, Table II. Copyright 1986 Chapman and Hall.

	A	B	C	D	E	F
2						
3	SUMMARY	WQ	OQ	AB	FC	Total
4	*0*					
5	Count	3	3	3	3	12
6	Sum	55.4	66.2	61	59.5	242.1
7	Average	18.466667	22.066667	20.333333	19.833333	20.175
8	Variance	0.9233333	4.6433333	5.5033333	0.2033333	3.8602273
9						
10	*3*					
11	Count	3	3	3	3	12
12	Sum	57.1	61.3	66.7	59.8	244.9
13	Average	19.033333	20.433333	22.233333	19.933333	20.408333
14	Variance	0.2033333	0.2033333	0.3733333	0.9033333	1.7917424
15						
16	*5*					
17	Count	3	3	3	3	12
18	Sum	62.3	64.4	64.2	56.5	247.4
19	Average	20.766667	21.466667	21.4	18.833333	20.616667
20	Variance	2.0133333	1.5633333	2.59	4.6433333	3.2033333
21						
22	*10*					
23	Count	3	3	3	3	12
24	Sum	48.9	53.5	50.2	51.6	204.2
25	Average	16.3	17.833333	16.733333	17.2	17.016667
26	Variance	0.97	1.7433333	0.6633333	0.48	1.0542424
27						
28	*Total*					
29	Count	12	12	12	12	
30	Sum	223.7	245.4	242.1	227.4	
31	Average	18.641667	20.45	20.175	18.95	
32	Variance	3.5244697	4.3445455	6.4620455	2.4481818	

E	A	B	C	D	E	F	G
34							
35	ANOVA						
36	*Source of Variation*	*SS*	*df*	*MS*	*F*	*P-value*	*F crit*
37	Sample	104.1942	3	34.73139	20.11713	1.64E-07	4.459423
38	Columns	28.6275	3	9.5425	5.527211	0.003571	4.459423
39	Interaction	25.13083	9	2.792315	1.617366	0.152259	3.02083
40	Within	55.24667	32	1.726458			
41							
42	Total	213.1992	47				

Problem 11.65

a. Construct an ANOVA summary table for the experiment.

b. Conduct a test to determine whether the two factors, amount of antimony and cooling method, interact. Use $\alpha = .01$.

c. Interpret the result obtained in part **b**.

d. If appropriate, conduct the tests for main effects. Use $\alpha = .01$.

e. Use Bonferroni's multiple comparisons procedure to compare the mean shear strengths of the four cooling methods and the four levels of antimony. Use EER $= .06$.

11.66 A trade-off study regarding the inspection and test of transformer parts was conducted by the quality department of a major defense contractor. The investigation was structured to examine the effects of varying inspection levels and incoming test times in detecting early part failure or fatigue. The levels of inspection selected were full military inspection (A), reduced military specification level (B), and commercial grade (C). Operational burn-in test times chosen for this study were at 1-hour increments from 1 hour to 9 hours. The response was failures per 1,000 pieces obtained from samples taken from lot sizes inspected to a specified level and burned in over a prescribed time length. Three replications were randomly sequenced under each condition making this a complete 3×9 factorial experiment (a total of 81 observations). The data for the study, shown in the accompanying table, were subjected to an ANOVA using Excel. The Excel printouts are reproduced on pages 729–730. Analyze and interpret the results. Be sure to check that the ANOVA assumptions are satisfied.

TRANSFORM.XLS

			INSPECTION LEVELS					
Burn-in (hours)	**Full Mil. Spec. A**			**Reduced Mil. Spec B**			**Commercial C**		
1	7.60	7.50	7.67	7.70	7.10	7.20	6.16	6.13	6.21
2	6.54	7.46	6.84	5.85	6.15	6.15	6.21	5.50	5.64
3	6.53	5.85	6.38	5.30	5.60	5.80	5.41	5.45	5.35
4	5.66	5.98	5.37	5.38	5.27	5.29	5.68	5.47	5.84
5	5.00	5.27	5.39	4.85	4.99	4.98	5.65	6.00	6.15
6	4.20	3.60	4.20	4.50	4.56	4.50	6.70	6.72	6.54
7	3.66	3.92	4.22	3.97	3.90	3.84	7.90	7.47	7.70
8	3.76	3.68	3.80	4.37	3.86	4.46	8.40	8.60	7.90
9	3.46	3.55	3.45	5.25	5.63	5.25	8.82	9.76	8.52

Source: Danny La Nuez, College of Business Administration, graduate student, University of South Florida, 1989–1990.

E

	A	B	C	D	E
3	SUMMARY	A	B	C	Total
4	*1*				
5	Count	3	3	3	9
6	Sum	22.77	22	18.5	63.27
7	Average	7.59	7.3333333	6.1666667	7.03
8	Variance	0.0073	0.1033333	0.0016333	0.459675
9					
10	*2*				
11	Count	3	3	3	9
12	Sum	20.84	18.15	17.35	56.34
13	Average	6.9466667	6.05	5.7833333	6.26
14	Variance	0.2201333	0.03	0.1414333	0.37645
15					
16	*3*				
17	Count	3	3	3	9
18	Sum	18.76	16.7	16.21	51.67
19	Average	6.2533333	5.5666667	5.4033333	5.7411111
20	Variance	0.1276333	0.0633333	0.0025333	0.2009611
21					
22	*4*				
23	Count	3	3	3	9
24	Sum	17.01	15.94	16.99	49.94
25	Average	5.67	5.3133333	5.6633333	5.5488889
26	Variance	0.0931	0.0034333	0.0344333	0.0639611
27					
28	*5*				
29	Count	3	3	3	9
30	Sum	15.66	14.82	17.8	48.28
31	Average	5.22	4.94	5.9333333	5.3644444
32	Variance	0.0399	0.0061	0.0658333	0.2247028

E

	A	B	C	D	E
33					
34	*6*				
35	Count	3	3	3	9
36	Sum	12	13.56	19.96	45.52
37	Average	4	4.52	6.6533333	5.0577778
38	Variance	0.12	0.0012	0.0097333	1.5154444
39					
40	*7*				
41	Count	3	3	3	9
42	Sum	11.8	11.71	23.07	46.58
43	Average	3.9333333	3.9033333	7.69	5.1755556
44	Variance	0.0785333	0.0042333	0.0463	3.5888028
45					
46	*8*				
47	Count	3	3	3	9
48	Sum	11.24	12.69	24.9	48.83
49	Average	3.7466667	4.23	8.3	5.4255556
50	Variance	0.0037333	0.1047	0.13	4.7510278
51					
52	*9*				
53	Count	3	3	3	9
54	Sum	10.46	16.13	28.1	54.69
55	Average	3.4866667	5.3766667	9.3666667	6.0766667
56	Variance	0.0030333	0.0481333	0.2385333	6.83075
57					
58	*Total*				
59	Count	27	27	27	
60	Sum	140.54	141.7	182.88	
61	Average	5.2051852	5.2481481	6.7733333	
62	Variance	2.1516798	1.0184387	1.7908231	

E

Amount → 67
Method → 68

65	ANOVA						
66	*Source of Variation*	*SS*	*df*	*MS*	*F*	*P-value*	*F crit*
67	Sample	27.9744	8	3.4968	54.62907	3.78E-23	2.115222
68	Columns	43.08412	2	21.54206	336.5427	3.25E-31	3.168246
69	Interaction	97.55355	16	6.097097	95.25244	1.13E-33	1.834628
70	Within	3.456533	54	0.06401			
71							
72	Total	172.0686	80				
73							

Problem 11.66

11.67   In an *Ecology* study of the lifespan of a certain species of predatory protozoan, 166 individuals of this species were captured from the Sacramento River (California) and randomly divided into five groups. All individuals were fed on a predetermined schedule; however, the food level varied among groups. The first group was fed 3 paramecia per day; the second group, 1 paramecium per day; the third group, $\frac{1}{2}$ paramecium per day; the fourth, $\frac{1}{4}$ paramecium per day; and the fifth group was starved. The lifespan, in days, was recorded for each individual. An analysis of variance was performed with the results shown in the table.

Source	df	SS	MS	F
Food levels	4	—	196.14	—
Error	111	—	59.12	
TOTAL	115	—		

Source: Kent, E. B. "Life history responses to resource variation in a sessile predator, the ciliate protozoan *Tokophyra Lemnarum* Stein." *Ecology,* Apr. 1981, Vol. 62, pp. 296–302. Copyright © 1981, the Ecological Society of America.

a.   Complete the ANOVA summary table.

b.   Is there evidence of a difference (at $\alpha = .025$) among the mean lifespans for the five food levels?

11.E.1 Using Microsoft Excel for a One-Way ANOVA

Solution Summary:

Use the Data Analysis Anova: Single Factor tool to perform a one-way analysis of variance.

Example: Grade point average analysis of Examples 11.3 and 11.4 on pages 676–679

Solution:

Use the Data Analysis Anova: Single Factor tool to determine whether there is evidence of a difference in grade point average based on socioeconomic class. To determine whether or not there is evidence of a significant difference in the grade point averages for three different socioeconomic classes, do the following:

1. Open the GPACLASS.XLS workbook and click the Data sheet tab. Verify that the grade point averages for the three socioeconomic classes of Table 11.3 on page 676 have been entered into columns A through C.

2. Select Tools | Data Analysis. Select Anova: Single Factor from the Analysis Tools list box in the Data Analysis dialog box. Click the OK button.

3. In the Anova: Single Factor dialog box (see Figure 11.E.1):

   a Enter A1:C8 in the Input Range: edit box.

   b Select the Columns option button.

   c Select the Labels in First Row check box.

   d Enter 0.05 in the Alpha: edit box.

   e Select the New Worksheet Ply option button and enter ANOVA as the name of the new sheet.

   f Click the OK button.

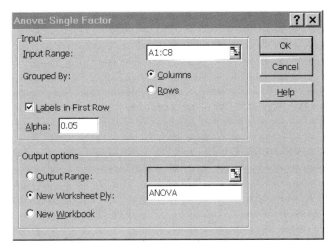

E Figure 11.E.1
Data Analysis Anova:
Single Factor Tool
Dialog Box

The Data Analysis Anova tool add-in inserts a worksheet containing the analysis of variance for the grade point averages for the three socioeconomic classes, similar to the one shown in Figure 11.6 on page 679. This worksheet is *not* dynamically changeable, so any changes made to the underlying grade point average data would require using the Data Analysis Anova tool a second time to produce updated test results.

Using Stacked Data

The Data Analysis Anova tool requires that the data for each group be in separate columns, an arrangement known as unstacked data. Some data sets are "stacked," in which the data for both groups appear in a single column, identified by a grouping variable in another column. To use such stacked data with the Data Analysis Anova: Single Factor tool, select the Data Preparation | Unstack Data choice of the PHStat add-in to unstack the data as part of Step 1.

11.E.2 Using Microsoft Excel for a Two-Way ANOVA

Solution Summary:

Use the Data Analysis Anova: Two Factors with Replication tool to perform a two-way analysis of variance.

 Example: Gasket manufacturing analysis of Examples 11.5–11.8 on pages 687–695

Solution:

Use the Data Analysis Anova: Two Factors with Replication tool. As an example, consider the gasket manufacturing problem of Section 11.4. To determine whether or not there is evidence of a significant difference in the average number of gaskets manufactured for two stamping machines and three materials, do the following:

1. Open the GASKETS.XLS workbook and click the Data sheet tab. Verify that the data of Table 11.4 on page 687 appears as follows: the machine values appear in column A (beginning in row 2), category labels for the three gasket materials have been entered into row 1 of columns B through D, and the data have been entered in rows 2 through 7 of columns B through D, with the three values for machine 1 and gasket material C in cells B2:B4, the three values for machine 2 and gasket material C in cells B5:B7, and so on, with the three values for machine 2 and gasket material P in cells D5:D7.

2. Select Tools | Data Analysis. Select Anova: Two Factors with Replication from the Analysis Tools list box in the Data Analysis dialog box. Click the OK button.

3. In the Anova: Two-Factor With Replication dialog box (see Figure 11.E.2):
   - **a** Enter A1:D7 in the Input Range: edit box.
   - **b** Enter 3 in the Rows per sample: edit box.
   - **c** Enter 0.05 in the Alpha: edit box.
   - **d** Select the New Worksheet Ply option button and enter Two-Way ANOVA as the name of the new sheet.
   - **e** Click the OK button.

E Figure 11.E.2
Data Analysis Anova:
Two Factors with
Replication Tool
Dialog Box

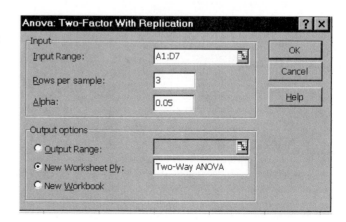

The Data Analysis Anova tool add-in inserts a worksheet containing calculations for the analysis of variance similar to the one shown in Figure 11.9 on pages 691–692. This worksheet is *not* dynamically changeable, so any changes made to the underlying manufacturing data would require using the Data Analysis Anova tool a second time to produce an updated analysis.

11.E.3 Using Microsoft Excel for Multiple Comparisons

Solution Summary:

Obtain the sample means and sample sizes of each group, the degrees of freedom within groups, and the MSE value from the Data Analysis Anova: Single Factor output.

Example: Grade point average analysis of Examples 11.3 and 11.4 on pages 676–679

Solution:

To perform the Bonferroni procedure for making multiple comparisons between all possible pairs of c means, first obtain the necessary statistics for the procedure. As an example, consider the grade point average examples of Section 11.3. To generate multiple comparisons between all possible pairs of the means of grade point average data for the three socioeconomic classes, do the following:

1 Generate a worksheet containing analysis of variance statistics using the instructions of Section 11.E.1.

2 From the worksheet generated by the Data Analysis tool in Step 1:

    a Obtain the sample sizes and means from Count and Average columns, respectively, of the SUMMARY table that begins in row 4.

    b Obtain the degrees of freedom within groups from the cell in the "df" column and "Within Groups" row of the ANOVA table.

    c Obtain the MSE value from the cell in the "MS" column and "Within Groups" row of the ANOVA table.

Implementing a Worksheet to Perform the Bonferroni Procedure

BONFERRONI.XLS

We can use arithmetic and logical formulas and the values obtained in Steps 1 and 2 as the basis for performing the Bonferroni procedure for making multiple comparisons between all possible pairs of c means. Tables 11.E.1 and 11.E.2 present a Bonferroni sheet design that performs the Bonferroni procedure using the grade point averages data of Table 11.3. Similar to the worksheet designs presented in Section 10.E.3, this design also includes a formula that uses the IF function (see Section 8.E.4) to display a message informing the user whether or not a pair of means significantly differs.

To implement the Tables 11.E.1 and 11.E.2 design, do the following:

1 Select File | New to open a new workbook (or open the existing workbook into which the Bonferroni worksheet is to be inserted).

2 Select an unused worksheet (or select Insert | Worksheet if there are none) and rename the sheet Bonferroni.

3 Enter the title, headings, and labels for column A as shown in Table 11.E.1.

4 Enter the sample size and sample mean for each group, the MSE statistic value, the experimentwise error rate, and the number of groups in column B as shown in Table 11.E.1.

5 Enter the headings, labels, and formulas for columns C and D as shown in Table 11.E.2.

6 Select the cell range A3:B3 and click the Merge and Center button on the formatting toolbar (see Section EP.2.9). Repeat this step for the cell ranges A6:B6, A9:B9, and for the two-cell ranges in columns C and D in rows 3, 7, 9, 13, 15, and 19.

The completed worksheet will be similar to the one shown in Figure 11.11 on page 704.

Table 11.E.1 Bonferroni Procedure Sheet Design for Columns A and B for the Grade Point Average Examples of Section 11.3

	A	B
1	Bonferroni Procedure for Grade Point Average	
2		
3	Group 1	
4	Sample Mean	xx
5	Sample Size	xx
6	Group 2	
7	Sample Mean	xx
8	Sample Size	xx
9	Group 3	
10	Sample Mean	xx
11	Sample Size	xx
12	MSE	xx
13	Error DF	=(B5-1)+(B8-1)+(B11-1)
14	Experiment Error Rate	0.xx
15	Number of Groups	3
16	Treatment pairs	=B15*(B15-1)/2
17	α value for comparisons	=B14/B16
18	t statistic for comparisons	=TINV(B17, B13)

Table 11.E.2 Bonferroni Procedure Sheet Design for Columns C and D
for the Grade Point Average Examples of Section 11.3

	C	D
1		
2		
3	Group 1 to Group 2 Comparison	
4	Absolute Difference	=ABS(B4-B7)
5	Standard Error of Difference	=SQRT((B12)*((1/B5)+(1/B8)))
6	Bonferroni Critical Difference	=B18*D5
7	=IF(D4>D6,"Means are different","Means are not different")	
8		
9	Group 1 to Group 3 Comparison	
10	Absolute Difference	=ABS(B4-B10)
11	Standard Error of Difference	=SQRT((B12)*((1/B5)+(1/B11)))
12	Bonferroni Critical Difference	=B18*D11
13	=IF(D10>D12,"Means are different","Means are not different")	
14		
15	Group 2 to Group 3 Comparison	
16	Absolute Difference	=ABS(B7-B10)
17	Standard Error of Difference	=SQRT((B12)*((1/B8)+(1/B11)))
18	Bonferroni Critical Difference	=B18*D17
19	=IF(D16>D18,"Means are different","Means are not different")	

Chapter 12

Nonparametric Statistics

CONTENTS

12.1 Distribution-Free Tests

12.2 Testing for Location of a Single Population

12.3 Comparing Two Populations: Independent Random Samples

12.4 Comparing Two Populations: Matched-Pairs Design

12.5 Comparing Three or More Populations: Completely Randomized Design

12.6 Testing for Rank Correlation

EXCEL TUTORIAL

12.E.1 Using Microsoft Excel for the Sign Test

12.E.2 Using Microsoft Excel to Perform the Wilcoxon Rank Sum Test for Differences in Two Medians

12.E.3 Using Microsoft Excel to Perform the Kruskal-Wallis H Test for Differences in c Medians

REAL-WORLD APPLICATION
Do Women Really Understand the Benefit of a Mammography?

Women are often faced with a decision of when to screen for breast cancer with a mammography. This decision is often based on quantitative information provided by physicians on the risks of dying from breast cancer. Do most women understand this quantitative information? To answer this question, a group of medical researchers collected data on female veterans' assessment of the risks of breast cancer. An analysis reveals that these data are not normally distributed. Consequently, one of the assumptions required to conduct the parametric tests described in earlier chapters is violated. In this chapter we learn how to use nonparametric (distribution-free) statistical tests to analyze data of this type. We will apply these techniques to the risk assessment data collected on the female veterans in Real-World Revisited.

12.1 DISTRIBUTION-FREE TESTS

The confidence interval and testing procedures developed in Chapters 7–11 all involve making inferences about population parameters. Consequently, they are often referred to as *parametric statistical tests*. Many of these parametric methods (e.g., the small sample *t* test of Chapter 8 or the ANOVA *F* test of Chapter 11) rely on the assumption that the data are sampled from a normally distributed population. When the data are normal, these tests are *most powerful*. That is, the use of these parametric tests maximizes the chance of the researcher correctly rejecting the null hypothesis.

Consider a population of data that is decidedly nonnormal. For example, the distribution might be very flat, peaked, or strongly skewed to the right or left (see Figure 12.1). Applying the small-sample *t* test to such a data set may result in serious consequences. Since the normality assumption is clearly violated, the results of the *t* test are unreliable: (1) the probability of a Type I error (i.e., rejecting H_0 when it is true) may be larger than the value of α selected; and (2) the power of the test (see box at left) may be different from what we expect.

A variety of *nonparametric techniques* are available for analyzing data that do not follow a normal distribution. Nonparametric tests do not depend on the distribution of the sampled population; thus, they are called *distribution-free tests*. Also, nonparametric methods focus on the location of the probability distribution of the population, rather than specific parameters of the population, such as the mean (hence, the name "nonparametrics").

Power of a Test

The probability of rejecting H_0 when H_0 is false is called the *power* of the test.

Definition 12.1

Distribution-free tests are statistical tests that do not rely on any underlying assumptions about the probability distribution of the sampled population.

Definition 12.2

The branch of inferential statistics devoted to distribution-free tests is called **nonparametrics**.

Nonparametric tests are also appropriate when the data are nonnumerical in nature, but can be ranked. For example, when taste-testing foods or in other types of consumer product evaluations, we can say we like product A better than product B, and B better than C, but we cannot obtain exact quantitative values for the respective measurements. Nonparametric tests based on the ranks of measurements are called *rank tests*.

Figure 12.1 Some Nonnormal Distributions for which the *t* Statistic Is Invalid

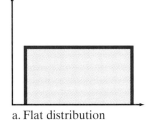

a. Flat distribution

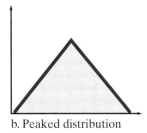

b. Peaked distribution

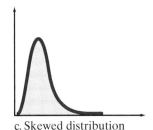

c. Skewed distribution

> ### Definition 12.3
>
> Nonparametric statistics (or tests) based on the ranks of measurements are called **rank statistics** (or **rank tests**).

In the following sections, we present several useful nonparametric methods. Keep in mind that these nonparametric tests are more powerful than their corresponding parametric counterparts in those situations where either the data are nonnormal or the data are ranked.

12.2 TESTING FOR LOCATION OF A SINGLE POPULATION

Recall that small-sample procedures for estimating a population mean or for testing a hypothesis about a population mean (Section 8.4) require that the population have an approximately normal distribution. For situations in which we collect a small sample from a nonnormal population, the t test may not be valid, and we should use a *nonparametric procedure*. The simplest nonparametric technique to apply in this situation is the *sign test*. The sign test is specifically designed for testing hypotheses about the median of any continuous population. Like the mean, the median is a measure of the center, or location, of the distribution; consequently, the sign test is sometimes referred to as a **test for location**.

Example 12.1   Distribution of the Sign Test Statistic

Consider a population with unknown median η and suppose we want to test the null hypothesis $H_0\colon \eta = 100$ against the one-sided alternative $H_a\colon \eta > 100$. From Definition 3.2 (page 151) we know that the **median** is a number such that half the area under the probability distribution lies to the left of η and half lies to the right (see Figure 12.2).

Therefore, the probability that an x value selected from the population is larger than the population median η is .5, i.e., $P(x_i > \eta) = .5$. If, in fact, the null hypothesis is true, then we would expect to observe approximately half the sample x values greater than $\eta = 100$.

The sign test utilizes the test statistic S, where

S = Number of sample observations (x's) that exceed the hypothesized
     population median of 100

Find the probability of the test statistic, S.

Figure 12.2 Location of
the Population Median η

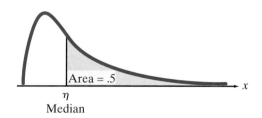

Solution

Notice that S depends only on the *sign* (positive or negative) of the difference between each sample value x_i and 100. That is, we are simply counting the number of positive ($+$) signs among the sample differences ($x_i - 100$). If S is "too large" (i.e., if we observe an unusual number of x_i's exceeding 100), then we will reject H_0 in favor of the alternative $H_a: \eta > 100$.

The probability distribution of the test statistic is derived as follows. Let each sample difference ($x_i - 100$) denote the outcome of a single trial in an experiment consisting of n identical trials. If we call a positive difference a "success" and a negative difference a "failure," then S is the number of successes in n trials. Under H_0, the probability of observing a success on any one trial is

$$\pi = P(\text{Success}) = P(x_i - 100 > 0)$$
$$= P(x_i > 100)$$
$$= .5$$

Since the trials are independent, the properties of a binomial experiment, listed in Section 5.3 (page 268), are satisfied. Therefore, S has a binomial distribution with parameters n and $\pi = .5$. We can use this fact to calculate the observed significance level (p-value) of the sign test. ❏

Example 12.2   Applying the Sign Test

The Environmental Protection Agency (EPA) sets certain pollution guidelines for major industries. For a particular company that discharges waste into a nearby river, the EPA criterion is that the median amount of pollution in water samples collected from the river may not exceed 5 parts per million (ppm). Responding to numerous complaints, the EPA takes 10 water samples from the river at the discharge point and measures the pollution level in each sample. The results (in ppm) are given here:

WATER.XLS

| 5.1 | 4.3 | 5.3 | 6.2 | 5.6 | 4.7 | 8.4 | 5.9 | 6.8 | 3.0 |

Do the data provide sufficient evidence to indicate that the median pollution level in water discharged at the plant exceeds 5 ppm? Test using $\alpha = .05$.

Solution

Letting η represent the population median pollution level, we want to test

$$H_0: \eta = 5$$
$$H_a: \eta > 5$$

using the sign test. The test statistic is

$S =$ Number of sample observations that exceed the hypothesized median 5
$\quad = 7$

where S has a binomial distribution with parameters $n = 10$ and $\pi = .5$.

Recall (Section 8.5) that the observed significance level (p-value) of the test is the probability that we observe a value of the test statistic S that is at least as contradictory to the null hypothesis as the computed value. For this one-sided case,

E Figure 12.3 PHStat Add-in for Excel Printout of Binomial Probabilities, $n = 10$, $\pi = .5$

	A	B	C	D	E	F	G
1	Binomial Probabilities						
2							
3	Sample size	10					
4	Probability of success	0.5					
5	Mean	5					
6	Variance	2.5					
7	Standard deviation	1.581139					
8							
9	Binomial Probabilities Table						
10			X	P(X)	P(<=X)	P(<X)	P(>X)	P(>=X)
11			0	0.000977	0.000977	0	0.999023	1
12			1	0.009766	0.010742	0.000977	0.989258	0.999023
13			2	0.043945	0.054688	0.010742	0.945313	0.989258
14			3	0.117188	0.171875	0.054688	0.828125	0.945313
15			4	0.205078	0.376953	0.171875	0.623047	0.828125
16			5	0.246094	0.623047	0.376953	0.376953	0.623047
17			6	0.205078	0.828125	0.623047	0.171875	0.376953
18			7	0.117188	0.945313	0.828125	0.054687	0.171875
19			8	0.043945	0.989258	0.945313	0.010742	0.054687
20			9	0.009766	0.999023	0.989258	0.000977	0.010742

E Figure 12.4 Excel Output of Sign Test for Data of Example 12.2

	A	B
1	**Sign Test**	
2		
3	**Hypothesized Median**	5
4	**Sample Size**	10
5	**Number greater than Median (S_1)**	7
6	**Number less than or equal to Median (S_2)**	3
7	**Level of Significance (α)**	0.05
8		
9	Two-Tailed Test	
10	Maximum of S_1 and S_2	7
11	p-value	0.34375
12	Do not reject the null hypothesis	
13		
14	Lower Tail Test	
15	Probability greater than or equal to S_2	0.945313
16	Do not reject the null hypothesis	
17		
18	Upper Tail Test	
19	Probability greater than or equal to S_1	0.171875
20	Do not reject the null hypothesis	

the p-value is the probability that we observe a value of S greater than or equal to 7. From the PHStat add-in for Excel printout, Figure 12.3, we find $P(S \geq 7) = .171875$. Therefore,

$$p\text{-value} = P(S \geq 7) = .1719$$

The p-value is also indicated (highlighted) on an Excel printout of the sign test shown in Figure 12.4. Since the p-value .1719 is larger than $\alpha = .05$, we cannot reject the null hypothesis. That is, there is insufficient evidence to indicate that the median pollution level of water discharged from the plant exceeds 5. ❑

A summary of the **sign test** for one-sided and two-sided alternatives is provided in the box on page 741.

Sign Test for the Median η of a Single Population

Upper-Tailed Test	Two-Tailed Test	Lower-Tailed Test
$H_0: \eta = \eta_0$	$H_0: \eta = \eta_0$	$H_0: \eta = \eta_0$
$H_a: \eta > \eta_0$	$H_a: \eta \neq \eta_0$	$H_a: \eta < \eta_0$
Test statistic:	*Test statistic:*	*Test statistic:*
S = Number of sample observations greater than η_0	S = Larger of S_1, S_2, where S_1 = Number of sample observations greater than η_0 S_2 = Number of sample observations less than η_0	S = Number of sample observations less than η_0
Observed significance level:	*Observed significance level:*	*Observed significance level:*
p-value $= P(x \geq S)$	p-value $= 2P(x \geq S)$	p-value $= P(x \geq S)$

where x has a binomial distribution with parameters n and $\pi = .5$. [*Note:* Reject H_0 if p-value $< \alpha$.]

Assumption: The sample is randomly selected from a continuous probability distribution. [*Note:* No assumptions have to be made about the shape of the probability distribution.]

Self-Test

12.1

Consider the following sample of five measurements:

185   150   210   344   192

a. Find the number of measurements greater than 200.
b. Find the *p*-value for testing the hypothesis that the median of the population exceeds 200.

PROBLEMS
12.1–12.11

Using the Tools

12.1 For each of the following, find the test statistic S for testing $H_0: \eta = 20$:
  a. $H_a: \eta \neq 20$; sample data: 14, 22, 21, 34, 12, 28, 25
  b. $H_a: \eta < 20$; sample data: 14, 22, 21, 34, 12, 28, 25
  c. $H_a: \eta > 20$; sample data: 8, 13, 22, 25, 19, 30, 37, 41, 23

12.2 Suppose you want to use the sign test to test the null hypothesis that the population median equals 75, i.e., $H_0: \eta = 75$. Use Excel to find the *p*-value of the test for each of the following situations:

a. $H_a: \eta > 75, n = 5, S = 2$ **b.** $H_a: \eta \neq 75, n = 20, S = 16$

c. $H_a: \eta < 75, n = 10, S = 8$

12.3   A random sample of six observations from a continuous population resulted in the following:

18.2   21.3   20.5   19.4   19.6   17.7

Is there sufficient evidence to indicate that the population median differs from 20? Test using $\alpha = .05$.

Applying the Concepts

12.4   A rare species of wildflower, native to Sumatra, was recently studied (*Lindleyana,* Mar. 1995). A clump of wildflowers was observed on each of four randomly selected days in November. The total number of insects and spiders visiting the wildflowers was recorded each day at 8 A.M. The data for the four days are listed below.

26   34   29   19

WILDFLOWER.XLS

a.   Is there evidence to indicate that the median number of insects and spiders that visit the wildflower clump daily in November exceeds 20? Test using $\alpha = .10$.

b.   Show that the null hypothesis tested in part **a** will be rejected in favor of the alternative hypothesis only if all four data values exceed 20.

12.5   Based on an analysis of automobile insurance claims, the Highway Loss Data Institute (HLDI) compiles injury and collision-loss data for popular cars, station wagons, and vans. The data in the table, reported in *Consumer's Research* (Nov. 1993), are the HLDI collision-damage ratings of large station wagons and minivans. The collision-damage rating reflects how much, compared to other cars, is paid out by insurance companies to the model's owners for collision damage repairs. The higher the rating, the greater the amount paid for repairs.

HLDI.XLS

Vehicle Model	Collision-Damage Rating
Chevrolet Astro 4-wheel drive	50
Plymouth Voyager	59
Chevrolet Caprice	77
Oldsmobile Silhouette	72
Dodge Caravan	60
GMC Safari	60
Mazda MPV 4-wheel drive	121
Toyota Previa	77
Chevrolet Lumina APV	71
Ford Aerostar	74
Chevrolet Astro	59
Mazda MPV	114
Pontiac Trans Sport	72

a.   If exactly half of all station wagons and minivans on the highway have collision-damage ratings that exceed 80, what is the median of the relevant population?

b.   Set up the null and alternative hypotheses for testing whether the population median is less than the value specified in part **a**.

c. The accompanying Excel printout gives the results of the sign test for testing the hypotheses of part **b**. Interpret the test results.

E	A	B
1	**Sign Test**	
2		
3	Hypothesized Median	80
4	Sample Size	13
5	Number greater than Median (S_1)	2
6	Number less than or equal to Median (S_2)	11
7	Level of Significance (α)	0.05
8		
9	Two-Tailed Test	
10	Maximum of S_1 and S_2	11
11	p-value	0.022461
12	Reject the null hypothesis	
13		
14	Lower Tail Test	
15	Probability less than or equal to S	0.01123
16	Reject the null hypothesis	
17		
18	Upper Tail Test	
19	Probability greater than or equal to S	0.998291
20	Do not reject the null hypothesis	

12.6   The data on retail prices of a random sample of 22 VCR (videocassette recorder) models, Problem 2.21 (page 80), is reproduced below. Is there sufficient evidence to say that the median retail price of VCRs differs from $200? Test using $\alpha = .10$.

VCRPRICE.XLS

350	300	340	220	320	450	270	265	210	250	180
300	190	170	190	170	170	200	180	220	200	250

Source: Adapted from "VCRs," *Consumer Reports,* Nov. 1996, pp. 36–38.

12.7   To investigate the transitory migration of guppy populations, 40 adult female guppies were placed into the left compartment of an experimental aquarium tank that was divided in half by a glass plate. After the plate was removed, the number of fish passing through the slit from the left compartment to the right one, and vice versa, was monitored every minute for 30 minutes (*Zoological Science,* Vol. 6, 1989). If an equilibrium is reached, the researchers would expect the median number of fish remaining in the left compartment to be 20. The data for the 30 observations (i.e., number of fish in left compartment at the end of the minute interval) are shown here. Use the sign test to determine whether the median is less than 20. Test using $\alpha = .05$.

GUPPY.XLS

16	11	12	15	14	16	18	15	13	15
14	14	16	13	17	17	14	22	18	19
17	17	20	23	18	19	21	17	21	17

Source: Terami, H., and Watanabe, M. "Excessive transitory migration of guppy populations. III. Analysis of perception of swimming space and a mirror effect." *Zoological Science,* Vol. 6, 1989, p. 977 (Figure 2).

12.8   Refer to the *Risk Management* study on victims who attempted to evacuate fires at compartmented fire-resistive buildings, described in Problem 3.84 (page 202). The number of victims who died attempting to evacuate for a sample of 14 recent fires is shown in the accompanying table.

FIRE.XLS

Fire	Number of Victims
Las Vegas Hilton (Las Vegas)	5
Inn on the Park (Toronto)	5
Westchase Hilton (Houston)	8
Holiday Inn (Cambridge, OH)	10
Conrad Hilton (Chicago)	4
Providence College (Providence)	8
Baptist Towers (Atlanta)	7
Howard Johnson (New Orleans)	5
Cornell University (Ithaca, New York)	9
Westport Central Apartments (Kansas City, MO)	4
Orrington Hotel (Evanston, IL)	0
Hartford Hospital (Hartford, CT)	16
Milford Plaza (New York, NY)	0
MGM Grand (Las Vegas)	36

Source: Macdonald, J. N. "Is evacuation a fatal flaw in fire-fighting philosophy?" *Risk Management,* Vol. 33, No. 2, Feb. 1986, p. 37.

a. In part **a** of Problem 3.84 you constructed a stem-and-leaf display for the data. Does it appear that the sample data are from a normally distributed population?

b. Based on your answer to part **a**, why is a nonparametric test for location preferred over a parametric test? Explain.

c. Conduct a test (at $\alpha = .01$) to determine whether the median number of victims who attempt to evacuate fires at compartmented fire-resistive buildings differs from 6. Use the accompanying Excel printout to make the proper inference.

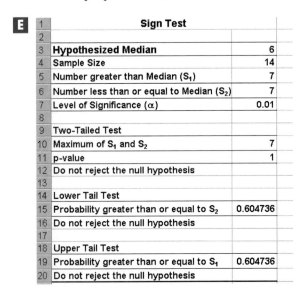

E 1	**Sign Test**	
2		
3	**Hypothesized Median**	6
4	Sample Size	14
5	Number greater than Median (S_1)	7
6	Number less than or equal to Median (S_2)	7
7	Level of Significance (α)	0.01
8		
9	Two-Tailed Test	
10	Maximum of S_1 and S_2	7
11	p-value	1
12	Do not reject the null hypothesis	
13		
14	Lower Tail Test	
15	Probability greater than or equal to S_2	0.604736
16	Do not reject the null hypothesis	
17		
18	Upper Tail Test	
19	Probability greater than or equal to S_1	0.604736
20	Do not reject the null hypothesis	

12.9 Refer to the *Journal of Applied Behavior Analysis* (Summer 1997) study of a mentally retarded boy playing with toys, Problem 7.23 (page 381).

The number of destructive responses (e.g., head banging, hair pulling, pinching, etc.) per minute observed for 10 sessions is listed below:

| 3 | 2 | 4.5 | 5.5 | 4.5 | 3 | 2.5 | 1 | 3 | 3 |

Determine whether the median number of destructive responses observed per minute exceeds 2. Test using $\alpha = .05$.

12.10 The biting rate of a particular species of fly was investigated in a study reported in the *Journal of the American Mosquito Control Association* (Mar. 1995). The biting rate was defined as the number of flies biting a volunteer during 15 minutes of exposure. This species of fly is known to have a median biting rate of five bites per 15 minutes on Stanbury Island, Utah. However, it is theorized that the median biting rate is higher in bright, sunny weather. To test this theory, 122 volunteers were exposed to the flies during a sunny day on Stanbury Island. Of these volunteers, 95 experienced biting rates greater than 5.

a. Set up the null and alternative hypotheses for the test.

b. Use Excel to find the *p*-value of the test.

12.11 An important property of certain manufactured products that are in powder or granular form is their particle-size distribution. For example, refractory cements are adversely affected by too high a proportion of coarse granules, which can lead to weakness because of poor packing. The following data, extracted from the *Journal of Quality Technology* (July 1985), represent the percentages of coarse granules for a random sample of eight refractory cement specimens from a large lot:

| 1.7 | .9 | 3.4 | 2.5 | 3.1 | .6 | 1.0 | 2.1 |

Is there sufficient evidence to indicate that the median percentage of coarse granules in the lot is less than 2%? Test using $\alpha = .05$.

12.3 COMPARING TWO POPULATIONS: INDEPENDENT RANDOM SAMPLES

In Section 9.1 we presented *parametric* tests (tests about population parameters based on the z and t statistics) to test for a difference between two population means μ_1 and μ_2. Recall that both the mean and median of a population measure the *location* of the population distribution. If the two population distributions have the same shape, the equivalent nonparametric test is a test about the difference between population medians η_1 and η_2. The test, based on independent random samples of n_1 and n_2 observations from the respective populations, is known as the *Wilcoxon rank sum test*.

To use the Wilcoxon rank sum test, we first rank all $(n_1 + n_2)$ observations, assigning a rank of 1 to the smallest, 2 to the second smallest, and so on. The sum of the ranks, called a **rank sum**, is then calculated for each sample. If the two distributions are identical, we would expect the sample rank sums, designated as T_1 and T_2, to be nearly equal. In contrast, if one rank sum—say, T_1—is much larger than the other, T_2, then the data suggest that the distribution for population 1 is shifted to the right of the distribution for population 2. The procedure for conducting a **Wilcoxon rank sum test** is summarized in the box and illustrated in Example 12.3.

> ## Wilcoxon Rank Sum Test for a Difference between Population Medians: Independent Random Samples
>
> Let η_1 and η_2 represent the medians for populations 1 and 2, respectively.
>
> | Upper-Tailed Test | Two-Tailed Test | Lower-Tailed Test |
> |---|---|---|
> | $H_0: \eta_1 = \eta_2$ | $H_0: \eta_1 = \eta_2$ | $H_0: \eta_1 = \eta_2$ |
> | $H_a: \eta_1 > \eta_2$ | $H_a: \eta_1 \neq \eta_2$ | $H_a: \eta_1 < \eta_2$ |
>
> *Test statistic:* T_1, if $n_1 < n_2$; T_2, if $n_2 < n_1$
>
> Either rank sum can be used if $n_1 = n_2$ (denote this rank sum as T for the two-tailed test)
>
> | *Rejection region:* | *Rejection region:* | *Rejection region:* |
> |---|---|---|
> | If T_1: $T_1 \geq T_U$ | Either $T \leq T_L$ or $T \geq T_U$ | If T_1: $T_1 \leq T_L$ |
> | If T_2: $T_2 \leq T_L$ | | If T_2: $T_2 \geq T_U$ |
>
> where T_L and T_U are obtained from Table B.6 of Appendix B. [*Note:* Tied observations are assigned ranks equal to the average of the ranks that would have been assigned to the observations had they not been tied.]
>
> *Assumption:* The shapes of the two population distributions are approximately the same.

Example 12.3   Applying the Rank Sum Test

In a comparison of visual acuity of deaf and hearing children, eye movement rates are taken on 10 deaf and 10 hearing children. (See Table 12.1.) A clinical psychologist believes that deaf children have greater visual acuity than hearing children. Test the psychologist's claim by using the data in Table 12.1. (The larger a child's eye movement rate, the more visual acuity the child possesses.) Use $\alpha = .05$.

Solution
The psychologist believes that the visual acuity of deaf children is greater than that of hearing children. If so, the median visual acuity of deaf children, η_1, will exceed the median for hearing children, η_2. Therefore, we will test

$$H_0: \eta_1 = \eta_2$$
$$H_a: \eta_1 > \eta_2$$

To conduct the test, we need to rank all 20 visual acuity observations in the combined samples. The ranks are shown in Table 12.2. Tied ranks receive the average of the ranks the scores would have received if they had not been tied.

Since the sample sizes n_1 and n_2 are equal, the Wilcoxon rank sum test statistic is (according to the box) either T_1 (the sum of the ranks for deaf children) or T_2

DEAF.XLS

Table 12.1 Visual Acuity of Children

Deaf Children		Hearing Children	
2.75	1.95	1.15	1.23
3.14	2.17	1.65	2.03
3.23	2.45	1.43	1.64
2.30	1.83	1.83	1.96
2.64	2.23	1.75	1.37

Table 12.2 Calculation of Rank Sums for Example 12.3

DEAF CHILDREN		HEARING CHILDREN	
Visual Acuity	**Rank**	**Visual Acuity**	**Rank**
2.75	18	1.15	1
3.14	19	1.65	6
3.23	20	1.43	4
2.30	15	1.83	8.5
2.64	17	1.75	7
1.95	10	1.23	2
2.17	13	2.03	12
2.45	16	1.64	5
1.83	8.5	1.96	11
2.23	14	1.37	3
$T_1 = 150.5$		$T_2 = 59.5$	

Table 12.3 A Portion of the Wilcoxon Rank Sum Table, Table B.6

n_2 \ n_1	3		4		5		6		7		8		9		10	
	T_L	T_U	T_L	T_U	T_L	T_U	T_L	T_U	T_L	T_U	T_L	T_U	T_L	T_U	T_L	T_U
3	6	15	7	17	7	20	8	22	9	24	9	27	10	29	11	31
4	7	17	12	24	13	27	14	30	15	33	16	36	17	39	18	42
5	7	20	13	27	19	36	20	40	22	43	24	46	25	50	26	54
6	8	22	14	30	20	40	28	50	30	54	32	58	33	63	35	67
7	9	24	15	33	22	43	30	54	39	66	41	71	43	76	46	80
8	9	27	16	36	24	46	32	58	41	71	52	84	54	90	57	95
9	10	29	17	39	25	50	33	63	43	76	54	90	66	105	69	111
10	11	31	18	42	26	54	35	67	46	80	57	95	69	111	83	127

(the sum of the ranks for hearing children). Arbitrarily, we choose T_1 and our test statistic is $T_1 = 150.5$.

Table B.6 of Appendix B gives lower- and upper-tailed critical values of the rank sum distribution, denoted T_L and T_U, respectively, for values $n_1 \leq 10$ and $n_2 \leq 10$. The portion of Table B.6 for a one-tailed test with $\alpha = .05$ and for a two-tailed test with $\alpha = .10$ is reproduced in Table 12.3. Values of n_1 are given across the top of the table; values of n_2 are given at the left.

Examining Table 12.3, you will find that the critical values (highlighted) corresponding to $n_1 = n_2 = 10$ are $T_L = 83$ and $T_U = 127$. Therefore, for a one-tailed (upper-tailed) test at $\alpha = .05$, we will reject H_0 if $T_1 \geq T_U$, i.e.,

$$\text{Reject } H_0 \text{ if } T_1 \geq 127$$

Since the observed value of the test statistic, $T_1 = 150.5$, is greater than 127, we reject H_0 and conclude (at $\alpha = .05$) that the median of the visual acuity levels of deaf children (η_1) exceeds the median of the visual acuity levels of hearing children (η_2).

The same conclusion can be obtained by using the PHStat add-in for Excel. The Excel output of the Wilcoxon rank sum analysis is displayed in Figure 12.5. Both the test statistic (150.5) and p-value of the test (≈ 0) are indicated on the printout. Since the p-value is less than $\alpha = .05$, we again have evidence to reject H_0 and to support the psychologist's belief. ❑

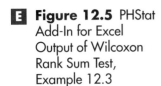

Figure 12.5 PHStat Add-In for Excel Output of Wilcoxon Rank Sum Test, Example 12.3

	A	B	C	D	E	F	G
1	Visual Acuity Comparison						
2							
3	Level of Significance	0.05					
4	Population 1 Sample						
5	Sample Size	10					
6	Sum of Ranks	150.5					
7	Population 2 Sample						
8	Sample Size	10					
9	Sum of Ranks	59.5					
10	Warning: Large-scale approximation formula not designed for small sample sizes.						
11	Total Sample Size n	20					
12	$T1$ Test Statistic	150.5					
13	$T1$ Mean	105					
14	Standard Error of $T1$	13.22876					
15	Z Test Statistic	3.439477					
16							
17	Upper-Tail Test						
18	Upper Critical Value	1.644853					
19	p-value	0.000291					
20	Reject the null hypothesis						

Self-Test

12.2

The following sample data represent independent random samples selected from two populations.

Sample 1:	31	40	27	22	25
Sample 2:	50	32	20	48	45

a. Rank the data for the combined samples.

b. Compute the rank sums T_1 and T_2.

c. Find the critical value for testing whether the median of population 1 is less than the median of population 2. Use $\alpha = .05$.

d. Give the conclusion of the test, part **c**.

Many nonparametric test statistics have sampling distributions that are approximately normal when n_1 and n_2 are large. For these situations we can test hypotheses using the large-sample z test of Chapter 8. The procedure is outlined in the following box.

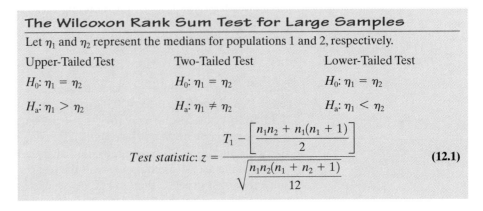

The Wilcoxon Rank Sum Test for Large Samples

Let η_1 and η_2 represent the medians for populations 1 and 2, respectively.

Upper-Tailed Test	Two-Tailed Test	Lower-Tailed Test
$H_0: \eta_1 = \eta_2$	$H_0: \eta_1 = \eta_2$	$H_0: \eta_1 = \eta_2$
$H_a: \eta_1 > \eta_2$	$H_a: \eta_1 \neq \eta_2$	$H_a: \eta_1 < \eta_2$

$$\text{Test statistic: } z = \frac{T_1 - \left[\dfrac{n_1 n_2 + n_1(n_1 + 1)}{2}\right]}{\sqrt{\dfrac{n_1 n_2(n_1 + n_2 + 1)}{12}}} \qquad (12.1)$$

Example 12.4   Large-Sample Rank Sum Test

Refer to Example 12.3 and the Wilcoxon rank sum test statistic $T_1 = 149.5$. Show that the large-sample Wilcoxon rank sum z test gives the same result as the exact test performed in Example 12.3.

Solution

According to equation 12.1 in the box, the value of the large-sample Wilcoxon rank sum z test statistic is

$$z = \frac{T_1 - \left[\dfrac{n_1 n_2 + n_1(n_1 + 1)}{2}\right]}{\sqrt{\dfrac{n_1 n_2(n_1 + n_2 + 1)}{12}}} = \frac{150.5 - \left[\dfrac{(10)(10) + (10)(11)}{2}\right]}{\sqrt{\dfrac{(10)(10)(10 + 10 + 1)}{12}}}$$

$$= \frac{150.5 - 105}{\sqrt{2,100/12}}$$

$$= 3.44$$

(This value is shown on the Excel printout, Figure 12.5, as "Z Test Statistic.")

   Since we want to detect if the median for population 1 (deaf children) exceeds the median for population 2 (hearing children), we will reject H_0 for values of z in the upper tail of the z distribution. For $\alpha = .05$, the rejection region is $z > z_{.05}$, or

$$\text{Reject } H_0 \text{ if } z > 1.645 \text{ (see Figure 12.6)}$$

Also, the p-value of the large-sample test is

$$p \text{ value} = P(z \geq 3.44) \approx 0$$

Using either the rejection region or p-value approach, our conclusion is the same as that for Example 12.3—reject H_0 and conclude that the distribution of the visual acuity levels for deaf children (population 1) is shifted to the right of the distribution of the visual acuity levels of hearing children (population 2). ❑

Figure 12.6 Rejection Region for Example 12.4

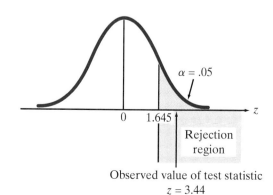

PROBLEMS
12.12–12.23

Using the Tools

12.12   Specify the rejection region for the Wilcoxon rank sum test for independent samples in each of the following situations. Assume the test statistic is T_1.

    a. $H_0: \eta_1 = \eta_2, H_a: \eta_1 \neq \eta_2, n_1 = 5, n_2 = 10, \alpha = .05$

    b. $H_0: \eta_1 = \eta_2, H_a: \eta_1 > \eta_2, n_1 = 8, n_2 = 8, \alpha = .05$

    c. $H_0: \eta_1 = \eta_2, H_a: \eta_1 < \eta_2, n_1 = 5, n_2 = 7, \alpha = .05$

12.13   Repeat Problem 12.12, but assume the test statistic is T_2.

12.14   For each of the following situations, find the large-sample Wilcoxon rank sum test statistic and corresponding rejection region:

    a.   Lower-tailed test, $T_1 = 71, n_1 = 10, n_2 = 14, \alpha = .05$

    b.   Upper-tailed test, $T_1 = 750, n_1 = 25, n_2 = 25, \alpha = .10$

    c.   Two-tailed test, $T_1 = 430, n_1 = 20, n_2 = 15, \alpha = .05$

12.15   Independent random samples were selected from two populations. The data are shown in the table below. Suppose you want to determine whether the median of population 1 is greater than the median of population 2.

Sample from Population 1		Sample from Population 2	
15	12	5	7
17	16	9	4
12		13	5
14		10	10

    a.   Compute the rank sums T_1 and T_2.

    b.   Give the rejection region for the test using $\alpha = .05$. (Use T_1 as a test statistic.)

    c.   State the appropriate conclusion.

Applying the Concepts

12.16   A study was conducted to compare the self-perceived assertiveness of women with traditional and nontraditional (i.e., less stereotypical) gender attitudes (*Small Group Research,* Nov. 1997). Forty female college undergraduates (20 with traditional and 20 with nontraditional attitudes) were selected to be group leaders. Each group was asked to complete a card sort task in 25 minutes. After completing the task, the assertiveness of each leader was measured using the Rathus Assertiveness Schedule (RAS). The RAS scores (simulated, based on information provided in the journal article) are listed on page 751. (Higher scores indicate higher levels of self-perceived assertiveness.) Use the Excel printout on page 751 to compare the medians of the RAS scores of the two groups of leaders. Do nontraditional leaders appear to be more self-assertive than traditional leaders? Test using $\alpha = .10$.

RAS.XLS

Nontraditional Leader			Traditional Leader		
17	28	25	6	9	7
28	24	37	4	8	11
27	23	29	10	8	2
20	26	30	10	9	6
9	23	26	7	13	7
43	42	21	4	13	8
40	28		6	21	

	A	B
1	Assertiveness Comparison	
2		
3	Level of Significance	0.1
4	Population 1 Sample	
5	Sample Size	20
6	Sum of Ranks	600.5
7	Population 2 Sample	
8	Sample Size	20
9	Sum of Ranks	219.5
10		
11	Total Sample Size n	40
12	$T$1 Test Statistic	600.5
13	$T$1 Mean	410
14	Standard Error of $T$1	36.96846
15	Z Test Statistic	5.153042
16		
17	Upper-Tail Test	
18	Upper Critical Value	1.281551
19	p-value	1.28E-07
20	Reject the null hypothesis	

12.17 Type II collagen is a candidate drug for suppressing the symptoms of rheumatoid arthritis. Medical researchers at Harvard University conducted a clinical trial to test the ability of collagen to reduce swollen joints in rheumatoid arthritis sufferers (*Science,* Sept. 24, 1993.) Each of 59 patients with severe, active rheumatoid arthritis was randomly assigned to receive either a daily dose of collagen or an indistinguishable placebo over a 30-day period. (Twenty-eight received collagen, 31 received the placebo.) The variable of interest was change in the number of swollen joints (after treatment minus before treatment) in each patient. The means and standard deviations for the two samples are reported here.

	CHANGE IN NUMBER OF SWOLLEN JOINTS (AFTER − BEFORE)	
	Collagen	Placebo
Sample size	28	31
Mean	−2.7	2.0
Std. deviation	.5	1.4

Source: Trentham, D. E., et al. "Effects of oral administration of Type II collagen on rheumatoid arthritis." *Science,* Vol. 261, No. 5129, Sept. 24, 1993, p. 1727 (Table 1).

a. Although no information on the distribution of the variable of interest for the two patient groups was provided in the *Science* article, the researchers used the Wilcoxon rank sum test rather than a *t* test to compare the two groups. Give two reasons why a nonparametric test may be more appropriate in this study.

b. Give the null hypothesis tested by the Wilcoxon rank sum procedure in this study.

c. The observed significance level of the test was reported to be smaller than .05 (i.e., *p*-value < .05). Make the appropriate conclusion.

12.18 Refer to the *Journal of Geography* (May/June 1997) study of the influence of computer software on learning geography, Problem 2.27 (page 84). Recall that two groups of seventh graders were given a test on the regions of Africa. One group used computer resources (e.g., electronic atlases and interactive maps) to research the region, while the other group used conventional methods (e.g., notes, worksheets). The test scores are reproduced in the table. Is there evidence that the median test scores of the computer resource group is larger than the median of the conventional group? Conduct the appropriate nonparametric test using $\alpha = .01$.

GEO.XLS

Group 1 (Computer Resources) Student Scores											
41	53	44	41	66	91	69	44	31	75	66	69
75	53	44	78	91	28	69	78	72	53	72	63
66	75	97	84	63	91	59	84	75	78	88	59
97	69	75	69	91	91	84					

Group 2 (No Computer) Student Scores											
56	59	9	59	25	66	31	44	47	50	63	19
66	53	44	66	50	66	19	53	78	78		

Source: Linn, S. E. "The effectiveness of interactive maps in the classroom: A selected example in studying Africa." *Journal of Geography,* Vol. 96, No. 3, May/June 1997, p. 167 (Table 1).

12.19 The Harris Corporation/University of Florida data on the voltage readings of production runs at two locations, Problem 9.19 (page 493) are reproduced below. If the median voltage reading at the new location is significantly higher (at $\alpha = .05$) than the median at the old (remote) location, the production process will be permanently shifted to the new location. What decision should Harris Corporation make?

VOLTAGE.XLS

Old Location			New Location		
9.98	10.12	9.84	9.19	10.01	8.82
10.26	10.05	10.15	9.63	8.82	8.65
10.05	9.80	10.02	10.10	9.43	8.51
10.29	10.15	9.80	9.70	10.03	9.14
10.03	10.00	9.73	10.09	9.85	9.75
8.05	9.87	10.01	9.60	9.27	8.78
10.55	9.55	9.98	10.05	8.83	9.35
10.26	9.95	8.72	10.12	9.39	9.54
9.97	9.70	8.80	9.49	9.48	9.36
9.87	8.72	9.84	9.37	9.64	8.68

Source: Harris Corporation, Melbourne, Fla.

12.20   In an epidemiological study, 55 female patients were diagnosed with full syndrome *bulimia nervosa* (an eating disorder). Each of these patients was asked to rate her satisfaction with life in general on a 7-point scale, where 1 = extremely dissatisfied and 7 = extremely satisfied. Their responses were compared to those of an independent sample of 4,208 normal female subjects (*American Journal of Psychiatry,* July 1995). The researchers hypothesized that female bulimia nervosa patients will have lower satisfaction levels than normal females.

   a.   The two groups' satisfaction levels were compared with the large-sample Wilcoxon rank sum test. State the null hypothesis of this test.

   b.   State the alternative hypothesis of interest to the researchers.

   c.   The large-sample test statistic was calculated to be $z = 5.48$. Use this result to carry out the test. What is your conclusion if $\alpha = .01$?

12.21   *Environmental Science & Technology* (Oct. 1993) reported on a study of insecticides used on dormant orchards in the San Joaquin Valley, California. Ambient air samples were collected and analyzed daily at an orchard site during the most intensive period of spraying. The oxon/thion ratios (in ng/m^3) in the air samples are recorded in the table. Compare the median oxon/thion ratio on foggy days to the median ratio on clear/cloudy days at the orchard using a nonparametric test. Use $\alpha = .05$.

OXONTHION.XLS

Date	Condition	Oxon/Thion Ratio
Jan. 15	Fog	.270
17	Fog	.241
18	Fog	.205
19	Fog	.523
20	Clear	.618
21	Fog	.112
21	Clear	.591
22	Fog	.330
23	Fog	.270
23	Cloudy	.225
25	Fog	.239
25	Clear	.375

Source: Selber, J. N., et al. "Air and fog deposition residues of four organophosphate insecticides used on dormant orchards in the San Joaquin Valley, California." *Environmental Science & Technology,* Vol. 27, No. 10, Oct. 1993, p. 2240 (Table V).

12.22   The Customer Satisfaction Index, computed by J. D. Powers & Associates, is designed to measure customer satisfaction with new automobiles and automakers. The results of one of these customer satisfaction surveys are shown in the table on page 754. Determine whether the median customer satisfaction index for foreign automakers is greater than the median for domestic automakers. Use $\alpha = .05$.

12.23   Consider a study of the relationship between job performance and turnover at a large national oil company (*Academy of Management Journal,* Mar. 1982). The company's employees were divided into two groups: "stayers"—those employees who stayed with the company over a 15-year period, and "leavers"—those former employees who left the

CUSTSAT.XLS

Auto (manufacturer)	Foreign (F) or Domestic (D)	Customer Satisfaction Index
Lexus (Toyota)	F	179
Infiniti (Nissan)	F	167
Saturn (GM)	D	160
Acura (Honda)	F	148
Mercedes-Benz	F	145
Toyota	F	144
Audi (VW)	F	139
Cadillac (GM)	D	138
Honda	F	138
Jaguar (Ford)	D	137

Source: J. D. Powers and Associates, 1992. Customer Satisfaction Study.

Problem 12.22

Stayers		Leavers	
3	4	4	5
5	3	3	5
2	2	3	3
3	5	2	4
4		3	3

Problem 12.23

STAYLEAV.XLS

company at varying points during the 15-year period. The company's annual performance appraisals corresponding to the initial years of service were used to form an initial performance rating for each employee. The performance ratings for a sample of 9 stayers and 10 leavers (simulated based on information provided in the article) are shown in the table at left. [*Note:* Smaller values of performance rating correspond to better performance.] Is there evidence that leavers are better performers than stayers at the oil company? Test using $\alpha = .05$.

12.4 COMPARING TWO POPULATIONS: MATCHED-PAIRS DESIGN

Recall from Section 9.2 that the parametric matched-pairs t test for a difference in population means is based on the differences within the matched pairs of observations. The nonparametric alternative is the *Wilcoxon signed ranks test* for the median difference. To perform the test, we assign ranks to the absolute values of the differences and then base the comparison on the rank sums of the negative (T^-) and the positive (T^+) differences. Differences equal to 0 are eliminated, and the number n of differences is reduced accordingly. Tied absolute differences receive ranks equal to the average of the ranks they would have received had they not been tied. The **Wilcoxon signed ranks test** is summarized in the box and is illustrated in Example 12.5.

The Wilcoxon Signed Ranks Test for Median Difference: Matched Pairs

Let η_D represent the median of the differences of paired observations from populations 1 and 2, where the difference is population 1's measurement minus population 2's measurement.

Upper-Tailed Test	Two-Tailed Test	Lower-Tailed Test
H_0: $\eta_D = 0$	H_0: $\eta_D = 0$	H_0: $\eta_D = 0$
H_a: $\eta_D > 0$	H_a: $\eta_D \neq 0$	H_a: $\eta_D < 0$

Calculate the difference within each of the n matched pairs of observations. Then rank the absolute values of the n differences from the smallest (rank 1) to the highest (rank n) and calculate the rank sum T^- of the negative differences and the rank sum T^+ of the positive differences.

Test statistic:	*Test statistic:*	*Test statistic:*
T^-, the rank sum of the negative differences	T, the smaller of T^- or T^+	T^+, the rank sum of the positive differences
Rejection region:	*Rejection region:*	*Rejection region:*
$T^- \le T_0$	$T \le T_0$	$T^+ \le T_0$

where T_0 is given in Table B.7 of Appendix B.

[*Note:* Differences equal to 0 are eliminated and the number n of differences is reduced accordingly. Tied absolute differences receive ranks equal to the average of the ranks they would have received had they not been tied.]

Example 12.5   Applying the Signed Ranks Test

In Example 9.8 (page 496), we used the parametric t test to determine whether the mean reading achievement test scores for first graders taught by two methods are different. First graders were matched according to IQ, socioeconomic status, and learning ability and one member of each pair was taught by method 1 and the other by method 2. The data for the $n = 10$ matched pairs are reproduced in Table 12.4. Test the hypothesis of no median difference in the reading achievement test scores for the two teaching methods. Use the Wilcoxon signed ranks test with $\alpha = .05$.

Solution

Let η_D represent the population median of the differences in test scores, where each difference is calculated as test score for method 1 minus test score for method 2. Then we want to test

$$H_0: \eta_D = 0$$

$$H_a: \eta_D \ne 0$$

Table 12.4 Reading Test Scores and Differences, Example 12.5

First-Grader Pair	TEACHING METHOD 1	TEACHING METHOD 2	Difference	Absolute Value of Difference	Rank of Absolute Value
1	78	71	7	7	3
2	63	44	19	19	10
3	72	61	11	11	5
4	89	84	5	5	2
5	91	74	17	17	9
6	49	51	−2	2	1
7	68	55	13	13	6
8	76	60	16	16	7.5
9	85	77	8	8	4
10	55	39	16	16	7.5

Table 12.5 A Portion of the Wilcoxon Signed Ranks Table, Table B.7

One-Tailed	Two-Tailed	$n = 5$	$n = 6$	$n = 7$	$n = 8$	$n = 9$	$n = 10$
$\alpha = .05$	$\alpha = .10$	1	2	4	6	8	1
$\alpha = .025$	$\alpha = .05$		1	2	4	6	8
$\alpha = .01$	$\alpha = .02$			0	2	3	5
$\alpha = .005$	$\alpha = .01$				0	2	3
		$n = 11$	$n = 12$	$n = 13$	$n = 14$	$n = 15$	$n = 16$
$\alpha = .05$	$\alpha = .10$	14	17	21	26	30	36
$\alpha = .025$	$\alpha = .05$	11	14	17	21	25	30
$\alpha = .01$	$\alpha = .02$	7	10	13	16	20	24
$\alpha = .005$	$\alpha = .01$	5	7	10	13	16	19

To apply the Wilcoxon signed ranks test, we need to rank the absolute values of the differences between the matched pairs of observations. The differences and ranks are shown in Table 12.4. The rank of the only negative difference (highlighted on Table 12.4) is 1. Consequently, the rank sums of the positive and negative differences are $T^+ = 54$ and $T^- = 1$, respectively. Since we want to conduct a two-tailed test, we use the smaller of T^- and T^+, i.e., T^-, as the test statistic and reject H_0 if $T^- \leq T_0$.

The critical values of the Wilcoxon signed ranks statistic are provided in Table B.7 of Appendix B, which gives the value of T_0 for one-tailed tests for values of α equal to .05, .025, .01, and .005, and for two-tailed tests for values of α equal to .10, .05, .02, and .01. A portion of Table B.7 is reproduced in Table 12.5.

Since we want to conduct a two-tailed test, we move down the two-tailed test column to the desired value of $\alpha = .05$. We find the value $T_0 = 8$ (highlighted in Table 12.5). Thus, our rejection region is:

$$\text{Reject } H_0 \text{ if } T^- \leq 8$$

Clearly, our test statistic, $T^- = 1$, is less than $T_0 = 8$. Therefore, we reject H_0 and conclude that there is sufficient evidence of a median difference in the reading achievement test scores for first graders taught by the two methods. (This is the same conclusion we obtained by using the t test in Example 9.8.) ❏

Self-Test

12.3

A random sample of $n = 5$ pairs of measurements is shown below.

	PAIR NUMBER				
	1	2	3	4	5
Measurement 1	600	510	622	331	308
Measurement 2	721	615	580	422	295

a. Compute and rank the difference between the two measurements.

b. Find the rank sums T^+ and T^-.

c. Find the critical value for testing for a difference in the locations of the distributions for the two measurements. Use $\alpha = .05$.

d. Give the conclusion of the test, part c.

The Wilcoxon signed ranks statistic also has a sampling distribution that is approximately normal when the number n of pairs is large—say, $n \geq 20$. This large-sample nonparametric matched-pairs test is summarized in the following box.

Wilcoxon Signed Ranks Test for Large Samples

Let η_D represent the population median difference.

Upper-Tailed Test	Two-Tailed Test	Lower-Tailed Test
$H_0: \eta_D = 0$	$H_0: \eta_D = 0$	$H_0: \eta_D = 0$
$H_a: \eta_D > 0$	$H_a: \eta_D \neq 0$	$H_a: \eta_D < 0$

$$\text{Test statistic: } z = \frac{T^+ - [n(n+1)/4]}{\sqrt{[n(n+1)(2n+1)]/24}} \qquad (12.2)$$

Rejection region:	Rejection region:	Rejection region:
$z > z_\alpha$	$|z| > z_{\alpha/2}$	$z < -z_\alpha$

Assumptions: The sample size n is greater than or equal to 20. Differences equal to 0 are eliminated and the number n of differences is reduced accordingly. Tied absolute differences receive ranks equal to the average of the ranks they would have received had they not been tied.

PROBLEMS
12.24–12.35

Using the Tools

12.24 Specify the test statistic and rejection region for the Wilcoxon signed ranks test for the matched-pairs design in each of the following situations:

   a. $H_0: \eta_D = 0$, $H_a: \eta_D \neq 0$, $n = 19$, $\alpha - .05$

   b. $H_0: \eta_D = 0$, $H_a: \eta_D > 0$, $n = 36$, $\alpha = .01$

   c. $H_0: \eta_D = 0$, $H_a: \eta_D < 0$, $n = 50$, $\alpha = .005$

12.25 Suppose you want to test the hypothesis that two treatments, A and B, are equivalent against the alternative that the responses for A tend to be larger than those for B.

   a. If $n = 8$ and $\alpha = .01$, give the rejection region for a Wilcoxon signed ranks test.

   b. Suppose you want to determine if the population median difference differs from 0. If $n = 7$ and $\alpha = .10$, give the rejection region for the Wilcoxon signed ranks test.

12.26 For each of the following sample differences in a matched-pairs experiment, compute T^+ and T^-.

   a. $-4, -1, 16, -2, 3, 1, 2, -5$

   b. $21.2, 1.7, -4.0, 3.3, 11.5, 6.6, -15.7, 20.0, 31.5$

   c. $1,700, 1,334, -45, 911, 726, -1,271$

12.27   A random sample of nine pairs of measurements is shown in the table.

Pair	Sample Data from Population 1	Sample Data from Population 2
1	8	7
2	10	1
3	6	4
4	10	10
5	7	4
6	8	3
7	4	6
8	9	2
9	8	4

a.   Use the Wilcoxon signed ranks test to determine whether observations from population 1 tend to be larger than those from population 2. Test using $\alpha = .05$.

b.   Use the Wilcoxon signed ranks test to determine whether a median difference exists. Test using $\alpha = .05$.

Applying the Concepts

12.28   One of the most critical aspects of a new atlas design is its thematic content. In a survey of atlas users (*Journal of Geography,* May/June 1995), a large sample of high school teachers in British Columbia ranked 12 thematic atlas topics for usefulness. The consensus rankings of the teachers (based on the percentage of teachers who responded they "would definitely use" the topic) are given in the table. These teacher rankings were compared to the rankings made by a group of university geography alumni. Compare the theme rankings for the two groups with an appropriate nonparametric test. Use $\alpha = .05$. Interpret the results.

ATLAS.XLS

| Theme | RANKINGS | |
	High School Teachers	Geography Alumni
Tourism	10	2
Physical	2	1
Transportation	7	3
People	1	6
History	2	5
Climate	6	4
Forestry	5	8
Agriculture	7	10
Fishing	9	7
Energy	2	8
Mining	10	11
Manufacturing	12	12

Source: Keller, C. P., et al. "Planning the next generation of regional atlases: Input from educators." *Journal of Geography,* Vol. 94, No. 3, May/June 1995, p. 413 (Table 1).

12.29   Eleven prisoners of the war in Croatia were evaluated for neurological impairment after their release from a Serbian detention camp (*Collegium Antropologicum,* June 1997). All eleven released POWs received blows to the head and neck and/or loss of consciousness during imprisonment. Neurological impairment was assessed by measuring the amplitude of the visual evoked potential (VEP) in both eyes at two points in time: 157 days and 379 days after release from prison. (The higher the VEP value, the greater the neurological impairment.) The data for the 11 POWs are shown in the table. Determine whether the VEP measurements of POWs 379 days after their release tend to be greater than the VEP measurements of POWs 157 days after their release. Test using $\alpha = .05$.

POWVEP.XLS

POW	157 Days after Release	379 Days after Release
1	2.46	3.73
2	4.11	5.46
3	3.93	7.04
4	4.51	4.73
5	4.96	4.71
6	4.42	6.19
7	1.02	1.42
8	4.30	8.70
9	7.56	7.37
10	7.07	8.46
11	8.00	7.16

Source: Vrca, A., et al. "The use of visual evoked potentials to follow-up prisoners of war after release from detention camps." *Collegium Antropologicum,* Vol. 21, No. 1, June 1997, p. 232. (Data simulated from information provided in Table 3.)

12.30 Researchers at Purdue University compared human real-time scheduling in a processing environment to an automated approach that utilizes computerized robots and sensing devices (*IEEE Transactions,* Mar. 1993). The experiment consisted of eight simulated scheduling problems. Each task was performed by a human scheduler and by the automated system. Performance was measured by the *throughput* rate, defined as the number of good jobs produced weighted by product quality. The resulting throughput rates are shown in the accompanying table. Compare the throughput rates of tasks scheduled by a human and by the automated method with a nonparametric test. Use $\alpha = .01$.

THRUPUT.XLS

Task	Human Scheduler	Automated Method	Task	Human Scheduler	Automated Method
1	185.4	180.4	5	240.0	269.3
2	146.3	248.5	6	253.8	249.6
3	174.4	185.5	7	238.8	282.0
4	184.9	216.4	8	263.5	315.9

Source: Yih, Y., Liang, T., and Moskowitz, H. "Robot scheduling in a circuit board production line: A hybrid OR/ANN approach." *IEEE Transactions,* Vol. 25, No. 2, Mar. 1993, p. 31 (Table 1).

12.31 Dental researchers have developed a new material for preventing cavities—a plastic sealant that is applied to the chewing surfaces of teeth. To determine whether the sealant is effective, it was applied to half of the teeth of each of 12 school-age children. After five years, the numbers of cavities in the sealant-coated teeth and untreated teeth were counted.

The results are given in the accompanying table. Is there sufficient evidence to indicate that sealant-coated teeth are less prone to cavities than are untreated teeth? Test using $\alpha = .05$.

CAVITIES.XLS

Child	Sealant-Coated	Untreated	Child	Sealant-Coated	Untreated
1	3	3	7	1	5
2	1	3	8	2	0
3	0	2	9	1	6
4	4	5	10	0	0
5	1	0	11	0	3
6	0	1	12	4	3

12.32   Refer to the *Journal of STAR Research* (July 1994) study of traffic counts at the 30th and 100th highest hourly volume at Florida highway stations, Problem 9.30 (page 502). The data are reproduced below.

DOTHOUR.XLS

Station	30th Highest Hour	100th Highest Hour
0117	1,890	1,736
0087	2,217	2,069
0166	1,444	1,345
0013	2,105	2,049
0161	4,905	4,815
0096	2,022	1,958
0145	594	548
0149	252	229
0038	2,162	2,048
0118	1,938	1,748
0047	879	811
0066	1,913	1,772
0094	3,494	3,403
0105	1,424	1,309
0113	4,571	4,425
0151	3,494	3,359
0159	2,222	2,137
0160	1,076	989
0164	2,167	2,039
0165	3,350	3,123

Source: Ewing, R. "Roadway levels of service in an era of growth management." *Journal of STAR Research,* Vol. 3, July 1994, p. 103 (Table 2).

a.   Compare the traffic counts at the 30th and 100th highest hours with a nonparametric test. Use $\alpha = .10$.

b.   Compare the results of the nonparametric test, part **a**, to the results of the parametric test you conducted in Problem 9.30.

12.33   Refer to the Merck Research Labs experiment to evaluate a new drug using rats in a swim maze, Problem 9.32 (page 503). Recall that 19 impregnated dam rats were given a dosage of the drug, then one male and one female rat pup from each resulting litter were selected to perform in the swim maze. The number of swims required by each rat pup to escape three times is reproduced in the table on page 761.

SWIMMAZE.XLS

Litter	Male	Female	Litter	Male	Female
1	8	5	11	6	5
2	8	4	12	6	3
3	6	7	13	12	5
4	6	3	14	3	8
5	6	5	15	3	4
6	6	3	16	8	12
7	3	8	17	3	6
8	5	10	18	6	4
9	4	4	19	9	5
10	4	4			

Source: Thomas E. Bradstreet, Merck Research Labs, BL 3-2, West Point, Penn. 19486.

a. Compare the number of swim attempts for male and female pups using the Wilcoxon signed ranks test. Use $\alpha = .10$.

b. Compare the results of the nonparametric test, part **a**, to the results of the parametric test you conducted in Problem 9.32.

12.34 A study was conducted to test the significance of the measurement errors in the U.S. Department of Treasury's bond yield series (*Quarterly Journal of Business and Economics,* Autumn 1986). The sample data consisted of 164 new government debt offerings that were issued at exact maturities (e.g., 2-year, 3-year, etc.). The actual yields for these bond issues were compared to the estimated yields from the Treasury's published yield series using the Wilcoxon signed ranks test for matched pairs. [*Note:* For each issue, the difference was calculated as Difference = (Actual yield − Estimated yield).] The results of the tests for both 2-year and 30-year bonds are shown in the accompanying table.

	BOND MATURITY	
	2-Year	30-Year
Number of bonds	92	15
Number of positive differences	56	8
Rank Sum, T^+	2,549.68	73.52
Number of negative differences	36	7
Rank Sum, T^-	1,728.36	73.01

a. Test the hypothesis (at $\alpha = .05$) of no median difference between the exact 2-year bond yields and estimated 2-year bond yields. (Use the large-sample version of the test.)

b. Repeat part **a** for the 30-year bonds.

12.35 Refer to the *Environmental Science & Technology* (Oct. 1993) study of air deposition residues of the insecticide diazinon used on dormant orchards, Problem 12.21 (page 753). Ambient air samples were collected and analyzed at an orchard site for each of 11 days during the most intensive period of spraying. The levels of diazinon residue (in ng/m^3) during the day and at night are recorded in the table on page 762. The researchers want to know whether the diazinon residue levels tend to differ from day to night. Perform the appropriate nonparametric analysis at $\alpha = .05$ and interpret the results.

DIAZINON.XLS

| Date | DIAZINON RESIDUE | | Date | DIAZINON RESIDUE | |
	Day	Night		Day	Night
Jan. 11	5.4	24.3	Jan. 17	6.1	104.3
12	2.7	16.5	18	7.7	96.9
13	34.2	47.2	19	18.4	105.3
14	19.9	12.4	20	27.1	78.7
15	2.4	24.0	21	16.9	44.6
16	7.0	21.6			

Source: Selber, J. N., et al. "Air and fog deposition residues for organophosphate insecticides used on dormant orchards in the San Joaquin Valley, California." *Environmental Science & Technology,* Vol. 27, No. 10, Oct. 1993, p. 2240 (Table IV).

12.5 COMPARING THREE OR MORE POPULATIONS: COMPLETELY RANDOMIZED DESIGN

In Section 12.3, we showed how the Wilcoxon rank sum test can be used in place of the *t* test for comparing the medians of two populations. The advantage of the Wilcoxon rank sum test over the *t* test is that we do not need to make restrictive assumptions (e.g., normality, equal variances) about the sampled populations.

The *Kruskal-Wallis H test* provides a nonparametric alternative to the one-way analysis of variance *F* test (Section 11.3) for comparing the medians of more than two populations based on independent random samples (i.e., the completely randomized design). As with the Wilcoxon rank sum test, no assumptions regarding the normality or variances of the sampled populations are required.

The sample observations are ranked from the smallest to the largest and the rank sums are calculated for each sample. For example, if you had three samples with $n_1 = 8, n_2 = 6$, and $n_3 = 7$, you would rank the $n_1 + n_2 + n_3 = 21$ observations from the smallest (rank 1) to the largest (rank 21) and then calculate the rank sums, T_1, T_2, and T_3, for the three samples. The Kruskal-Wallis *H* test uses these rank sums to calculate an *H* statistic that possesses an approximate *chi-square* sampling distribution. The elements of the **Kruskal-Wallis *H* test** are summarized in the box and illustrated in Example 12.6.

The Kruskal-Wallis *H* Test for Comparing *k* Population Medians: Completely Randomized Design

Let η_i represent the median for population *i*.

$$H_0: \eta_1 = \eta_2 = \cdots = \eta_k$$

H_a: At least two of the population medians are different

$$Test\ statistic:\ H = \frac{12}{n(n+1)} \sum_{i=1}^{k} \frac{T_i^2}{n_i} - 3(n+1) \tag{12.3}$$

where
  n_i = number of observations in sample *i*
  T_i = rank sum of sample *i*
  n = total sample size = $n_1 + n_2 + \cdots + n_k$

[*Note:* Tied observations are assigned ranks equal to the average of the ranks that would have been assigned had they not been tied.]

Rejection region: $H > \chi_\alpha^2$ where χ_α^2 is based on $(k - 1)$ degrees of freedom

Assumptions:
1. The k samples have been independently and randomly selected from their respective populations.
2. For the chi-square approximation to be adequate, there should be five or more observations in each sample.

Example 12.6   Applying the Kruskal-Wallis H Test

In Examples 11.3 and 11.4 (pages 676 and 678), we used an analysis of variance to compare the mean grade point averages of college freshmen from three different socioeconomic backgrounds. Independent random samples of $n_1 = n_2 = n_3 = 7$ freshmen were selected from each of the three populations. The data are reproduced in Table 12.6. Use the Kruskal-Wallis H test to determine whether the data provide sufficient evidence to indicate that the median grade point average of freshmen depends on the students' socioeconomic backgrounds. Test using $\alpha = .05$.

Solution

Let η_1, η_2, and η_3 represent the population median GPAs of lower class, middle class, and upper class college freshmen, respectively. For this problem, we want to test:

$$H_0: \eta_1 = \eta_2 = \eta_3$$

$$H_a: \text{At least two population medians differ}$$

The first step in finding the Kruskal-Wallis H test statistic is to rank the $n_1 + n_2 + n_3 = 7 + 7 + 7 = 21$ observations from the smallest to the largest. Thus, we give the smallest observation (1.36) a rank of 1, the next smallest (1.80) a rank of 2, ..., and the largest observation (3.77) a rank of 21. The original data, their associated ranks, and the sample rank sums T_1, T_2, and T_3 are shown in Table 12.7 (page 764).
From equation 12.3,

$$H = \frac{12}{n(n + 1)} \sum_{i=1}^{3} \frac{T_i^2}{n_i} - 3(n + 1)$$

GPACLASS.XLS

Table 12.6 Grade Point Averages for Three Socioeconomic Groups

Lower Class	Middle Class	Upper Class
2.87	3.23	2.25
2.16	3.45	3.13
3.14	2.78	2.44
2.51	3.77	3.27
1.80	2.97	2.81
3.01	3.53	1.36
2.16	3.01	2.53

Table 12.7 Rank Sums for Grade Point Average Data

LOWER CLASS		MIDDLE CLASS		UPPER CLASS	
GPA	Rank	GPA	Rank	GPA	Rank
2.87	11	3.23	17	2.25	5
2.16	3.5	3.45	19	3.13	15
3.14	16	2.78	9	2.44	6
2.81	7	3.77	21	3.27	18
1.80	2	2.97	12	2.81	10
3.01	13.5	3.53	20	1.36	1
2.16	3.5	3.01	13.5	2.53	8
$T_1 = 56.5$		$T_2 = 111.5$		$T_3 = 63$	

Figure 12.7 Rejection Region for the Kruskal-Wallis *H* Test of Example 12.6

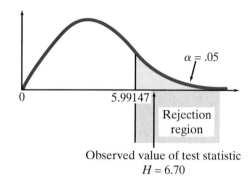

where $n_1 = n_2 = n_3 = 7, n = 21, T_1 = 56.5, T_2 = 111.5$, and $T_3 = 63$. Substituting these values into the formula for *H*, we obtain

$$H = \frac{12}{21(21 + 1)}\left[\frac{(56.5)^2}{7} + \frac{(111.5)^2}{7} + \frac{(63)^2}{7}\right] - 3(21 + 1)$$

$$= \frac{12}{462}\left(\frac{3192.25}{7} + \frac{12,432.25}{7} + \frac{3969}{7}\right) - 66$$

$$= 6.70$$

The rejection region for the *H* test is $H > \chi_\alpha^2$, where χ_α^2 is based on $(k - 1)$ degrees of freedom. Since we have selected $\alpha = .05$ and we want to compare $k = 3$ population relative frequency distributions, we need the value of $\chi_{.05}^2$ for $(k - 1) = (3 - 1) = 2$ degrees of freedom. This value is given in Table B.4 as $\chi_{.05}^2 = 5.99147$. Therefore, the rejection region for the test includes all values of *H* larger than 5.99147 (see Figure 12.7).

The final step in conducting the test is to determine whether the value of the test statistic falls in the rejection region. Since the value computed for the grade point data, $H = 6.70$, is larger than $\chi_{.05}^2 = 5.99147$, we reject the null hypothesis, i.e., there is sufficient evidence to indicate differences in median freshmen grade point averages among the three socioeconomic populations.

The same conclusion can be reached using the PhStat add-in for Excel. The Excel printout of the analysis is shown in Figure 12.8 (page 765). The test statistic (rounded) is indicated, as is the *p*-value of the test. Note that the *p*-value (.035) is less than $\alpha = .05$, resulting in our conclusion of "reject H_0." ❑

E **Figure 12.8** PHStat
Add-In for Excel
Output, Example 12.6

	A	B
1	GPA Comparison	
2		
3	Level of Significance	0.05
4	Group 1	
5	Sum of Ranks	56.5
6	Sample Size	7
7	Group 2	
8	Sum of Ranks	111.5
9	Sample Size	7
10	Group 3	
11	Sum of Ranks	63
12	Sample Size	7
13	Sum of Squared Ranks/Sample Size	2799.071
14	Sum of Sample Sizes	21
15	Number of groups	3
16	H Test Statistic	6.703154
17	Critical Value	5.991476
18	p-Value	0.035029
19	Reject the null hypothesis	

Self-Test

12.4

The rank sums for samples of five measurements selected from each of four populations (a total of 20 measurements) are $T_1 = 84$, $T_2 = 15$, $T_3 = 50$, and $T_4 = 61$.

a. For each sample, find T_i^2/n_i.

b. Compute the Kruskal-Wallis H test statistic for the data.

c. Find the critical value (at $\alpha = .10$) for testing whether the distributions of the four populations have the same location.

d. Give the conclusion for the test, part c.

PROBLEMS
12.36–12.47

Using the Tools

12.36 For each part, compute the rank sums of the three samples.

a. Sample 1: 6, 18, 20, 2; Sample 2: 7, 11, 6, 5; Sample 3: 19, 40, 3, 3

b. Sample 1: 100, 123, 194; Sample 2: 215, 208, 137; Sample 3: 204, 256, 245

c. Sample 1: −16, 41, 57, 8, 0; Sample 2: 11, 32, 24, 16, 15;
Sample 3: 20, 14, 10, 8; Sample 4: −3, −12, 15, 4

12.37 Consider a test to determine whether three distributions differ in location. Independent random samples of size $n = 5$ were collected from each distribution and all 15 observations ranked. For each set of rank sums, compute the Kruskal-Wallis H test statistic.

a. $T_1 = 60$, $T_2 = 40$, $T_3 = 20$ b. $T_1 = 55$, $T_2 = 45$, $T_3 = 20$

c. $T_1 = 70$, $T_2 = 15$, $T_3 = 35$ d. $T_1 = 30$, $T_2 = 30$, $T_3 = 60$

12.38 Find the rejection region for comparing k medians with the Kruskal-Wallis H test if:

a. $k = 3, n = 50, \alpha = .01$ b. $k = 5, n = 30, \alpha = .10$

c. $k = 4, n = 100, \alpha = .05$ d. $k = 10, n = 75, \alpha = .05$

12.39 Suppose you want to use the Kruskal-Wallis H test to compare the medians of three populations. The following are independent random samples selected from the three populations:

PROB12_39.XLS

```
I:    45   33   55   88   58
II:   22   31   16   25   30   33
III:  91   96  102   75   88
```

a. What type of experimental design was used?

b. Specify the null and alternative hypotheses you would test.

c. Specify the rejection region that would be used for your hypothesis test at $\alpha = .05$.

d. Calculate the rank sums for the three samples.

e. Calculate the test statistic.

f. State the appropriate conclusion of the test.

Applying the Concepts

12.40 Refer to the *Journal of the American Mosquito Control Association* study of biting flies, Problem 12.10 (page 745). The effect of wind speeds in kilometers per hour (kph) on the biting rate of the fly on Stanbury Island, Utah, was investigated by exposing samples of volunteers to one of six wind speed conditions. The distributions of the biting rates for the six wind speeds were compared using the Kruskal-Wallis H test. The rank sums of the biting rates for the six conditions are shown in the accompanying table.

Wind Speed (kph)	Number of Volunteers (n_i)	Rank Sum of Biting Rates (T_i)
<1	11	1,804
1–2.9	49	6,398
3–4.9	62	7,328
5–6.9	39	4,075
7–8.9	35	2,660
9–20	21	1,388
TOTALS	217	23,653

Source: Strickman, D., et al. "Meteorological effects on the biting activity of *Leptoconops americanus* (Diptera: Ceratopogonidae)." *Journal of the American Mosquito Control Association,* Vol. II, No. 1, Mar. 1995, p. 17 (Table 1).

a. The researchers reported the test statistic as $H = 35.2$. Verify this value.

b. Find the rejection region for the test using $\alpha = .01$.

c. Make the proper conclusions.

d. The researchers reported the p-value of the test as $p < .01$. Does this value support your inference in part c? Explain.

12.41 A field experiment was conducted to determine if the prevalence of nematodes (roundworms) in sedimentation in the Baltic Sea depends on

January	May	August
35.0	224.5	217.4
149.8	104.9	202.3
71.0	216.9	203.0
105.0	312.0	10.1
1.2	268.3	155.9

Problem 12.41

NEMATODE.XLS

season (*Estuarine, Coastal and Shelf Science,* Aug. 1997). Five sediment samples were randomly selected during one month in each of three seasons—January, May, and August—and the biomass (dry weight) of nematodes in each sediment sample was determined. The data (recorded in grams) for a particular nematode species are provided in the table. (The data are simulated based on summary statistics from the journal article.)

a. The researchers used the nonparametric Kruskal-Wallis H test to compare the median dry weights of the nematode species across the three seasons. Explain why this test is more appropriate than the ANOVA F test of Chapter 11.

b. Using Excel, analyze the data using the Kruskal-Wallis H test. Interpret the results at $\alpha = .05$.

12.42 In a study to evaluate the effectiveness of performance appraisal training, each in a sample of middle-level managers was randomly assigned to one of three training conditions: standard (in-class) training, Internet-assisted training, or Internet-assisted training plus a behavior modeling workshop. After the formal training, the managers were administered a 25-question multiple-choice test of managerial knowledge and the number of correct answers was recorded for each. The data are provided in the table. Is there sufficient evidence of differences in the median scores for the three types of performance appraisal training? Test using $\alpha = .01$.

APPRTRAIN.XLS

Standard Training	Internet-Assisted Training	Internet Training Plus Workshop
16	19	12
18	22	19
11	13	18
14	15	22
23	20	16
	18	25
	21	

FISH.XLS

12.43 Refer to the U.S. Army Corps of Engineers data on contaminated fish captured in the Tennessee River. Recall that three species of fish were investigated: channel catfish, smallmouth buffalo, and largemouth bass. Suppose you want to compare the median levels of the pollutant DDT in the three species of fish.

a. Use Excel to graphically check whether the DDT distribution for each species is approximately normal.

b. Explain why the graphs, part **a**, suggest that a nonparametric procedure is more appropriate for analyzing the data.

c. Use Excel to rank the data. Then perform the appropriate nonparametric test at $\alpha = .05$. Interpret the results.

12.44 Organic chemical solvents are used for cleaning fabricated metal parts in industries such as aerospace, electronics, and automobiles. These solvents, when disposed of, have the potential to become hazardous waste. The *Journal of Hazardous Materials* (July 1995) published the results of a study of the chemical properties of three different types of hazardous organic solvents used to clean metal parts: aromatics, chloroalkanes, and esters. One variable studied was sorption rate, measured as mole percentage. Independent samples of solvents from each type were tested and their sorption rates recorded in the table. What do the results imply about the median sorption rates for the three solvents? Use $\alpha = .01$.

SORPRATE.XLS

Aromatics		Chloroalkanes		Esters		
1.06	.95	1.58	1.12	.29	.43	.06
.79	.65	1.45	.91	.06	.51	.09
.82	1.15	.57	.83	.44	.10	.17
.89	1.12	1.16	.43	.61	.34	.60
1.05				.55	.53	.17

Source: Reprinted from *Journal of Hazardous Materials,* Vol. 42, No. 2, J. D. Ortego et al., "A review of polymeric geosynthetics used in hazardous waste facilities," p. 142 (Table 9), July 1995, Elsevier Science-NL, Sara Burgerhartstraat 25, 1055 KV Amsterdam, The Netherlands.

12.45 Refer to the sale prices of residential properties in seven Tampa (Florida) neighborhoods, Problem 11.7 (page 683). The data (in thousands of dollars) is reproduced in the table.

TAMSALES.XLS

			NEIGHBORHOOD			
A	**B**	**C**	**D**	**E**	**F**	**G**
191.5	100.5	35.3	86.0	69.9	46.0	14.5
208.5	147.5	50.0	140.0	159.5	42.0	50.0
255.0	115.0	160.0	120.0	84.5	48.0	39.0
375.0	210.0	59.9	72.9	57.7	57.0	20.0
205.0	155.0	38.5	72.0	45.4	34.5	32.2
191.5	127.5	105.9	68.7	43.9	74.5	54.0
189.0	115.0	312.0	90.0	205.0	66.9	21.5
139.0	110.0	84.5	100.0	59.9	54.5	55.0
138.0	183.0	68.5	97.0	37.0	46.8	43.0
162.0	107.5	160.0	84.0	68.0	6.9	13.3

Source: Hillsborough County (Fla.) property appraiser's office.

a. In Problem 11.7 you conducted an analysis of variance F test to compare the mean sale prices of properties in the seven neighborhoods. Why might this parametric test be invalid?

b. Conduct a nonparametric test to determine whether the median sale prices differ for the seven neighborhoods. Test using $\alpha = .01$. Compare your result to the findings in Problem 11.7.

12.46 Refer to the study of the relationship between hair color and pain threshold, Problem 11.57 (page 722). The pain threshold scores for people of assorted hair colors are reproduced in the table. Determine whether the median pain threshold scores differ for the four hair color types. Test using $\alpha = .05$.

HAIRPAIN.XLS

	HAIR COLOR		
Light Blond	**Dark Blond**	**Light Brunette**	**Dark Brunette**
62	63	42	32
60	57	50	39
71	52	41	51
55	41	37	30
48	43		35

12.47   Phosphoric acid is chemically produced by reacting phosphate rock with sulfuric acid. An important consideration in the chemical process is the length of time required for the chemical reaction to reach a specified temperature. An experiment was conducted to compare the reactivity of phosphate rock mined in north, central, south, and the panhandle of Florida. Rock samples were collected from each location and placed in vacuum bottles with a sulfuric acid solution. The time (in seconds) for the chemical reaction to reach 200°F was recorded for each sample. Do the data provide sufficient evidence to indicate differences among the median reaction times of phosphoric rock mined at the four locations? Test using $\alpha = .05$.

PHOSROCK.XLS

South	Central	North	Panhandle
40	41	25	20
42	38	36	22
37	40	28	51
38	33	31	37
41	35	29	
		22	
		27	

12.6   TESTING FOR RANK CORRELATION

We learned in Chapter 10 how to use a parametric t test to conduct a test for the slope β_1 of a simple linear regression model or the population coefficient of correlation ρ. (Recall that the tests are equivalent.) These tests are suspect, however, if the assumptions about the random error term ε in the model are violated. In this situation, an alternative procedure is to conduct a nonparametric test for linear correlation ρ in the population.

The nonparametric test is based on **Spearman's rank correlation coefficient**, denoted r_s. The formula for computing r_s and the nonparametric test of hypothesis for rank correlation are shown in the box. Example 12.7 illustrates the procedure.

Spearman's Nonparametric Test for Rank Correlation

Upper-Tailed Test	Two-Tailed Test	Lower-Tailed Test
$H_0: \rho = 0$	$H_0: \rho = 0$	$H_0: \rho = 0$
$H_a: \rho > 0$	$H_a: \rho \neq 0$	$H_a: \rho < 0$

$$\text{Test statistic: } r_s = 1 - \frac{6\sum_{i=1}^{n} d_i^2}{n(n^2 - 1)} \tag{12.4}$$

where d_i is the difference between the y rank and x rank for the ith observation. [*Note:* In the case of ties, calculate r_s by substituting the ranks of the y's and the ranks of the x's for the actual y values and x values in the formula for r given in Section 3.9 (page 190).]

Rejection region:	Rejection region:	Rejection region:
$r_s > r_\alpha$	$|r_s| > r_{\alpha/2}$	$r_s < -r_\alpha$

where the values of r_α and $r_{\alpha/2}$ are given in Table B.8 of Appendix B.

Example 12.7 Applying Spearman's Test for Rank Correlation

An experiment was designed to learn whether eye pupil size is related to a person's attempt at deception. Fifteen college students were asked to respond verbally to a series of questions. Before questioning began, the pupil size of each student was measured and the students were instructed to answer some of the questions dishonestly. During questioning, the percentage increase in pupil size was recorded. After answering the questions, each student was given a deception score based on the proportion answered dishonestly. (High scores indicate a large number of deceptive responses.) The data are shown in Table 12.8.

a. Calculate Spearman's rank correlation coefficient as a measure of the strength of the relationship between deception score and percentage increase in pupil size.

b. Is there sufficient evidence to indicate that percentage increase in pupil size is positively correlated with deception score? Test using $\alpha = .01$.

Solution

a. Spearman's rank correlation coefficient is found by first ranking the values of each variable separately. (Ties are treated by averaging the tied ranks.) Then r_s is computed in exactly the same way as the Pearson correlation coefficient r; the only difference is that the values of x and y that appear in the formula for r are replaced by their ranks. That is, the *ranks* of the raw data are used to compute r_s rather than the raw data themselves. When there are no (or few) ties in the ranks, this formula reduces to the simple expression given in equation 12.4:

$$r_s = 1 - \frac{6 \sum_{i=1}^{n} d_i^2}{n(n^2 - 1)}$$

where d_i is the difference between the rank of y and x for the ith observation.

Table 12.8 Data for Example 12.7

DECEPT.XLS

Student	Deception Score y	Percent Increase in Pupil Size x
1	87	10.5
2	63	6.2
3	95	1.1
4	50	7.4
5	43	0.8
6	89	15.2
7	33	4.4
8	55	5.9
9	80	9.3
10	59	7.5
11	45	8.8
12	72	10.0
13	23	0.5
14	91	6.5
15	85	13.4

Table 12.9 Calculation Table for Example 12.7

Student	Deception Score y	Rank	Percent Increase in Pupil Size x	Rank	d_i	d_i^2
1	87	12	10.5	13	-1	1
2	63	8	6.2	6	2	4
3	95	15	1.1	3	12	144
4	50	5	7.4	8	-3	9
5	43	3	0.8	2	1	1
6	89	13	15.2	15	-2	4
7	33	2	4.4	4	-2	4
8	55	6	5.9	5	1	1
9	80	10	9.3	11	-1	1
10	59	7	7.5	9	-2	4
11	45	4	8.8	10	-6	36
12	72	9	10.0	12	-3	9
13	23	1	0.5	1	0	0
14	91	14	6.5	7	7	49
15	85	11	13.4	14	-3	9
					$\sum d_i^2 = 276$	

The ranks of y and x, the differences between the ranks, and the squared differences for each of the 15 students are shown in Table 12.9. Note that the sum of the squared differences is $\sum d_i^2 = 276$. Substituting this value into equation 12.4, we obtain

$$r_s = 1 - \frac{6 \sum_{i=1}^{n} d_i^2}{n(n^2 - 1)} = 1 - \frac{6(276)}{15(224)} = .507$$

This positive value of r_s implies that a moderate positive linear correlation exists between deception score y and percent increase in pupil size x in the sample.

b. To determine whether a positive correlation exists in the population, we test

$$H_0: \rho = 0$$

$$H_a: \rho > 0$$

using r_s as a test statistic. As you would expect, we reject H_0 for large values of r_s. Upper-tailed critical values of Spearman's r_s are provided in Table B.8 of Appendix B. This table is partially reproduced in Table 12.10 (page 772). For $\alpha = .01$ and $n = 15$, the critical value (highlighted in Table 12.10) is $r_{.01} = .623$. Thus, the rejection region for the test is

$$\text{Reject } H_0 \text{ if } r_s > .623$$

Since the test statistic, $r_s = .507$, does not fall in the rejection region, there is insufficient evidence (at $\alpha = .01$) of positive linear correlation between deception score y and percent increase in pupil size x in the population. ❑

Table 12.10 A Portion of the Spearman's r_s Table, Table B.8

	$\alpha = .10$	$\alpha = .05$	$\alpha = .02$	$\alpha = .01$	← Two-Tailed
n	$\alpha = .05$	$\alpha = .025$	$\alpha = .01$	$\alpha = .005$	← One-Tailed
15	.441	.525	➤ .623	.689	
16	.425	.507	.601	.666	
17	.412	.490	.582	.645	
18	.399	.476	.564	.625	
19	.388	.462	.549	.608	
20	.377	.450	.534	.591	
21	.368	.438	.521	.576	
22	.359	.428	.508	.562	
23	.351	.418	.496	.549	
24	.343	.409	.485	.537	
25	.336	.400	.475	.526	

Self-Test

12.5

Data for a random sample of $n = 5$ (x, y) values are listed below:

x	.2	.4	.8	1.1	1.5
y	30	25	20	21	13

a. Rank the five x values.
b. Rank the five y values.
c. Compute the differences between the ranks.
d. Find Spearman's rank correlation coefficient r_s.

Spearman's rank correlation coefficient r_s has a sampling distribution that is approximately normal when the sample size n is large—say, $n > 30$. For large samples, the critical values of r_s are obtained using the formulas shown in the box. The test is then conducted in the usual manner.

Critical Values of r_s for Large Samples

$$\text{One-tailed test: } r_\alpha = \frac{z_\alpha}{\sqrt{n-1}} \qquad (12.5)$$

$$\text{Two-tailed test: } r_{\alpha/2} = \frac{z_{\alpha/2}}{\sqrt{n-1}} \qquad (12.6)$$

where z_α and $z_{\alpha/2}$ are obtained from the standard normal table, Table B.2 in Appendix B.

Example 12.8

Spearman's Large-Sample Test for Rank Correlation

For a large sample of $n = 40$ (x, y) pairs, Spearman's rank correlation is calculated to be $r_s = -.38$. Test the null hypothesis $H_0: \rho = 0$ against the alternative hypothesis $H_a: \rho \neq 0$ using $\alpha = .05$.

Solution

The test statistic is $r_s = -.38$ (as demonstrated in the previous example). For this two-tailed, large-sample test, we use equation 12.6 to find the critical value. Since $\alpha = .05, \alpha/2 = .025$, and $z_{.025} = 1.96$. Substituting these values into equation 12.6, we obtain

$$r_{\alpha/2} = r_{.025} = \frac{z_{.025}}{\sqrt{n-1}} = \frac{1.96}{\sqrt{40-1}} = .31385$$

Consequently, we reject H_0 if $|r_s| > r_{.025} = .31385$. Since $|r_s| = |-.38| = .38$ exceeds the critical value, we reject H_0 and conclude (at $\alpha = .05$) that the population rank correlation ρ differs from 0. ❑

Nonparametric tests are also available for the general multiple regression model. These tests are very sophisticated, however, and require the use of specialized statistical software. Consult the Chapter 12 references if you want to learn more about these nonparametric techniques.

PROBLEMS 12.48–12.58

Using the Tools

12.48   Specify the rejection region for Spearman's nonparametric test for rank correlation in each of the following situations:

a. $H_a: \rho > 0, n = 15, \alpha = .05$ **b.** $H_a: \rho < 0, n = 20, \alpha = .01$

c. $H_a: \rho \neq 0, n = 50, \alpha = .05$ **d.** $H_a: \rho \neq 0, n = 10, \alpha = .05$

e. $H_a: \rho > 0, n = 100, \alpha = .05$

Pair	x	y
1	65	59
2	57	61
3	55	58
4	38	23
5	29	34
6	43	38
7	49	37

Problem 12.49

12.49   A random sample of seven pairs of observations is recorded on two variables, x and y. The data are shown in the table at left.

a.   Rank the values of each variable, x and y. (Note that there are no tied ranks.)

b.   Compute the test statistic r_s.

c.   Do the data provide sufficient evidence to conclude that the ranked pairs are correlated? Test using $\alpha = .05$.

Pair	x	y
1	185	16
2	188	20
3	210	13
4	243	10
5	277	7

Problem 12.50

12.50   A random sample of five pairs of observations is recorded on two variables, x and y. The data are shown in the table at left.

a.   Rank the values of each variable, x and y. (Note that there are no tied ranks.)

b.   Compute the test statistic, r_s.

c.   Do the data provide sufficient evidence to conclude that the ranked pairs are correlated? Test using $\alpha = .05$.

Applying the Concepts

12.51   Universities receive gifts and donations from corporations, foundations, friends, and alumni. It has long been argued that universities rise or fall depending on the level of support of their alumni. The table on page 774 reports the total dollars raised during 1994–95 by a sample of major U.S. universities. In addition, it reports the percentage of that total donated by alumni.

FUNDRAISE.XLS

University	Total Funds Raised	Alumni Contribution
Harvard	$323,406,242	47.5%
Yale	199,646,606	54.6
Cornell	198,736,229	56.2
Wisconsin	164,349,458	17.4
Michigan	145,757,642	45.4
Pennsylvania	135,324,761	34.3
Illinois	116,578,975	36.6
Princeton	103,826,392	53.2
Brown	102,513,437	34.7
Northwestern	101,041,213	27.3

Source: The Chronicle of Higher Education, Sept. 2, 1996, p. 27.

a. Do these data indicate that total fundraising and alumni contributions are correlated? Test using Spearman's rank correlation at $\alpha = .05$.

b. What assumptions must hold to ensure the validity of your test?

12.52 The *Journal of Financial Planning* (Jan. 1993) offered advice on selecting stocks and bonds for the 1990s. One of the many exhibits presented in the article involved a comparison of 5-year net returns and expense ratios (operating expenses as a percentage of net assets) for 18 insured AAA-rated municipal bond funds. The 18 funds were ranked according to both of these variables; the rankings are displayed in the accompanying table.

AAABONDS.XLS

Municipal Bond Fund	Expense Ratio Rank	Net Return Rank	Municipal Bond Fund	Expense Ratio Rank	Net Return Rank
1	1	1	10	10	18
2	2	6	11	11	15
3	3	3	12	12	17
4	4	10	13	13	2
5	5	11	14	14	7
6	6	9	15	15	14
7	7	8	16	16	13
8	8	5	17	17	16
9	9	4	18	18	12

Source: Bogle, J. C. "A crystal ball look at U.S. markets in the 1990s." *Journal of Financial Planning,* Jan. 1993, p. 19 (Exhibit 15).

a. Give a reason why nonparametric statistics should be used to analyze the data.

b. Use Spearman's rank correlation method to analyze the data. Interpret the results.

12.53 Amorphous alloys have been found to have superior corrosion resistance. *Corrosion Science* (Sept. 1993) reported on the resistivity of an amorphous iron-boron-silicon alloy after crystallization. Five alloy specimens were annealed at 700°C, each for a different length of time. The passivation potential—a measure of resistivity of the crystallized alloy—was then measured for each specimen. The experimental data are shown on page 775.

ALLOY.XLS

Annealing Time x (minutes)	Passivation Potential y (mV)
10	-408
20	-400
45	-392
90	-379
120	-385

Source: Chattoraj, I., et al. "Polarization and resistivity measurements of post-crystallization changes in amorphous Fe–B–Si alloys." *Corrosion Science,* Vol. 49, No. 9, Sept. 1993, p. 712 (Table 1).

> **a.** Calculate Spearman's correlation coefficient between annealing time x and passivation potential y. Interpret the result.
>
> **b.** Use the result, part **a**, to test for a significant correlation between annealing time and passivation potential. Use $\alpha = .05$.

12.54 Refer to the *Journal of the American Mosquito Control Association* (Mar. 1995) study of the relationship between temperature and mosquito catch ratio, Problem 10.9 (page 567). The data are reproduced in the accompanying table. Use Spearman's method to test for a positive rank correlation between the two variables. Use $\alpha = .05$.

MOSQUITO.XLS

Date	Average Temperature	Catch Ratio
July 24	16.8	.66
25	15.0	.30
26	16.5	.46
27	17.7	.44
28	20.6	.67
29	22.6	.99
30	23.3	.75
31	18.2	.24
Aug. 1	18.6	.51

Source: Petric, D., et al. "Dependence of CO_2-baited suction trap captures on temperature variations." *Journal of the Mosquito Control Association,* Vol. 11, No. 1, Mar. 1995, p. 8.

12.55 Refer to the study of the variables related to the number of wins by a professional basketball (NBA) team, Problem 10.11 (page 569). The data (measured using z scores) are reproduced in the table on page 776. The table also gives the value of Spearman's rank correlation coefficient relating standardized wins y to each of the three independent variables.

> **a.** Interpret each of the three values of r_s.
>
> **b.** Is there evidence of rank correlation between standardized wins y and standardized 2-point field goals x_1? Test using $\alpha = .01$.
>
> **c.** Is there evidence of rank correlation between standardized wins y and standardized rebounds x_2? Test using $\alpha = .01$.
>
> **d.** Is there evidence of rank correlation between standardized wins y and standardized turnovers x_3? Test using $\alpha = .01$.

12.56 Refer to the *Proceedings of the Institute of Civil Engineers* (Mar. 1990) study of the relationship between shear strength of masonry joints and precompression stress, Problem 10.12 (page 570). The stress results for seven brick triplets are reproduced in the table on page 776. Conduct a

NBA.XLS

Team	Wins y	Two-Point Field Goals x_1	Rebounds x_2	Turnovers x_3
Chicago	2.21	0.77	2.14	2.02
Seattle	1.64	1.70	0.29	0.67
Orlando	1.35	1.28	−0.34	0.69
San Antonio	1.28	1.02	−0.23	0.54
Utah	1.00	1.49	1.11	1.61
LA Lakers	0.85	1.10	−0.65	1.50
Indiana	0.78	0.73	0.86	−0.67
Houston	0.50	0.50	−0.99	−0.18
New York	0.43	0.81	−0.56	0.18
Cleveland	0.43	0.18	−0.47	1.84
Detroit	0.36	0.29	0.36	−0.54
Atlanta	0.36	−0.89	−0.06	1.56
Portland	0.21	0.58	1.93	−1.63
Miami	0.07	0.92	0.78	−0.93
Phoenix	0.00	−0.05	0.49	−0.14
Charlotte	0.00	−0.60	−0.27	−0.23
Washington	−0.14	0.54	−0.86	0.31
Sacramento	−0.14	−0.29	0.36	−0.76
Golden State	−0.36	−0.90	0.21	0.28
Denver	−0.43	−0.23	0.89	−1.19
Boston	−0.57	−0.93	−0.60	0.11
New Jersey	−0.78	−1.14	1.86	−0.93
LA Clippers	−0.85	−0.10	−0.90	0.02
Dallas	−1.07	−2.26	−0.10	0.69
Minnesota	−1.07	−0.44	−0.38	−0.73
Milwaukee	−1.14	−0.31	−0.66	−0.73
Toronto	−1.42	−0.44	−0.35	−1.92
Philadelphia	−1.64	−1.66	−1.78	−1.11
Vancouver	−1.85	−1.65	−2.07	0.67

	r_s
Two-point field goals x_1	.8399
Rebounds x_2	.3798
Turnovers x_3	.5218

Problem 12.55

Problem 12.55

nonparametric test for positive rank correlation between the variables. Test using $\alpha = .05$.

TRIPLETS.XLS

	TRIPLET TEST						
	1	2	3	4	5	6	7
Shear Strength y	1.00	2.18	2.24	2.41	2.59	2.82	3.06
Precompression Stress x	0	.60	1.20	1.33	1.43	1.75	1.75

Source: Riddington, J. R., and Ghazali, M. Z. "Hypothesis for shear failure in masonry joints." *Proceeding of the Institute of Civil Engineers,* Part 2, Mar. 1990, Vol. 89, p. 96 (Figure 7).

12.57   Refer to the *Wetlands* (Mar. 1995) study of the relationship between the accuracy of a land survey determined by a Global Positioning System and the percentage of readings obtained when three satellites are used, Problem 10.44 (page 590). The data are reproduced in the table on page 777. Spearman's rank correlation relating distance error y to percent of readings with three satellites x is $r_S = .5956$. Give a full interpretation of this result (using $\alpha = .10$).

12.58   The *Journal of Archaeological Science* (Oct. 1997) reported on two indexes developed for classifying whale bones. The architectural utility index (AUI) is based on bone length, shape, and weight. The meat utility

GPS.XLS

Survey Location	Distance Error y	Percent of Readings with Three Satellites x
1	7.8	94
2	4.2	83
3	4.5	23
4	2.8	58
5	1.5	78
6	3.0	100
7	1.4	100
8	14.9	100
9	0.4	28
10	18.9	100
11	5.8	75
12	0.6	2
13	30.9	100
14	30.3	100
15	31.4	100
16	7.5	53

Source: Hook, D. D., et al. "Locating delineated wetland boundaries in coastal South Carolina using Global Positioning Systems." *Wetlands,* Vol. 15, No. 1, Mar. 1995, p. 34 (Table 2).

Problem 12.57

index (MUI) is based on estimated flesh weight per skeletal portion. Thirteen different skeletal body parts of a prehistoric bowhead whale were ranked from low (1) to high (9) according to the AUI. In addition, the MUI (measured as a percentage) was determined for each part. The data are listed below.

WHALES.XLS

Whale Part	AUI Ranking	MUI (%)
Cranium	9	3.8
Maxillae/premaxillae	8	7.55
Mandible	9	7.55
Hyoid	2	88.8
Cervical vert.	8	4.3
Thoracic vert.	6	49.2
Lumbar vert.	7	100.0
Caudal vert.	3	91.2
Rib	7	39.7
Sternum	1	2.1
Scapula	6	4.8
Humerus	5	1.9
Radius/ulna	4	3.9

Source: Savelle, J. M. "The role of architectural utility in the formation of zooarchaeological whale bone assemblages." *Journal of Archaeological Science,* Vol. 24, No. 10, Oct. 1997, pp. 872–873 (Tables 2 and 3).

a. Find the rank correlation between AUI rank and MUI percent. Interpret the result.

b. Is there evidence (at $\alpha = .10$) of a significant rank correlation between AUI and MUI?

c. The research archaeologist also ranked the whale body parts according to how they were distributed among whaling crews. (The portion of the whale retained by the boat's captain—the edible material—is rated the highest, while the portion distributed to others who did not participate in the kill—the skin or blubber—is rated the lowest.) The rank correlation between MUI and whaling crew distribution rating is $r_s = .9118$ for the 13 whole body parts. Interpret this value.

REAL-WORLD REVISITED
Do Women Really Understand the Benefit of a Mammography?

Each year approximately 45,000 women in the United States die from breast cancer. Studies have shown that women can dramatically reduce their chances of dying of the disease with a regular screening mammography. The National Institutes of Health (NIH) recommends that women aged 50 years or older be screened for breast cancer once a year. Recently, an NIH Consensus Panel decided not to make the same recommendation for women aged 40 to 49 years, but to advocate that these women make their own decisions about screening based on their personal evaluation of the risk of breast cancer.

The NIH decision is based on the assumption that patients understand readily available quantitative information on the risks of breast cancer and benefits of a mammography. Do they? This question was posed by medical researchers Lisa Schwartz, M.D., Steven Woloshin, M.D., Gilbert Welch, M.D. (all of the Department of Veterans Affairs Outcomes Group), and William Black, M.D. (of Dartmouth Medical School). The researchers hypothesized that this quantitative information may only be meaningful to patients who have some understanding of basic probability and numerical concepts. To investigate this issue, they conducted a study of women's comprehension of quantitative-based messages about mammography and published their results in the Dec. 1, 1997, issue of the *Annals of Internal Medicine.**

*Schwartz, L. M., Woloshin, S., Black, W. C., and Welch, H. G. "The role of numeracy in understanding the benefit of screening mammography." *Annals of Internal Medicine,* Vol. 127, Dec. 1, 1997, pp. 996–972.

A random sample of 500 women was selected from a registry of female veterans kept at a VA Medical Center in Vermont. Each veteran was mailed one of four questionnaires that differed only in how the same information on average risk reduction with mammography was presented (see Table 12.11 on page 779). A total of 230 complete survey responses were returned.

To assess the respondents' ability to understand basic probability and numerical concepts—called *numeracy*—the following three questions were asked:

1. Imagine that we flip a fair coin 1,000 times. What is your best guess about how many times the coin would come up heads in 1,000 flips?

2. In the BIG BUCKS LOTTERY, the chance of winning a $10 prize is 1%. What is your best guess about how many people would win a $10 prize if 1,000 people each buy a single ticket to BIG BUCKS?

3. In ACME PUBLISHING SWEEPSTAKES, the chance of winning a car is 1 in 1,000. What percent of tickets to ACME PUBLISHING SWEEPSTAKES win a car?

The number of correct answers (0, 1, 2, or 3) was then recorded as the numeracy score for each respondent.

Table 12.11 Information on Risk of Dying from Breast Cancer
Provided on Four Questionnaires

Survey Group	Information Provided
1) Relative Risk Reduction with Baseline Risk	33% risk reduction from 12 in 1,000 in the next 10 years
2) Relative Risk Reduction without Baseline Risk	33% risk reduction in the next 10 years
3) Absolute Risk Reduction with Baseline Risk	4 in 1,000 risk reduction from 12 in 1,000 in the next 10 years
4) Absolute Risk Reduction without Baseline Risk	4 in 1,000 risk reduction in the next 10 years

To assess the female veterans' perceived risk of death from breast cancer, the following two questions were asked:

QA. Imagine 1,000 women just like you. How many will die from breast cancer without mammography?

QB. Imagine 1,000 women just like you. How many will die from breast cancer with mammography?

The difference between the responses to QA and QB—called *perceived absolute risk reduction*—enabled the researchers to determine how well the respondents applied the information on breast cancer given in the survey questionnaire.

MAMMO.XLS

The data for the 230 females who returned fully completed questionnaires are stored in the Excel workbook MAMMO.XLS. Three variables are included in the data set: GROUP (1, 2, 3, or 4), NUMERACY score (0, 1, 2, or 3), and RISKRED (i.e., perceived absolute risk reduction).

Study Questions

a. One of the objectives of the research was to compare the numeracy scores across the four survey groups. Use Excel to rank the numeracy scores. Then apply the appropriate nonparametric test to the data. Interpret the results.

b. Another objective was to compare the perceived absolute risk reduction values across the four survey groups. Use Excel to rank the risk reduction values. Then conduct the appropriate nonparametric test and interpret the results.

c. Explain why a nonparametric test is more appropriate for analyzing the data than a parametric test.

d. The researchers also wanted to investigate whether numeracy score impacts the respondents' ability to assess risk reduction. For each survey group, the rank correlation between numeracy score and perceived absolute risk reduction is shown at left. Which groups (if any) show a significant correlation between the two variables?

e. Discuss the importance of the results, parts **a**, **b**, and **d**, in making recommendations about the type of quantitative information provided to women about the risks of death from breast cancer and benefits of screening mammography.

Group	Sample Size	Rank Correlation
1	59	−.1899
2	52	−.3904
3	58	−.2576
4	61	−.1956

Key Terms

Distribution-free tests 737	Rank sum 745	Wilcoxon rank sum test 745
Kruskal-Wallis H test 762	Sign test 740	Wilcoxon signed ranks test 754
Median 738	Spearman's rank correlation	
Nonparametrics 737	coefficient 769	
Rank statistics (tests) 738	Test for location 738	

Key Formulas

Test	Test Statistic	Large Sample Approximation
Sign	S = number of sample measure- 738 ments greater than (or less than) hypothesized median, η_0	—
Wilcoxon rank sum	T_1 = rank sum of sample 1 or 745 T_2 = rank sum of sample 2	$z = \dfrac{T_1 - \dfrac{n_1(n_1 + n_2 + 1)}{2}}{\sqrt{\dfrac{n_1 n_2(n_1 + n_2 + 1)}{12}}}$ **(12.1)**, 748
Wilcoxon signed ranks	T^- = negative rank sum or 755 T^+ = positive rank sum	$z = \dfrac{T^+ - \dfrac{n(n + 1)}{4}}{\sqrt{\dfrac{n(n + 1)(2n + 1)}{24}}}$ **(12.2)**, 757
Kruskal-Wallis	$H = \dfrac{12}{n(n + 1)} \displaystyle\sum_{i=1}^{k} \dfrac{T_i^2}{n_i} - 3(n + 1)$ **(12.3)**, 762	—
Spearman rank correlation (shortcut formula)	$r_s = 1 - \dfrac{6 \sum d_i^2}{n(n^2 - 1)}$ **(12.4)**, 769 where d_i = difference in ranks of ith observation for samples 1 and 2	$r_\alpha = \dfrac{z_\alpha}{\sqrt{n - 1}}$ **(12.5)**, 772 $r_{\alpha/2} = \dfrac{z_{\alpha/2}}{\sqrt{n - 1}}$ **(12.6)**, 772

Key Symbols

Symbol	Description
η	Population median
S	Test statistic for sign test
T_1	Sum of ranks of observations in sample 1
T_2	Sum of ranks of observations in sample 2
T_L	Critical lower Wilcoxon rank sum value
T_U	Critical upper Wilcoxon rank sum value
T^+	Sum of ranks of positive differences of paired observations
T^-	Sum of ranks of negative differences of paired observations
T_0	Critical value of Wilcoxon signed ranks test
T_i	Rank sum of observations in sample i

H Test statistic for Kruskal-Wallis H test
r_s Spearman's rank correlation coefficient
ρ Population correlation coefficient

Supplementary Problems 12.59–12.70

BIOFEED.XLS

Subject	Before	After
1	136.9	130.2
2	201.4	180.7
3	166.8	149.6
4	150.0	153.2
5	173.2	162.6
6	169.3	160.1

Problem 12.59

12.59 The blood pressure data for the biofeedback experiment, Problem 9.26 (page 499), are reproduced at left. The "before" and "after" measurements are blood pressure readings (in millimeters) of human subjects before and after biofeedback training, respectively. Conduct a nonparametric test to determine whether biofeedback training reduces blood pressure in humans. Test using $\alpha = .05$.

12.60 An educational psychologist claims that the order in which test questions are asked affects a student's ability to answer correctly. To investigate this assertion, a professor randomly divides a class of 13 students into two groups—seven in one group and six in the other. The professor prepares one set of test questions but arranges the questions in two different orders. On test A, the questions are arranged in order of increasing difficulty (that is, from easiest to most difficult), whereas on test B the order is reversed. One group of students is given test A, the other test B, and the test score is recorded for each student. The results are as follows:

Test A:	90	71	83	82	75	91	65
Test B:	66	78	50	68	80	60	

Do the data provide sufficient evidence to indicate a difference between the median scores of the two tests? Test using $\alpha = .05$.

TESTAB.XLS

12.61 *Scram* is the term used by nuclear engineers to describe a rapid emergency shutdown of a nuclear reactor. The nuclear industry has made a concerted effort to significantly reduce the number of unplanned scrams each year. The number of unplanned scrams at each of a random sample of 20 nuclear reactor units in a recent year is given below. Test the hypothesis that the median number of unplanned scrams that occur each year at nuclear reactor plants is less than 5. Use $\alpha = .10$.

1	8	0	3	3	9	1	2	4	5	4	3
1	2	7	10	2	6	3	0				

SCRAM.XLS

12.62 Refer to the *Chemosphere* (Vol. 20, 1990) study on the TCDD levels of 20 Massachusetts Vietnam veterans who were possibly exposed to Agent Orange, Problem 3.56 (page 180). The amounts of TCDD (measured in parts per trillion) in blood plasma and fat tissue drawn from each veteran are shown on page 782. Is there sufficient evidence of a median difference between the TCDD levels in plasma and fat tissue for Vietnam veterans exposed to Agent Orange? Test using $\alpha = .05$.

12.63 Many successful corporations issue bonds to investors to raise capital for expansion. The sale of the bonds is usually handled by a bond underwriter at an investment bank. The table on page 782 gives the changes in bond prices (in dollars) over a 12-month period for independent random

TCDDFAT.XLS

Veteran	TCDD Levels in Plasma	TCDD Levels in Fat Tissue	Veteran	TCDD Levels in Plasma	TCDD Levels in Fat Tissue
1	2.5	4.9	11	6.9	7.0
2	3.1	5.9	12	3.3	2.9
3	2.1	4.4	13	4.6	4.6
4	3.5	6.9	14	1.6	1.4
5	3.1	7.0	15	7.2	7.7
6	1.8	4.2	16	1.8	1.1
7	6.0	10.0	17	20.0	11.0
8	3.0	5.5	18	2.0	2.5
9	36.0	41.0	19	2.5	2.3
10	4.7	4.4	20	4.1	2.5

Source: Schecter, A., et al. "Partitioning of 2,3,7,8-chlorinated dibenzo-p-dioxins and dibenzofurans between adipose tissue and plasma lipid of 20 Massachusetts Vietnam veterans." *Chemosphere,* Vol. 20, Nos. 7–9, 1990, pp. 954–955 (Tables I and II).

Problem 12.62

samples of bonds underwritten by each of four firms—Morgan Stanley, First Boston, Goldman Sachs, and Merrill Lynch. (Six bonds were randomly selected from each firm.)

BONDPRICE.XLS

Morgan Stanley	First Boston	Goldman Sachs	Merrill Lynch
.037	−.128	.025	−.047
−.016	−.054	−.080	.010
−.132	.007	−.031	−.003
−.148	−.011	.049	−.104
.022	.031	−.019	−.082
−.049	−.042	−.027	−.039

a. Give the null and alternative hypotheses appropriate for comparing the median change in bond prices for the four underwriters.

b. Calculate the nonparametric test statistic.

c. Make the appropriate conclusion at $\alpha = .01$.

12.64 In the early 1960s, the air-conditioning systems of a fleet of Boeing 720 jet airplanes came under investigation. The accompanying table presents the lifelengths (in hours) of the air-conditioning systems in two different Boeing 720 planes. Assuming the data represent random samples from the respective populations, is there evidence of a difference between the median lifelengths of the air-conditioning systems for the two Boeing 720 planes? Test using $\alpha = .05$.

BOEING720.XLS

Plane 1			Plane 2		
23	156	76	59	66	67
118	49	62	32	230	34
90	10		14	54	
29	310		102	152	

Source: Hollander, M., Park, D. H., and Proschan, F. "Testing whether *F* is 'more NBU' than is *G*." *Microelectronics and Reliability,* Vol. 26, No. 1, 1986, p. 43, Table 1, Pergamon Press, Ltd.

12.65   Refer to the Real-World Revisited problem in Chapter 7 (page 401). The data on scallop weights (measured as a multiple of $\frac{1}{36}$ of a pound) for a sample of 18 bags of scallops are reproduced below. Recall that the harbormaster wants to know if the "average" weight per bag of the ship's catch of 11,000 bags of scallops is less than $\frac{1}{36}$ of a pound. Apply the appropriate nonparametric test to the data. (Test using $\alpha = .05$.) What do you conclude?

SCALLOPS.XLS

.93	.88	.85	.91	.91	.84	.90	.98	.88
.89	.98	.87	.91	.92	.99	1.14	1.06	.93

Source: Bennett, A. "Misapplications review: Jail terms." *Interfaces,* Vol. 25, No. 2, March–April 1995, p. 20.

12.66   Housing planners, insurance companies, and credit institutions have shown great interest in measuring factors related to the risk of default in home mortgages. Typically, the default rate (i.e., the probability of defaulting on a home loan) is used as the standard measure of risk. Another, possibly superior, measure of risk is the expected loss for a home loan (i.e., the dollar loss expected on a default). A study was conducted to compare the two measures of mortgage default risk. The default rate and expected loss (measured as a percentage of loan value) for a sample of 16 different FHA categories of home mortgages are recorded in the accompanying table.

FHALOAN.XLS

Loan Type	Default Rate (%)	Expected Loss (%)
1	3.44	.948
2	5.20	1.181
3	12.50	3.178
4	16.95	2.696
5	3.89	1.256
6	4.85	1.578
7	13.40	6.453
8	14.67	4.025
9	2.15	.558
10	2.66	.630
11	5.60	1.448
12	7.95	2.304
13	2.41	.674
14	2.60	.897
15	10.27	4.836
16	5.66	1.579

Source: Evans, R. D., Maris, B. A., and Weinstein, R. I. "Expected loss and mortgage default risk." *Quarterly Journal of Business and Economics,* Vol. 24, No. 1, Winter 1985, pp. 75–92.

a.   Calculate Spearman's rank correlation coefficient between the two measures of mortgage default risk.

b.   Is there sufficient evidence to indicate that default rate and expected loss are positively correlated? Test using $\alpha = .01$.

12.67   A study was conducted to obtain information on the background levels of the toxic substance polychlorinated biphenyl (PCB) in soil samples in the United Kingdom (*Chemosphere,* Feb. 1986). The accompanying table contains the measured PCB levels of soil samples taken at 14 rural and 15 urban locations in the United Kingdom (PCB concentration is measured in .0001 gram per kilogram of soil). From these preliminary results, the researchers reported "a significant difference between (the PCB levels) for rural areas ... and for urban areas." Do the data support the researchers' conclusions? Test using $\alpha = .05$.

PCBSOIL.XLS

Rural			Urban		
3.5	23.0	12.0	24.0	94.0	22.0
8.1	1.5	8.2	29.0	141.0	13.0
1.8	5.3	9.7	16.0	11.0	18.0
9.0	9.8	1.0	21.0	11.0	12.0
1.6	15.0		107.0	49.0	18.0

Source: Badsha, K., and Eduljee, G. "PCB in the U.K. environment—a preliminary survey." *Chemosphere,* Vol. 15, No. 2, Feb. 1986, p. 213, Table 1, Pergamon Press, Ltd.

12.68   One method of lowering drilling costs is to increase drilling speed. Researchers at Cities Service Co. have developed a drill bit, called the PD-1, that they believe penetrates rock at a faster rate than any other bit on the market. It is decided to compare the speed of the PD-1 with the two fastest drill bits known, the IADC 1-2-6 and the IADC 5-1-7, at 15 drilling locations in Texas. Five drilling sites were randomly assigned to each bit, and the rate of penetration in feet per hour was recorded after drilling 3,000 feet at each site. The data are given below. Based on this information, can Cities Service Co. conclude that the median rate of penetration for the three drill bits differs? Test at the $\alpha = .05$ level of significance.

DRILLBIT.XLS

PD-1	IADC 1-2-6	IADC 5-1-7
35.2	25.8	14.7
30.1	29.7	28.9
37.6	26.6	23.3
34.3	30.1	16.2
33.2	27.4	20.6

12.69   A political scientist wants to determine if a voter's image of a particular liberal political candidate is correlated with the age of the voter. Each of 12 randomly selected voters rated the candidate on a scale of 1 to 20 (the higher the rating, the more favorable the candidate). The data are presented in the table below.

VOTERS.XLS

Voter	Rating	Age of Voter
1	9	56
2	18	30
3	3	47
4	8	50
5	15	22
6	4	61
7	12	49
8	7	56
9	5	43
10	19	35
11	17	28
12	12	41

a. Calculate Spearman's rank correlation coefficient r_s. Interpret its value in the context of this problem.

b. Does it appear that a voter's image of the liberal political candidate is correlated with the voter's age? Test using $\alpha = .01$.

TBAD.XLS

12.70 The case management practices of a group of HMO physicians, the Tampa Bay Area Doctors (TBAD), were investigated in *Medical Interfaces* (Oct. 1992). A case manager authorizes any out-of-ordinary testing, referral to physicians other than the primary care physician, and referral to a hospital for admission. The question of interest to the researchers is whether some TBAD case managers are more cost-effective than others. The Excel workbook TBAD.XLS contains the following variables measured on 186 TBAD physicians: (1) secondary specialty (yes or no); (2) country of medical school (USA or foreign); (3) years of experience; and (4) total accrued cost (per patient, per month).

a. Use Excel and a nonparametric test to determine whether primary care TBAD physicians (i.e., those with no secondary specialty) are more cost-effective than physicians with a secondary specialty.

b. Use Excel and a nonparametric test to determine whether foreign medical school graduates are less cost-effective than domestic medical school graduates.

c. Spearman's rank correlation between experience and total accrued cost is $r_s = -.119$. Fully interpret this result.

Using Microsoft Excel

12.E.1 Using Microsoft Excel for the Sign Test

Solution Summary:

Implement a worksheet that uses the COUNTIF and BINOMDIST functions to perform a sign test.

 Example: Water pollution data of Example 12.2 on page 739

Solution:

To perform the sign test, implement the two-worksheet design of Tables 12.E.1 and 12.E.2. This design uses the COUNTIF function to count the number of positive differences and the BINOMDIST function to compute binomial probabilities. The format of these functions is

$$COUNTIF(\textit{cell range, expression})$$

where
    cell range is the cell range of the values to be possibly counted and
    expression is the algebraic expression that determines which value(s) in the *cell range* to count.

$$BINOMDIST(X, n, p, \textit{cumulative})$$

where
    X is the number of successes
    n is the sample size
    p is the probability of success in the binomial distribution
    cumulative is a True or False value. If true, the function returns the probability of X or fewer successes; if false, the probability of exactly X successes.

SIGN.XLS

As an example, to use the sign test to determine whether there is evidence that the median pollution level of water discharged from the plant is greater than 5 ppm, implement rows 1 through 20 of the Table 12.E.1 design and rows 1 through 11 of the Table 12.E.2 design. To implement these rows, do the following:

1.   Select File | Open to open a new workbook (or open the existing workbook into which the Sign worksheet is to be inserted).
2.   Select an unused worksheet (or select Insert | Worksheet if there are none) and rename the sheet Sign.
3.   Enter the title, labels, and formulas for column A as shown in Table 12.E.1. Enter the formulas in cells A12, A16, and A20, which have been typeset as two lines, as one continuous line.
4.   Enter the hypothesized median in cell B3 and the level of significance in cell B7.

The resulting worksheet is shown in Table 12.E.1.

Table 12.E.1 Worksheet Design for the Sign Test

	A	B
1	Sign Test	
2		
3	Hypothesized Median	5
4	Sample Size	=COUNT(Data!B:B)
5	Number greater than Median (S1)	=COUNTIF(Data!B:B, ">0")
6	Number less than or equal to Median (S2)	=B4-B5
7	Level of Significance (α)	0.05
8		
9	Two-Tailed Test	
10	Maximum of S1 and S2	=MAX(B5:B6)
11	p-value	=IF(B5-B6, 1, 2*(1-BINOMDIST((B10-1), B4, 0.5, TRUE)))
12	=IF(B11<B7, "Reject the null hypothesis", "Do not reject the null hypothesis")	
13		
14	Lower Tail Test	
15	Probability greater than or equal to S2	=1-BINOMDIST((B6-1), B4, 0.5, TRUE)
16	=IF(B15<B7, "Reject the null hypothesis", "Do not reject the null hypothesis")	
17		
18	Upper Tail Test	
19	Probability greater than or equal to S1	=1-BINOMDIST((B5-1), B4, 0.5, TRUE)
20	=IF(B19<B7, "Reject the null hypothesis", "Do not reject the null hypothesis")	

Once Table 12.E.1 has been implemented, do the following:

5 Select an unused worksheet (or select Insert | Worksheet if there are none) and rename the sheet Data.

6 Copy the data to be analyzed to column A of the Data sheet.

7 Enter the formula =A2-Sign!$B$3 to cell B2, and copy this down through the last row of data.

The resulting worksheet is shown in Table 12.E.2.

Table 12.E.2 Design for the Data Worksheet for the Sign Test

	A	B
1	PPM	Difference
2	5.1	=A2-Sign!$B$3
3	4.3	=A3-Sign!$B$3
4	5.3	=A4-Sign!$B$3
5	6.2	=A5-Sign!$B$3
6	5.6	=A6-Sign!$B$3
7	4.7	=A7-Sign!$B$3
8	8.4	=A8-Sign!$B$3
9	5.9	=A9-Sign!$B$3
10	6.8	=A10-Sign!$B$3
11	3	=A11-Sign!$B$3

The results obtained will be similar to those of Figure 12.4 on page 740.

12.E.2 Using Microsoft Excel to Perform the Wilcoxon Rank Sum Test for Differences in Two Medians

Solution Summary:

Use the PHStat add-in to perform the Wilcoxon rank sum test for differences in two medians.

 Example: Visual acuity data of Example 12.3 on page 746

Solution:

To perform the Wilcoxon rank sum test for differences in two medians, use the Two-Sample Tests | Wilcoxon Rank Sum Test choice of the PHStat add-in. As an example, consider the visual acuity data of Example 12.3. To determine whether or not there is evidence of a significant difference in the eye movement rates of deaf and hearing children, do the following:

1. If the PHStat add-in has not been previously loaded, load the add-in using the instructions of Section EP.3.2.
2. Open the DEAF.XLS workbook and click the Data sheet tab. Verify that the data for the deaf children and the hearing children have been entered into columns A and B, respectively.
3. Select PHStat | Two-Sample Tests | Wilcoxon Rank Sum Test.
4. In the Wilcoxon Rank Sum Test dialog box (see Figure 12.E.1):
   - a. Enter 0.05 in the Level of Significance: edit box.
   - b. Enter A1:A11 in the Population 1 Sample Cell Range: edit box.
   - c. Enter B1:B11 in the Population 2 Sample Cell Range: edit box.
   - d. Select the First cells in both ranges contain label check box.

e Select the Upper-Tail Test option button.

f Enter Visual Acuity Comparison in the Output Title: edit box.

g Click the OK button.

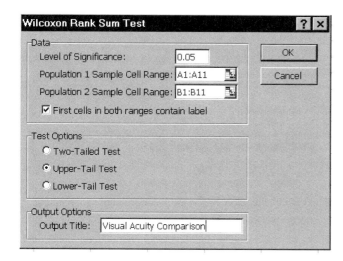

E Figure 12.E.1
PHStat Wilcoxon Rank
Sum Test Dialog Box

The PHStat add-in inserts a worksheet containing calculations for the Wilcoxon test similar to the one shown in Figure 12.5 on page 748.

Using Stacked Data

The PHStat Wilcoxon choice requires that the data for each group be in separate columns, an arrangement known as unstacked data. Some data sets are "stacked," in which the data for both groups appear in a single column, identified by a grouping variable in another column. To use such stacked data with the PHStat Wilcoxon choice, select the Data Preparation | Unstack Data choice of the PHStat add-in to unstack the data as part of Step 2 of the procedure listed above.

12.E.3 Using Microsoft Excel to Perform the Kruskal-Wallis *H* Test for Differences in *c* Medians

Solution Summary:

Use the PHStat add-in to perform the Kruskal-Wallis *H* test, or rank test, for differences in *c* medians. The add-in assumes that the sample sizes are the same for all groups.

 Grade point average analysis for the data of Example 12.6 on page 763

Solution:

To perform the Kruskal-Wallis rank test for differences in *c* medians, use the c-Sample Tests | Kruskal-Wallis Rank Test choice of the PHStat add-in. As an example, consider the grade point average data of Example 12.6. To determine whether or not there is evidence of a significant difference in the median grade point average for three different socioeconomic backgrounds, do the following:

1 If the PHStat add-in has not been previously loaded, load the add-in using the instructions of Section EP.3.2.

2 Open the GPACLASS.XLS workbook and click the Data sheet tab. Verify that the grade point averages for the three socioeconomic classes of Table 12.6 on page 763 have been entered into columns A through C.

3 Select PHStat | c-Sample Tests | Kruskal-Wallis Rank Test.

4 In the Kruskal-Wallis Rank Test dialog box (see Figure 12.E.2):

  a Enter .05 in the Level of Significance: edit box.

  b Enter A1:C8 in the Sample Data Cell Range: edit box.

  c Select the First cells contain label check box.

  d Enter GPA Comparison in the Output Title: edit box.

  e Click the OK button.

E **Figure 12.E.2**
PHStat Kruskal-Wallis
Rank Test Dialog Box

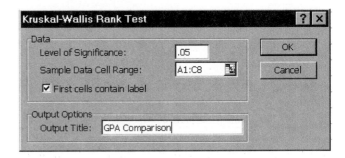

The PHStat add-in inserts a worksheet containing calculations for the Kruskal-Wallis test similar to the one shown in Figure 12.8 on page 765.

Using Stacked Data

The PHStat Kruskal-Wallis choice requires that the data for each group be in separate columns, an arrangement known as unstacked data. Some data sets are "stacked," in which the data for both groups appear in a single column, identified by a grouping variable in another column. To use such stacked data with the PHStat Kruskal-Wallis choice, select the Data Preparation | Unstack Data choice of the PHStat add-in to unstack the data as part of Step 2 of the procedure listed above.

Appendix A

Review of Arithmetic and Algebra

In writing this text, we realize that there are wide differences in the mathematical background of students taking a basic statistics course. Some students may have taken various courses in calculus and matrix algebra, while other students may not have taken any mathematics courses since high school. Since the emphasis in this text is on the concepts of statistical methods and the interpretation of output from Microsoft Excel, no mathematical prerequisite beyond elementary algebra is needed, since no formal mathematical proofs are derived. However, a proper foundation in basic arithmetic and algebraic skills will enable the student to focus on understanding the concepts of statistics rather than the mechanics of computing results.

With this goal in mind, we should make it clear that the object of this appendix is to review arithmetic and algebra for those students whose basic skills have become rusty over a period of time. If students need to actually learn arithmetic and algebra, then this review will probably not provide enough depth.

In order to assess their arithmetic and algebraic skills, and increase their confidence, we suggest that students take the following quiz before studying the text.

Part I: Fill in the correct answer.

1. $\dfrac{1}{\frac{2}{3}} =$

2. $.4^2 =$

3. $1 + \frac{2}{3} =$

4. $\left(\frac{1}{3}\right)^4 =$

5. $1/5 =$ (in decimals)

6. $1 - (-.3) =$

7. $4 \times .2 \times (-8) =$

8. $\left(\frac{1}{4} \times \frac{2}{3}\right) =$

9. $(1/100) + (1/200) =$

10. $\sqrt{16} =$

Part II: Select the correct answer.

1. If $a = bc$, then $c =$
   a. ab
   b. b/a
   c. a/b
   d. None of the above

2. If $x + y = z$, then $y =$
   a. z/x
   b. $z + x$
   c. $z - x$
   d. None of the above

3. $(x^3)(x^2) =$
   a. x^5
   b. x^6
   c. x^1
   d. None of the above

4. $x^0 =$
   a. x
   b. 1
   c. 0
   d. None of the above

5. $x(y - z) =$

   a. $xy - xz$

   b. $xy - z$

   c. $(y - z)/x$

   d. None of the above

6. $(x + y)/z =$

   a. $(x/z) + y$

   b. $(x/z) + (y/z)$

   c. $x + (y/z)$

   d. None of the above

7. $x/(y + z) =$

   a. $(x/y) + (1/z)$

   b. $(x/y) + (x/z)$

   c. $(y + z)/x$

   d. None of the above

8. If $x = 10, y = 5, z = 2$, and $w = 20$, then $(xy - z^2)/w =$

   a. 5

   b. 2.3

   c. 46

   d. None of the above

9. $(8x^4)/(4x^2) =$

   a. $2x^2$

   b. 2

   c. $2x$

   d. None of the above

10. $\sqrt{x/y}$

   a. $\sqrt{y}/\sqrt{x}$

   b. $\sqrt{1}/\sqrt{xy}$

   c. $\sqrt{x}/\sqrt{y}$

   d. None of the above

The answers to both parts of this quiz appear at the end of this appendix. Most students will find the review of arithmetic and algebraic operations presented in the following sections to be helpful. A student who gets fewer than ten of the twenty items correct might need a more extensive review of basic algebra.

Symbols

Each of the four basic arithmetic operations—addition, subtraction, multiplication, and division—is indicated by an appropriate symbol.

$+$	add	$\times$ or $\cdot$	multiply
$-$	subtract	$\div$ or $/$	divide

In addition to these operations, the following symbols are used to indicate equality or inequality:

$=$	equals	$\neq$	not equal	$\approx$	approximately equal to
$>$	greater than	$<$	less than		
$\geq$	greater than or equal to	$\leq$	less than or equal to		

Addition

The process of addition refers to the summation or accumulation of a set of numbers. In adding numbers together, there are two basic principles or laws, the commutative law and the associative law.

The *commutative law* states that the order in which numbers are added is irrelevant. This can be seen in the following two examples:

$$1 + 2 = 3 \qquad 2 + 1 = 3$$

$$x + y = z \qquad y + x = z$$

In each example, it did not matter which number was added first and which was added second, the result was the same.

The *associative law* of addition states that in adding several numbers, any sub-grouping of the numbers can be added first, last, or in the middle. This can be seen in the following examples:

1. $2 + 3 + 6 + 7 + 4 + 1 = 23$
2. $(5) + (6 + 7) + 4 + 1 = 23$
3. $5 + 13 + 5 = 23$
4. $5 + 6 + 7 + 4 + 1 = 23$

In each of these cases, the order in which the numbers have been added has no effect on the result.

Subtraction

The process of subtraction is the opposite or inverse of addition. The operation of subtracting 1 from 2 (i.e., $2 - 1$) means that one unit is to be taken away from two units, leaving a remainder of one unit. In contrast to addition, the commutative and associative laws do not hold for subtraction. Therefore, as indicated in the following examples, we have

$$8 - 4 = 4 \quad \text{but} \quad 4 - 8 = -4$$
$$3 - 6 = -3 \quad \text{but} \quad 6 - 3 = 3$$
$$8 - 3 - 2 = 3 \quad \text{but} \quad 3 - 2 - 8 = -7$$
$$9 - 4 - 2 = 3 \quad \text{but} \quad 2 - 4 - 9 = -11$$

When subtracting negative numbers, we must remember that the same result occurs when subtracting a negative number as adding a positive number. Thus we would have

$$4 - (-3) = +7 \quad 4 + 3 = 7$$
$$8 - (-10) = +18 \quad 8 + 10 = 18$$

Multiplication

The operation of multiplication is actually a shortcut method of addition when the same number is to be added several times. For example, if 7 is to be added three times $(7 + 7 + 7)$, we could just multiply 7 by 3 to obtain the product of 21.

In multiplication as in addition, the commutative laws and associative laws are in operation so that:

$$a \times b = b \times a$$
$$4 \times 5 = 5 \times 4 = 20$$
$$(2 \times 5) \times 6 = 10 \times 6 = 60$$

A third law of multiplication, the *distributive law*, applies to the multiplication of one number by the sum of several other numbers. Thus,

$$a(b + c) = ab + ac$$
$$2(3 + 4) = 2(7) = 2(3) + 2(4) = 14$$

Here the resulting product is the same regardless of whether b and c are summed and multiplied by a, or a is multiplied by b and by c and the two products are added together. We also need to remember that when multiplying negative numbers that a negative number multiplied by a negative number equals a positive number. Thus,

$$(-a) \times (-b) = ab$$
$$(-5) \times (-4) = +20$$

Division

Just as subtraction is the opposite of addition, division is the opposite or inverse of multiplication. When we discussed multiplication, we viewed it as a shortcut to addition in certain situations. Similarly, division can be viewed as a shortcut to subtraction. When we divide 20 by 4, we are actually determining the number of times that 4 can be subtracted from 20. In general, however, the number of times one number can be divided by another does not have to be an exact integer value, since there could be a remainder. For example, if 21 rather than 20 were divided by 4, the answer would be 5 with a remainder of 1, or $5\frac{1}{4}$.

As in the case of subtraction, neither the commutative nor associative law of addition and multiplication holds for division. Thus

$$a \div b \neq b \div a$$
$$9 \div 3 \neq 3 \div 9$$
$$6 \div (3 \div 2) = 4$$
$$(6 \div 3) \div 2 = 1$$

The distributive law will only hold when the numbers to be added are contained in the numerator, not the denominator. Thus,

$$\frac{a+b}{c} = \frac{a}{c} + \frac{b}{c} \qquad \text{but} \qquad \frac{a}{b+c} \neq \frac{a}{b} + \frac{a}{c}$$

For example,

$$\frac{6+9}{3} = \frac{6}{3} + \frac{9}{3} = 2 + 3 = 5$$

$$\frac{1}{2+3} = \frac{1}{5} \qquad \text{but} \qquad \frac{1}{2+3} \neq \frac{1}{2} + \frac{1}{3}$$

The last important property of division is that if the numerator and the denominator are each multiplied or divided by the same number, the resulting quotient will not be affected. Therefore, if we have

$$\frac{80}{40} = 2$$

then

$$\frac{5(80)}{5(40)} = \frac{400}{200} = 2$$

and

$$\frac{80 \div 5}{40 \div 5} = \frac{16}{8} = 2$$

Fractions

A fraction is a number that consists of a combination of whole numbers and/or parts of whole numbers. For instance, the fraction $\frac{1}{6}$ consists of only a portion of a number, while the fraction $\frac{7}{6}$ consists of the whole number 1 plus the fraction $\frac{1}{6}$. Each of the operations of addition, subtraction, multiplication, and division can be applied to fractions. When adding and subtracting fractions, one must obtain the lowest common denominator for each fraction prior to adding or subtracting them. Thus, in adding $\frac{1}{3} + \frac{1}{5}$, the lowest common denominator is 15 so that we have

$$\frac{5}{15} + \frac{3}{15} = \frac{8}{15}$$

In subtracting $\frac{1}{4} - \frac{1}{6}$, the same principle can be applied so that we would have a lowest common denominator of 12, producing a result of

$$\frac{3}{12} - \frac{2}{12} = \frac{1}{12}$$

The operations of multiplication of fractions and division of fractions do not have the lowest common denominator requirement associated with the addition and subtraction of fractions. Thus, if a/b is multiplied by c/d we obtain

$$\frac{ac}{bd}$$

The resulting numerator, ac, is the product of the numerators a and c, while the denominator, bd, is the product of the two denominators b and d. The resulting fraction can sometimes be reduced to a lower term by dividing numerator and denominator by a common factor. For example, taking

$$\frac{2}{3} \times \frac{6}{7} = \frac{12}{21}$$

and dividing numerator and denominator by 3 produces a result of $\frac{4}{7}$.

Division of fractions can be thought of as the inverse of multiplication, so that the divisor can be inverted and multiplied by the original fraction. Thus,

$$\frac{9}{5} \div \frac{1}{4} = \frac{9}{5} \times \frac{4}{1} = \frac{36}{5}$$

Finally, the division of a fraction can be thought of as a way of converting the fraction to a decimal number. For example, the fraction $\frac{2}{5}$ can be converted to a decimal number by dividing its numerator, 2, by its denominator, 5, to produce the decimal number 0.40.

Exponents and Square Roots

The process of exponentiation provides a shortcut in writing out numerous multiplications. For example, if we have $2 \times 2 \times 2 \times 2 \times 2$, then we may also write this as $2^5 = 32$. The 5 represents the exponent of the number 2, telling us that 2 is to be multiplied by itself five times.

There are several rules that can be applied for multiplying or dividing numbers that contain exponents.

Rule 1 $\qquad\qquad\qquad\qquad x^a \cdot x^b = x^{(a+b)}$

If two numbers involving powers of the same number are multiplied, the product is that same number raised to the sum of the powers. Thus

$$4^2 \cdot 4^3 = (4 \cdot 4)(4 \cdot 4 \cdot 4) = 4^5$$

Rule 2 $$(x^a)^b = x^{ab}$$

If we take the power of a number that is already taken to a power, the result will be a number that is raised to the product of the two powers. For example,

$$(4^2)^3 = (4^2)(4^2)(4^2) = 4^6$$

Rule 3 $$\frac{x^a}{x^b} = x^{(a-b)}$$

If a number raised to a power is divided by the same number raised to a power, the quotient will be the number raised to the difference of the powers. Thus

$$\frac{3^5}{3^3} = \frac{3 \cdot 3 \cdot 3 \cdot 3 \cdot 3}{3 \cdot 3 \cdot 3} = 3^2$$

If the denominator has a higher power than the numerator, the resulting quotient will be a negative power. Thus

$$\frac{3^3}{3^5} = \frac{3 \cdot 3 \cdot 3}{3 \cdot 3 \cdot 3 \cdot 3} = \frac{1}{3^2} = 3^{-2}$$

If the difference between the powers of the numerator and denominator is 1, the result will be the actual number itself, so that $x^1 = x$. For example,

$$\frac{3^3}{3^2} = \frac{3 \cdot 3 \cdot 3}{3 \cdot 3} = 3^1 = 3$$

If, however, there is no difference in the power of the numbers in the numerator and denominator, the result will be 1. Thus,

$$\frac{x^a}{x^a} = x^{a-a} = x^0 = 1$$

Therefore, any number raised to the zero power will equal 1. For example,

$$\frac{3^3}{3^3} = \frac{3 \cdot 3 \cdot 3}{3 \cdot 3 \cdot 3} = 3^0 = 1$$

The square root represents a special power of a number, the $\frac{1}{2}$ power. It indicates the value that when multiplied by itself will produce the original number. It is represented by the symbol $\sqrt{\ }$.

Equations

In statistics, many formulas are expressed as equations where one unknown value is a function of some other value. Therefore, it is extremely useful that we know how to manipulate equations into various forms. The rules of addition, subtraction, multiplication, and division can be used to work with equations. For example, if we have the equation $x - 2 = 5$, we can solve for x by adding 2 to each side of the equation. Thus, we would have $x - 2 + 2 = 5 + 2$ and, therefore, $x = 7$. If we had $x + y = z$, we could solve for x by subtracting y from both sides of the equation so that $x = z - y$.

If we have the product of two variables equal to the third, such as $x \cdot y = z$, we can solve for x by dividing both sides of the equation by y. Thus $x = z/y$. On the other hand, if $x/y = z$, we can solve for x by multiplying both sides of the equation by y. Thus, x would equal yz. Therefore, the various operations of addition, subtraction, multiplication, and division can be applied to equations as long as the same operation is performed on each side of the equation, thereby maintaining the equality.

Answers to Quiz

Part I

1. $\frac{3}{2}$
2. 0.16
3. $\frac{5}{3}$
4. $\frac{1}{81}$
5. 0.20
6. 1.30
7. -6.4
8. $+\frac{1}{6}$
9. 3/200
10. 4

Part II

1. c
2. c
3. a
4. b
5. a
6. b
7. d
8. b
9. a
10. c

Appendix B

Statistical Tables

Contents

Table B.1	Random Numbers
Table B.2	Normal Curve Areas
Table B.3	Critical Values for Student's t
Table B.4	Critical Values for the χ^2 Statistic
Table B.5	Critical Values of the F Statistic
Table B.5a	$\alpha = .10$
Table B.5b	$\alpha = .05$
Table B.5c	$\alpha = .025$
Table B.5d	$\alpha = .01$
Table B.6	Critical Values for the Wilcoxon Rank Sum Test
Table B.7	Critical Values for the Wilcoxon Signed Ranks Test
Table B.8	Critical Values of Spearman's Rank Correlation Coefficient

Table B.1 Random Numbers

Row	1	2	3	4	5	6	7	8	9	10	11	12	13	14
1	10480	15011	01536	02011	81647	91646	69179	14194	62590	36207	20969	99570	91291	90700
2	22368	46573	25595	85393	30995	89198	27982	53402	93965	34095	52666	19174	39615	99505
3	24130	48360	22527	97265	76393	64809	15179	24830	49340	32081	30680	19655	63348	58629
4	42167	93093	06243	61680	07856	16376	39440	53537	71341	57004	00849	74917	97758	16379
5	37570	39975	81837	16656	06121	91782	60468	81305	49684	60672	14110	06927	01263	54613
6	77921	06907	11008	42751	27756	53498	18602	70659	90655	15053	21916	81825	44394	42880
7	99562	72905	56420	69994	98872	31016	71194	18738	44013	48840	63213	21069	10634	12952
8	96301	91977	05463	07972	18876	20922	94595	56869	69014	60045	18425	84903	42508	32307
9	89579	14342	63661	10281	17453	18103	57740	84378	25331	12566	58678	44947	05585	56941
10	85475	36857	53342	53988	53060	59533	38867	62300	08158	17983	16439	11458	18593	64952
11	28918	69578	88231	33276	70997	79936	56865	05859	90106	31595	01547	85590	91610	78188
12	63553	40961	48235	03427	49626	69445	18663	72695	52180	20847	12234	90511	33703	90322
13	09429	93969	52636	92737	88974	33488	36320	17617	30015	08272	84115	27156	30613	74952
14	10365	61129	87529	85689	48237	52267	67689	93394	01511	26358	85104	20285	29975	89868
15	07119	97336	71048	08178	77233	13916	47564	81056	97735	85977	29372	74461	28551	90707
16	51085	12765	51821	51259	77452	16308	60756	92144	49442	53900	70960	63990	75601	40719
17	02368	21382	52404	60268	89368	19885	55322	44819	01188	65255	64835	44919	05944	55157
18	01011	54092	33362	94904	31273	04146	18594	29852	71585	85030	51132	01915	92747	64951
19	52162	53916	46369	58586	23216	14513	83149	98736	23495	64350	94738	17752	35156	35749
20	07056	97628	33787	09998	42698	06691	76988	13602	51851	46104	88916	19509	25625	58104
21	48663	91245	85828	14346	09172	30168	90229	04734	59193	22178	30421	61666	99904	32812
22	54164	58492	22421	74103	47070	25306	76468	26384	58151	06646	21524	15227	96909	44592
23	32639	32363	05597	24200	13363	38005	94342	28728	35806	06912	17012	64161	18296	22851
24	29334	27001	87637	87308	58731	00256	45834	15398	46557	41135	10367	07684	36188	18510
25	02488	33062	28834	07351	19731	92420	60952	61280	50001	67658	32586	86679	50720	94953
26	81525	72295	04839	96423	24878	82651	66566	14778	76797	14780	13300	87074	79666	95725
27	29676	20591	68086	26432	46901	20849	89768	81536	86645	12659	92259	57102	80428	25280
28	00742	57392	39064	66432	84673	40027	32832	61362	98947	96067	64760	64584	96096	98253
29	05366	04213	25669	26422	44407	44048	37937	63904	45766	66134	75470	66520	34693	90449
30	91921	26418	64117	94305	26766	25940	39972	22209	71500	64568	91402	42416	07844	69618
31	00582	04711	87917	77341	42206	35126	74087	99547	81817	42607	43808	76655	62028	76630
32	00725	69884	62797	56170	86324	88072	76222	36086	84637	93161	76038	65855	77919	88006
33	69011	65795	95876	55293	18988	27354	26575	08625	40801	59920	29841	80150	12777	48501
34	25976	57948	29888	88604	67917	48708	18912	82271	65424	69774	33611	54262	85963	03547
35	09763	83473	73577	12908	30883	18317	28290	35797	05998	41688	34952	37888	38917	88050

continued

Table B.1 Continued

Row	1	2	3	4	5	6	7	8	9	10	11	12	13	14
36	91576	42595	27958	30134	04024	86385	29880	99730	55536	84855	29080	09250	79656	73211
37	17955	56349	90999	49127	20044	59931	06115	20542	18059	02008	73708	83517	36103	42791
38	46503	18584	18845	49618	02304	51038	20655	58727	28168	15475	56942	53389	20562	87338
39	92157	89634	94824	78171	84610	82834	09922	25417	44137	48413	25555	21246	35509	20468
40	14577	62765	35605	81263	39667	47358	56873	56307	61607	49518	89656	20103	77490	18062
41	98427	07523	33362	64270	01638	92477	66969	98420	04880	45585	46565	04102	46880	45709
42	34914	63976	88720	82765	34476	17032	87589	40836	32427	70002	70663	88863	77775	69348
43	70060	28277	39475	46473	23219	53416	94970	25832	69975	94884	19661	72828	00102	66794
44	53976	54914	06990	67245	68350	82948	11398	42878	80287	88267	47363	46634	06541	97809
45	76072	29515	40980	07391	58745	25774	22987	80059	39911	96189	41151	14222	60697	59583
46	90725	52210	83974	29992	65831	38857	50490	83765	55657	14361	31720	57375	56228	41546
47	64364	67412	33339	31926	14883	24413	59744	92351	97473	89286	35931	04110	23726	51900
48	08962	00358	31662	25388	61642	34072	81249	35648	56891	69352	48373	45578	78547	81788
49	95012	68379	93526	70765	10592	04542	76463	54328	02349	17247	28865	14777	62730	92277
50	15664	10493	20492	38391	91132	21999	59516	81652	27195	48223	46751	22923	32261	85653
51	16408	81899	04153	53381	79401	21438	83035	92350	36693	31238	59649	91754	72772	02338
52	18629	81953	05520	91962	04739	13092	97662	24822	94730	06496	35090	04822	86774	98289
53	73115	35101	47498	87637	99016	71060	88824	71013	18735	20286	23153	72924	35165	43040
54	57491	16703	23167	49323	45021	33132	12544	41035	80780	45393	44812	12512	98931	91202
55	30405	83946	23792	14422	15059	45799	22716	19792	09983	74353	68668	30429	70735	25499
56	16631	35006	85900	98275	32388	52390	16815	69290	82732	38480	73817	32523	41961	44437
57	96773	20206	42559	78985	05300	22164	24369	54224	35083	19687	11052	91491	60383	19746
58	38935	64202	14349	82674	66523	44133	00697	35552	35970	19124	63318	29686	03387	59846
59	31624	76384	17403	53363	44167	64486	64758	75366	76554	31601	12614	33072	60332	92325
60	78919	19474	23632	27889	47914	02584	37680	20801	72152	39339	34806	08930	85001	87820
61	03931	33309	57047	74211	63445	17361	62825	39908	05607	91284	68833	25570	38818	46920
62	74426	33278	43972	10110	89917	15665	52872	73823	73144	88662	88970	74492	51805	99378
63	09066	00903	20795	95452	92648	45454	09552	88815	16553	51125	79375	97596	16296	66092
64	42238	12426	87025	14267	20979	04508	64535	31355	86064	29472	47689	05974	52468	16834
65	16153	08002	26504	41744	81959	65642	74240	56302	00033	67107	77510	70625	28725	34191
66	21457	40742	29820	96783	29400	21840	15035	34537	33310	06116	95240	15957	16572	06004
67	21581	57802	02050	89728	17937	37621	47075	42080	97403	48626	68995	43805	33386	21597
68	55612	78095	83197	33732	05810	24813	86902	60397	16489	03264	88525	42786	05269	92532
69	44657	66999	99324	51281	84463	60563	79312	93454	68876	25471	93911	25650	12682	73572
70	91340	84979	46949	81973	37949	61023	43997	15263	80644	43942	89203	71795	99533	50501

Table B.1 Continued

Row	1	2	3	4	5	6	7	8	9	10	11	12	13	14
71	91227	21199	31935	27022	84067	05462	35216	14486	29891	68607	41867	14951	91696	85065
72	50001	38140	66321	19924	72163	09538	12151	06878	91903	18749	34405	56087	82790	70925
73	65390	05224	72958	28609	81406	39147	25549	48542	42627	45233	57202	94617	23772	07896
74	27504	96131	83944	41575	10573	08619	64482	73923	36152	05184	94142	25299	84387	34925
75	37169	94851	39117	89632	00959	16487	65536	49071	39782	17095	02330	74301	00275	48280
76	11508	70225	51111	38351	19444	66499	71945	05422	13442	78675	84081	66938	93654	59894
77	37449	30362	06694	54690	04052	53115	62757	95348	78662	11163	81651	50245	34971	52924
78	46515	70331	85922	38329	57015	15765	97161	17869	45349	61796	66345	81073	49106	79860
79	30986	81223	42416	58353	21532	30502	32305	86482	05174	07901	54339	58861	74818	46942
80	63798	64995	46583	09785	44160	78128	83991	42865	92520	83531	80377	35909	81250	54238
81	82486	84846	99254	67632	43218	50076	21361	64816	51202	88124	41870	52689	51275	83556
82	21885	32906	92431	09060	64297	51674	64126	62570	26123	05155	59194	52799	28225	85762
83	60336	98782	07408	53458	13564	59089	26445	29789	85205	41001	12535	12133	14645	23541
84	43937	46891	24010	25560	86355	33941	25786	54990	71899	15475	95434	98227	21824	19585
85	97656	63175	89303	16275	07100	92063	21942	18611	47348	20203	18534	03862	78095	50136
86	03299	01221	05418	38982	55758	92237	26759	86367	21216	98442	08303	56613	91511	75928
87	79626	06486	03574	17668	07785	76020	79924	25651	83325	88428	85076	72811	22717	50585
88	85636	68335	47539	03129	65651	11977	02510	26113	99447	68645	34327	15152	55230	93448
89	18039	14367	64337	06177	12143	46609	32989	74014	64708	00533	35398	58408	13261	47908
90	08362	15656	60627	36478	65648	16764	53412	09013	07832	41574	17639	82163	60859	75567
91	79556	29068	04142	16268	15387	12856	66227	38358	22478	73373	88732	09443	82558	05250
92	92608	82674	27072	32534	17075	27698	98204	63863	11951	34648	88022	56148	34925	57031
93	23982	25835	40055	67006	12293	02753	14827	23235	35071	99704	37543	11601	35503	85171
94	09915	96306	05908	97901	28395	14186	00821	80703	70426	75647	76310	88717	37890	40129
95	59037	33300	26695	62247	69927	76123	50842	43834	86654	70959	79725	93872	28117	19233
96	42488	78077	69882	61657	34136	79180	97526	43092	04098	73571	80799	76536	71255	64239
97	46764	86273	63003	93017	31204	36692	40202	35275	57306	55543	53203	18098	47625	88684
98	03237	45430	55417	63282	90816	17349	88298	90183	36600	78406	06216	95787	42579	90730
99	86591	81482	52667	61582	14972	90053	89534	76036	49199	43716	97548	04379	46370	28672
100	38534	01715	94964	87288	65680	43772	39560	12918	86537	62738	19636	51132	25739	56947

Source: Abridged from W. H. Beyer (ed.), *CRC Standard Mathematical Tables*, 24th edition. (Cleveland: The Chemical Rubber Company), 1976.

Table B.2 Normal Curve Areas

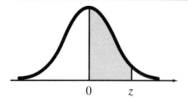

z	.00	.01	.02	.03	.04	.05	.06	.07	.08	.09
.0	.0000	.0040	.0080	.0120	.0160	.0199	.0239	.0279	.0319	.0359
.1	.0398	.0438	.0478	.0517	.0557	.0596	.0636	.0675	.0714	.0753
.2	.0793	.0832	.0871	.0910	.0948	.0987	.1026	.1064	.1103	.1141
.3	.1179	.1217	.1255	.1293	.1331	.1368	.1406	.1443	.1480	.1517
.4	.1554	.1591	.1628	.1664	.1700	.1736	.1772	.1808	.1844	.1879
.5	.1915	.1950	.1985	.2019	.2054	.2088	.2123	.2157	.2190	.2224
.6	.2257	.2291	.2324	.2357	.2389	.2422	.2454	.2486	.2517	.2549
.7	.2580	.2611	.2642	.2673	.2704	.2734	.2764	.2794	.2823	.2852
.8	.2881	.2910	.2939	.2967	.2995	.3023	.3051	.3078	.3106	.3133
.9	.3159	.3186	.3212	.3238	.3264	.3289	.3315	.3340	.3365	.3389
1.0	.3413	.3438	.3461	.3485	.3508	.3531	.3554	.3577	.3599	.3621
1.1	.3643	.3665	.3686	.3708	.3729	.3749	.3770	.3790	.3810	.3830
1.2	.3849	.3869	.3888	.3907	.3925	.3944	.3962	.3980	.3997	.4015
1.3	.4032	.4049	.4066	.4082	.4099	.4115	.4131	.4147	.4162	.4177
1.4	.4192	.4207	.4222	.4236	.4251	.4265	.4279	.4292	.4306	.4319
1.5	.4332	.4345	.4357	.4370	.4382	.4394	.4406	.4418	.4429	.4441
1.6	.4452	.4463	.4474	.4484	.4495	.4505	.4515	.4525	.4535	.4545
1.7	.4554	.4564	.4573	.4582	.4591	.4599	.4608	.4616	.4625	.4633
1.8	.4641	.4649	.4656	.4664	.4671	.4678	.4686	.4693	.4699	.4706
1.9	.4713	.4719	.4726	.4732	.4738	.4744	.4750	.4756	.4761	.4767
2.0	.4772	.4778	.4783	.4788	.4793	.4798	.4803	.4808	.4812	.4817
2.1	.4821	.4826	.4830	.4834	.4838	.4842	.4846	.4850	.4854	.4857
2.2	.4861	.4864	.4868	.4871	.4875	.4878	.4881	.4884	.4887	.4890
2.3	.4893	.4896	.4898	.4901	.4904	.4906	.4909	.4911	.4913	.4916
2.4	.4918	.4920	.4922	.4925	.4927	.4929	.4931	.4932	.4934	.4936
2.5	.4938	.4940	.4941	.4943	.4945	.4946	.4948	.4949	.4951	.4952
2.6	.4953	.4955	.4956	.4957	.4959	.4960	.4961	.4962	.4963	.4964
2.7	.4965	.4966	.4967	.4968	.4969	.4970	.4971	.4972	.4973	.4974
2.8	.4974	.4975	.4976	.4977	.4977	.4978	.4979	.4979	.4980	.4981
2.9	.4981	.4982	.4982	.4983	.4984	.4984	.4985	.4985	.4986	.4986
3.0	.4987	.4987	.4987	.4988	.4988	.4989	.4989	.4989	.4990	.4990

Source: Abridged from Table I of A. Hald, *Statistical Tables and Formulas* (New York: John Wiley & Sons, Inc.), 1952. Reproduced by permission of A. Hald and the publisher.

Table B.3 Critical Values for Student's t

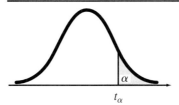

Degrees of Freedom	$t_{.100}$	$t_{.050}$	$t_{.025}$	$t_{.010}$	$t_{.005}$	$t_{.001}$	$t_{.0005}$
1	3.078	6.314	12.706	31.821	63.657	318.31	636.62
2	1.886	2.920	4.303	6.965	9.925	22.326	31.598
3	1.638	2.353	3.182	4.541	5.841	10.213	12.924
4	1.533	2.132	2.776	3.747	4.604	7.173	8.610
5	1.476	2.015	2.571	3.365	4.032	5.893	6.869
6	1.440	1.943	2.447	3.143	3.707	5.208	5.959
7	1.415	1.895	2.365	2.998	3.499	4.785	5.408
8	1.397	1.860	2.306	2.896	3.355	4.501	5.041
9	1.383	1.833	2.262	2.821	3.250	4.297	4.781
10	1.372	1.812	2.228	2.764	3.169	4.144	4.587
11	1.363	1.796	2.201	2.718	3.106	4.025	4.437
12	1.356	1.782	2.179	2.681	3.055	3.930	4.318
13	1.350	1.771	2.160	2.650	3.012	3.852	4.221
14	1.345	1.761	2.145	2.624	2.977	3.787	4.140
15	1.341	1.753	2.131	2.602	2.947	3.733	4.073
16	1.337	1.746	2.120	2.583	2.921	3.686	4.015
17	1.333	1.740	2.110	2.567	2.898	3.646	3.965
18	1.330	1.734	2.101	2.552	2.878	3.610	3.922
19	1.328	1.729	2.093	2.539	2.861	3.579	3.883
20	1.325	1.725	2.086	2.528	2.845	3.552	3.850
21	1.323	1.721	2.080	2.518	2.831	3.527	3.819
22	1.321	1.717	2.074	2.508	2.819	3.505	3.792
23	1.319	1.714	2.069	2.500	2.807	3.485	3.767
24	1.318	1.711	2.064	2.492	2.797	3.467	3.745
25	1.316	1.708	2.060	2.485	2.787	3.450	3.725
26	1.315	1.706	2.056	2.479	2.779	3.435	3.707
27	1.314	1.703	2.052	2.473	2.771	3.421	3.690
28	1.313	1.701	2.048	2.467	2.763	3.408	3.674
29	1.311	1.699	2.045	2.462	2.756	3.396	3.659
30	1.310	1.697	2.042	2.457	2.750	3.385	3.646
40	1.303	1.684	2.021	2.423	2.704	3.307	3.551
60	1.296	1.671	2.000	2.390	2.660	3.232	3.460
120	1.289	1.658	1.980	2.358	2.617	3.160	3.373
∞	1.282	1.645	1.960	2.326	2.576	3.090	3.291

Table B.4 Critical Values for the χ^2 Statistic

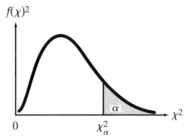

Degrees of Freedom	$\chi^2_{.995}$	$\chi^2_{.990}$	$\chi^2_{.975}$	$\chi^2_{.950}$	$\chi^2_{.900}$
1	.0000393	.0001571	.0009821	.0039321	.0157908
2	.0100251	.0201007	.0506356	.102587	.210720
3	.0717212	.114832	.215795	.351846	.584375
4	.206990	.297110	.484419	.710721	1.063623
5	.411740	.554300	.831211	1.145476	1.61031
6	.675727	.872085	1.237347	1.63539	2.20413
7	.989265	1.239043	1.68987	2.16735	2.83311
8	1.344419	1.646482	2.17973	2.73264	3.48954
9	1.734926	2.087912	2.70039	3.32511	4.16816
10	2.15585	2.55821	3.24697	3.94030	4.86518
11	2.60321	3.05347	3.81575	4.57481	5.57779
12	3.07382	3.57056	4.40379	5.22603	6.30380
13	3.56503	4.10691	5.00874	5.89186	7.04150
14	4.07468	4.66043	5.62872	6.57063	7.78953
15	4.60094	5.22935	6.26214	7.26094	8.54675
16	5.14224	5.81221	6.90766	7.96164	9.31223
17	5.69724	6.40776	7.56418	8.67176	10.0852
18	6.26481	7.01491	8.23075	9.39046	10.8649
19	6.84398	7.63273	8.90655	10.1170	11.6509
20	7.43386	8.26040	9.59083	10.8508	12.4426
21	8.03366	8.89720	10.28293	11.5913	13.2396
22	8.64272	9.54249	10.9823	12.3380	14.0415
23	9.26042	10.19567	11.6885	13.0905	14.8479
24	9.88623	10.8564	12.4011	13.8484	15.6587
25	10.5197	11.5240	13.1197	14.6114	16.4734
26	11.1603	12.1981	13.8439	15.3791	17.2919
27	11.8076	12.8786	14.5733	16.1513	18.1138
28	12.4613	13.5648	15.3079	16.9279	18.9392
29	13.1211	14.2565	16.0471	17.7083	19.7677
30	13.7867	14.9535	16.7908	18.4926	20.5992
40	20.7065	22.1643	24.4331	26.5093	29.0505
50	27.9907	29.7067	32.3574	34.7642	37.6886
60	35.5346	37.4848	40.4817	43.1879	46.4589
70	43.2752	45.4418	48.7576	51.7393	55.3290
80	51.1720	53.5400	57.1532	60.3915	64.2778
90	59.1963	61.7541	65.6466	69.1260	73.2912
100	67.3276	70.0648	74.2219	77.9295	82.3581
150	109.142	112.668	117.985	122.692	128.275
200	152.241	156.432	162.728	168.279	174.835
300	240.663	245.972	253.912	260.878	269.068
400	330.903	337.155	346.482	354.641	364.207
500	422.303	429.388	439.936	449.147	459.926

Source: From C. M. Thompson, "Tables of the Percentage Points of the χ^2-Distribution," *Biometrika,* 1941, Vol. 32, pp. 188–189. Reproduced by permission of the *Biometrika* Trustees and Oxford University Press.

Table B.4 Continued

Degrees of Freedom	$\chi^2_{.100}$	$\chi^2_{.050}$	$\chi^2_{.025}$	$\chi^2_{.010}$	$\chi^2_{.005}$
1	2.70554	3.84146	5.02389	6.63490	7.87944
2	4.60517	5.99147	7.37776	9.21034	10.5966
3	6.25139	7.81473	9.34840	11.3449	12.8381
4	7.77944	9.48773	11.1433	13.2767	14.8602
5	9.23635	11.0705	12.8325	15.0863	16.7496
6	10.6446	12.5916	14.4494	16.8119	18.5476
7	12.0170	14.0671	16.0128	18.4753	20.2777
8	13.3616	15.5073	17.5346	20.0902	21.9550
9	14.6837	16.9190	19.0228	21.6660	23.5893
10	15.9871	18.3070	20.4831	23.2093	25.1882
11	17.2750	19.6751	21.9200	24.7250	26.7569
12	18.5494	21.0261	23.3367	26.2170	28.2995
13	19.8119	22.3621	24.7356	27.6883	29.8194
14	21.0642	23.6848	26.1190	29.1413	31.3193
15	22.3072	24.9958	27.4884	30.5779	32.8013
16	23.5418	26.2962	28.8454	31.9999	34.2672
17	24.7690	27.5871	30.1910	33.4087	35.7185
18	25.9894	28.8693	31.5264	34.8053	37.1564
19	27.2036	30.1435	32.8523	36.1908	38.5822
20	28.4120	31.4104	34.1696	37.5662	39.9968
21	29.6151	32.6705	35.4789	38.9321	41.4010
22	30.8133	33.9244	36.7807	40.2894	42.7956
23	32.0069	35.1725	38.0757	41.6384	44.1813
24	33.1963	36.4151	39.3641	42.9798	45.5585
25	34.3816	37.6525	40.6465	44.3141	46.9278
26	35.5631	38.8852	41.9232	45.6417	48.2899
27	36.7412	40.1133	43.1944	46.9630	49.6449
28	37.9159	41.3372	44.4607	48.2782	50.9933
29	39.0875	42.5569	45.7222	49.5879	52.3356
30	40.2560	43.7729	46.9792	50.8922	53.6720
40	51.8050	55.7585	59.3417	63.6907	66.7659
50	63.1671	67.5048	71.4202	76.1539	79.4900
60	74.3970	79.0819	83.2976	88.3794	91.9517
70	85.5271	90.5312	95.0231	100.425	104.215
80	96.5782	101.879	106.629	112.329	116.321
90	107.565	113.145	118.136	124.116	128.299
100	118.498	124.342	129.561	135.807	140.169
150	172.581	179.581	185.800	193.208	198.360
200	226.021	233.994	241.058	249.445	255.264
300	331.789	341.395	349.874	359.906	366.844
400	436.649	447.632	457.305	468.724	476.606
500	540.930	553.127	563.852	576.493	585.207

Table B.5a Critical Values of the *F* Statistic, $\alpha = .10$

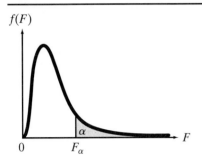

ν_1	NUMERATOR DEGREES OF FREEDOM								
ν_2	1	2	3	4	5	6	7	8	9
1	39.86	49.50	53.59	55.83	57.24	58.20	58.91	59.44	59.86
2	8.53	9.00	9.16	9.24	9.29	9.33	9.35	9.37	9.38
3	5.54	5.46	5.39	5.34	5.31	5.28	5.27	5.25	5.24
4	4.54	4.32	4.19	4.11	4.05	4.01	3.98	3.95	3.94
5	4.06	3.78	3.62	3.52	3.45	3.40	3.37	3.34	3.32
6	3.78	3.46	3.29	3.18	3.11	3.05	3.01	2.98	2.96
7	3.59	3.26	3.07	2.96	2.88	2.83	2.78	2.75	2.72
8	3.46	3.11	2.92	2.81	2.73	2.67	2.62	2.59	2.56
9	3.36	3.01	2.81	2.69	2.61	2.55	2.51	2.47	2.44
10	3.29	2.92	2.73	2.61	2.52	2.46	2.41	2.38	2.35
11	3.23	2.86	2.66	2.54	2.45	2.39	2.34	2.30	2.27
12	3.18	2.81	2.61	2.48	2.39	2.33	2.28	2.24	2.21
13	3.14	2.76	2.56	2.43	2.35	2.28	2.23	2.20	2.16
14	3.10	2.73	2.52	2.39	2.31	2.24	2.19	2.15	2.12
15	3.07	2.70	2.49	2.36	2.27	2.21	2.16	2.12	2.09
16	3.05	2.67	2.46	2.33	2.24	2.18	2.13	2.09	2.06
17	3.03	2.64	2.44	2.31	2.22	2.15	2.10	2.06	2.03
18	3.01	2.62	2.42	2.29	2.20	2.13	2.08	2.04	2.00
19	2.99	2.61	2.40	2.27	2.18	2.11	2.06	2.02	1.98
20	2.97	2.59	2.38	2.25	2.16	2.09	2.04	2.00	1.96
21	2.96	2.57	2.36	2.23	2.14	2.08	2.02	1.98	1.95
22	2.95	2.56	2.35	2.22	2.13	2.06	2.01	1.97	1.93
23	2.94	2.55	2.34	2.21	2.11	2.05	1.99	1.95	1.92
24	2.93	2.54	2.33	2.19	2.10	2.04	1.98	1.94	1.91
25	2.92	2.53	2.32	2.18	2.09	2.02	1.97	1.93	1.89
26	2.91	2.52	2.31	2.17	2.08	2.01	1.96	1.92	1.88
27	2.90	2.51	2.30	2.17	2.07	2.00	1.95	1.91	1.87
28	2.89	2.50	2.29	2.16	2.06	2.00	1.94	1.90	1.87
29	2.89	2.50	2.28	2.15	2.06	1.99	1.93	1.89	1.86
30	2.88	2.49	2.28	2.14	2.05	1.98	1.93	1.88	1.85
40	2.84	2.44	2.23	2.09	2.00	1.93	1.87	1.83	1.79
60	2.79	2.39	2.18	2.04	1.95	1.87	1.82	1.77	1.74
120	2.75	2.35	2.13	1.99	1.90	1.82	1.77	1.72	1.68
∞	2.71	2.30	2.08	1.94	1.85	1.77	1.72	1.67	1.63

Source: From M. Merrington and C. M. Thompson, "Tables of Percentage Points of the Inverted Beta (*F*)-Distribution," *Biometrika,* 1943, 33, 73–88. Reproduced by permission of the *Biometrika* Trustees and Oxford University Press.

Table B.5a Continued

ν_1	NUMERATOR DEGREES OF FREEDOM									
ν_2	10	12	15	20	24	30	40	60	120	∞
1	60.19	60.71	61.22	61.74	62.00	62.26	62.53	62.79	63.06	63.33
2	9.39	9.41	9.42	9.44	9.45	9.46	9.47	9.47	9.48	9.49
3	5.23	5.22	5.20	5.18	5.18	5.17	5.16	5.15	5.14	5.13
4	3.92	3.90	3.87	3.84	3.83	3.82	3.80	3.79	3.78	3.76
5	3.30	3.27	3.24	3.21	3.19	3.17	3.16	3.14	3.12	3.10
6	2.94	2.90	2.87	2.84	2.82	2.80	2.78	2.76	2.74	2.72
7	2.70	2.67	2.63	2.59	2.58	2.56	2.54	2.51	2.49	2.47
8	2.54	2.50	2.46	2.42	2.40	2.38	2.36	2.34	2.32	2.29
9	2.42	2.38	2.34	2.30	2.28	2.25	2.23	2.21	2.18	2.16
10	2.32	2.28	2.24	2.20	2.18	2.16	2.13	2.11	2.08	2.06
11	2.25	2.21	2.17	2.12	2.10	2.08	2.05	2.03	2.00	1.97
12	2.19	2.15	2.10	2.06	2.04	2.01	1.99	1.96	1.93	1.90
13	2.14	2.10	2.05	2.01	1.98	1.96	1.93	1.90	1.88	1.85
14	2.10	2.05	2.01	1.96	1.94	1.91	1.89	1.86	1.83	1.80
15	2.06	2.02	1.97	1.92	1.90	1.87	1.85	1.82	1.79	1.76
16	2.03	1.99	1.94	1.89	1.87	1.84	1.81	1.78	1.75	1.72
17	2.00	1.96	1.91	1.86	1.84	1.81	1.78	1.75	1.72	1.69
18	1.98	1.93	1.89	1.84	1.81	1.78	1.75	1.72	1.69	1.66
19	1.96	1.91	1.86	1.81	1.79	1.76	1.73	1.70	1.67	1.63
20	1.94	1.89	1.84	1.79	1.77	1.74	1.71	1.68	1.64	1.61
21	1.92	1.87	1.83	1.78	1.75	1.72	1.69	1.66	1.62	1.59
22	1.90	1.86	1.81	1.76	1.73	1.70	1.67	1.64	1.60	1.57
23	1.89	1.84	1.80	1.74	1.72	1.69	1.66	1.62	1.59	1.55
24	1.88	1.83	1.78	1.73	1.70	1.67	1.64	1.61	1.57	1.53
25	1.87	1.82	1.77	1.72	1.69	1.66	1.63	1.59	1.56	1.52
26	1.86	1.81	1.76	1.71	1.68	1.65	1.61	1.58	1.54	1.50
27	1.85	1.80	1.75	1.70	1.67	1.64	1.60	1.57	1.53	1.49
28	1.84	1.79	1.74	1.69	1.66	1.63	1.59	1.56	1.52	1.48
29	1.83	1.78	1.73	1.68	1.65	1.62	1.58	1.55	1.51	1.47
30	1.82	1.77	1.72	1.67	1.64	1.61	1.57	1.54	1.50	1.46
40	1.76	1.71	1.66	1.61	1.57	1.54	1.51	1.47	1.42	1.38
60	1.71	1.66	1.60	1.54	1.51	1.48	1.44	1.40	1.35	1.29
120	1.65	1.60	1.55	1.48	1.45	1.41	1.37	1.32	1.26	1.19
∞	1.60	1.55	1.49	1.42	1.38	1.34	1.30	1.24	1.17	1.00

DENOMINATOR DEGREES OF FREEDOM

Table B.5b Critical Values of the F Statistic, $\alpha = .05$

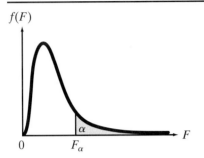

ν_1		NUMERATOR DEGREES OF FREEDOM							
ν_2	1	2	3	4	5	6	7	8	9
1	161.4	199.5	215.7	224.6	230.2	234.0	236.8	238.9	240.5
2	18.51	19.00	19.16	19.25	19.30	19.33	19.35	19.37	19.38
3	10.13	9.55	9.28	9.12	9.01	8.94	8.89	8.85	8.81
4	7.71	6.94	6.59	6.39	6.26	6.16	6.09	6.04	6.00
5	6.61	5.79	5.41	5.19	5.05	4.95	4.88	4.82	4.77
6	5.99	5.14	4.76	4.53	4.39	4.28	4.21	4.15	4.10
7	5.59	4.74	4.35	4.12	3.97	3.87	3.79	3.73	3.68
8	5.32	4.46	4.07	3.84	3.69	3.58	3.50	3.44	3.39
9	5.12	4.26	3.86	3.63	3.48	3.37	3.29	3.23	3.18
10	4.96	4.10	3.71	3.48	3.33	3.22	3.14	3.07	3.02
11	4.84	3.98	3.59	3.36	3.20	3.09	3.01	2.95	2.90
12	4.75	3.89	3.49	3.26	3.11	3.00	2.91	2.85	2.80
13	4.67	3.81	3.41	3.18	3.03	2.92	2.83	2.77	2.71
14	4.60	3.74	3.34	3.11	2.96	2.85	2.76	2.70	2.65
15	4.54	3.68	3.29	3.06	2.90	2.79	2.71	2.64	2.59
16	4.49	3.63	3.24	3.01	2.85	2.74	2.66	2.59	2.54
17	4.45	3.59	3.20	2.96	2.81	2.70	2.61	2.55	2.49
18	4.41	3.55	3.16	2.93	2.77	2.66	2.58	2.51	2.46
19	4.38	3.52	3.13	2.90	2.74	2.63	2.54	2.48	2.42
20	4.35	3.49	3.10	2.87	2.71	2.60	2.51	2.45	2.39
21	4.32	3.47	3.07	2.84	2.68	2.57	2.49	2.42	2.37
22	4.30	3.44	3.05	2.82	2.66	2.55	2.46	2.40	2.34
23	4.28	3.42	3.03	2.80	2.64	2.53	2.44	2.37	2.32
24	4.26	3.40	3.01	2.78	2.62	2.51	2.42	2.36	2.30
25	4.24	3.39	2.99	2.76	2.60	2.49	2.40	2.34	2.28
26	4.23	3.37	2.98	2.74	2.59	2.47	2.39	2.32	2.77
27	4.21	3.35	2.96	2.73	2.57	2.46	2.37	2.31	2.25
28	4.20	3.34	2.95	2.71	2.56	2.45	2.36	2.29	2.24
29	4.18	3.33	2.93	2.70	2.55	2.43	2.35	2.28	2.22
30	4.17	3.32	2.92	2.69	2.53	2.42	2.33	2.27	2.21
40	4.08	3.23	2.84	2.61	2.45	2.34	2.25	2.18	2.12
60	4.00	3.15	2.76	2.53	2.37	2.25	2.17	2.10	2.04
120	3.92	3.07	2.68	2.45	2.29	2.17	2.09	2.02	1.96
∞	3.84	3.00	2.60	2.37	2.21	2.10	2.01	1.94	1.88

Source: From M. Merrington and C. M. Thompson, "Tables of Percentage Points of the Inverted Beta (F)-Distribution." *Biometrika,* 1943, 33, 73–88. Reproduced by permission of the *Biometrika* Trustees and Oxford University Press.

Table B.5b Continued

ν_2 \ ν_1	NUMERATOR DEGREES OF FREEDOM									
	10	12	15	20	24	30	40	60	120	∞
1	241.9	243.9	245.9	248.0	249.1	250.1	251.1	252.2	253.3	254.3
2	19.40	19.41	19.43	19.45	19.45	19.46	19.47	19.48	19.49	19.50
3	8.79	8.74	8.70	8.66	8.64	8.62	8.59	8.57	8.55	8.53
4	5.96	5.91	5.86	5.80	5.77	5.75	5.72	5.69	5.66	5.63
5	4.74	4.68	4.62	4.56	4.53	4.50	4.46	4.43	4.40	4.36
6	4.06	4.00	3.94	3.87	3.84	3.81	3.77	3.74	3.70	3.67
7	3.64	3.57	3.51	3.44	3.41	3.38	3.34	3.30	3.27	3.23
8	3.35	3.28	3.22	3.15	3.12	3.08	3.04	3.01	2.97	2.93
9	3.14	3.07	3.01	2.94	2.90	2.86	2.83	2.79	2.75	2.71
10	2.98	2.91	2.85	2.77	2.74	2.70	2.66	2.62	2.58	2.54
11	2.85	2.79	2.72	2.65	2.61	2.57	2.53	2.49	2.45	2.40
12	2.75	2.69	2.62	2.54	2.51	2.47	2.43	2.38	2.34	2.30
13	2.67	2.60	2.53	2.46	2.42	2.38	2.34	2.30	2.25	2.21
14	2.60	2.53	2.46	2.39	2.35	2.31	2.27	2.22	2.18	2.13
15	2.54	2.48	2.40	2.33	2.29	2.25	2.20	2.16	2.11	2.07
16	2.49	2.42	2.35	2.28	2.24	2.19	2.15	2.11	2.06	2.01
17	2.45	2.38	2.31	2.23	2.19	2.15	2.10	2.06	2.01	1.96
18	2.41	2.34	2.27	2.19	2.15	2.11	2.06	2.02	1.97	1.92
19	2.38	2.31	2.23	2.16	2.11	2.07	2.03	1.98	1.93	1.88
20	2.35	2.28	2.20	2.12	2.08	2.04	1.99	1.95	1.90	1.84
21	2.32	2.25	2.18	2.10	2.05	2.01	1.96	1.92	1.87	1.81
22	2.30	2.23	2.15	2.07	2.03	1.98	1.94	1.89	1.84	1.78
23	2.27	2.20	2.13	2.05	2.01	1.96	1.91	1.86	1.81	1.76
24	2.25	2.18	2.11	2.03	1.98	1.94	1.89	1.84	1.79	1.73
25	2.24	2.16	2.09	2.01	1.96	1.92	1.87	1.82	1.77	1.71
26	2.22	2.15	2.07	1.99	1.95	1.90	1.85	1.80	1.75	1.69
27	2.20	2.13	2.06	1.97	1.93	1.88	1.84	1.79	1.73	1.67
28	2.19	2.12	2.04	1.96	1.91	1.87	1.82	1.77	1.71	1.65
29	2.18	2.10	2.03	1.94	1.90	1.85	1.81	1.75	1.70	1.64
30	2.16	2.09	2.01	1.93	1.89	1.84	1.79	1.74	1.68	1.62
40	2.08	2.00	1.92	1.84	1.79	1.74	1.69	1.64	1.58	1.51
60	1.99	1.92	1.84	1.75	1.70	1.65	1.59	1.53	1.47	1.39
120	1.91	1.83	1.75	1.66	1.61	1.55	1.50	1.43	1.35	1.25
∞	1.83	1.75	1.67	1.57	1.52	1.46	1.39	1.32	1.22	1.00

DENOMINATOR DEGREES OF FREEDOM

Table B.5c Critical Values of the F Statistic, $\alpha = .025$

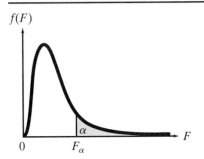

ν_1	NUMERATOR DEGREES OF FREEDOM								
ν_2	1	2	3	4	5	6	7	8	9
1	647.8	799.5	864.2	899.6	921.8	937.1	948.2	956.7	963.3
2	38.51	39.00	39.17	39.25	39.30	39.33	39.36	39.37	39.39
3	17.44	16.04	15.44	15.10	14.88	14.73	14.62	14.54	14.47
4	12.22	10.65	9.98	9.60	9.36	9.20	9.07	8.98	8.90
5	10.01	8.43	7.76	7.39	7.15	6.98	6.85	6.76	6.68
6	8.81	7.26	6.60	6.23	5.99	5.82	5.70	5.60	5.52
7	8.07	6.54	5.89	5.52	5.29	5.12	4.99	4.90	4.82
8	7.57	6.06	5.42	5.05	4.82	4.65	4.53	4.43	4.36
9	7.21	5.71	5.08	4.72	4.48	4.32	4.20	4.10	4.03
10	6.94	5.46	4.83	4.47	4.24	4.07	3.95	3.85	3.78
11	6.72	5.26	4.63	4.28	4.04	3.88	3.76	3.66	3.59
12	6.55	5.10	4.47	4.12	3.89	3.73	3.61	3.51	3.44
13	6.41	4.97	4.35	4.00	3.77	3.60	3.48	3.39	3.31
14	6.30	4.86	4.24	3.89	3.66	3.50	3.38	3.29	3.21
15	6.20	4.77	4.15	3.80	3.58	3.41	3.29	3.20	3.12
16	6.12	4.69	4.08	3.73	3.50	3.34	3.22	3.12	3.05
17	6.04	4.62	4.01	3.66	3.44	3.28	3.16	3.06	2.98
18	5.98	4.56	3.95	3.61	3.38	3.22	3.10	3.01	2.93
19	5.92	4.51	3.90	3.56	3.33	3.17	3.05	2.96	2.88
20	5.87	4.46	3.86	3.51	3.29	3.13	3.01	2.91	2.84
21	5.83	4.42	3.82	3.48	3.25	3.09	2.97	2.87	2.80
22	5.79	4.38	3.78	3.44	3.22	3.05	2.93	2.84	2.76
23	5.75	4.35	3.75	3.41	3.18	3.02	2.90	2.81	2.73
24	5.72	4.32	3.72	3.38	3.15	2.99	2.87	2.78	2.70
25	5.69	4.29	3.69	3.35	3.13	2.97	2.85	2.75	2.68
26	5.66	4.27	3.67	3.33	3.10	2.94	2.82	2.73	2.65
27	5.63	4.24	3.65	3.31	3.08	2.92	2.80	2.71	2.63
28	5.61	4.22	3.63	3.29	3.06	2.90	2.78	2.69	2.61
29	5.59	4.20	3.61	3.27	3.04	2.88	2.76	2.67	2.59
30	5.57	4.18	3.59	3.25	3.03	2.87	2.75	2.65	2.57
40	5.42	4.05	3.46	3.13	2.90	2.74	2.62	2.53	2.45
60	5.29	3.93	3.34	3.01	2.79	2.63	2.51	2.41	2.33
120	5.15	3.80	3.23	2.89	2.67	2.52	2.39	2.30	2.22
∞	5.02	3.69	3.12	2.79	2.57	2.41	2.29	2.19	2.11

Source: From M. Merrington and C. M. Thompson, "Tables of Percentage Points of the Inverted Beta (F)-Distribution," *Biometrika*, 1943, 33, 73–88. Reproduced by permission of the *Biometrika* Trustees and Oxford University Press.

Table B.5c Continued

ν_2 \ ν_1	NUMERATOR DEGREES OF FREEDOM									
	10	12	15	20	24	30	40	60	120	∞
1	968.6	976.7	984.9	993.1	997.2	1,001	1,006	1,010	1,014	1,018
2	39.40	39.41	39.43	39.45	39.46	39.46	39.47	39.48	39.49	39.50
3	14.42	14.34	14.25	14.17	14.12	14.08	14.04	13.99	13.95	13.90
4	8.84	8.75	8.66	8.56	8.51	8.46	8.41	8.36	8.31	8.26
5	6.62	6.52	6.43	6.33	6.28	6.23	6.18	6.12	6.07	6.02
6	5.46	5.37	5.27	5.17	5.12	5.07	5.01	4.96	4.90	4.85
7	4.76	4.67	4.57	4.47	4.42	4.36	4.31	4.25	4.20	4.14
8	4.30	4.20	4.10	4.00	3.95	3.89	3.84	3.78	3.73	3.67
9	3.96	3.87	3.77	3.67	3.61	3.56	3.51	3.45	3.39	3.33
10	3.72	3.62	3.52	3.42	3.37	3.31	3.26	3.20	3.14	3.08
11	3.53	3.43	3.33	3.23	3.17	3.12	3.06	3.00	2.94	2.88
12	3.37	3.28	3.18	3.07	3.02	2.96	2.91	2.85	2.79	2.72
13	3.25	3.15	3.05	2.95	2.89	2.84	2.78	2.72	2.66	2.60
14	3.15	3.05	2.95	2.84	2.79	2.73	2.67	2.61	2.55	2.49
15	3.06	2.96	2.86	2.76	2.70	2.64	2.59	2.52	2.46	2.40
16	2.99	2.89	2.79	2.68	2.63	2.57	2.51	2.45	2.38	2.32
17	2.92	2.82	2.72	2.62	2.56	2.50	2.44	2.38	2.32	2.25
18	2.87	2.77	2.67	2.56	2.50	2.44	2.38	2.32	2.26	2.19
19	2.82	2.72	2.62	2.51	2.45	2.39	2.33	2.27	2.20	2.13
20	2.77	2.68	2.57	2.46	2.41	2.35	2.29	2.22	2.16	2.09
21	2.73	2.64	2.53	2.42	2.37	2.31	2.25	2.18	2.11	2.04
22	2.70	2.60	2.50	2.39	2.33	2.27	2.21	2.14	2.08	2.00
23	2.67	2.57	2.47	2.36	2.30	2.24	2.18	2.11	2.04	1.97
24	2.64	2.54	2.44	2.33	2.27	2.21	2.15	2.08	2.01	1.94
25	2.61	2.51	2.41	2.30	2.24	2.18	2.12	2.05	1.98	1.91
26	2.59	2.49	2.39	2.28	2.22	2.16	2.09	2.03	1.95	1.88
27	2.57	2.47	2.36	2.25	2.19	2.13	2.07	2.00	1.93	1.85
28	2.55	2.45	2.34	2.23	2.17	2.11	2.05	1.98	1.91	1.83
29	2.53	2.43	2.32	2.21	2.15	2.09	2.03	1.96	1.89	1.81
30	2.51	2.41	2.31	2.20	2.14	2.07	2.01	1.94	1.87	1.79
40	2.39	2.29	2.18	2.07	2.01	1.94	1.88	1.80	1.72	1.64
60	2.27	2.17	2.06	1.94	1.88	1.82	1.74	1.67	1.58	1.48
120	2.16	2.05	1.94	1.82	1.76	1.69	1.61	1.53	1.43	1.31
∞	2.05	1.94	1.83	1.71	1.64	1.57	1.48	1.39	1.27	1.00

DENOMINATOR DEGREES OF FREEDOM

Table B.5d Critical Values of the F Statistic, $\alpha = .01$

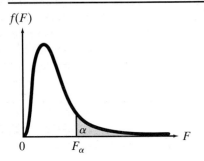

$f(F)$

α

F_α

0

F

ν_1	NUMERATOR DEGREES OF FREEDOM								
ν_2	1	2	3	4	5	6	7	8	9
1	4,052	4,999.5	5,403	5,625	5,764	5,859	5,928	5,982	6,022
2	98.50	99.00	99.17	99.25	99.30	99.33	99.36	99.37	99.39
3	34.12	30.82	29.46	28.71	28.24	27.91	27.67	27.49	27.35
4	21.20	18.00	16.69	15.98	15.52	15.21	14.98	14.80	14.66
5	16.26	13.27	12.06	11.39	10.97	10.67	10.46	10.29	10.16
6	13.75	10.92	9.78	9.15	8.75	8.47	8.26	8.10	7.98
7	12.25	9.55	8.45	7.85	7.46	7.19	6.99	6.84	6.72
8	11.26	8.65	7.59	7.01	6.63	6.37	6.18	6.03	5.91
9	10.56	8.02	6.99	6.42	6.06	5.80	5.61	5.47	5.35
10	10.04	7.56	6.55	5.99	5.64	5.39	5.20	5.06	4.94
11	9.65	7.21	6.22	5.67	5.32	5.07	4.89	4.74	4.63
12	9.33	6.93	5.95	5.41	5.06	4.82	4.64	4.50	4.39
13	9.07	6.70	5.74	5.21	4.86	4.62	4.44	4.30	4.19
14	8.86	6.51	5.56	5.04	4.69	4.46	4.28	4.14	4.03
15	8.68	6.36	5.42	4.89	4.56	4.32	4.14	4.00	3.89
16	8.53	6.23	5.29	4.77	4.44	4.20	4.03	3.89	3.78
17	8.40	6.11	5.18	4.67	4.34	4.10	3.93	3.79	3.68
18	8.29	6.01	5.09	4.58	4.25	4.01	3.84	3.71	3.60
19	8.18	5.93	5.01	4.50	4.17	3.94	3.77	3.63	3.52
20	8.10	5.85	4.94	4.43	4.10	3.87	3.70	3.56	3.46
21	8.02	5.78	4.87	4.37	4.04	3.81	3.64	3.51	3.40
22	7.95	5.72	4.82	4.31	3.99	3.76	3.59	3.45	3.35
23	7.88	5.66	4.76	4.26	3.94	3.71	3.54	3.41	3.30
24	7.82	5.61	4.72	4.22	3.90	3.67	3.50	3.36	3.26
25	7.77	5.57	4.68	4.18	3.85	3.63	3.46	3.32	3.22
26	7.72	5.53	4.64	4.14	3.82	3.59	3.42	3.29	3.18
27	7.68	5.49	4.60	4.11	3.78	3.56	3.39	3.26	3.15
28	7.64	5.45	4.57	4.07	3.75	3.53	3.36	3.23	3.12
29	7.60	5.42	4.54	4.04	3.73	3.50	3.33	3.20	3.09
30	7.56	5.39	4.51	4.02	3.70	3.47	3.30	3.17	3.07
40	7.31	5.18	4.31	3.83	3.51	3.29	3.12	2.99	2.89
60	7.08	4.98	4.13	3.65	3.34	3.12	2.95	2.82	2.72
120	6.85	4.79	3.95	3.48	3.17	2.96	2.79	2.66	2.56
∞	6.63	4.61	3.78	3.32	3.02	2.80	2.64	2.51	2.41

The leftmost axis label reads: DENOMINATOR DEGREES OF FREEDOM

Source: From M. Merrington and C. M. Thompson, "Tables of Percentage Points of the Inverted Beta (*F*)-Distribution," *Biometrika,* 1943, 33, 73–88. Reproduced by permission of the *Biometrika* Trustees and Oxford University Press.

Table B.5d Continued

ν_2 \ ν_1	NUMERATOR DEGREES OF FREEDOM									
	10	12	15	20	24	30	40	60	120	∞
1	6,056	6,106	6,157	6,209	6,235	6,261	6,287	6,313	6,339	6,366
2	99.40	99.42	99.43	99.45	99.46	99.47	99.47	99.48	99.49	99.50
3	27.23	27.05	26.87	26.69	26.60	26.50	26.41	26.32	26.22	26.13
4	14.55	14.37	14.20	14.02	13.93	13.84	13.75	13.65	13.56	13.46
5	10.05	9.89	9.72	9.55	9.47	9.38	9.29	9.20	9.11	9.02
6	7.87	7.72	7.56	7.40	7.31	7.23	7.14	7.06	6.97	6.88
7	6.62	6.47	6.31	6.16	6.07	5.99	5.91	5.82	5.74	5.65
8	5.81	5.67	5.52	5.36	5.28	5.20	5.12	5.03	4.95	4.86
9	5.26	5.11	4.96	4.81	4.73	4.65	4.57	4.48	4.40	4.31
10	4.85	4.71	4.56	4.41	4.33	4.25	4.17	4.08	4.00	3.91
11	4.54	4.40	4.25	4.10	4.02	3.94	3.86	3.78	3.69	3.60
12	4.30	4.16	4.01	3.86	3.78	3.70	3.62	3.54	3.45	3.36
13	4.10	3.96	3.82	3.66	3.59	3.51	3.43	3.34	3.25	3.17
14	3.94	3.80	3.66	3.51	3.43	3.35	3.27	3.18	3.09	3.00
15	3.80	3.67	3.52	3.37	3.29	3.21	3.13	3.05	2.96	2.87
16	3.69	3.55	3.41	3.26	3.18	3.10	3.02	2.93	2.84	2.75
17	3.59	3.46	3.31	3.16	3.08	3.00	2.92	2.83	2.75	2.65
18	3.51	3.37	3.23	3.08	3.00	2.92	2.84	2.75	2.66	2.57
19	3.43	3.30	3.15	3.00	2.92	2.84	2.76	2.67	2.58	2.49
20	3.37	3.23	3.09	2.94	2.86	2.78	2.69	2.61	2.52	2.42
21	3.31	3.17	3.03	2.88	2.80	2.72	2.64	2.55	2.46	2.36
22	3.26	3.12	2.98	2.83	2.75	2.67	2.58	2.50	2.40	2.31
23	3.21	3.07	2.93	2.78	2.70	2.62	2.54	2.45	2.35	2.26
24	3.17	3.03	2.89	2.74	2.66	2.58	2.49	2.40	2.31	2.21
25	3.13	2.99	2.85	2.70	2.62	2.54	2.45	2.36	2.27	2.17
26	3.09	2.96	2.81	2.66	2.58	2.50	2.42	2.33	2.23	2.13
27	3.06	2.93	2.78	2.63	2.55	2.47	2.38	2.29	2.20	2.10
28	3.03	2.90	2.75	2.60	2.52	2.44	2.35	2.26	2.17	2.06
29	3.00	2.87	2.73	2.57	2.49	2.41	2.33	2.23	2.14	2.03
30	2.98	2.84	2.70	2.55	2.47	2.39	2.30	2.21	2.11	2.01
40	2.80	2.66	2.52	2.37	2.29	2.20	2.11	2.02	1.92	1.80
60	2.63	2.50	2.35	2.20	2.12	2.03	1.94	1.84	1.73	1.60
120	2.47	2.34	2.19	2.03	1.95	1.86	1.76	1.66	1.53	1.38
∞	2.32	2.18	2.04	1.88	1.79	1.70	1.59	1.47	1.32	1.00

DENOMINATOR DEGREES OF FREEDOM

Table B.6 Critical Values for the Wilcoxon Rank Sum Test

Test statistic is the rank sum associated with the smaller sample (if equal sample sizes, either rank sum can be used).

a. $\alpha = .025$ one-tailed; $\alpha = .05$ two-tailed

n_2	n_1 3 T_L	T_U	4 T_L	T_U	5 T_L	T_U	6 T_L	T_U	7 T_L	T_U	8 T_L	T_U	9 T_L	T_U	10 T_L	T_U
3	5	16	6	18	6	21	7	23	7	26	8	28	8	31	9	33
4	6	18	11	25	12	28	12	32	13	35	14	38	15	41	16	44
5	6	21	12	28	18	37	19	41	20	45	21	49	22	53	24	56
6	7	23	12	32	19	41	26	52	28	56	29	61	31	65	32	70
7	7	26	13	35	20	45	28	56	37	68	39	73	41	78	43	83
8	8	28	14	38	21	49	29	61	39	73	49	87	51	93	54	98
9	8	31	15	41	22	53	31	65	41	78	51	93	63	108	66	114
10	9	33	16	44	24	56	32	70	43	83	54	98	66	114	79	131

b. $\alpha = .05$ one-tailed; $\alpha = .10$ two-tailed

n_2	n_1 3 T_L	T_U	4 T_L	T_U	5 T_L	T_U	6 T_L	T_U	7 T_L	T_U	8 T_L	T_U	9 T_L	T_U	10 T_L	T_U
3	6	15	7	17	7	20	8	22	9	24	9	27	10	29	11	31
4	7	17	12	24	13	27	14	30	15	33	16	36	17	39	18	42
5	7	20	13	27	19	36	20	40	22	43	24	46	25	50	26	54
6	8	22	14	30	20	40	28	50	30	54	32	58	33	63	35	67
7	9	24	15	33	22	43	30	54	39	66	41	71	43	76	46	80
8	9	27	16	36	24	46	32	58	41	71	52	84	54	90	57	95
9	10	29	17	39	25	50	33	63	43	76	54	90	66	105	69	111
10	11	31	18	42	26	54	35	67	46	80	57	95	69	111	83	127

Source: From F. Wilcoxon and R. A. Wilcox, "Some Rapid Approximate Statistical Procedures," 1964, 20–23. Courtesy of Lederle Laboratories Division of American Cyanamid Company, Madison, NJ.

Table B.7 Critical Values for the Wilcoxon Signed Ranks Test

One-Tailed	Two-Tailed	n = 5	n = 6	n = 7	n = 8	n = 9	n = 10
$\alpha = .05$	$\alpha = .10$	1	2	4	6	8	11
$\alpha = .025$	$\alpha = .05$		1	2	4	6	8
$\alpha = .01$	$\alpha = .02$			0	2	3	5
$\alpha = .005$	$\alpha = .01$				0	2	3
		n = 11	n = 12	n = 13	n = 14	n = 15	n = 16
$\alpha = .05$	$\alpha = .10$	14	17	21	26	30	36
$\alpha = .025$	$\alpha = .05$	11	14	17	21	25	30
$\alpha = .01$	$\alpha = .02$	7	10	13	16	20	24
$\alpha = .005$	$\alpha = .01$	5	7	10	13	16	19
		n = 17	n = 18	n = 19	n = 20	n = 21	n = 22
$\alpha = .05$	$\alpha = .10$	41	47	54	60	68	75
$\alpha = .025$	$\alpha = .05$	35	40	46	52	59	66
$\alpha = .01$	$\alpha = .02$	28	33	38	43	49	56
$\alpha = .005$	$\alpha = .01$	23	28	32	37	43	49
		n = 23	n = 24	n = 25	n = 26	n = 27	n = 28
$\alpha = .05$	$\alpha = .10$	83	92	101	110	120	130
$\alpha = .025$	$\alpha = .05$	73	81	90	98	107	117
$\alpha = .01$	$\alpha = .02$	62	69	77	85	93	102
$\alpha = .005$	$\alpha = .01$	55	61	68	76	84	92
		n = 29	n = 30	n = 31	n = 32	n = 33	n = 34
$\alpha = .05$	$\alpha = .10$	141	152	163	175	188	201
$\alpha = .025$	$\alpha = .05$	127	137	148	159	171	183
$\alpha = .01$	$\alpha = .02$	111	120	130	141	151	162
$\alpha = .005$	$\alpha = .01$	100	109	118	128	138	149
		n = 35	n = 36	n = 37	n = 38	n = 39	
$\alpha = .05$	$\alpha = .10$	214	228	242	256	271	
$\alpha = .025$	$\alpha = .05$	195	208	222	235	250	
$\alpha = .01$	$\alpha = .02$	174	186	198	211	224	
$\alpha = .005$	$\alpha = .01$	160	171	183	195	208	
		n = 40	n = 41	n = 42	n = 43	n = 44	n = 45
$\alpha = .05$	$\alpha = .10$	287	303	319	336	353	371
$\alpha = .025$	$\alpha = .05$	264	279	295	311	327	344
$\alpha = .01$	$\alpha = .02$	238	252	267	281	297	313
$\alpha = .005$	$\alpha = .01$	221	234	248	262	277	292
		n = 46	n = 47	n = 48	n = 49	n = 50	
$\alpha = .05$	$\alpha = .10$	389	408	427	446	466	
$\alpha = .025$	$\alpha = .05$	361	379	397	415	434	
$\alpha = .01$	$\alpha = .02$	329	345	362	380	398	
$\alpha = .005$	$\alpha = .01$	307	323	339	356	373	

Source: From F. Wilcoxon and R. A. Wilcox, "Some Rapid Approximate Statistical Procedures," 1964, p. 28. Courtesy of Lederle Laboratories Division of American Cyanamid Company, Madison, NJ.

Table B.8 Critical Values of Spearman's Rank Correlation Coefficient

The α values correspond to a one-tailed test of H_0: $\rho = 0$. The value should be doubled for two-tailed tests.

n	$\alpha = .05$	$\alpha = .025$	$\alpha = .01$	$\alpha = .005$	n	$\alpha = .05$	$\alpha = .025$	$\alpha = .01$	$\alpha = .005$
5	.900	—	—	—	18	.399	.476	.564	.625
6	.829	.886	.943	—	19	.388	.462	.549	.608
7	.714	.786	.893	—	20	.377	.450	.534	.591
8	.643	.738	.833	.881	21	.368	.438	.521	.576
9	.600	.683	.783	.833	22	.359	.428	.508	.562
10	.564	.648	.745	.794	23	.351	.418	.496	.549
11	.523	.623	.736	.818	24	.343	.409	.485	.537
12	.497	.591	.703	.780	25	.336	.400	.475	.526
13	.475	.566	.673	.745	26	.329	.392	.465	.515
14	.457	.545	.646	.716	27	.323	.385	.456	.505
15	.441	.525	.623	.689	28	.317	.377	.448	.496
16	.425	.507	.601	.666	29	.311	.370	.440	.487
17	.412	.490	.582	.645	30	.305	.364	.432	.478

Source: From E. G. Olds, "Distribution of Sums of Squares of Rank Differences for Small Samples," *Annals of Mathematical Statistics,* 1938, 9. Reproduced with the permission of the Institute of Mathematical Statistics.

Appendix C

Documentation for Microsoft Excel Diskette Files

C.1 CD-ROM Overview

The CD-ROM that accompanies this text contains all the data files used in this text. The data file icon ▣ in the margin of the text names the appropriate file. In addition, the CD-ROM contains the PHStat Add-In for Microsoft Excel.

C.2 File Contents

The CD-ROM that accompanies this text contains two types of files. Files with the extension .XLS are Microsoft Excel workbook files. Files with the extension .TXT are text files that contain the raw data sets. Presented below are an alphabetical listing and description of the files.

Name	Description of Variables
AAABONDS	Expense ratio rank and net return rank
ALCOHOL	Age and years of excessive drinking
ALLOY	Annealing time and passivation potential
APPRAISAL	Property, sales price, improvement value, area
APPRTRAIN	Training and performance score
ATLAS	Theme, high school teacher ranking, geography alumni ranking
ATTENT	Attention time
AUDITOR	Years of experience and salary
BACKPACKS	Price, volume, number of books
BEBR	County, refusal rate, per capita income
BEERCAL	Brand and number of calories per 12-ounce serving
BIDRIG	Type of contract and ratio of low-to-fair bid
BILE	Belief in lunar effects scale score
BIOFEED	Subject, before biofeedback and after biofeedback blood pressure readings
BLACKSOX	Black Sox player, player, number player game percentage (PGP), slugging average
BLCKJACK	Profit and strategy
BOEING720	Plane and life length of air-conditioning systems
BOILER	Man-hours, boiler capacity, boiler design pressure, boiler type, drum type
BOND	Length and whether the metal compound is copper-phosphorus
BONDPRICE	Underwriter and change in bond price
BONES	Length-to-width ratio of humerus bones
BONFERRONI	Excel workbook for obtaining Bonferroni multiple comparisons
BOSNIA	Order-to-delivery times
BOY16	Number of destructive responses
BRAIN	Veteran, verbal memory retention, right hippocampal volume

CANS	Amount of fill in ounces
CAVITIES	Child and cavities for sealant-coated teeth and untreated teeth
CELLPHON	Case number, hazard interval, control interval
CHKOUT	Checkout times in seconds
COLOGNE	Brand and intensity rating
COMMIT	Employee number, age, gender, length of organizational tenure, job teure, commitment scale, affective commitment scale, moral commitment scale, quitting scale, search scale, intent to leave scale, leave the organization
CONFINTVAR	Confidence interval for a variance workbook
CONICOTINE	Nicotine level and carbon monoxide level
COOKIES	Calories
COPEPOD	Percent of shallow-living copepods in diet of 35 species of myctophid fish
CORRTEST	Excel workbook for a test for a correlation coefficient
CRACK	Crack size and presence of flaw in specimens
CRASH	Class, make, model, doors, weight, driver star rating, passenger star rating, presence of driver air bag, presence of passenger air bag, driver head injury rating, passenger head injury rating, driver chest injury rating, passenger chest injury rating, driver left femur injury rating, driver right femur injury rating, passenger left femur injury rating, passenger right femur injury rating
CREATIVITY	Child, flexibility score, creativity scores
CROPWT	Crop weight
CUSTSAT	Country of origin and customer satisfaction index
DARTS	Distance from target and frequency
DAYSABSENT	Number of days absent
DECEPT	Student, deception score, percent increase in pupil size
DECODE	Group and trials required for decoding
DEFAULT	Student loan default rate
DENTAL	Discomfort level of novocaine and new anesthetic
DESTRUCT	Number of destructive responses
DIAZINON	Day, level of diazinon residue during day, level of diazinon residue at night
DISCRIM	Hiring status and gender
DOTHOUR	Station, traffic counts in 30th highest hour, 100th highest hour
DRILL	Drilling depth and time to drill 5 feet
DRILLBIT	Drill bit and penetration
EEG	Age and average peak EEG frequency
EERAC	Energy efficiency rating (EER)
ELECUTIL	Rate of return of electric utilities
ENDOW	Endowments in millions of dollars
ENTRAP	Points awarded
FACTORS	Number of factors and length of stay
FASTFOOD	City and weekly sales at stores
FERTIL	Country, contraceptive prevalence, fertility rate

FHALOAN	Loan type, default rate percent, expected loss percent
FIRE	Deaths from fire
FIREFRAUD	Month, invoice number, selling price, profit, profit margin
FISH	Location, species, length, weight, amount of DDT
FOREST	Breast-height diameter of trembling aspen trees
FOREST1	Breast-height diameter and height of white spruce trees
FTC	Brand, length, menthol, filter, light, pack, tar content, nicotine, carbon monoxide
FUNDRAISE	University, total funds raised, alumni contributions
GASKETS	Gaskets produced for machines and materials
GEM	Retail price per carat and demand
GEO	Scores for group 1 and scores for group 2
GEOSCORE	Student, score on test 1, score on test 2
GOLD	Production of gold
GPACLASS	Grade point averages for lower class, middle class, upper class
GPS	Location, distance error, percent of readings with three satellites
GRANULES	Percent of coarse granules
GUPPY	Number of guppies migrating
HAIRPAIN	Pain threshold score for light blond, dark blond, light brunette, dark brunette
HEAT	Unflooded area ratio and heat transfer ratio
HELIUM	Proportion of impurity passing through helium and temperature
HEMATO	White blood cell and lymphocyte counts
HLDI	Collision damage rating
IDTWIN	Twin A IQ and twin B IQ
IGG	Subject, amount of immunoglobulin, maximal oxygen uptake
IODINE	Iodine concentration measurement
IOUNITS	Input-output units utilized
KIDNEY	Weight for the factors of size and diet of rats
KNEESURG	Number of post-surgical hospitalization days for the factors of type of knee surgery and age group
LANDSALE	Pair, land sale ratio, downward price adjustment
LIQCO2	Amount of liquid carbon monoxide
LONGJUMP	Jumper, best jumping distance, average takeoff error
MAMMO	Information group, numeracy score, perceived risk reduction of mammography
MANHRS	Man-hours required
MATHCPU	Time to solve programming problems
MONRENTS	Monthly rents for two-bedroom apartments
MOSQUITO	Date, average temperature, catch ratio
MOVIES	Movie, rating by Siskel, rating by Ebert
MUCK	Depth of muck
MUM	Time between end of test and delivery of feedback for the factors of subject visibility and confederate success

NBA	Wins, standardized difference between team and opponent two-point field goal percentage, standardized difference between team and opponent rebounds, standardized difference between team and opponent turnovers
NEMATODE	Dry weight and month
NOSHOW	Number of cancellations
OILSPILL	Spillage amount and cause of puncture
ORING	Flight number, temperature, and O-ring damage index
OTRATIO	Day, condition, thion, oxon, oxon/thion ratios
OXONTHION	Day, weather condition, oxon/thion ratios
PCB	PCB level
PCBSOIL	Location and PCB level
PCBUK	Location and PCB level
PCBWATER	PCB 1984, PCB 1985, name of bay
PH	Food and pH values
PHOSROCK	Location and time for chemical reaction
PHYSICS	Student, inventory score, baseline score
PIPES	Breaking strength of sewer pipe
PMI	Postmortem interval
POWVEP	POW and VEP measurements 157 and 379 days after release
PRINTER	Brand, price, text speed
PROB12_39	Group and values
PROTEIN	Food, calories, protein, fat, saturated fat, cholesterol
RADIUM	Mean radium-226 level in soil samples
RAIN	Observation, station, direction facing, average annual precipitation, altitude, latitude, distance from coast
RAS	RAS scores of nontraditional and traditional gender attitudes
REACTIME	Driver, restrained reaction time, unrestrained reaction time
READING	Test scores in reading for method 1 and method 2
REFRIG	Brand, price, energy cost
ROOMBOARD	College and room and board costs
RUBIDIUM	Trace amount
SAL50	Starting salaries
SAMPLE50	Sample number and five measurements
SANIT	Sanitation inspection scores
SATSCORE	Student, junior year SAT score, senior year SAT score
SCALLOPS	Scallop weights
SCRAM	Number of unplanned scrams
SIGN	Excel workbook for the sign test
SILICA	Silica determination in parts per million
SMALLGRP	Group and grade

SMOKE	Weight percentile and number of cigarettes smoked
SMOKING	Brand, carbon monoxide level, nicotine
SOCIOLOGY	Student, verbal ability test score, final sociology grade
SORPRATE	Sorption rate for aromatics, chloroalkanes, and esters
SPEAKER	Interruption number, speaker gender, interrupter gender
SPL	Sound pressure level
STAYLEAV	Employment status and performance rating
SUICIDE	Days in jail before committing suicide
SUPRMKT	Winn-Dixie price and Publix price
SWIMMAZE	Litter, number of swims for males, number of swims for females
TAMSALES	Sales price and neighborhood
TAXRATE	Software program and satisfaction score
TBAD	Secondary specialty, country of medical school, years of experience, total accrued cost per patient per month
TBEAMS	Cracking torsion moments of 70-cm slab widths and 100-cm slab widths
TCDD	TCDD level in plasma
TCDDFAT	TCDD level in plasma and TCDD level in fat tissue
TELEPHON	Country, charge per minute, minutes expended
TESTAB	Test and test score
TESTVARIANCE	Test for a variance workbook
THRUPUT	Task and throughput rates for a human scheduler and an automated system
TINLEAD	Shear strength with the factors of antimony amount and cooling method
TIRES	Pressure and mileage of tires
TRAIN	Arrival time lateness
TRAINING	Time and type of training
TRANSFORM	Failures per 1,000 pieces for factors of hours and levels of inspection
TRIPLETS	Shear strength and pre-compression stress
TWINS	Pair number, twin A IQ score, twin B IQ score
USFRAISE	Administrator, raise, average rating
VANADIUM	Vanadium trace amount in oyster tissue, citrus leaves, bovine liver, and human serum
VCR	Repair time for factors of service center and brand
VCRPRIC	Retail price of VCRs
VIDEOS	Movie, box office gross, home video units sold
VOLTAGE	Voltage reading at old location and new location
VOTERS	Voter, rating, age
WAREHOUS	Number of vehicles and congestion time
WATER	Parts per million
WHALES	Whale part, AUI ranking, MUI percent
WILDFLOWER	Insects visiting wildflower clump

Appendix D

Installation Instructions for the PHStat Add-In for Microsoft Excel and the Data Files on the CD-ROM

The CD-ROM packaged with this text contains Microsoft Excel workbook files and the files necessary to set up and install PHStat, the Prentice Hall statistical add-in for Microsoft Excel.

Excel workbook files are located in the Excel directory on the CD-ROM. This directory contains workbook files (.XLS) for each of the data sets used in this text for which an icon is displayed next to the data set along with the data set name. In addition, a separate directory contains these files in text (.TXT) format.

Files necessary to set up and install PHStat are located in the PHStat directory. To use PHStat, you must first run the setup program (Setup.exe) in that directory. The setup program installs the files that are required by PHStat and creates Desktop and Start icons that facilitate its use. Review the PHStat readme file (PHStat readme.doc) before running the setup program to learn all about the setup process and the technical requirements and limitations of PHStat.

Answers to Self-Test Problems

Chapter 1

1.1a. qualitative   **b.** quantitative   **c.** quantitative   **d.** qualitative

1.2a. all U.S. school children   **b.** 1,000 children that were included in the survey   **c.** number of cavities; quantitative

1.3 length of maximum span: quantitative; number of vehicle lanes: quantitative; structural condition: qualitative; observational study; population

1.4 begin at row 30, column 6: 25940 (use 2); 35126 (use 3); 88072 (use 8)

1.5 selection bias

Chapter 2

2.1 potential growth deficiency, reduced head circumference, and lowered IQ

2.2a. 1, 2, 3, 4, and 5

b.

1	2 7 9
2	1 2 3 5 6 7 8
3	0 2 4 6 8 9 9
4	1 2
5	6

c. .15, .35, .35, .10, .05

2.3a. 20   **b.** .2667

2.4a. 22% of the males sampled indicated that they reached puberty earlier.   **b.** Male and female attitudes about reaching puberty appear to be very similar.

2.5 negative

2.6 bars increase in both height and width; distorts magnitude of differences

Chapter 3

3.1a. 12   **b.** 40   **c.** 7   **d.** 21

3.2 $\bar{x} = 5.7$; $M = 4.5$; mode $= 4$

3.3 range $= 8$; $s^2 = 7.556$; $s = 2.749$

3.4a. approximately 70%   **b.** approximately 95%   **c.** approximately 100%

3.5a. 20%, 80%   **b.** 50%, 50%   **c.** 86%, 14%   **d.** 25%, 75%   **e.** 75%, 25%   **f.** 10%, 90%

3.6 IQR $= 70$

3.7a. $-.667$; not an outlier   **b.** 6; outlier

3.8 $-.9163$

Chapter 4

4.1a. 0, 1, 2, and 3   **b.** the likelihood that a family with three children has exactly 3 boys

4.2 (First does not use coupons and second does not use coupons), (First does not use coupons and second does use coupons)

4.3 $P(A) = 1/10$; $P(B) = 6/10$; $P(C) = 3/10$

4.4 9/10

4.5 120

4.6 $\frac{1}{3}$

4.7 1/36

4.8a. .7   **b.** .75

Chapter 5

5.1a. continuous   **b.** discrete   **c.** discrete   **d.** continuous

5.2a. .10   **b.** 1.3   **c.** 1.345

5.3a. .009   **b.** .000000025   **c.** .00145

5.4 yes, $P(x < 14) = .00239$

5.5 $\mu = 200$; $\sigma = 6.325$

5.6a. .250   **b.** .5578   **c.** 1.5

5.7a. 20/35   **b.** 4/35

Chapter 6

6.1a. .2734   **b.** .4873   **c.** .0737   **d.** .84

6.2 .0080

6.3a. no, IQR/$s = 4$ **b.** yes   **c.** yes   **d.** no

6.4c. small number of samples (20)

6.5a. 5   **b.** .10

6.6 .2266

Chapter 7

7.1a. proportion of intervals that enclose target parameter   **b.** value such that $P(z > z_{\alpha/2}) = \alpha/2$

7.2 485 ± 15.56

7.3 1.943

7.4 485 ± 150.21

7.5a. .4   **b.** $.4 \pm .033$

7.6 $.00896 \pm .00584$

7.7 68

7.8 26.1190

7.9 (64.86, 300.96)

Chapter 8

8.1 $H_0: \pi = .65; H_a: \pi < .65$

8.2a.   conclude $\mu < 16$ when $\mu = 16$ **b.** conclude $\mu = 16$ when $\mu < 16$ **c.** Type II error   **d.** Type I error

8.3 $z < -1.28$

8.4a. μ **b.** $H_0: \mu = 60; H_a: \mu > 60$

8.5 $z = -3.37$, reject H_0

8.6 $t = -1.066$, do not reject H_0

8.7a.   do not reject H_0 **b.** reject H_0 **c.** reject H_0

8.8 $z = -2.35$, reject H_0

8.9 $\chi^2 = 6.975$, do not reject H_0

Chapter 9

9.1a. $\mu_1 - \mu_2$ **b.** σ_1^2/σ_2^2

9.2a. 2.5 ± 3.47 **b.** $t = 1.41$, do not reject H_0

9.3a. $t = .298$ **b.** $t > 1.701$

9.4a.   yes, $z = 6.51$ **b.** 115 ± 34.65

9.5a. $z = -2.53$, reject H_0 **b.** $-.08 \pm .062$

9.6

		Column	
		1	**2**
Row	1	18.75	11.25
	2	81.25	48.75

9.7a.   .27   **b.** $\chi^2 > 6.6349$ **c.** do not reject H_0

9.8 $F > 2.19$

9.9 $F > 4.77$

Chapter 10

10.1a.   dependent = sales price; independent = living area, assessed value, rooms
b. $y = \beta_0 + \beta_1 x_1 + \beta_2 x_2 + \beta_3 x_3 + \varepsilon$

10.2 y-intercept = -100; slope = 67

10.3b.   Liquid H: 102.5; Solid H: 120.6

10.4a.   6.124

10.5 $t = 1.18$, do not reject H_0

10.6 $.26 \pm .36$

10.7 $t = 3.55$

10.8   model explains 62% of sample variation in salary

10.9a.   32,000   **b.** 30,000

10.10a.   95% confident that mean salary for all workers with 8 years of experience is between $27,500 and $32,500   **b.** 95% confident that the salary for an individual worker with 8 years of experience is between $23,000 and $37,000

10.11 $\hat{y} = .7 + 1.5x$

10.12   SSE = .3; $s = .316$

10.13   15.0

10.14 $1.5 \pm .32$

10.15   .987

10.16a. $(-.45, 1.85)$ **b.** $(3.15, 4.25)$

Chapter 11

11.1   response = distance; factor = brand; levels = 4 brands; treatments = 4 brands; experimental units = driver hits

11.2a. $H_0: \mu_A = \mu_B = \mu_C = \mu_D$ **b.** 43.99   **c.** reject H_0

11.3a.   response = distance; factors = brands and clubs; treatments = $(A, \text{driver}), (A, 5\text{-iron}), (B, \text{driver}), (B, 5\text{-iron}), (C, \text{driver}), (C, 5\text{-iron}), (D, \text{driver}), (D, 5\text{-iron})$ **b.** 4

11.4a.   interaction   **b.** no   **c.** yes; club main effect significant, no brand main effect

11.5a.   6   **b.** .01   **c.** 5.6   **d.** C; D or A   **e.** overall confidence level of .94

11.6   yes

Chapter 12

12.1a.   2   **b.** .8125

12.2a.   20, 22, 25, 27, 31, 32, 40, 45, 48, 50   **b.** $T_1 = 21$; $T_2 = 34$ **c.** $T_2 = 19$ **d.** do not reject H_0

12.3a.   differences: $-721, -105, 42, -91, 13$; ranks: 5, 4, 2, 3, 1
b. $T^+ = 3; T^- = 12$ **c.** $T^+ \le 1$
d. do not reject H_0

12.4a.   1,411.2, 45, 500, 744.2   **b.** 14.15   **c.** 7.81473
d. reject H_0

12.5a.   1, 2, 3, 4, 5   **b.** 5, 4, 2, 3, 1   **c.** $-4, -2, 1, 1, 4$
d. $-.9$

Answers to Selected Problems

Chapter 1

1.1a. quantitative **b.** qualitative **c.** qualitative
d. quantitative **e.** quantitative **1.3a.** individual property
b. quantitative; quantitative; qualitative; qualitative
1.5a. quantitative **b.** quantitative **c.** qualitative
d. qualitative **e.** qualitative **f.** quantitative **g.** qualitative **h.** quantitative **1.7a.** quantitative **b.** quantitative
c. qualitative **d.** qualitative **e.** qualitative **f.** qualitative **g.** quantitative **1.9a.** qualitative **b.** qualitative
c. quantitative **1.11a.** sample **b.** sample **c.** sample
d. sample **1.13a.** 1 **b.** individual question **c.** qualitative **d.** sample **1.15a.** U.S. federal court case **b.** 3,854
U.S. federal court cases that occurred between 1850 and
1859 **c.** whether or not the court case involved tyranny;
qualitative **d.** population **1.17a.** all adult Americans
b. no **1.19a.** diameters of all the stones found in the delta
region of the Amazon **b.** diameters of the 50 stones
collected by the geographer **c.** extremely unlikely
1.21 observational **1.23** designed experiment
1.25a. designed experiment **b.** productivity measurements of all the students falling in the four groups **c.** the
20 students in each of the four groups **d.** yes **e.** sample
data **1.27a.** 034 **b.** 635 **c.** 409 **d.** 03 **1.29** beginning
at row 4, column 3: 624, 785, 1637, 3944, and 84 **1.31** beginning at row 14, column 1: 103, 482, 15, 163, and 202
1.33 method B **1.35** probably not be representative of
business employees **1.37** nonresponse **1.39** begin at row
30, column 6: 25,940; 35,126; 27,354; 18,317; 17,032; 25,774;
38,857; 24,413; 34,072; and 4,542 **1.41a.** quantitative
b. qualitative **c.** qualitative **d.** quantitative **1.43a.** all
sea turtle hatchlings traveling under the three conditions
b. set of directions that the 60 sea turtle hatchlings travel
c. qualitative **d.** experimental design **1.45a.** high school
junior **b.** change in the SAT score **c.** differences in SAT
scores of all students taking a beta blocker one hour before
their second attempt **d.** experimental design **e.** 22 SAT
differences **f.** yes

Chapter 2

2.1a. A: .26; B: .56; C: .18 **2.5a.** vast majority of scrapped
automobile tires (77.6%) are dumped **2.7c.** 46%
2.9a. experimental unit: artifact found at excavation site;
variable: type of artifact found (drillpoint, bead, or grinding
slab) **c.** drillpoints; 57% **2.11d.** support Jerry Jones's
claim **2.13** Anomic type of Aphasia is the most common

2.15

3	3
4	0 2
5	0 1 4
6	1 3 6 6 7 9
7	2 5
8	1 3 7
9	0
10	2 5

2.17a. 21, 22, 23, 24, 25, 26, 27, 28, 29, 30, 31, and 32

b.

21	3 5
22	4 6 7 8
23	4 7
24	1
25	4
26	5 6 7 8
27	0 4
28	5 8
29	1
30	3 3
31	6 9
32	0

2.19a. 8.6 **b.** 1.72 **c.** $(1.10 - 2.82); (2.82 - 4.54)$;
$(4.54 - 6.26); (6.26 - 7.98); (7.98 - 9.70)$

2.21a.

1	7 7 7 8 8 9 9
2	0 0 1 2 2 5 5 6 7
3	0 0 2 4 5
4	5

b. under $200 **c.** no

2.23a. frequency histogram **b.** approximately 250
c. very slightly skewed to the right

2.27a.

2	8
3	1
4	1 1 4 4 4
5	3 3 3 9 9
6	3 3 6 6 6 9 9 9 9 9
7	2 2 5 5 5 5 5 8 8 8
8	4 4 4 8
9	1 1 1 1 1 7 7

b.

0	9
1	9 9
2	5
3	1
4	4 4 7
5	0 0 3 3 6 9 9
6	3 6 6 6 6
7	8 8

c. group 1 tended to perform better

2.29a. see below **b.** yes **2.31a.** (X,A), (X,B), (Y,A),
(Y,B), (Z,A), and (Z,B) **b.** 1; 3; 3; 2; 1; 2 **c.** .2; .4286; .6;
.2857; .2; .2857 **d.** .25; .75; .6; .4; .3333; .6667

Problem 2.29a

0	0 0 1 1 2 2 2 2 3 3 3 3 3 3 3 4 4 4 4 4 4 4 5 5 5 5 6 6 6 6 6 6 7 7 7 7 7 8 8 8 8 8 9
1	0 0 0 0 0 0 0 1 2 2 3 6
2	
3	1 3
4	4

	North	East	South	West	Total
A	0.1053	0.2105	0.4211	0.2632	1.0000
B	0.4000	0.3200	0.0800	0.2000	1.0000
C	0.3529	0.2353	0.1176	0.2941	1.0000
Total	0.2500	0.2500	0.2500	0.2500	1.0000

Problem 2.33d

	North	East	South	West	Total
A	0.2000	0.4000	0.8000	0.5000	0.4750
B	0.5000	0.4000	0.1000	0.2500	0.3125
C	0.3000	0.2000	0.1000	0.2500	0.2125
Total	1.0000	1.0000	1.0000	1.0000	1.0000

Problem 2.33e

2.33a. 400   **b.** 100   **c.** 25   **d.** see above   **e.** see above
2.35a. nationality (British, German, or American); humorous (yes or no)

b.

	Yes	No	Total
British	0.2562	0.7438	1.0000
German	0.2292	0.7708	1.0000
American	0.2074	0.7926	1.0000
Total	0.2286	0.7714	1.0000

2.37a. complication type (redo surgery, post-op infection, both, or none); drug (yes or no)

b.

	Yes	No	Total
Redo Surgery	0.1228	0.0877	0.1053
Infection	0.1228	0.0702	0.0965
Both	0.0526	0.0175	0.0351
None	0.7018	0.8246	0.7632
Total	1.0000	1.0000	1.0000

d. possibly, but difference between the two groups is small
2.39c. yes   **2.41a.** positive   **b.** none   **c.** negative
2.43b. no   **2.45b.** no   **2.47b.** yes; negative   **c.** yes
2.49b. positive   **2.55** very little chart junk   **2.57** considerable chart junk; bars distort the data   **2.63b.** yes; slightly
2.65a. quantitative   **b.** relative frequency histogram; approximately 64%

2.67a.

```
22 | 3 3 4 5 6 6 6
23 | 7 8 8 9 9 9
24 | 0 0 0 0 1 1 6 6 6 6 6 7 8
25 | 0 0 3 5 6 6 6 9
26 | 1 1 2 5 6 8
27 | 2 4
28 | 4
29 | 1
```

b. skewed to the right   **c.** no trend   **2.69b.** yes; positive
c. yes

2.73a.

```
0 | 1 2 2 2 4 5 8 8 9
1 | 0 0 2 5
2 | 3
```

b.

```
 1 | 1 1 2 3 6 8 8
 2 | 1 2 4 9
 3 |
 4 | 9
 5 |
 6 |
 7 |
 8 |
 9 | 4
10 | 7
11 |
12 |
13 |
14 | 1
```

c. yes;

```
 0 | 1 2 2 2 4 5 8 8 9
 1 | 0 0 1 1 2 2 3 5 6 8 8
 2 | 1 2 3 4 9
 3 |
 4 | 9
 5 |
 6 |
 7 |
 8 |
 9 | 4
10 | 7
11 |
12 |
13 |
14 | 1
```

2.75 graphic represents the data fairly well

Chapter 3

3.1a. 58   **b.** 690   **c.** 3,364   **d.** -7 **e.** 27   **3.3a.** 17.2
b. 30   **c.** 680   **3.5a.** 6   **b.** 50   **c.** 42.8   **3.7a.** 6   **b.** 122
c. 0.25   **3.9** mean $= 4.6$; median $= 4$ **3.11a.** 5; 5; 5
b. 12; 5; 5   **c.** 0; 0; 0   **d.** 0; 4.5; 9   **3.13a.** mode $= 9$
b. mean $= 7.6$ **c.** median $= 8$ **3.15a.** mean $= 3$;
mode $= 4$ **b.** mean $= 14.3$ **c.** use median due to
skewness   **d.** mean will decrease, but median and mode
unchanged   **3.17** mean $= 91.044$; median $= 92.000$; mode
$= 89$ **3.19** mean $= 4.246$; median $= 4.2$; mode $= 4.4$
3.21 mean $= 75.575$; median $= 75.15$; no mode
3.23a. mean $= 68.86$; median $= 69$; mode $= 91$ **b.** mean $=$
50.045; median $= 53$; mode $= 66$ **c.** group 1:
approximately symmetric; group 2: skewed to the left
d. yes   **3.25a.** 10, 14, 3.742   **b.** 10, 10.4, 3.225   **c.** 0, 0, 0
3.27 $\bar{x} = 4.64$; $s = 1.912$ **3.29** $\bar{x} = 12.75$; $s = 3.51$
3.31a. 280   **b.** 5,367.8   **c.** 73.27   **d.** 245.23   **e.** (98.70,
391.76)   **3.33a.** 4.5   **b.** 1.317   **c.** (0.566, 5.834)
3.35a. (3.95, 12.03)   **b.** no   **3.37a.** 74.31   **b.** 20.94
c. (32.43, 116.18)   **3.39a.** $\bar{x} = 84.91$; $s^2 = 261.26$; $s = 16.16$
b. (52.59, 117.24)   **c.** 34   **3.41a.** $\bar{x} = 6.03$; $s^2 = 66.10$,
$s = 8.13$ **b.** approximately 95%   **c.** 19   **d.** decrease;
decrease   **3.43a.** -1.6 **b.** .6   **c.** 1.6   **3.45a.** $M = 6.1$;
$Q_1 = 3.3$; $Q_3 = 8.2$ **b.** 1.2   **3.47a.** 40%   **b.** 5%   **c.** 67%
d. 85%   **3.49** 75% of the checkout times fall below 65
seconds   **3.51** $Q_1 = 89$; $M = 92$; $Q_3 = 95$ **3.53a.** $z = -.76$
b. $z = 1.87$ **3.55a.** $Q_1 = 76.5$; $M = 92$; $Q_3 = 99$ **b.** 100
c. $z = -0.242$ **3.57a.** 98   **b.** 122   **c.** 24   **d.** 75   **e.** 143
g. none   **3.59** $x_{\text{smallest}} = 213$; $Q_1 = 228$; $M = 266.5$; $Q_3 =$
291, $x_{\text{largest}} = 320$ **3.61a.** .5   **b.** 7.5   **c.** -3.9 **d.** -2
3.63a. Five-number summary: 9, 13.25, 15, 18, 20
b. skewed left   **c.** no outliers   **d.** no outliers   **3.65a.** no
skewness   **b.** yes, 4 of the 500 nicotine contents fall above

the value 1.7. **3.67a.** 2.8, 4.05, 5.65, 7.8, 82 **b.** yes, 82 in city C **c.** $z = 4.65$ **d.** $\bar{x} = 5.99$; $s = 2.28$; Five-number summary: 2.8, 4.05, 5.65, 7.75, 10.6; no outliers **3.69a.** .20 **b.** $-.70$ **c.** .625 **d.** -1.00 **3.71a.** .993 **b.** .174 **c.** $-.989$ **3.73** .2509 **3.75** positive **3.77** .7937 **3.79** moderate positive relationship **3.81** $-.865$ **3.83** decreasing order of skewness: Tampa Palms, Carrollwood Village, Northdale, Ybor City, and Town & Country **3.85a.** mean = 37; median = 40 **b.** mean = 6.71; median = 6 **c.** median **d.** mean **3.87a.** calories: 193.4, 190, 98, 299, 4799.5, 69.28, 157, 239, 82; protein: 26.52, 27, 27, 17, 14.09, 3.75, 16, 29, 5; fat: 36.56, 37, 37, 71, 407.4, 20.18, 6, 51, 33; saturated fat: 12.72, 14, 12, 27, 70.71, 8.41, 0, 20, 15; cholesterol: 102.2, 89, 89, 443, 6992.67, 83.62, 39, 94, 23 **c.** calories and cholesterol: skewed to the right; protein: skewed to the left; fat and saturated fat: symmetric **d.** calories: (124.12, 262.68) 64%, (54.84, 331.96) 96%, $(-14.44, 401.24)$ 100%; protein: (22.77, 30.27) 80%, (19.01, 34.03) 96%, (15.26, 37.78) 100%; fat: (16.38, 56.74) 60%, $(-3.81, 76.93)$ 96%, $(-23.99, 97.11)$ 100%; saturated fat: (4.31, 21.13) 60%, $(-4.10, 29.54)$ 100%, $(-12.51, 37.95)$ 100%; cholesterol: (18.58, 185.82) 96%, $(-65.04, 269.44)$ 96%, $(-148.67, 353.07)$ 96% **e.** 482 mg for cholesterol **f.** calorie, protein: .4644; calorie, fat: .8509; calorie, saturated fat: .8888; calorie, cholesterol: .1777; protein, fat: .4606; protein, saturated fat: .4984; protein, cholesterol: .1417; fat, saturated fat: .9299; fat, cholesterol: .1088; saturated fat, cholesterol: .1083 **3.89a.** 14.682 **b.** $s^2 = 199.974$; $s = 14.141$ **c.** approximately 95% **d.** .91 **e.** decrease; decrease **f.** $\bar{x} = 13.74$; $s = 11.961$ **g.** approximately 95%; (.91) **3.91a.** with ants: $(-1.9, 7.7)$; without ants: $(-1.2, 16.4)$ **b.** approximately 95% **3.93a.** 117.82, 117.5, 97 **b.** 62, 225.3343, 15.01 **c.** (102.809, 132.831): 31, .62; (87.798, 147.842): 49, .98; (72.787, 162.853): 50, 1.00; no outliers **3.95a.** $\bar{x} = 29.45$; $s = 9.33$ **b.** 7, 23.75, 30, 35, 57 **c.** (20.12, 38.78): 28, .70; (10.79, 48.11): 38, .95; (1.46, 57.44): 40, 1.00 **d.** 18 **e.** slightly positively skewed **f.** 57 inches

Chapter 4

4.1a. 1/6 **4.3a.** .08 **b.** .43 **c.** .71 **d.** .95 **4.5** .540 **4.7** no; .70 **4.9a.** $P(A) = .60$; $P(B) = .60$, $P(C) = .75$ **b.** $A' = \{S_3, S_5\}$; $B' = \{S_1, S_4\}$; $C' = \{S_4\}$ **c.** $P(A') = .40$; $P(B') = .40$; $P(C') = .25$ **4.11a.** $\frac{1}{4}$ **b.** $\frac{1}{4}$ **c.** $\frac{3}{4}$ **d.** $\frac{3}{4}$ **4.15a.** Huggies, Pampers, Private, Luvs, and Other **b.** .317, .265, .210, .142, .066 **c.** .407 **d.** .683 **4.17a.** .1739 **b.** 0 **c.** Yes **d.** .6522 **4.19a.** .2685 **b.** .1128 **c.** .8794 **4.21a.** .5694 **b.** .3873 **c.** .0433 **d.** .6127 **e.** .6127 **4.23a.** .0476 **b.** .2857 **4.25a.** $A, B, C, D, E, AB, AC, AD,$ $AE, BC, BD, BE, CD, CE, DE, ABC, ABD, ABE, ACD,$ $ACE, ADE, BCD, BCE, BDE, CDE, ABCD, ABCE,$ $ABDE, ACDE, BCDE, ABCDE,$ and none **b.** .28125 **4.27a.** 1 **b.** 50 **c.** 50 **d.** 1 **4.29a.** 10 **b.** $ABC, ABD,$ $ABE, ACD, ACE, ADE, BCD, BCE, BDE, CDE$ **4.31** 28 **4.33** 1,820 **4.35a.** 2,598,960 **b.** 4 **c.** .0000015 **4.37** .18

4.39a. no **b.** no **c.** no **4.41a.** .40 **b.** .60 **4.43a.** .4833 **b.** .2250 **4.45a.** $P(S) = .34$; $P(I) = .32$; $P(S \mid I) = .58$; $P(S \mid I') = .23$ **b.** no **4.47a.** 1/181,795,000; 1/200,000,000; 1/84,000,000 **b.** 3.27×10^{-25} **c.** .9999999776 **d.** .0000000224 **4.51** .6 **4.53** .06 **4.55a.** (1,6), (2,5), (3,4), (4,3), (5,2), and (6,1); 6/36 **b.** (1,4), (2,4), (3,4), (4,4), (5,4), (6,4), (4,1), (4,2), (4,3), (4,5), and (4,6); 11/36 **c.** (3,4) and (4,3); 2/36 **d.** (1,6), (2,5), (3,4), (4,3), (5,2), (6,1), (1,4), (2,4), (4,4), (5,4), (6,4), (4,1), (4,2), (4,5), and (4,6); 15/36 **4.57** .24 **4.59a.** .000025 **b.** .00002 **4.61a.** .50 **b.** .7632 **c.** .3509 **d.** .7017 **e.** .8246 **4.63a.** .53 **b.** .05 **c.** .89 **d.** .41 **e.** .0205 **f.** B and C **4.65a.** $P(A \mid B)$ is unknown **b.** .035 **4.67a.** .1221 **b.** .7187 **c.** .6825 **4.69a.** Basic browns, True-blues, Greenback greens, Sprouts, and Grousers **b.** .28, .11, .11, .26, .24 **c.** .52 **d.** .48 **4.71** all are mutually exclusive **4.73a.** (40, 300), (40, 350), (40, 400), (45, 300), (45, 350), (45, 400), (50, 300), (50, 350), (50, 400) **b.** probably not **4.75a.** .46 **b.** .68 **c.** .20 **4.77a.** 142,506 **b.** .000007 **4.79a.** .0612 **b.** .1400 **c.** .1826 **d.** .0902 **e.** .0397 **f.** .9671 **4.81a.** .0256 **b.** .0001049 **c.** independence **4.83** 635,013,559,600 **4.85a.** .4927 **b.** .5073 **c.** .1169 **4.87b.** P(woman has two boys | woman has at least one boy) $= \frac{1}{3}$; P(man has two boys | older child is a boy) $= \frac{1}{2}$

Chapter 5

5.1a. $-5, 0, 2, 5$ **b.** 2 **c.** .5 **d.** .2 **f.** $\mu = .3$; $\sigma = 3.002$ **5.3a.** .8, .16, .032, .0064, .00128 **b.** .99968; no **c.** .96 **5.5b.** .1875 **c.** .5875 **d.** 12.425 **e.** 2.235 **f.** (7.955, 16.895) **g.** .95 **5.7a.** $\mu = 15.14$; $\sigma = 1.114$ **b.** (12.912, 17.368) **c.** .29 **5.9** $p(1) = .09$; $p(2) = .42$; $p(3) = .49$ **5.11a.** $p(0) = .09091$; $p(1) = .48485$; $p(2) = .42424$ **b.** $\mu = 1.333$; $\sigma = .6356$ **5.13a.** 24 **b.** 4 **c.** 10 **d.** .064 **e.** .2304 **5.15** no **5.17a.** .8192 **b.** .1808 **c.** complements **d.** sum to 1 **5.19a.** .9983 **b.** .9877 **c.** .00168 **5.21a.** $\mu = 297$; $\sigma = 1.723$ **b.** $\mu = 160$; $\sigma = 5.657$ **c.** $\mu = 50$; $\sigma = 5$ **d.** $\mu = 20$; $\sigma = 4$ **e.** $\mu = 10$; $\sigma = 3.146$ **5.23a.** (1) .3874, (2) .7361, (3) .6513, (4) .3487, (5) .2639 **b.** $\mu = 9$ **5.25** yes **5.27a.** .0287 **b.** .0333 **c.** $\pi = .2$ suspect **5.29a.** $p(0) = .125$; $p(1) = .375$; $p(2) = .375$; $p(3) = .125$ **b.** .875 **c.** no **5.31** no **5.33a.** 24 **b.** support; $P(x \le 20) = .1978$ **5.35a.** .0012 **b.** .0166 **c.** 41.6 **d.** 3.816 **e.** policy effective **5.37a.** $p(x) = (2.5)^x(.082085)/x!$ **b.** $\mu = 2.5$; $\sigma = 1.58$ **c.** .958 **5.39a.** .981 **b.** .857 **c.** .537 **d.** .265 **e.** .059 **5.41a.** .1295 **b.** .3397 **c.** .1468 **d.** .7953 **e.** .384; .125; .040; .722 **5.43a.** .205 **b.** .544 **c.** no, $P(x \ge 7) = .0142$ **5.45a.** .3491 **b.** .0303 **c.** $\sigma = 4.243$; (9.514, 26.486) **d.** independence assumption probably invalid **5.47a.** 350/792 **b.** 246/792 **c.** 21/792 **d.** 196/792 **e.** 2.92 **f.** .88 **5.49a.** 0 **b.** 21/252 **c.** 1/3 **d.** 1/6 **5.51a.** .0000000387 **b.** .00001115 **c.** .000655 **d.** .01339 **e.** .47514 **f.** .0000002605, .00005315, .00219, .03118, .35038

5.53a. .3   **b.** 1   **5.55** .291, yes   **5.57a.** .2734   **b.** .3633
c. .03516   **5.59a.** .6   **b.** .24   **c.** .096   **d.** .0384
5.61a. 30,800   **b.** 166.71   **c.** percentage understated,
$z = 25.19$ **5.63a.** .8333   **b.** .5   **c.** .0333   **d.** 1.2 ± 1.496
5.65 no   **5.67a.** probably not   **b.** no   **c.** probably not
5.69a. .0282   **b.** .9984   **c.** .000144

Chapter 6

6.1a. .3849   **b.** .4319   **c.** .1844   **d.** .4147   **e.** .0918
6.3a. .25   **b.** .92   **c.** 1.28   **d.** 1.645   **e.** 1.96   **6.5a.** .75
b. -1 **c.** -1.625 **d.** 2   **e.** -2 **6.7a.** -1.28 **b.** -1.04
c. $-.52$ **d.** 0   **6.9a.** .0668   **b.** .8664   **c.** .1336   **d.** .0668
e. 146.8 seconds   **6.11a.** .2033   **b.** no   **c.** 0   **d.** no
e. 23.24   **6.13a.** 9%   **b.** 11.01%   **c.** 13.92%
6.15a. .0228   **b.** .9544   **c.** 38.58 seconds   **6.17a.** .2655
b. .0143   **c.** no   **6.19a.** .9406   **b.** .0068   **6.21a.** .6826
b. .9544   **c.** .9974   **6.23** approximately normal   **6.25** no
6.27 no, skewed right   **6.29** immature: approximately
normal; mature: slight skewness   **6.31** no, skewed right
6.33a. see below   **c.** .05   **6.35c.** mean = 4.74, standard
deviation = 1.275

6.39c.

Stem	Leaves
6	0 3
6	5 5 7 7 7 8 8
7	0 0 2 2 3 3 3
7	5 5 7
8	0
8	

d. $\mu_{\bar{x}} = 70; \sigma_{\bar{x}} = 4.96$ **6.43a.** $\mu_{\bar{x}} = 60; \sigma_{\bar{x}} = 3.16$
b. $\mu_{\bar{x}} = 60; \sigma_{\bar{x}} = 2$ **c.** $\mu_{\bar{x}} = 60; \sigma_{\bar{x}} = 1.41$ **d.** $\mu_{\bar{x}} = 60;$
$\sigma_{\bar{x}} = 1.15$ **e.** $\mu_{\bar{x}} = 60; \sigma_{\bar{x}} = 1$ **f.** $\mu_{\bar{x}} = 60; \sigma_{\bar{x}} = .45$
g. $\mu_{\bar{x}} = 60; \sigma_{\bar{x}} = .32$ **6.45a.** .9932   **b.** .2061   **c.** .0159
6.47a. $\mu_{\bar{x}} = 400; \sigma_{\bar{x}} = 10$, approximately normal   **b.** .1587
c. .3023   **6.49a.** skewed right   **b.** Central Limit Theorem
c. .2119   **6.51a.** $\mu_{\bar{x}} = 15; \sigma_{\bar{x}} = 1.118$, approximately
normal   **b.** 0   **c.** $\mu < 15$ **6.53a.** .0166   **b.** $\mu > 3$
6.55a. $\mu_{\bar{x}} = 5.5; \sigma_{\bar{x}} = .783$ **b.** no, approximately normal
c. yes, $P(\bar{x} \leq 5.3) = .3974$ **6.57a.** .0838   **b.** 1   **c.** .1793
6.59a. .0018   **b.** yes   **6.61a.** .0548   **b.** 97   **6.63a.** .7642
b. .2037   **c.** 54,175   **d.** 1   **6.65** all variables are approximately normal   **6.67a.** .5199   **b.** 4.39   **6.69** .6915
6.71a. .0179   **b.** yes   **6.73a.** $\mu_{\bar{x}} = 121.74; \sigma_{\bar{x}} = 4.86$,
approximately normal   **b.** .7348

Chapter 7

7.1 proportion of times interval encloses target parameter
7.3a. 81 ± 1.97 **b.** 81 ± 2.35 **c.** 81 ± 3.09 **7.5a.** $5.7 \pm .66$
b. 5.7 ± 1.03 **7.7** approximately 95% of similarly
constructed intervals will enclose μ **7.9a.** $\mu; 2.8 \pm .11$
b. $\mu; .1 \pm .04$ **c.** sons more desired than daughters

7.11a. $(90.07, 92.01)$ **b.** 90% confident that μ falls
between 90.07 and 92.01   **c.** random sample
7.13a. $.044 \pm .104$ **b.** possible since $\mu = 0$ falls within
interval   **7.15a.** Late: $3.04 \pm .011$; Massive: $2.83 \pm .015$;
Cumberlandite: $3.05 \pm .044$ **7.17** random sample from
normal population   **7.19a.** 5 ± 2.02 **b.** 5 ± 4.37
7.21 both symmetric, mound-shaped, mean of 0; t is flatter
(more variable)   **7.23** $(2.26, 4.14)$ **7.25a.** $(128.50, 142.59)$
b. 90% confident that μ falls between 128.5 and 142.59
calories   **c.** normal population   **d.** no, distribution skewed
left   **7.27a.** 74.31 ± 15.57 **b.** yes   **7.29a.** .031   **b.** .016
c. .043   **7.31a.** $.2 \pm .078$ **b.** $.2 \pm .035$ **7.33a.** .075
b. $.075 \pm .028$ **c.** 90% confident that π falls between .047
and .103   **7.35a.** $.257 \pm .003$ **b.** $.241 \pm .003$ **7.37a.** $.10 \pm$
$.013$ **b.** $.06 \pm .032$ **c.** yes   **7.39a.** .833   **b.** $.833 \pm .133$
7.41 $.137 \pm .079$ **7.43a.** 153   **b.** 423   **c.** 6,766
7.45a. 385   **b.** 139   **7.47** 1,068   **7.49a.** 27,061   **b.** 3,150
7.51 271   **7.53a.** $(4.54, 8.81)$ **b.** $(.00024, .00085)$
c. $(641.86, 1,809.09)$ **d.** $(.95, 12.66)$ **7.55** $(4.274, 65.988)$
7.57 $(.0028, .0105)$ **7.59** $(.81, .95)$ **7.61a.** $(4.727, 9.445)$
b. no   **7.63a.** 95% confident that μ falls between 13.6 and
26.4 seconds   **b.** 95% of similarly constructed intervals will
enclose the true value of μ **7.65a.** 9.9 ± 4.813 **b.** $6.7 \pm$
6.188 **c.** $(8.73, 14.32)$ **7.67a.** $3.39 \pm .0466$ **b.** $\mu > 2.50$
7.69 $.43 \pm .012$ **7.71a.** $(178.88, 299.52)$ **b.** $(231.0,$
$16,584.4)$ **7.73a.** 49.3 ± 8.60 **b.** 99% confident that μ
falls between 40.70% and 57.90%   **c.** normal population
7.75a. $(.163, 29.142)$ **b.** 85.65 ± 1.33 **7.77a.** .0129
b. yes $(.025 \approx .03)$

Chapter 8

8.3 P(Type II error) is unknown   **8.5** $H_0: \pi = .52$;
$H_a: \pi \neq .52$ **8.7a.** $H_0: \mu = 22; H_a: \mu < 22$ **b.** conclude
$\mu < 22$ when $\mu = 22$ **c.** conclude $\mu = 22$ when $\mu < 22$
8.9a. $H_0: \mu = 4; H_a: \mu > 4$ **b.** conclude $\mu > 4$ when $\mu = 4$
c. conclude $\mu = 4$ when $\mu > 4$ **8.11a.** $|z| > 2.33$
b. $z > 1.645$ **c.** $z < -2.33$ **d.** $z < -1.28$ **8.13a.** reject
H_0 **b.** reject H_0 **c.** reject H_0 **d.** do not reject H_0
8.15a. $H_0: \mu = 500; H_a: \mu \neq 500$ **b.** $|z| > 1.645$
8.17a. $H_0: \mu = 675; H_a: \mu < 675$ **b.** $z > -1.645$
8.19a. $H_0: \mu = 16; H_a: \mu < 16$ **b.** $z = -4.31$
c. $z < -2.33$ **d.** reject H_0 **8.21a.** $z = .33$ **b.** $z = -10.64$
c. $z = -1.49$ **8.23a.** $|t| > 2.145$ **b.** $|t| > 2.977$
c. $t < -1.761$ **d.** $t > 1.533$ **e.** $t > 1.318$ **8.25a.** $t = 1.13$,
do not reject H_0 **b.** $t = 1.13$, do not reject H_0
8.27 $z = .87$, do not reject H_0 **8.29** $z = 5.03$, reject H_0
8.31 $t = -5.16$, reject H_0 **8.33a.** $t = -3.46$, reject H_0
b. normal population   **8.35a.** .0022   **b.** .0571   **c.** .0139
d. .003   **8.37a.** do not reject H_0 **b.** reject H_0
c. reject H_0 **d.** do not reject H_0 **e.** do not reject H_0
8.39a. $H_0: \mu = 30; H_a: \mu > 30$ **b.** .1277   **c.** do not

Problem 6.33a

$\bar{x}$	1	1.5	2	2.5	3	3.5	4	4.5	5
$p(\bar{x})$	.04	.12	.17	.20	.20	.14	.08	.04	.01

reject H_0 **8.41** $z = 3.30; p \approx .001$; reject H_0
8.43 $t = -3.898; p = .0003$; reject H_0 **8.45a.** $|z| > 1.96$
b. $z < -1.645$ **c.** $z > 1.28$ **d.** $z < -2.33$ **e.** $|z| > 2.58$
8.47a. $z = -1$, do not reject H_0 **b.** $z = 2.10$, reject H_0
c. $z = 1.02$, do not reject H_0 **8.49** $z = -3.10$, reject H_0
8.51 no; $z = 2.30$, reject H_0 **8.53a.** $z = 4.59$, reject H_0
b. $z = 1.33; p = .092$; do not reject H_0 **8.55** $z = 11.17$,
reject H_0 **8.57a.** $\chi^2 = 19.2$ **b.** $\chi^2 = 18.7$
c. $\chi^2 = 30.38$ **d.** $\chi^2 = 396$ **8.59** no; do not reject H_0,
$\chi^2 = 8.44$ **8.61** yes; reject H_0, $\chi^2 = 127.86$
8.63 $\chi^2 = 21.95$, reject H_0 **8.65** $\chi^2 = 14.13$, do not reject
H_0 **8.67** $\chi^2 = .97$, reject H_0 **8.71** reject H_0; do not reject
H_0 **8.73** when you reject H_0; when you accept H_0
8.75a. $|z| > 2.575$ **b.** $|z| > 2.33$ **c.** $|z| > 2.055$
8.77a. $\alpha > .07$ **b.** $\alpha \le .07$ **8.79a.** $z = .77$ **b.** $z > 1.28$
c. do not reject H_0 **e.** .2206   **8.81** $z = 9.82$, reject H_0
8.83 yes; $z = 1.85$, reject H_0 **8.85** $z = 11.0; p \approx 0$; reject H_0
8.87 $z = 3.77$, reject H_0 **8.89** yes; $z = -5.81$, reject H_0

Chapter 9

9.1a. 10; 1.31   **b.** 13; 2   **9.3** $z = -7.87$ **9.5a.** $|t| > 2.101$
b. $t > 1.372$ **c.** $t < -2.821$ **9.7a.** 9.8 ± 15.46 **b.** $9.8 \pm$
18.71   **c.** 9.8 ± 25.58 **9.9** normal with equal variances;
independent and random   **9.11a.** $z = -.41$, do not reject
H_0 **b.** $-.33 \pm 1.59$ **9.13a.** no; $z = -5.82$; reject H_0 at
$\alpha = .05$ **b.** yes; $t = -1.16$; do not reject H_0 at $\alpha = .05$
9.15a. $t = .80$, do not reject H_0 **b.** $t = 0$, do not reject H_0
9.17 yes, $z = 3.53$ **9.19a.** $t = 2.89$, reject H_0: $\mu_{\text{old}} = \mu_{\text{new}}$
b. yes; evidence that $\mu_{\text{new}} < \mu_{\text{old}}$ **9.21a.** $|z| > 2.58$;
$z = 9.20$; reject H_0 **b.** $t > 2.132$; $t = 13.42$; reject H_0
c. $t < -1.533$; $t = -2.98$; reject H_0 **9.23a.** $z = 26.24$
b. approximately 0   **c.** reject H_0 **d.** 19.3 ± 1.44
9.25a. random sample of differences   **b.** random sample of
differences from a normal population   **9.27b.** $t = -7.18$;
$p \approx 0$; reject H_0 **9.29** $t = -1.31$, do not reject H_0
9.31a. yes   **b.** yes   **c.** $t = -.69$, do not reject H_0
9.33a. yes, $t = -2.49$ **b.** differences from a normal
population   **9.35a.** $z = -2.74$ **b.** $|z| > .96$ **c.** reject H_0
d. $-.1 \pm .06$ **e.** $-.1 \pm .071$ **f.** $-.1 \pm .093$ **9.37a.** $.05 \pm$
.099   **b.** $.05 \pm .045$ **c.** $.28 \pm .075$ **9.39a.** $.18 \pm .058$
b. $z = 5.795$, reject H_0 **9.41a.** $z = 2.59$, reject H_0 **b.** .152
$\pm .140$; yes   **9.43** $.147 \pm .069$; infer $\pi_M > \pi_A$ **9.45** $z =$
3.21, reject H_0 **9.47** $z = -3.03$, reject H_0

9.49a.

		Column		
		1	2	3
Row	1	15.7	30.95	27.36
	2	19.3	38.05	33.64

b. 3.74   **c.** yes; do not reject H_0 **9.51** $\chi^2 = 1.12$, do not
reject H_0 **9.53a.** $\chi^2 = 5.61$, reject H_0 **b.** yes   **c.** $.087 \pm$
.062; yes   **9.55** $\chi^2 = 1.56$, do not reject H_0 **9.57a.** H_0:
AMD incidence and alcohol type are independent
b. $\chi^2 = 25.65; p \approx 0$; reject H_0 **9.59a.** $p \approx 0$ **b.** reject H_0
9.61a. 3.18   **b.** 2.62   **c.** 2.10   **9.63a.** $F = 1.42$

b. $F = 3.88$ **c.** $F = 1.77$ **9.65a.** no, $F = 1.09$ **b.** random
samples from normal distributions   **9.67a.** $F = 1.60$, do not
reject H_0 **b.** $F = 2.16$, do not reject H_0
9.69b. H_0: $\sigma_{\text{E}}^2/\sigma_{\text{U}}^2 = 1$; H_a: $\sigma_{\text{E}}^2/\sigma_{\text{U}}^2 \neq 1$ **c.** $F = 2.45$,
reject H_0 **d.** random samples from normal distributions
9.71 $F = 2.03$, do not reject H_0 **9.73a.** yes, $z = -2.21$
b. $-.027 \pm .024$ **9.75a.** mean differences in all 4 variables
significant at $\alpha = .05$ **b.** exertion: $t = 1.15$, do not reject
H_0; heart rate: $t = .42$, do not reject H_0; WAT: $t = -1.02$, do
not reject H_0; fatigue: $t = .07$, do not reject H_0 **9.77a.** -12
± 25.62 **b.** no evidence of a difference   **c.** no, $F = 1.29$
9.79a. $z = -6.41$, reject H_0 **b.** $\pi_{\text{control}} < \pi_{\text{E2F1}}$
9.81a. $t = -7.06$, reject H_0 **b.** normal populations with
equal variances   **c.** $F = 3.41$, reject H_0 at $\alpha = .02$; test results
are suspect   **9.83a.** $.2 \pm .066$, $\mu_{\text{fast}} > \mu_{\text{slow}}$ **b.** $F = 8.82$,
reject H_0 **c.** inference may not be valid   **9.85** yes,
$z = -9.51$ **9.87a.** yes, $\chi^2 = 313.15$ **b.** $.181 \pm .069$

9.89a.

		Gender	
		Male	Female
Status	Hired	6	3
	Not hired	7	12

b. $\chi^2 = 2.18$, do not reject H_0

Chapter 10

10.1a. 1.5   **b.** 2   **c.** increase 2   **d.** decrease 2   **e.** 1.5
10.5 $\beta_0 = y$-intercept; $\beta_1 = $ slope   **10.7a.** $y = \beta_0 + \beta_1 x + \varepsilon$
b. $b_0 = 1,238.9$; $b_1 = 2.628$ **c.** $\hat{y} = 1,238.9 + 2.628x$
f. 1,449.14   **10.9a.** yes   **b.** $\hat{y} = -.643 + .064x$ **10.11a.** x_1
b. $\hat{y} = -.000215 + .8116x_1$ **d.** $\hat{y} = -.000184 + .4671x_2$
e. $\hat{y} = -.0174 + .5148x_3$ **10.13a.** yes
b. $y = \beta_0 + \beta_1 x + \varepsilon$ **c.** $\hat{y} = 4.44 + .26x$ **g.** 6.01
10.15a. .0275   **b.** .1658   **10.17a.** 2.75   **b.** 1.658
10.19 SSE $= 289,875$; $s^2 = 41,410.76$; $s = 203.5$ **10.21** .596;
.901; .895; x_1 **10.23** SSE $= 2,290.98$; $s^2 = 91.64$; $s = 9.57$
10.25a. $|t| > 3.182$ **b.** $t > 1.638$ **c.** $t < -3.365$
10.27a. $t = -1.55$, reject H_0 **b.** -300.7 ± 390.9 **10.29** yes,
$t = 5.05$ **10.31** yes, $t = 10.25$ **10.33a.** $\hat{y} = 6.73 - .055x$
b. $t = -8.62$, reject H_0 **c.** $(-.068, -.042)$ **d.** yes
10.35a. $\hat{y} = 25.796 + .371x$ **b.** $t = 3.72$, reject H_0
10.37a. $t > 2.896$; do not reject H_0, $t = 2.62$ **b.** $t < -2.415$;
reject H_0, $t = -6.56$ **c.** $|t| > 1.86$; reject H_0, $t = -4.38$
10.39 yes, $t = 3.88$ **10.41a.** H_0: $\rho = 0$; H_a: $\rho \neq 0$
10.43a. reject H_0: $\rho = 0$ in favor of H_a: $\rho > 0$ at $\alpha = .05$
b. do not reject H_0: $\rho = 0$ at $\alpha = .10$ **c.** reject H_0: $\rho = 0$ at
$\alpha = .01$ **10.45a.** $y = \beta_0 + \beta_1 x + \varepsilon$ **b.** H_0: $\beta_1 = 0$;
H_a: $\beta_1 > 0$ **c.** weak negative linear relationship   **d.** no
10.47 .772   **10.49** .1225   **10.51** .8899   **10.53** .3863
10.55 .5776   **10.57a.** 10.6; $E(y)$ and y when $x = 2.5$
b. $10.6 \pm .22$ **c.** $8.9 \pm .32$ **d.** $12.3 \pm .32$ **e.** wider
f. 10.6 ± 1.02 **10.59a.** 39 ± 5.76 **b.** 39 ± 7.73 **c.** PI for y
10.61 $(964.8, 1,986.1)$ **10.63a.** $(6.01, 6.53)$ **b.** $(5.31, 7.23)$
10.65a. $(3.51, 17.20)$ **b.** $x = 32$ is outside range of sample
data   **10.67** $(-6.97, 37.36)$ **10.69b.** $\hat{y} = .35 + 1.05x$

d. 7.00   **e.** reject H_0 **f.** .942   **g.** $(-.48, 1.18)$
10.71b. $\hat{y} = 3.343 + .576x$ **d.** 7.99   **e.** reject H_0 **f.** .927
g. $(-.38, 2.46)$ **10.73a.** curvilinear term omitted from
model   **b.** constant error variance assumption violated
c. outlier   **d.** independent error assumption violated
e. normal errors assumption violated   **10.75a.** 3.197, 3.215,
$-2.258, .269, -1.713, -7.186, -2.659, 3.359, -2.132, 5.904$
b. yes; add curvature term   **c.** no outliers
d. yes; add curvature term   **10.77** add curvature term to
model   **10.79a.** $\hat{y} = 40.99 - .26x$ **d.** no
10.81a. $\hat{y} = 99.82 + .045x$ **c.** yes; add curvature to model
10.83a. $\hat{y} = 85.01 + .04x; R^2 = .383; s = 145.7$ **b.** yes,
$t = 4.67$ **c.** no   **d.** $\hat{y} = 4.01 + .99x; R^2 = .897$;
$s = 49.12$; yes   **10.85** $y = \beta_0 + \beta_1 x_1 + \beta_2 x_2 + \varepsilon$
10.87 $y = \beta_0 + \beta_1 x + \beta_2 x^2 + \varepsilon$ **10.89a.** $t = 1.35$, do not
reject H_0 **b.** $t = 3.41$, reject H_0 **c.** 3.1 ± 3.95
d. $.92 \pm .76$ **10.91a.** $y = \beta_0 + \beta_1 x_1 + \beta_2 x_2 + \varepsilon$
b. $\hat{y} = -36.29 + .0028x_1 + 1.471x_2$ **d.** 42.85;
$(21.67, 64.04)$ **e.** yes, $F = 13.31$ **f.** 9.99   **g.** .4592
h. do not reject H_0: $\beta_1 = 0$; reject H_0: $\beta_2 = 0$
10.93a. $y = \beta_0 + \beta_1 x_1 + \beta_2 x_2 + \beta_3 x_3 + \varepsilon$ **b.** model
explains 70% of sample variation in depression   **c.** yes,
reject H_0: $\beta_1 = \beta_2 = \beta_3 = 0$ **d.** do not reject H_0: $\beta_2 = 0$
10.95a. quantitative: x_4 and x_5; qualitative: x_1, x_2, and x_3
b. yes, $p = .0146$ **c.** for every 1-month increase in x_4,
y decreases by .0017   **d.** no, $p = .2501$ **10.97a.** model
explains 82% of sample variation in y **b.** reject
H_0: $\beta_1 = \beta_2 = \cdots = \beta_9 = 0$ **c.** for every 1-year increase
in x_1, y increases by 1.98%   **d.** reject H_0 **e.** do not reject
H_0 **10.99a.** reject H_0: $\beta_1 = \beta_2 = \beta_3 = 0$ **b.** model
explains 16.5% of the sample variation in score y **c.** score
increases .341 for every 1-unit increase in x_1; score increases
.070 for every 1-unit increase in x_2; score increases .05 for
every 1-unit increase in x_3 **d.** reject H_0: $\beta_1 = 0$ **e.** do not
reject H_0: $\beta_3 = 0$ **10.101a.** for every percentile increase in
rental price, number of homeless increases by 2.87
b. reject H_0: $\beta_4 = 0$ in favor of H_a: $\beta_4 < 0$ **c.** inflated
overall Type I error rate   **d.** model explains 83% of sample
variation in number of homeless   **10.103** $\hat{y} = .21 + 2.43x$;
$R^2 = .837; s = .454; t = 10.63; p \approx 0$; constant variance
assumption may be violated   **10.105a.** .081   **b.** no, $t = .26$
10.107a. $\hat{y} = -13.62 - .053x; R^2 = .852; s = .134; t =$
$-6.80; p \approx 0$; constant variance assumption may be
violated; add curvature to model   **b.** no, outside range of
sample data   **10.109a.** $x_1, x_2, x_3, x_4, x_6, x_7$, and x_8
b. H_0: $\beta_1 = \beta_2 = \cdots = \beta_8 = 0$ **c.** reject $H_0, p < .001$
d. model explains 29% of sample variation in y **e.** estimated difference between mean length of story for female
victims and male victims is .26   **f.** for every 1-unit increase
in per capita income, length of story decreases .02
10.111a. for every 1-day increase in TV news exposure,
support level increases by .02   **b.** reject H_0: $\beta_1 = 0$ in favor
of H_a: $\beta_1 > 0$ **c.** yes, reject H_0: $\beta_8 = 0$ **d.** no, do not reject
H_0: $\beta_9 = 0$ **e.** model explains 19.4% of the sample varia-

tion in support level   **f.** reject H_0: $\beta_1 = \beta_2 = \cdots = \beta_9 = 0$
10.113a. reject H_0: $\beta_1 = \beta_2 = \cdots = \beta_{27} = 0$ **b.** for every
1-sq.-ft. increase in G.L.A., 95% confident that price increases
between \$15.13 and \$29.67 (holding other x's fixed)   **c.** no;
inflated overall Type I error rate   **10.115a.** $\hat{y} = -102.36 +$
$.0041x_1 + 3.451x_2 - .143x_3$; yes, $F = 13.02$ **b.** possible
violation of constant variance assumption   **d.** $\hat{y} = -97.88$
$+ .0022x_1 + 3.453x_2 - .054x_3 - 15.85x_4$; yes, $t = -3.63$

Chapter 11

11.1b. 2.38   **c.** 7.50   **d.** 3.15

e.

Source	df	SS	MS	F
Treatments	1	7.50	7.50	3.15
Error	11	26.19	2.38	
Total	12	33.69		

f. $F > 4.84$ **g.** do not reject H_0 **11.3a.** 3.11
b. 1.38   **c.** 2   **d.** 6   **e.** $F = 2.795$

f.

Source	df	SS	MS	F
Treatments	2	7.73	3.86	2.795
Error	6	8.29	1.38	
Total	8	16.02		

g. $F > 5.14$ **h.** do not reject H_0 **i.** .1387; yes
11.5a. age at onset of bulimia nervosa   **b.** three age
groups   **c.** H_0: $\mu_1 = \mu_2 = \mu_3$ **d.** reject H_0 at $\alpha = .01$
11.7a. completely randomized   **b.** response = sale price;
treatments = 7 neighborhoods
c. H_0: $\mu_A = \mu_B = \mu_C = \mu_D = \mu_E = \mu_F = \mu_G$

d.

Source	df	SS	MS	F
Neighborhoods	6	199,341	33,223.5	13.34
Error	63	156,931	2,490.97	
Total	69	356,272		

e. yes, reject H_0

11.9

Source	df	SS	MS	F
Species	2	23,454.5	11,727.2	1.22
Error	141	1,360,549	9,649.28	
Total	143	1,384,003		

do not reject H_0: $\mu_1 = \mu_2 = \mu_3$ at $\alpha = .10$
11.11a. H_0: $\mu_1 = \mu_2 = \mu_3 = \mu_4$ **b.** reject H_0
c. H_0: $\mu_1 = \mu_2 = \mu_3 = \mu_4$ **d.** do not reject H_0

11.13

Source	df
A	2
B	3
$A \times B$	6
Error	12
Total	23

11.15a. yes, $F = 3.27$ **b.** no

11.17a.

Source	df	SS	MS	F
Size	1	8.068	8.068	141.18
Diet	1	.012	.012	.22
Size × Diet	1	.036	.036	.64
Error	24	1.372	.057	
Total	27	9.488		

b. no size × diet interaction; no diet main effect; significant size main effect   **11.19a.** factors = confirmation and verification; confirm levels = completed, not completed; verify levels = completed, not completed; treatments = $(C, C), (C, NC), (NC, C), (NC, NC)$ **c.** yes

11.21a.

Source	df
Photoperiod	1
Gender	1
$P \times G$	1
Error	120
Total	123

b. no evidence of $P \times G$ interaction   **c.** $\mu_{16 \text{ less}}$ differs from $\mu_{16 \text{ more}}$; μ_{male} differs from μ_{female} **11.23a.** factors = setting and story; setting levels = Yale, Princeton; story levels = true, false; treatments = (Yale, true), (Yale, false), (Princeton, true), (Princeton, false)   **b.** agreement rating   **c.** evidence of setting × story interaction   **d.** no   **e.** yes   **11.25a.** yes   **11.27a.** 15   **b.** 6   **c.** 28   **d.** 15   **11.29a.** .0033   **b.** .0083   **c.** .0018   **d.** .0033   **11.31a.** D **b.** B and C **11.33** $\underline{A \quad \overline{C \quad D} \quad B}$ **11.35a.** reject H_0 at $\alpha = .01$ **b.** $\mu_{\text{Time Shifter}}$ less than all other means; no differences among other four means   **11.37** $\mu_{\text{Dealer}} < (\mu_{5\text{count}}, \mu_{\text{Basic10}}, \mu_{\text{Adv10}})$ **11.39** $\mu_{\text{USAnglo}} < (\mu_{\text{MexHisp}}, \mu_{\text{CAHisp}}); \mu_{\text{USHisp}} < \mu_{\text{CAHisp}}$ **11.41a.** overall confidence level of .95   **b.** $\mu_{\text{min}} < (\mu_{\text{med}}, \mu_{\text{sub}}, \mu_{\text{ext}})$ **c.** $(\mu_{\text{min}}, \mu_{\text{med}}, \mu_{\text{ext}}) < \mu_{\text{sub}}$ **11.43** $(\mu_{\text{Oct}}, \mu_{\text{Jul}}, \mu_{\text{Sep}}) < \mu_{\text{Jun}}; \mu_{\text{Oct}} < \mu_{\text{Sep}}$ **11.45** equal variance assumption appears to be violated   **11.47** normality and equal variance assumptions appear to be violated   **11.49** see answer to Problem 11.3f   **11.51** see answer to Problem 11.7d   **11.53** see answer to Problem 11.17a   **11.55a.** job satisfaction   **b.** career stage; time on road   **c.** (early, short), (early, long), (mid, short), (mid, long), (late, short), (late, long)   **d.** no evidence of interaction; no RT main effect; significant CS main effect   **e.** no   **f.** yes   **11.57a.** yes, $F = 9.46$ **b.** dark blond and light blond   **11.59** $(\mu_{\text{Rest}}, \mu_{\text{A}}) < \mu_{\text{B}} < (\mu_{\text{C}}, \mu_{\text{D}}, \mu_{\text{E}})$ **11.61a.** no $H \times S$ interaction; H main effect and S main effect are significant   **b.** yes   **c.** $(\mu_{\text{PLC}}, \mu_{\text{LGN}}) < (\mu_{\text{QMD}}, \mu_{\text{MTZ}})$

11.63a.

Source	df	SS	MS	F
Material	3	16.80	5.60	12.33
Error	8	3.63	.45	
Total	11	20.43		

b. yes, $F = 12.33$ **c.** $\mu_{\text{citrus}} < (\mu_{\text{bovine}}, \mu_{\text{human}})$; all other pairs not significantly different

11.65a.

Source	df	SS	MS	F
Amount (A)	3	104.194	34.731	20.12
Method (C)	3	28.627	9.542	5.53
$A \times C$	9	25.131	2.792	1.62
Error	32	55.247	1.726	
Total	47	213.199		

b. do not reject H_0 **c.** no evidence of interaction   **d.** both A and C main effects are significant   **e.** amount: $\mu_{10} > (\mu_0, \mu_3, \mu_5)$; method: $(\mu_{\text{FC}}, \mu_{\text{WQ}}) < (\mu_{\text{OQ}}, \mu_{\text{AB}})$

11.67a.

Source	df	SS	MS	F
Food Levels	4	784.56	196.14	3.32
Error	111	6,562.32	59.12	
Total	115	7,346.88		

b. yes

Chapter 12

12.1a. 5   **b.** 2   **c.** 6   **12.3** no, $S = 4, p = .6875$ **12.5a.** 80   **b.** $H_0: \eta = 80; H_a: \eta < 80$ **c.** $S = 11; p = .011$; reject H_0 **12.7** $S = 25; p = .00016$; reject H_0 **12.9** $S = 8; p = .055$; do not reject H_0 **12.11** no, $S = 4, p = .637$ **12.13a.** $T_2 \leq 24$ or $T_2 \geq 56$ **b.** $T_2 \leq 52$ **c.** $T_2 \leq 43$ **12.15a.** $T_1 = 67; T_2 = 38$ **b.** $T_1 \geq 58$ **c.** reject H_0 **12.17b.** $H_0: \eta_{\text{after}} = \eta_{\text{before}}$ **c.** reject H_0 **12.19** shift to new location; $T_{\text{old}} = 1,134$, $T_{\text{new}} = 697, z = 3.23$ **12.21** $T_{\text{fog}} = 43; T_{\text{clear}} = 35; z = 1.44$; do not reject H_0 **12.23** no, do not reject H_0; $T_{\text{stay}} = 88.5$, $p = .948$ **12.25a.** $T \leq 2$ **b.** $T \leq 4$ **12.27a.** $T^- = 2.5$, do not reject H_0 **b.** $T^- = 2.5$, do not reject H_0 **12.29** $T^+ = 9; p = .016$; reject H_0 **12.31** no, $T^+ = 11$, $p = .053$ **12.33a.** do not reject $H_0, T^+ = 83, z = .28$ **12.35** reject $H_0, T^+ = 1, p < .01$ **12.37a.** 8   **b.** 6.5   **c.** 15.5   **d.** 6   **12.39a.** completely randomized   **b.** $H_0: \eta_1 = \eta_2 = \eta_3$ **c.** $\chi^2 > 5.99147$ **d.** $T_1 = 46, T_2 = 21.5, T_3 = 68.5$ **e.** 12.47   **f.** reject H_0 **12.41b.** reject $H_0, \chi^2 = 6.86, p = .032$ **12.43a.** nonnormal distributions   **c.** reject $H_0, \chi^2 = 37.65$, $p \approx 0$ **12.45b.** reject $H_0, \chi^2 = 45.86, p \approx 0$ **12.47** no, $\chi^2 = 7.32, p = .062$ **12.49a.** x-ranks: 7, 6, 5, 2, 1, 3, 4; y-ranks: 6, 7, 5, 1, 2, 4, 3   **b.** .893   **c.** yes, reject H_0 **12.51a.** yes, $r_s = .491$ **12.53a.** $r_s = .9$ **b.** reject H_0 **12.55b.** yes   **c.** no   **d.** yes   **12.57** reject H_0 **12.59** $T^+ = 20; p = .03$; reject H_0 **12.61** $S = 14; p = .058$; reject H_0 **12.63a.** $H_0: \eta_1 = \eta_2 = \eta_3 = \eta_4$ **b.** $\chi^2 = 1.17$ **c.** do not reject H_0 **12.65** $S = 16; p = .006$; reject H_0 **12.67** yes, $T_{\text{urban}} = 21, z = 3.88$ **12.69a.** $-.705$ **b.** no, do not reject H_0

References

Excel Primer

Grauer, R., and Barber, M. *Exploring Microsoft Excel 97.* Englewood Cliffs, N.J.: Prentice-Hall, 1998.

Microsoft Excel 97. Redmond, Wash.: Microsoft Corp., 1996.

Wells, E. *Developing Microsoft Excel 5 Solutions.* Redmond, Wash.: Microsoft Corp., 1995.

Chapter 1

Careers in Statistics. American Statistical Association, Biometric Society and the Institute of Mathematical Statistics, 1995.

Cochran, W. G. *Sampling Techniques,* 3rd ed. New York: Wiley, 1977.

Deming, W. E. *Sample Design in Business Research.* New York: Wiley & Sons, Inc., 1963.

Ethical Guidelines for Statistical Practice. American Statistical Association, 1995.

Hansen, M. H., Hurwitz, W. N., and Madow, W. G. *Sample Survey Methods and Theory,* Vol. 1. New York: Wiley & Sons, Inc., 1953.

Kirk, R. E., ed. *Statistical Issues: A Reader for the Behavioral Sciences.* Monterey, Calif.: Brooks/Cole, 1972.

Kish, L. *Survey Sampling.* New York: Wiley & Sons, Inc., 1965.

Milgram, S. *Obedience to Authority.* New York: Harper & Row, 1974.

Scheaffer, R., Mendenhall, W., and Ott, R. L. *Elementary Survey Sampling,* 2nd ed. Boston: Duxbury, 1979.

Tanur, J. M., Mosteller, F., Kruskal, W. H., Link, R. F., Pieters, R. S., and Rising, G. R., eds. *Statistics: A Guide to the Unknown,* 3rd ed. San Francisco: Holden-Day, 1989.

What Is a Survey? Section on Survey Research Methods, American Statistical Association, 1995.

Yamane, T. *Elementary Sampling Theory,* 3rd ed. Englewood Cliffs, N.J.: Prentice-Hall, 1967.

Chapter 2

Huff, D. *How to Lie with Statistics.* New York: Norton, 1954.

Mendenhall, W. *Introduction to Probability and Statistics,* 9th ed. North Scituate, Mass.: Duxbury, 1994.

Microsoft Excel 97. Redman, Wash.: Microsoft Corp., 1997.

Tanur, J. M., Mosteller, F., Kruskal, W. H., Link, R. F., Pieters, R. S., and Rising, G. R. (eds.). *Statistics: A Guide to the Unknown,* 2nd ed. San Francisco: Holden-Day, 1978.

Tufte, E. R. *Envisioning Information.* Cheshire, Conn.: Graphics Press, 1990.

Tufte, E. R. *Visual Display of Quantitative Information.* Cheshire, Conn.: Graphics Press, 1983.

Tufte, E. R. *Visual Explanations.* Cheshire, Conn.: Graphics Press, 1997.

Tukey, J. *Exploratory Data Analysis.* Reading, Mass.: Addison-Wesley, 1977.

Chapter 3

Mendenhall, W. *Introduction to Probability and Statistics,* 8th ed. Boston: Duxbury, 1990.

Microsoft Excel 97. Redman, Wash.: Microsoft Corp., 1997.

Tukey, J. *Exploratory Data Analysis.* Reading, Mass.: Addison-Wesley, 1977.

Chapter 4

Epstein, R. A. *The Theory of Gambling and Statistical Logic,* revised ed. New York: Academic Press, 1977.

Feller, W. *An Introduction to Probability Theory and Its Applications,* 3rd ed. Vol. I. New York: Wiley, 1968.

Mendenhall, W., Wackerly, D., and Scheaffer, R. *Mathematical Statistics with Applications,* 4th ed. Boston: PWS-Kent, 1990.

Mosteller, F., Rourke, R., and Thomas, G. *Probability with Statistical Applications,* 2nd ed. Reading, Mass.: Addison-Wesley, 1970.

Parzen, E. *Modern Probability Theory and Its Applications.* New York: Wiley, 1960.

Wright, G., and Ayton, P., eds. *Subjective Probability.* New York: John Wiley and Sons, 1994.

Chapter 5

Feller, W. *An Introduction to Probability Theory and Its Applications,* Vol. I, 3rd ed. New York: Wiley, 1968.

Hogg, R. V., and Craig, A. T. *Introduction to Mathematical Statistics,* 4th ed. New York: Macmillan, 1978.

Mendenhall, W., Wackerly, D., and Scheaffer, R. L. *Mathematical Statistics with Applications,* 4th ed. Boston: PWS-Kent, 1990.

Chapter 6

Hogg, R. V., and Craig, A. T. *Introduction to Mathematical Statistics,* 4th ed. New York: Macmillan, 1978.

Mendenhall, W., and Sincich, T. *Statistics for Engineering and the Sciences,* 4th ed. Upper Saddle River, N.J.: Prentice Hall, 1995.

Mendenhall, W., Wackerly, D., and Scheaffer, R. *Mathematical Statistics with Applications,* 4th ed. Boston: PWS-Kent, 1990.

Ramsey, P. P., and Ramsey, P. H. "Simple tests of normality in small samples." *Journal of Quality Technology,* Vol. 22, 1990, pp. 299–309.

Snedecor, G. W., and Cochran, W. G. *Statistical Methods,* 7th ed. Ames, Iowa: Iowa State University Press, 1980.

Chapter 7

Agresti, A., and Coull, B. A. "Approximate is better than 'exact' for interval estimation of binomial proportions." *The American Statistician,* Vol. 52, No. 2, May 1998, pp. 119–126.

Cochran, W. G. *Sampling Techniques,* 3rd ed. New York: Wiley, 1977.

Kish, L. *Survey Sampling.* New York: Wiley, 1965.

Mendenhall, W., Wackerly, D., and Scheaffer, R. *Mathematical Statistics with Applications,* 4th ed. Boston: PWS-Kent, 1990.

Snedecor, G. W., and Cochran, W. G. *Statistical Methods,* 7th ed. Ames, Iowa: Iowa State University Press, 1980.

Wilson, E. B. "Probable inference, the law of succession, and statistical inference." *Journal of the American Statistical Association,* Vol. 22, 1927, pp. 209–212.

Chapter 8

Maclure, M. "The case-crossover design: A method for studying transient effects on the risk of acute events," *American Journal of Epidemiology,* Vol. 133, 1991, p. 144–153.

Mendenhall, W., Wackerly, D., and Scheaffer, R. *Mathematical Statistics with Applications,* 4th ed. Boston: PWS-Kent, 1990.

Snedecor, G. W., and Cochran, W. G. *Statistical Methods,* 7th ed. Ames, Iowa: Iowa State University Press, 1980.

Chapter 9

Agresti, A., and Agresti, B. F. *Statistical Methods for the Social Sciences,* 2d ed. San Francisco: Dellen, 1986.

Cochran, W. G. "The χ^2 test of goodness of fit." *Annals of Mathematical Statistics,* 1952, p. 23.

Daniel, W. W. *Applied Nonparametric Statistics,* 2d ed. Boston: PWS-Kent, 1990.

Hollander, M., and Wolfe, D. A. *Nonparametric Statistical Methods.* New York: Wiley, 1973.

Lawontin, R. C., and Felsenstein, J. "Robustness of homogeneity tests in $2 \times n$ tables." *Biometrics,* Vol. 21, Mar. 1965, pp. 19–33.

Marascuilo, L. A., and McSweeney, M. *Nonparametric and Distribution-Free Methods for the Social Sciences.* Monterey, Calif.: Brooks/Cole, 1977.

Mendenhall, W. *Introduction to Probability and Statistics,* 8th ed. Boston: PWS-Kent, 1991.

Satterthwaite, F. E. "An approximate distribution of estimates of variance components." *Biometrics Bulletin,* Vol. 2, 1946, pp. 110–114.

Snedecor, G. W., and Cochran, W. *Statistical Methods,* 7th ed. Ames: Iowa State University Press, 1980.

Chapter 10

Barnett, V., and Lewis, T. *Outliers in Statistical Data.* New York: Wiley, 1978.

Belsley, D. A., Kuh, E., and Welsch, R. E. *Regression Diagnostics: Identifying Influential Data and Sources of Collinearity.* New York: Wiley, 1980.

Berenson, M., and Levine, D. *Basic Business Statistics,* 7th ed. Upper Saddle River, N.J.: Prentice Hall, 1999.

Chatterjee, S., and Price, B. *Regression Analysis by Example,* 2d ed. New York: Wiley, 1991.

Draper, N., and Smith, H. *Applied Regression Analysis,* 2d ed. New York: Wiley, 1981.

Mendenhall, W., and Sincich, T. *A Second Course in Statistics: Regression Analysis,* 5th ed. Upper Saddle River, N. J.: Prentice Hall, 1996.

Montgomery, D. C., and Peck, E. A. *Introduction to Linear Regression Analysis.* New York: Wiley, 1982.

Mosteller, F., and Tukey, J. W. *Data Analysis and Regression: A Second Course in Statistics.* Reading, Mass.: Addison-Wesley, 1977.

Neter, J., Kutner, M., Nachtsheim, C., and Wasserman, W. *Applied Linear Statistical Models,* 4th ed. Homewood, Ill.: Richard Irwin, 1996.

Rousseeuw, P. J., and Leroy, A. M. *Robust Regression and Outlier Detection.* New York: Wiley, 1987.

Weisberg, S. *Applied Linear Regression,* 2d ed. New York: Wiley, 1985.

Chapter 11

Box, G. E. P., Hunter, W. G., and Hunter, J. S. *Statistics for Experimenters.* New York: Wiley, 1978.

Cochran, W. G., and Cox, G. M. *Experimental Designs,* 2nd ed. New York: Wiley, 1957.

Gujarati, D. N. *Basic Econometrics,* 3rd ed. New York: McGraw-Hill, 1995 (Chapter 11).

Hsu, J. C. *Multiple Comparisons: Theory and Methods.* London: Chapman & Hall, 1996.

Johnson, N., and Leone, F. *Statistics and Experimental Design in Engineering and the Physical Sciences,* Vol. II, 2nd ed. New York: Wiley, 1977.

Kirk, R. E. *Experimental Design,* 2nd. ed. Monterey, Calif.: Brooks/Cole, 1982.

Kramer, C. Y. "Extension of multiple range tests to group means with unequal number of replications." *Biometrics,* Vol. 12, 1956, pp. 307–310.

Mason, R. L., Gunst, R. F., and Hess, J. L. *Statistical Design and Analysis of Experiments.* New York: Wiley, 1989.

Mendenhall, W. *Introduction to Linear Models and the Design and Analysis of Experiments.* Belmont, Calif.: Wadsworth, 1968.

Miller, R. G., Jr. *Simultaneous Statistical Inference.* New York: Springer-Verlag, 1981.

Neter, J., Kutner, M., Nachtsheim, C., and Wasserman, W. *Applied Linear Statistical Models,* 4th ed. Homewood, Ill.: Richard Irwin, 1996.

Scheffé, H. *The Analysis of Variance.* New York: Wiley, 1959.

Snedecor, G. W., and Cochran, W. G. *Statistical Methods,* 7th ed. Ames, Iowa: Iowa State University Press, 1980.

Steel, R. G. D., and Torrie, J. H. *Principles and Procedures of Statistics: A Biometrical Approach,* 2nd ed. New York: McGraw-Hill, 1980.

Tukey, J. W. "Comparing individual means in the analysis of variance." *Biometrics,* Vol. 5, 1949, pp. 99–114.

Chapter 12

Conover, W. J. *Practical Nonparametric Statistics,* 3rd ed. New York: Wiley, 1999.

Daniel, W. W. *Applied Nonparametric Statistics,* 2nd ed. Boston: PWS-Kent, 1990.

Friedman, M. "The use of ranks to avoid the assumption of normality implicit in the analysis of variance." *Journal of the American Statistical Association,* Vol. 32, 1937.

Hollander, M., and Wolfe, D. A. *Nonparametric Statistical Methods.* New York: Wiley, 1973.

Kruskal, W. H., and Wallis, W. A. "Use of ranks in one-criterion variance analysis." *Journal of the American Statistical Association,* Vol. 47, 1952.

Lehmann, E. L. *Nonparametrics: Statistical Methods Based on Ranks.* San Francisco: Holden-Day, 1975.

Siegel, S. *Nonparametric Statistics for the Behavioral Sciences.* New York: McGraw-Hill, 1956.

Wilcoxon, F., and Wilcox, R. A. "Some rapid approximate statistical procedures." The American Cyanamid Co., 1964.

Appendix A

Bashaw, W. L. *Mathematics for Statistics.* New York: Wiley, 1969.

Lanzer, P. *Video Review of Arithmetic.* East Hills, NY: Video Aided Instruction, 1990.

Levine, D. *Video Review of Statistics.* East Hills, NY: Video Aided Instruction, 1989.

Shane, H. *Video Review of Elementary Algebra.* East Hills, NY: Video Aided Instruction, 1990.

Sobel, M., and Lerner, N. *College Algebra,* 4th ed. Englewood Cliffs, NJ: Prentice Hall, 1995.

Microsoft Excel Index

Abs function, 735
Absolute cell reference, 12
Add-in, 23–24
Arithmetic operators, 9

Binomdist function, 786
Bin range, 128
Bonferroni procedure, 733–735
Box-and-whisker plot, 209–210
Buttons, 2

Cancel button, 7
CD-ROM
   File contents, 819
   installation, 824
   overview, 819
   workbook files, 819–823
Cell range, 9
Cells, 9
Chart Wizard
   Histogram, 128–134
   Scatterplot, 141–143, 661–663
   Side-by-side bar chart, 138–141
Check box, 7
Chidist function, 475
Chiinv function, 414
Click, 2
Closing button, 4
Confidence interval for the
      population variance, 414–415
Configuring Microsoft Excel, 7–9
Copying, 11–12
Correcting errors, 11
Correl function, 210
Countif function, 786

Data analysis installation, 23–24
Data analysis tool used for
   Descriptive statistics, 209
   F test for the difference between
      two variances, 552–553
   Histogram, 128–130
   One-way ANOVA, 731–732
   Two-sample t test assuming equal
      variances, 543–544
   Two-sample t test for the
      difference between means
      (paired samples), 545–546
   Two-way ANOVA, 732–733

Dialog boxes, 5
   Add-ins, 24
   Add trendline, 662–663
   Anova: Single factor, 731
   Anova: Two factor with replication,
      732
   Descriptive statistics, 209
   File | open, 6
   File | print, 6
   F-test: Two sample for variance, 552
   Histogram, 128–130
   Macro virus, 24
   Page Setup
      Header/footer tab, 16
      Margins tab, 16
      Page tab, 15
      Sheet tab, 16
   t-test: Paired two sample for
      means, 546
   t-test: Two Sample assuming equal
      variances, 543
   Text Import Wizard, 22–23
   Tools | Options
      View tab, 8
      General tab, 8
Double-click, 2
Drag, 2
Drop-down list box, 6
Dummy variables, 668

Edit box, 6
Enhancing appearance of a
      worksheet, 17–18
Entering values, 10–11

File menu, 5
File | open dialog box, 6
File | page setup dialog box, 16
File | print dialog box, 6
Formatting toolbar, 4, 18
Formula bar, 4
Formulas, 9
Functions, 9
   Abs, 735
   Binomdist, 786
   Chidist, 475
   Chiinv, 414
   Correl, 210
   Countif, 786

Sum, 299
Sumproduct, 299
Tdist , 665
Tinv, 665

Gap width edit box, 130

Header/footer tab settings, 15
Histogram, 128–130

Icons, 2

Kruskal-Wallis test, 789–790

List box, 6

Menu bar, 4
Mouse, 1
Mouse pointer, 1
Microsoft Office, 1

Normal probabilities, 355–356
Normal probability plot, 356–357

Office Assistant, 19–20
Ok button, 7
One-way ANOVA, 731–732
Operators, 9
Option buttons, 6

Page tab settings, 15
PHStat add-in, 25
   Binomial distribution, 300–301
   Box-and-whisker plot, 209–210
   Confidence interval
      for the mean (σ known),
        408–409
      for the mean (σ unknown),
        409–410
      for the proportion, 410–411
   Hypergeometric distribution, 302
   Menu, 25
   Multiple regression, 667
   Normal distribution, 355–356
   Normal probability plot, 356–357
   One way tables and charts, 125–126
   Poisson distribution, 301

PHStat add-in, *continued*
Random sample without replacement, 60
Regression, 663–664
Residual plots, 664
Sample size determination
for the mean, 412
for the proportion, 413–414
Sampling distributions, 357–360
Stem-and-leaf display, 127
Test of hypothesis
For independence in the $r \times c$ table, 548–550
For the difference between c medians, 789–790
For the difference between two means for summary data, 544–545
For the difference between two medians, 788–789
For the difference between two proportions, 546–548
For the difference between two variances for summary data, 550–552
For the existence of correlation, 665–666
for the mean (σ known), 472–473
for the mean (σ unknown), 473–474
for the proportion, 474–475
Two way tables and charts, 141
Pivot tableWizard used for Contingency tables, 134–138
Print-outs
Customizing, 15

Question mark buttons, 19

Referring to other sheets, 9
Regression, 661–668
Relative cell reference, 12
Renaming sheets, 10
Resizing button, 4
Right-click, 2

Sample size determination
for the mean, 412
for the proportion, 413–414
Saving workbooks, 17
Scroll bar, 4
Select, 1
Selecting a simple random sample, 60
Setting up the application window, 4
Sheet tab, 4, 10
Sheet tab settings, 15
Shortcut menu, 2
Spinner buttons, 6
Stacked data, 544, 546, 553, 789, 790
Standard toolbar, 4
Start button, 2
Status bar, 4
Stem-and-leaf display, 127
Sum function, 299
Sumproduct function, 299

Tdist function, 665
Tests of hypothesis
Chi-square tests
For contingency tables, 548–550
For the variance, 475–477
for a proportion, 474–475

for the difference between two means, 543–545
for the difference between two means (paired samples), 545–546
for the difference between two variances, 550–553
Kruskal-Wallis test, 789–790
One sample Z-test for the mean (σ known), 472–473
One sample t-test for the mean (σ unknown), 473–474
One-way ANOVA, 731–732
Pooled variance t-test, 543–545
Sign test, 786–788
Two-way ANOVA, 732–733
Wilcoxon rank sum test, 788–789
Z-test for the difference between two proportions, 546–548
Text files, 22
Text Import Wizard, 22–23
Tinv function, 665
Toolbars, 2
Tool tips, 19
Two-way ANOVA, 732–733

Wilcoxon rank sum test, 788–789
Windows, 4
Wizards, 21
Chart, 128
Text Import, 22–23
Workbook, 1
Worksheet, 1

Subject Index

Additive Law, 240–241
Additive Rule for Mutually Exclusive Events, 218–220
Adjusted coefficient of determination, 630–632
Agent Orange, 584
Agoraphobia, 390
Alloys, amorphous, 774
Alternative (research) hypothesis, 418–420
American Association for Marriage and Family Therapy (AAMFT), 225
Analysis of variance (ANOVA), 669–735
  calculation formulas for, 712–718
  checking assumptions of, 710–712
  concept of, 673–675
  experimental design and, 670–673
  follow-up analysis to, 700–710
    Bonferroni multiple comparisons procedure, 701–705
  one-way, 675–686, 713, 714–716
    conducting, 678–680
    F statistic for, 677–678
    setting up, 676–677
  two-way, 686–700, 713–714, 716–718
    ANOVA table for, 690
    factor interaction in, 688–689
    F tests for, 690–693
    main effects, 689–690, 693–694
    2×3 design, 686–688
Analysis of variance (ANOVA) table, 678, 690
Aphasia, 71
Areas, probabilities and, 304–305
Arithmetic mean (average), 148–149
Association
  linear (straight-line), 192
  measures of, 145, 146
  negative, 96–97
  positive, 96–97

Bar graphs, 63–67
  constructing, 65
  side-by-side, 87–90
Beck Depression Inventory (BDI), 405
Behavior, emotional, 207–208

Benefit bundle, 230–231
Bennett, Arnold, 401–402
Beta blockers, 535
Between-sample variation, 674
Bias(es)
  in data collection, eliminating, 462
  nonresponse, 51
  selection, 51
Binomial experiment, 268–269
Binomial probability distribution, 268–280
  computing, 271–272
  mean, variance, and standard deviation for, 274–275
  summing, 272–273
Bins, 76
Bioethics, 33
Biofeedback, 499
Bonferroni multiple comparisons procedure, 701–705
Box-and-whisker plots, 180–182
Bulimia nervosa, 682, 753
Business organizations, commitment to, 258, 291–292

Categorical (qualitative) data, 29
Causality, correlation and, 192
Cause-and-effect relationship, 642
Cell count, expected, 513
Cellular telephones, 416, 464–465
Central Limit Theorem, 339–342, 375
Central tendency measures, 145, 146, 148–156
  mean, 148–149, 152, 160–161
  median, 149–151, 152, 738
  mode, 151
Chart junk, 105
Chewing cycle, 318
Chi-square (χ^2) distribution, 396, 762
Citrus mealybug, 697
Class(es), 62
  in histogram, 75, 78
  modal, 151
Class boundary, 76
Class intervals, 75
Class percentage, 62
Class relative frequency, 62
Clonidine, 502
Cluster sampling, 49

Coefficient(s)
  confidence, 365
  of correlation, 192
    for linear regression, 585–592
    Pearson product moment, 192
  of determination (R^2), 592–596, 608–609, 630–632
    adjusted, 630–632
Cogeners, 584
Collagen, Type II, 751
Combinatorial Rule, 228–231
Commitment to business organizations, 258, 291–292
Complements Rule, 221–222
Completely randomized design, 675–676. *See also* One-way ANOVA
  ANOVA calculation formulas for, 713
  confidence intervals for, 675–676
Conditional probability, 231–239
Confidence coefficient, 365
Confidence interval(s)
  for ANOVA, 680
  for comparing two means, 482, 483–485
  for difference between two proportions, 506
  for estimation of population mean, 364–371, 377–378
    Excel-generated, 379–380
    90%, 367–368
    95%, 364–367
    99%, 368–369
    100%, 368
    small-sample, 375–376, 378–379
    width of, 369–371
  hypothesis testing and, 417–418
  inference about proportion (π) from, 417–418
  inferences about mean (μ) from, 417
  for matched pair experiment, 496
  for proportion, 385–387
  for slope of linear regression, 579–585, 608
  for Student's t, 377–380
  for variance (σ^2), 397–398
Consent, informed, 463

Constant error variance, 614–615
Contingency tables, 511–523
  computing expected numbers in, 512–513
  independence of two qualitative proportions, 511–512
  rejection region for, 514–515
  test of hypothesis for, 515–516
  test statistic for, 514
  2×3, 517–518
Continuous random variables, 259, 304–305
Control, 499
Control group, 41
Correlation, 190–197
  causality and, 192
  computing, 191–192
  defined, 190–191
  interpreting, 192–194
Correlation coefficient, 192
  for linear regression, 585–592
    zero correlation, 586–588
  Pearson product moment, 192
Cost estimation, 353
Critical value, 427, 679
Cross-classification tables, 85–87, 90
Cross-product model, 631
Cumulative probabilities, 273–274
Curvilinear relationship, 99
Customer Satisfaction Index, 753
Cycle time, 345

Data, 28
  collecting, 41–44. *See also* Sampling
    eliminating biases in, 462
    methods of, 42
  statistics as science of, 28–29
  types of, 29
Data description. *See also* Graph(s); Table(s)
  objective of, 62
  for single qualitative variable, 62–72
  for single quantitative variable, 72–85
  for two qualitative variables, 85–98
  for two quantitative variables, 95–102
Data-ink, 105
Data set, features of, 29–30
Data snooping, 463
Data variations, measures of, 145, 146, 157–167
  Empirical Rule, 164–167
  range, 157

standard deviation. *See* Standard deviation (σ)
variance. *See* Variance (σ^2)
Dear enemy recognition, 522–523
Deep hole drilling, 442
Degrees of freedom, 375–376, 377
  for estimating variance (σ^2), 574
Depakote, 510
Dependent (response) variable, 555, 670
Depression, postpartum, 39
Descriptive statistics, 34–36, 38. *See also* Data description; Graph(s); Numerical descriptive measures; Table(s)
  for assessing normality, 321–330
    graphs, 321, 322–323
    interquartile range (IQR), 321, 323
    normal probability plots, 321, 323
Design, experimental
  ANOVA and, 670–673
  unethical, 53
Designed experiments, 41, 642
Detection probability, 354
Deviation, 157. *See also* Standard deviation (σ)
  negative, 158
  positive, 158
Disagreement probability. *See* p-values
Discrete probability distributions, 258–300
  binomial, 268–280
  hypergeometric, 286–292
  Poisson, 281–286
  properties of, 261–263
  random variables and, 259–260
Discrete random variable, 259, 262–263
Dispersion, measures of (measures of variation), 145
Distribution(s). *See also* Binomial probability distribution; Discrete probability distributions; Normal probability distributions; Sampling distribution(s)
  chi-square (χ^2), 396
  geometric, 294
  hypergeometric, 286–292
  nonnormal, 737
  nonskewed (symmetric), 77

Poisson, 281–286
  cumulative, 281–283
  mean, variance, and standard deviation of, 283–284
sampling, 331
skewed, 145, 183, 737
  left (negatively), 77
  right (positively), 76
Distribution-free tests, 737–738
Double blind experiment, 509
Drug screening, 425
Dummy variable model, 631, 632, 633

Effect size, 351
Electroencephalogram, 570
Emotional behavior, 207–208
Empirical Rule, 164–167, 185, 264–268
Endowment, 405
Error(s)
  independent, 618
  mean error of zero, 612–613
  mean square for (MSE), 674
  measurement, 52
  nonresponse, 51–52
  normal, 616–617
  sampling, 52, 392
    of least squares line, 598
  standard, 339, 363
  sum of squared (SSE), 674, 675
  Type I, 420–421
  Type II, 420–421
Error variance
  constant, 614–615
  nonconstant, 615–616
Ethical issues
  in hypothesis testing, 461–464
  in linear regression analysis, 644
  in statistical applications, 50
Ethylene oxide (ETO), 526–527
Euthanasia, 40
Event(s), 213
  complementary, 221–222
  independent, 233–234
  mutually exclusive, 218–220
  probability of, 214
  rare, 235, 273
  simple, 218, 228–231
Expected cell count, 513
Expected value. *See* Mean (μ)
Experiment(s), 211–212, 670. *See also* Matched pairs experiment
  binomial, 268–269
  designed, 41, 642
  double blind, 509

Experimental design
    ANOVA and, 670–673
    unethical, 53
Experimental (sampling) unit, 670
Experimentwise error rate (EER),
    701
Explanatory (independent) variable,
    555
Exploratory data analysis, 180, 463

Factorial, 228
Factorial designs, 686. *See also* Two-
    way ANOVA
    ANOVA calculation formulas for,
        713–714
Factor level, 670–671
Factor main effects, 689–690
Factors, 670–671
False alarm, probability of, 354
False negative results, 252, 425
False positive results, 252, 425
Fertility rate, 196–197
Fetal alcohol syndrome, 66–67
Fire insurance fraud, detecting,
    347–348
1st quartile, 174
First-order regression model, 626–628
Five-number summary, 180, 181
Fixed significance level, 444
Flat distribution, 737
Frame, 48
Fraud, fire insurance, 347–348
Frequency, 62
    relative, 213–214
Frequency histogram, 75–77
Frequency table, 63
    for stem-and-leaf displays, 74
F statistic, 524–526, 528
    for ANOVA, 690–693
    in multiple regression, 628–629

Game selection, 248
Gauss, C.F., 305
General regression model, 555,
    625, 626
Geometric distribution, 294
Global Positioning System (GPS), 590
Gosset, W.S., 375
Graph(s)
    bar graphs, 63–67, 87–90
    histograms, 75–79
    of normal distribution, 321,
        322–323
    pie charts, 63–67

proper presentation of, 104–108
    graphical excellence principles,
        105
    scatterplots, 95–104, 557–558
    stem-and-leaf displays, 72–75
Gross Profit Factor (GPF), 347

Hermaphrodites, 530
Highway contracting industry, bid-
    collusion in, 144, 198–200
Histogram(s), 75–79
    classes in, 75, 78
    constructing, 75–77
    frequency, 75–77
    interpreting, 78–79
Horizontal bar chart, side-by-side, 87
Hotspots, 290
Hypergeometric probability
    distribution, 286–292
    computing, 287–288
    mean, variance, and standard
        deviation for, 288–289
Hypothesis. *See also* Test(s) of
    hypothesis
    alternative (research), 418–420
    formulating, 418–425
    null, 418–420
    statistical, 418

Identical twins, IQ comparison of, 478,
    531–533
Independent errors, 618
Independent events, 233–234
Independent (explanatory) variable,
    555
Independent samples, drawback of
    using, 494–495
Inference(s), 234–239, 273–274
    about mean (μ) from confidence
        interval, 417
    about proportion (π) from
        confidence interval, 417–418
    rare event approach to making, 235
    reliability of, 211
    about slope of linear regression,
        576–579
Inferential statistics, 37–38
Informed consent, 463
Interaction between factors, 688–689
Interaction model, 631, 632, 633
Interquartile range (IQR), 180–181,
    321, 323
Intersection of *A* and *B*, 239
Interval data, 29

Intervals
    around mean, 160–161
    class, 75
Interventions, 280

Jackpot, 224
Jitter, 400–401

Kruskal-Wallis *H* test, 762–769

Learning, Earning, and Parenting
    (LEAP) program, 227
Least squares, method of, 562
Least squares line (least squares
    prediction equation), 562
    sampling errors of, 598
    variability around, 574–576
Left (negatively) skewed distribution,
    77
Lemmings, collared, 698
Level of significance. *See* Significance
    level
Lie factor, 106
Linear regression, 554–668
    assumptions of, 572–573, 612, 641
    coefficient of determination (R^2)
        for, 592–596, 608–609,
        630–632
    computations in, 604–612
        estimating β_0 and β_1, 605–606
        estimation and prediction,
            609–610
        of R^2, 608–609
        of SSE and s, 606–607
    correlation coefficient for, 585–592
        zero correlation, 586–588
    defined, 555
    equation for, 558–559
    for estimating mean value,
        597–598, 609–610
    ethical issues in, 644
    identifying variables for, 556
    parameters in
        estimating and interpreting,
            560–565
        identifying, 559
    pitfalls in, 641–644
    prediction with, 596–604
    residual analysis, 612–624
        constant error variance, 614–615
        detecting model
            misspecification, 613–615
        independent errors, 618
        mean error of zero, 612–613

Linear regression, *continued*
    nonconstant error variance, 615–616
    normal errors, 616–617
    outliers, 617–618
    steps in, 619
    robust, 616–617
    slope of
      confidence intervals for, 579–585, 608
      inferences about, 576–579
      testing, 607–608
    steps in, 557
    variability around least squares line, 574–576
Linear relationship, 99
Linear (straight-line) association, 192
Lottery Buster, 210, 248–249
Lower-tailed test, rejection regions for, 427–428, 430

Machiavellianism, 589
Machining conditions, 226
Magnetotherapy, 493
Main effects, 689–690, 693–694
Mammography, 736, 778–779
Matched pairs experiment, 494–503.
    See also Wilcoxon signed ranks test
    assumptions with small sample, 496
    confidence interval for, 496
    hypothesis test for, 496–498
Mean (μ), 148–149, 152, 160–161, 262–263
    arithmetic (average), 148–149
    for binomial probability distribution, 274–275
    comparison of two, 480–503
      confidence intervals for, 482, 483–485
      hypothesis testing for, 482–483, 485–488
      matched pairs experiment, 494–503
      sampling distribution of point estimate, 481–482
      selecting point estimate for, 480–481
    estimating, 362–383
      linear regression for, 597–598, 609–610
      normal (z) statistic, 362–374
      sample size for, 391–393
      Student's t statistic, 375–383

    for hypergeometric probability distribution, 288–289
    hypothesis testing for, 435–441
      large-sample, 435–438
      small-sample, 438–441
    inferences about, from confidence interval, 417
    intervals around, 160–161
    multiple comparisons of, 700–710
      Bonferroni multiple comparisons procedure, 701–705
    of Poisson distribution, 283–284
    sampling distribution of, 363
Mean error of zero, 612–613
Mean square for error (MSE), 674
Mean square for treatments (MST), 675
Measurement error, 52
Measures of central tendency. *See* Central tendency measures
Median, 149–151, 152, 738
    comparing two. *See* Wilcoxon rank sum test
Melatonin, 344
Meta-analysis, 330
Method of least squares, 562
Mid-quartile, 174
Milgram experiment, 27, 55–56
Minnesota Multiphasic Personality Inventory (MMPI), 685
Modal class, 151
Mode, 151
Model building, 644
Muck, 207
Multicollinearity, 642–643
Multiple regression, 624–646
    dummy variable model, 631, 632, 633
    first-order model, 626–628
    F test in, 628–629
    general model, 555, 625, 626
    interaction model, 631, 632, 633
    interpreting x and R^2, 630–632
    quadratic (second-order) model, 631, 632
    steps in, 625
    testing β parameters, 630
    test of overall adequacy of, 628
Multiplicative Law, 239–240
MUM effect, 669, 718–720

Negative association, 96–97
Negatively (left) skewed, 77

Nominal data, 29
Nominal variable, 64
Nonconstant error variance, 615–616
Nondestructive evaluation (NDE), 84
Nonparametric statistics, 380, 736–790
    distribution-free tests, 737–738
    Kruskal-Wallis H test, 762–769
    sign test, 738–745
      applying, 739–740
      rejection region of, 738–739
    Spearman's rank correlation coefficient, 769–778
      applying, 770–772
      for large samples, 772–773
    Wilcoxon rank sum test, 745–754
      applying, 746–748
      for large samples, 748–750
    Wilcoxon signed ranks test, 754–762
      applying, 755–756
      for large samples, 757
Nonresponse bias, 51
Nonresponse error, 51–52
Nonskewed (symmetric) distribution, 77
Normal curve. *See* Normal probability distributions
Normal errors, 616–617
Normalizing transformation, 710
Normal probability distributions, 303–360
    descriptive methods for assessing normality, 321–330
      graphical, 321, 322–323
      interquartile range (IQR), 321, 323
      normal probability plots, 321, 323
    finding, 311–315
    sampling distributions, 330–348
      approximating, 332–333
      concept of, 331–332
      defined, 331
      impact of n on, 334–335
      of sample mean, 332, 333, 338–342
    standard, 307–311
Normal probability plot, 321, 323
Normal random variable, 314–315
Normal (z) statistic, 362–374. *See also* Confidence interval(s)
Null hypothesis, 418–420
Numerical descriptive measures, 144–209

box-and-whisker plots, 180–182
central tendency measures, 145, 146, 148–156
  mean, 148–149, 152, 160–161
  median, 149–151, 152, 738
  mode, 151
correlation, 190–197
  causality and, 192
  computing, 191–192
  defined, 190–191
  interpreting, 192–194
data variations, 145, 146, 157–167
  Empirical Rule, 164–167
  range, 157
  standard deviation. *See* Standard deviation (σ)
  variance. *See* Variance (σ^2)
outlier detection, 184–190
  using quartiles, 185–186
  using z scores, 184–185, 186
for populations, 198–200
relative standing, 145, 146, 173–180
  percentiles, 173–176
  z scores, 176–177, 184–185, 186
summation notation, 146–147
types of, 145–146

Observational study, 42
Observed significance level, 446
One-tailed test of hypothesis, 420, 463
One-way ANOVA, 675–686, 713
  calculation formulas for, 714–716
  conducting, 678–680
  F statistic for, 677–678
  setting up, 676–677
Ordinal data, 29
Ordinal variable, 64
Outcomes, listing, 212
Outlier(s)
  detection of, 184–190
    using quartiles, 185–186
    using *z* scores, 184–185, 186
  discarding, 463–464
  identifying, 186–187
  in linear regression, 617–618

Parameter(s), 198, 330
  estimation of, 361–415
    choosing sample size for, 390–402
    mean, 362–383
    proportion, 383–390
    variance, 395–401

target
  for comparing samples, 479–480
  for hypothesis testing, 433–435
Parametric statistical tests, 737
Partitioning of sum of squares (SS), 675, 689
Peaked distribution, 737
Pearson product moment coefficient of correlation, 192
Percentage, 62
Percentiles, 173–176
  defined, 174
  finding, 175–176
  interpreting, 174–175
Perfectionists, 635
Pheromones, 227
Pie charts, 63–67
  constructing, 65–66
  interpreting, 66
Point estimates, 62
  for comparing two means, 480–481
    sampling distribution of, 481–482
  for difference between two proportions, 504
    sampling distribution of, 505
  of mean value of *y*, 597–598
Poisson, Siméon D., 281
Poisson probability distribution, 281–286
  cumulative, 281–283
  mean, variance, and standard deviation of, 283–284
Poisson random variable, 281
Pooled estimate of variance, 485
Population(s), 34, 35–38
  comparing two or more, 762–769
  detecting nonnormal, 710
  numerical descriptive measures for, 198–200
  parameter estimation. *See under* Parameter(s)
  testing for location of single. *See* Sign test
Positive association, 96–97
Positively (right) skewed distribution, 76
Postmortem interval (PMI), 72–73
Postpartum depression, 39
Power of tests of hypothesis, 737
Predicting outside the experimental region, 643–644
Probability(ies), 210–257
  Additive Law of, 240–241

Additive Rule of, 218–220
  areas and, 304–305
  Combinatorial Rule for counting simple events, 228–231
  Complements Rule of, 221–222
  conditional, 231–239
  cumulative, 273–274
  detection, 354
  of an event, 214
  of false alarm, 354
  for independent events, 233–234
  inferences using, 234–239
  interpreting, 213–214
  Multiplicative Law of, 239–240
  Probability Rule of, 220–221
  properties of, 221
  role in statistics, 210
  summing, 217–218
  unconditional, 231
Probability distributions. *See* Distribution(s)
Proportion (π)
  difference between two, 504–523
    contingency tables, 511–523
    large-sample confidence interval for, 506
    sampling distribution of point estimate, 505
    selecting point estimate for, 504
    test of hypothesis for, 505, 506–507
  estimation of, 383–390
    point estimate selection, 383–384
    sample size selection for, 393–395
    sampling distribution of, 384
    Wilson's adjustment in, 387
  hypothesis testing for, 452–457
    large-sample, 452–454
    p-value for, 454
  inferences about, from confidence interval, 417–418
Propranolol, 535
Published source, data collection using, 41
p-values, 444–452
  for one-tailed test, 446–447
  for population proportion, 454
  for two-tailed test, 446, 447–448

Quadratic (second-order) regression model, 631, 632
Quadratic (squared) model, 631, 632
Qualitative (categorical) data, 29

Qualitative variable(s), 511–512
    data description for single, 62–72
    data description for two, 85–98
    independence of two, 511–512
    nominal, 64
    ordinal, 64
Quantitative data, 29
Quantitative variable
    data description for single, 72–85
    data description for two, 95–102

Random error component, 558
Random number generators, 45
Random sample, simple, 44
Random sampling, 44–46
    stratified, 48–49
Random variable(s)
    continuous, 259, 304–305
    discrete, 259, 262–263
    discrete probability distributions
        and, 259–260
    normal, 314–315
    Poisson, 281
    standard normal, 308
Range, 157
    interquartile (IQR), 180–181,
        321, 323
Rank correlation, testing for, 769–778
    applying, 770–772
    for large samples, 772–773
Rank sum, 745
Rank tests, 737–738. See also
        Spearman's rank correlation
        coefficient; Wilcoxon rank sum
        test; Wilcoxon signed ranks test
Rare event, 235, 273
Ratio data, 29
Regression analysis. See Linear
        regression; Multiple regression
Regression models, 555
Rejection region(s), 425–431
    for contingency tables, 514–515
    guidelines for specifying, 429
    for lower-tailed test, 427–428, 430
    of sign test, 738–739
    for two-tailed test, 428–429, 430
    for upper-tailed test, 426–427,
        429–430
Relative frequency, 62, 213–214
Relative standing, measures of, 145,
        146, 173–180
    percentiles, 173–176
        defined, 174
        finding, 175–176

interpreting, 174–175
z scores, 176–177, 184–185, 186
Reliability of inference, 211
Replacement
    sampling with, 268
    sampling without, 286
Replication of factorial experiment,
        686
Reporting, pitfalls and ethical issues
        in, 464
Representative sample, 43
Research (alternative) hypothesis,
        418–420
Residual analysis, 612–624
    constant error variance, 614–615
    detecting model misspecification,
        613–615
    independent errors, 618
    mean error of zero, 612–613
    nonconstant error variance,
        615–616
    normal errors, 616–617
    outliers, 617–618
    steps in, 619
Residuals, 573, 612
Response (dependent) variable, 555,
        670
Right (positively) skewed
        distribution, 76
Robust linear regression, 616–617
Royal flush, 231
Running distance, 346

Sample(s), 35–38
    independent, drawback of using,
        494–495
    representative, 43
Sample space, 218
Sample statistic, 331
Sample variance, 157
Sampling
    cluster, 49
    random, 44–46
        stratified, 48–49
    with replacement, 268
    without replacement, 286
    survey, 48, 394
    systematic, 50
Sampling distribution(s), 330–348
    approximating, 332–333
    chi-square, 396, 762
    concept of, 331–332
    defined, 331
    impact of n on, 334–335

of the mean, 363
of proportion, 384
of sample mean, 332, 333, 338–342
Sampling error, 52, 392
    of least squares line, 598
Sampling (experimental) unit, 670
Scale break, 108
Scalelessness, 192
Scallop harvesting laws, 361, 401–402
Scatterplots, 95–104
    constructing, 95–97
    interpreting, 98–99, 557–558
Screening, drug, 425
Second-order (quadratic) regression
        model, 631, 632
Selection bias, 51
Side-by-side bar graphs, 87–90
Significance level, 421–422
    choice of, 463
    fixed, 444
    observed, 446
Sign test, 738–745
    applying, 739–740
    rejection region of, 738–739
Simple events, 218, 228–231
Simple linear regression. See Linear
        regression
Simple random sample, 44
Simulated test marketing, 53
Skewed distribution, 145, 183, 737
    box-and-whisker plot to detect, 183
    left (negatively), 77
    right (positively), 76
Slope, 558. See also under Linear
        regression
Spearman's rank correlation
        coefficient, 769–778
    applying, 770–772
    for large samples, 772–773
SSR, 594
Standard deviation (σ), 159, 262
    for binomial probability
        distribution, 274–275
    estimating, 575
    for hypergeometric probability
        distribution, 288–289
    interpretation of, 575
    of Poisson distribution, 283–284
Standard error, 339, 363
Standard normal distribution, 307–311
Standard normal random variable, 308
Statistical applications, ethical issues
        in, 50
Statistical hypothesis, 418

Statistics
   defined, 28–29
   descriptive, 34–36, 38
   inferential, 37–38
   objective of, 38
Stem-and-leaf displays, 72–75
Straight-line model. *See* Linear
   regression
Stratified random sampling, 48
Stratum/strata, 49
Student's *t* statistic, 375–383
   characteristics of, 376–377
   confidence intervals for, 377–380
   finding, 377
Summation notation, 146–147
Sum of squared errors (SSE),
   674, 675
Sum of squares, partitioning of (SS),
   675, 689
Sum of squares table, 191
Supercooling temperature, 345
Supremism, 443
Survey, 41–42
   misleading results from, 52–53
   unethically designed, 53
Survey sampling, 48, 394
Symmetric (nonskewed) distribution,
   77
Systematic sampling, 50

Table(s)
   analysis of variance (ANOVA),
      678, 690
   contingency. *See* Contingency
      tables
   cross-classification, 85–87, 90
   frequency, 63
   sum of squares, 191
Tail probability associated with x_0,
   314–315
Taylor Manifest Anxiety Scale
   (TMAS), 349
Tchebysheff's Theorem, 165, 185
Telephones, cellular, 416, 464–465
Temperature, supercooling, 345
Test marketing, simulated, 53
Test(s) of hypothesis, 416–477
   choosing target parameter, 433–435
   for comparing two means, 482–483,
      485–488
   for comparing two variances,
      523–528
   confidence intervals and, 417–418
   for contingency tables, 515–516

   for difference between two
      proportions, 505, 506–507
   for linear correlation, 585–588
   logic of, 422–423
   for matched pair experiment,
      496–498
   methodology, 418–433
      formulating hypotheses,
         418–425
      rejection regions, 425–431
      significance level, 421–422
      test statistics, 425–426
      Type I and Type II errors,
         420–421
   one-tailed, 420, 463
   of population mean (μ), 435–441
      large sample, 435–438
      small sample, 438–441
   of population proportion (π),
      452–457
      large sample, 452–454
      p-value for, 454
   of population variance (σ^2),
      457–461
   potential pitfalls and ethical issues
      in, 461–464
   power of, 737
   reporting results of (*p*-values),
      444–452
      for one-tailed test, 446–447
      for two-tailed test, 446, 447–448
   for slope of straight-line model,
      577–578
   two-tailed, 420, 463
Test statistics, 425–426
   for contingency tables, 514
3rd quartile, 174
Throughput rate, 759
Time series, 618
Trail Making Test (TMT), 169, 319
Transformation(s)
   normalizing, 710
   variance-stabilizing, 616, 711
Treatment, 670–671
Treatment group, 41
Twins, IQ comparison of identical,
   478, 531–533
Two-number summary, 180
Two-tailed test of hypothesis, 420
   rejection regions for, 428–429, 430
   selecting, 463
Two-way ANOVA, 686–700, 713–714
   ANOVA table for, 690
   calculation formulas for, 716–718

   factor interaction in, 688–689
   F statistic for, 690–693
   main effects, 689–690, 693–694
   2×3 design, 686–688
Type I error, 420–421
Type II error, 420–421

Uncertainty, 210, 211
Unconditional probability, 231
Unethical survey design, 53
Union of *A* and *B*, 239
United Faculty of Florida (UFF),
   644–646
Upper-tailed test, rejection regions
   for, 426–427, 429–430

Variable(s)
   classifying, 31–32
   dependent (response), 555, 670
   dummy, 631, 632, 633
   independent (explanatory), 555
   qualitative. *See* Qualitative
      variable(s)
   quantitative
      data description for single,
         72–85
      data description for two, 95–102
   random. *See* Random variable(s)
   regression, 556
Variance (σ^2), 157–159, 262–263. *See
   also* Analysis of variance
   (ANOVA)
   for binomial probability
      distribution, 274–275
   comparing two, 523–528
   confidence intervals for (σ^2),
      397–398
   detecting unequal, 710–711
   estimating, 395–401, 575
      degrees of freedom for, 574
   for hypergeometric probability
      distribution, 288–289
   hypothesis testing of, 457–461
   of Poisson distribution, 283–284
   pooled estimate of, 485
   sample, 157
Variance-stabilizing transformations,
   616, 711
Variation
   between-sample, 674
   measures of (measures of
      dispersion), 145
   within-sample, 674
Venn diagram, 239

Wayfinding problem, 699
Wilcoxon rank sum test, 745–754
    applying, 746–748
    for large samples, 748–750
Wilcoxon signed ranks test, 754–762
    applying, 755–756

    for large samples, 757
Wilson's adjustment, 387
Within-sample variation, 674
"Work measurement" analyses, 206

y-intercept, 558

Zero correlation, test for, 586–588
z scores, 176–177, 184–185, 186
    outlier detection using, 184–185, 186
z value, computing, 306–307